Semiconductor Process Reliability in Practice

About the Authors

Zhenghao Gan, Ph.D., is a Senior Manager of Reliability in the Technology Development Center at the Semiconductor Manufacturing International Corporation (SMIC) in Shanghai, China. He has extensive technical and management experience in research and development for semiconductor reliability improvement, testing/characterization, problem solving, project management, modeling, and analysis. Dr. Gan has published one book and more than 50 research articles, and holds more than 60 patents filed.

Waisum Wong, Ph.D., is a Senior Director at SMIC. He was in charge of process and device reliability in SMIC's Technology Development Center and is currently responsible for device modeling and process design kit development. Dr. Wong has worked with companies such as Intel and Silconix and has extensive experience in device design, device reliability, device modeling, and characterization.

Juin J. Liou, Ph.D., is the Pegasus Distinguished Professor and UCF–Analog Devices Fellow in the Department of Electrical Engineering and Computer Science at the University of Central Florida. He has published eight books and 500 research articles, and has been awarded more than $12 million in research contracts and grants from federal agencies, state governments, and industry sectors. Dr. Liou has directed several research labs in the United States and Asia specializing in electrostatic discharge protection design, modeling, and simulation.

Semiconductor Process Reliability in Practice

Zhenghao Gan, Ph.D.
Waisum Wong, Ph.D.
Juin J. Liou, Ph.D.

New York Chicago San Francisco
Lisbon London Madrid Mexico City
Milan New Delhi San Juan
Seoul Singapore Sydney Toronto

1 2 3 4 5 6 7 8 9 0 LPI/LPI 1 8 7 6 5 4 3 2

ISBN 978-0-07-175427-9
MHID 0-07-175427-X

Sponsoring Editor
Michael Penn

Editing Supervisor
Stephen M. Smith

Production Supervisor
Pamela A. Pelton

Acquisitions Coordinator
Bridget L. Thoreson

Project Manager
Sheena Uprety,
Cenveo Publisher Services

Copy Editor
James K. Madru

Proofreader
Carol Shields

Indexer
Michael Ferreira

Art Director, Cover
Jeff Weeks

Composition
Cenveo Publisher Services

This book is printed on acid-free paper.

Contents

PART 1

General

CHAPTER 1

Introduction

1.1 Background

Reliability is an important requirement for almost all users of integrated circuits (ICs).[1] Scaling for enhanced performance and cost reduction has pushed existing complementary metal-oxide-semiconductor (CMOS) materials much closer to their intrinsic physical and reliability limits.[2] Reliability challenges have become more and more concerned with continuous scaling for 65-nm technology node and beyond also owing to the introduction of new materials and device architectures with unknown reliability behaviors. For example, the application of stress-memorization technique (SMT), high-*k*/metal gate in front-end-of-line (FEOL), and ultralow-*k* porous dielectrics in back-end-of-line (BEOL) has a serious impact on the reliability performance.

A newly developed semiconductor technology node cannot be released to designers without going through a rigorous process and product-reliability qualification. The reliability qualification of a developing node usually starts in parallel with every phase of the process and device development and ends only when the full node of the process development is completed. In other words, any process and/or tool change during development should be confirmed to meet reliability requirements. After the process is released for mass production, the reliability still should be monitored periodically to ensure that there are no process/tool variations. The reliability also refines the design rule; for instance, the minimum channel length L_{min} for a 2.5-V input/output (IO) device overdriven to 3.3 V is defined as the L_{min} that meets the hot-carrier reliability requirement.

This book covers the areas of process reliability and its qualification of a deep submicron semiconductor process node from theory to the very practical industry practice for each area of reliability concern. It starts with initial spec definition and moves through test-structure design to analysis of the silicon data taken from the test structures and final qualification of the process. Many practical examples, including test-structure designs to qualify the front-end-of-line devices or the back-end-of-line of the interconnects, have been included to strengthen the reader's understanding of reliability theory and how the theory is used to resolve issues in the standard practice. In addition, this book includes some fundamental process flow and device physics when necessary to help in explaining process and device reliability concerns.

1.2 Process Reliability Items

In general, process integration engineers (PIEs) divide a semiconductor process into two parts, the FEOL and BEOL. The FEOL includes the process steps up to polysilicon, where as BEOL covers the process after the polysilicon. FEOL defines the metal-oxide-semiconductor (MOS) device and other commonly used active and passive devices, whereas BEOL defines interconnects.

The five major significant device-level failure items in IC technology FEOL are *hot-carrier injection* (HCI), *gate-oxide time-dependent dielectric breakdown* (TDDB), *negative-bias temperature instability* (NBTI), *plasma-induced damage* (PID) and *electrostatic discharge* (ESD). The three major interconnect failure modes considered most commonly in BEOL are *electromigration* (EM), *stress migration* (SM), and *TDDB of intermetal dielectrics*. Note that PID is drawing more and more attention because both FEOL and BEOL processes can lead to PID. There is another device-level failure mode, PID, that is drawing more and more attention because it affects the reliability of both FEOL and BEOL.

1.2.1 FEOL

Hot-Carrier Injection

HCI is a phenomenon of short-channel devices in which high field in the channel near the drain region leads to impact ionization, followed by injection of energetic carrier into the gate dielectrics and degradation in the MOS *I-V* characteristics and parameter shift. Since the 1980s, HCI degradation in N-channel metal-oxide-semiconductor (NMOS) nodes has become a critical issue because of the continuous scaling of transistor dimensions without accompanying scaling of the power-supply voltage V_{dd} (maintained at 5 V).[3] HCI immunity was achieved for scaling devices from 1- to ~0.5-μm nodes by new drain and gate architectures derived from the lightly doped drain (LDD) structures. The corresponding worst-case HCI degradation of NMOS nodes was clearly identified in correlation with the large amount of energetic carriers at the maximum of I_{sub} condition at $V_{gs} \cong V_{ds}/2$ corresponding to the maximum of interface traps generated.[4]

From the mid-1990s, together with the scaling of channel length and oxide thickness, V_{dd} was forced to be reduced considering both reliability requirements and power-consumption reduction. Hence HCI damage became less of an issue for CMOS nodes down to 120 nm.

However, with CMOS node continuous scaling down to 100 nm and beyond, HCI degradation has become a concern again because V_{dd} scaling is slowing or stopping, whereas gate length is continuing to scale down to 40 nm and beyond. In addition, the requirement for stressing an MOS device and passing HCI has been changed from $V_{gs} = V_{ds}/2$ to $V_{gs} = V_{ds}$, and this leads to a higher substrate current and worse HCI, whereas the temperature for stressing an MOS device and passing HCI also has shifted from room temperature to high temperature.[5] Chapter 5 details the HCI phenomenon with emphasis on the 90-nm and beyond technology node, where temperature/oxide-thickness dependence and degradation mechanisms of thin oxide are somehow different from those of previous technologies. The HCI improvement methodology through design and process optimization will be explained.

Gate-Oxide Integrity and Time-Dependent Dielectric Breakdown

With the gate-oxide physical thickness becoming thinner and thinner (~1.2 nm at the 65-nm node with only a few monolayers of Si-O bonds), the oxide becomes more and more susceptible to gate current leakage (~100 A/cm^2 at 1.0 V),[6] defect creation, and ultimately, dielectric breakdown. With thinner oxide, the electrical field across the gate oxides has been increased with each technology node and will increase further at future technologies, which leads to further challenges to reliability assessment and control. The International Technology Roadmap for Semiconductor (ITRS) predicts that the oxide electrical fields under normal operation can be up to 10 MV/cm.[7] The TDDB behaviors of the gate oxide also have been changed and mixed with hard breakdown (HBD), soft breakdown (SBD), and stress-induced leakage current (SILC), which lead to harder interpretation of the testing data and lifetime extrapolation. The lifetime-extrapolation methodology and model also have been evolved from the $1/E$ model for thick oxide to E model for intermediate-thickness oxide and to the current-power-law model for thin oxide. Chapter 6 will introduce the gate-oxide integrity (GOI) and TDDB phenomena, mechanisms, and models, with focus on the most advanced technology. The improvement methodology through process optimization also will be elaborated.

Negative-Bias Temperature Instability

NBTI on P-channel metal-oxide-semiconductor (PMOS) nodes causes a shift in threshold voltage V_t and a decrease in the mobility of the inversion channel. The basic understanding is that the Si-H bonds at the Si-SiO_2 interface, which are needed for interface-state passivation, can become broken under NBTI stressing. NBTI has become more and more important today[2] because electrical fields in the gate oxide are higher and devices are operating at higher temperatures owing to higher power consumption. In addition, the voltage headroom (difference between V_g and V_t) becomes much smaller than in the past, and a smaller V_t shift (ΔV_t) can lead to a larger I_{dsat} shift ($\Delta I_{dsat}/I_{dsat}$) [details are given in Chap. 2, Eq. (2.50)]. The NBTI has a larger impact on the ring oscillators at lower operating voltages.[8] Thus the NBTI impact on $V_{\min}$ circuit operation can be quite significant and an important reliability concern. Chapter 7 will provide a comprehensive summary for the negative-bias temperature instability phenomenon, which is observed more clearly in thinner oxides. The corresponding mechanisms, models, and improvement methodology through process optimization will be detailed.

Plasma-Induced Damage

PID can be a serious reliability concern for state-of-the-art ICs of modern process generations. Although in recent years the publications on PID have been much reduced, PID is still detectable for the most advanced gate oxides, for example, with an oxide thickness of 1.6 nm.[9] In general, specially designed test structures are used to monitor the severity of PID during manufacturing. However, it is hard to ensure that the test structures and the corresponding measurement methods are carefully designed so that the PID effects are properly detected. In addition, measurements for evaluating failure rates in the parts-per-million (ppm) regime are not possible because the size of test structures for PID investigation employed in one test vehicle is limited.[10] Investigations in the literature demonstrate that dielectrics of a wide thickness range can degrade because of PID, leading to significant MOS transistor parameter shift such as threshold voltage V_t, transconductance G_m, and gate-oxide leakage currents I_{g_leak}. Different gate-oxide thickness indicates different PID sensitivity. To prevent IC failures because of PID, design rules (DRs) have been established as guidelines to ensure the layout robustness against a large amount of PID that typically occurs during processing. However, even with design rules in place, PID damage to gate dielectrics is still possible, attributed to some hidden layout weaknesses with respect to the PID DRs. Protection schemes such as protection diodes are widely employed to reduce the PID impact, but sometimes they are not properly placed. Chapter 8 details the PID mechanism, antenna ratio, test structures, characterization methods, and PID control/improvement methodology through design and process changes.

Electrostatic Discharge

The ESD protection for CMOS devices has become more and more challenging because there is a general downward trend in ESD robustness with scaling, including the use of silicide, channel-length, and gate-oxide thickness reduction; the use of fin field effect transistor (FinFET) and silicon on insulator (SOI); and so on.[2] Chapter 9 provides give a summary for ESD phenomena and how to prevent ESD-induced damage.

1.2.2 BEOL

The transition from aluminum- to copper-based interconnect technology has been an important milestone for the continued scaling of micro-/nanoelectronics. The technological benefits of integrating Cu as the conductor material for interconnect metallization of advanced ultra-large-scale integration (ULSI) include, for example, the reduced resistance-capacitance (RC) delay and the improved EM resistance, which are quite accepted by the industry. For example, the EM lifetime for Cu is much greater than that for Al, by more than 100 times,[11] owing to the lower diffusivity of Cu compared with Al. Copper interconnects will continue to be used for the 32- and 22-nm technology nodes.

To accompany the application of Cu interconnects, the dual-damascene (DD) process flow has been developed, which requires the use of electrolytic Cu plating and chemical mechanical polishing (CMP) techniques. Major concerns with Cu were its extremely high diffusivity in silica or polymer dielectrics and its vulnerability to corrosion. A sidewall barrier layer, usually a derivative of tantalum, titanium, or tantalum nitride, thus is proposed to effectively prevent Cu diffusion at temperatures below 873 K.[12] All these approaches have created a complex electromigration, stress migration, and Cu/low-*k* dielectric breakdown behaviors.[13]

Electromigration

EM is the migration of metal atoms in a conductor owing to an electric current and can lead to the formation of voids in the interconnects and finally failure of the circuits. The aggressive shrinking of Cu lines and vias, together with the usage of (ultra) low-*k* dielectric materials for newer technologies, has exacerbated EM challenges. With the smaller vias, the critical void size for circuit failure also was reduced. The interconnect failure also becomes more sensitive to void shape and location. Therefore, the process variation in the new technologies will lead to a broader EM failure time distribution, in particular, for interconnections without sufficient liner/via redundancy,[14] imposing a larger EM challenge.

In DD interconnects, a high-aspect-ratio via has been recognized as the most complicated region for integration and also has been reported as the weakest link for reliability.[15,16] Although it is hard, it is very critical to form a thin but conformal barrier in the via. Otherwise, a via with a nonuniform barrier with poor adhesion/integrity/mechanical strength may result in serious extrinsic EM reliability issues.[17,18] Thus via-relevant processes such as via barrier coverage, via barrier etch, and via cleaning definitely will have an impact on EM performance.

Another critical aspect of DD interconnects is the Cu–cap-layer interface, which has been found as the fastest diffusion path during

EM stressing. In general, Cu interconnect reliability can be much improved by good liner and cap adhesion and barrier properties and integration practices that prevent contamination or corrosion of the metal films and their interfaces.

Accompanying the continuously decreasing *k*-value of the low-*k* dielectric with shrinkage is a higher level of porosity, which further leads to a poorer thermal conductivity. The exponential dependence of thermal conductivity of low-*k* dielectric on its *k* value is given in Eq. (1.1)[19]:

$$\lambda \propto \exp(1.05 * \kappa) \tag{1.1}$$

where λ is the thermal conductivity in mW/cm·K, and κ is the dielectric constant.

In contrast, the operating current density inevitably increases owing to the scaling trend. This increase in current density causes not only an acceleration of EM degradation but also a temperature rise owing to joule heating. The application of the low-*k* dielectric makes the situation even worse. Chapter 10 details the EM phenomenon, including mechanisms, lifetime prediction methodology, process effect, and test-structure effects. This chapter also will show how the joule heating affects the interconnect EM performance and how to control the joule heating effect from a design point of view.

Stress Migration

SM can result in void formation under and/or within vias for advanced interconnect systems and can be accelerated by high-temperature baking (150–200°C).[20] Generally, this has been an issue when a single minimum-size via is connected to a wide Cu lead. In this case, the high Cu diffusivity path at the Cu–cap-layer interface and stress gradients under the via enable voids to nucleate below the via and easily grow to a fatal size, which becomes even worse when there is no conductive shunt layer at the via-bottom–Cu–cap-layer intersection. On the other hand, SM also has been reported more recently for narrow metal leads.[21] With continued scaling of the via, via voiding with narrow metal leads indeed becomes an issue simply because fewer vacancies will be required to form a "killer" void for abrupt resistance rise and failure. Chapter 11 will present the SM mechanisms, finite element analysis as a tool for SM understanding, and then how to reduce SM damage through process modification and optimized interconnect design.

Intermetal Dielectric Time-Dependent Dielectric Breakdown

With a high electrical field continuously applied to the BEOL intermetal dielectrics (IMDs), similar to FEOL gate dielectrics, damage can be induced in IMDs and the interfaces between the interconnects and the IMDs, that will lead to the formation of a conduction path and finally dielectric breakdown—commonly referred as TDDB. In older technologies, when the spacing between metal interconnects was relatively large, the IMD TDDB is a much less critical reliability concern because the electrical field across the BEOL dielectric is low. However, current-interconnect low-*k* dielectric minimum spacing can be 70 to 80 nm or less, which is similar to gate-oxide thickness approximates 20 years ago. In other words, the electrical field across the IMD increases continuously with continuous scaling, as shown in Fig. 1.1. Although the electrical field across the BEOL dielectric is still about one order lower than that across the gate dielectrics for 32-nm technology nodes and below, the breakdown strength of the BEOL dielectric is much less because the intrinsic quality of these low-*k* dielectrics is much less perfect than that of gate oxides in terms of both electrical[22] and mechanical strength.[23] There are a number of reasons for the low breakdown strength of BEOL dielectrics.[13] First of all, there is a much higher defect density in the IMDs, especially for low-*k* dielectrics.[22] Such preexisting defects serve as precursors for traps during TDDB stressing. Second, damage or contamination of the dielectric can be induced during processing such as CMP. Furthermore, Cu diffusion into the dielectric can take place if the integrity of the barrier layers is not well ensured during processing.[24] Last but not least, patterning problems such as line-width roughness (LWR) or via misalignment can result in a locally high electrical field. As a result, IMD TDDB becomes a more and more critical reliability concern

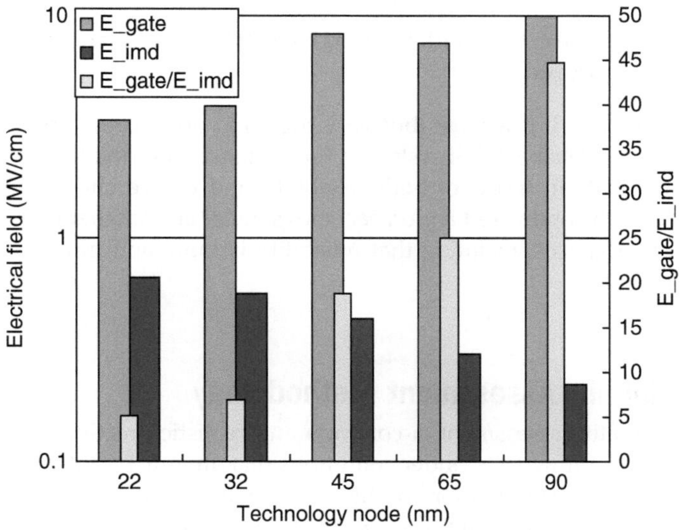

FIGURE 1.1 A plot of the electrical field across FEOL gate dielectrics and BEOL dielectrics for minimum pitch of interconnects in terms of the technology node.

when device dimensions are scaled and (ultra) low-*k* materials with high porosity are used as BEOL insulators. Chapter 12 summarizes the most updated knowledge and understanding of the intermetal dielectric breakdown mechanisms and how to control reliability through design and process optimization, with especial focus on low-*k* materials.

1.3 Process-Dependent Reliability

Reliability is definitely process dependent. Different process steps and conditions potentially can affect reliability and its related parameters. It is imperative for process engineers to understand the process-reliability correlation so as to make prudent process changes for reliability robustness. *How process parameters affect reliability is one of the key focuses of this book.* An example is the hydrogen-related processes in forming-gas annealing (FGA), which is a common practice to passivate the dangling bonds in the gate-oxide–channel interface—an approach to recover the plasma-induced damage.[25] For FGA, the key parameter is the thermal budget, which can be characterized by the annealing temperature and time, deciding how much the hydrogen will be incorporated into the device. In general, the higher the thermal budget, the more hydrogen will be introduced. When considering the effects of reducing thermal budget (i.e., less hydrogen is incorporated), it is seen that

- The retention time of dynamic random-access memory (DRAM) is degraded, which is not desirable.
- On the other hand, the NBTI and HCI performance could be improved.

Therefore, it is a must (but very tricky) to optimize the processes for a highly reliable ULSI system in future generations in order to meet requirements in terms of both reliability and device characteristics. Note that the optimized region becomes smaller and smaller with technology advances, meaning that reliability is more and more process dependent.

1.4 Reliability Assessment Methodology

For reliability assessment, a common and realistic practice is to perform accelerating tests under conditions that are much more "severe" than those under operation conditions to obtain failure statistics within a reasonable testing time. The so-called severe conditions in accelerating tests in general mean a much higher stressing temperature or stressing current/voltage or both than the operation conditions. After obtaining the time-to-failure (TTF) and failure distributions on specially designed test structures from the accelerating tests, the assessments of reliability under normal operation for a larger chip area then is realized through the relevant physical/mathematical models by extrapolation. There are three extrapolations from the accelerating test to the normal operation as follows:

1. *Toward lower percentiles (typically 0.1 percent or 1 ppm).* This extrapolation is done based on the assumed statistical distribution (Weibull or log-normal, etc.) because we always have a limited number of devices under test (DUTs).
2. *Toward larger chip areas (sometimes referred to as area scaling).* TTF of a smaller-area (A_1) DUT is tested, which is extrapolated to a much larger chip area (A_2) based on the following Poisson relationship[26]

$$\frac{TTF_1}{TTF_2} = \left(\frac{A_2}{A_1}\right)^{1/\beta} \tag{1.2}$$

   where β is an area scaling factor.
3. *Toward operating conditions.* This extrapolation is done based on acceleration models for the specific failure mechanism from the stress conditions (i.e., high voltage/current and/or high temperature) to operating conditions (i.e., relatively low voltage/current and/or low temperature). Table 1.1 shows

Reliability Item	Acceleration Model
Gate-oxide TDDB	$\text{MTF} = AV^{-n}\exp\left(\frac{E_a}{k_B T}\right)$
Hot-carrier injection	$\text{MTF} = A(I_{\text{sub}})^{-m}\exp\left(\frac{E_a}{k_B T}\right)$
Negative-bias temperature instability	$\text{MTF} = A\exp(-B \cdot V_g)\exp\left(\frac{E_a}{k_B T}\right)$
Electromigration	$\text{MTF} = Aj^{-n}\exp\left(\frac{E_a}{k_B T}\right)$
Stress migration	$\text{MTF} = A(T_o - T)^{-n}\exp\left(\frac{E_a}{k_B T}\right)$
Intermetal-dielectric TDDB	$\text{MTF} = A\exp(-\gamma E)\exp\left(\frac{E_a}{k_B T}\right)$

TABLE 1.1 Typical Acceleration Models for Various Reliability Items

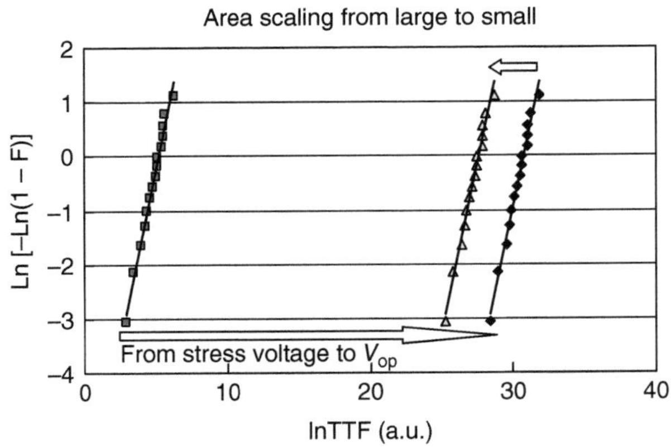

FIGURE 1.2 An example of a gate-oxide TDDB lifetime projection from stress conditions (voltage 32 V, 0.01 m) to operating conditions (voltage 3.63 V, 100 m).

some typical acceleration models for various reliability items. Therefore, the acceleration models are very critical for an accurate projection of lifetime from accelerated stress data to operating conditions. Indeed, these models, in conjunction with their reliability physics, have migrated along with the migration of the technology nodes because the models should properly reflect the physics of the reliability failure process. *The migration of the models for each reliability item is, thus, one of the focuses of this book.*

Figure 1.2 provides an example of IMD TDDB lifetime projection from stress conditions (voltage 32 V, 0.01 m) to operating conditions (voltage 3.63 V, 100 m). The voltage acceleration is based on the *E* model, whereas the area/length projection is based on the area scaling given in Eq. (1.2). *In this book, the test structures, the accelerating tests, and the corresponding physical/mathematical models will be elaborated for the reliability items in respective chapters.*

1.5 Organization of This Book

This book consists of three parts as shown in Fig. 1.3. Part 1 includes some background knowledge for understanding semiconductor process reliability. Chapter 1 provides an introduction to semiconductor process reliability, its importance, the reliability items, and an overall view of the book. Chapter 2 discusses the basic device physics that serve as base for understanding process-reliability mechanisms. Chapter 3 presents the basic semiconductor manufacturing process flow, including FEOL and BEOL flows, knowledge of which will assist readers in understanding the process effects on reliability performance. Chapter 4 includes some measurement methodologies to

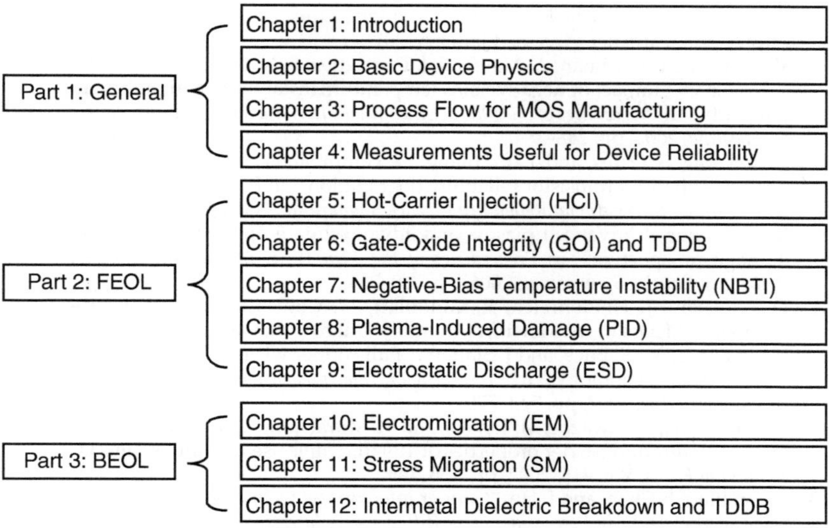

FIGURE 1.3 Organization of this book.

identify the interface trap and oxide trap in a gate-oxide dielectrics induced by reliability stressing and degradation, information about which is a prerequisite to understanding reliability mechanisms.

Part 2 discusses in detail FEOL reliability items, including hot-carrier injection (Chap. 5), gate-oxide integrity and TDDB (Chap. 6), negative-bias temperature instability (Chap. 7), plasma-induced damage (Chap. 8), and electrostatic discharge (Chap. 9).

Part 3 presents BEOL reliability items including electromigration, (Chap. 10), stress migration (Chap. 11), and intermetal dielectric breakdown and TDDB (Chap. 12).

1.6 References

1. "Process integration, devices, and structures," in *International Technology Roadmap for Semiconductors*, 2009 ed., p.19.
2. J. W. McPherson, "Reliability challenges for 45 nm and beyond," in *Proceedings of the IEEE 43rd Annual Design Automation Conference*, 2006, pp. 176–181.
3. C. Hu, S. C. Tam, F. C. Hsu, P.-K. Ko, T.-Y. Chan, and K. W. Terrill, "Hot-electron-induced MOSFET degradation-model, monitor and improvement," *IEEE Transactions on Electron Devices*, Vol. 48, No. 4, 1985, pp. 375–385.
4. P. Heremans, R. Bellens, G. Groeseneken, and H. E. Maes, "Consistent model for the hot-carrier degradation in *n*-channel and *p*-channel MOSFETs," *IEEE Transactions on Electron Devices*, Vol. 35, No. 12, 1988, pp. 2194–2209.
5. W. Wang, J. Tao, and P. Fang, "Dependence of HCI mechanism on temperature for 0.18-μm technology and beyond," *IEEE International Integrated Reliability Workshop Final Report*, 1999, pp. 66-68.
6. Y.-H. Lee, N. Mielke, M. Agostinelli, S. Gupta, R. Lu, and W. McMahon, "Prediction of logic product failure due to thin gate oxide breakdown," in *Proceedings of the 44th IEEE Annual International Reliability Physics Symposium*, 2006, pp. 18–28.

7. *International Technology Roadmap for Semiconductors*, 2006 ed.; available at http://public.itrs.net. Accessed January, 2011.
8. V. Reddy, A. T. Krishnan, A. Marshall, J. Rodriguez, S. Natarajan, T. Rost, and S. Krishnan, "Impact of negative bias temperature instability on digital circuit reliability," in *Proceedings of the 40th IEEE Annual International Reliability Physics Symposium*, 2002, pp. 248–254.
9. W. T. Weng, S. Oates, and T.-Y. Huang, "A comprehensive model for plasma damage enhanced transistor reliability degradation," in *Proceedings of the 45th IEEE Annual International Reliability Physics Symposium*, 2007, pp. 364–369.
10. K. P. Cheung, "Advanced plasma and advanced gate dielectric: A charging damage prospective," in *Proceedings of the 44th IEEE Annual International Reliability Physics Symposium*, 2006, pp. 360–364.
11. D. Edelstein, J. Heidenreich, R. Goldblatt, W. Cote, C. Uzoh, N. Lustig, P. Roper, T. McDevitt, W. Motsiff, A. Simon, J. Dukovic, R. Wachnik, H. Rathore, R. Schulz, L. Su, S. Luce, and J. Slattery, "Full copper wiring in a sub-0.25-μm CMOS ULSI technology," in *Proceedings of the International Electron Devices Meeting (IEDM)*, 1997, pp. 773–776.
12. M. Moriyama, T. Kawazoe, M. Tanaka, and M. Murakami, "Correlation between microstructure and barrier properties of TiN thin films used Cu interconnects," *Thin Solid Films*, Vol. 416, 2002, pp. 136–144.
13. J. Gambino, F. Chen, and J. He, "Copper interconnect technology for the 32-nm node and beyond," in *Proceedings of the IEEE Custom Intergrated Circuits Conference (CICC)*, 2009, pp. 141–148.
14. B. Li, J. Gill, C. J. Christiansen, T. D. Sullivan, and P. S. McLaughlin, "Impact of via-line contact on Cu interconnect EM performance," in *Proceedings of the 43rd IEEE Annual International Reliability Physics Symposium*, 2005, pp. 24–30.
15. E. T. Ogawa, K.-D. Lee, H. Matsuhashi, K.-S. Ko, P. R. Justison, A. N. Ramamurthi, A. J. Bierwag, P. S. Ho, V. A. Blasche, and R. H. Havemann, "Statistics of electromigration early failures in Cu/oxide dual-damascene interconnects," in *Proceedings of the 39th IEEE Annual International Reliability Physics Symposium*, 2001, pp. 341–349.
16. J. Gill, T. Sullivan, S. Yankee, H. Barth, and A. von Glasow, "Investigation of via-dominated multimodal electromigration failure distributions in dual damascene Cu interconnects with a discussion of the statistical implications," in *Proceedings of the 40th IEEE Annual International Reliability Physics Symposium*, 2002, pp. 298–304.
17. A. H. Fisher, A. von Glasow, S. Penka, and F. Ungar, "Process optimization: The key to obtain highly reliable Cu interconnects," in *Proceedings of the IEEE International Interconnect Technology Conference*, 2003, pp. 253–255.
18. J. W. Pyun, X. Lu, S. Yoon, N. Henis, K. Neuman, K. Pfeifer, and P.S. Ho, "Scaling effect on electromigration reliability for CU/low-*k* interconnects," in *Proceedings of the 43rd IEEE Annual International Reliability Physics Symposium*, 2005, pp. 191–194.
19. S. Yokogawa, Y. Kakuhara, and H. Tsuchiya, "Joule heating effects on electromigration in Cu/low-*k* interconnects," in *Proceedings of the 47th IEEE Annual International Reliability Physics Symposium*, 2009, pp. 837–843.
20. E. T. Ogawa, J. W. Mcpherson, J. A. Rosal, K.J. Dickerson, T.-C. Chiu, L. Y. Tsung, M. K. Jain, T. D. Bonifield, J. C. Ondrusek, and W. R. McKee, "Stress-induced voiding under vias connected to wide Cu metal leads," in *Proceedings of the 40th IEEE Annual International Reliability Physics Symposium*, 2002, pp. 312–321.
21. T. Kouno, I. T. Suzuki, S. Otsuka, T. Hosoda, I. T. Nakamura, I. Y. Mizushima, M. Shiozu, H. Matsuyama, K. Shono, H. Watatani, Y. Ohkura, M. Sato, S. Fukuyama, and M. Miyajima, "Stress-induced voiding under vias connected to "narrow" copper lines," in *Proceedings of the International Electron Devices Meeting (IEDM)*, 2005, pp. 187–190.
22. E. T. Ogawa, J. Kim, G. S. Haase, H. C. Mogul, and J. W. McPherson, "Leakage, breakdown, and TDDB characteristics of porous low-*k* silica-based interconnect dielectrics," in *Proceedings of the 41st IEEE Annual International Reliability Physics Symposium*, 2003, pp. 166–172.

23. Y. Zhou, G. Xu, T. Scherban, J. Leu, G. Kloster, and C.-I Wu, "Correlation of surface and film chemistry with mechanical properties," in *AIP Conference Proceedings of the International Conference on Characterization and Metrology for ULSI Technology*, Vol. 683, pp. 455–461.
24. F. Chen, O. Bravo, K. Chanda, P. McLaughlin, T. Sullivan, J. Gill, J. Lloyd, R. Kontra, and J. Aitken, "A comprehensive study of low-*k* SiCOH TDDB phenomena and its reliability lifetime model development," in *Proceedings of the 44th IEEE Annual International Reliability Physics Symposium*, 2006, pp. 46–53.
25. E. Morifuji, T. Kumamori, M. Muta, K. Suzuki, M. S. Krishnan, T. Brozek, X. Li, W. Asano, M. Nishigori, N. Yanagiya, S. Yamada, K. Miyamoto, T. Noguchi, and M. Kakumu, "New guideline for hydrogen treatment in advanced system LSI," *Symposium on VLSI Technology Digest of Technical Papers*, 2002, pp. 218–219.
26. T. Nigam, R. Degraeve, G. Groeseneken, M. M. Heyns, and H. E. Maes, "Constant current charge-to-breakdown: Still a valid tool to study the reliability of MOS structures?" in *Proceedings of the 36th IEEE Annual International Reliability Physics Symposium*, 1998, pp. 62–69.

CHAPTER 2

Basic Device Physics

This chapter briefly introduces the basic metal-oxide-semiconductor (MOS) theory. We start with basic semiconductor material property, cover the junction diode, which is part of the MOS transistor, and discuss the physics of the MOS capacitor and the metal-oxide-semiconductor field-effect transistor (MOSFET). Main physical effects are shown in this chapter. The reader is also encouraged to reference other textbooks to gain insight into basic semiconductor material properties, MOS devices, and MOS modeling for further understanding of the topics of this book.

2.1 Basic Material Property Introduction

2.1.1 Conductors, Semiconductors, and Insulators

For conductors (Fig. 2.1*a*), the conduction band is partially filled with electrons. When an electrical field is applied, the electrons in the partially filled bands are accelerated and gain energy with the capability to conduct electric current. On the other hand, for both semiconductors and insulators, there is a band gap between the conduction band (which is empty) and the valence band (which is full). In insulators, the band-gap energy E_g is larger than 3 eV (Fig. 2.1*b*). The electrons in the valence band have very little chance to jump to the conduction band via thermal excitation. In contrast, the band-gap energy in semiconductors is lower ($E_g < 2$ eV) (Fig. 2.1*c*), and a limited number of electrons in the valence band can gain enough energy to jump to the conduction band via thermal excitation (sometimes referred as an *intrinsic process*).

In an intrinsic semiconductor with no impurities added, the intrinsic carrier density (n for electron and p for hole) is

$$n = N_C \exp\left[-\frac{(E_C - E_F)}{k_B T}\right] \tag{2.1}$$

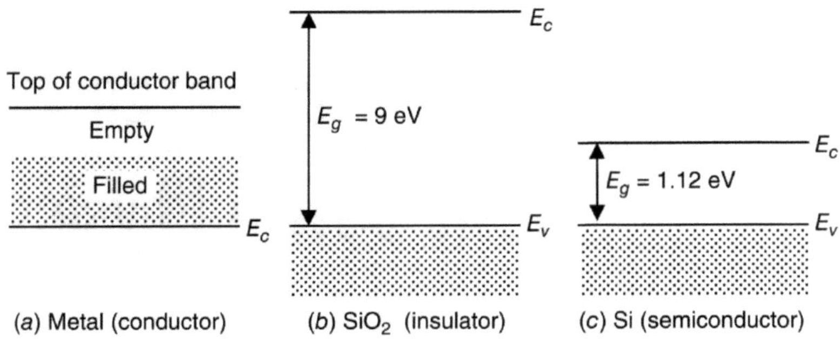

FIGURE 2.1 Schematic energy bands in (a) a conductor (metal as an example), (b) an insulator (SiO_2 as an example), and (c) a semiconductor (Si as an example).

$$p = N_V \exp\left[-\frac{(E_F - E_V)}{k_B T}\right] \tag{2.2}$$

$$N_C = 2\left(\frac{2\pi m_n k_B T}{h^2}\right)^{3/2} \tag{2.3}$$

$$N_V = 2\left(\frac{2\pi m_p k_B T}{h^2}\right)^{3/2} \tag{2.4}$$

where N_C and N_V are effective density of states of the conduction and valence bands, respectively. For Si at 300 K, N_C and N_V are 2.9×10^{19} and 2.7×10^{19} cm^{-3}, respectively. m_n and m_p are effective masses of electrons and holes, respectively.

Considering that $n = p$ in the intrinsic semiconductor, the intrinsic Fermi level E_i is expressed as

$$E_i = E_F = \frac{E_C + E_V}{2} + \frac{k_B T}{2}\ln\left(\frac{N_V}{N_C}\right) \tag{2.5}$$

where the second term in Eq. (2.5) is much smaller than the first term, which is defined as the midgap (E_{midgap}). In other words, in intrinsic semiconductors, E_F is close to the midgap [E_{midgap}, Eq. (2.6)]:

$$E_i = E_F \approx E_{\text{midgap}} \equiv \frac{E_C + E_V}{2} \tag{2.6}$$

The intrinsic carrier density n_i is defined when n and p are equal; using Eqs. (2.1) and (2.2), we have

$$n_i = n = p = \sqrt{N_C N_V}\exp\left(-\frac{E_g}{2k_B T}\right) \tag{2.7}$$

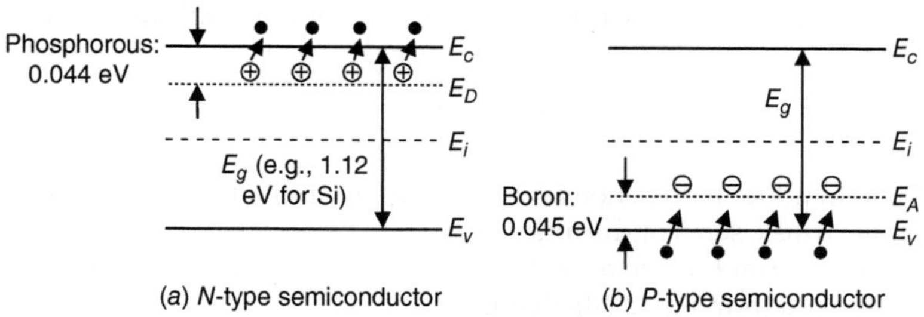

FIGURE 2.2 Schematic energy bands for (*a*) an *N*-type semiconductor with a donor dopant (e.g., phorphorous) and (*b*) a *P*-type semiconductor with an acceptor dopant (e.g., boron).

The *mass-action law* indicates that

$$np = n_i^2 \tag{2.8}$$

For Si at 300 K, the intrinsic carrier density n_i is 9.65×10^9 cm^{-3}.[1] Equation (2.7) shows that the intrinsic carrier density depends exponentially on the band-gap energy E_g and the temperature.

Semiconductors become useful only after special impurities (dopants) are added, and then they are referred as *extrinsic semiconductors*. There are basically two types of extrinsic semiconductors:

- *N-type semiconductors* (Fig. 2.2*a*) are obtained when elements from Column V of the periodic table are added (e.g., phosphorous, arsenic, and antimony). These elements have five valence electrons in their outmost orbit and are known as *donor impurities* because they can provide one electron after forming the covalent bonding. The donor impurities have an energy E_D located between E_i and E_C.
- *P-type semiconductors* (Fig. 2.2*b*) are obtained when elements from Column III of the periodic table are added (e.g., boron, gallium, and indium). These atoms have three electrons in their outermost orbit and are known as *acceptor impurities* because the dopant atom can accept an electron. The acceptor impurities have an energy E_A located between E_i and E_V.

For an extrinsic *N*-type semiconductor, the electron density n is

$$n = N_C \exp\left[-\frac{(E_C - E_F)}{k_B T}\right] = N_C \exp\left[-\frac{(E_C - E_i)}{k_B T}\right] \exp\left[\frac{(E_F - E_i)}{k_B T}\right] \tag{2.9}$$

$$\Rightarrow n = n_i \exp\left[\frac{(E_F - E_i)}{k_B T}\right] \tag{2.10}$$

Similarly, for an extrinsic *P*-type semiconductor, the hole density p is

$$p = n_i \exp\left[\frac{(E_i - E_F)}{k_B T}\right] \tag{2.11}$$

Note that the mass-action law [Eq. (2.8)] is applied to both intrinsic and extrinsic semiconductors.

The Fermi level in an *N*-type semiconductor (where $n \approx N_D$) can be derived from Eq. (2.10), that is,

$$E_F \cong E_i + k_B T \ln\left(\frac{N_D}{n_i}\right) \tag{2.12}$$

Similarly, the Fermi level in a *P*-type semiconductor (where $p \approx N_A$) can be derived from Eq. (2.11), that is,

$$E_F \cong E_i - k_B T \ln\left(\frac{N_A}{n_i}\right) \tag{2.13}$$

Note that the intrinsic carrier density n_i is on the order of 1×10^{10} cm^{-3}, whereas in practice, the dopant concentration N_D and N_A can be on the order of 1×10^{13} to 1×10^{16} cm^{-3}. Therefore, the Fermi level in an *N*-type semiconductor is higher than E_i and closer to the conduction band, whereas the Fermi level in a *P*-type semiconductor is lower than E_i and closer to the valence band.

2.1.2 Electron and Hole Energies

An electron-hole pair can be generated owing to, for example, hot-electron-induced impact ionization (see Chap. 5) in a semiconductor. In the energy-band diagram, the generated electron may gain enough energy (larger than the band gap E_g) and then jump into the conduction band. Therefore, the generated electron and hole are separated as shown schematically in Fig. 2.3*a*.

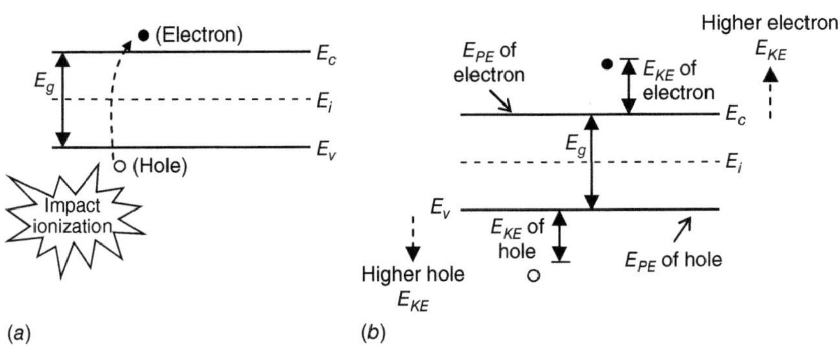

FIGURE 2.3 (*a*) After an electron-hole pair is generated, the energetic electron likely jumps into the conduction band. (*b*) Schematic illustration of the kinetic energy E_{KE} and potential energy E_{PE} of an electron and a hole and their energy directions.

The energy of an electron (or hole) can be divided into two parts, (i.e., potential energy E_{PE} and kinetic energy E_{KE}), as shown schematically in Fig. 2.3*b*. The E_{PE} of an electron is the conduction-band energy E_C, and the E_{KE} of an electron is the difference between the energy location and E_C. Similarly, the E_{PE} of a hole is the valence-band E_V and the E_{KE} of a hole is the difference between the energy location and E_V. When an electron moves upward (e.g., caused by electrical field), its E_{KE} increases, whereas its E_{KE} stays at E_C if there is no band bending. Similarly, when hole moves downward, its E_{KE} increases, whereas its E_{PE} stays at E_V if there is no band banding. If there is band bending, the E_{PE} of an electron or hole will change, following the band bending, as shown in Figs. 2.4 and 5.1.

2.1.3 Collision and Energy Exchange in Semiconductors

Figure 2.4*a*, *b* schematically illustrates the energy transfer in a vacuum with length *L*. An electron located at point *A* has a potential energy (PE) of qV owing to the electrical field applied. When this electron is released, it will be accelerated by the electrical field, during which it will gain kinetic energy (KE). Since the electron is traveling within a vacuum, no collision takes place along the way, and there is no energy loss. According to the law of energy conservation, during travel, the electron's potential energy decreases, whereas its kinetic energy increases. In other words, it is the potential energy of the electron that converts to the kinetic energy along the way. At point *B*, all the potential energy is converted to kinetic energy when the electron travels from point *A* to point *B*. However, the electron at point *B* will lose all

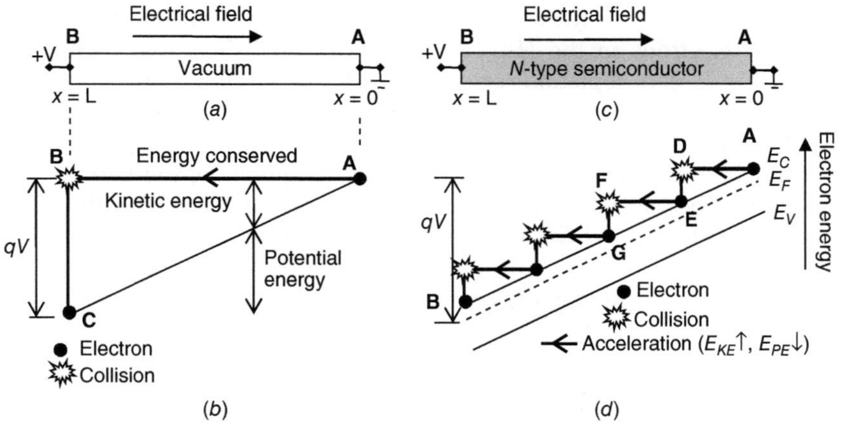

FIGURE 2.4 Schematic illustration of conduction in a vacuum: (*a*) vacuum with voltage applied; (*b*) the motion of the electron and collision within an energy-band diagram. Schematic illustration of conduction in an *N*-type semiconductor: (*c*) *N*-type semiconductor with voltage applied; (*d*) the motion of the electron and collision within an energy-band diagram.

its kinetic energy, which is, in turn, converted into heat attributed to the collision at this terminal. The energy loss is shown schematically in Fig. 2.4*b*, where energy drops from point *B* to point *C*.

In contrast, Fig. 2.4*c*, *d* schematically shows the situation for an *N*-type semiconductor with +V applied to point *B* and grounded to point *A*, where E_C, E_F and E_V are shown. In Fig. 2.4*d*, the bottom of the conduction band at E_C is the level of the potential energy of an electron whose kinetic energy is zero. Therefore, similar to the case for vacuum, the energy-band diagram of a semiconductor for an electron in Fig. 2.4*d* shows a decreasing potential energy and hence a decreasing level of E_C (and thus E_F and E_V) from point *A* to point *B*.

However, the difference between the vacuum and the semiconductor is clear: The accelerated electron in a semiconductor (e.g., at point *D*) may lose its kinetic energy in due course owing to the lattice collision, and thus the electron drops to the corresponding level E_C at point *E*. It is accelerated again by the electrical field applied, collides with the lattice at point *F*, and drops back to the lower E_C level at point *G*. This process continues until the electron reaches point *B*, where it will lose all the qV potential energy that it had at point *A*. Note that in this case (Fig. 2.4*d*), the potential energy has been converted into heat in the lattice along the line of travel rather than being converted into heat at point *B* in the vacuum.

2.2 *PN* Junctions

2.2.1 The Energy Band of *PN* Junctions

When *N*-type and *P*-type semiconductors are connected to each other to form a *PN* junction, the holes in the *P*-type semiconductor and the electrons in the *N*-type semiconductor diffuse into each other (Fig. 2.5). A depletion region then is formed in the connection region, and a built-in electrical field is established that will block further interdiffusion of the holes and electrons. Note that in equilibrium, where no current flows through the *PN* junction, the Fermi level must be constant throughout the *PN* junction. Therefore, the energy bands (E_V and E_C) are bent in the *PN* junction to meet this requirement (Fig. 2.5*c*).

The built-in voltage V_{bi}, also known as the *barrier voltage*, is derived as

$$V_{bi} = \frac{k_B T}{q} \ln\left(\frac{N_A N_D}{n_i^2}\right) \tag{2.14}$$

It is clear that V_{bi} depends on the doping of the *P* and *N* regions (N_A and N_D) and on temperature and bandgap through n_i^2 [Eq. (2.7)]. As an example, the barrier voltage V_{bi} for an abrupt silicon junction varies between 0.7 and 1.0 V.

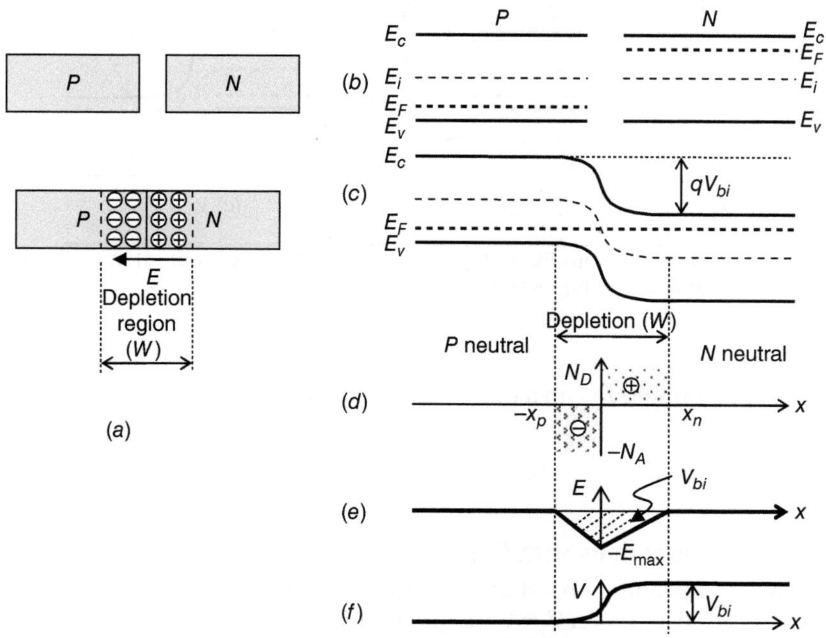

FIGURE 2.5 Energy band when *N*-type and *P*-type semiconductors are connected to each other to form a *PN* junction. (*a*) Schematic for the connection under balanced conditions; (*b*) energy band before connection; (*c*) band bending after connection; (*d*) space charge; (*e*) distribution of electrical field; and (*f*) distribution of potential.

The width of the depletion W at equilibrium is

$$W = \sqrt{\frac{2\varepsilon}{q}\left(\frac{1}{N_A}+\frac{1}{N_D}\right)V_{bi}} = \sqrt{\frac{2\varepsilon k_B T}{q^2}\left(\frac{1}{N_A}+\frac{1}{N_D}\right)\ln\left(\frac{N_A N_D}{N_C N_V \exp\left(-\frac{E_g}{k_B T}\right)}\right)} \tag{2.15}$$

The maximum electrical field E_{max} (Fig. 2.5*e*) at equilibrium owing to the built-in voltage is

$$E_{max} = \frac{2V_{bi}}{W} \tag{2.16}$$

For a *one-side abrupt junction* (e.g., a p^+-n junction where $N_A >> N_D$), $x_p << x_n$, and W in Eq. (2.15) can be simplified as

$$W \approx x_n = \sqrt{\frac{2\varepsilon}{q}\frac{V_{bi}}{N_D}} \tag{2.17}$$

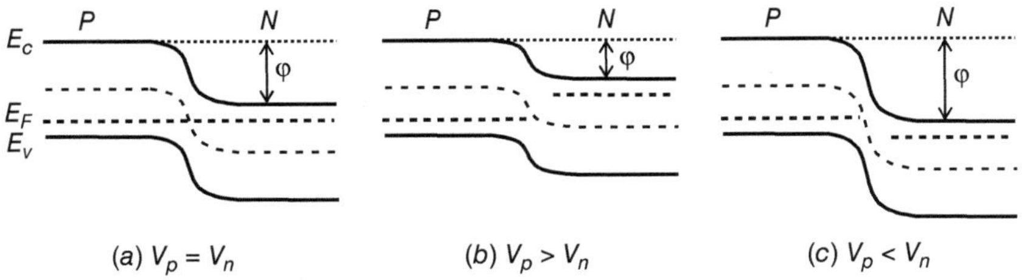

FIGURE 2.6 Energy-band diagram for a *PN* junction with voltage applied: (*a*) at equilibrium; (*b*) forward biased; (*c*) reverse biased.

The corresponding E_{max} in a one-side abrupt junction is expressed as

$$E_{\max} = \frac{qN_D W}{\varepsilon} \tag{2.18}$$

2.2.2 *PN* Junctions with Bias

Figure 2.6 is an energy-band diagram for a *PN* junction with different voltages applied. When $V_p = V_n$ (Fig. 2.6*a*), the *PN* junction is balanced with V_{bi} on the depletion region, and no current flows through it. However, with an externally applied voltage V_a (= $V_p - V_n$), the voltage across the depletion region V_j turns to be

$$V_j = V_{\text{bi}} - (V_p - V_n) = V_{\text{bi}} - V_a \tag{2.19}$$

Note that V_a is positive for forward bias and negative for reverse bias. Thus the depletion region width *W* in Eq. (2.15) is modified by replacing V_{bi} with V_j as

$$W = \sqrt{\frac{2\varepsilon}{q}\left(\frac{1}{N_A} + \frac{1}{N_D}\right)V_j} = \sqrt{\frac{2\varepsilon}{q}\left(\frac{1}{N_A} + \frac{1}{N_D}\right)(V_{bi} - V_a)} \tag{2.20}$$

Therefore, when $V_a > 0$ (i.e., $V_p > V_n$) (Fig. 2.6*b*), the barrier voltage is lowered by the forward-biased voltage, and the depletion region is narrower, resulting in more electrons and holes diffusing through the junction. When $V_a < 0$ (i.e., $V_p < V_n$) (Fig. 2.6*c*), the barrier voltage is higher owing to the reverse-biased voltage. Reverse current forms as a result of to the drift of the minority electrons in the *P* region to the *N* region, and the drift of the minority holes in the *N* region to the *P* region. However, the reverse current is small owing to the minority characteristics of the electrons and holes.

2.2.3 Junction Capacitance

For *PN* junctions, the charges in the *P* and *N* regions [or the depletion region-width *W* in Eq. (2.20)] are changed with the change in biased

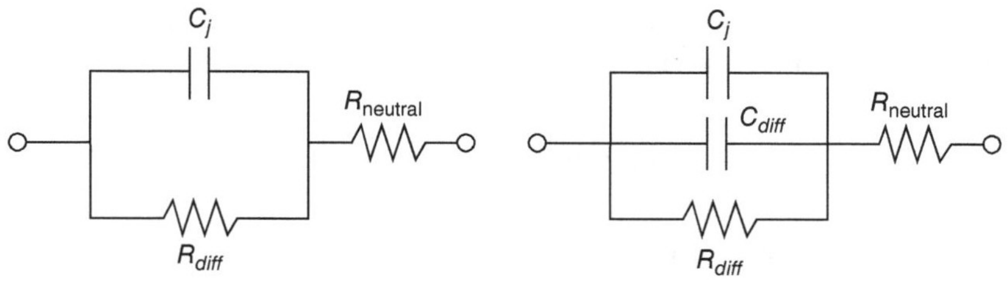

FIGURE 2.7 Equivalent circuits of a *PN* junction considering both parasitic resistance and capacitance under (*a*) reverse bias and (*b*) forward bias.

voltage. The *PN* junction thus acts as a capacitor. The capacitance is defined as

$$C = \frac{dQ}{dV} \tag{2.21}$$

Figure 2.7 gives the small-signal equivalent circuits of a *PN* junction considering both parasitic resistance and capacitance. For a reverse-biased junction ($V_a < 0$) (Fig. 2.7*a*), the change in Q comes from the change in the depletion region, which gives the junction capacitance C_j or transition capacitance. The physical characteristics of C_j are similar to those of a parallel-plate capacitor. Therefore, C_j is expressed simply as

$$C_j = \frac{\varepsilon A}{W} \tag{2.22}$$

where A is the cross-sectional area of the junction. C_j is inversely correlated with the depletion width W, which is a function of the concentrations of the N and P regions, as given in Eq. (2.20). Thus a higher concentration has a thinner depletion width and gives a higher C_j. In Fig. 2.7*a*, R_{neutral} is the bulk resistance of the neutral region. R_{diff} is the differential resistance (sometimes referred to as *incremental resistance* or *slope resistance*). In the reverse direction, R_{diff} is very high and often may be assumed to be infinite. However, R_{diff} in forward bias (Fig. 2.7*b*) is a very small quantity.

On the other hand, for a forward-biased junction ($V_a > 0$) (Fig. 2.7*b*), the depletion region changes slightly, and the change in Q comes mainly from the diffusion of electrons and holes, resulting in the so-called diffusion capacitance C_{diff}, which is sometimes referred as *storage capacitance*. The reason for this is that with forward bias, the reduction in the depletion width leads to an injection of majority carriers across the depletion region into the opposite region, where they are stored as excess minority carriers. The density of the excess stored minority carriers increases with an increase in the forward bias. The diffusion capacitance is expressed as

$$C_{diff} \propto \exp\left(\frac{qV_a}{k_B T}\right) \tag{2.23}$$

That is, C_{diff} increases exponentially with forward bias. Note that in forward bias, C_{diff} is significantly larger than C_j.

2.3 Basic Physics of Metal-Oxide-Semiconductor Capacitors

2.3.1 The Energy Band of Metal-Oxide-Semiconductor Capacitors

Before digging into the complex MOS transistor, let's first introduce the basic MOS capacitor, as illustrated schematically in Fig. 2.8. The basic MOS capacitor is comprised of an insulator layer (such as SiO_2) sandwiched between the gate (either polysilicon or metal) and the silicon substrate.

Figure 2.9 is an energy-band diagram for an ideal basic MOS capacitor. When a voltage is applied to the MOS capacitor, the energy levels E_C, E_i, and E_V are bent because some of the applied voltage is dropped in the semiconductor at the interface with the oxide. On the other hand, *since there is no current in the MOS capacitor, the position of the Fermi level in silicon does not change.*

- *Flat band:* When the gate voltage V_g is at the flat-band voltage V_{fb}, the energy band is flat.

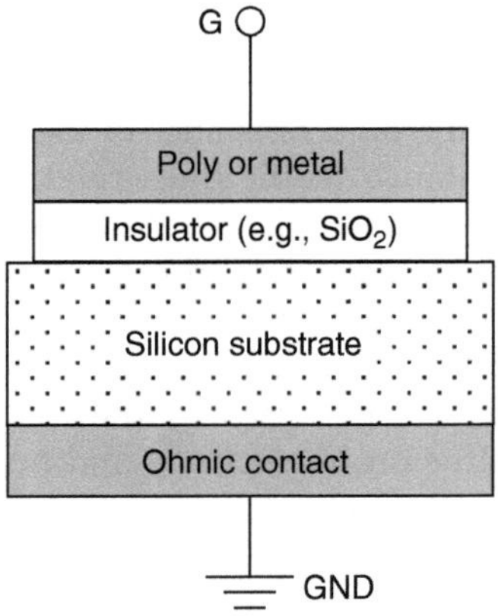

FIGURE 2.8 Schematic of basic MOS capacitor.

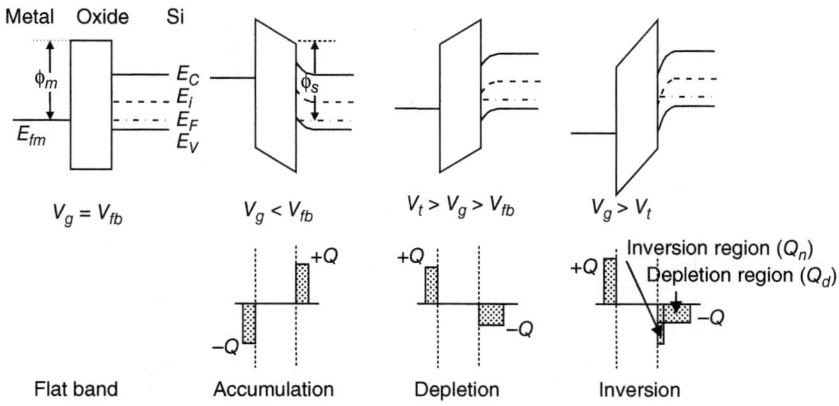

FIGURE 2.9 Energy-band diagram for an ideal basic MOS capacitor.

- *Accumulation:* When $V_g < V_{fb}$, the majority holes accumulate at the interface between the oxide and the Si.
- *Depletion:* When $V_t > V_g > V_{fb}$, the majority holes are driven far away from the interface. At the interface, a depletion layer is formed with E_i at the interface still be above E_F.
- *Inversion:* When $V_g > V_t$, more band bending takes place, and E_i at the interface goes below E_F. The majority carriers turn out to be electrons.

The flatband voltage V_{fb} is a voltage applied between the metal and the silicon to achieve flattening of the bands in order to compensate for the work-function difference ϕ_{ms} and the oxide charges Q_i:

$$V_{fb} = \phi_{ms} - \frac{Q_i}{C_{ox}} \tag{2.24}$$

where Q_i is the trapped charge in the gate oxide (a positive quantity), including fixed charge Q_f, interface trap Q_{it}, mobile charge Q_m, and oxide trap Q_{ot}. ϕ_{ms} $(= \phi_m - \phi_s)$ is the work-function difference depending on the type of metal or poly used and on the density/type of the dopant atoms in the silicon.

In the inversion mode, the gate voltage V_g applied can be divided into the following three components (both V_{ox} and ψ_s are shown in Fig. 2.10):

$$V_g = V_{ox} + \psi_s + V_{fb} \tag{2.25}$$

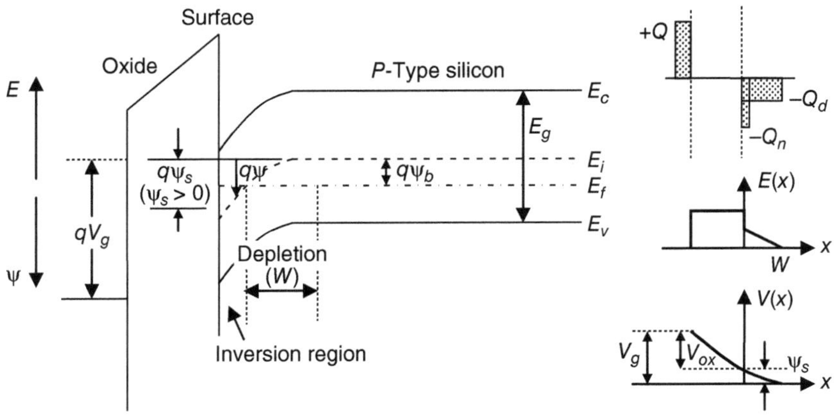

FIGURE 2.10 Band diagram at the surface between oxide and *p*-type silicon when a positive bias is applied to the gate.

The threshold voltage V_t is defined as the turn-on voltage that is applied to the gate to cause strong inversion, with the following quantities defined as

$$\psi_s = 2\psi_b \tag{2.26}$$

$$V_{ox} = \frac{-Q_s}{C_{ox}} = \frac{-Q_{dm} - Q_n}{C_{ox}} \approx \frac{-Q_{dm}}{C_{ox}} \tag{2.27}$$

$$Q_{dm} = -qN_A W_m = -\sqrt{2\varepsilon_{Si} q N_A (2\psi_b)} \tag{2.28}$$

where W_m and Q_{dm} are the maximum values of W (depletion width) and Q_d (charge density in the depletion layer) under strong inversion, respectively. Q_n is the charge density in the inversion layer, which is small and is assumed to be zero. Therefore, the V_t is derived when there is no substrate bias based on Eqs. (2.25) to (2.28):

$$V_t = V_{fb} + 2\psi_b + V_{ox} = \phi_{ms} - \frac{Q_i}{C_{ox}} + 2\psi_b + \frac{\sqrt{2\varepsilon_{Si} q N_A (2\psi_b)}}{C_{ox}} \tag{2.29}$$

2.3.2 Metal-Oxide-Semiconductor Capacitor Capacitance-Voltage Curve

Figure 2.11 gives the capacitance-voltage CV characteristics of a *P*-type-substrate MOS capacitor. The MOS capacitance is given by

$$C = \frac{dQ_m}{dV_g} \tag{2.30}$$

$$Q_m = -Q_s = -(Q_n + Q_d) \tag{2.31}$$

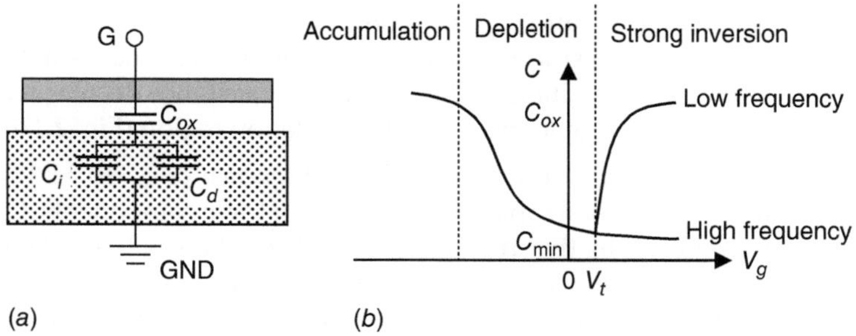

FIGURE 2.11 (*a*) The capacitance components of a MOS capacitor. (*b*) Capacitance-voltage characteristics of a *P*-type-substrate MOS capacitor.

Thus, combining Eqs. (2.30), (2.31), and (2.25) gives

$$C = \frac{dQ_m}{d(V_{ox} + \psi_s)} = \frac{1}{\dfrac{1}{\dfrac{dQ_m}{dV_{ox}}} + \dfrac{1}{\dfrac{dQ_m}{d\psi_s}}} = \frac{1}{\dfrac{1}{C_{ox}} + \dfrac{1}{C_s}} \tag{2.32}$$

$$C_s = \frac{dQ_n}{d\psi_s} + \frac{dQ_d}{d\psi_s} = C_i + C_d \tag{2.33}$$

where C_{ox} and C_s are the capacitances of the oxide and the semiconductor, respectively. C_s consists of the depletion-layer capacitance C_d and the inversion-layer capacitance C_i. The capacitance does vary when the gate voltage V_g changes, as shown in Fig. 2.11*b*. The details are given below:

- Under the *accumulation mode*, the accumulated holes in the interface Q change exponentially with the ψ_s change. According to Eq. (2.32), the corresponding capacitance C_s is very large. Therefore, the total C is approximately C_{ox}.
- Under the *depletion mode*, the change in the charge in the depletion layer varies, and thus the corresponding C_s plays an important role, as described below:

$$Q_s = Q_d = -qN_A W = -\sqrt{2\varepsilon q N_A \psi_s} \tag{2.34}$$

$$C_s = \frac{-dQ_s}{d\psi_s} = \sqrt{\frac{\varepsilon q N_A}{2\psi_s}} = \frac{\varepsilon}{W} \tag{2.35}$$

$$\frac{1}{C} = \frac{1}{C_s} + \frac{1}{C_{ox}} = \frac{W}{\varepsilon} + \frac{1}{C_{ox}} \tag{2.36}$$

where the C_s and C_{ox} are comparable. An increase in V_g increases ψ_s, resulting in a wider depletion layer and a smaller depletion capacitance. Thus the total capacitance decreases as the gate voltage increases in the depletion region (Fig. 2.11*b*).

- Under the *inversion mode with low-frequency* (e.g., 10 Hz) V_g applied, the total capacitance consists of three components, C_{ox}, C_i, and C_d. In this region, the width of the depletion layer reaches its maximum W_m, which will not change with V_g. Therefore, C_d in this region is a constant. On the other hand, the density of the inversion electrons is proportional to exp $(K \cdot \psi_s)$, resulting a very large C_i (note that this is similar to the situation for the accumulation holes in the accumulation mode). Since C_i and C_d are in parallel, the resulting C_s $(= C_d + C_i)$ is also very large compared with C_{ox}. Therefore, the total C under the inversion mode with low frequency is again proximately C_{ox}.
- Under the *inversion mode with high-frequency* (e.g., 1 MHz) V_g applied, owing to the slow generation-recombination process (at a rate of 10^{-5} to 10^{3} s) not responding fast enough to the alternating-current (ac) signal whose period is 1 μs, the density of the inversion electrons is fixed at the direct-current (dc) value, and the depletion layer width will increase and decrease about its dc value. As a result, the total capacitance with high frequency is given by $C_{\min}$ as

$$\frac{1}{C_{\min}} = \frac{1}{C_{ox}} + \frac{W_m}{\varepsilon} \tag{2.37}$$

2.4 Physics of Metal-Oxide-Semiconductor Field-Effect Transistors

2.4.1 The Current-Voltage Characteristics of the Metal-Oxide-Semiconductor Field-Effect Transistor

The derivation of the $I_d - V_d$ analytical relationship for the MOSFET can let us better understand MOSFET physics. For simplification, assumptions are generally made as follows:

- The source and the substrate are grounded (i.e., $V_s = V_b = 0$).
- The surface mobility is a constant along the channel.
- V_t is defined as the gate voltage applied for generation of an inversion layer at $V_d = 0$.
- *Gradual channel approximation* (GCA) for the channel carrier density is determined by the gate-to-channel potential based on the assumption that the potential gradient from drain to source is much smaller than that from gate to channel.

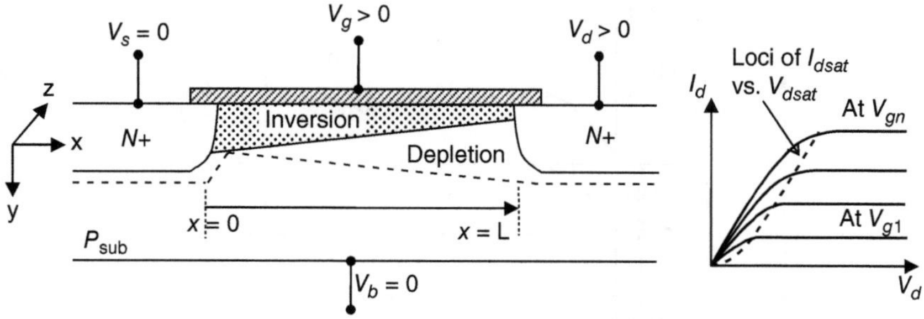

FIGURE 2.12 Schematic ideal NMOS with positive bias voltages at gate and drain.

Figure 2.12 schematically shows an ideal NMOSFET with bias voltages at gate and drain for the analysis. Following Eq. (2.25), the gate voltage V_g applied can be divided into three components as

$$V_g = V_{ox} + \psi_s(x) + V_{fb} \tag{2.38}$$

where V_{ox} is the voltage across the oxide given by $|Q_s|/C_{ox}$, and Q_s $(= Q_n + Q_{dm})$ is the negative charge density for NMOSFETs in silicon and is the summation of the electron surface charge Q_n in the channel and the negative depletion-layer charge Q_{dm}, V_{fb} is the flat-band voltage, and $\psi_s(x)$ is the surface potential in silicon, which is $2\psi_b$ (as defined in Fig. 2.10) at the source under strong inversion. Note that this is based on the Fermi level E_F at the substrate in the bulk and $\psi_b = (E_i - E_F)/q$. However, when moving along the channel from the source ($x = 0$) to the drain ($x = L$), the E_F in the channel moves down owing to the channel voltage $V(x)$, which represents the voltage drop V_{cb} $(= V_{\text{channel}} - V_{\text{bulk}})$ from channel to bulk (ground here). The condition for strong inversion becomes $(2\psi_b + V_{cb})$, where $V_{cb} = V_{cs} + V_{sb} = V_{cs}$ because $V_{sb} = 0$. Therefore, as illustrated in Fig. 2.13, we get

$$\psi_s(x) = 2\psi_b + V(x) \tag{2.39}$$

and then Eq. (2.38) becomes

$$V_g = -\frac{Q_n(x) + Q_{dm}(x)}{C_{ox}} + V_{fb} + 2\psi_b + V(x) \tag{2.40}$$

and the electron surface charge Q_n in the channel is

$$Q_n(x) = -C_{ox}\left[V_g - V(x) - V_{fb} - 2\psi_b\right] - Q_{dm}(x) \tag{2.41}$$

$$\Rightarrow Q_n(x) = -C_{ox}\left[V_g - V(x) - V_t\right] \tag{2.42}$$

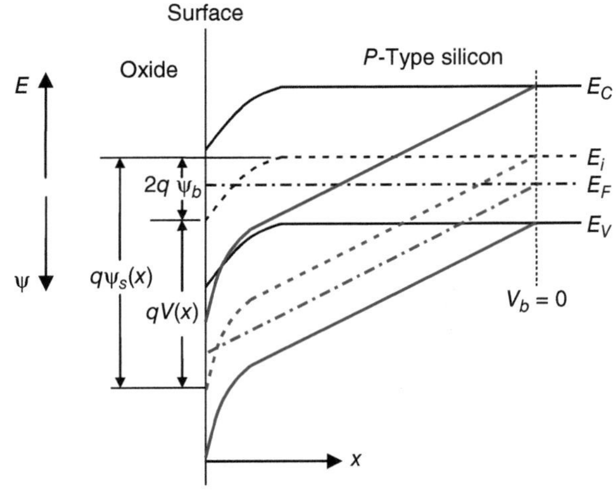

Figure 2.13 Band diagram at position x with channel bulk voltage $V(x)$ in Fig. 2.12 for the surface between oxide and p-type silicon.

According to Eq. (2.42), a larger $|V_t|$ will lead to a smaller carrier population in the inversion channel ($|Q_n|$). In case $V_d = V_s = 0$ [e.g., for the negative-bias temperature instability (NBTI) stress condition], $V(x)$ thus is 0. Therefore, the carrier population in the inversion channel ($|Q_n|$) is proportional to C_{ox} and $|V_g - V_t|$:

$$|Q_n| = C_{ox} \cdot |V_g - V_t| \tag{2.43}$$

The drain current I_d in the linear region at a specific V_g is approximated as

$$I_d \approx \mu_{eff} C_{ox} \frac{Z}{L}\left[(V_g - V_t)V_d - \frac{V_d^2}{2}\right] \tag{2.44}$$

$$\approx \mu_{eff} C_{ox} \frac{Z}{L}\left[(V_g - V_t)V_d\right], \quad V_d << (V_g - V_t) \tag{2.45}$$

where Z is the channel width. The empirical equation for the effective surface mobility (μ_{eff}) is

$$\mu_{eff} = \frac{\mu_0}{1 + \theta(V_g - V_t)} \tag{2.46}$$

where μ_0 is the bulk mobility, and θ is a constant. It is seen that a larger V_g and a smaller V_t will lead to a lower effective surface mobility.

The transconductance g_m in the linear region is

$$g_m \equiv \left.\frac{\partial I_d}{\partial V_g}\right|_{Vd=\text{constant}} \approx \mu_{eff} C_{ox} \frac{Z}{L} V_d \tag{2.47}$$

The saturation drain current I_{dsat} is given at $V_d = V_{dsat} = V_g - V_t$ and is expressed as

$$I_{dsat} \approx \mu_{eff} C_{ox} \frac{Z}{2L} (V_g - V_t)^2 \tag{2.48}$$

Inputting Eq. (2.46) into Eq. (2.48), we get

$$I_{dsat} \approx \mu_0 C_{ox} \frac{Z}{2L} \frac{(V_g - V_t)^2}{1 + \theta(V_g - V_t)} \tag{2.49}$$

$$\frac{\Delta I_{dsat}}{I_{Idsat}} = \left[1 + \frac{1}{1 + \theta(V_g - V_t)}\right] \frac{\Delta V_t}{V_g - V_t} \tag{2.50}$$

It is seen that with smaller voltage headroom (defined as the difference between V_g and V_t),[3] a *smaller V_t shift (ΔV_t) can lead to larger I_{dsat} shift ($\Delta I_{dsat}/I_{dsat}$)*. This is very critical when the operating voltages of devices are continuously scaled down for advanced technology nodes.

In terms of wafer acceptance test (WAT) characterization, some common practices are listed below. The I_{dsat} is the drain current measured at $V_d = V_g = V_{op}$ (i.e., operating voltage), with source and bulk grounded. For V_t measured at constant current, V_{t_lin} (or V_{thi}) is the gate voltage when $I_d = 0.1{*}W/L$ µA at $V_d = 0.1$ V (or 0.05 V), and V_{t_sat} is the gate voltage when $I_d = 0.1{*}W/L$ µA at $V_d = V_{op}$. The off-state current I_{off} is the I_d measured under $V_g = V_s = V_b = 0$ and $V_d = V_{op}$ (or $1.1{*}V_{op}$).

2.4.2 Long-Channel V_t of the Metal-Oxide-Semiconductor Field-Effect Transistor

Equation (2.29) is the expression for the long channel V_t for a MOSFET without substrate bias (i.e., $V_b = 0$ V). Therefore, the following factors can affect V_t:

- Adjusting channel implantation, providing the additional charge needed for formation of the inversion layer [Eq. (2.51)]. A higher implantation dosage Q_{imp} gives a higher V_t.

$$V_t = V_{fb} + 2\psi_b + \frac{\sqrt{2\varepsilon_{Si} q N_A (2\psi_b)}}{C_{ox}} + \frac{Q_{imp}}{C_{ox}} \tag{2.51}$$

	ϕ_{ms}	$2_{\Psi b}$	$-Q_{dm}/C_{ox}$	V_t
NMOS (with n + poly/p-Si substrate)	−	+	+	+
PMOS (with p + poly/n-Si substrate)	+	−	−	−

TABLE 2.1 A Qualitative Comparison for the Dual-Doped Polysilicon Gate Process

- Adjusting C_{ox} through gate-oxide thickness and the k value.
- Adjusting V_{fb} by adjusting ϕ_{ms} through poly predoping. Poly predoping also can adjust the C_{ox}. A higher predoping dosage leads to a lower V_t.

Table 2.1 gives a qualitative comparison for the dual-doped polysilicon gate process, where the oxide charge Q_i and charge caused by channel implantation Q_{imp} are ignored.

2.4.3 The Capacitances in a Metal-Oxide-Semiconductor Field-Effect Transistor

The capacitances that are associated with an MOS transistor can be categorized into intrinsic and extrinsic, as shown in Fig. 2.14. These capacitances reduce the voltage gain at high frequencies and should be well understood and well evaluated for controlling device performance. A simple MOSFET model includes intrinsic capacitance of gate to source C_{gs}, gate to drain C_{gd}, gate to body C_{gb}, and the extrinsic junction capacitances of drain to body C_{db}, source to body C_{sb}, and the overlap capacitances of C_{gdo}, C_{gso}, and C_{gbo}.

It is relatively easy to understand the definition of extrinsic capacitance and how to formulate it. However, it may take more steps to gain insights into the physics of intrinsic capacitance and its modeling.

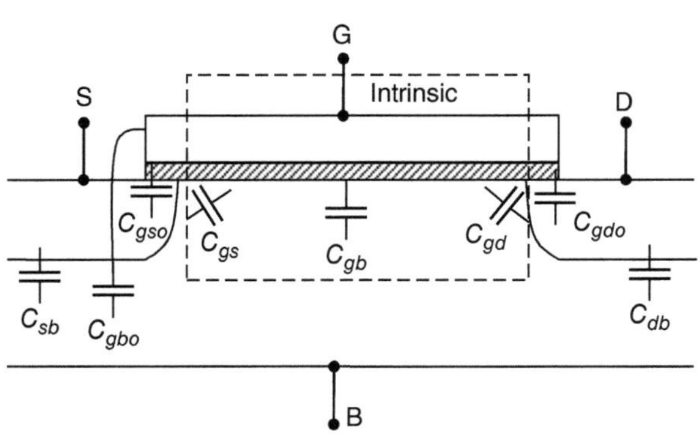

FIGURE 2.14 Schematic illustration of intrinsic and extrinsic capacitances in an MOS transistor.

A MOSFET capacitance will be the sum of the intrinsic and extrinsic capacitances put together.

Extrinsic Capacitance and Its Modeling

As indicated earlier, extrinsic capacitance consists of gate overlap capacitance in the drain/source region and drain/source junction capacitance. In general, an *overlap capacitance* is defined as

$$C_{ov} = \frac{\varepsilon \cdot d}{t_{ox}} \tag{2.52}$$

where t_{ox} is the oxide thickness and d is the width of the overlap region. Since for each particular technology, d is a fixed value, therefore we can simplify the equation as

$$\begin{aligned} C_{ov,gd} &= \frac{\varepsilon \cdot d}{t_{ox}} = C_{gdo} \cdot W \\ C_{ov,gs} &= \frac{\varepsilon \cdot d}{t_{ox}} = C_{gso} \cdot W \end{aligned} \tag{2.53}$$

There is an additional but smaller parasitic capacitance component in MOSFETs, namely, the gate-to-bulk capacitance, which is caused by the MOSFET end-cap regions overlaping the gate extension to the bulk and is defined in the same way as above:

$$C_{ov,gb} = C_{gbo} \cdot L \tag{2.54}$$

As the technology feature size shrinks smaller, there are two other extrinsic parasitic capacitances that cannot be ignored. These are (1) the inner fringing capacitance between the gate and the sidewall of the drain and source and (2) the outer fringing capacitance between the gate and drain/source area (excluding the gate-to-contact coupling capacitance). The equations for these capacitances are defined.

$$C_{fo} = \frac{\varepsilon}{\alpha} \ln\left(1 + \frac{X_p}{t_{ox}}\right) \tag{2.55}$$

$$C_{fi} = \frac{\varepsilon}{\beta} \ln\left(1 + \frac{X_j \sin\beta}{t_{ox}}\right) \tag{2.56}$$

where X_p is the polysilicon thickness, X_j is the junction depth, and α is the slope angle of the gate.

Other extrinsic components in a MOSFET are junction capacitances of drain to body and drain to source; the detailed equations for these capacitances can be referred to the *PN* junction (Sec. 2.2.3).

Intrinsic Capacitance and Its Modeling

The intransic capacitances C_{gd} and C_{gs} are the gate-to-drain and gate-to-source capacitances, and C_{gb} is the gate-to-bulk capacitance. The capacitances are distributed in nature. Since a detailed descrption of intrinsic capacitance is beyond the scope of this book, readers are encourage to refer to textbooks such as the Berkeley Short-channel Insulated-gate-field-effect-transistor Model (BSIM) user's guide for further information.

2.5 The Secondary Effects of Metal-Oxide-Semiconductor Field-Effect Transistors

2.5.1 Short-Channel Effect

Charge Sharing and Drain-Induced Barrier Lowering

Figure 2.15*a* shows a charge-sharing model for a short-channel device. The dotted-triangle depletion regions are not caused by V_g (as denoted by the shaded trapezoidal area) but rather are caused by source/drain (S/D) junction depletion. For a long-channel device, these parts are negligible. However, for short-channel devices, they are comparable with the trapezoidal area that was caused by V_g. Therefore, for a short-channel device, V_t becomes lower and lower with shorter channel length, which is commonly referred as V_t *roll-off*. Figure 2.15*b* shows V_t roll-off for both linear (at $V_d = 0.1$ V) and saturation (at $V_d = 1.1{*}V_{op}$) regions. It is clear that the saturation–V_t roll-off is much larger. Note that V_t roll-off also increases off-state leakage current (or standby current) for short-channel devices.

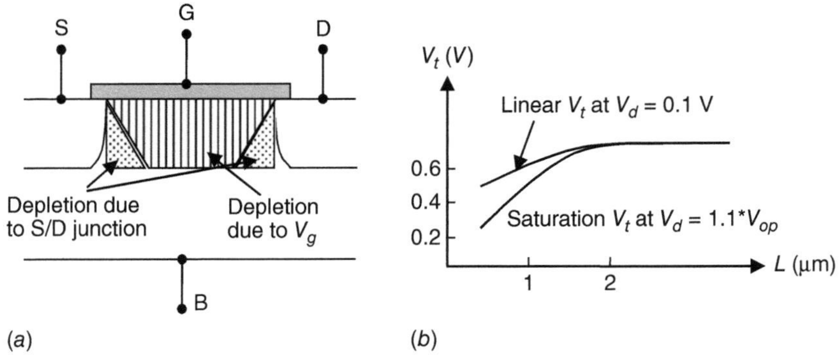

FIGURE 2.15 (*a*) Charge-sharing model for a short channel device. (*b*) Short-channel threshold-voltage roll-off.

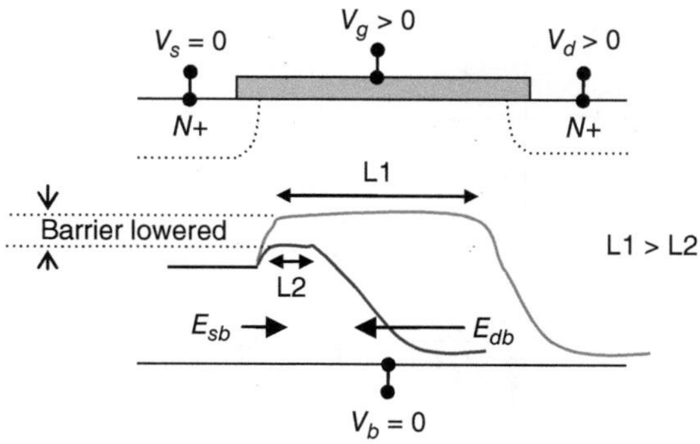

FIGURE 2.16 Schematic illustration of drain-induced barrier lowering (DIBL).

The larger saturation–V_t roll-off is caused by drain-induced barrier lowering (DIBL), as shown schematically in Fig. 2.16. When V_d is high and the channel length is short enough (e.g., L2), the high-drain electrical field E_{db} can compensate the source electrical field E_{sb}, which reduces the source/bulk (S/B) barrier. More carriers thus are injected from the source, resulting in higher I_d with a V_d increase at the same V_g. DIBL will increase the subthreshold slope and subthreshold leakage current. *Punch-through* may be induced when the drain-side depletion region touches the source-side depletion region, and the gate totally loses control of the channel. DIBL is quantitatively characterized in WAT as

$$\Delta V_t = V_t(\text{linear}) - V_t(\text{saturation}) \tag{2.57}$$

Short-Channel Effect Control

Since the short-channel effect is caused mainly by charge sharing and DIBL, there are some methods to reduce it:

- Use a shallower S/D junction (i.e., reduce x_j). There will be less depletion charge in the triangle portion in Fig. 2.15*a* and thus a smaller V_t lowering.
- Increase channel doping. This is done to increase the depletion charge in the trapezoidal portion of Fig. 2.15*a* and thus result in a smaller V_t lowering.
- Use halo (pocket) implantation. This is done to increase the concentration of the channel adjacent to the S/D and to reduce the thickness of the triangle depletion region.
- Use thinner gate oxide. In this way, the vertical electrical field is increased, thus raising the gate-control capability to the channel.

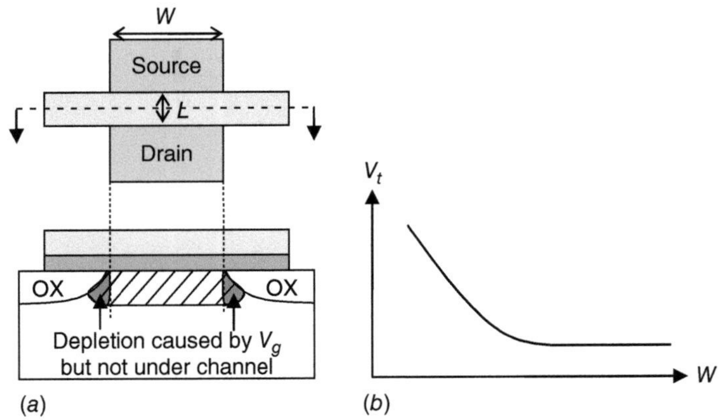

Figure 2.17 Schematic illustrations of (*a*) narrow-width effect for LOCOS; (*b*) V_t versus channel width for narrow-width effect.

2.5.2 Width Effect

Narrow-Width Effect Local Oxidation of Silicon

Figure 2.17*a* provides a schematic illustration of the narrow-width effect (NWE) for a local oxidation of silicon (LOCOS) structure. It is clear that there is a depletion charge near the field oxide that is caused by the V_g but not under the channel, which also will contribute to the V_t. In other words, the V_g gives rise not only to the channel-region depletion but also to some other region besides the channel. If the channel is wide enough, these additional depletion charges $Q_{\text{additional}}$ are negligible. However, when the channel becomes narrower, this portion ($Q_{\text{additional}}$) turns out to be comparable with the depletion charges under the channel. As indicated in Eq. (2.58) [which is rewritten for Eq. (2.29)], the V_t will be higher when the channel width is smaller (Fig. 2.17*b*):

$$V_t = V_{fb} + 2\psi_b + \frac{\sqrt{2\varepsilon_{Si}qN_A(2\psi_b)}}{C_{ox}} + \frac{Q_{\text{additional}}}{C_{ox}} \tag{2.58}$$

Inverse Narrow-Width Effect: Shallow-Trench Isolation

For a shallow-trench isolation (STI) structure, there is a negligible parallel-channel Si out of poly, so V_g only contributes to the channel depletion charge. However, at the STI corner, the electrical field is much stronger than that under the channel (Fig. 2.18*a*). As a result, when *W* becomes narrower, V_t becomes lower (Fig. 2.18*b*). An example of the inverse narrow-width effect (INWE) for the STI isolation technology is given in Ref. 6.

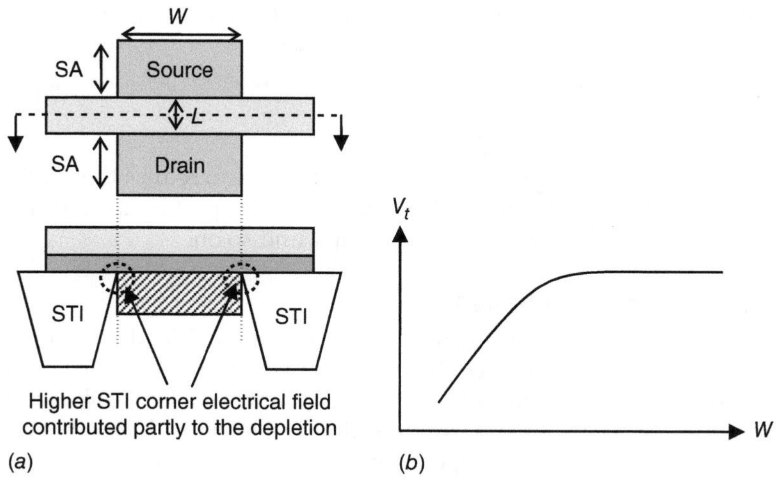

FIGURE 2.18 Schematic illustration of (*a*) inverse narrow-width effect for STI; (*b*) V_t versus channel width for inverse narrow-width effect.

2.5.3 Gate-Induced Drain Leakage

Figure 2.19*a* is a schematic illustration of the mechanism of gate-induced drain leakage (GIDL) current with NMOS as an example. When the gate is grounded and the drain is biased with a high voltage, the gate-to-drain overlap region will have an inversion layer with a hole as the minority carrier. However, since the substrate is also grounded, the hole will be drained laterally to the substrate. Therefore, no inversion layer is formed. Simultaneously, there is band-to-band tunneling-electron flow into the drain, as shown in Fig. 2.19*b*. The drain current I_d under the bias condition shown in Fig. 2.19*a* can be expressed as follows:

$$I_d = AE_s e^{-B/E_s} \tag{2.59}$$

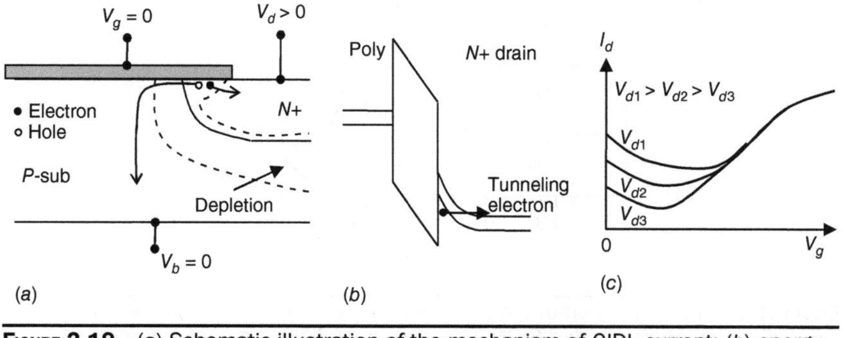

FIGURE 2.19 (*a*) Schematic illustration of the mechanism of GIDL current; (*b*) energy-band diagram of the tunneling electron contributing to the GIDL current; (*c*) the drain current in terms of gate voltage for varied V_d. With a higher V_d applied, the GIDL current is larger.

where $B = 21.3$ MV/cm, E_s is the normal electrical field in the Si, and

$$E_s \approx \frac{\varepsilon_{ox}(V_{dg} - E_g)}{\varepsilon_{Si}t_{ox}} \tag{2.60}$$

Therefore, smaller V_{dg} (Fig. 2.19c) and thicker t_{ox} can reduce the GIDL. The other factors affecting GIDL include drain profile, gate/drain overlap, gate-edge profile, interface traps, and so on.

2.5.4 Boron Penetration

Boron penetration is not desirable because it will lead to out-of-control device performance. In case there is boron penetration into the channel, it will neutralize the channel dopant (e.g., *P* in *N* substrate) and thus reduce its dopant concentration and lead to a lower threshold voltage V_t. Higher nitrogen concentration in gate oxide is more effective in suppressing the boron penetration, thus resulting in a higher V_t. Figure 2.20[7] clearly shows that when peak nitrogen concentration increases from 9 (S1) to 15 (S3), V_t increases from 0.58 to 0.73 V for devices with gate lengths of 2 μm and width of 10 μm. However, one should be very careful that the concentration of nitrogen in the gate dielectric should be well controlled for desired NBTI performance, which is elaborated in Chap. 7.

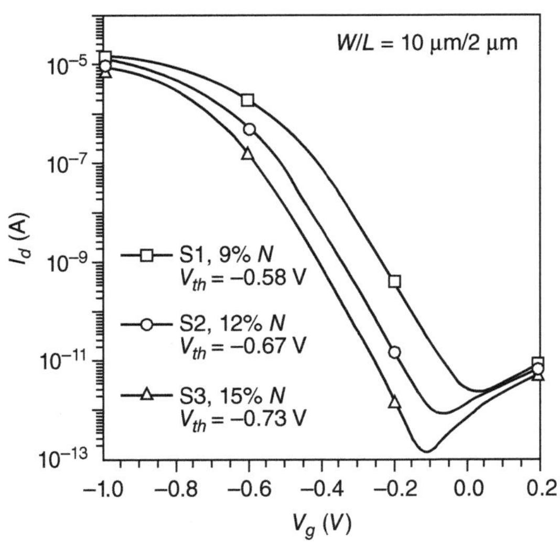

FIGURE 2.20 Room-temperature drain-current–gate-voltage ($I_d - V_g$) curves for a PMOS with plasma-nitrided SiON gate dielectrics having a peak nitrogen concentration of 9 (S1), 12 (S2), and 15 (S3). The $I_d - V_g$ curves were measured at $V_d = -0.1$ V. (*Used with permission from the American Institute of Physics.*[7])

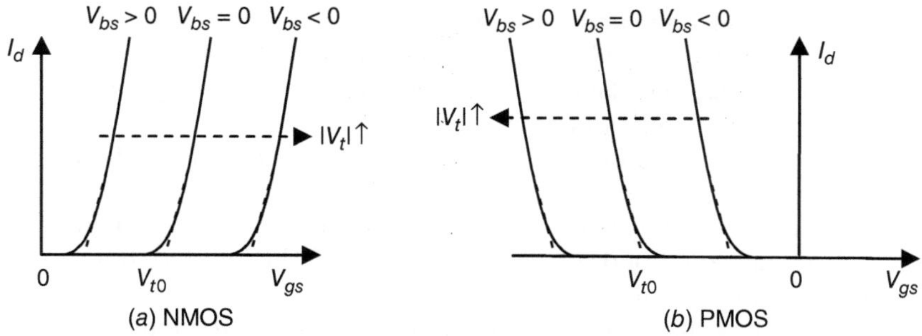

FIGURE 2.21 Substrate-bias V_{bs} effect on V_t modulation for (*a*) an NMOS and (*b*) a PMOS.

2.5.5 Effect of Substrate Bias

It is possible to modulate V_t by applying a voltage to the substrate. For a MOSFET with substrate bias (i.e., $V_{bs} \neq 0$), the expression for V_t is

$$V_t = V_{fb} + 2\psi_b + \frac{\sqrt{2\varepsilon_{Si} q N_A (2\psi_b - V_{bs})}}{C_{ox}} \tag{2.61}$$

$$\Rightarrow V_t = V_{t0} + \gamma\left(\sqrt{2\psi_b - V_{bs}} - \sqrt{2\psi_b}\right) \tag{2.62}$$

where V_{t0} is given in Eq. (2.29), and the *body effect* γ is defined as

$$\gamma = \frac{\sqrt{2\varepsilon_{Si} q N_A}}{C_{ox}} \tag{2.63}$$

Based on Eq. (2.61), positive V_{bs} will reduce and increase $|V_t|$ for N- and P-channel metal-oxide-semiconductors (NMOSs and PMOSs), respectively, as shown schematically in Fig. 2.21.

From Eq. (2.63) it is seen that the ways to minimize the body effect γ are to reduce the gate-oxide thickness (i.e., increase C_{ox}) and reduce the N_A.

2.6 Interface Traps and Oxide Traps

Device reliability hot-carrier injection (HCI), gate-oxide integrity (GOI), NBTI and, plasma-induced damage (PID) concerns the generation of an oxide trap Q_{ot} and an interface trap Q_{it}, which leads to a reduction in the mobility and V_t shift and then I_{dsat} reduction. As a result, the switching speed of the circuit is slowed down owing to a longer charging time to the interconnect or load capacitances. The trapped charges also shift the subthreshold gate voltage and decrease the subthreshold slope of the $I_d - V_g$ curve, which means that the transistor off-state leakage and standby power consumption will increase, and the signal/noise margin will decrease.

The polarity and number of interface traps measured depend on the device type and the voltage applied during measurement. Interface traps are commonly believed to be acceptor-like in the upper half of the band gap and donor-like in the lower half.[8] The interface traps are neutral when the upper half of the band gap is totally empty and the lower half is fully occupied. A specific case is that the interface traps are neutral when the Si surface Fermi level E_{Fs} is at the middle energy E_i,[9] namely, at $V_g = V_{mg}$. In case where E_{Fs} (or E_F) $\neq E_i$, if the upper half of the band gap is occupied (fully or partially), it is negatively charged. On the contrary, if the lower half of the band gap is empty (fully or partially), it is positively charged. In contrast, as shown in Fig. 2.2, for doping atoms, donor impurities are in the upper half and acceptor impurities are in the lower half of the band gap. Detailed descriptions of the interface traps for both *N*-channel and *P*-channel devices are given below.[9]

Figure 2.22*a* shows the case for a *P*-type substrate (for an *N*-channel device):

- At flat band, electrons occupy all the interface states below the Fermi energy E_F. The interface states filled with electrons below the Fermi energy are within the lower half of the band gap, so they are neutral and marked here by 0. Those between midgap E_i and the Fermi energy E_F are empty, but within the lower half of the band gap, so they are positively charged and marked here by +. Those above E_i are neutral because they are not occupied, and again they are marked by 0.
- On the other hand, when the channel is inverted, the fraction of interface traps between midgap and the Fermi level is now occupied by acceptors, leading to negatively charged interface traps and marked here by –. Hence interface traps in *N*-channel devices in inversion are negatively charged, leading to positive threshold voltage shifts according to Eq. (2.24).

Figure 2.22*b* shows the case for an *N*-type substrate (for a *P*-channel device):

- At flat band, electrons occupy all the interface states below the Fermi energy E_F. The interface states in the lower half of the band gap are fully occupied by donor states, so they are neutral and marked here by 0. Those between midgap E_i and the Fermi energy E_F are occupied by acceptor states, so they are negatively charged and marked here by –. Those above

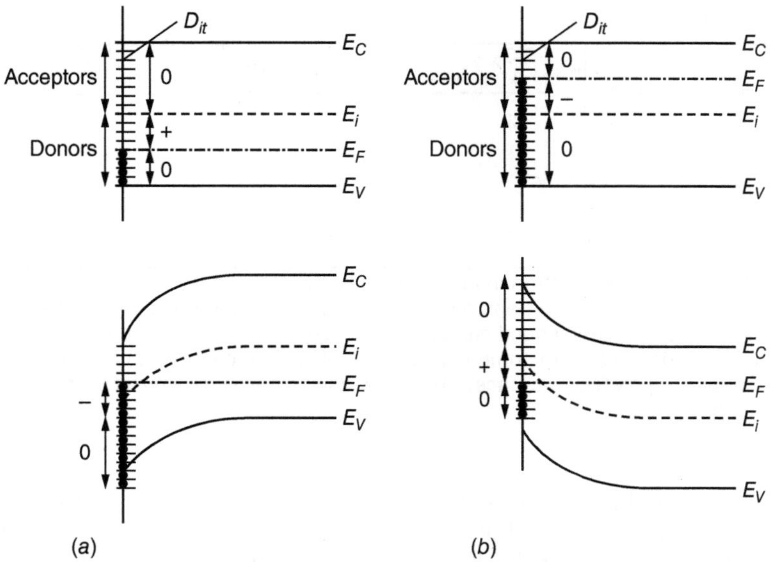

FIGURE 2.22 Band diagrams of the Si substrate showing the occupancy of interface traps and the various charge polarities: (*a*) *P*-type substrate (*N*-channel) with positive interface trap charge at flatband and negative interface trap charge at inversion; (*b*) *N*-type substrate (*P*-channel) with negative interface trap charge at flatband and positive interface trap charge at inversion. (*Used with permission from the American Institute of Physics.*[9])

E_F are neutral because they are not occupied and again are marked by 0.

- On the other hand, when the channel is inverted, the fraction of interface traps between midgap and the Fermi level is now unoccupied donors, leading to positively charged interface traps marked here by +. Hence interface traps in *P*-channel devices in inversion are positively charged, leading to negative threshold voltage shifts according to Eq. (2.24).

In contrast, oxide traps are always positive. However, their impact on the flat band and the V_t shift depends on where the charges are located in the dielectric film (gate oxide). For a random distribution $[\rho(x)]$ of the oxide traps, the corresponding flat-band change can be expressed as

$$\Delta V_{fb} = -\frac{\int_{x=0}^{x=t_{ox}} x \cdot \rho(x) dx}{t_{ox} \cdot C_{ox}} \tag{2.64}$$

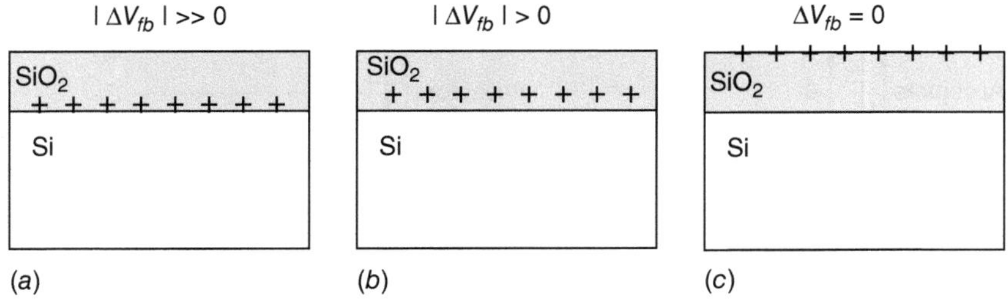

FIGURE 2.23 Schematic illustration of the effect of the location of oxide traps on flat-band change: (*a*) the oxide traps are close to the Si-SiO_2 interface, which has the largest impact on the flat-band change; (*b*) the oxide traps are located in the middle of the oxide, leading to some impact on the flat-band change; (*c*) the oxide traps are near the poly-SiO_2 interface, which may have a negligible impact on the flat-band change.

where t_{ox} and C_{ox} are gate-oxide thickness and capacitance, respectively. Equation (2.64) clearly shows that oxide traps near the Si/SiO_2 interface will lead to a larger flat-band change and thus a larger V_t shift, as illustrated in Fig. 2.23. The oxide traps near the poly-SiO_2 interface may have negligible impact on the flat band.

2.7 References

1. S. M. Sze, *Semiconductor Devices Physics and Technology*, 2nd ed., Chap. 2, Hoboken, NJ: Wiley, 2002.
2. http://en.wikipedia.org/wiki/Electrical_resistance_and_conductance. Accessed January, 2011.
3. J. W. McPherson, "Reliability challenges for 45 nm and beyond," in *Proceedings of the 43rd annual IEEE Design Automation Conference*, 2006, pp. 176–81.
4. R. Shrivasrava and K. Fitzpatrick, "A simple mothod for the overlap capacitance of a VLSI MOS device," *IEEE Transactions on Electron Devices*, Vol. 29, 1995, pp. 229–232.
5. Y. Cheng and C. Hu, *MOSFET Modeling & BSIM3 User's Guide*, KAP, Boston, 1999.
6. W. Lee, S. Lee, T. Ahn, and H. Hwang, "Degradation of hot carrier lifetime for narrow width MOSFET with shallow trench isolation," in *Proceedings of the 37th IEEE Annual International Reliability Physics Symposium*, 1999, pp. 259–262.
7. S. Zhu, A. Nakajima, T. Ohashi, and H. Miyake, "Interface trap and oxide charge generation under negative bias temperature instability of *p*-channel metal-oxide-semiconductor field-effect transistors with ultrathin plasma-nitrided SiON gate dielectrics," *Journal of Applied Physics*, Vol. 98, 2005, p. 114504.
8. P. V. Gray and D. M. Brown, "Density of SiO_2-Si Interface States," *Applied Physics Letters*, Vol. 8, 1966, pp. 31–33.
9. D. K. Schroder and J. A. Babcock, "Negative bias temperature instability: Road to cross in deep submicron silicon semiconductor manufacturing," *Journal of Applied Physics*, Vol. 94, No.1, 2003, pp. 1–18.

CHAPTER 3

Process Flow for Metal-Oxide-Semiconductor Manufacturing

This chapter provide the basic semiconductor manufacturing process flow, including front-end-of-line (FEOL) and back-end-of-line (BEOL) flows. This knowledge will assist readers in understanding process effects on reliability performance.

3.1 Front-End-of-Line

The commonly used starting material/wafer is a *P*-substrate silicon with a (100) surface orientation doped with boron. The (100) provides a much better Si-SiO_2 interface than other orientations. There is an option to use (110) substrate to boost PMOSFET performance for advanced technology: A doubled mobility can be achieved.[1] However, the reliability degradation should be taken care. Figure 3.1 provides a schematic step-by-step FEOL process flow for modern complementary metal-oxide-semiconductor (CMOS) technology, which is detailed as follows:

1. *P-Substrate wafer with pad oxide growth and nitride deposition.* This is the preparation step for forming the active-area (AA) for devices. The silicon nitride (Si_3N_4) is usually deposited by low-pressure chemical vapor deposition (LPCVD) at around 800°C. The nitride (~1,000 Å thick) serves as a hard mask to define the device pattern. The pad oxide (~100 Å thick) in between the nitride and the substrate is a thermal oxide grown in O_2 at around 1,000°C. The main purpose of the pad oxide under the nitride is to compensate its high tensile stress because the thermal oxide deposited on silicon is under compressive stress.

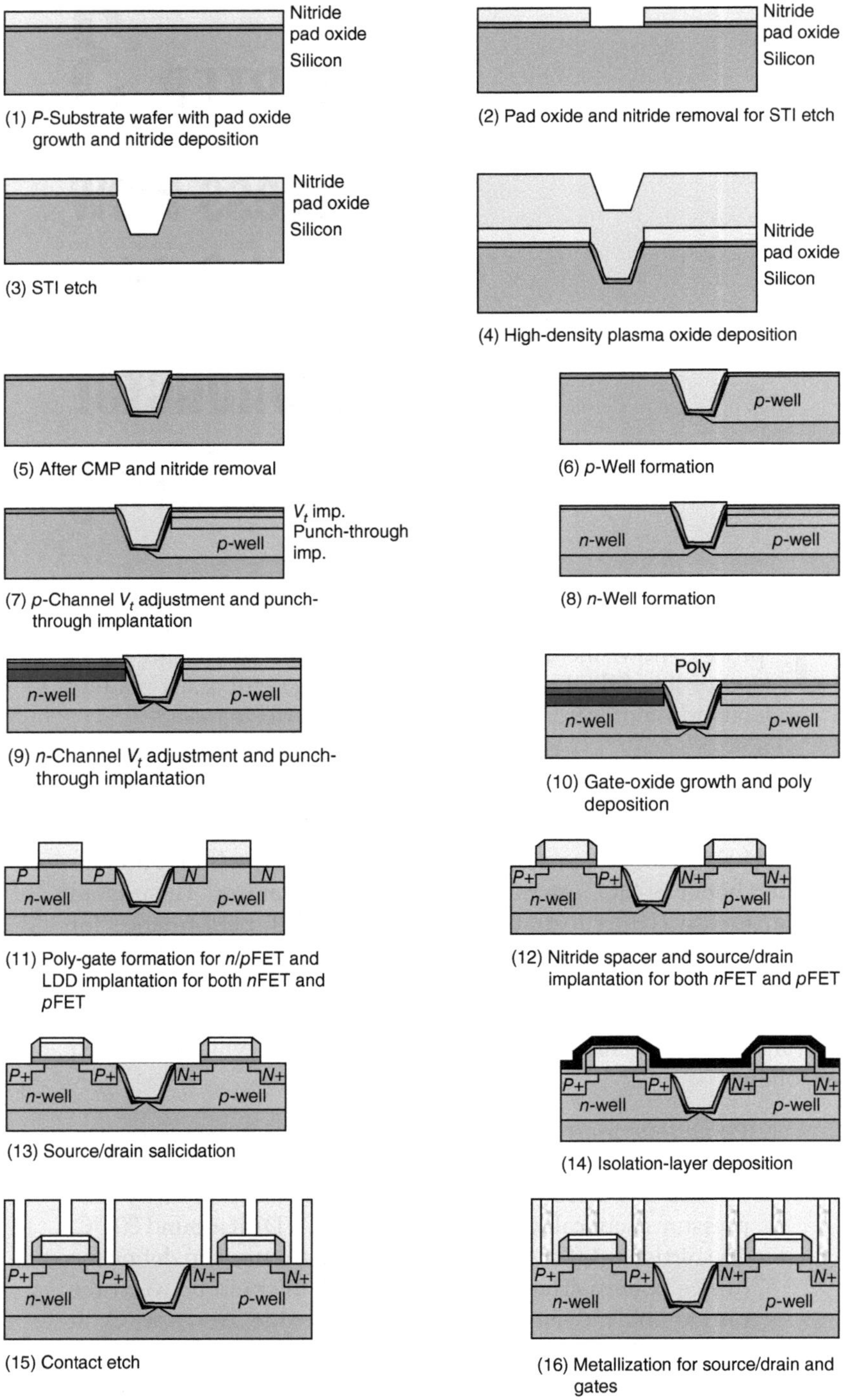

FIGURE 3.1 Schematic step-by-step process flow for modern CMOS technology (FEOL).

2. *Pad oxide and nitride removal for shallow-trench isolation (STI) etch.* The AA portion of the silicon substrate is exposed by etching away the relevant pad oxide and nitride using a fluorine plasma.
3. *STI etch.* The purpose of STI (~0.5 μm deep) is for device isolation. The nitride serves as the hard mask for the successive etch to form the trench in the silicon, whose walls typically are tapered to avoid the potential issues with a perfectly vertical wall. To improve the packing density of integrated circuits, scaling down of the isolation region is necessary. STI is usually used in order to avoid the subthreshold hump, bird's beak, and field oxide thinning effect in local oxidation of silicon (LOCOS). Compared with LOCOS isolation, STI has various additional advantages, such as perfect surface planarity, scalability, and latchup immunity, and has become the most widely used isolation scheme in advanced technology nodes.
4. *High-density plasma oxide deposition.* The trench then is refilled with SiO_2 by high-density plasma (HDP) CVD. Before that, a thermally grown "lining" oxide (~100 Å thick) is placed on the trench sidewalls and bottom to improve corner rounding as well as to provide a better Si-SiO_2 interface.
5. *Chemical-mechanical polishing (CMP) planarization and nitride removal.* CMP is used to planarize the wafer surface by removing the excess SiO_2 from the top surface. The nitride layer serves as a polishing stop. Once the CMP is finished, the nitride is chemically removed in hot phosphoric acid, which is highly selective between SiN and SiO_2.
6. *P-Well formation.* The *P*-well for the N-channel metal-oxide-semiconductor (NMOS) device is first formed by ion implantation, for example, boron ions with typically 150 to 200 keV of energy and a dose on the order of 10^{13} cm^{-2}. The implantation energy is large enough so that the ions not only penetrate through the thin gate oxide but also penetrate through the STI oxide to avoid any potential parasitic inversion layer in Si beneath the STI oxide. A furnace anneal then is applied to recover the damage caused by the ion implantation.
7. *N-Channel V_t adjustment and punch-through implantation.* One of the most important device parameters is the threshold voltage V_t as described in Chap. 2. One typical way for V_t adjustment to meet the design target is through channel implantation. With more ions implanted, the V_t will be higher. In modern CMOS circuits, the target V_t for both NMOS and P-channel metal-oxide-semiconductor (PMOS) is generally around 0.5 to 0.8 V. For example, boron implantation with a dose around 1×10^{13} cm^{-2} at an energy of 10 to 30 keV may be applied for *N*-channel V_t adjustment. A higher energy with a similar-dosage implantation

is applied to prevent the so-called punch-through that appeared in ad-vanced CMOS devices as a result of the short-channel effect.

8. *N-Well formation.* Similarly, an *N*-well is formed to accommodate the PMOS device by ion implantation, for example, phosphorous. The implantation energy of phosphorous ions is around 300 to 400 keV, which is much larger than that for boron ions because B^+ is much lighter than P^+.

9. *P-Channel V_t adjustment and punch-through implantation.* This is the identical process applied to the *P*-channel. The ion species is again phosphorus with higher implantation energy.

10. *Gate-oxide growth and poly deposition.* Prior to growing the gate oxide, the preexisting oxide is first stripped in a dilute HF solution. Note that HF etch of SiO_2 is highly selective, which can effectively stop at the Si surface. The reason for this stripping is to well control the final oxide thickness and to improve the quality of the oxide and the Si-SiO_2 interface because the preexisting oxide has been subjected to several steps of ion implantation, as described earlier. The oxide can be formed either in a dry O_2 environment (e.g., at 800°C for 2 hours) or in an H_2O in-situ steam-generation (ISSG) chamber (e.g., at 800°C for 25 minutes). Then the polysilicon gate electrode is deposited by LPCVD technique at about 600°C by the thermal decomposition of the silane (SiH_4). For advanced CMOS, the polysilicon is 1,000 Å thick or thinner. The polysilicon then is predoped to the desired type if needed. The purpose of the predoping is to reduce the poly depletion, which might degrade device performance.

11. *Poly-gate formation for N/PMOS and lightly doped drain implantation for both NMOS and PMOS.* The poly gate is formed by etching away the portion not needed. A chlorine- or bromine-based plasma chemistry typically is used to achieve good selectivity to SiO_2. This is followed by an anneal at around 1,000°C, called *rapid-thermal oxidation* (RTO). The purpose of the anneal is to repair the damage in the gate oxide caused by the etch. Lightly doped drain (LDD) implantation then is applied to achieve a graded doping region between the channel and the highly doped drain. The purpose of the LDD is to reduce the electrical field in between the drain and channel and thus to resolve the hot-carrier injection (HCI) issue, as detailed in Chap. 5. HCI has been known as one of the key reliability issues in advanced devices with smaller and smaller dimensions. Therefore, LDD is critical for the control of HCI issue. LDD sometimes is referred as the *extension region* of the source/drain.

12. *Nitride spacer and source/drain implantation for both NMOS and PMOS.* A conformal dielectric spacer consisting of SiO_2 and Si_3N_4 layers is deposited by LPCVD. Note that the thickness of this spacer layer determines the dimension of the extension region and thus is critical for device characteristics and reliability control. The source/drain regions then are formed by ion implantation of the corresponding species, for example, P or As or both for the source/drain of NMOS and B for that of PMOS. Source/drain implantation is followed by rapid thermal anneal (RTA). implant activation with a temperature greater than 1,000°C for about 1 minute. The RTA also can repair the implant damage and drive the junction to the desired depth.
13. *Source/drain salicidation.* The purpose of this step is to reduce the sheet resistance of the gate and source/drain when connecting to the contact. *Salicide* is shorten form of *self-aligned silicide.* $TiSi_2$ was used at the 250-nm technology node, $CoSi_2$ at the 180- and 130-nm technology nodes, and NiSi at the 90-nm and below node.[2] A monosilicide such as NiSi consumes less of the silicon in the sallow junctions and generally translates to less junction leakage. The steps include the deposition of a salicide block (SAB) with a composition of silicon-rich oxide (SRO), SAB etch, salicide Co (or Ni) deposition, and an RTA anneal at about 300°C for 30 minutes for NiSi to form the salicide.
14. *Isolation-layer deposition (ILD).* The following steps, including the BEOL description in the next section, are to provide a means of interconnecting the active devices to form circuits and a means of connection to the input/output (I/O) pads. The ILD is to isolate the FEOL devices and the BEOL interconnects. First, a layer of silicon oxynitride (SiON) of around 150 Å is deposited, and this is followed by a layer of high-tensile nitride (SiN) of around 300 Å by PECVD technique. The high-tensile nitride serves as a contact etch stop layer (CESL). It also helps to improve the carrier mobility and thus to improve the device drive current. A fairly thick SiO_2 layer then is deposited by high-density plasma CVD. This layer is known as *phosphosilicate glass* (PSG) because it often is doped with phosphorus, which provides some protection against mobile ions (such as Na^+) originating from BEOL process. The mobile ions can lead to extra fixed charge in the gate oxide and thus adversely affect device performance. The ILD thus is planarized by CMP. Note that the PSG layer is around 1 μm thick, which is thick enough that after CMP the metal interconnect layers can be deposited directly on a highly planarized topography. *In these steps, because a plasma process is involved in*

the deposition of both the high-tensile nitride and the PSG layers, they are critical for the control of plasma-induced damage (PID), as detailed in Chap. 8.

15. *Contact etch.* This step is to define the regions where contact will be made between metal level 1 and the underlying structures (e.g., source/drain). The contact hole is obtained by plasma etch through the ILD/PSG layers.
16. *Metallization for source/drain and gates.* This step is to form the contact. A thin glue layer such as a TiN layer or Ti/TiN bilayer is first deposited by sputtering or CVD. This layer typically is on the order of a few nanometers thick. The next step is to deposit a W layer by CVD using WF_6 as a precursor, which is followed again by CMP planarization. The steps described here, in which contact holes are etched, filled, and planarized, is known as the *damascene process.*

Some key steps that have a great impact on device reliability are highlighted here:

- One key step is the *gate-oxide nitridation to replace the gate oxide with* SiO_xN_y, which is common in advanced devices to reduce boron penetration, to increase the gate-oxide dielectric constant, and thus to reduce the equivalent oxide thickness (EOT) and the gate-oxide leakage current. Nitridation can reduce the leakage by about an order of magnitude.[2] The nitridation is applied in step 10 above in between gate-oxide growth and poly deposition. Tables 3.1 and 3.2 compare different gate-oxide nitridation processes which attect both nitrogen concentration and profile in the gate oxide. Note that nitrogen concentration and profile highly affect the negative-bias temperature instability (NBTI) performance, which will be elaborated in Chap. 7. Among the nitridation processes, decoupled plasma nitridation (DPN) followed by postnitridation anneal (PNA) has become the most desirable process for today's advance devices.
- *A dual-gate-oxide (DGO) or triple-gate-oxide (TGO) technique is currently widely accepted to manufacture high-voltage I/O drivers and high-performance core circuit blocks within one process flow.*[11] For example, in a typical 65-nm DGO process flow for 2.5-V I/O and 1.2-V core devices, the thick gate oxide for I/O devices is grown first. Then the region corresponding to the core thin oxide is etched away by a heavy or diluted HF solution, followed by growth of the thin oxide in this region (Fig. 3.2). The HF etching step may affect core device performance and reliability, such as gate-oxide integrity (GOI), as detailed in Chap. 6, and NBTI, as detailed in Chap. 7. In a typical TGO technique, a 2.5-V I/O, 1.8-V I/O, and 1.2-V core can be manufactured within one process flow.

Nitridation Method	Characteristic	Advantages and Disadvantages
Rapid thermal nitridation (RTN)	Nitridation using an NH_3, N_2O, or NO gas anneal around 1,000°C.	A relatively low concentration on nitrogen, of the order of 10^{15} cm^{-2}, in SiO_2.[3,4] a nitrogen-rich region near the SiO_2-Si interface,[5] will worsen NBTI performance.
Remote plasma nitridation (RPN)	Nitrogen can be incorporated into thin SiO_2 gate dielectrics[6,7] by RPN, where plasma is created away from the wafer or chamber.	Compared with RTN, a higher atomic concentration of nitrogen can be incorporated.[8] The plasma nitridation process will generate a nitrogen-rich layer at the top of the silicon dioxide gate dielectric[9] that is more efficient in preventing boron penetration than having one near the SiO_2-Si interface and beneficial to NBTI resistance.[10]
Decoupled plasma nitridation (DPN)	The DPN system is not fully remote and is actually quasi-remote because the plasma is created in the chamber at the dome position using an inductive radio frequency (RF) coupling system away from the wafer. DPN is followed by a PNA.	Compared with RPN, higher plasma density can be achieved by DPN because there is no need for the active nitrogen species to travel a long distance, and thus there is much less possibility for them to recombine. However, with a high-density plasma system, the potential risk of plasma-induced charging damage to the gate dielectric may be increased.

Table 3.1 Comparison of Different Gate-Oxide Nitridation Processes

Nitridation Method	Nitrogen Profile		Advantages	Disadvantages
	Schematic	Description		
NO/O_2 mixture growth	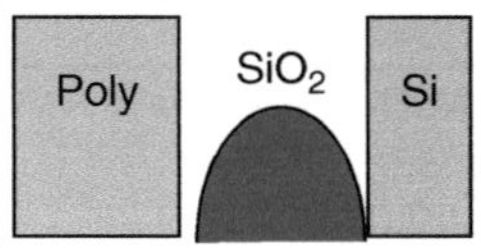	*N* distributed throughout gate oxide	High *N* concentration; good barrier to boron	Uniformity hard to control; worse NBTI
N_2O anneal	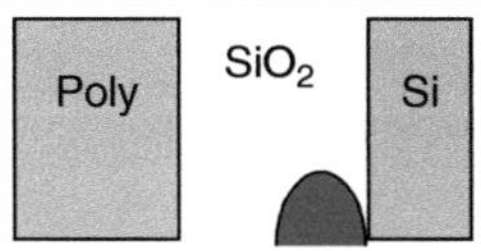	*N* close to SiO_2/Si interface	Release stress; better electrical properties	*N* Concentration too low; high thermal budget
NO anneal	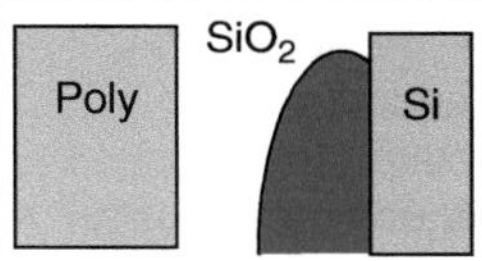	*N* peak at SiO_2/Si interface	High *N* concentration; better barrier to boron	Traps boron in SiO_2; poor Si/SiO_2 interface properties; worse NBTI
Plasma nitridation	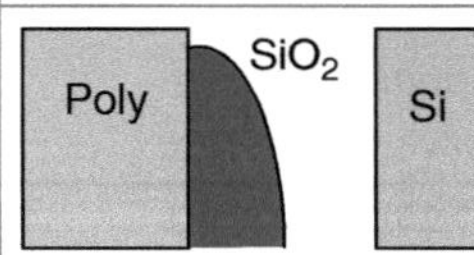	*N* peak at poly/SiO_2 surface	High *N* concentration; better barrier to boron. Desired.	Not applicable

TABLE 3.2 Comparison Between Thermal Anneal and Plasma Nitridation

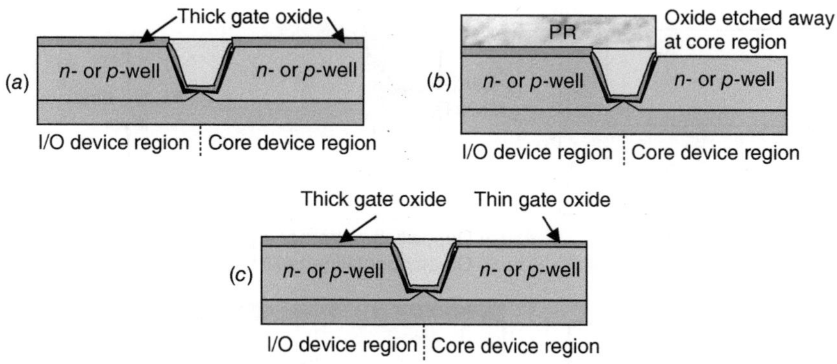

FIGURE 3.2 Typical process flow for manufacturing of DGO: (*a*) deposition of thick oxide for I/O devices; (*b*) oxide in the core region is etched away, for example, by HF solution; (*c*) deposition of thin oxide for core devices.

- *Sacrificial oxide before well implantation.* In step 5 of Fig. 3.1, after CMP and nitride removal, normally the pad oxide is stripped at this point.[12] A sacrificial oxide layer about 20 nm thick then is grown to protect the exposed Si surface of the AA from contamination during the subsequent implant process. The stripping of the pad oxide and growing of the sacrificial oxide also are effective to reduce the Kooi effect, which leads to gate-oxide thinning near the edge of an AA in an STI structure,[13] which will worsen the GOI performance, as detailed in Chap. 6. The Kooi effect is caused by a series of reactions explained as follows. When the STI oxide is annealed in an O_2 ambient, the O_2 will react with residual H_2 in the STI to form H_2O, which, in turn, will react with SiN if there is any and then form NH_3. A reaction between NH_3 and Si will take place and results in nitridation of Si and a thinner gate oxide grown in the subsequent step. This is the reason why the thinner gate oxide is located at the AA corner close to the STI. Note that the sacrificial oxide is etched away after well implant and before gate-oxide growth. This is the reason it is called *sacrificial.*

3.2 Back-End-of-Line of the Cu Dual Damascene Processes

As ultra-large-scale integration (ULSI) devices shrink in size and multilevel integration becomes imperative, copper has successfully replaced Al in advanced interconnects for continued improvements in integrated circuit (IC) performance. However, Cu cannot be easily patterned by reactive ion etching (RIE) owing to the low volatility of Cu chlorides and Cu fluorides. The *dual damascene process* (Fig. 3.3) is

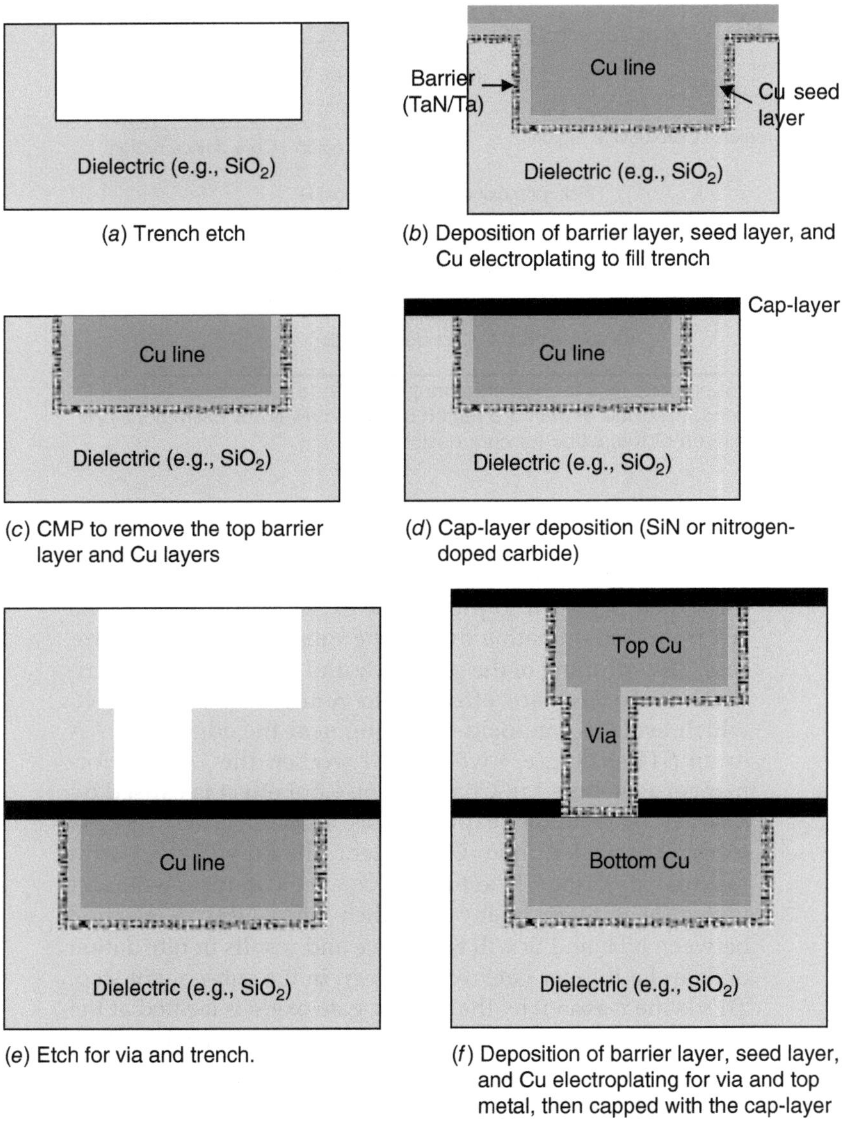

FIGURE 3.3 Illustration of Cu dual damascene processes.

presently recognized as a standard Cu interconnect fabrication technique.[14] A general sequence of this technique is detailed below:

- The via holes and trenches are etched first (Fig. 3.3*a*), followed by deposition of barrier layers (TaN/Ta, for example) and Cu seed layers (Fig. 3.3*b*) to the walls of via holes and trenches by sputtering, a physical vapor deposition (PVD) process. The barrier layers can effectively prevent Cu from diffusing into the surrounding dielectric. After the via/trench etch, the Cu surface is exposed at a clean-room ambient, and a surface layer containing Cu_2O, CuO, $Cu(OH)_2$, and $CuCO_3$ is formed.[15] Prior to the barrier-layer deposition, conventionally, an ionized argon (Ar^+) plasma is employed to clean the Cu surfaces under the via bottoms by physical bombardment in order to reduce via resistance and improve adhesion and reliability.

- Cu interconnects then are filled into the via holes and trenches using an electrochemical deposition method (ECD) or electrochemical plating method (ECP) (Fig. 3.3*b*).
- Then CMP is used to remove the excess Cu and metal barrier and thus to achieve a planarized topography (Fig. 3.3*c*). Blanket dielectric is deposited for the next level in succession (Fig. 3.3*d*). In multilevel interconnect structures, successive levels of interconnects are connected by vias (Fig. 3.3*e,f*).

In advanced technology (130 nm and onwards), the dual TaN/Ta barriers with an additional resputter process are introduced to realize adequate barriers in via holes with high aspect ratios so as to cover and compensate preexisting microdefects to a certain degree and improve reliability performance.[16,17] Two different resputter deposition techniques are proposed,[18] and they differ in terms of resputter sequences. The first method (ex situ) contains a separate argon-sputter clean between the TaN and Ta PVD resputter deposition, which runs in a separate chamber. The other method (in situ) takes advantage of an argon-sputter clean running in one and the same chamber during the TaN depoition.

Figure 3.4 provides a schematic step-by-step BEOL process flow for modern CMOS technology with Cu dual damascene processes. In conjunction with the FEOL steps, the BEOL steps in Fig. 3.4 are numbered in succession as follows:

17. *Metal 1 (M1) isolation layer deposition.* A cap-layer around tens of nanometers thick is deposited first by PECVD at around 400°C. For an advanced interconnect, an example of the cap-layer is with a composition of nitrogen-doped carbide (NDC). The cap-layer serves as an etch stop layer for the following metal etch, which also protects the Cu from oxidation. A layer of intermetal dielectric (IMD1 here, e.g., SiCOH) around hundreds of nanometers thick then is deposited by PECVD, again at about 400°C. There are several commercially available IMDs on the market, which will be elaborated in Chap. 12. A nother cap layer of tetraethylorthosilicate (TEOS) about 30 nm thick then is deposited by PECVD to protect the IMD from the damage caused by the subsequent photolithography and etch processes.

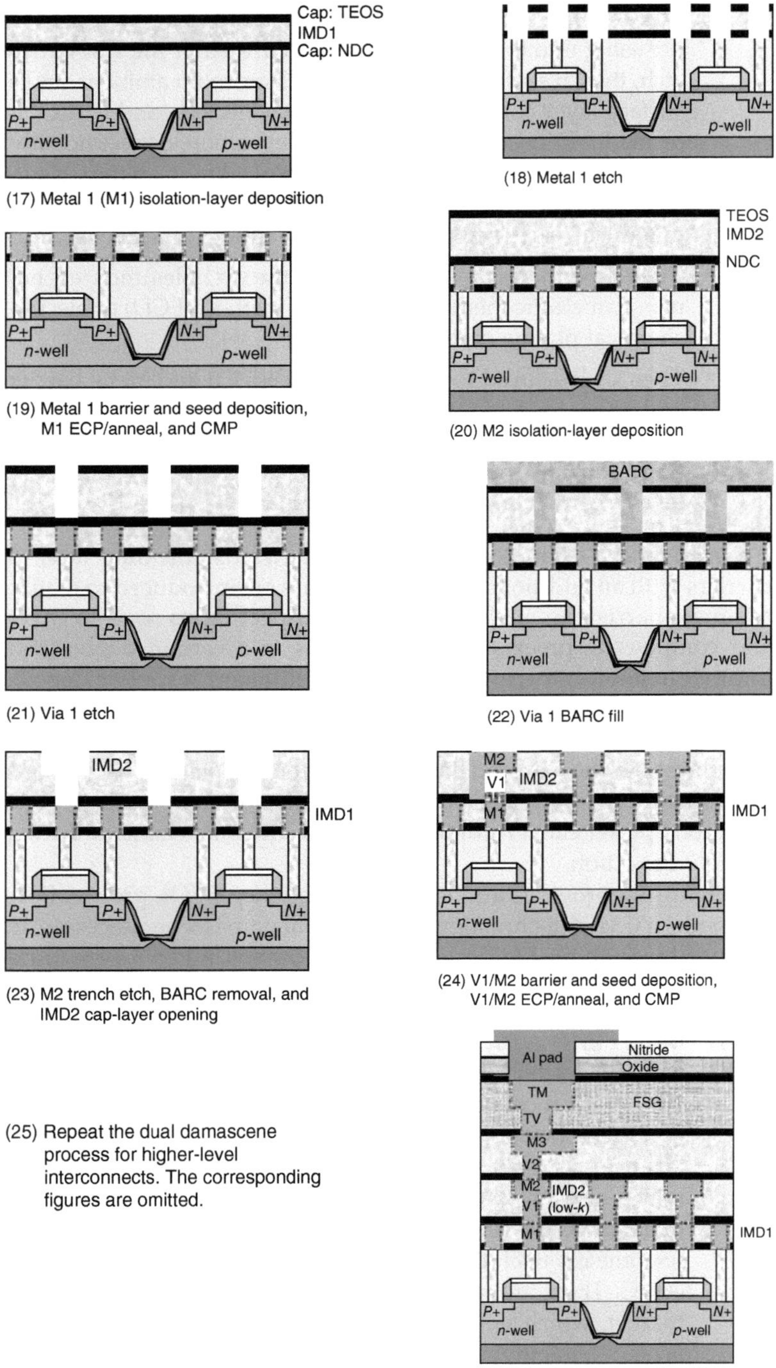

(17) Metal 1 (M1) isolation-layer deposition

(18) Metal 1 etch

(19) Metal 1 barrier and seed deposition, M1 ECP/anneal, and CMP

(20) M2 isolation-layer deposition

(21) Via 1 etch

(22) Via 1 BARC fill

(23) M2 trench etch, BARC removal, and IMD2 cap-layer opening

(24) V1/M2 barrier and seed deposition, V1/M2 ECP/anneal, and CMP

(25) Repeat the dual damascene process for higher-level interconnects. The corresponding figures are omitted.

(26) Top-layer metal deposition, passivation-layer deposition, and pad opening

FIGURE 3.4 Schematic step-by-step process flow for modern CMOS technology (BEOL) with Cu dual damascene processes.

18. *Metal 1 (M1) etch.* This step is to define the regions for M1.
19. *Metal 1 barrier and seed deposition, M1 ECP/anneal and CMP.* This step is to fill the M1 holes with ECP Cu based on the damascene process. The details have already been illustrated in Fig. 3.3. Note that the previously deposited TEOS is here removed by CMP because it has a much higher dielectric constant.
20. *M2 isolation layer deposition.* Similar to M1 isolation layers, the deposition of cap-layer (NDC)/IMD (BD)/cap layer (TEOS) in sequence is done again for IMD2, within which both via 1 holes and M2 trenches will be etched. The NDC deposition is a plasma-involved process, which affects the device PID performance. In addition, the NDC-Cu interface quality has been demonstrated to be very critical for BEOL reliability, including electromigration (Chap. 10), stress migration (Chap. 11), and IMD dielectric breakdown (Chap. 12).
21. *Via 1 etch.* This step is to define the regions for V1.
22. *Via 1 bottom anti-reflective coatings (BARC) fill.* In this step, BARC is filled into the V1 holes by spun-on technique. BARC is a sacrificial material for the subsequent M2 trench etch.
23. *M2 trench etch, BARC removal, and IMD2 cap-layer opening.* The regions for M2 trench then are etched, followed by removal of the BARC materials. And then it is the time to open the IMD2 cap-layer on top of the M1 so as to expose the Cu interconnect there.
24. *V1/M2 barrier and seed deposition, V1/M2 ECP/anneal, and CMP.* Following the same procedure applied to M1 deposition, first the barrier and seed layer are deposited, and then the vias and trenches are filled with ECP Cu, followed by CMP planarization. An extra process step prior to barrier deposition is to do a reactive-plasma clean (RPC) for tens of seconds to remove the native oxide on the Cu surface, thus improving the M1-V1 interface quality. *Note that the RPC is a plasma involved process, which potentially may degrade device performance by PID.*
25. *Repeat the dual damascene process for higher-level interconnects.* The corresponding figures are omitted here because they are just repetitious.
26. *Top-layer metals deposition, passivation layer deposition, and pad opening.* This is a schematic for three layers of intermetal, one layer of top metal (TM), and passivation layers. Since the top metal is connected to the Al pad, it is generally thicker and wider than the intermetal layers. In addition, the top-layer dielectric is not a low-k material but is replaced by a normal oxide with better rigidity. A typical material used is fluorosilicate glass (FSG).

Together with the passivation layers, including oxide and nitride shown in the figure, they can provide better protection for the chip because there is mechanical handling during packaging. The chip also may be subjected to mechanical stress during operation. The passivation layers also protect the chip from ambient contamination (Na^+ or K^+) and moisture adsorption.

27. After the final processing steps, the wafers are *annealed/alloyed in conventional forming gas* (a typical composition is 10 percent H_2 in N_2) at a relatively low temperature (400 to 450°C) for about 30 minutes. The main purpose of this step is to reduce some of the electric charges associated with the Si-SiO_2 interface and thus improve gate-oxide reliability and PID performance. This will be detailed in Chaps. 6 and 8. Note also that the *composition of the forming gas, anneal temperature, and time may affect the NBTI performance* because hydrogen is involved in this process. This will be elaborated in Chap. 7.

3.3 References

1. S. S. Chung, "Reliability issues for high-performance nanoscale CMOS technologies with channel mobility enhancing schemes," in *Proceedings of the International Workshop on Nano CMOS*, 2006, pp. 128–131.
2. J. W. McPherson, "Reliability challenges for 45 nm and beyond," in *Proceedings of the 43rd Annual IEEE Design Automation Conference*, San Francisco, CA, 2006, pp. 176–181.
3. M. L. Green, D. Brasen, K. W. Evans-Lutterodt, L. C. Feldman, K. Krish, W. Lennard, H. T. Tang, L. Manchanda and M. T. Tang, "Rapid thermal oxidation of silicon in N_2O between 800 and 1200°C: Incorporated nitrogen and interfacial roughness," *Applied Physics Letters*, Vol. 65, 1994, pp. 848–850.
4. E. P. Gusev, H. C. Lu, T. Gustafsson, E. Galfunkel, M. L. Green, and D. Brasen, "The composition of ultrathin oxynitrides thermally grown in nitric oxide," *Journal of Applied Physics*, Vol. 82, No. 2, 1997, pp. 896–898.
5. E. P. Gusev, H. C. Lu, E. L. Garfunkel, T. Gustafsson, and M. L. Green, "Growth and characterization of ultrathin nitrided silicon oxide films," *IBM Journal of Research and Development*, Vol. 43, 1999, pp. 265–286.
6. D. Kapila, S. Hattangady, M. Douglas, R. Kraft, and M. Gribelyuk, "Modeling and optimization of oxynitride gate dielectrics formation by remote plasma nitridation of silicon dioxide," *Journal of the Electrochemical Society*, Vol. 146, 1999, pp. 1111–1116.
7. S. V. Hattangady, R. Kraft, D. T. Grider, M. A. Douglas, G. A. Brown, P. A. Tiner, J. W. Kuehne, P. E. Nicollian, and M. F. Pas, "Ultrathin nitrogen-profile engineered gate dielectric films," in *Proceedings of the International Electron Devices Meeting (IEDM)*, 1996, pp. 495–498.
8. E. Cartier, D. A. Buchanan, and G. I. Dunn, "Atomic hydrogen-induced interface degradation of reoxidised nitrided silicon dioxide on silicon," *Applied Physics Letters*, Vol. 64, 1994, pp. 901–903.
9. P. C. Chen, K. Y. J. Hsu, J. Y. Lin, and H. L. Hwang," *Characterization of ultrathin* dielectrics grown by microwave afterglow oxygen, and N_2O plasma," *Japanese Journal of Applied Physics*, Vol. 34, 1995, pp. 973–977.
10. Y. Wu, G. Lucovsky, and H. Z. Massoud, "Improvement of gate dielectric reliability for *p*+ poly MOS devices using remote PECVD top nitride deposition on thin gate oxides," in *Proceedings of the 36th Annual IEEE International Reliability Physics Symposium*, 1998, pp. 70–75.

11. N. Bhat, P. Chen, P. Tsui, A. Das, M. Foisy, Y. Shiho, J. Higman, J.-Y. Nguyen, S. Gonzales, S. Collins, and D. Workman, "Hot carrier reliability considerations in the integration of dual gate oxide transistor process on a sub-0.25-pm CMOS technology for embedded applications," in *Proceedings of the International Electron Devices Meeting (IEDM)*, 1998, pp. 931–934.
12. S. Wolf, Deep-Submicron Process Technology, Vol. 4: *Silicon Processing for the VLSI Era*, Lattice Press, Sunset Beach, CA, 2002.
13. N. Balasubramanian, "Gate oxide thinning in MOS structures with shallow trench isolation," in *Meeting Abstract of Electrochemistry Society*, 2000, Abstract 454.
14. M. Quirk, and J. Serda, *Semiconductor Manufacturing Technology*, Prentice-Hall, Upper Saddle River, NJ, 2001.
15. M. R. Baklanov, D. G. Shamiryan, Z. Tökei, G. P. Beyer, T. Conard, S. Vanhaelemeersch, and K. Maex, "Characterization of Cu surface cleaning by hydrogen plasma," *Journal of Vacuum Science and Technology B*, Vol. 19, No. 4, 2001, pp. 1201–1211.
16. G. B. Alers, R.T. Rozbicki, G. J. Harm, S. K. Kailasam, G.W. Ray, and M. Danek, "Barrier first integration for improved reliability in copper dual damascene interconnects," in *Proceedings of the IEEE International Interconnect Technology Conference*, 2003, pp. 27–29.
17. K. D. Lee, Y. J. Park, T. Kim, and W. R. Hunter, "Via processing effects on electromigration in 65 nm technology," in *Proceedings of the 44th IEEE Annual International Reliability Physics Symposium*, 2006, pp. 103–106.
18. A. H. Fischer, O. Aubel, J. Gill, T. C. Lee, B. Li, C. Christiansen, F. Chen, M. Angyal, T. Bolom, and E. Kaltalioglu, "Reliability challenges in copper metallizations arising with the PVD resputter liner engineering for 65 nm and beyond," in *Proceedings of the 45th IEEE Annual International Physics Reliability Symposium*, 2007, pp. 511–515.

CHAPTER 4

Measurements Useful for Device Reliability Characterization

As described in Sec. 2.6, describing front-end-of-line (FEOL) device reliability characteristics, it is the interface traps and/or oxide traps that lead to the V_t/I_{dsat} shift of devices. For example, ΔV_t becomes negative, indicating a positive charge buildup in the gate dielectric, where a positive charge can result from either hole trapping in the dielectric or the creation of positively charged interface states at the dielectric interface. This chapter summarizes the characterization methodologies used to distinguish between bulk and interface traps, although it is hard to do so in the case of ultra-thin dielectrics.[1]

4.1 Capacitance-Voltage Measurement

The flat-band voltage shift ΔV_{fb} between pre- and poststressing from high-frequency (e.g., 1 MHz) capacitance-voltage (*C-V*) curves is an indication of oxide traps caused by the stressing, as shown in Fig. 4.1. The ΔV_{fb} induced by the oxide trap charge Q_{ot} can be expressed as follows[2]:

$$\Delta V_{fb} = -\frac{\alpha Q_{ot}}{C_{ox}} \tag{4.1}$$

where C_{ox} is the oxide capacitance, and α is a factor to account for the charge distribution throughout the oxide. For a simplified analysis, $\alpha = 1$ is used by assuming that all the charges are located at the Si-SiO_2 interface, which gives

$$Q_{ot} = -\Delta V_{fb} \cdot C_{ox} \tag{4.2}$$

Interface trap density D_{it} can be estimated from *C-V* measurement at a sufficiently high frequency that interface traps are

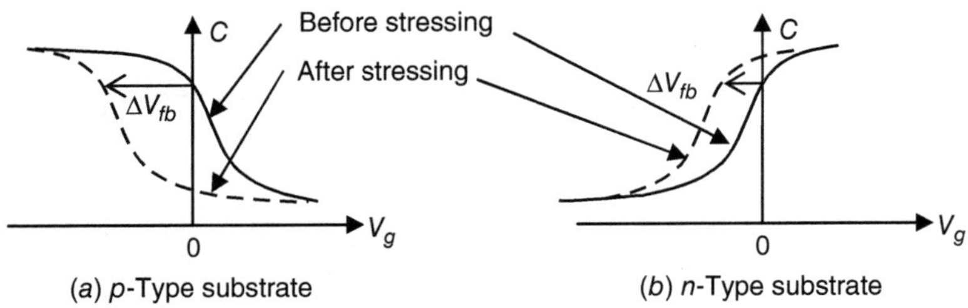

FIGURE 4.1 Schematic of C-V curve shift along the voltage axis owing to positive oxide charge indicated by the V_{fb} shift (ΔV_{fb}).

assumed not to respond using the Terman method.[2] In this method, a surface-potential ψ_s versus gate-voltage V_g curve is constructed first as follows:

Step 1: From the ideal high-frequency (HF) *C-V* curve, find ψ_s for a given capacitance at high frequency (C_{HF}).
Step 2: Find V_g on the experimental curve for the same C_{HF}, giving one point of a ψ_s-V_g curve.
Step 3: Repeat for other points to get a full ψ_s-V_g curve.

The experimental ψ_s-V_g curve is a stretched-out version of the theoretical curve, and the interface trap density is determined by the following equation:

$$D_{it} = \frac{C_{ox}}{q}\frac{d(\Delta V_g)}{d(\psi_s)} \tag{4.3}$$

where $\Delta V_g = V_g - V_g$ (ideal) is the voltage shift of the experimental from the ideal curve, V_g is the gate voltage, and ψ_s is the surface potential of Si at a specific gate voltage.

4.2 Direct-Current Current-Voltage

Direct-current current-voltage (DCIV)[3,4,5] measurements provide interface trap information based on DCIV peak height and oxide trap information from the peak position shift. However, for an ultrathin gate dielectric, modification of the method is necessary to reduce the gate tunneling effect, as detailed below.

Figure 4.2*a* is a schematic diagram of the DCIV method, by which the well current I_{well} at a forward bias of the source/drain well junction (with $V_e = V_s = V_d > 0$, $V_{well} = 0$) is measured as a function of V_g. Theoretically, there are three peaks in a DCIV curve for a device, as shown in Fig. 4.2*b*. As illustrated in Fig. 4.2*c*, the first peak occurs in the channel region (region I), the second peak occurs at the near source-substrate or drain-substrate junction (region II), and the third peak is

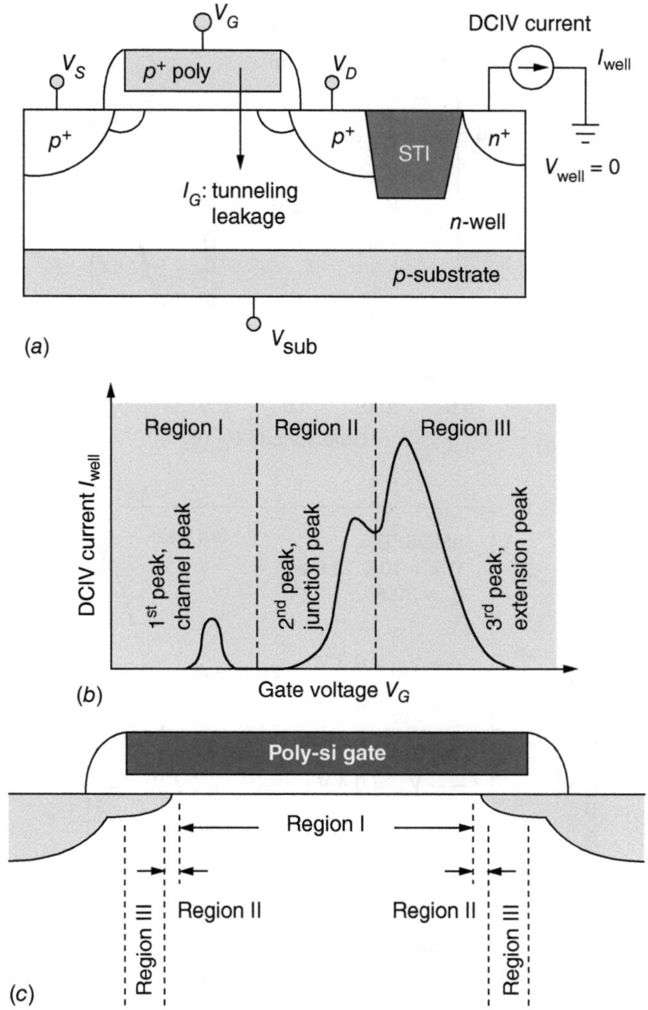

FIGURE 4.2 (*a*) Schematic diagram of the DCIV method for the interface trap/oxide trap measurement technique; (*b*) schematic three peaks of DCIV current as a measure of N_{it} or Q_{ot}; and (c) the corresponding three-peak regions. (*d*) The DCIV measured results for ultrathin gate oxide. Note that in the 20-Å gate oxide, direct tunneling leakage I_G overwrites the DCIV current I_{well} such that the first peak is not observable. (*e*) An improved DCIV by removing the direct tunneling leakage of the ultrathin gate oxide device under various V_{pn} biases ($V_{pn} = V_D - V_{well}$, and $V_{well} = 0$ V).[5] (*Used with permission from IEEE.*)

inside the source/drain extension region (region III). From the peak values and shift of the measured well current, oxide damage in their respective regions can be detected.

However, in reality, for an ultrathin (e.g., 20 Å thick) gate oxide, there is no peak in the DCIV curves owing to the large gate leakage I_G

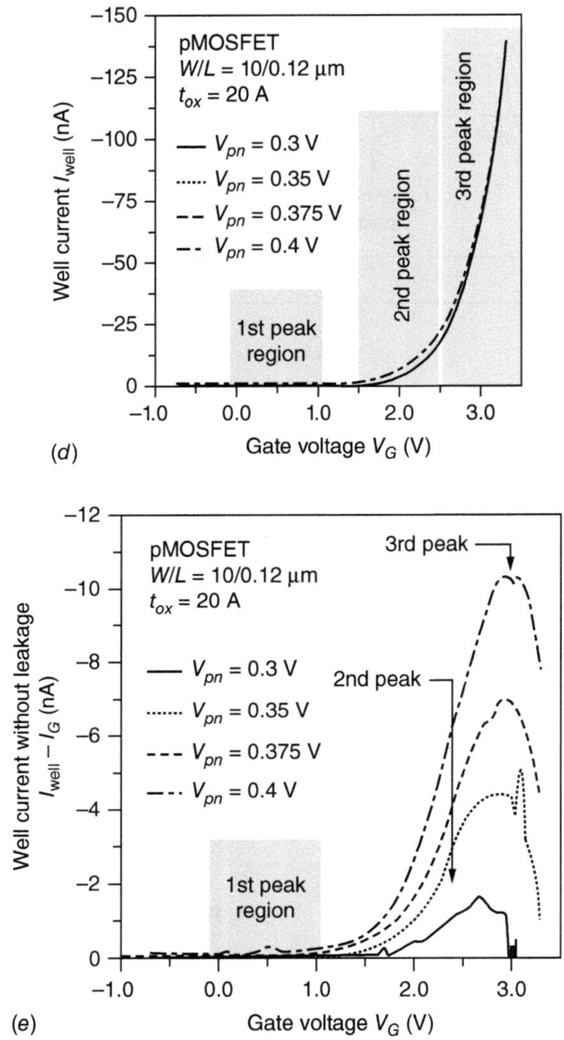

FIGURE 4.2 *(Continued)*

in ultrathin gate dielectrics, which makes the peak of the well current I_{well} unobservable, as shown in Fig. 4.2*d*. At least two solutions are provided in literature as follows:

- Chung et al.[5] proposed to remove the leakage current owing to a very thin gate oxide by also measuring gate leakage and using $I_{well} - I_G$ as an index instead of just I_{well} values, as shown in Fig. 4.2*e*. Under a larger V_{pn} bias (where $V_{pn} = V_D - V_{well}$, and $V_{well} = 0$ V), the peaks can be identified clearly. The $I_{well} - I_G$

curve in terms of V_g can be used to obtain a low-leakage DCIV measurement with forward bias (V_d = 0.3 to 0.4 V)

- Zhu et al.[6] modified the conventional DCIV method by measuring the well current at both $V_e > 0$ and subsequently at zero bias ($V_e = 0$) as a function of V_g. The modified DCIV current I_{DCIV} is defined as $I_{DCIV} = I_{well}(@V_e > 0) - I_{well}(@V_e = 0)$ to reduce the direct-tunneling-current effect of the ultrathin gate dielectric. Figure 4.3 shows the modified DCIV curves of a sample before and after negative-bias temperature instability (NBTI) stress from 10 to 31,200 s, with a clear peak at a certain V_g (denoted as V_{peak}).

4.2.1 Extraction of Interface Trap from Direct-Current Current-Voltage Measurement

The DCIV peak height above the baseline ΔI_{DCIV} is approximately proportional to the effective interface trap generated N_{it}[3,6]:

$$\Delta I_{DCIV} = \frac{1}{2} \cdot q \cdot N_{it} \cdot W \cdot L \cdot n_i \cdot \sigma \cdot v_{th} \cdot \left[\exp\left(\frac{qV_e}{nk_BT}\right) - 1\right] \tag{4.4}$$

where W and L are channel width and channel length, respectively. $\sigma = \sqrt{\sigma_n \sigma_p}$, σ_n, and σ_p are the electron and hole capture cross sections, q is the electronic charge, n_i is the intrinsic carrier concentration, v_{th} is

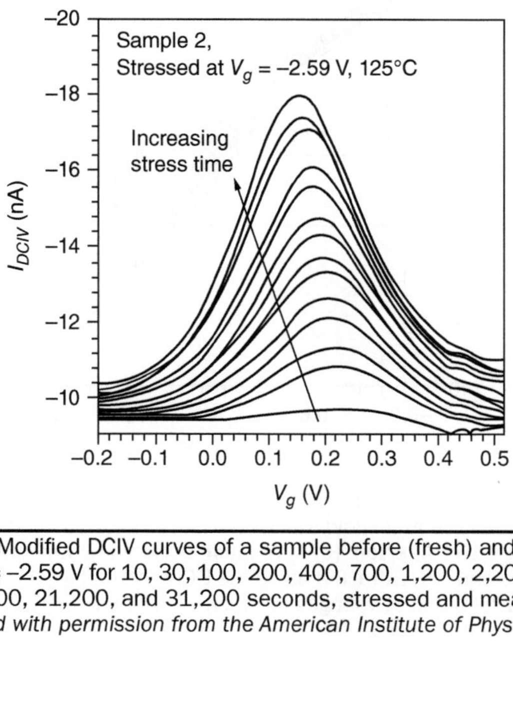

FIGURE 4.3 Modified DCIV curves of a sample before (fresh) and after NBTI stress at V_g = −2.59 V for 10, 30, 100, 200, 400, 700, 1,200, 2,200, 4,200, 7,200, 13,200, 21,200, and 31,200 seconds, stressed and measured at 125°C. (*Used with permission from the American Institute of Physics.*[6])

the thermal velocity, k_B is Boltzmann's constant, T is the temperature in kelvin, and n is a factor around 1.7 to 1.8. It has been confirmed that N_{it} extracted from the preceding formula is independent of the measurement temperature.[7] The interface trap generation induced by stressing such as hot-carrier insertion (HCI) or NBTI then can be calculated as follows:

$$\Delta N_{it} = N_{it}(\text{stress}) - N_{it}(\text{fresh}) \tag{4.5}$$

The stress-induced ΔN_{it} also can be related to the change in the subthreshold swing ΔS by the following formula[8,6]:

$$\frac{\Delta S}{S_0} = \frac{S(\text{stressed}) - S(\text{fresh})}{S(\text{fresh})} = \frac{q\Delta N_{it}/\Delta E_{it}}{C_{ox} + C_D} \tag{4.6}$$

where $S_0 = S(\text{fresh})$ is the subthreshold swing before stress, C_{ox} is the gate-oxide capacitance, C_D is the depletion capacitance (see Fig. 2.11*a*), and ΔE_{it} is the energy width of the interface trap distribution. Figure 4.4*b* shows a strong linear correlation between ΔN_{it} and $\Delta S/S_0$ that is independent of the stress voltage. For example, the S values for before and after NBTI stress from 10 to 31,200 seconds are extracted by linearly fitting the $I_d - V_g$ curves (Fig. 4.4*a*) at a range of I_d between approximately 2×10^{-9} and 2×10^{-8} Å.[6]

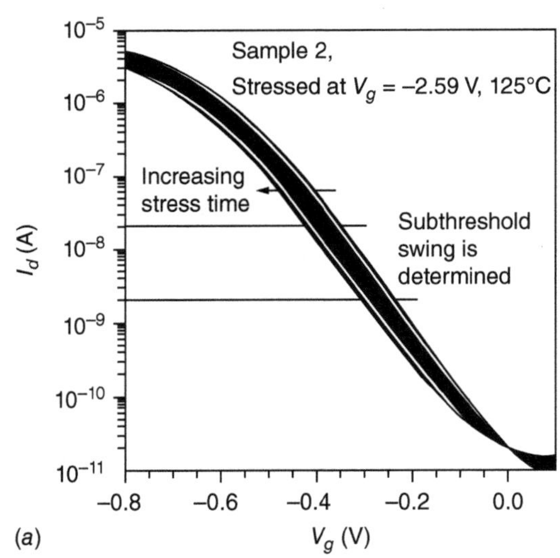

FIGURE 4.4 (*a*) An example of I_d-V_g curves of a sample before (fresh) and after NBTI stress. Subthreshold swings are extracted from linear fitting in the 2×10^{-9} to 2×10^{-8} Å region. (*b*) Correlation between subthreshold swing change $\Delta S/S_0$ and ΔN_{it} extracted using the DCIV method for the sample stressed at 125°C and at voltages between −2.09 and −3.09 V. (*Used with permission from the American Institute of Physics.*[6])

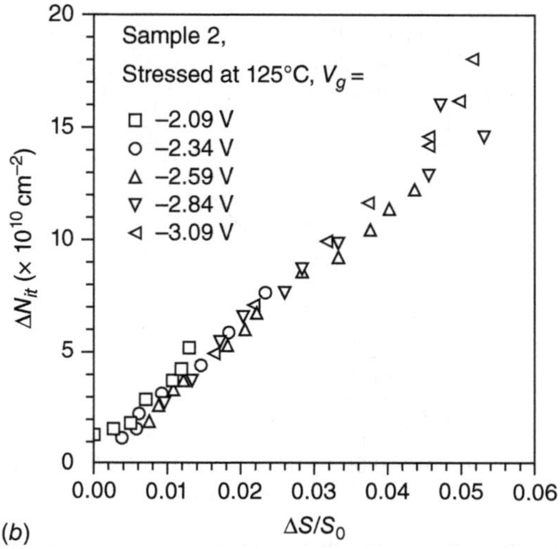

FIGURE 4.4 *(Continued)*

Equation (4.6) can be rewritten as Eq. (4.7), for which the slope of the correlation between ΔN_{it} and $\Delta S/S_0$ [(Fig. 4.4*b*) is $(\Delta E_{it}/q)(C_{ox}+C_D)$]. Thus, C_{ox} can be calculated based on Eq. (4.7) with the following assumptions: The ΔE_{it} simply can be assumed to be $E_g/2$ (where E_g is the Si energy gap), and C_D is independent of stress. Interestingly, equivalent oxide thickness (EOT) can be further estimated from C_{ox}. A very good agreement between this methodology and those extracted from the *C-V* measurement can be achieved, indicating that the ΔN_{it} value extracted from the modified DCIV measurement has a relatively high degree of accuracy[6]:

$$\Delta N_{it} = \frac{\Delta E_{it}}{q}(C_{ox}+C_D)\cdot\frac{\Delta S}{S_0} \qquad (4.7)$$

4.2.2 Extraction of Oxide Traps from the Direct-Current Current-Voltage Measurement

The preceding is talking about how to extract the ΔN_{it} from the DCIV measurement. On the other hand, ΔN_{ot} also can be estimated, which is discussed below.[6] In the DCIV measurement, the electron-hole recombination at the SiON-Si interface reaches a maximum, at which $\sigma_n N_s \approx \sigma_p N_p$, where N_s and N_p are, respectively, the surface concentrations of the electrons and holes.[4,9] Because $\sigma_n \approx \sigma_p$, V_{peak} corresponds to the case where $N_s \approx N_p$, namely, E_{Fs} is close to E_i at V_{peak}. Therefore, similar to the midgap method discussed in Sec. 4.5:, the shift of ΔV_{peak} can be considered an indicator of stress-induced oxide charges.

However, it is noted that V_{peak}, which depends on V_e, is different from V_{mg}, which corresponds to the case where $V_e = (V_s = V_d) = 0$ because

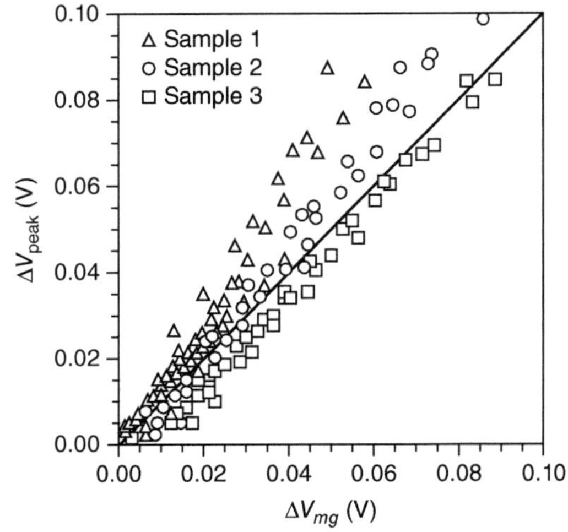

FIGURE 4.5 Correlation between midgap voltage shift ΔV_{mg} extracted from $I_d - V_g$ curves and DCIV peak-position shift ΔV_{peak} for different samples under NBTI stressed at 125°C at different voltages. (*Used with permission from the American Institute of Physics.*[6])

the channel potential is modified by the source/drain bias V_e during the DCIV measurement. As an example, the relationship between ΔV_{peak} and ΔV_{mg} is shown in Fig. 4.5. The relatively large deviation arises from the uncertainty of I_{mg} in the midgap method and from the broad DCIV peak, as shown in Fig. 4.3*b*, owing to the interface-trap spatial and energy-level distributions.[4] Nevertheless, ΔV_{peak} can be used as a rough, simple, and direct indicator of the stress-induced oxide charges.[6]

4.3 Gated-Diode Method

The gated-diode (GD) method[5,10,11] is shown schematically in Fig. 4.6. There are some similarities between the DCIV method and the GD method. The differences between these methods include (1) the DCIV method measures the well current (Fig. 4.3*a*), whereas the GD method measures the drain current (I_D, Fig. 4.6) and (2) the DCIV method needs four terminals, whereas the GD needs only three terminals.

Similar to DCIV method, the conventional GD method also suffers from large gate tunneling leakage in a device with thin gate oxide (T_{ox} < 30 Å). The low-gate-leakage gate-diode (L^2-GD) method thus was proposed.[5] Figure 4.7*a* shows that in the conventional GD measurement for the ultrathin-gate-oxide device, tunneling leakage overwrites the GD current, especially for V_g > 2 V. On the other hand,

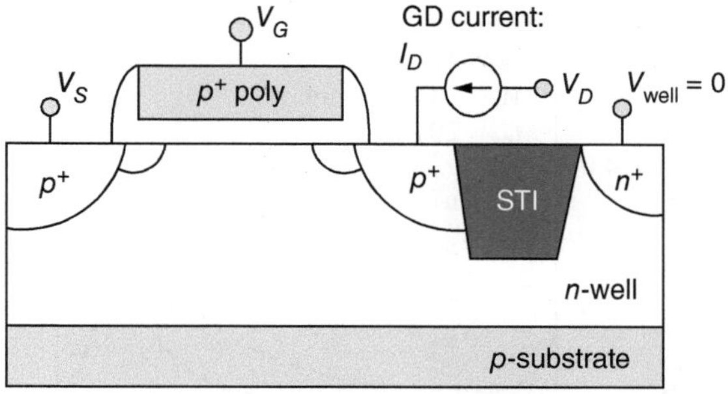

FIGURE 4.6 Schematic diagram of the gated-diode (GD) method for the interface-trap/oxide-trap measurement technique.[5] (*Used with permission from IEEE.*)

with the L²-GD measurement, second/third peaks appear at high V_{pn} bias ($V_{pn} = V_D - V_{well}$, and $V_{well} = 0$ V) for the same ultrathin-gate-oxide device, as shown in Fig. 4.7*b*.

Table 4.1 compares DCIV, conventional GD, and the new L²-GD methods.[5] In short, (1) DCIV makes use of the bipolar structure such that four terminals are required for the measurement, (2) the conventional GD method uses a small reverse bias at the drain/substrate junction such that usually only one peak can be detected, and (3) the L²-GD method combines the merits of the other two methods, providing two/three peaks.

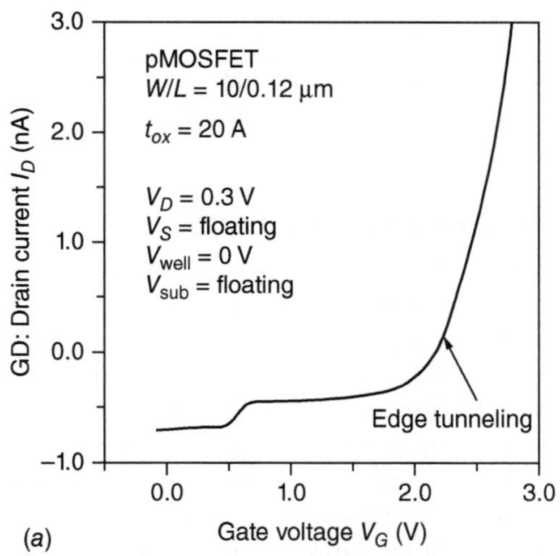

FIGURE 4.7 (*a*) Conventional GD measurement for the ultrathin-gate-oxide device. Tunneling leakage overwrites the GD current for $V_G > 2$ V. (*b*) L²-GD measurement for the ultrathin-gate-oxide device under various V_{pn} biases. Second/third peaks appear at high V_{pn} bias.[5] (*Used with permission from IEEE.*)

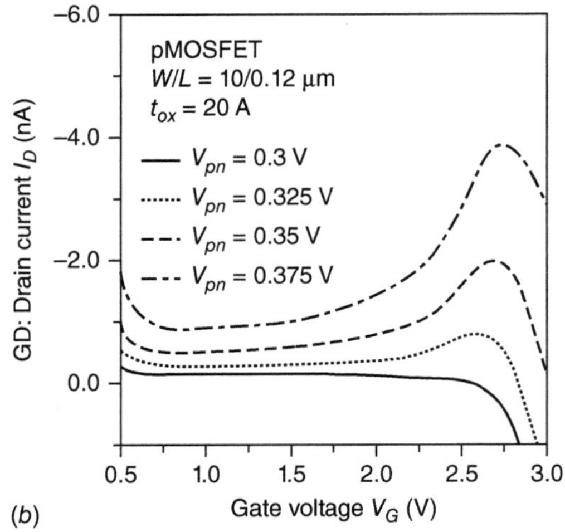

Figure 4.7 *(Continued)*

DCIV Method	Gated-Diode Method	L²-GD Method
Four terminals: gate, drain, well, substrate	Three terminals: gate, drain, well	Three terminals: gate, drain, well
V_d = 0–0.7 V	V_d < 0.3 V or reverse bias	V_d = 0–0.7 V
Two/three peaks	One peak	Two/three peaks
I_{well}-V_g curve	I_d-V_g curve	I_d-V_g curve

Table 4.1 Comparison among the DCIV Method, Gated-Diode (GD) Method, and L²-GD Method.[5] *(Used with permission from IEEE.)*

4.4 Charge-Pumping Measurement

Charge-pumping (CP)[12] measurements (typically at 0.01 to 1 MHz) can be used to characterize the interface trap (N_{it}) on small-geometry metal-oxide-semiconductor field-effect transistors (MOSFETs) instead of large-dimension MOS capacitors.[2] The measurement configuration is given in Fig. 4.8 for an N-channel metal-oxide semiconductor (NMOS), with the source/drain of the MOS connected together at a reverse-bias voltage V_r and the gate under pulse-biased voltage. The magnitude of the reverse-bias voltage generally is small, for example, 0.1 V. The amplitude of the pulse should be large enough to drive the gate into inversion and accumulation alternatively. The pulse applied to gate can be square, triangular, or trapezoidal.

In theory, the explanation for the CP method is as follows:

- In Fig. 4.8*a*, the gate pulse is with positive bias, and the NMOS is under strong inversion. The corresponding band diagram is shown in Fig. 4.8*c*, where the interface traps are more occupied, as represented by ⊗. The corresponding trapping state is negative according to the analysis from Fig. 2.24.

- When the gate pulse falls from its positive to its negative values (Fig. 4.8*b*), the NMOS turns to be under flat band (Fig. 4.8*d*) first and then strong accumulation (Fig. 4.8*e*). During this transient time, electrons in the inversion layer drift to source/drain, and the interface is less trapped. The corresponding trapping state then is positive.
- In the case of channel transition from inversion (Fig. 4.8*c*) to flat band (Fig. 4.8*d*), the energy level involved in the trap

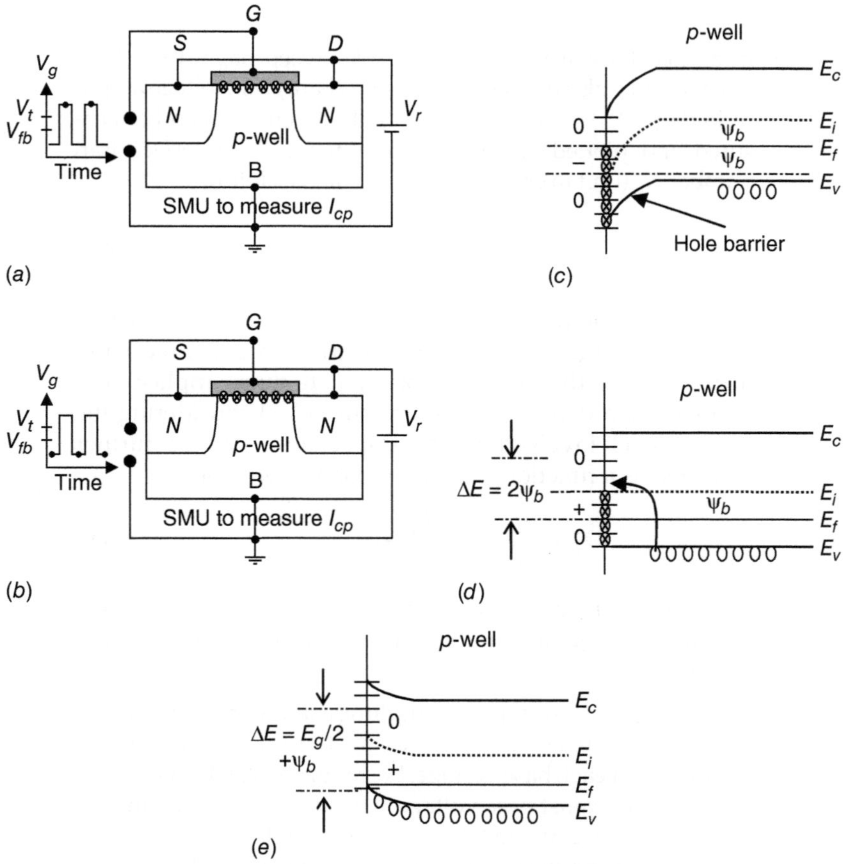

FIGURE 4.8 The measurement setup and band diagram for charge-pumping measurement. (*a*) Gate pulse under inversion (interface traps more filled). (*b*) Gate pulse under accumulation (interface traps less filled). (*c*) Band diagram for inversion with interface negatively trapped. (*d*) Band diagram for flat band with interface positively trapped. (e) Band diagram for strong accumulation (assuming that E_f touches E_v at interface) with interface positively trapped.

change ΔE is $2\psi_b$. In the case of channel transition from inversion (Fig. 4.8*c*) to further to strong accumulation, the flat band assuming E_f touches E_v at the interface (Fig. 4.8*e*), and the energy level involved in the trap change ΔE is $E_g/2 + \psi_b$.

- The process of alternative interface traps getting occupied and emptied will repeat when the gate bias is switched from positive to negative potential alternatively.

In practice, the basic CP technique involves measuring the substrate current while applying voltage pulses of fixed amplitude, rise time, fall time, and frequency to the gate of the transistor, with the source, drain, and body tied to ground. The pulse can be applied with a fixed-amplitude voltage base sweep or a fixed-base variable-amplitude sweep.[13]

- In a fixed base, a variable-amplitude sweep has a fixed base voltage and pulse frequency with step changes in voltage amplitude (Fig. 4.9*a*). At each amplitude, body current can be measured and plotted against the amplitude. The interface trap number N_{it} (unit in cm^{-2}) can be calculated as follows:

$$N_{it} = \frac{I_{cp}}{q \cdot A \cdot f} \tag{4.8}$$

where I_{cp} is the measured CP current, q is the fundamental electronic charge, A is the gate dielectric area under measurement, and f is the frequency of the gate pulse applied. The interface trap density D_{it} (unit in cm^{-2} eV^{-1}) considering the energy level involved in the trap change can be further extracted as a function of ΔE based on this equation:

$$D_{it} = \frac{N_{it}}{\Delta E} = \frac{I_{cp}}{q \cdot A \cdot f \cdot \Delta E} \tag{4.9}$$

ΔE is the difference between the inversion Fermi level and the accumulation (or flat-band) Fermi level, as depicted in Fig. 4.8*e* (Fig. 4.8*d*).

- In a voltage base sweep, the amplitude and period (width) of the pulse are fixed while sweeping the pulse base voltage (Fig. 4.9*b*). At each base voltage, body current can be measured and plotted against base voltage. The information obtained is similar to that extracted from an amplitude sweep. Typical curves of CP versus base voltage are given in Fig. 4.10.

Again, for ultrathin gate dielectrics, the CP method has suffered from high leakage current owing to the direct tunneling through the gate. As a result, the traditional CP technique is limited because the CP current is covered by the leakage current. The incremental-frequency

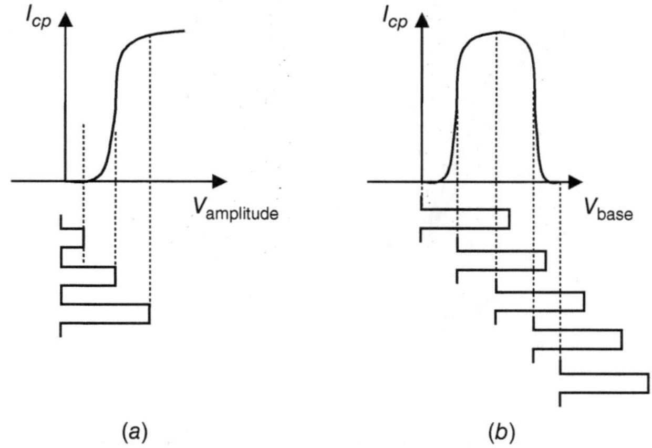

FIGURE 4.9 Examples of pulse waveforms applied in CP measurements and the corresponding CP current I_{cp}: (*a*) for amplitude sweep, base voltage is constant; (*b*) for base voltage sweep, pulse amplitude is constant.

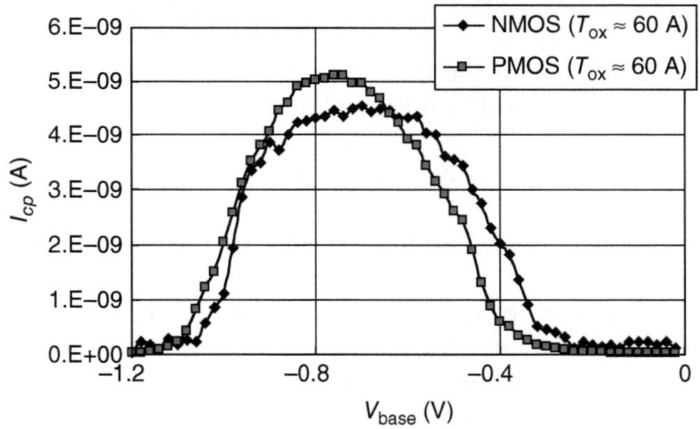

FIGURE 4.10 Typical CP current I_{cp} plotted against the base voltage using the voltage base sweep method. The pulse frequency used is 0.5 MHz.

charge-pumping (IFCP) method,[14,15] an improved CP method, thus was proposed for determination of the interface traps in ultrashort-gate-length CMOS devices with ultrathin gate oxide in the direct tunneling regime. For a MOSFET with an ultrathin gate dielectric, the CP current I_{cp} to the gate consists of both interface-trap and tunneling currents. The interface-trap current depends on the frequency of the gate bias pulse, whereas the tunneling current does not. Therefore, the current owing to tunneling electrons can be eliminated by measuring I_{cp} at two different frequencies and subtracting the results. Figure 4.11 gives an example of CP current for the IFCP measurement with the frequencies of the gate pulse 10 and 1,000 kHz. The interface-trap

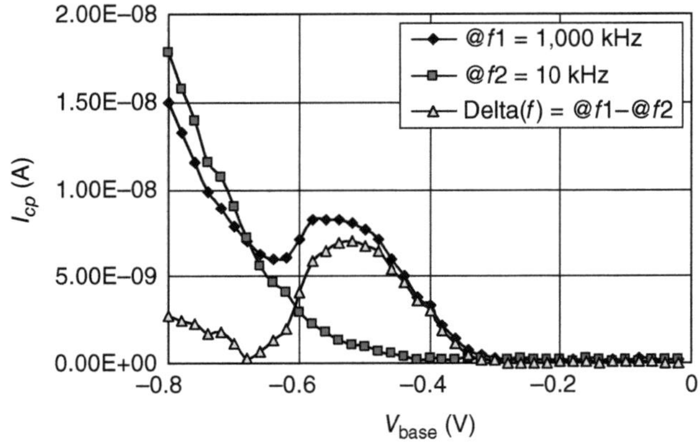

FIGURE 4.11 Typical I_{cp} at both high frequency ($f1$ = 1,000 kHz) and low frequency ($f2$ = 10 kHz) for ultrathin gate oxide. The I_{cp} difference between the two frequencies is also plotted, from which the peak can be clearly identified.

density is calculated after subtracting the measured I_{cp} at 1,000 kHz from the one at 10 kHz.

The peak value of the IFCP currents ΔI_{peak} is related to ΔN_{it} by Eq. (4.10), which is quite similar to Eq. (4.8).

$$N_{it} = \frac{\Delta I_{IFCP}^{\text{peak}}}{q \cdot A \cdot \Delta f} \tag{4.10}$$

where Δf is the frequency difference between the two IFCP measurements.

A derivative of the conventional CP method discussed earlier is the VT²CP method[16] proposed recently, which is very useful to probe the depth-dependent traps in the dielectric, especially for traps in the interfacial-layer (SiO_2)/high-k (HfO_2)/metal-gate dielectric stacks.

4.5 Midgap Measurement for Interface and Oxide Trap Separation

Interface traps are commonly believed to be acceptor-like in the upper half of the band gap and donor-like in the lower half, so all interface traps are neutral when the Si surface Fermi level E_{Fs} is at the middle energy E_i,[17] namely, at $V_g = V_{mg}$. Therefore, the V_{mg} shift (ΔV_{mg}) is only due to the stress-induced (either by HCI, NBTI, or whatever) oxide charges (ΔN_{ot}) as

$$\Delta N_{ot} = C_{ox} \cdot \Delta V_{mg} / q \tag{4.11}$$

where the midgap voltage shift ΔV_{mg} is defined as

$$\Delta V_{mg} = (V_{mg})_{\text{poststress}} - (V_{mg})_{\text{prestress}} \tag{4.12}$$

which can be extracted from I-V[18] or C-V[19] measurements. Using the C-V measurement requires a relatively large area and is adversely affected by the large gate leakage current of the ultrathin gate dielectric, resulting in a large experimental error, especially for an ultrathin gate dielectric.[6] On the other hand, the I-V measurement and data interpretation can be performed more easily. In this method, the I_d-V_g curves are measured first. Then the midgap current I_{mg} is calculated by substituting the surface potential $\psi_s = (k_BT/q)\ln(N_Dn_i)$ into the subthreshold current equation[6,20]:

$$I_{mg} = \mu_{eff}\left(\frac{W}{L}\right)\frac{aC_{ox}}{2(q/k_BT)^2}\left(\frac{n_i}{N_D}\right)(1-e^{-V_Dq/k_BT})\left(\ln\frac{N_D}{n_i}\right)^{-1/2} \tag{4.13}$$

where μ_{eff} is the effective hole mobility, a is a parameter, and N_D is the channel doping concentration. When I_{mg} is calculated, the V_{mg} value then can be extracted from the extrapolation of the linear fitting of the log (I_d) versus V_g curves (e.g., Fig. 4.4*a*). One precaution is that the μ_{eff} value depends on the interface trap density N_{it}, which makes quantitative analysis difficult. However, for a first-order estimation, an ideal μ value (e.g., 400 cm^2/V·s)[6] can be used by ignoring the N_{it} effect.

To derive the interface trap from the midgap measurement, the stretchout voltage V_{so} is defined as the voltage difference between the midgap and threshold points as follows[20]:

$$V_{so} = V_t - V_{mg} \tag{4.14}$$

The V_t shift owing to interface traps thus is the difference in the stretchout voltage on respective subthreshold-current curves as follows:

$$\Delta V_{Nit} = (V_{so})_{\text{poststress}} - (V_{so})_{\text{prestress}} \tag{4.15}$$

The interface traps (ΔN_{it}) thus are determined by

$$\Delta N_{it} = C_{ox}\cdot\Delta V_{Nit}/q \tag{4.16}$$

4.6 Carrier-Separation Measurement

For the gate oxide in an MOS under electrical stress, the total gate leakage current I_g basically can be separated into its electron I_e and hole I_h components. This separation can be realized by the carrier separation measurement technique.[21] The purpose of separating the electron and hole components is to identify the major contributor to the leakage current under various polarities of stress conditions. With such information at hand, the effect of substrate and gate injection of electrons or holes on the breakdown and electrical degradation of the gate stack and devices then can be characterized.

Figure 4.12 provides the schematic illustrations of the carrier-separation measurement for NMOS and P-channel metal-oxide

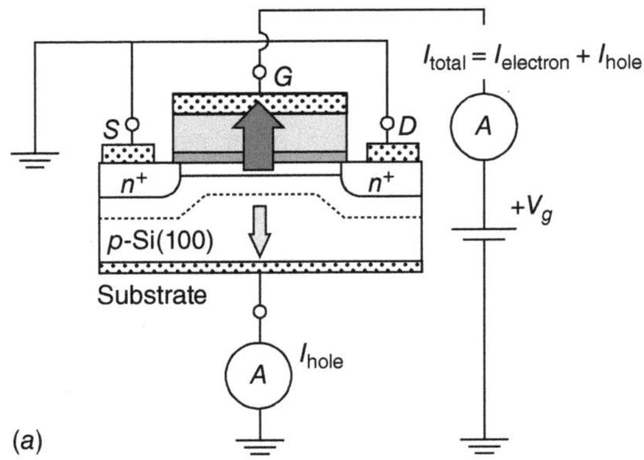

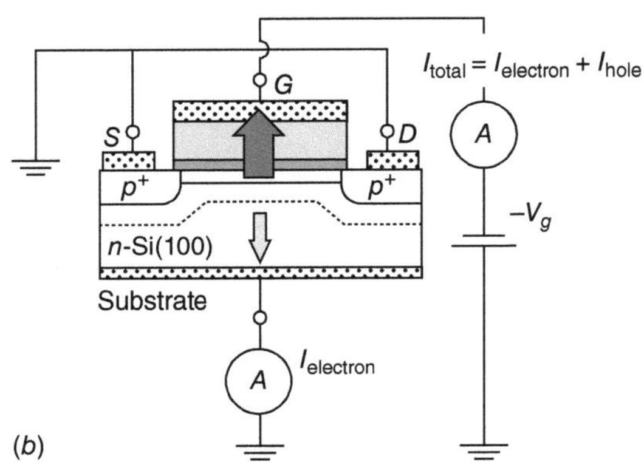

FIGURE 4.12 Schematic illustrations of the carrier-separation measurement for (*a*) NMOS and (*b*) PMOS devices, respectively. (*Used by permission from the Institute of Electronics, Information and Communication Engineers.*[22])

semiconductor (PMOS) devices, respectively.[22] For both cases, the inversion condition is reached. That is, a positive voltage is applied to the gate of the NMOS with substrate injection of electrons, whereas a negative voltage is applied to the gate of the PMOS with gate injection of electrons. For the NMOS case in Fig. 4.12*a*, the substrate current is the hole component, whereas for the PMOS in Fig. 4.12*b*, the substrate current is the electron component. Therefore, with such experimental setups, the contribution of electron and hole components to the gate leakage current can be properly separated.

4.7 Current-Voltage Characteristics

Current-voltage (*I-V*) measurement is the most basic method to characterize device performance. With justifiable interpretation, I_d-V_g measurement can provide interface-trap and oxide-trap information. To characterize the HCI or NBTI degradation, a gradual degradation in

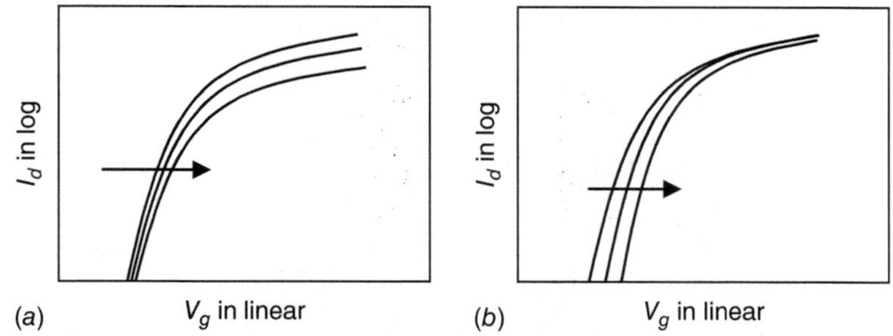

FIGURE 4.13 Schematic of two types of I_d-V_g shift during reliability stressing.

the subthreshold slope with increasing stress time is an indication of the interface trap, as shown in Fig. 4.13*a*. On the other hand, the parallel shift of the I_d-V_g curves to higher gate voltages with increasing stress time indicates the trapping of electrons in the oxide, as shown in Fig. 4.13*b*.

4.8 References

1. I. Polishchuk, Y. C. Yeo, Q. Lu, T. J. King, and C. Hu, "Hot carrier reliability of P-MOSFET with ultrathin silicon nitride gate dielectric," in *Proceedings of the 39th Annual International Reliability Physics Symposium*, 2001, pp. 425–430.
2. D. K. Schroder, "Oxide and interface trapped charge," in *Semiconductor Material and Device Characterization*, Wiley, New York, 1990, p. 278.
3. A. Neugroschel, C. T, Sah, K. M. Han, M. S. Caroll, T. Nishida, I. T. Kavalieros and Y. Yu, "Direct-current measurements of oxide and interface traps on oxidized silicon," *IEEE Transactions on Electron Devices*, Vol. 42, 1995, pp. 1657–1662.
4. B. B. Jie, W. K. Chim, M. F. Li, and K. F. Lo, "Analysis of the DCIV peaks in electrically stressed pMOSFETs," *IEEE Transactions on Electron Devices*, Vol. 48, No. 5, 2001, pp. 913–920.
5. S. S. Chung, D. K. Lo, J. J. Yang, and T. C. Lin, "Localization of NBTI-induced oxide damage in direct tunneling regime gate oxide pMOSFET using a novel low gate-leakage gated-diode (L^2-GD) method," in *Proceedings of the International Electron Devices Meeting (IEDM)*, 2002, pp. 513–516.
6. S. Zhu, A. Nakajima, T. Ohashi, and H. Miyake, "Interface trap and oxide charge generation under negative bias temperature instability of *p*-channel metal-oxide-semiconductor field-effect transistors with ultrathin plasma-nitrided SiON gate dielectrics," *Journal of Applied Physics*, Vol. 98, 2005, p. 114504.
7. S. Y. Zhu, A. Nakajima, T. Ohashi, and H. Miyake, "Modified direct-current current-voltage method for interface trap density extraction in metal-oxide-semiconductor field-effect-transistor with tunneling gate dielectrics at high temperature," *Japanese Journal of Applied Physics*, Vol. 44, 2005, pp. L60–L62.
8. S. M. Sze, *Physics of Semiconductor Devices*, 2nd ed., Wiley, New York, 1981, Chap. 8.
9. J. Cai and C. T. Sah, "Monitoring interface traps by DCIV method," *IEEE Electron Device Letters*, Vol. 20, No. 1, 1999, pp. 60–63.
10. P. Seckbacher, J. Berger, A. Asenov, F. Koch, and W. Weber, "The 'gated-diode' configuration in MOSFET's, a sensitive tool for characterizing hot-carrier degradation," *IEEE Transactions on Electron Devices*, Vol. 42, No. 7, 1995, pp. 1287–1296.

11. S. S. Chung, C. M. Yih, S. M. Cheng, and M. S. Liang, "A new oxide damage characterization technique for evaluating hot carrier reliability of flash memory cell after P/E cycles," in *Proceedings of the Symposium on VLSI Technology Digest of Technical Papers*, 1997, pp. 111–112.
12. G. Groeseneken, H. E. Maes, N. Beltran, and R. F. De Keersmaecker, "A reliable approach to charge-pumping measurements in MOS transistors," *IEEE Translations on Electron Devices*, Vol. 31, No. 1, 1984, pp. 42–53.
13. Keithley Application Note Series, Number 2457, "Making charge-pumping measurements with the Model 4200-SCS semiconductor characterization system," www.keithley.com, Access date: December 2010.
14. S. S. Chung, S. J. Chen, C. K. Yang, S. M. Cheng, S. H. Lin, Y. C. Sheng, H. S. Lin, K. T. Hung, D. Y. Wu, T. R. Yew, S. C. Chien, F. T. Liou, and F. Wen, "A novel and direct determination of the interface traps in sub-100-nm CMOS devices with direct tunneling regime (12 to16 Å) gate oxide," *VLSI Technical Digest*, 2002, pp. 74–75.
15. S. U. Han, H. S. Kang, and B. K. Kang, "Time-dependent degradation due to negative bias temperature instability of *p*-MOSFET with an ultrathin SiON gate dielectric," *Microelectronic Engineering*, Vol. 83, 2006, pp. 520–527.
16. M. B. Zahid, R. Degraeve, L. Pantisano, J. F. Zhang, and G. Groeseneken, "Defects generation in SiO_2/HfO_2 studied with variable T_{charge}-$T_{discharge}$ charge pumping (VT^2CP)," in *Proceedings of the 45th Annual International Reliability Physics Symposium*, 2007, pp. 55–60.
17. D. K. Schroder and J. A. Babcock, "Negative bias temperature instability: Road to cross in deep submicron silicon semiconductor manufacturing," *Journal of Applied Physics*, Vol. 94, 2003, pp. 1–18.
18. S. S. Tan, T. P. Chen, C. H. Ang, and L. Chan, "Mechanism of nitrogen-enhanced negative bias temperature instability in pMOSFET," *Microelectronics Reliability*, Vol. 45, No. 1, 2005, pp. 19–30.
19. S. Tsujikawa and J. Yugami, "Positive charge generation due to species of hydrogen during NBTI phenomenon in pMOSFETs with ultra-thin SiON gate dielectrics," *Microelectronics Reliability*, Vol. 45, 2005, pp. 65–69.
20. P. J. McWhorter and P. S. Winokur, "Simple technique for separating the effects of interface traps and trapped-oxide charge in metal-oxide-semiconductor transistors," *Applied Physics Letters*, Vol. 48, 1996, pp. 133–135.
21. C. Chang, C. Hu, and R. W. Brodersen, "Quantum yield of electron impact ionization in silicon," *Journal of Applied Physics*, Vol. 57, No. 2, 1985, pp. 302–309.
22. J. Molina, K. Kakushima, P. Ahmet, K. Tsutsui, N. Sugii, T. Hattori, and H. Iwai, "Carrier separation and V_{th} measurements of W-La2O3-gated MOSFET structures after electrical stress," *IEICE Electronics Express*, Vol. 4, No. 6, 2007, pp. 185–191.

PART 2

Front-End-of-Line

CHAPTER 5

Hot-Carrier Injection

Hot-carrier injection (HCI) is one of the most important reliability issues accompanying the reduction of gate sizes to micrometer and submicrometer dimensions in metal-oxide-semiconductor field-effect transistor (MOSFET) devices. This effect basically arises from the high field in the channel near the drain region, which causes the inversion-layer electrons [in the case of N-channel metal oxide semiconductors (NMOS)] to be accelerated and thus to gain sufficient energy to overcome the Si-SiO_2 barrier resulting in damage to the gate oxide, which results in a degradation in the MOS current-voltage (*I-V*) characteristics and parameter shift (e.g., shifts of drive current I_{dsat}, threshold voltage V_t, transconductance G_m, and so on). HCI is a well-known phenomenon for short-channel devices. However, in the 90-nm and beyond technology nodes, temperature/oxide-thickness dependence and degradation mechanisms of thin oxide are somehow different from those of previous nodes.

Several trap-generation mechanisms have been proposed to contribute to hot-carrier degradation as follows:

- Oxide trap Q_{ot},[1] in which electron-hole defect recombination and the corresponding energy release are likely the root cause. The traps in the gate dielectric can lead to a shift of $V_t/I_{dsat}/G_m$[2] as well as a gate leakage current increase owing to trap-assisted tunneling. This increase in gate leakage current can become a reliability concern for logic transistors with ultrathin gate dielectrics because it can result in increased power consumption.
- Interface trap Q_{it} at the Si-SiO_2 interface.[3]
- Interface trap Q_{it} at the SiO_2-polysilicon interface.[4] In both interface-trap cases, the release of hydrogen and the creation of dangling bonds have been identified as root causes.[4,5]

Hot-carrier-induced degradation in the MOS *I-V* characteristics and parameter shift can be explained in the following two ways.

On the one hand, hot carrier can lead to a mobility reduction of the carriers in the channel underneath the damaged region, which increases the channel resistance and then decreases the drive current. On the other hand, hot carrier also leads to an increase in threshold voltage V_t according to Eq. (5.1) because the Q_i (including both the oxide trap and the interface trap) will increase owing to hot-carrier stressing. An increased V_t means a lower carrier density in the inversion layer for a fixed applied gate voltage [Eq. (2.43)]. Note that using the MOS capacitor structure, V_t can be written as follows [Eq. (2.29)]:

$$V_t = V_{fb} + 2\psi_b + V_{ox} = \phi_{ms} - \frac{Q_i}{C_{ox}} + 2\psi_b + \frac{\sqrt{2\varepsilon_{\text{Si}} q N_A (2\psi_b)}}{C_{ox}} \tag{5.1}$$

where $\psi_b = V_{\text{thermal}} \ln(N_A/n_i)$, n_i is the intrinsic electron concentration, $V_{\text{thermal}} (= k_B T/q)$ is the thermal voltage, N_A is the bulk dopant concentration, q is the electron charge, ε_{Si} is the silicon permittivity, C_{ox} is the gate-oxide capacitance per unit area, ϕ_{ms} is the metal-Si work-function difference, Q_i includes Q_{ot} and Q_{it} for the oxide and interface traps, respectively.

In the 1980s, HCI degradation in NMOS devices was a critical issue because the transistor became smaller and smaller when the device channel length evolved from micron to 0.5- to 0.4-μm nodes, but the power-supply voltage V_{dd} was kept the same (~5 V).[6,3] At that time, the hot-carrier-induced reliability issue was resolved through new drain and gate architectures, for example, double-diffusion drains (DDDs),[7] lightly doped drain (LDD) structure such as the inverse-T LDD (ITLDD), [8] gate-to-drain overlapped devices (GOLDs),[9] and large-angle-tilt implanted drains (LATIDs). [10] The corresponding worst case of NMOS degradation was clearly identified to occur for the maximum-substrate-current (I_{submax}) condition at $V_{gs} \approx V_{ds}/2$ because under such conditions the maximum number of interface traps (ΔN_{it}) were induced by the hot (energetic) carriers.[11] Correspondingly, the channel-hot-carrier (CHC) generation rate was well related to the substrate-to-drain-current ratio (i.e., I_{sub}/I_d).

From the mid-1990s, not only the channel length L_g and gate-oxide thickness t_{ox} were scaled down, but also the power-supply voltage V_{dd} started to drop, considering both reliability improvement and power-consumption reduction. As a result, HCI damage became less of critical for CMOS nodes down to 120 nm owing to reductions in both the lateral field and HCI generation rate ($\propto I_{sub}/I_d$).[12] In this period, the worst-case hot-carrier-induced damage still corresponded to the I_{submax} condition.

With the continuous scaling down of V_{dd} and T_{ox} and the shortening of the gate length L_g, one would expect a large reduction in classic HCI effects owing to the large reduction in the channel field and thus a reduction in the hot-carrier generation rate. Also, the channel carriers might not be able to gain enough energy to climb over the electron (hole) barrier height at the SiO_2-Si interface. This was observed in the 0.13-μm technology, in which NMOS HCI performance is better for thinner oxides when gate oxide thicknesses 18 and 22 Å are compared.[13] However, the story is different when the gate oxide becomes thinner and thinner. With the CMOS continuous scaling down to 100 nm and beyond, CHC degradation became a concern again as V_{dd} scaling was slowed down or stopping because of the nonscalability of the subthreshold slope, whereas L_g continued to scale down to 40 nm and beyond. This induced an increase in lateral electrical field again. Other highlights include that the worst-case stressing conditions have changed from $V_g \approx V_d/2$ for input-output (I/O) devices (corresponding to thick oxides) to $V_g = V_d$ for the core devices (corresponding to thin oxides).[14–15,16] The worst temperature also has shifted from room temperature to high temperature.[17] All these topics will be addressed in this chapter.

In this chapter, an introduction to HCI is given first. Then the physical mechanisms of HCI are compared between the old technology nodes (0.13 μm and above) and advanced nodes (90 nm and below). The HCI characterization methodology then is detailed. The other HCI phenomena, such as screening effect, degradation saturation, temperature effect, and bulk bias effect, are also emphasized. This is followed by the detailed explanation of HCI improvement methodologies through structural design and process optimization. Finally, the HCI qualification requirements and practice are given.

5.1 Maximum Channel Electrical Field

This section describes the maximum channel electrical field E_{max}, where the largest impact ionization is located. Figure 5.1 is a schematic illustration of the CHC phenomenon using NMOS device as an example. At the off-state with $V_g = 0$ V (Fig. 5.1*b*), electrons from the source side cannot flow to the drain because there is a potential barrier at the source side. However, with a larger V_g applied (e.g., $V_g = V_d$ for the on-state) (Fig. 5.1*c*), the potential barrier at the source side is "pulled down," and the electrons can flow from source to drain. At the pinch-off region, the channel electrons are accelerated by the electrical field, which is just the derivative of the potential in terms of *x*, as shown in Fig. 5.1*d*. The maximum electrical field E_{max} is located near the drain for long-channel devices.

The accelerated channel electrons then gain kinetic energy that is higher than the thermal energy of the surrounding lattice. Referring to Fig. 2.3*b*, the electron kinetic energy E_{KE} is expressed as follows:

$$E_{KE} = E_{\text{total}} - E_c = k_B T_e > k_B T_{\text{Si}} \tag{5.2}$$

where T_e is the electron temperature and T_{Si} is the lattice temperature. With the kinetic energy gained, the carriers turn to be "hot" because they possess a temperature T_e that is much larger than the surrounding lattice temperature T_{Si}. As a result, impact ionization in silicon takes place, and then the electron-hole pairs are created (Fig. 5.1*a*). Attributed to the local electrical field, the generated electrons likely inject into the gate oxide, creating damage in the form of oxide charge

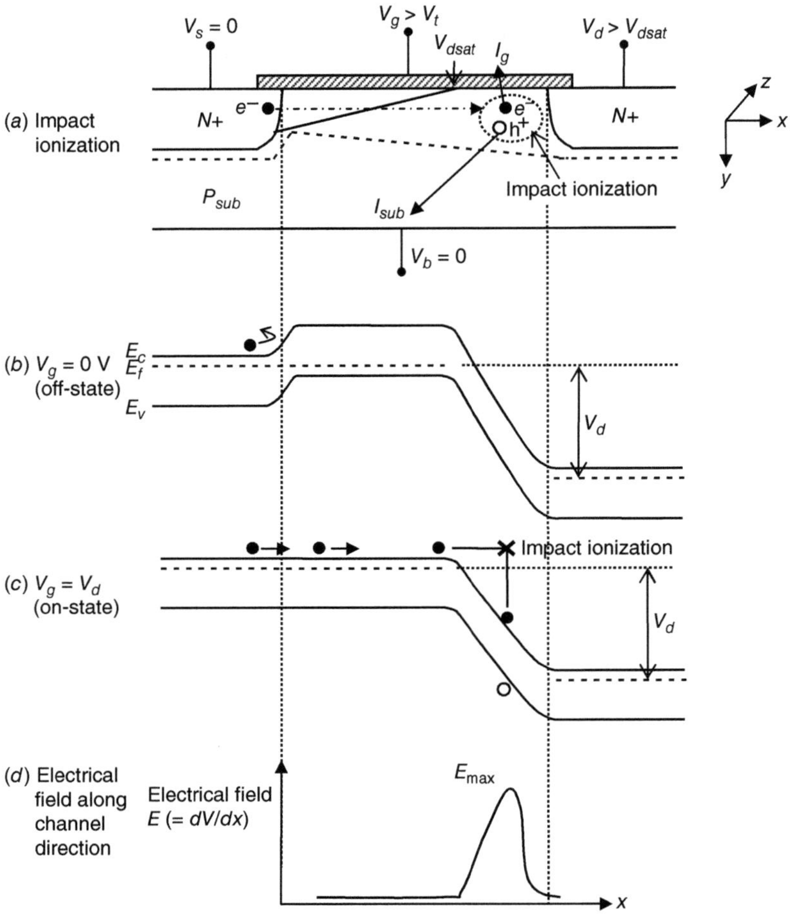

FIGURE 5.1 (*a*) Schematic illustration of channel hot-carrier (CHC) impact ionization. (*b*) Band diagram along the channel direction for $V_g = 0$ V (off-state). (*c*) Band diagram along the channel direction for $V_g = V_d$ (on-state). (*d*) Electrical field along the channel direction.

trapping and/or interface traps, which, in turn, cause the transistor *I-V* characteristics to degrade (e.g., V_t, I_{dlin}, I_{dsat}, G_m, etc.). Hole current through the substrate I_{sub} is also observed. Note that I_{sub} itself does not cause the damage, but it has been found to be a strong indicator of the number of hot electrons.

The substrate current I_{sub} and the gate current I_g induced by the hot carriers are expressed as follows:

$$I_{sub} = C_1 \cdot I_d \cdot \exp\left(-\frac{\phi_{ii}}{\lambda \cdot E_{\max}}\right) \tag{5.3}$$

$$I_g = C_2 \cdot I_d \cdot \exp\left(-\frac{\phi_b}{\lambda \cdot E_{\max}}\right) \tag{5.4}$$

where ϕ_{ii} is the minimum energy for a hot carrier to create impact ionization (i.e., the band gap for Si is 1.1 eV), ϕ_b is the barrier height from Si to SiO_2 (3.2 eV), C_1 and C_2 are constants, I_d is the drain current, λ is the channel carrier mean free path, and $\lambda * E_{max}$ is the average carrier energy. Note that electron mean free path is much larger than the hole. Thus NMOS HCI is generally much worse than PMOS HCI. Significant hot-carrier effects occur when the channel electrical field exceeds the critical field for velocity saturation (~10^4 V/cm),[18] which happens in the pinch-off region (Fig. 5.1*a*).

A simplified analytic expression for E_{max} is given as follows:

$$E_{\max} \approx \frac{V_d - V_{dsat}}{l} \tag{5.5}$$

where l is the characteristic length expressed as follows:

$$l \propto \sqrt{t_{ox} \cdot x_j} \tag{5.6}$$

$$V_{dsat} = \frac{E_{sat} \cdot L \cdot (V_g - V_t)}{E_{sat} \cdot L + (V_g - V_t)} \tag{5.7}$$

where t_{ox} is the oxide thickness, x_j is the junction depth, V_{dsat} is the channel voltage beyond which the carrier velocity saturates, E_{sat} is the electrical field for carrier velocity saturation (~10^4 V/cm), and L is the channel length. For long channels, Eq. (5.7) can be simplified as follows:

$$V_{dsat} \approx V_g - V_t \tag{5.8}$$

By inputting Eqs. (5.8) and (5.6), Eq. (5.5) can be rewritten as follows for long-channel devices:

$$E_{\max} \approx \frac{V_d - V_g + V_t}{\sqrt{t_{ox} \cdot x_j}} \tag{5.9}$$

Based on Eq. (5.9), it is seen that the E_{max} reduces with increasing channel length.

It is well known that the I_{sub} has a bell-shaped characteristic in terms of V_g, as shown in Fig. 5.2. This can be explained as follows based on Eq. (5.3). In a low V_g region, I_d is low and increases linearly with increases in V_g according to Eq. (2.45) in Chap. 2. Therefore, I_{sub} also increases with increasing V_g. However, in a high V_g region, I_d becomes saturated, and E_{max} is much lower because E_{max} decreases with larger V_g according to Eq. (5.9). As a result, it is generally observed that the I_{sub} peaks at around $V_g = V_d/2$.

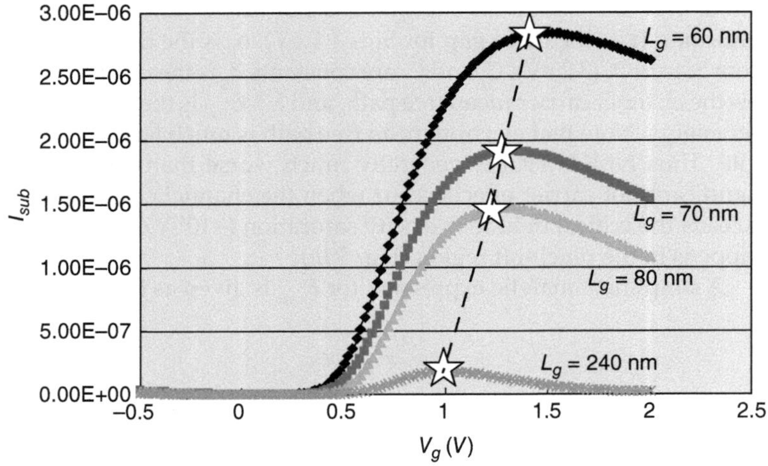

FIGURE 5.2 NMOS substrate current I_{sub} is plotted in terms of V_g for various channel lengths. V_g at I_{sub} decreases with increases in channel length.

5.2 HCI Physical Mechanisms

5.2.1 Field-Driven CHC Mechanisms

In this mode of CHC, it is the channel lateral electrical field that changes the carrier effective temperature and serves as the driving force for the hot-carrier-injection-induced damage, which includes the widely accepted drain-avalanche hot-carrier (DAHC) injection model and the lucky-electron model (LEM). The models detailed in this section well explain CHC behaviors until the mid-1990s for CMOS nodes down to 120 nm with long channels and high energies (low gate voltage and high drain voltage).

Drain-Avalanche Hot-Carrier Injection

DAHC injection is detailed in Fig. 5.1. DAHC injection is caused by hot electrons and holes that are generated by impact ionization and avalanche multiplication at the drain. DAHC injection occurs at the maximum channel electrical field E_{max} with V_g around $V_d/2$. It is well known that the mechanism responsible for this kind of HCI degradation is interface state generation. The peak of the substrate current I_{sub} appears under DAHC stress conditions ($V_g \approx V_d/2$) and can be located the same as that for interface traps N_{it} and G_m degradation.[19] The generated N_{it} was observed to increase in a power-law relation with stress time and is modeled as follows:

$$\Delta N_{it} = \left[t \cdot \frac{I_d}{W} \cdot \exp\left(-\frac{\phi_{it}}{q\lambda E_{\max}} \right) \right] \tag{5.10}$$

where t is stress time, I_d is drain current, W is gate width, ϕ_{it} is the minimum energy for a hot carrier to create an interface trap, q is electron charge, and λ is mean free path of the charge carrier.

The CHC mechanism is applicable to low drain current. In this case, the channel is long, and the carriers can gain quite high energy owing to a local high electrical field, as shown in Fig. 5.1*d*. The corresponding HCI lifetime is inversely proportional to the drain current [Eq. (5.11)]. With this mechanism, it is the impact ionization induced by the high-energy carriers that generates electron-hole pairs so that the electrons are injected into the Si-SiO_2 interface and gate oxide and further break the Si-H bonds at the interface. As a result, interface traps are induced.

$$\text{Time-to-failure (TTF)} = A \cdot \left(\frac{I_{sub}}{I_d}\right)^{-m} \cdot \left(\frac{I_d}{W}\right)^{-1} \tag{5.11}$$

where the exponent m is described by ϕ_{it}/ϕ_{ii}, with ϕ_{it} being the critical electron energy to create an interface state and ϕ_{ii} being the minimum electron energy to create an impact ionization.

Channel-Hot-Electron or Lucky-Electron Model

In channel-hot-electron (CHE) injection, the hot electrons or holes in the channel experience collisions near the drain (or sometimes even in the middle of the channel for short-channel devices in deep-submicron technology nodes) and acquire a momentum in the direction of the oxide thickness. Some of these are "lucky electrons" or "lucky holes" that do not suffer energy-losing collisions and thus get injected over the Si-SiO_2 barrier into the oxide conduction band because of their favorable momentum and direction (Fig. 5.3).

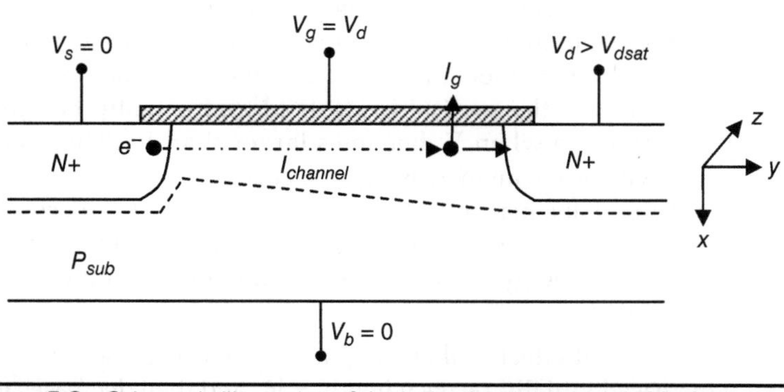

FIGURE 5.3 Schematic illustration of the lucky-electron model (LEM). Sometimes this is referred as CHC injection.

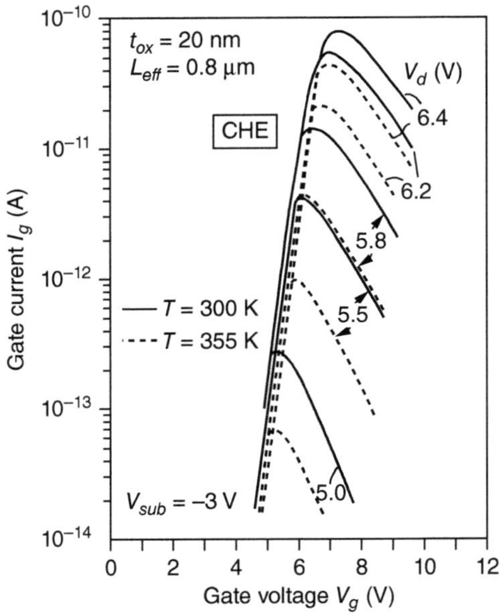

Figure 5.4 Typical gate currents owing to CHE injection. t_{ox} = 20 nm; L_{eff} = 0.8 μm. (*Used by permission from Elsevier.*[20])

CHE generally takes place at $V_g \approx V_d$, where channel hot electrons are injected into the gate oxide at a maximum number, and thus the gate current peaks (Fig. 5.4).[20] This results predominantly in electron trapping in the oxide. The reason for this I_g peak can be explained as follows:

- When V_g is small and less than V_d, the maximum channel electrical field E_{max} exists very near the drain. However, the vertical oxide field repels the injected hot electrons as $V_g < V_d$.
- As V_g increases, there are two competing trends:
  - First, the vertical oxide field at the drain becomes more favorable in attracting the injected hot electrons; that is, the ability to repel the electrons at the drain end decreases owing to the vertical oxide field and actually becomes attractive when V_g becomes larger than V_d. This trend gives rise to an increasing gate current.
  - Second, the maximum channel electrical field E_{max} decreases with increasing V_g [Eq. (5.9)]. This has the effect of decreasing the energy of the hot carriers and decreasing the gate current.
- The overall effect is that the gate current increases as V_g is increased initially, peaks when $V_g = V_d$, and then decreases for further increases of V_g beyond V_d, as shown in Fig. 5.4.

5.2.2 Energy-Driven Channel-Hot-Carrier Mechanism: Electron-Electron Scattering

With the continuous reduction in the channel length, the supply voltage (or operating voltage) for the deep submicron technology nodes does not reduce correspondingly, but the gate-oxide field continues to increase owing to a thinner application of gate oxide (in order to achieve a higher drive current). *For a given V_d, the drain current is increased with V_g* [Eq. (2.49)], *whereas the energy is lowered with V_g because E_{max} reduces with* V_g [Eq. (5.9)]. The reduction in E_{max} reduces the possibility for impact ionization. Therefore, the carriers may not be able to acquire enough energy to break the bonds for trap formation. It is found that the preceding mechanisms (DAHC or LEM) are not valid for this range.[21,22]

One of the mechanisms that was proposed recently is the electron-electron-scattering (EES) model.[23–25] Through carrier-carrier interactions, some of the low-energy carriers can gain enough energy to break the bonds. Therefore, the driving force for device degradation in this case is energy-driven, where interactions between carriers play a key role in the acquisition of electron energy and the distribution function. The underlying principle is detailed below.

As discussed in Sec. 5.2.1, the hot-carrier impact is often described as the interaction of a high-energy channel carrier with either the silicon substrate [leading to impact ionization (ii)] or an interface defect [creating interface states (it)]. The probability of interaction depends mainly on the energy of the carrier and is described by a scattering-rate function $S_i(E)$. The hot carriers are distributed in energy according to the electron-energy-distribution function (EEDF) $f(E)$. Thus, the total probability of interaction is described as[21]

$$R_i = \int f(E) \cdot S_i(E)\, d(E) \tag{5.12}$$

It is found that the interaction rate R_i peaks at one preferential energy referred as the *dominant energy E_{dom}*. In the classic LEM picture, the EEDF has no significant peak, and the dominant energy E_{dom} is determined by the peak in the scattering-rate function $S_i(E)$, which is close to the energy threshold of the scattering-rate function (ϕ_{it} or ϕ_{ii}) and depends slightly on the bias conditions. On the other hand, in the energy-driven framework, the dominant energy is determined by the EEDF peak. The interaction rate R_i can be described as energy-driven because it depends primarily on the energy dependence of the scattering-rate function $S_i(E)$, and the dominant energy is a strong function of the bias conditions.

Basically, the EES mechanism is applicable to moderate drain current. The corresponding HCI lifetime is a quadratic function of the inverse of the drain current, as expressed below:

$$\text{TTF} = A' \cdot \left(\frac{I_{sub}}{I_d}\right)^{-m} \cdot \left(\frac{I_d}{W}\right)^{-2} \tag{5.13}$$

5.2.3 Multiple-Vibrational-Excitation Mechanism

Compared with the EES model, *when the channel electrons have even lower energy (i.e., higher drain current and V_g)*, the other mechanism—the *multiple-vibrational-excitation* (MVE) model[26,27]—is proposed and is considered to be the main contributor to hot-carrier degradation. In this low-energy channel carrier region, the damage strongly depends on the drain current (i.e., the number of electrons "hitting" the bond per second), but no longer on the device geometry. Varying the gate voltage and/or the channel length is just equivalent to a change in the drain current; that is, a single lifetime corresponds to a single drain-current value. In this case, it is *not the energy but the number of the hot electrons* that causes the hot-carrier degradation. The electrons that are injected into the oxide may not have enough energy to break the Si-H bonds at the interface, and it is the electrons remaining in the channel (not overcoming the oxide-Si barrier) that break the Si-H bonds.[28] As a result, hot-electron degradation is still an important reliability issue even if the drain voltage is scaled down to below 1 V[27] because the drive current has become larger and larger with the continuous scaling down of the channel length. The International Technology Roadmap for Semiconductor (ITRS)[29] shows a continuous historical 17 percent per year increase in saturation drive-current values in order to maintain the performance-improvement scaling trend of devices.

The basic understanding of this mechanism is the hot-electron-induced vibrational excitations of hydrogen that passivate silicon (or polysilicon) at an interface. These vibrations and collisions with electrons having *a few electronvolts of energy* then can lead to desorption of the hydrogen, creating atomic hydrogen and a dangling bond—an interface state thus is generated and the gate dielectric is degraded.[30] The major effect of the MVE model may be to lower the energy needed for a single carrier to generate an interface state. This is demonstrated by scanning tunneling microscope (STM) experiments dealing with passivation and depassivation of silicon surfaces in ultrahigh vacuum (UHV)[31] and dedicated experiments to intentionally inject holes and electrons deep into the oxide and also to have them present only at the interface.[28]

The MVE mechanism explains the hot-carrier performance improvement induced by D_2 anneal replacing the conventional H_2 anneal. In this scenario, the low-energy electrons are able to break the Si-H/D bonds in MOS devices because of the unique nature of multiple-vibrational-excitation—a heating effect. Based on STM experiments, the desorption of the Si-H/D occurs for STM current densities in the range of ~10^5 A/cm^2, a value that is comparable with the possible current density at the Si-SiO_2 interface in an MOS.[30] The desorption efficiency for deuterium from Si(100) is about one-fiftieth to one-hundredth of that of hydrogen for energies above 5 eV. This large isotope effect is explained by the electronic excitation and vibrational heating mechanism.[30]

The *MVE mechanism* is applicable to high drain current (i.e., low energy and high V_g). The HCI lifetime is found to be a cubic function of the inverse of the drain current. This HCI mechanism becomes more and more important in advance technologies because the driving currents are higher and higher. However, this mechanism is not well understood yet, and more effort is necessary. On the other hand, the HCI lifetime also can be described empirically as a function of both I_d and V_d. This is justifiable because for the same I_d, a single carrier with higher energy (i.e., higher V_d) will have a higher probability of generating an interface state and thus will have a lower lifetime.

$$\mathrm{TTF} = A'' \cdot V_d^{-\gamma} \cdot \left(\frac{I_d}{W}\right)^{-\alpha} \tag{5.14}$$

where α is the I_d exponent. $\alpha = 6.7$ and $\gamma = 15.5$ were obtained empirically.[21]

5.2.4 NMOS Hot-Carrier-Injection Mechanisms/Models

Table 5.1 summarizes NMOS HCI injection and degradation mechanisms, for which the worst stress condition is at $V_g = V_d/2$, and the interface trap (ΔN_{it}) is the main CHC degradation mechanism.[12]

Stress Condition	Injection Mechanism	Degradation Mechanism	Remarks
$V_t \le V_g \le V_d/4$	Hot-hole injection (DAHC)	Positive charge trapping	
$V_g \approx V_d/2$	Hot-electron injection (CHE) + some hot-hole injection (DAHC)	• Compensation of negative and positive charge trapping • Acceptor-like interface traps (negative charges) dominate.	Mode for long-channel and/or thick-oxide devices (e.g., I/O devices) and lower temperature worse degradation. I_{sub} model is used for lifetime prediction.
$V_g \approx V_d$	Hot-electron injection (CHE)	Negative charge trapping	Mode for short-channel and/or thin-oxide devices (e.g., core devices) and higher temperature worse degradation; $1/V_d$ model is used for lifetime prediction.

TABLE 5.1 Summary for NMOS Injection and Degradation Mechanisms

Stress Condition	Injection Mechanism	Degradation Mechanism
$V_t < V_g < V_d/2$	Hot-hole injection (DAHC)	• Negative charge trapping • $I_{g,max}$ model
$V_g \approx V_d/2$	Hot-hole injection (CHH) + some hot-hole injection (DAHC) (especially near the drain)	• Positive charge trapping • Interface trap generation • I_{sub} model
$V_g \approx V_d$	Hot-hole injection (CHH)	• Positive charge trapping; generation of donor-like interface traps • $1/V_d$ model • The worst stress condition with $V_g = V_d$

Table 5.2 Summary for PMOS Injection and Degradation Mechanisms

On the other hand, for lower V_g ($V_t \leq V_g \leq V_d/4$), HCI is caused mainly by oxide charge trapping under hole injection, whereas in high V_g regions ($\sim V_d$), the main mechanism is electron injection. Both mechanisms can be monitored by the injected gate current I_g.

5.2.5 PMOS Hot-Carrier-Injection Mechanisms/Models

Table 5.2 summarizes the PMOS HCI injection and degradation mechanisms. It has long been believed that hot-carrier reliability of PMOS devices is less serious than that of NMOS devices for the following reasons:

- The mean free path of holes in silicon is about one-half of that of the electrons.[32] Therefore, the holes scatter more frequently, and fewer holes can gain high enough energy (about 3.7 eV) to create interface states.[33]
- The barrier height between the valence-band edges of Si and SiO_2 (i.e., barrier-to-hot-hole injection) is 4.7 eV, whereas the barrier between the conduction-band edges of Si and SiO_2 (i.e., barrier-to-hot-electron injection) is only 3.2 eV.

However, as the transistor channel length has been scaled down into the deep submicron regime, hot-carrier-induced degradation of PMOS devices has been approaching that of NMOS devices.[34] Three hot-carrier degradation mechanisms in PMOS devices have been identified[35,36]:

- *Negative oxide charge trapping*. Electron trapping near the drain region leads to a reduction in the threshold voltage and to the effective channel shortening. As a result, PMOS drive current increases. This mechanism is sometimes called *hot-electron-induced punch-through* (HEIP). This mechanism is most important in longer-channel PMOS devices, and gate current I_g has been used as a predictor of the device lifetime. In this case, V_g is stressed at the $I_{g,\max}$ point, which is generally $1/5V_d$ (see Table 5.2). Li et al.[37] reported that for PMOS devices with oxide thicknesses of around 6.4 nm and gate lengths of 0.27 μm, at the V_g stressed at $I_{g,\max}$, the I_{dsat} moves in the positive direction, indicating that electron trapping is the dominant degradation mechanism. However, this mechanism becomes less significant and unobservable as the channel length becomes shorter than 0.25 μm.[35,36,38]

- *Generation of interface states by hot holes.* In this case, channel mobility will degrade. Similar to the NMOS, the substrate current I_{sub} is generally used to predict device lifetime. In this case, V_g is stressed at the I_{sub} point, which is generally between $1/2V_d$ and $1/3V_d$ (see Table 5.2). This mechanism has been shown to be the dominant degradation mechanism for 0.25-μm surface-channel PMOS devices,[35] and this is also true for 100-nm device as well.[2] In this case, the hot-carrier-induced V_t shift becomes a power-law function of time.[35]
- *Positive oxide charge trapping.* In this case, V_g is stressed at between $2/3V_d$ and V_d (see Table 5.2). However, one should be cautious that the stress at $V_g = V_d$ may lead to rapid device degradation under Fowler-Nordheim (rather than hot-carrier) stress because one is applying such a high gate voltage V_g, especially at high temperature. It has found that for PMOS devices, Fowler-Nordheim (FN) stressing causes the same order of I_{dsat} degradation as hot-carrier stressing, although I_{dsat} degrades more under hot-carrier stressing than under FN stressing.[13] A higher FN stressing temperature will lead to a larger V_t shift, and a thinner oxide corresponds to a higher degradation.[13] In other words, such a stressing condition is something like the coupling between HCI and negative-bias temperature instability (NBTI). This mechanism is important with a nitrided gate oxide.

Figure 5.5 Illustrates I_{dsat} degradation of three test conditions for PMOS hot-carrier stressing in the 65-nm technology node, including the peak-gate-current condition (V_g @ $I_{g,\max}$), peak-substrate-current condition (V_g @ $I_{g,\max}$ I_{sub}), and $V_g = V_d$. It is seen that for all three stressing conditions, I_{dsat} shifts to a negative direction, indicating hole trapping in the gate oxide. Although at V_g stressed at $I_{g,\max}$ the I_{dsat} no longer shifts to the positive direction, the I_{dsat} shift value is much lower than for the other two stress conditions. It is clear that with V_g stressing at a higher value approaching V_d, the I_{dsat} degradation is more serious.

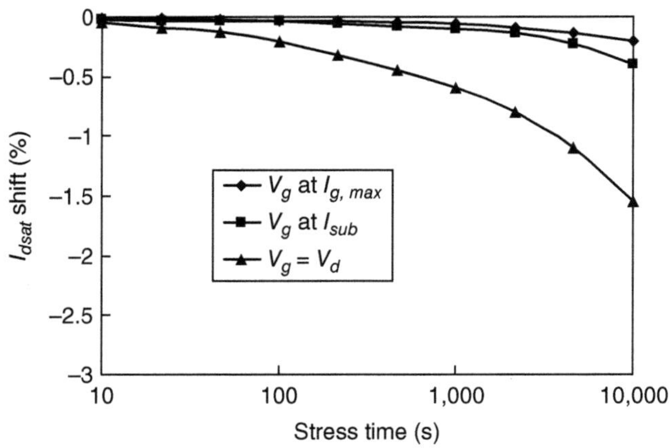

FIGURE 5.5 The saturation-drain-current degradation at three stress conditions for PMOS hot-carrier stressing in the 65-nm technology node.

5.3 Hot-Carrier-Injection Characterization Methodology

5.3.1 Device Parameters to Monitor

Generally, hot-carrier degradation is described as the degradation of macroscopic transistor parameters such as saturation drain current I_{dsat}, linear drain current I_{dlin}, threshold voltage V_t, peak transconductance G_m, and subthreshold swing S. These parameters are monitored when the hot-carrier stress is interrupted periodically at desired preset intervals.

In HCI characterization, I_{dlin} degradation follows exactly the same time dependence as G_m degradation, whereas I_{dsat} changes only half as much as or less than G_m changes. This is expected because both I_{dlin} and G_m are directly proportional to carrier (electron or hole) channel mobility, whereas I_{dsat} depends on the carrier saturation velocity as well as mobility. A much larger I_{dlin} degradation than I_{dsat} degradation also can be explained more explicitly by using a band diagram as follows. Figure 5.6 shows the band diagram in the vertical direction (x direction) near the drain for both linear and saturation regions. The I_{dlin} is measured at the linear region, where the acceptor-type interface traps are all filled. According to Fig. 5.6*a*, this means that negative charges are detected, corresponding to a positive V_t shift [Eq. (5.1)]. On the other hand, I_{dsat} is measured at the saturation region (Fig. 5.6*b*). It is seen that the acceptor-type interface traps are empty, indicating that the charges are neutral. As a result, the interface traps near the drain do not contribute to the I_{dsat} shift measured in the saturation regon.

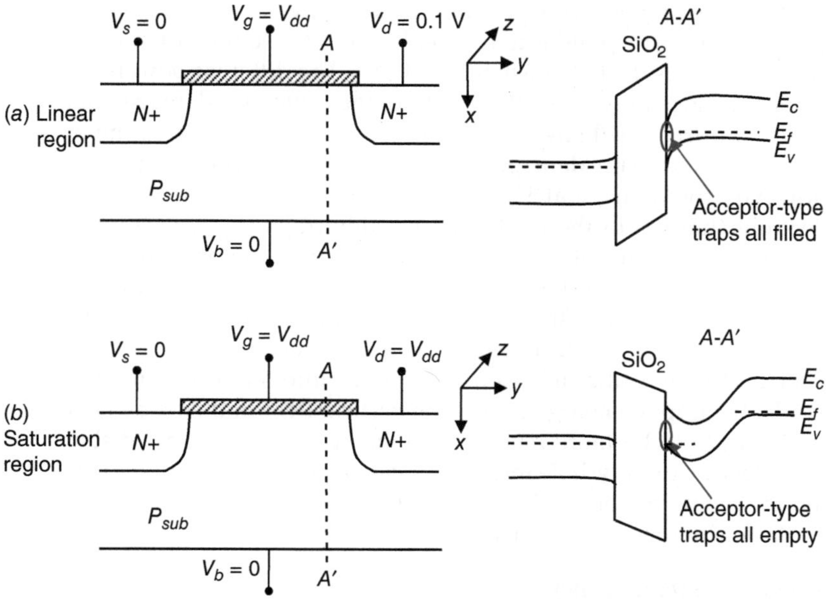

Figure 5.6 Band diagram in the vertical direction (*x* direction) near the drain for (*a*) the linear region, and (*b*) the saturation region.

5.3.2 Hot-Carrier-Injection Degradation Models

Time Power-Law Dependency

Generally, time-dependent degradation induced by HCI is well described by the power-law dependence as follows[39]:

$$\Delta P\left(\text{or } \frac{\Delta P}{P_0}\times 100\%\right) = A\cdot t^n \tag{5.15}$$

where ΔP is the shift in the device parameter of interest (V_t, G_m, I_{dsat}, etc.), P_0 is the initial value of the device parameter, A is a material-dependent parameter, t is the stress time, and n is an empirically determined time power-law exponent, which is a function of stress voltage, temperature, and effective transistor channel length. A simple exponential relationship between A and n was derived as follows[40]:

$$A = a\cdot \exp(-b\cdot n) \tag{5.16}$$

which was found to be independent of the processing technology (e.g., buried versus surface channel, nonsalicide versus salicide).

The time power-law exponent n is generally 0.5 to 0.6 when the gate voltage correspondes to I_{sub} (i.e, $V_g \approx V_d/2$). On the other hand, the value of n is 0.2 to 0.3 when $V_g = V_d$.[41,42] The different power-law exponents generally are attributed to the different damage that occurred during HCI stress, as discussed earlier. When stressing at $V_g \approx V_d/2$, the main damage is related to the interface trap (N_{it}). However, the damage at $V_g = V_d$ is caused by the injection of hot electrons/holes into the oxide, when oxide traps (N_{ot}) are created. This has been demonstrated by the significant reduction in the gate current with stress time because the electrons are trapped in the oxide in a manner identical to the PMOS case.[43] The lower exponent n in the latter case ($V_g = V_d$) is attributed to the trapped charges, which act as a barrier to the further injection of electrons into the oxide. In other words, the trapped charge causes the vertical field in the oxide near the interface to be more repulsive. Sinha et al.[44] also reported that only-electron or only-hole injection causes a power law with $n = 0.25$ and that combination of electron and hole injection leads to a value $n = 0.5$ for interface state generation.

Temperature Dependency

Temperature-dependent time-to-failure (TTF) for HCI is expressed by following the Arrhenius model:

$$\mathrm{TTF} = B \cdot \exp[E_a/(k_B T)] \tag{5.17}$$

where k_B is Boltzmann's constant, T is temperature in kelvins, B is a fitted constant, and E_a is the apparent activation energy, which is device-dependent, as detailed in Table 5.3.

MOS Type		E_a (eV)	Remark
NMOS		–0.2 to +0.4 eV	The apparent activation energy can be negative or positive depending on channel length and voltage.[17,45,46] For the I/O devices, E_a is negative, whereas for core devices, especially in the 65-nm technology node and below, E_a is positive.
PMOS[47]	L <0.25 μm	+0.1 to +0.4 eV	TTF is lower at a higher temperature.
	L >0.25 μm	–0.1 to –0.2 eV	TTF is higher at a higher temperature.

Table 5.3 Apparent Activation Energy E_a for Different Types of Devices

Voltage and/or Current Models

Substrate/Drain-Current-Ratio I_{sub}/I_d Model The widely used I_{sub}/I_d model for NMOS HCI lifetime (TTF) is related to the impact ionization rate r_{ii} (characterized by I_{sub}/I_d), as given in Eq. (5.11),[3] which is rewritten as follows:

$$\mathrm{TTF}\cdot\left(\frac{I_d}{W}\right)=C\cdot\left(\frac{I_{sub}}{I_d}\right)^{-m} \tag{5.18}$$

with

$$m=\frac{\phi_{it}}{\phi_{ii}} \tag{5.19}$$

where C is an arbitrary fitted scale factor (strong function of proprietary factors such as doping profiles, sidewall spacing dimensions, etc.), I_d is the drain current, W is the device gate width, I_{sub} is the peak substrate current during stressing, m is the acceleration exponent depending on ϕ_{it} and ϕ_{ii}, ϕ_{it} is the minimum energy for generation of interface traps (which is ~ 3.7 eV), and ϕ_{ii} is the minimum energy for creation of an impact ionization (which varies from 1.12 to ~ 3 eV).[3] Therefore, simple calculation gives $m \approx 3.3$ to 1.12.

Substrate-Current I_{sub} Model Since I_d is proportional to W and does not vary much, Eq. (5.18) is sometimes simplified to the following approximate form:

$$\mathrm{TTF}=C\cdot(I_{sub})^{-m} \tag{5.20}$$

The I_{sub} model is applicable to both NMOS and PMOS devices. For NMOS devices, m is around 3. However, for PMOS HCI, the power-law-dependent slope under V_g at I_{sub} stressing is much lower (e.g., ~1.5, as shown in Polishchuk et al.).[2]

Drain-Source-Voltage V_d Model The hot-carrier lifetime (TTF) also can be plotted against the reciprocal of the stressing drain voltage ($1/V_d$)—this is the so-called $1/V_d$ model:

$$\mathrm{TTF}=C\cdot e^{\beta/V_d} \tag{5.21}$$

where β is a fitted factor. A linear relationship between ln(TTF) and $1/V_d$ is observed because the maximum lateral electrical field E_{max} is an approximate linear function of drain voltage [Eq. (5.9)], and I_{sub} and I_g are exponents that depend on $1/E_{max}$ [Eqs. (5.3) and (5.4)]. The $1/V_d$ model is currently widely applied to HCI lifetime extrapolation for PMOS core and I/O devices and NMOS core devices.

Gate-Current I_g Model The I_g model generally is applicable to PMOS devices with channel lengths $L \geq 0.25$ µm[47]:

$$\mathrm{TTF}=C\cdot(I_g)^{-m} \tag{5.22}$$

where C is again an arbitrary scale factor (strong function of proprietary factors, such as doping profiles, sidewall spacing dimensions, etc.), I_g is the peak gate current during stressing, and $m = 2$ to 4.[47] However, $m = 1.5$ also was reported.[48] The gate current may not be a good variable for modeling in the case for deep-submicron devices with thin gate oxides because the increased direct tunneling leakage current likely misleads the data interpretation.

5.3.3 Lifetime Extrapolation

Joint Electron Device Engineering Council (JEDEC) Standard 28A (JESD28A) provides the hot-carrier stress-test procedure.[49] The first step is to select the desired devices to be tested, and then their initial unstressed parameter values (e.g., I_{dsat}, V_t, G_m, etc.) are characterized. After obtaining the initial device parameters, we need to make a decision as to whether they are suitable for further hot-carrier stressing. If the device is determined to be "good," the stress cycle begins. During the stress cycle, the device is biased using the selected stress bias condition (e.g., $V_g = V_d$ or V_g at I_{sub}). Since parameter degradations typically exhibit a power-law dependence of the stress time, as given in Eq. (5.15), the recommended stress intervals are 1/3 (or 1/4, or 1/5) decade time steps. During each stress interval, the device parameters are characterized, recorded, and compared with the initial values. If the parameter degradation exceeds the termination criterion (e.g., 10 percent shift) or the total stressing time (e.g., 10,000 seconds) is met, testing ends. Otherwise, another stress cycle is initiated.

Figure 5.7*a* provides a schematic for degradation of the device parameters $\Delta P/P_0$ with stressing time under three different stress conditions A, B, and C. As a usual practice, the TTF for a particular stress condition is defined by 10 percent degradation in the saturation drain current I_{dsat}. According to Eq. (5.15), both the stress time and the parameter shift are plotted with log scale. The HCI lifetime then is extrapolated to the operating voltage (e.g., $1.1 * V_{dd}$ to ensure a 10 percent safety margin) according to the applicable models discussed earlier; that is, the I_{sub}/I_d model (Fig. 5.7*b*), the I_{sub} model (Fig. 5.7*c*), and the $1/V_d$ model (Fig. 5.7*d*). For high confidence in the estimation of operating device lifetime, an R^2 of at least 0.98 is necessary.[50]

5.4 Screening Effect on Hot-Carrier-Injection Characterization

Figure 5.8 provides a schematic of the drain current I_d versus drain-source-voltage V_{ds} characteristics before and after HCI stress. It is expected that the I_d decreases after stress owing to the HCI effect. One interesting point is that the I-V characteristics of the transistor after stressing are quite different if the drain and source are switched

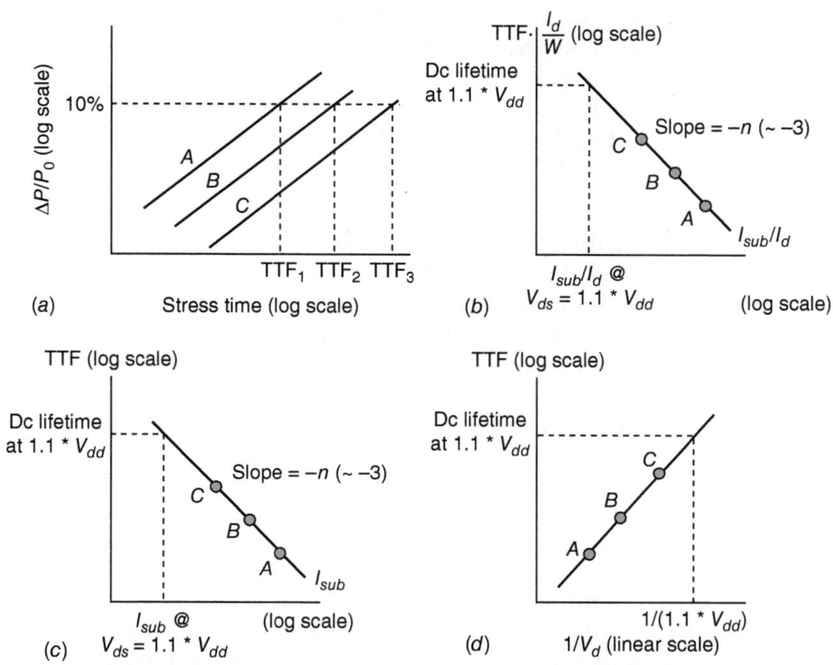

Figure 5.7 Schematic for (*a*) degradation of device parameters (e.g., I_{dsat}) with stressing time under three different stress conditions *A*, *B*, and *C*; lifetime extrapolation to operating voltage (1.1 * V_{dd}) with (*b*) the I_{sub}/I_d model; (*c*) the I_{sub} model; and (*d*) the $1/V_d$ model.

("forward" connection versus "reverse" connection) when measuring $I_d - V_{ds}$. A strong asymmetry with respect to "forward" and "reverse" connection is always observed in NMOS HCI *I-V* characteristic, which is sometimes referred as the *screening effect*, as elaborated below.

- *"Forward" connection.* The damaged area of an NMOS device caused by hot carriers is located at the Si-SiO_2 interface near the drain region (Fig. 5.1*a*). When measuring $I_d - V_{ds}$ curve with low V_g ($>V_t$) and V_{ds} values, the transistor is in the linear region covering the entire length of the channel (including the damaged region) in inversion state. Since the electrical characteristics of MOSFETs are determined by the inverted region of the channel, the effect of hot-carrier-induced damage can be fully reflected in the *I-V* characteristic, and the I_d degradation is clearly observed, as shown in Fig. 5.8. However, as V_{ds} increases, the device moves into the saturation region, and a pinch-off region (noninverted) appears near the drain, which includes part of the damaged region. A larger V_{ds}

means more pinch-off region extending toward the source, and a greater and greater part of the damaged region will be excluded to the inversion state. As a result, the effect of damage near the drain decreases with higher V_{ds}, and thus the I_d increases with a larger slope and approaches its value before the stress.

- *"Reverse" connection.* In this case, the $I_d - V_{ds}$ characteristic is measured with the drain and the source interchanged compared with the "forward" connection. When V_{ds} is low with the device in the linear region, the *I-V* characteristic is the same as that of the "forward" connection. However, when the device moves into the saturation region, the damaged region is always located in the inverted part of the channel, even though V_{ds} increases further, meaning that the effect of the damaged region to the device *I-V* characteristic will not change (increase or decrease) with a larger applied V_{ds}. This is the reason that a saturated *I-V* is observed for the "reverse" connection, as shown in Fig. 5.8.

It should be mentioned that the screening effect is always observed in NMOS degradation owing to the hot-carrier-induced damage located around the drain. However, for PMOS HCI, the difference in I_{dsat} degradation between the forward and reverse modes is much less significant,[13] indicating that the damaged region is less localized than in NMOS devices and more uniformly distributed in the channel. Therefore, the screening effect is less important in PMOS devices than in NMOS devices. In other words, the screening effect also can be used to qualitatively locate the HCI-induced damage in the device.

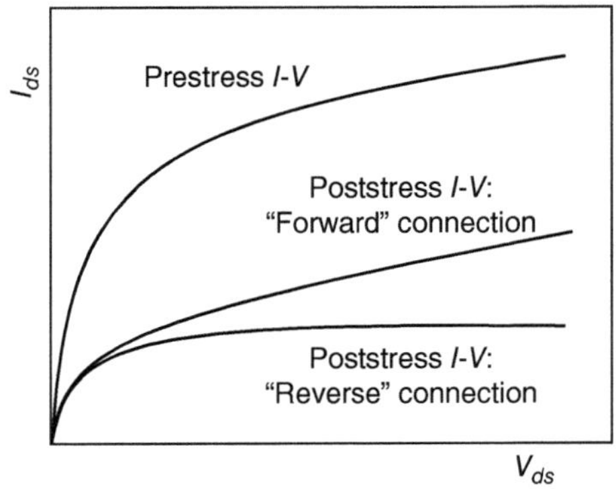

FIGURE 5.8 Comparison of *I-V* characteristic before and after hot-carrier stressing.

When comparing the *I-V* characteristic of the forward connection before and after stress, as shown in Fig. 5.8, I_{dsat} is reduced, whereas the slope of the curve in the saturation region increase substantially. The slope is the output conductance of the transistor g_{ds} in the small signal model of the transistor. The increase in g_{ds} means degradation of the performance of the analog circuits made of MOSFETs.

5.5 Hot-Carrier-Injection Degradation Saturation

Sometimes HCI degradation saturation is observed after exceeding a certain threshold value, leading to a smaller time-exponent slope n in Eq. (5.15).[51–56] An example of the exponent n change is shown in Fig. 5.9. Figure 5.9*b* shows that the degradation time-exponent n value reduces as stressing continues. Several models have been proposed for explaining the saturation behavior of HCI degradation as summarized below.

- *Reduction of barrier height.*[51] In this model, the saturation of degradation is correlated with the increased potential barrier

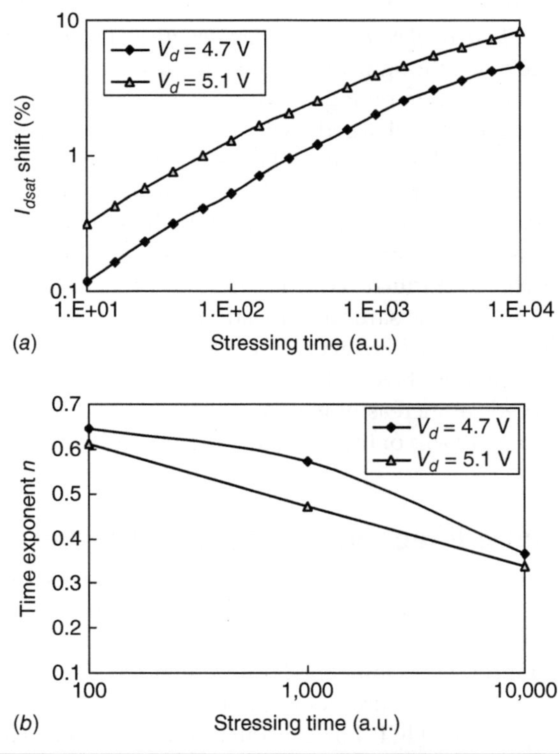

FIGURE 5.9 (*a*) Typical hot-carrier degradation in NMOS devices. (*b*) The degradation time exponent n of part *a* showing the high n value of the initial degradation and the low n value after a long stressing time.

height owing to the accumulation of hot-carrier-induced oxide-trapped charges and interface charges at the gate edge region, which repel subsequent hot carriers from causing further damage.

- *Modification of lateral maximum electrical field maxima* E_{max}. The accumulation of interface charges in the gate edge region can reduce E_{max}, thus leading to the degradation saturation.[52] In addition, the location of the peak electrical field also may shift toward the spacer region so that the degradation rate in the channel and gate-overlapped LDD regions will be reduced.[53]
- *Mobility change.*[54] In this scenario, the observed saturation in the degradation time dependence is attributed to a combination of an increase in the series resistance in the LDD region and a reduction of the carrier mobility in the channel and subdiffusion regions.
- *Logarithmic growth of traps.*[55] The saturation behavior of the degradation of PMOS transistors is assumed to originate from the logarithmic growth of a region of filled traps from the drain junction toward the source.
- *Saturation of defect propagation length.*[56] In this model, the defects are thought to be first produced in the spacer-LDD interface region owing to hot-carrier stressing. These defects are negatively charged, which basically corresponds to the location of the maximum lateral electrical field E_{max}. Defect sites may saturate with electrons. However, the profile of the lateral electrical field should not change much. In other words, the lateral electrical field of the location next to E_{max} is not changed much. With further hot-carrier stress, the location next to the saturated location also will become filled up with electrons. However, this takes longer owing to the lower electrical field below it. Thus, growth of the degraded region proceeds at increasingly slower rates with time. Hence, a saturating growth of the degraded region with time is observed.

5.6 Temperature Effect on Hot-Carrier Injection

Quite different temperature-dependent HCI behaviors are reported with the continuous scaling down of semiconductor technology. As discussed earlier, substrate current I_{sub} is one of the indicators of HCI degradation, especially for NMOS devices. Therefore, people always discuss temperature-dependent I_{sub} when talking about temperature-dependent HCI. For the thick gate oxides of early days, I_{sub} was higher (i.e., HCI was worse) at lower temperature,[57,58] and this was ascribed to the longer mean free path λ between scattering events at low temperature. In the very low temperature range from 77 to 300 K, it was shown that I_{sub} was not sensitive to temperature owing to the insensitivity of the carrier mean free path to temperature.[59]

With technology migration, an observation is that higher temperature gives a larger I_{sub} for short-channel devices. One interesting explanation is detailed below.[17] The widely used I_{sub} model is given in Eq. (5.3). As the gate length scales down, device operating voltage decreases, whereas the operating temperature increases. The thermal kinetic energy becomes more important for low drain bias, and thus the preceding model could be modified by also including the thermal kinetic energy:

$$I_{sub} = C_1 \cdot I_d \cdot \exp\left[-\frac{\phi_{ii}}{\lambda E_{\max} + \beta\left(\frac{3}{2}k_B T\right)}\right] \tag{5.23}$$

where β is the fitting coefficient (~3). A quantitative analysis is given as follows: To estimate the magnitude of $\lambda E_{\max}$, the following values are used: $\lambda = 7.8$ nm,[3] $E_{\max}$ is 1.6×10^6 V/cm and 1×10^6 V/cm for a long device (e.g., $L = 1$ μm) and a short device (e.g., $L = 0.07$ μm), respectively (see Fig. 5.12). Therefore, $\lambda E_{\max}$ is 1.25 eV and 0.78 eV for long and short devices, respectively. On the other hand, when assuming $T = 300$ K, $\beta(\frac{3}{2}k_B T)$ is 0.12 eV. In other words, for long-channel devices, the thermal kinetic energy $\beta(\frac{3}{2}k_B T)$ is one order less than the $\lambda E_{\max}$, whereas for short channel devices, $\beta(\frac{3}{2}k_B T)$ is within the same order of magnitude as $\lambda E_{\max}$. Therefore, the thermal kinetic energy should not be ignored when considering hot-carrier-induced damage for devices with thin gate oxides.

Bravaix et al.[12] also reported that the CHC number increases with temperature as compared between room temperature (RT) and 125°C. On the other hand, the value of the lateral electrical field peak does not change at high temperature. This could be explained by the contribution from the thermal kinetic energy indicated in Eq. (5.23).

5.7 Effect of Bulk Bias on Hot-Carrier Injection

The use of bulk bias V_{bs} for low-power (LP) management has been proposed[60,61] to extend the scalability of bulk MOSFETs. However, the use of reverse V_{bs} clearly can worsen the hot-carrier-induced damage owing to the stronger carrier injections from the channel to the gate caused by the so-called second-impact ionization (2I), as indicated by the increase in the gate leakage correlated with electron trapping I_{ge} at smaller V_g.[62–64] 2I-phenomenon-induced interface traps and electron trapping most likely take place in thicker gate oxides (with $T_{ox} > 3.2$ nm).

In short, I/O devices are much more sensitive to 2I effects than L_G = 45- to 60-nm core devices. The lifetime of I/O devices may have more than four orders of magnitude difference in linear mode when V_{bs} is changed from 0 to $-V_{dd}$. Data have shown that to meet a 10-year lifetime specification based on 10 percent reduction of linear parameters [ΔI_{ds} or $\Delta(1/G_m)$], the dc reverse back bias V_{bs} should not exceed $-V_{dd}/2$.[12]

More interesting, in I/O devices, although the reverse V_{bs} exacerbates HCI degradation, high temperature (HT) can relax CHC to some extent. However, the reverse V_{bs} effect is larger than the HT effect. As a consequence, $V_{bs} = -V_{dd}$ gives the worst case of device lifetime at RT in I/O devices. In contrast, core devices (e.g., channel length 40 nm) show mixed temperature effects versus V_{bs}, where both acceleration factors tend to compensate each other, and forward V_{bs} does not improve lifetime performance.

5.8 Effects of Structure on Hot-Carrier Injection

5.8.1 Effect of Channel Width on Hot-Carrier Injection

Normally, hot-carrier reliability is evaluated on wide devices to avoid unknown edge effects. The dependence of channel width of hot-carrier degradation has received attention from those who are interested in SRAMs. There are contradictory reports showing that compared with wide-width NMOS devices, narrow-width devices show more[65–71] or less[72,73] degradation. Some work reported that local oxidation of silicon (LOCOS) and shallow-trench isolation (STI) devices show opposite effects on device performance,[74] whereas others[75] showed that there is no clear difference between LOCOS and STI devices in terms of impact on channel-width-dependent characteristics of hot-carrier reliability. A summary is given for the channel-width-dependent HCI characteristics in this section.

LOCOS Isolation Technology

For the LOCOS isolation technology of the early days, enhanced degradation was attributed to the higher electrical field near the LOCOS region owing to diffusion of the channel-stop implantation[76] or the electron trapping in the gate oxide owing to the mechanical constraints arising from the proximity of the bird's-beak structure.[66] The narrow width effect could be suppressed by using split-well drive-in.[72] A 0.25-μm-technology NMOS with recessed LOCOS shows no enhanced degradation on narrow devices.[73]

For the LOCOS isolation technology, it is found that narrow devices (W = 1.4 μm) degrade predominantly by oxide charge trapping, whereas wider devices (W = 49 μm) show interface-state creation,[66] which is demonstrated in the following way: First, both the wide and

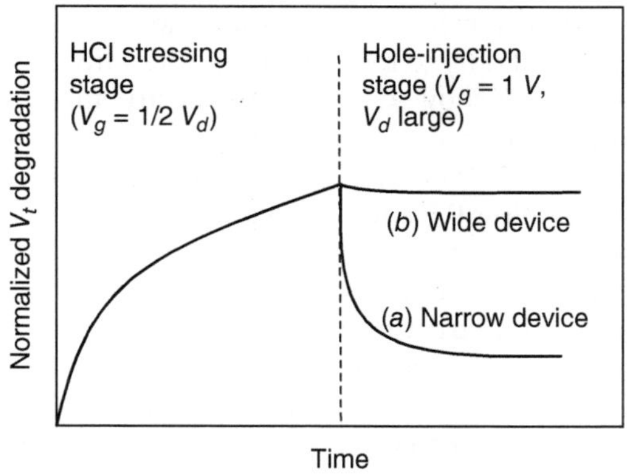

FIGURE 5.10 Schematic of V_t degradation versus time for narrow and wide devices stressed initially at $V_g = V_d/2$ and then subjected to a hole-injection phase (V_d high, $V_g = 1$ V).

narrow devices were stressed for certain time at $V_g = V_d/2$, corresponding to the maximum HCI damage. The two devices then were stressed under conditions of high V_d and $V_g = 1$ V in order to inject holes into the gate dielectric. Such conditions are far enough away from the point of maximum damage to avoid possible additional damage. It was supposed that if, an electron trap was induced by the HCI stressing, the injected holes could neutralize the trapped electrons. Figure 5.10 shows that for narrow devices, the V_t degradation was reduced in the hole-injection phase, indicating the HCI-induced electron traps are neutralized by the injected holes. On the other hand, there is not much V_t shift change in the hole-injection phase for the wide devices, indicating that the V_t shift induced by HCI stressing should be interface-state-related, whose occupancy depends solely on the Fermi level in the silicon.

Shallow-Trench Isolation Technology: NMOS

STI is currently used in ultra-large-scale integration (ULSI) to meet the high-density requirement for field isolation. STI-induced mechanical stress can result in an apparent impact on device performance, attributed to the so-called piezoresistance effect.[77,78] That is, when longitudinal and compressive stresses are increased along current flow, saturation current I_{dsat} will be decreased and increased for NMOS and PMOS devices, respectively. On the contrary, increasing tensile stress will enhance and degrade I_{dsat} for NMOS and PMOS devices, respectively. There are also many reports about the effects of mechanical stress from different process steps on device reliability and gate-oxide integrity.[79,80] However, contrary conclusions existed in these reports. One considered that compressive stress had a serious effect on electron trapping in SiO_2.[79] However, others indicated that externally imposed mechanical stress had no effect on carrier trapping and ascribed it only to the piezoresistance effect.[80]

Nishigohri et al.[68] showed that in the STI structure, as the channel width becomes narrower, the time-dependent slope [i.e., n value in Eq. (5.15)] for hot-carrier degradation drastically increases when channel width is below 0.5 μm. The n value is ~0.2 for a channel width of 1 to 10 μm, whereas it is ~0.45 when channel width is 0.28 μm. The results indicate that acceleration of drain-current degradation is well correlated to the STI structure in the narrow-channel region. Intuitively, one may attribute the increase in impact-ionization rate to the acceleration of drain-current degradation. However, on the contrary, Nishigohri et al.[68] reported that that the impact-ionization rate (characterized by I_{sub}/I_d) of narrow devices is reduced by electrical field modulation at the field edge in the STI structure[81] because the STI edge shows a thicker gate oxide.[71]

The larger HCI-induced I_{dsat} shift in narrow devices is demonstrated to be strongly correlated with the larger interface state (N_{it}) measured by the charge-pumping method.[71] Device degradation is controlled not only by the number of hot carriers (as indexed by the impact ionization rate) but also by their flow direction. Only energetic carriers bombarding the Si-SiO_2 interface cause device degradation. It is thus straightforward that the higher degradation in narrower NMOS devices is attributed to the electrical field modulation at the STI edge caused by the interface traps.[68] On the other hand, some others' work attributed it to the mechanical stress near the STI edge and higher degradation rate at the channel region adjacent to the isolation edge.[69–71] A summary is given below.

Effect of the Lateral Electrical Field The width effect on HCI could be explained by the interaction between the lateral electrical field and channel current path in the STI structure, which could be quite different for the isolation edge and center of the channel. For both cases, the simulation from Nishigohri et al. [68] showed that after HCI stressing, the current path goes deeper into the substrate (i.e., to be farther away from the interface) on the drain side compared with that before the stress. This is caused by the electron traps on the drain side. However, the maximum electrical field is closer to the interface for the isolation-edge case than that for center channel. In other words, on the isolation edge, the current path enters the maximum field location after HCI stressing, making HCI performance worse than for the center of channel. Detailed description is given in Table 5.4.

Effect of the Vertical Electrical Field The vertical electrical field at the Si-SiO_2 interface will affect the hot-electron injection rate because such a vertical electrical field always pushes the current away from the interface. Simulations show that the vertical-channel electrical field is less at the isolation edge than at with the center of the channel.[73,82]

Location	Schematic	Current Path	Maximum Electrical Field	Remark
Isolation edge	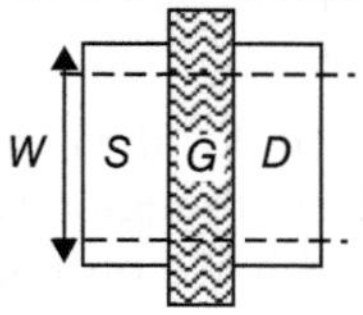	After HCI stressing, the current path goes deeper	Away from the interface	The current path enters the maximum field location after HCI stressing, making HCI worse.
Center of channel	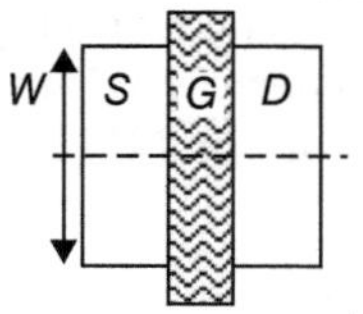	Away from the interface	Closer to the interface	The current path is away from the maximum field location after HCI stressing.

Table 5.4 Interaction Between Current Path and Maximum Electrical Field and Its Impact on HCI Performance

Therefore, it is proposed that it is the two competing factors, that is, the number of hot carriers (as indexed by the impact ionization rate I_{sub}/I_d) and the vertical electrical field, that explain the enhanced degradation on narrow-width devices.[75]

- When stressing under I_{sub}, the I_{sub}/I_d value decreases with a decrease in device width, which will improve HCI performance. However, the effect from the lower vertical channel electrical field at the isolation edge is the dominant factor, which then can compensate the fewer number of energetic carriers and cause enhanced degradation on narrow-width devices. As a result, there is enhanced HCI for narrow-width devices.
- When stressing under $V_g = V_d$, the I_{sub}/I_d value increases with a decrease in device width. Different from the case with stressing under I_{sub}, the vertical field is small (which is an order of magnitude smaller than that under I_{sub}), which will not have much impact the HCI performance. As a result, the HCI degradation follows the trend of I_{sub}/I_d, that is, there is also enhanced HCI for narrow-width devices.

Effect of the Mechanical Stress In this scenario, it is the interface-trap (N_{it}) generation during HCI stressing that leads to the degradation. For the device after hot-electron (I_{sub}) stress, the generated N_{it} at the edge region is higher than that in the center region because of mechanical stress near the STI edge.[71] Let's consider a simple linear-dependent model as follows:

$$\frac{N_{it}}{W} = \alpha \cdot \left(1 - \frac{\Delta W}{W}\right) + \beta \cdot \left(\frac{\Delta W}{W}\right) \tag{5.24}$$

where α and β are constants of N_{it} per unit width for the center and edge regions, respectlively. ΔW is the channel-width region with the STI edge effect, and N_{it}/W is the N_{it} per unit width. According to this mechanism, $\alpha < \beta$. It is thus expected that the effective N_{it} per unit width for a narrow-width device is larger than for a wide one. Effective N_{it} per unit width also can be generated much faster owing to the generated defects during the trench process in a narrow-width device. These interface traps are the dominant factor for NMOS drain-current degradation.

Shallow-Trench Isolation Technology: PMOS

For HCI of PMOS devices under $I_{g,max}$ stress condition, electron trapping dominates the transistor degradation (see Fig. 5.5). The higher degradation for narrow PMOS devices is attributed to the enhanced electron-trapping efficiency owing to the mechanical stress at the STI edge.[70,83]

However, Chung et al.[71] reported that the electron-trapping efficiency is not the dominant mechanism for the PMOS HCI drain current I_d degradation through an edge FN experiment. It is the channel-shortening effect[84,85] in narrow devices that leads to higher degradation. In this two-dimensional channel-shortening model, channel hot-electron injection in the PMOS device is localized in the channel near the drain-junction edge, which is demonstrated by the reverse S/D measurement showing a higher degradation than the forward S/D measurement.[82] As the electrons being injected into the oxide, they are trapped, and then the trapping of these negative charges causes the threshold voltage V_t of the injected part of the channel to become more positive (i.e., a lower $|V_t|$ for the PMOS). In other words, at voltages just below the threshold voltage of the undamaged part of the channel, the damaged part is already in inversion. Therefore, compared with the drawn channel length (L_{drawn}), the effective channel length L_{eff} is decreased by the size of the damaged area L_{damage} as follows:

$$L_{eff} = L_{\text{drawn}} - L_{\text{damage}} \tag{5.25}$$

Similar to the case shown in Table 5.4, after hot-carrier stress, the device can be divided into two devices in parallel, that is, one at the center and the other one at the isolation edge. The size of the damaged area for the center device $L_{\text{damage\_center}}$ is less than that for the isolation-edge device $L_{\text{damage\_edge}}$ owing to the STI-enhanced vertical field at the isolation edge. In other words, the effective channel length for the center device (L_{eff_center}) is larger than that for the isolation-edge device (L_{eff_edge}). Therefore, the effective channel-shortening length per unit width is larger for a narrow-width device, which causes an enhancement of the drain-current degradation.[71]

On the other hand, Lee et al.[82] reported that it is the electrical field difference between the STI edge and channel area (rather than the mechanical stress) that is responsible for the channel-width-dependent degradation in PMOSFET devices. It was found that electron trapping dominates for narrow devices, but hole trapping dominates for wide devices.

5.8.2 Effect of Channel Length on Hot-Carrier Injection

As discussed earlier, for long-channel NMOS devices, the maximum I_{dsat} degradation occurs around $V_g = V_d/2$ with I_{sub} for the nominal channel length. However, for short-channel transistors ($L < 0.15$ μm), the most damaging stress condition has switched from I_{sub} to $V_g = V_d$. Figure 5.11 gives an example. This transition of the damage model with channel length is applied to both SiO_2 and high-*k* transistors.[13,14,16] MEDICI simulations of short- ($L = 70$ nm) and long-channel ($L = 1$ μm) devices with high-*k* material as a dielectric show differences between the electrical field and impact ionization along the channel, as shown in Fig. 5.12.[16] While for long-channel devices impact ionization spikes mainly on the drain side (Fig. 5.12*a*), in short-channel devices large impact ionization occurs along almost the entire channel (Fig. 5.12*b*). Note that the maximum impact-ionization rate is larger in short-channel devices, whereas the maximum electrical field is

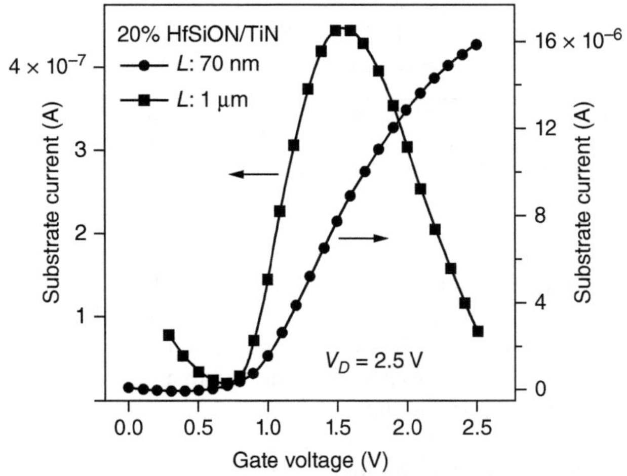

FIGURE 5.11 Substrate current I_{sub} is plotted in terms of gate voltage for various channel lengths. The gate voltage at I_{sub} peak becomes larger with shorter channel lengths. (*Used with permission from IEEE.*[16])

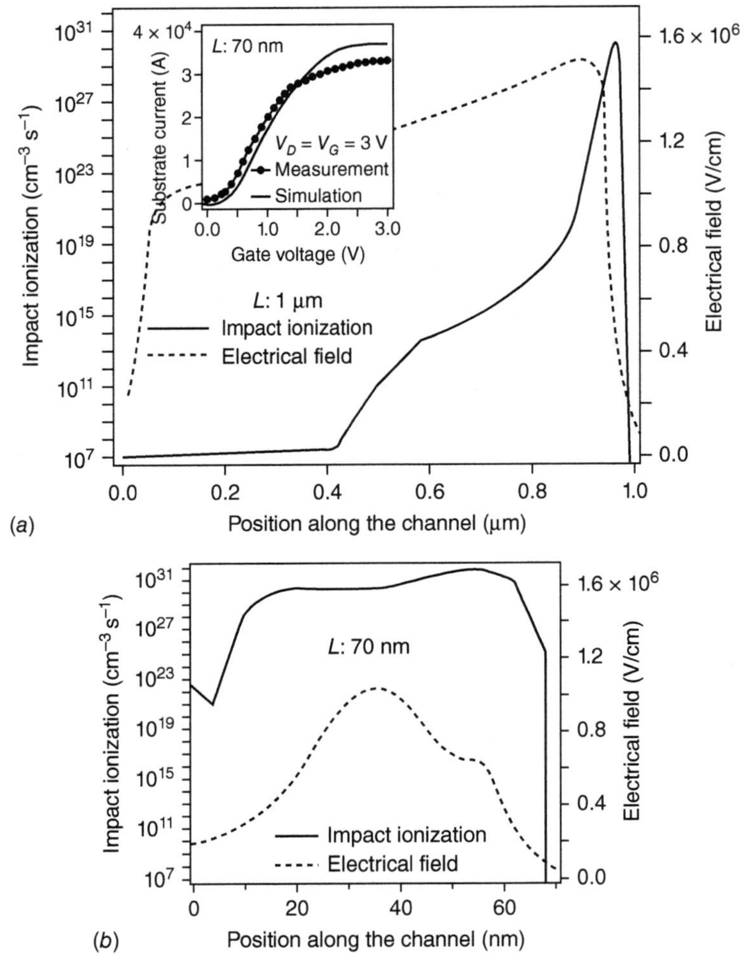

FIGURE 5.12 MEDICI simulation of the electrical field and impact-ionization rate for (*a*) long-channel ($L = 1\ \mu m$) and (*b*) short-channel ($L = 70$ nm) devices. (*Used with permission from IEEE.*[16])

greater in long-channel devices. On the other hand, the total electrical field (vertical plus horizontal) along the channel is also different for the two lengths analyzed. In short transistors the electrical field shows a peak at the center of the channel, whereas in a long channel devices this peak is located on the drain side.

There is a "competition" between the gate voltage in and the drain voltage in controlling the channel electrical field and consequently also the impact ionization because the impact ionization clearly depends on the lateral electrical field. For example, for short-channel transistors with a thicker oxide, the drain voltage has a larger influence on the channel field than the gate voltage, which can extend the pinch-off region from the drain side to the source side—to spread to almost the entire channel. In this case, large impact ionization occurs along almost the entire channel (Fig. 5.12*b*), and the "classical" I_{sub} (peak substrate current) disappears. However, when any lateral electrical field–related parameter (e.g., dielectric thickness or *S*/*D* depths) is reduced, the gate regains a larger influence, resulting in reduction of the pinch-off region. The I_{sub} then may appear again.

On the other hand, unlike NMOS devices, the maximum I_{dsat} degradation of PMOS devices always occurs at $V_g = V_d$, which is independent of channel length.[13]

5.8.3 Effect of Offset Spacers on Hot-Carrier Injection

Feng et al.[86] reported the effect of offset spacers on NMOS HCI. The offset spacer consisting of silicon oxide was formed by chemical-vapor deposition (CVD) after poly reoxidation, and followed by N-type lightly doped drain (N-LDD) arsenic/phosphorous implantation. In this process, the offset spacer was found to reduce the gate-to-drain overlap capacitance (Cgd0) as well as the short-channel effect (SCE). As shown in Fig. 5.13, compared with the device without an offset spacer, the Cgd0 of the devices with an offset spacer is reduced about 8 percent. Figure 5.14 is a schematic illustrating the interaction of hot electrons and the local electrical field near the drain under stress conditions for NMOS devices. V_d is the drain voltage applied, whereas V_g is the gate voltage at the maximum substrate current, which is generally about one-half of V_d, as shown in Fig. 5.2. The reverse vertical electrical field located on gate-to-drain overlap can "repel" the hot

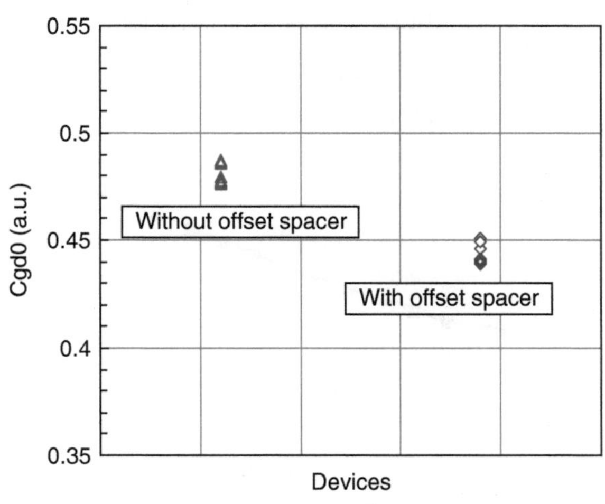

FIGURE 5.13 Cgd0 plot on an NMOS device. (*Reproduced with permission of the Electrochemical Society.*[86])

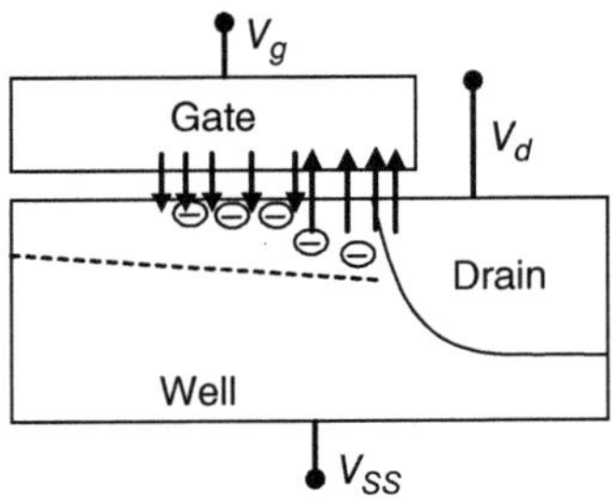

FIGURE 5.14 Vertical electrical field effect on hot carriers. *(Reproduced with permission of the Electrochemical Society.*[86]*)*

electrons from getting close to the Si-SiO_2 interface and creating damage. Therefore, intuitively, the reduction of Cgd0 on the device with offset spacer will worsen the hot carrier performance.

However, it is found that a device with an offset spacer has about four times the hot-carrier lifetime improvement in an I/O NMOSFET compared with one without an offset spacer, as shown in Fig. 5.15. A much decreased substrate current is seen in the process with an offset spacer, as indicated in Fig. 5.16, which shows that the maximum substrate current of NMOS devices with offset spacer is ~20 percent lower than in those without an offset spacer. Thus it is supposed that an additional offset spacer can reduce the number of hot electrons and thus reduce the HCI effect.

The common driving force of the HCI phenomenon is the channel electrical field. Specifically, control over hot-electron effects is exerted by the maximum channel electrical field E_{max}, which occurs near the drain end of the channel. To further explain the detail mechanism of

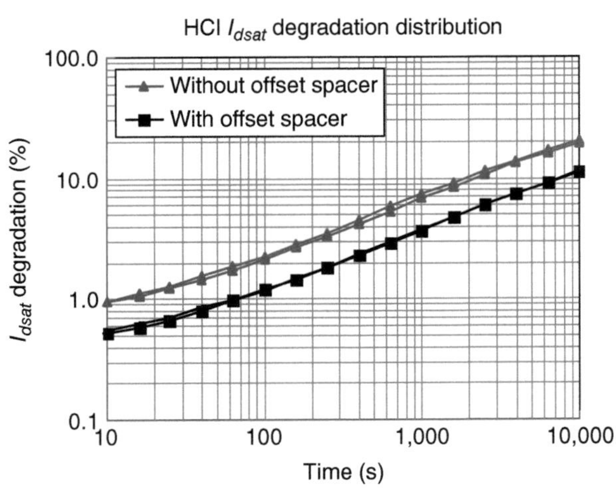

FIGURE 5.15 HCI I_{dsat} degradation distribution on NMOSFETs with and without offset spacers. *(Reproduced with permission of the Electrochemical Society.*[86]*)*

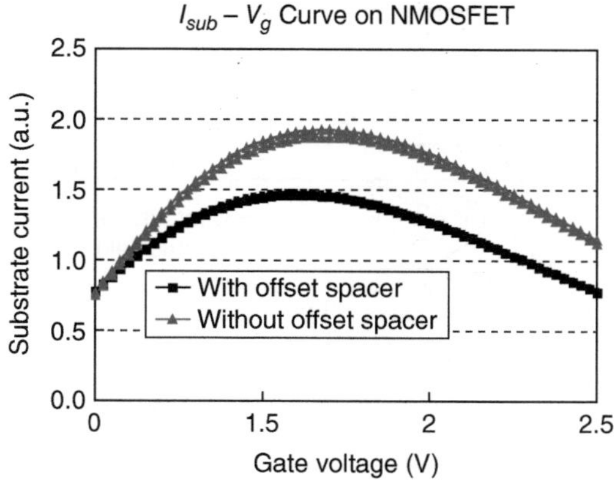

FIGURE 5.16 I_{sub}–V_g curve on NMOSFETs with and without offset spacers. (*Reproduced with permission of the Electrochemical Society.*[86])

lower I_{dsat} degradation and substrate current reduction on devices with an offset spacer, impact-ionization contour plots based on technology computer-aided design (TCAD) simulation are given in Fig. 5.17 for devices with and without offset spacers, respectively. The hot spot indicates where the strongest impact ionization occurred under stress conditions. It can be clearly seen that similar to the LDD effect, the hot-spot size corresponding to strong impact ionization for the device with an offset spacer in Fig. 5.17*b* is much smaller than that for the one without an offset spacer in Fig. 5.17*a*. This observation indicates that the existence of an offset spacer can effectively reduce the number of hot electrons that are generated by impact ionization and then could possibly inject into the gate oxide and degrade device performance. Correspondingly, the substrate current is reduced, as shown in Fig. 5.16.

Furthermore, when there is an offset spacer, the location of the hot spot indicating strong impact ionization shifts. The distance of the center of the hot spot (i.e., the maximum impact ionization) to the gate edge on the device with an offset spacer is about 5 nm, which is only one-third of that of a device without an offset spacer (i.e., 15 nm). According to Fig. 5.14, for hot electrons located away from the drain, the strong electrical field (directed from the gate oxide to the channel) can drive the electrons into the gate oxide, surpassing the interface barrier. This effect worsens HCI. However, it is a different story for electrons close to the drain (i..e, the gate-to-drain overlap region), where the local vertical electrical field is directed from the channel to the oxide because here the drain-voltage effect dominates, and $V_d > V_g$. As a result, a local reverse electrical field is formed in this region, which is repulsive for electron injection to the gate oxide. The hot electrons generated are thus injected into the gate oxide with much less

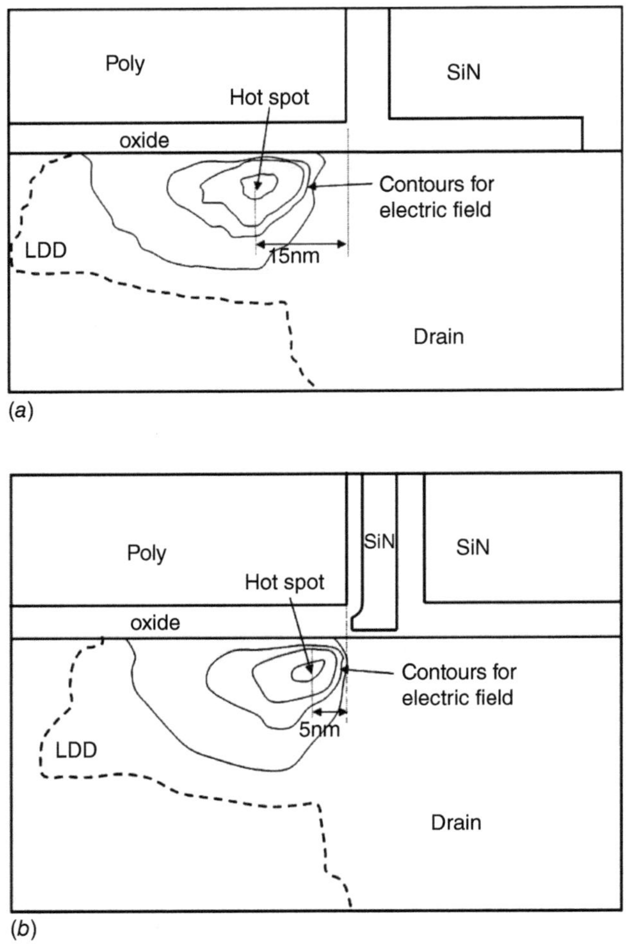

FIGURE 5.17 Impact ionization contours on (*a*) device without an offset spacer and (*b*) a device with an offset spacer. (*Reproduced with permission of the Electrochemical Society.*[86])

probability than those in the channel center. Under this condition, most of the hot electrons flow into the drain under the effect of drain voltage rather than into the gate. Therefore, it is justifiable that if the location of the strong impact ionization (hot spot), which generate hot carriers, is closer to the drain, fewer hot carriers will be injected into the gate. However, one should be careful that the location of the hot spot must not extend outside the gate edge. Otherwise, it would be detrimental to the device owing to hot-carrier damage from a lack of gate-to-drain electrical field on the drain-depletion region.

In short, TCAD simulation results show that with the application of an offset spacer, a much longer hot-carrier lifetime is achieved, contributed by the reduced E_{max} and optimized E_{max} location, even though the Cgd0 is reduced.

5.8.4 Effect of Spacing Between the Gate Edge and the Shallow-Trench Isolation Edge

In ULSI technology, not only are channel length and gate dielectrics scaled down to achieve higher performance, the *S*/*D* area also should be reduced to meet the requirement of ultrahigh-circuit density. The interaction of a smaller *S*/*D* area and the STI-induced mechanical stress and/or electrical field modulation may affect hot-carrier reliability. Shih et al.[87] have demonstrated that symmetric spacing (SA) (Fig. 2.18) between the gate edge and the STI edge may affect hot-carrier-induced degradation because the SA affects the intensity of STI-induced mechanical stress. Figure 5.18 shows schematically the SA-dependent HCI lifetime, applicable for both NMOS and PMOS devices with different channel widths. The observations are summarized below:

- When transistors have a large SA value (e.g., SA = 10 μm), transistors with small channel widths have a much worse lifetime (Fig. 5.18). For example, the HCI lifetime of a device with $W = 0.24$ μm is about one order less than that of a device with $W = 10$ μm, which is applicable to both NMOS and PMOS devices. This observation is attributed to mechanical stress–induced hot-carrier degradation[68] or the competition between impact ionization and vertical electrical field at the STI edge.[75]

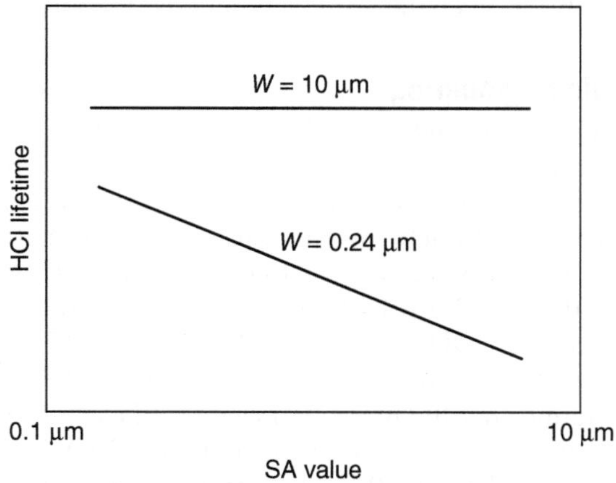

FIGURE 5.18 Schematic of SA-dependent HCI lifetime, applicable for both NMOS and PMOS devices with different channel widths.

- When the SA value is reduced, the HCI lifetime increases, especially for narrow devices ($W < 10$ μm). Furthermore, the HCI lifetime of narrow devices approaches that of a device with $W = 10$ μm when the SA value approaches zero. In other words, the lifetime degradation owing to channel-width reduction seems to be relieved to some extent if the SA value is reduced.

Shih et al.[87] reported that the preceding interesting results cannot be explained by the possible doping-profile change in transistors with small channel widths and SA values because the related impact ionization rate does not change for both NMOS and PMOS devices with different channel widths and SA values. Instead, a plausible mechanism was proposed, and the observations were attributed to the tensile pocket with the stress component normal to the trench side, whose intensity increases with decreasing dimensions. Because the topography of the gate oxide in small transistors is not completely uniform in the corner region owing to Si recesses during STI formation and divots after pad oxide and sacrificial oxide removal, degradation from this tensile-stress pocket will play a more important role. When the SA value is decreased, the impact of increasing compressive stress on HCI degradation owing to channel-width reduction will be relieved by the tensile-stress pocket. As a result, less hot-carrier lifetime degradation can be expected.

5.9 Process Effects on Hot-Carrier Injection Performance

To minimize and control HCI-induced degradation, two approaches are widely used, and both approaches will be elaborated in this section, namely, drain engineering and gate-oxide robustness.

5.9.1 Drain Engineering

The first approach for reducing HCI damage is *drain engineering* to reduce the population/energy of hot carriers by decreasing the maximum channel electrical field E_{max} near the drain region, which is achieved by manufacturing a space-charge transition region near the drain junction edge so as to provide some $I * R$ drop and reduce the effective drain bias. A short list of drain-engineering structures includes DDDs, LDDs,[88] GOLDs,[89] and LATIDs. Izawa et al.[89] compared the electrical field single drain, DDD, LDD, and GOLD and found that the maximum electrical field can be reduced-effectively from 0.48 MV/cm in a single drain to 0.28 MV/cm in a DDD, 0.24 MV/cm in an LDD, and 0.18 MV/cm in a GOLD. In addition, the

GOLD can spread the electrical field more widely than other drain-engineering techniques. However, the negative effect of this drain-engineering approach is that device performance (e.g., drive current I_{dsat}) will degrade owing to the reduction in the effective drain bias. There is always a tradeoff between device performance and reliability.

Large-Angle-Tilt Implanted Drain

It is well known that the LDD implant with a large tilt angle makes the dopant profile more gradient. As a result, a LATID (e.g., 45° angle)[10] can effectively reduce the maximum lateral electrical field E_{max} and thus reduce the substrate current. Correspondingly, HCI of devices with LATIDs can be improved to meet requirements. However, it should be noted that with the continuous scaling down of the technology node, the requirement to reduce the poly-line pitch limits the usable tilt angle of an LDD implant, which usually should be less than 30°.

As/P NLDD Co-Implant

To achieve better performance and hot carrier reliability, a hybrid arsenic/phosphorus (As/P) LDD junction is used in I/O transistors.[90] The phosphorus in-hybrid implant helps in grading the NLDD doping profile, thus reducing the maximum electrical field in the channel. For the normal process sequence, a rapid thermal anneal (RTA) is applied after the As/P LDD implant for I/O devices. To further improve the hot-carrier reliability of I/O devices, phosphorus transient-enhanced diffusion (TED) is used to achieve a more graded NLDD doping profile.[91] The phosphorus TED is achieved by reversing the process sequence, namely, by putting RTA before the As/P implant. As a result, substantial substrate current reduction and hot-carrier lifetime improvement for I/O devices are attained owing to a more graded NLDD doping profile.

5.9.2 Gate-Oxide Robustness

The second approach for HCI improvement is to increase the gate-oxide and interface "robustness" through fabrication-process modifications. Hot-carrier damage generally is due to interface traps, which are dangling bonds at the Si-SiO_2 interface. The dangling bonds can be passivated by several species, for example, H, F, Cl, and N. The Si bonds with the latter three species are much stronger than the Si-H bond. A high-temperature anneal is also found to be helpful for interface robustness but not for bulk SiO_2 trapping. Plasma-induced damage (PID) also should be reduced because it can damage gate-oxide integrity. Implantation through gate oxide should be avoided because a high-dose implant is known to create oxide traps.

Effect of Nitrogen on Hot-Carrier Injection

Nitrogen can be incorporated into the gate dielectric through a gate-oxide anneal or ion implantation. Nitrogen incorporation can prevent boron (B) penetration for PMOS devices,[92] and also improve hot-carrier performance for NMOS devices.[93] The details are given below.

Nitridation Through Gate-Oxide Anneal

Nitridation has been proven to increase hot-carrier performance of NMOS devices as long as the dominant degradation mechanism is the interface-trap generation.[94–96] Because nitrided gate oxides exhibit a high density of traps, a proper additional annealing step in an O_2 environment is necessary. Gate processing based on NO (or NO_2) nitridation rather than NH_3 is more common because the latter has H species.[97] PMOS devices with an NO anneal also showed less degradation than devices without an NO anneal (i.e., annealed under an O_2 environment) for HCI stress at $V_g = V_d$.[37]

In terms of the underlying mechanism for the NMOS HCI improvement induced by the nitridation process, it was found that the incorporation of nitrogen can strongly reduce the injected hole current. It also can reduce the injected electron current, although this is less pronounced.[96] Hence a net reduction in nitrided gate oxide was observed both in interface trap generation and in hole efficiency compared with the conventional gate oxide without nitridation.[12]

Lin et al.[13] demonstrated that NMOS device hot-carrier reliability can be improved with a higher nitrogen dose, whereas PMOS devices degraded with nitrogen doses, where NMOS devices are stressed at V_g with I_{sub} and PMOS devices are stressed at $V_g = V_d$. A device with $W/L = 10$ μm/0.12 μm and a 22-Å-thick gate oxide was manufactured using 0.13-μm technology. Compared with the samples without nitrogen doping, the NMOS V_t shift can be halved, whereas the PMOS V_t shift can be doubled if a median dose of nitrogen was introduced.

It is quite acceptable that nitridation can help to reduce NMOS HCI degradation. However, it is quite arguable in the literature how the nitridation affects PMOS HCI performance. Nitridation has been claimed to improve,[37] degrade,[13] or have no impact[2,98] on PMOS HCI. NO anneal helps PMOS HCI, whereas nitridation by jet vapor deposition (JVD) has no impact[98] These phenomena should be correlated with the nitrogen location in the gate dielectric. When the nitrogen is closer to the Si-SiON interface, it will help to strengthen the interface, thus leading to less interface-state generation during HCI stressing, which is the case for the NO anneal. However, for nitridation by JVD, it looks like the nitrogen is closer to the poly, which will have less impact on interface-state generation during HCI stressing. Therefore, it can be concluded that the effect of nitridation on HCI is quite process-dependent. In contrast, it is noted that PMOS devices with an NO anneal have shown more severe NBTI damage compared with those without NO anneal.[99] This is attributed basically to the nitrogen-assisted oxide trap in the gate oxide, especially when the nitrogen concentration peak is closer to the channel, which is elaborated in Chap. 7. In other words, there is a tradeoff between HCI and NBTI when employing nitridation for PMOS devices.

Nitrogen Introduction Through Implantation Nitrogen also can be introduced into the gate dielectric or source/drain through ion implantation. An example was given by Kuroi et al.,[100] where nitrogen ions were implanted into the source/drain region (at a dose of 4×10^{15} ions/cm^2) after gate-electrode definition was performed, and the HCI degradation was remarkably suppressed. The hot electrons are injected mainly into the sidewall in LDD structure. This drastic improvement in HCI degradation was attributed to the pileup/segregation of nitrogen at the interface between the substrate and the sidewall SiO_2 during the relatively low-temperature heat treatment, which further reduced the generation of interface states at the drain edges.

As another example, phosphorus (P_{31}) and nitrogen (N_{14}) LDDs are co-implanted through the tetraethylorthosilicate (TEOS) liner, followed by Si_3N_4 (SiN)/TEOS composite spacer formation, and an As S/D implant was suggested to improve HCI performance.[101] In this example, the N_{14} implant dose ranges between 1E14 and 5E14 atoms/cm^2. With this co-implantation, the substrate current I_{sub} can be reduced more than twice at the same drain current I_{dsat}. The HCI acceleration exponent m value increased from 2.31 without N_{14} implantation to 2.58 with N_{14} implantation, which also contributed to a longer HCI lifetime. However, the N_{14} implant did not show less interface-state (N_{it}) generation, indicating that this method does not improve the oxide-Si interface. On the other hand, the reduction in I_{sub} was attributed to modulation of the doping profiles of B_{11} and P_{31} at the LDD region caused by the N_{14} implantation, which enhanced the TED effects. As a result, a smaller lateral electrical field $E_{\max}$ was achieved with N_{14} implantation.

Effect of H_2/D_2 Anneal on Hot-Carrier Injection

Low-temperature postmetal anneals (350 to 450°C) in a hydrogen ambient atmosphere which usually are referred to as *forming-gas (FG) annealing*, have been used successfully in MOS fabrication technologies to passivate the silicon dangling bonds and consequently reduce Si-SiO_2 interface trap-charge density.[102–104] The passivation process is described by the following equation:

$$P_b + H_2 \rightarrow P_bH + H \tag{5.26}$$

where P_bH is the passivated dangling bond.[105]

FG annealing is also believed to be able to passivate the dangling bonds in the gate oxide caused by plasma damage. Morifuji et al.[106] showed that PMOS device hot-carrier lifetime (under the worst stress condition of $V_g = V_d$) is improved significantly by decreasing FG annealing time. PMOS device hot-carrier degradations at $V_g = V_d$ are caused mainly by hole injection and trapping[107] at precursor sites such as the Si-H in the oxide-Si interface and thus exhibit similar dependence on H as NBTI (Chap. 7).

When FG annealing was performed in a deuterium (D_2) environment rather than in a hydrogen (H_2) environment, it was found that CMOS transistor HCI lifetimes can be 10 to 50 times longer, which is referred as a *large H/D isotope effect*.[108,109] As detailed earlier, for long-channel devices, the interface degradation is due primarily to desorption of Si-H/D by high-energy electrons through the direct electronic desorption mechanism, whereas for short-channel devices in deep submicron devices, the MVE mechanism becomes important for electron-stimulated desorption of H/D, as detailed in Sec. 5.2.3.[110] Therefore, the large H/D isotope effect for hot electrons strongly depends on the density of hot electrons present at the interface attributed to the MVE. Furthermore, Hof et al.[111] reported that the interfacial oxide quality benefits from a deuterium isotope effect only when the deuterium is incorporated at the end of the process. This is to avoid any possible desorption of the deuterium during the subsequent high-temperature processing steps. The oxide interface is more stable only when deuterium is introduced in the postmetal anneal (PMA). In other words, there is not much effect with deuterium incorporated at the gate-oxide growth and postoxidation anneal stages.

D_2 anneal can improve HCI performance even for tensile-stressed NMOS devices.[112] In the plot of lifetime versus $1/V_d$ (i.e., the $1/V_d$ model; Fig. 5.7*d*), a tensile-strained NMOS device has a larger slope for lifetime extrapolation than a conventional NMOS device without strain. This can be attributed to an increased Si-to-SiO_2 barrier height for electrons in strained-Si devices, which is caused by the lowering of the conduction band in the strained Si.[113] The D_2 annealing can make the slope in both tensile-strained and unstrained devices even higher, which will ensure a better HCI lifetime under operating conditions. Devices with narrow channel widths showed a more significant effect than wide-channel devices.[114]

Cheng et al.[115] reported the dependence of device HCI reliability on deuterium anneal pressure and temperature. Several findings are listed below:

- *Deuterium-pressure effect.* HCI reliability improves initially as deuterium pressure is increased. However, the benefit reaches a maximum, and then HCI performance begins to degrade with further increases in deuterium

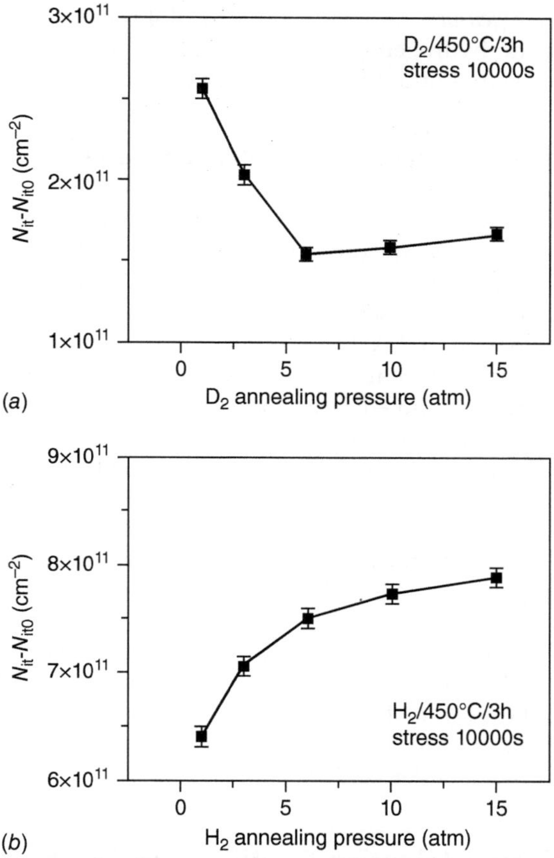

FIGURE 5.19 (*a*) Deuterium and (*b*) hydrogen pressure dependence of hot-carrier-induced interface-trap creation. Annealing condition: 450°C/3 h. Stress conditions: V_{ds} = 3.8 V, V_{gs} = 1.4 V for 10,000 s. (*Used with permission from Elsevier.*[115])

pressure (Fig. 5.19*a*). The benefit of high-pressure deuterium processing on device reliability is attributed to improved deuterium incorporation, whereas annealing-induced interface-trap creation can negate the benefit at extremely high pressures. The competition of these two effects suggests that process optimization on the pressure of the deuterium annealing is necessary to achieve the maximum device reliability improvement.

- *Processing-temperature effect*. The temperature can be lowered with high pressure while still maintaining the deuteration benefit.

- *Hydrogen-pressure effect*. Device reliability constantly degrades as the hydrogen pressure increases (Fig. 5.19*b*) caused by annealing-induced interface-trap creation.

Effect of F or Cl Incorporation on Hot-Carrier Injection

To manufacture an F-contained gate oxide, wafers can be cleaned by the Radio Corporation of America (RCA) cleaning process, followed by immersion in an aqueous HF solution (e.g., ≤3% concentration) for ~5 minutes before they are dried in N_2 and loaded into the dry oxidation furnace (e.g., set at 1,000°C).[116] Cl is introduced in SiO_2 by thermal oxidation at 1,000°C in a dry O_2 + trichloroethane (TCA) ambient.[116] The incorporation of F and Cl into the gate oxide can effectively suppress the hot-electron-induced interface traps, especially for long-channel devices. The underlying mechanism is that the bond strain distribution near the Si-SiO_2 interface may be modified by the presence of F or Cl in the SiO_2 network. However, the TCA amounts should be well controlled. Otherwise, excess amounts of TCA during oxidation may cause the oxide to be more susceptible to hot-electron damage, perhaps owing to the incorporation of hydrogen into the oxide.

The fluorine also can be introduced into the gate oxide through F ion implantation (e.g., 25 keV, dose 2×10^{15} cm^{-2}) into the surface region of the polycrystalline-Si gate, followed by annealing (e.g., 950°C, 10 minutes) to drive some of the F atoms into the gate SiO_2 toward the SiO_2-Si interface.[67,117]

5.10 Hot-Carrier Injection Qualification Practice

For HCI qualification, JEDEC JP-001[118] provides good guidance for practioners. However, different applications may require quite different tests to be performed. A few considerations are given below that are further elaborated in Table 5.5:

- The worst-case bias conditions for the HCI degradation should be well evaluated and understood. For example, it is necessary to demonstrate whether lower or higher temperatures will give a higher HCI degradation. Gate bias condition also should be carefully selected. In addition, the worst-case bias condition for HCI may vary for different transistor types.
- We generally determine device-level HCI under dc bias. However, it is ac in real applications. Circuit-level ac stress testing and/or simulation of HCI is highly recommended. Therefore, the challenge is how to establish the lifetime correlation between the device-level dc lifetime and the product-level ac lifetime so as to ensure product reliability at the early stage of process development.

	NMOS HCI	PMOS HCI
Failure criteria	• A device is considered to fail when a chosen device parameter changes by more than the specified failure criterion at worst-use condition that is temperature and voltage dependent. • A common practice is to use 10% I_{dsat} degradation as the failure criterion. • The selection of failure criterion for the monitored parameters will depend on the specific requirements of the circuit application. Other device parameters, such as G_m, I_{dlin}, and V_{tlin}, can be used to monitor hot-carrier degradation. For example, in some cases, we may use 50- or 100-mV V_t degradation as the criterion.	
Test structures	• Devices with the norminal geometry for each transistor type (NMOS, PMOS, native, or low/standard/high-V_t devices, etc.) are advised to be evaluated first because they are used mostly in product design. • A matrix of device lengths/widths as defined in the design rules may be necessary to cover all the application cases. • A safe-operation-area (SOA) matrix of devices in some cases should be provided by the foundry for IC designer reference.	
Sample size	For each type of device, test devices from 3 lots, 3 wafers per lot, minimum of 5 sites (typically 8 or 12 sites for advanced technology nodes) per bias condition.	
Stress conditions	• Stress temperature: Find the worst-case temperature (typically ranging from 25 to 125°C); –40°C may be necessary for some special applications. • Drain-source voltage @ stress: Test at 3 to 5 (typically 3) different V_{ds} stress conditions. • Gate-source voltage @ stress: Worst-case V_{gs} for the corresponding V_{ds}. • Readouts: Log time scale with at least 3 points per decade.	
Wafer/package level	Wafer level or package level.	
Acceptance criteria	TTF ≥ 10 years considering 2% duty at worst temperature/ $V_{ddmax} = 1.1\ V_{dd}$. Other duty (such as 1%) may also be considered based on specific application.	
Models	• Substrate/drain-current ratio model: Eq. (5.18); • Substrate-current model: Eq. (5.20); • Drain-source voltage ($1/V_d$) model: Eq. (5.21).	• Drain-source voltage ($1/V_d$) model: Eq. (5.21); • Gate-current model: Eq. (5.22).
Parameters	TTF (at 1.1 V_{dd}), n, or β	TTF (at 1.1 V_{dd}), m, or β

TABLE 5.5 Typical Process Qualification Practice for dc Device-Level HCI

5.11 References

1. T H. Ning, "Hot-electron emission from silicon into silicon dioxide," *Solid-State Electronics*, Vol. 21, 1973, pp. 273–282.
2. I. Polishchuk, Y. C. Yeo, Q. Lu, T. J. King, and C. Hu, "Hot carrier reliability of P-MOSFET with ultrathin silicon nitride gate dielectric," in *Proceedings of the 39th IEEE International Annual Reliability Physics Symposium*, 2001, pp. 425–430.
3. C. Hu, S. C. Tam, F. C Hsu, P. K. Ko, T. Y. Chan, and K.W. Terrill, "Hot electron induced MOSFET degradation: Model, monitor, and improvement," *IEEE Transactions on Electron Devices*, Vol. 48, No. 4, 1985, pp. 375–385.
4. D. J. DiMaria, E. Cartier, and D. Arnold, "Impact ionization, trap creation, degradation and breakdown in silicon dioxide films in silicon," *Journal of Applied Physics*, Vol. 73, 1993, pp. 3367–3384.
5. T. Matsuoka, S. Taguchi, H. Ohtsuka, K. Taniguchi, C. Hamaguchi, S. Kakimoto, and K. Uda, "Hot-carrier-induced degradation of N_2O oxynitrided gate oxide N-MOSFETs," *IEEE Transactions on Electron Devices*, Vol. 43, 1996, pp. 1364–1373.
6. E. Takeda, "Hot-carrier effects in submicrometre MOS VLSIs," *IEEE Proceedings: Solid-State and Electron Devices*, Vol. 131, No. 5, 1984, pp. 153–162.
7. T. I. Liou and C. S. Teng, "Double-diffused drain CMOS process using a conterdoping technique," U. S. Patent No. 4956311, 1990.
8. T. Y Huang, W. W. Yao, R. A. Martin, A. G. Lewis, M. Koyanagi, and J. Y. Chen, "A novel submicron LDD transistor with inverse-T gate structure," in *Proceedings of the International Electron Devices Meeting (IEDM)*, 1986, pp. 742–745.
9. T. Buti, S. Ogura, N. Rovedo, and K. Tobimatsu,"A new asymmetrical halo source GOLD drain (HS-GOLD) deep sub-half micrometer *n*-MOSFET design for reliability and performance," *IEEE Transactions on Electron Devices*, Vol. 38, No. 8, 1991, pp. 1757–1763.
10. T. Hori, J. Hirase, Y. Odake, and T. Yasui, "Deep-submicrometer large-angle-tilt implanted drain (LATID) technology," *IEEE Transactions on Electron Devices*, Vol. 39, No. 10, 1992, pp. 2312–2324.
11. P. Heremans, R. Bellens, G. Groeseneken, and H. E. Maes, "Consistent model for the hot-carrier degradation in *n*-channel and *p*-channel MOSFETs," *IEEE Transactions on Electron Devices*, Vol. 35, No. 12, 1988, pp. 2194–2209.
12. A. Bravaix, C. Guerin, V. Huard, D. Roy, J. M. Roux, and E. Vincent, "Hot-carrier acceleration factors for low power management in dc-ac stressed 40-nm NMOS node at high temperature,"in *Proceedings of the 47th IEEE International Annual Reliability Physics Symposium*, 2009, pp. 531–548.
13. C. Lin, S. Biesemans, L. K. Han, K. Houlihan, T. Schiml, K. Schruefer, C. Wann, J. Chen, and R. Mahnkopf, "Hot carrier reliability for 0.13-μm CMOS technology with dual gate oxide thickness," in *Proceedings of the International Electron Devices Meeting (IEDM)*, 2000, pp. 135–138.
14. J. H. Sim, L. Byoung Hun, C. Rino, S. Seung-Chul, and G. Bersuker, "Hot-carrier degradation of HfSiON gate dielectrics with TiN electrode," *IEEE Transactions on Device and Materials Reliability*, Vol. 5, 2005, pp. 177–182.
15. E. Amat, T. Kauerauf, R. Degraeve, R. Rodríguez, M. Nafría, X. Aymerich, and G. Groeseneken, "Competing degradation mechanisms in short-channel transistors under channel hot-carrier stress at elevated temperatures," *IEEE Transactions on Devices and Materials Reliability*, Vol. 9, No. 3, 2009, pp. 454–458.
16. E. Amat, T. Kauerauf, R. Degraeve, A. De Keersgieter, R. Rodriguez, M. Nafria, X. Aymerich, and G. Groeseneken, "Channel hot-carrier degradation under static stress in short channel transistors with high-k/metal gate stacks," in *Proceedings of the 9th International Conference on Ultimate Integration of Silicon*, 2008, pp. 103–106.
17. W. Wang, J. Tao, and P. Fang, "Dependence of HCI mechanism on temperature for 0.18-μm technology and beyond," in *IEEE International Integrated Reliability Workshop Final Report*, 1999, pp. 66–68.
18. S. M. Sze, *Semiconductor Devices: Physics and Technology*, Wiley, New York. 1985.

19. E. Takeda, A. Shimizu, and T. Hagiwara, "Role of hot-hole injection in hot-carrier effects and the small degraded channel region in MOSFETs," *IEEE Electron Device Letters*, Vol. 4, 1983, pp. 329–331.
20. E. Takeda, C. Y. Yang, and A. Miura-Hamada, *Hot-Carrier Effects in MOS Devices*, Academic Press, New York, 1995, p. 50.
21. C. Guerin, V. Huard, and A. Bravaix, "The energy-driven hot-carrier degradation modes of *n*MOSFETs," *IEEE Transactions on Device and Materials Reliability*, Vol. 7, 2007, pp. 225–235.
22. C. Guerin, V. Huard, and A. Bravaix, "The energy-driven hot carrier degradation modes,"in *Proceedings of the 45th Annual International Reliability Physics Symposium*, 2007, pp. 692–693.
23. S. E. Rauch III, G. La Rosa, and F. J. Guarin, "Role of E-E scattering in the enhancement of channel hot carrier degradation of deep-submicron NMOSFETs at high V_{GS} conditions," *IEEE Transactions on Device and Materials Reliability*, Vol. 1, 2001, pp. 113–119.
24. S. E. Rauch and G. La Rosa, "The energy-driven paradigm of NMOSFET hot-carrier effects," in *Proceedings of the 43rd Annual International Reliability Physics Symposium*, 2005, pp. 708–709.
25. S. E. Rauch and G. La Rosa, "The energy-driven paradigm of NFET hot carrier effects," *IEEE Transactions on Devices and Materials Reliability*, Vol. 5, No. 4, 2005, pp. 701–705.
26. K. Hess, B. Tuttle, F. Register, and D. K. Ferry, "Magnitude of the threshold energy for hot electron damage in metal-oxide-semiconductor field-effect transistors by hydrogen desorption," *Applied Physics Letters*, 1999, pp. 3147–3149.
27. Z. Chen, P. Ong, A. K. Mylin, V. Singh, and S. Chetlur, "Direct evidence of multiple vibrational excitation for the Si-H/D bond breaking in metal-oxide-semiconductor transistors," *Applied Physics Letters*, Vol. 81, No. 17, 2002, pp. 3278–3280.
28. Z. Chen, K. Hess, J. Lee, J. W. Lyding, E. Rosenbaum, I. Kizilyalli, S. Chetlur, and R. Huang, "On the mechanism for interface trap generation in MOS transistors due to channel hot carrier stressing," *IEEE Electron Device Letters*, Vol. 21, 2000, pp. 24–26.
29. "Process integration, devices, and structures," in *International Technology Roadmap for Semiconductors*, 2005, p. 15.
30. K. Hess, I. C. Kizilyalli, and J. W. Lyding, "Giant isotope effect in hot electron degradation of metal oxide silicon devices," *IEEE Transactions on Electron Devices*, Vol. 45, No. 2, 1998, pp. 406–416.
31. J. W. Lyding, T.-C. Shen, J. S. Hubacek, J. R. Tucker, and G. C. Abeln, "Nanoscale patterning and oxidation of H-passivated Si(100)-2X1 surfaces with an ultrahigh vacuum scanning tunneling microscope," *Applied Physics Letters*, Vol. 64, 1994, pp. 2010–2012.
32. T. C. Ong, P. K. Ko, and C. Hu, "Modeling of substrate current in *p*-MOSFETs," *IEEE Electron Device Letters*, Vol. 8, 1987, pp. 413–415.
33. T. C. Ong, P. K. Ko, and C. Hu, "Hot-carrier current modeling and device degradation in surface-channel *p*-MOSFET's," *IEEE Transactions on Electron Devices*, Vol. 37, 1990, pp. 1658–1666.
34. T. Tsuchiya, Y. Okazaki, M. Miyake, and T. Kobayashi, "New hot carrier degradation mode and lifetime prediction method in quarter-micrometer PMOSFET," *IEEE Transactions on Electron Devices*, Vol. 39, 1992, p. 404.
35. R. Woltjer, G. M. Paulzen, H. G. Pomp, H. Lifka, and P.H. Woerlee, "Three hot carrier degradation mechanisms in deep submicron PMOSFETs," *IEEE Transactions on Electron Devices*, Vol. 42, 1995, p. 109.
36. A. Bravaix, "Hot carrier degradation evolution in deep submicrometer CMOS technologies," in *IEEE International Integrated Reliability Workshop, Final Report*, 1999, pp. 174–183.
37. E. Li, S. Prasad, and L. Duong, "Process dependence of hot carrier degradation in PMOSFETS," in *IEEE International Integrated Reliability Workshop Final Report*, 2004, pp. 166–168.

38. A. Bravaix, D. Vuillaume, D. Goguenheim, V. Lasserre, and M. Haond, "Competing ac hot-carrier degradation mechanisms in surface-channel *p*-MOSFET's during pass transistor operation," in *Proceedings of the International Electron Devices Meeting (IEDM)*, 1996, pp. 873–876.
39. E. Takeda and N. Suzuki, "An empirical model for device degradation due to hot-carrier injection," *IEEE Electron Device Letters*, Vol. 4, 1983, pp. 111–113.
40. S. W. Sun, M. Orlowski, and K. Y. Fu, "Parameter correlation and modeling of the power-law relationship in MOSFET hot-carrier degradation," *IEEE Electron Device Letters*, Vol. 11, No. 7, 1990, pp. 297–299.
41. B. Doyle, M. Bourcerie, J.-C. Marchetaux, and A. Boudou, "Interface state creation and charge trapping in the medium-to-high gate voltage range ($V_d \geq V_g \geq V_d/2$) during hot-carrier stressing of *n*-MOS transistors," *IEEE Transactions on Electron Devices*, Vol. 37, 1990, pp. 744–754.
42. B. S. Doyle, K. R. Mistry, and J. Faricelli, "Examination of the time power law dependencies in hot carrier stressing of *n*-MOS transistors," *IEEE Electron Device Letters*, Vol. 18, No. 2, 1997, pp. 51–53.
43. B. S. Doyle and K. R. Mistry, "A lifetime predition method for hot-carrier degradation in surface channel p-MOS devices," *IEEE Transactions on Electron Devices*, Vol. 37, 1990, pp. 301–307.
44. S. P. Sinha, F. L. Duan, and D. E. Ioannou, "Time dependence power laws of hot-carrier degradation in SOI MOSFETs," in *Proceedings of the 1996 IEEE International SOI Conference*, 1996, pp. 18–19.
45. P. Aminzadeh, M. Alavi, and D. Scharfetter, "Temperature dependence of substrate current and hot carrier-induced degradation at low drain bias," in *Symposium on VLSI Technology Digest of Technical Papers*, 1998, pp. 178–179.
46. P. Su, K. Goto, T. Sugii, and C. Hu, "A thermal activation view of low voltage impact ionization in MOSFETs," *IEEE Electron Device Letters*, Vol. 23, No. 9, 2002, pp. 550–552.
47. JEP122E, "Failure Mechanisms and Models for Semiconductor Devices," *JEDEC Solid State Technology Association, Arlington*, VA, March 2009.
48. Y.-H. Lee, L. D. Yau, E. Hansen, R. Chau, B. Sabi, S. Hossaini, and B. Asakawa, "Hot-carrier degradation of submicrometer *p*-MOSFETs with thermal/LPCVD composite oxide," *IEEE Transactions on Electron Devices*, Vol. 40, No. 1, 1993, pp. 163–168.
49. JESD28A, "A procedure for measuring *N*-channel MOSFET hot-carrier-induced degradation under dc stress," *JEDEC Solid State Technology Association, Arlington*, VA, December 2001.
50. S. Rangan, S. Krishnan, A. Amerasckara, S. Aur, and S. Ashok, "A model for channel hot carrier reliability degradation due to plasma damage in MOS devices," in *Proceedings of the 37th IEEE International Annual Reliability Physics Symposium*, 1999, pp. 370–374.
51. C. Liang, H. Gaw, and P. Cheng, "An analytical model for self-limiting behavior of hot-carrier degradation in 0.25-μm *n*-MOSFETs," *IEEE Electron Device Letters*, Vol. 13, 1992, pp. 569–571.
52. K. M. Cham, J. Hui, P. V. Voorde, and H. S. Fu, "Self-limiting behavior of hot-carrier degradation and its implication on the validity of lifetime extraction by accelerated stress," in *Proceeding of the 25th IEEE International Annual Reliability Physics Symposium*, 1987, pp. 191–194.
53. Q. Wang, W. H. Krautschneider, M. Brox, and W. Weber, "Time dependence of hot-carrier degradation in LDD *n*MOSFETs," *Microelectronic Engineering*, Vol. 15, 1991, pp. 441–444.
54. V. H. Chan and J. E. Chung, "Two-stage hot-carrier degradation and its impact on submicrometer LDD NMOSFET lifetime prediction," *IEEE Transactions on Electron Devices*, Vol. 42, 1995, pp. 957–962.
55. Q. Wang, M. Brox, W. H. Krautschneider, and W. Weber, "Explanation and model for the logarithmic time dependence of P-MOSFET degradation," *IEEE Electron Device Letters*, Vol. 12, 1991, pp. 218–220.
56. A. Raychaudhuri, M. J. Deen, M. I. H. King, and W. Kwan, "Features and mechanisms of the saturating hot carrier degradation in LDD MOSFETs," *IEEE Transactions on Electron Devices*, Vol. 43, No. 7, 1996, pp. 1114–1122.

57. F.-C. Hsu, and K.-Y. Chiu, "Temperature dependence of hot-electron-induced degradation in MOSFETs," *IEEE Electron Device Letters*, Vol. 5, No. 5, 1984, pp. 148–150.
58. P. Heremans, G.V.D. Bosch, R. Bellens, G. Groeseneken, and H.E. Maes, "Temperature dependence of the channel hot-carrier degradation of *n*-channel MOSFETs," *IEEE Transactions on Electron Devices*, Vol. 37, No. 4, 1990, pp. 980–993.
59. J. H. Huang, G. B. Zhang, Z. H. Liu, J. Duster, S. J. Wann, P. Ko, and C. Hu, "Temperature dependence of MOSFET substrate current," *IEEE Electron Device Letters*, Vol. 14, No. 5, 1993, pp. 268–271.
60. T. Skotnicki, C. Fenouillet-Beranger, C. Gallon, F. Boeuf, S. Monfray, F. Payet, A. Pouydebasque, M. Szczap, A. Farcy, F. Arnaud, S. Clerc, M. Sellier, A. Cathignol, J.-P. Schoellkopf, E. Perea, R. Ferrant, and H. Mingam, "Innovative materials, devices, and CMOS technologies for low-power mobile multimedia," *IEEE Transactions on Electron Devices*, Vol. 55, No. 1, 2008, pp. 96–130.
61. A. Hokazono, S. Balasubramanian, K. Ishimaru, H. Ishiuchi, C. Hu, and T.-J. King Liu, "MOSFET hot-carrier reliability improvement by forward-body bias," *IEEE Electron Device Letters*, Vol. 27, No. 7, 2006, pp. 605–608.
62. D. Esseni, L. Selmi, A. Ghetti, and E. Sangiorgi, "Injection efficiency of CHISEL gate currents in short MOS devices: Physical mechanisms, device implications and sensitivity to technological parameters," *IEEE Transactions on Electron Devices*, Vol. 47, No. 11, 2000, pp. 2194–2199.
63. A. Bravaix, D. Goguenheim, N. Revil, and E. Vincent, "Injection mechanisms and lifetime prediction with the substrate voltage in 0.15-µm channel-length *N*-MOSFETs," *Microelectronics Reliability*, Vol. 41, 2001, pp. 1313–1318.
64. F. Driussi, D. Esseni, and L. Selmi, "Performance, degradation monitors, and reliability of the CHISEL injection regime," *IEEE Transactions on Electron Devices*, Vol. 4, No. 3, 2004, pp. 327–334.
65. V. Srinivasan and J. J. Barnes, "Small width effects on MOSFET hot electron reliability," in *Proceedings of the International Electron Devices Meeting (IEDM)*, 1980, pp. 740–743.
66. M. Bourcherie, B. S. Doyle, J. C. Marchetaux, A. Boudou, and H. Mingam, "Hot-carrier stressing damage in wide and narrow LDD NMOS transistors," *IEEE Electron Device Letters*, Vol. 10, 1989, pp. 132–134.
67. Y. Nishioka, K. Ohyu, Y. Ohji, and T.-P. Ma, "Channel length and width dependence of hot carrier hardness in fluorinated MOSFETs," *IEEE Electron Device Letters*, Vol. 10, 1989, pp. 540–542.
68. M. Nishigohri, K. Ishimaru, M. Takahashi, Y. Unno, Y. Okayama, F. Matsuoka, and M. Kinugawa, "Anomalous hot-carrier induced degradation in very narrow channel, nMOSFETs with STI structure," in *Proceedings of the International Electron Devices Meeting (IEDM)*, 1996, pp. 881–884.
69. H. Hwang, J. Lee, P. Fazan, and C. Dennison, "Hot-carrier reliability characteristics of narrow-width MOSFETs," *Solid-State Electronics*, Vol. 36, 1993, pp. 665–666.
70. W. Lee, S. Lee, T. Ahn, and H. Hwang, "Degradation of hot carrier lifetime for narrow width MOSFET with shallow trench isolation," in *Proceedings of the 37th IEEE International Annual Reliability Physics Symposium*, 1999, pp. 259–262.
71. S. S. Chung, S.-J. Chen, W.-J. Yang, and J.-J. Yang, "A new physical and quantitative width dependent hot carrier model for shallow-trench-isolated CMOS devices," in *Proceedings of the 39th IEEE International Annual Reliability Physics Symposium*, 2001, pp. 419–424.
72. C. Mazure, A. Lili, and C. Zeller, "Width-independent narrow nMOSFET reliability by split-well drive-in," *IEEE Electron Device Letters*, Vol. 11, 1990, pp. 224–226.
73. J. M. P. Yue, W. K. Chim, B. J. Cho, D. S. H. Chan, W. H. Qin, Y.-B. Kim, S.-A. Jang, and I.-S. Yeo, "Hot-carrier degradation mechanism in narrow- and wide-channel *n*-MOSFETs with recessed LOCOS isolation structure," *IEEE Electron Device Letters*, Vol. 21, 2000, pp. 130–132.
74. T. Oishi, K. Shiozawa, A. Furukawa, Y. Abe, and Y. Tokuda, "Isolation edge effect depending on gate length of MOSFETs with various isolation structures," *IEEE Transactions on Electron Devices*, Vol. 47, 2000, pp. 822–827.

75. E. Li and S. Prasad, "Channel width dependence of NMOSFET hot carrier degradation," *IEEE Transactions on Electron Devices*, Vol. 50, No. 6, 2003, pp. 1545–1548.
76. S. Sawada, Y. Matsumoto, S. Shinozaki, and O. Ozawa, "Effect of field boron dose on substrate current in narrow channel LDD MOSFETs" in *Proceedings of the International Electron Devices Meeting (IEDM)*, 1984, pp. 778–781.
77. G. Scott, J. Lutze, M. Rubin, F. Nouri, and M. Manley, "NMOS drive current reduction caused by transistor layout and trench isolation induced stress," in *Proceedings of the International Electron Devices Meeting (IEDM)*, 1999, pp. 827–830.
78. G. Dorda, " Piezoresistance in quantized conduction bands in silicon inversion layers," *Journal of Applied Physics*, Vol. 42, No. 5, 1971, pp. 2053–2060.
79. A. Hamada, T. Furusawa, N. Saito, and E. Takeda, "A new aspect of mechanical stress effects in scaled MOS devices," *IEEE Transactions on Electron Devices*, Vol. 38, No. 4, 1991, pp. 895–900.
80. R. Degraeve, G. Groeseneken, I. De Wolf, and H. E. Maes, "The effect of externally imposed mechanical stress on the hot-carrier-induced degradation of deep-sub micron *n*MOSFETs," *IEEE Transactions on Electron Devices*, Vol. 44, No. 6, 1997, pp. 943–950.
81. K. Hieda, F. Horiguchi, H. Watanabe, K. Sunouchi, I. Inoue, and T. Hamamoto, "New effects of bench isolated transistor using sidewall gates", in *Proceedings of the International Electron Devices Meeting (IEDM)*, 1987, pp. 736–739.
82. Y. H. Lee, K. Wu, T. Linton, N. Miekle, S. Hu, and B. Wallace, "Channel width dependent hot-carrier degradation of thin-gate pMOSFETs," in *Proceedings of the 38th IEEE International Annual Reliability Physics Symposium*, 2000, pp. 77–82.
83. J. F. Chen, K. lshimaru, and C. Hu, "Enhanced hot carrier induced degradation in shallow trench isolation narrow channel PMOSFETs" *IEEE Electron Device Letters*, Vol. 19, 1998, p. 332.
84. R. Woltjer, G. M. Paulzen, H. Lijka, and P. Woerlee, "Positive oxide-charge generation during 0.25-μm PMOSFET hot-carrier degradation," *IEEE Electron Device Letters*, Vol. 15, 1994, pp. 427–429.
85. B. S. Doyle, K. R. Mistry, and D. B. Jackson, "Examination of gradual-junction *p*-MOS structure for hot carrier control using a new lifetime extraction method," *IEEE Transactions on Electron Devices*, Vol. 39, 1992, pp. 2290–2297.
86. J. Feng, Z. Gan, L. Zhang, L. Chang, Z. Pan, X. Shi, H. Wu, B. Ye, and T. Yu, "The effects of offset spacer on *n*MOSFET hot-carrier lifetime," *ECS Transactions*, Vol. 44, No. 1, 2012, pp. 135–139.
87. J. R. Shih, R. Wang, Y. M. Sheu, H. C. Lin, J. J. Wang, and K. Wu, "Pattern density effect of trench isolation-induced mechanical stress on device reliability in sub-0.1μm technology," in *Proceedings of the 42nd IEEE International Annual Reliability Physics Symposium*, 2004, pp. 489–492.
88. S. Ogura, P. J. Tsang, W. W. Walker, D. L. Critchlow, and J. F. Shepard, "Design and characteristics of the lightly doped drain-source (LDD) insulated gate field-effect transistor," *IEEE Transactions on Electron Devices*, Vol. 27, No. 8, 1980, pp. 1359–1367.
89. R. Izawa, T. Kure, and E. Takeda, "Impact of the gate-drain overlapped device (GOLD) for deep submicrometer VLSI," *IEEE Transactions on Electron Devices*, Vol. 35, No. 12, 1988, pp. 2088–2093.
90. D. K. Nayak, M. Y. Hao, J. Umali, and R. Rakkhit, "A comprehensive study of performance and reliability of P, As, and hybrid As/P nLDD junctions for deep-submicron CMOS logic technology," *IEEE Electron Device Letters*, Vol. 18, 1997, pp. 281–283.
91. H. C. H. Wang, C. H. Diaz, B. K. Liew, J. Y. C. Sun, and T. Wang, "Hot carrier reliability improvement by utilizing phosphorus transient enhanced diffusion for input/output devices of deep submicron CMOS technology," *IEEE Electron Device Letters*, Vol. 21, No. 12, 2000, pp. 598–600.
92. C. T. Liu, Y. Ma, K. P. Cheung, C. P. Chang, L. Fritzinger, J. Becerro, H. Luftman, H. M. Vaidya, J. I. Colonell, A. Kamgar, J. F. Minor, R.G. Murray, W. Y. C. Lai, C. S. Pai, and S. J. Hillenius, "25-Å gate oxide without boron penetration for

0.25- and 0.3-pm PMOSFETs," *Symposium on VLSI Technology Digest of Technical Papers*, 1996, pp. 18–19.

93. C. T. Liu, E. J. Lloyd, Yi Ma, M. Du, R. L. Opila, and S. J. Hillenius, "High-performance 0.2-μm CMOS with 25-Å gate oxide grown on nitrogen implanted Si substrates," in *Proceedings of the International Electron Devices Meeting (IEDM)*, 1996, pp. 499–502.
94. B. J. Aniruddha and D. L. Kwong, "Hot carrier effects on analog performance of *N*- and *P*-MOSFETs with oxynitride gate dielectrics," *IEEE Transactions on Electron Devices*, Vol. 41, No. 8, 1994, pp. 1465–1467.
95. Y. Y. Chen, I. M. Liu, M. Gardner, J. Fulford, and D. L. Kwong, "Performance and reliability assessment of dual-gate CMOS devices with gate oxide grown on nitrogen implanted Si substrates," in *Proceedings of the International Electron Devices Meeting (IEDM)*, 1997, pp. 639–642.
96. T. Kaga and T. Hagiwara, "Short- and long-term reliability of nitrided oxide MOSFETs," *IEEE Transactions on Electron Devices*, Vol. 35, No. 3, 1988, pp. 929–934.
97. T. Hori, T. Yasui, and S. Akamatsu, "Hot-carrier effects in MOSFETs with nitrided-oxide gate dielectrics prepared by rapid thermal processing", *IEEE Transactions on Electron Devices*, Vol. 39, No. 1, 1992, pp. 134–147.
98. J. F. Zhang, H. K, Sii, G. Groeseneken, and R. Degraeve, "Degradation of oxides and oxynitrides under hot hole stress," *IEEE Transactions on Electron Devices*, Vol. 47, 2000, p. 378.
99. N. Kimizuka, K. Yamaguchi, K. Imai, T. Iizuka, C. T. Liu, R. C. Keller, and T. Horiuchi,"NBTI enhancement by nitrogen incorporation into ultrathin gate oxide for 0.10-μm gate CMOS generation," in *Symposium on VLSI Technology Digest of Technical Papers*, 2000, pp. 92–93.
100. T. Kuroi, S. Shimizu, A. Furukawa, S. Komori, Y. Kawasaki, S. Kusunoki, Y. Okumura, M. Inuishi, N. Tsubouchi, and K. Horie, "Highly reliable 0.15-μm MOSFETs with surface proximity gettering (SPG) and nitrided oxide spacer using nitrogen implantation," in *Symposium on VLSI Technology Digest of Technical Papers*, 1995, pp. 19–20.
101. J. R. Shih, M. C. Chiang, H. C. Lin, R. Y. Shiue, Y. Peng, and J. T. Yue, "*N*-FET HCI reliability improvement by nitrogen interstitialization and its mechanism," *40th Annual International Reliability Physics Symposium*, 2002, pp. 272–277.
102. G. Cheroff, F. Fang, and H. Hochberg, "Effect of low temperature annealing on the surface conductivity of Si in the Si-SiO2-Al system," *IBM Journal*, Vol. 8, 1964, pp. 416–421.
103. P. Balk, "Effects of hydrogen annealing on silicon surfaces," in *Proceedings of the Electrochemical Societies Spring Meeting*, Abstract No. 109, 1965, pp. 237–240.
104. N. M. Johnson, D. K. Biegelsen, and M. D. Moyer, "Low-temperature annealing and hydrogenation of defects at the Si-SiO2 interface," *Journal of Vacuum Science and Technology*, Vol. 19, 1981, pp. 390–394.
105. E. Cartier, J. H. Stathis, and D. A. Buchanan, "Passivation and depassivation of silicon dangling bonds at the Si–SiO_2 interface by atomic hydrogen," *Applied Physics Letters*, Vol. 63, 1993 pp. 1510–1512.
106. E. Morifuji, T. Kumamori, M. Muta, K. Suzuki, M. S. Krishnan, T. Brozek, X. Li, W. Asano, M. Nishigori, N. Yanagiya, S. Yamada, K. Miyamoto, T. Noguchi, and M. Kakumu, "New guideline for hydrogen treatment in advanced system LSI," in *Symposium on VLSI Technology Digest of Technical Papers*, 2002, pp. 218–219.
107. A. Hamada, T. Furusawa, N. Saito, and E. Takeda, "A new aspect of mechanical stress effects in scaled MOS devices," *IEEE Transactions on Electron Devices*, Vol. 38, No. 4, 1991, pp. 895–900.
108. J. W. Lyding, K. Hess, and I. C. Kizilyalli, "Reduction of hot electron degradation in MOS transistors by deuterium processing," *Applied Physics Letters*, Vol. 68, 1996, pp. 2526–2528.
109. I. C. Kizilyalli, J. W. Lyding and K. Hess, "Deuterium post-metal annealing of MOSFETs for improved hot carrier reliability," *IEEE Electron Device Letters*, Vol. 18, 1997, pp. 81–83.

110. K. Cheng, J. Lee, Z. Chen, S. A. Shah, K. Hess, J. P. Leburton, and J. W. Lyding, "Fundamental connection between hydrogen/deuterium desorption at silicon surfaces in ultrahigh vacuum and at oxide/silicon interfaces in metal-oxide-semiconductor devices," *Journal of Vacuum Science and Technology B*, Vol. 19, No. 4, 2001, pp. 1119–1123.
111. A. J. Hof, E. Hoekstra, A. Y. Kovalgin, R. van Schaijk, W. M. Baks, and J. Schmitz, "The impact of deuterated CMOS processing on gate oxide reliability," *IEEE Transactions on Electron Devices*, Vol. 52, No. 9, 2005, pp. 2111–2115.
112. S. M. Cho, J. H. Lee, M. Chang, M. S. Jo, H. S. Hwang, J. K. Lee, S. B. Hwang, and J. H. Lee, "High-pressure deuterium annealing effect on nanoscale strained CMOS devices," *IEEE Transactions on Device and Materials Reliability*, Vol. 8, No. 1, 2008, pp. 153–159.
113. D. Q. Kelly, D. Onsongo, S. Dey, R. Wise, R. Cleavelin, and S. K. Banerjee, "Enhanced hot-electron performance of strained Si NMOS over unstrained Si,"in *Proceedings of the 42th Annual International Reliability Physics Symposium*, 2004, pp. 455–462.
114. S. M. Cho, J. H. Lee, M. Chang, M. S. Jo, H.S. Hwang, J. K. Lee, S. B. Hwang, and J. H. Lee, "High pressure deuterium annealing effect on nano-scale CMOS devices with different channel width," in *Proceedings of the Nanotechnology Materials and Devices Conference (NMDC)*, 2006, pp. 98–99.
115. K. Cheng, J. Lee, Z. Chen, S. Shah, K. Hess, J. W. Lyding, Y. K. Kim, Y. W. Kim, and K. P. Suh, "Deuterium pressure dependence of characteristics and hot-carrier reliability of CMOS devices," *Microelectronic Engineering*, Vol. 56, 2001, pp. 353–358.
116. Y. Nishioka, E. F. Da Silva, Y. Wang, and T. P, Ma, "Dramatic improvement of hot-electron-induced interface degradation in MOS structures containing F or Cl in SiO_2," *IEEE Electron Device Letters*, Vol. 9, No. 1, 1988, pp. 38–40.
117. Y. Nishioka, K. Ohyu, Y. Ohji, N. Natuaki, K. Mukai, and T. P. Ma, "Hot-electron hardened Si-gate MOSFET utilizing F implantation," *IEEE Electron Device Letters*, Vol. 10, 1989, p. 141.
118. JP001.01, "Foundry process qualification guidelines (Wafer fabrication manufacturing sites)," JEDEC Solid State Technology Association and Fabless Semiconductor Association, May 2004.

CHAPTER 6

Gate-Oxide Integrity and Time-Dependent Dielectric Breakdown

6.1 Tunneling in Metal-Oxide-Semiconductor Structure

6.1.1 Gate Leakage Tunneling Mechanism

Direct Tunneling versus Fowler-Nordheim Tunneling

Figure 6.1 provides a schematic for oxide under direct tunneling (DT) through a trapezoidal barrier and Fowler-Nordheim (FN) tunneling through a triangular barrier into the insulator conduction band. For thick oxides, DT dominates for low applied voltage (V_{g1}) when the gate oxide voltage V_{ox} is less than the interfacial energy barrier ϕ_b (Fig. 6.1*a*), whereas FN tunneling takes place for high applied voltage (V_{g2}) when $V_{ox} > \phi_b$ (Fig. 6.1*b*). On the other hand, for thin oxides, DT is the dominant tunneling mechanism (Fig. 6.1*c*). At a low voltage (V_{g1}) with DT dominant, for the same voltage, direct tunneling allows more current than FN tunneling. In other words, for the same current, DT yields a lower voltage (and electrical field). The tunneling mechanisms are generally mutually exclusive, with only one tunneling mechanism occurring.[1] As a rough guideline, significant FN currents flow when the electrical field is greater than 6 MV/cm, whereas significant DT occurs with electrical fields of less than 5 MV/cm.

The FN tunneling current J_{FN} has been derived as a function of the applied field to the oxide as follows[2]:

$$J_{FN} = AE_{ox}^2 \exp\left(-\frac{B}{E_{ox}}\right) \tag{6.1}$$

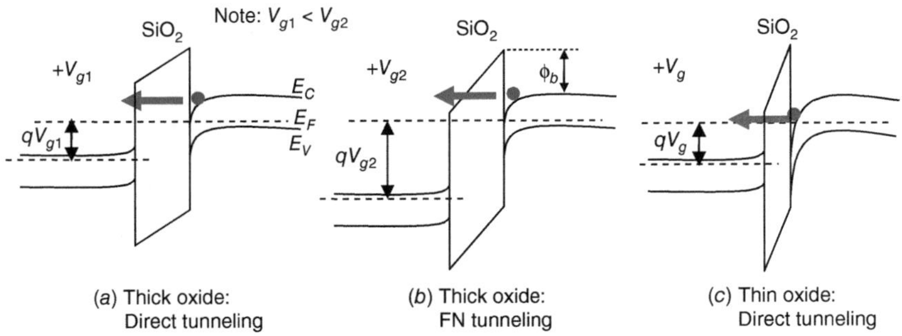

FIGURE 6.1 Schematics for DT and FN tunneling.

where
$$A = \frac{q^3}{8\pi h \phi_b} \tag{6.2}$$

and
$$B = \frac{8\pi}{3} \frac{\sqrt{2m_{eff}}\,\phi_b^{3/2}}{qh} \tag{6.3}$$

where q is the electron charge, h is Planck's constant, m_{eff} is the electron effective mass in the insulator, ϕ_b is the barrier height (Fig. 6.1*b*) at the semiconductor-oxide interface, and E_{ox} is the electrical field across the oxide. It is clear that a plot of the logarithm of $J_{FN}/E_{ox}^{\,2}$ versus $1/E_{ox}$ (known as the *Fowler-Nordheim plot*) yields a straight line whose slope is B. Therefore, the barrier height ϕ_b can be calculated. For metal-oxide-semiconductor (MOS) structures with a relatively thick oxide and metal gate, this model provides a very good fit to experimental data for a wide range of applied voltages.[3] It should be mentioned that this model does not consider the depletion effects on the polysilicon gate. Typical values of A and B are given as $A = 1.6$ MA/(MV)^2 and $B = 222$ MV/cm. When there is large deviation in the current density from the FN equation, it may indicate that the oxide has defects.

For thin oxides with thickness on the order of 1 or 2 nm, the DT current at low fields is no longer negligible. In this case, the barrier is approximately trapezoidal (Fig. 6.1*c*). The DT current density J_{DT} is expressed as[4]

$$J_{DT} = \left(\frac{e}{2\pi h d^2}\right)\left\{\begin{aligned}&\left(\phi_b - E_0 - \frac{eV}{2}\right)\exp\left[-\frac{4\pi(2m_{eff})^{1/2}}{h}\times\alpha\left(\phi_b - E_0 - \frac{eV}{2}\right)^{1/2} d\right]\\ &-\left(\phi_b - E_0 + \frac{eV}{2}\right)\times\exp\left[-\frac{4\pi(2m_{eff})^{1/2}}{h}\alpha\left(\phi_b - E_0 + \frac{eV}{2}\right)^{1/2} d\right]\end{aligned}\right\} \tag{6.4}$$

where d is the barrier width, E_0 is the ground-state energy, V is applied bias, and α is a unitless adjustable parameter.

Equations (6.1) and (6.4) are the basic equations for gate-current modeling of FN and DT components, respectively. It was demonstrated that the DT current is dominant at below 3 V for 30 to 50 Å, FN current is dominant at above 4 V for the rest.[5]

In this chapter, the tunneling mechanism in MOS structures is introduced first, followed by a description of gate-oxide integrity (GOI) and time-dependent dielectric-breakdown (TDDB) phenomena and mechanisms. The evolution of gate-oxide TDDB models for lifetime extrapolation from stress conditions to operating conditions then is elaborated, with a focus on the most advanced technology. Both test structures and processes have a large impact on gate-oxide TDDB performance, both of which are then discussed in detail for process optimization to meet process qualification requirements.

Tunneling-Current Components for Metal-Oxide Semiconductors

In dual-gate CMOS devices, for a simplified consideration, the work-function differences $|\phi_{ms}|$ in n^+-gate/NMOS and p^+-gate/PMOS devices can be assumed symmetric. Therefore, the NMOS and PMOS devices have the same absolute value of flat-band voltage as well as threshold voltage.[6] It is seen that except for a sign difference, the relationships between the gate voltage V_g and oxide voltage V_{ox} are the same for both PMOS and NMOS devices. Such $|V_{ox}|$ versus $|V_g|$ relationships are expressed in the following equations:

- Under strong accumulation for either a p^+/PMOS or an n^+/NMOS device,

$$|V_g| = |V_{ox}|_{acc} + E_g/q \tag{6.5}$$

  where $|V_{ox}|_{acc}$ is the oxide voltage under accumulation, and E_g/q is the silicon band gap in volts.

- Under strong inversion for either a p^+/PMOS or an n^+/NMOS device,

$$|V_g| = |V_{ox}|_{inv} + |V_{poly}| \tag{6.6}$$

  where $|V_{ox}|_{inv}$ is the oxide voltage under inversion, and $|V_{poly}|$ is polysilicon band bending.

Therefore, Eqs. (6.5) and (6.6) define the voltage applied to the gate oxide $|V_{ox}|$, which applies equally well to both p^+/PMOS and n^+/NMOS devices with the same $|V_g|$. However, this does not mean that these two devices should have similar tunneling current characteristics (I_g versus V_g) under the same polarity condition (either strong accumulation or strong inversion). Other factors such as the tunneling species (electron and/or hole) indeed have an impact, as detailed below.

With the same $|V_g|$, the difference in the voltage applied to the gate oxide $|V_{ox}|$ between strong accumulation and strong inversion is obtained simply by subtracting Eq. (6.5) from Eq. (6.6):

$$|V_{ox}|_{inv} - |V_{ox}|_{acc} = E_g/q - |V_{poly}| > 0 \tag{6.7}$$

Equation (6.7) indicates that the voltage imposed to the gate oxide $|V_{ox}|$ under strong inversion is generally larger than that under strong accumulation with the same $|V_g|$ applied.

The voltage applied to the gate oxide $|V_{ox}|$ leads to the gate-oxide tunneling current. The significant DT current components (Fig. 6.2) in an Si-SiO_2-Si structure include electron tunneling from the conduction band (ECB) with a barrier height 3.1 eV, electron tunneling from the valence band (EVB) with a barrier height 4.2 eV, and hole tunneling from the valence band (HVB) with a barrier height 4.5 eV.

Figure 6.3 shows schematically the gate-leakage components for an MOS system with source/drain/gate/bulk. The corresponding tunneling mechanisms are summarized in Table 6.1. Energy-band diagrams of NMOS and PMOS devices under inversion and accumulation modes with the tunneling components are illustrated in Fig. 6.4. EVB generates the substrate current in both NMOS and PMOS devices under inversion mode. However, there is no EVB component under accumulation because no energy state with the corresponding energy level is available in the band gap to receive the tunneling electrons. ECB tunneling dominates in NMOS devices for both bias polarities and in the accumulation regime in PMOS devices. In all these ECB cases, the injecting electrodes are well supplied with conduction-band

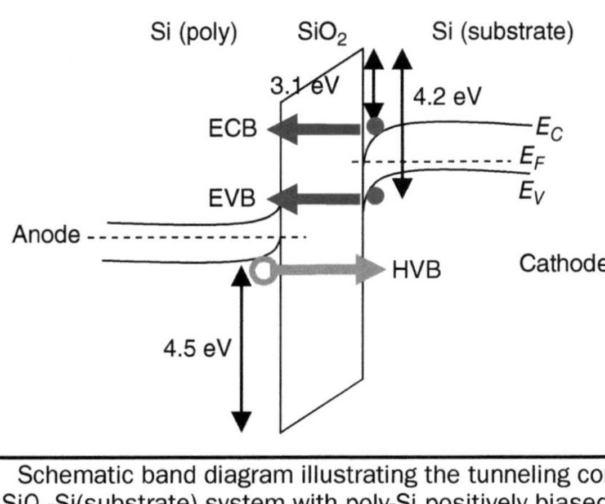

FIGURE 6.2 Schematic band diagram illustrating the tunneling components in an Si(poly)-SiO_2-Si(substrate) system with poly-Si positively biased, including electron tunneling from the conduction band (ECB), electron tunneling from the valence band (EVB), and hole tunneling from the valence band (HVB).

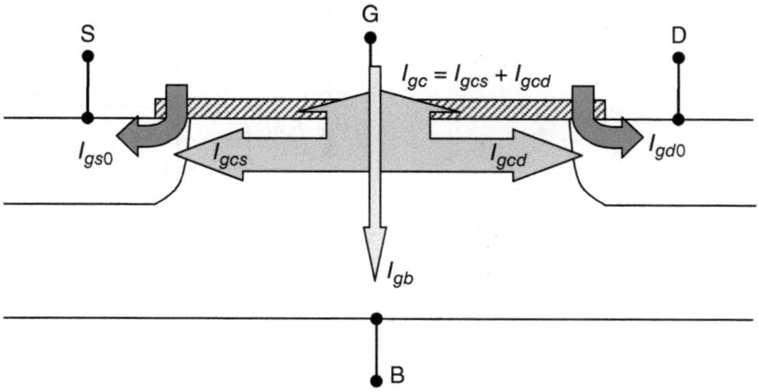

FIGURE 6.3 Components of tunneling current. I_{gb} flows between gate and substrate. I_{gs0} and I_{gd0} are tunneling currents through gate-source/drain-overlap regions. I_{gc} is the gate-to-channel tunneling current, which is partitioned into I_{gcs} (flowing to the source) and I_{gcd} (flowing to the drain).

	Inversion ($V_g > 0$ for NMOS, $V_g < 0$ for PMOS)			**Accumulation ($V_g < 0$ for NMOS, $V_g > 0$ for PMOS)**	
	I_{gc} ($I_{s/d}$)	I_{gb} (I_{sub})	I_{gs0}/I_{gd0}	I_{gb} (I_{sub})	I_{gs0}/I_{gd0}
NMOS	ECB (dominates at all voltage levels)	EVB	ECB	ECB	ECB
PMOS	HVB (dominates at low V_g)	EVB (dominates at high V_g)	HVB	ECB	HVB

TABLE 6.1 Dominant Gate-Leakage-Current Mechanisms for Each Component Shown in Fig. 6.4.

electrons at the surface. These conduction-band electrons can tunnel through the relatively small 3.1-eV barrier. HVB dominates the gate leakage of inverted PMOS devices in the low-bias range.[7]

Comparisons for different polarity biases and MOS types are detailed below:

- *n+/NMOS versus p+/PMOS in strong inversion:* Since both devices under inversion follow Eq. (6.6), the carrier density depending on $C_{ox} \cdot |V_{ox}|$ should be the same for both NMOS and PMOS devices. Figure 6.5 shows carrier-separation results (method is given in Fig. 4.12) of NMOS and PMOS devices with 2-nm-thick gate oxides under inversion conditions. The gate-leakage-current mechanism in a PMOS device is quite different from that in an NMOS device:
  - Referring to Fig. 6.4*b* and Fig. 6.5*a*, for NMOS devices, the dominate gate leakage is from the ECB supplied from the

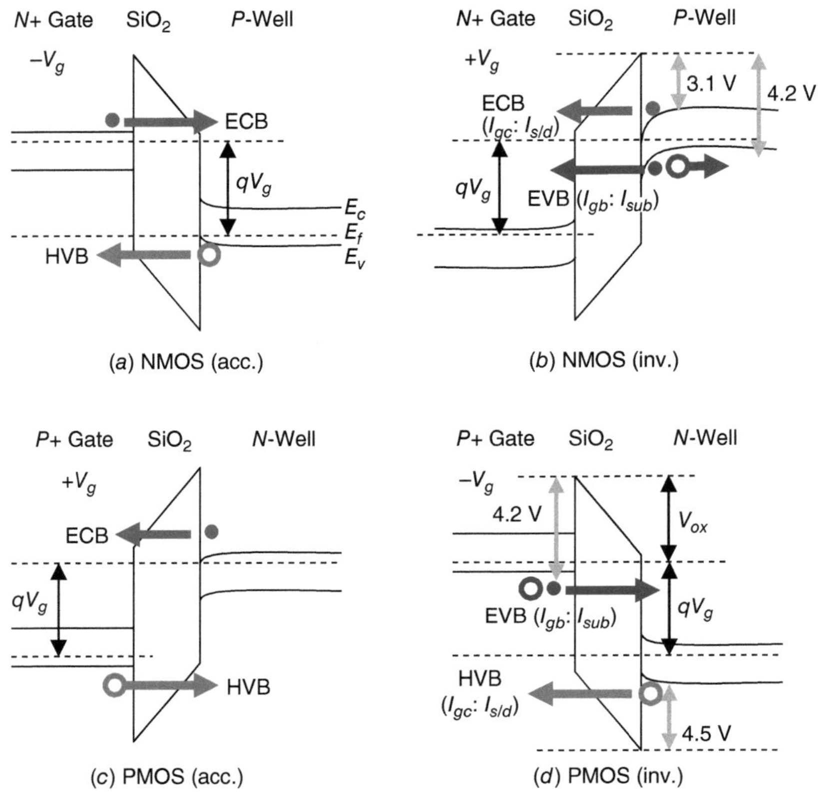

FIGURE 6.4 Energy-band diagrams of NMOS and PMOS devices under inversion and accumulation modes, and the corresponding gate-leakage-current components in Table 6.1.

source/drain (denoted as I_{gc} or $I_{s/d}$); the minor gate leakage is the substrate current (I_{gb} or I_{sub}), which is from EVB. Figure 6.5*a* shows that $I_{s/d}$ is always larger than I_{sub} regardless of the gate voltage applied. The reason is that the barrier height for the EVB is 4.2 eV, which is quite larger than that for the ECB, namely, 3.1 V. In addition, the EVB component is important only for relatively high voltages because at low voltages, receiving states would be required in the energy gap of the semiconductor, which is not acceptable.

- Referring to Fig. 6.4*d* and Fig. 6.5*b*, for PMOS devices, the gate leakage current is dominated by hole current at low gate voltages ($|V_g| < 2.1$V), which is from the HVB and is supplied from the source/drain (denoted as I_{gc} or $I_{s/d}$). For the HVB, the barrier height is 4.5 eV, which is much higher than for the ECB. On other hand, the gate leakage current

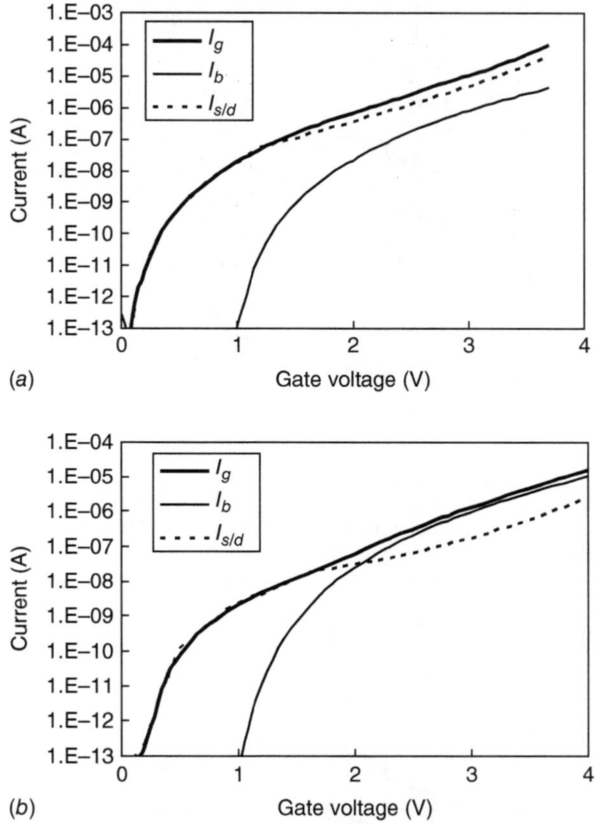

FIGURE 6.5 Carrier separation measurements on NMOS and PMOS devices with 2 nm of oxide under inversion conditions. In the NMOS device, the dominate gate leakage comes from then source/drain regardless of the gate-voltage magnitude. In the PMOS device the gate leakage current is dominated by the hole current at low gate voltage ($|V_g| < 2.1$ V).

begins to be dominated by electron current by valence-band (EVB) tunneling from the *p*+ poly-Si gate at high gate voltages (Fig. 6.4*d*), characterized as the substrate current (I_{gb} or I_{sub}). Although there is a high density of electrons in the valence band of the p+ polysilicon gate (EVB), these electrons see a lower barrier (4.2 eV) than those of the HVB. Therefore, the leakage current in inverted PMOS devices is always substantially lower than that in NMOS devices at both low and high V_g (Fig. 6.6) because in PMOS devices both the HVB at low V_g and the EVB at high V_g have higher barrier heights than the ECB in NMOS devices.

- For PMOS devices, it should be pointed out that the contribution of the electron tunnel from the negatively

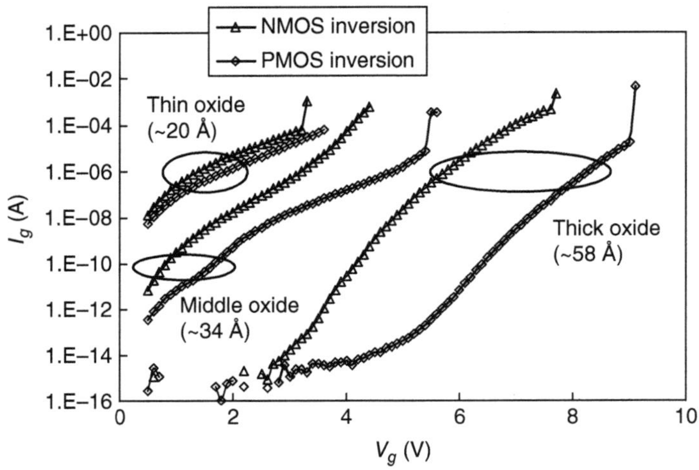

FIGURE 6.6 Gate tunneling current measured in inversion exhibits a strong polarity dependence in dual-gate CMOS devices. *P*+/PMOS devices show substantially lower tunneling current than *n*+/NMOS devices. The source/drain is grounded with the substrate during measurement.

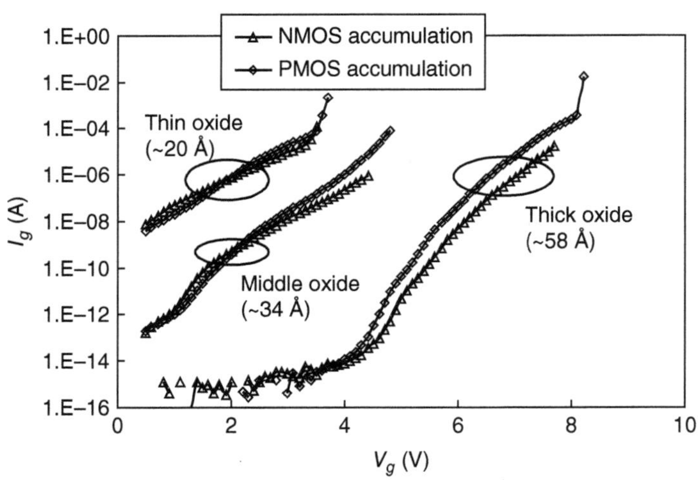

FIGURE 6.7 Gate tunneling currents in accumulated NMOS and PMOS devices show much less polarity dependence than is the case under inversion.

biased *p*+ polysilicon gate conduction band (i.e., the ECB) is very limited because the *p*+ polysilicon is not inverted. Thus the ECB component is not shown in Fig. 6.4*d*.

- *n+/NMOS versus p+/PMOS in accumulation:* Figure 6.7 shows that the gate tunneling currents in accumulated NMOS and PMOS capacitors show no polarity dependence, which indeed matches Eq. (6.5). Referring to Fig. 6.4*a* and *c*, the dominant carrier is the ECB for both NMOS and PMOS devices under accumulation.

- *Inversion versus accumulation in n+/NMOS:* As described by Eq. (6.7), the voltage imposed on the gate oxide is smaller under the accumulation mode. Therefore, the gate leakage under the accumulation mode is smaller than that under the inversion mode, as shown in Fig. 6.8.
- *Inversion versus accumulation in p+/PMOS:* Similarly, for PMOS devices, the voltage imposed on the gate oxide is smaller under the accumulation mode. However, the ECB with barrier height 3.1 eV is the main tunneling-current mechanism under accumulation, which has a lower barrier height than the HVB (4.5 eV) and the EVB (4.2 eV) under inversion. As a result, PMOS devices under accumulation are much worse than those under inversion. This is especially true for thick oxide as shown in Fig. 6.9.

6.1.2 Polarity-Dependent Q_{bd} and T_{bd}

Charge to breakdown Q_{bd} and *time to breakdown* T_{bd} are two critical parameters characterizing gate-oxide reliability. The relationship between Q_{bd} and T_{bd} will be elaborated in this section.

In NMOS devices, Q_{bd} at $-V$ (i.e., gate injection, accumulation mode) is consistently lower than that of $+V$ (i.e., substrate injection, inversion mode) roughly by a factor of 20× with various gate-oxide thicknesses

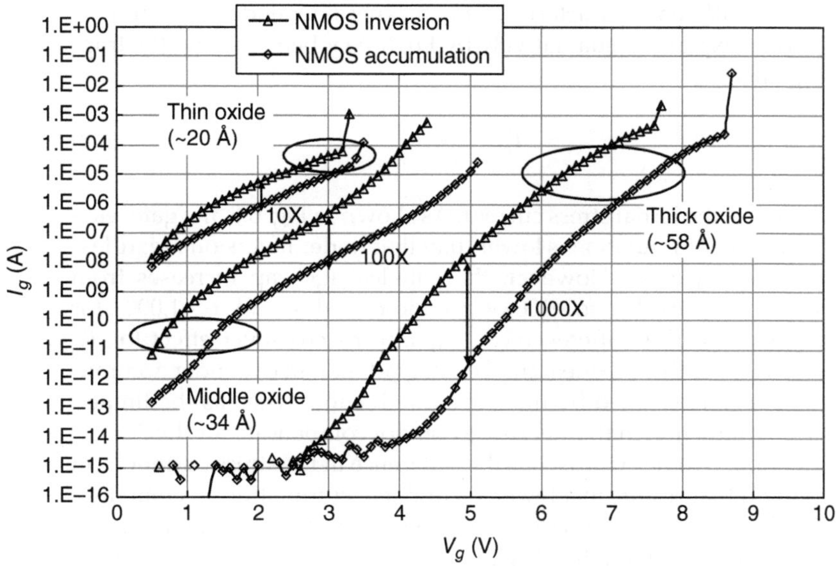

FIGURE 6.8 Polarity-dependent tunneling current for NMOS inversion and accumulation with varied oxide thicknesses.

Gate oxide thickness (Å)	19	35	50
Ratio of Q_{bd} (inversion/ accumulation)	~20 (>1)	~10–20 (>1)	~10–20 (>1)
Ratio of T_{bd} (inversion/ accumulation)	~5–10 (>1)	~0.1 (<1)	~0.3 (<1)

Source: Data extracted from Ref. 8.

TABLE 6.2 Ratio of Q_{bd} (Inversion/Accumulation) and T_{bd} (Inversion/Accumulation) for NMOS Devices with Various Gate-Oxide Thicknesses

(Table 6.2). The data are extracted from Ref. 8. The strong polarity dependence of Q_{bd} was attributed to the weakness of a structure transition layer (STL) located at the SiO_2-Si interface.[9] The injected energetic electrons may release energy at the STL interface and cause bond breaking, a precursor of dielectric breakdown, under gate injection.

However, Table 6.2 also shows that T_{bd} for the accumulation mode is higher than that for the inversion mode for the thicker oxide (50 Å thick) stressed at high voltages in the FN regime, which is consistent with the lower V_{ox} under accumulation, as indicated by Eq. (6.7). On the other hand, for thinner oxides (19 Å thick) often stressed in the DT regime, T_{bd} for the accumulation mode is lower than that for the inversion mode. Wu et al.[8] suggested a crossover behavior in T_{bd} between the accumulation mode and the inversion mode as t_{ox} is varied, and the crossover thickness is around 2.9 nm.

The different characteristics of thickness-dependent Q_{bd} and T_{bd} could be explained qualitatively by Eq. (6.8) when ignoring the trapping effect:

$$T_{bd} = \frac{Q_{bd}}{J_0} \tag{6.8}$$

where J_0 is the initial stress current. As shown in Fig. 6.8, the gate leakage under accumulation is lower than that under inversion regardless of oxide thickness. However, the gate-leakage gap increases from ~10× for thin oxides, to ~100× for middle oxides, and to ~1,000× for thick oxides. This difference can explain the crossover effect for T_{bd} versus t_{ox}. For thick oxides, an ~20× degradation in Q_{bd} under accumulation is compensated by the current reduction, whereas the limited current reduction in thinner oxides cannot compensate for the 20× Q_{bd} reduction. In other words, it is the difference in the stress current [J_0 in Eq. (6.8)] that results in a lower T_{bd} under accumulation as compared with the inversion mode as t_{ox} is reduced. In short, T_{bd} is a parameter for gate oxide from the reliability point of view, whereas Q_{bd} is a fundamental breakdown parameter directly related to the defect-generation process.

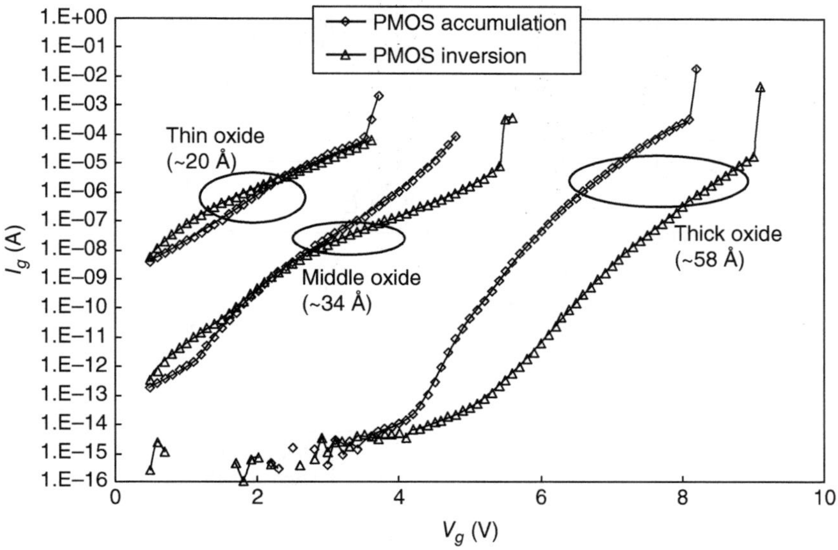

FIGURE 6.9 Polarity dependent tunneling current for PMOS inversion and accumulation with varied oxide thicknesses.

Chen et al.[10] reported that Q_{bd} values for NMOS devices with a 3-nm-thick N_2O-nitrided oxide under substrate-injection polarity (i.e., +1 A/cm^2, gate positive bias, inversion mode) are much higher than those under gate-injection polarity (i.e., –1 A/cm^2, gate negative bias, accumulation mode). Shi et al.[11] and Hasegawa et al.[12] showed the same polarity-dependent Q_{bd} for PMOS devices. However, Q_{bd} values for PMOS devices show the opposite trend: Under substrate injection (i.e., +1 A/cm^2, gate positive bias, accumulation mode), they are slightly lower than those under gate injection (i.e., –1 A/cm^2, gate negative bias, inversion mode). It is likely that this is caused by the nitrogen location in the dielectric because nitrogen indeed plays a critical role in PMOS gate-oxide reliability, as detailed later. Boron segregation at the grain boundary of the polysilicon gate is also known to degrade Q_{bd} under substrate injection.[12]

Figure 6.10 shows a comparisons of TDDB lifetime versus stress voltage between inversion and accumulation modes for both NMOS and PMOS devices with 5-nm-thick oxides.[13] The T_{bd} under inversion is worse than that under accumulation. Indeed, it is seen in Fig. 6.10 that for NMOS devices, the power-law voltage-acceleration exponent n value of the accumulation mode is much better than that of the inversion mode. In contrast, the inversion mode is better in lifetime and n values for PMOS devices. Note that the *n value is an indicator of oxide-interface quality,* and its predicted value is around 35 to 40 for a 50-Å-thick oxide.[14] The results in Fig. 6.10 imply that *in terms of n value, gate injection of electrons is better than substrate injection of*

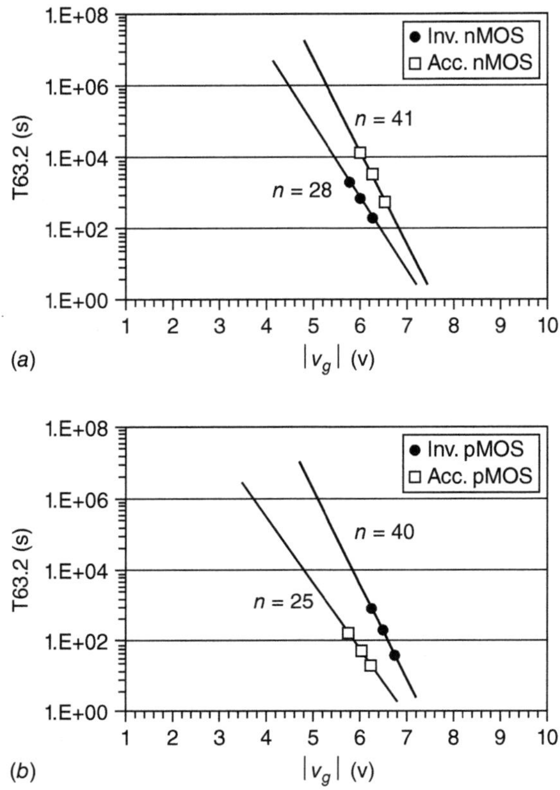

FIGURE 6.10 TDDB lifetime (*a*) inversion versus accumulation for NMOS and (*b*) inversion versus accumulation for PMOS. (*Used with permission from* IEEE.[13])

electrons. Note that for $|V_g| > 5$ V, the electron current is dominant in gate leakage.[6]

6.1.3 Relationship Between Gate Leakage Current and V_{bd}/T_{bd}

It is interesting to see that *gate current is a sensitive parameter to detect the TDDB lifetime characteristics* (Fig. 6.11).[5] Gate-oxide thickness can be extracted based on the gate current and the FN and DT current modeling. Figure 6.11 shows that the oxide lifetime is proportional to t_{ox} and inversely proportional to gate current.

Other independent work shows *a strong power-law dependence between the breakdown voltage V_{bd} (and T_{bd}) and the gate leakage* (Fig. 6.12).[13,15] The relationship between the gate leakage and the inversion t_{ox} looks more complicated when both intrinsic and effective thinning components exist, as shown in Fig. 6.12*b*. Figure 6.12*c* shows that the power-law

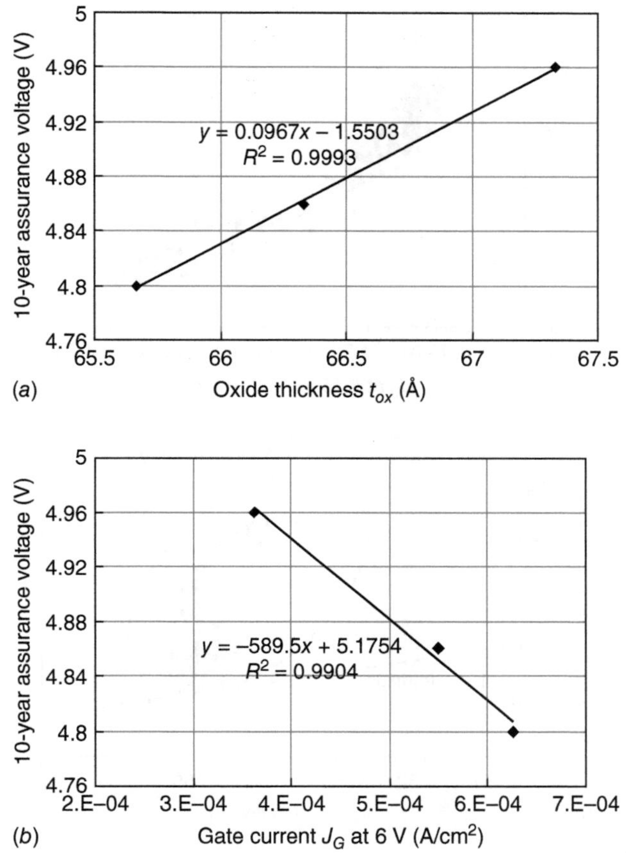

FIGURE 6.11 Correlation between electrical parameters and TDDB lifetime. (*a*) The oxide lifetime is proportional to t_{ox}; (*b*) the oxide lifetime is inversely proportional to gate current. (*Redrawn from raw data from Park et al.*[5])

exponent m increases rapidly with thinner oxides (<5 nm). This is attributed to fewer traps needed for the conductive breakdown path in the ultrathin gate oxide. In addition, as indicated in Fig. 6.5, two components [i.e., both the majority from source/drain ($J_{s/d}$) and minority from substrate carrier tunneling (J_{sub})] dominate the gate leakage current in thin oxides when the gate bias is greater than 2 V rather than the majority carrier tunneling ($J_{s/d}$) only in thick oxides. The contribution of substrate current J_{sub} to T_{bd} cannot be neglected in ultrathin oxides, leading to a much higher DT current, larger m value, and shorter T_{bd} lifetime.

Figure 6.13 separates the exponent m for the components from the source/drain ($J_{s/d}$) and substrate (J_{sub}) of 1.2-nm nitrided oxide with various physical oxide thicknesses and nitrogen concentrations, where $J_{s/d}$ is the current density of the source/drain region at a

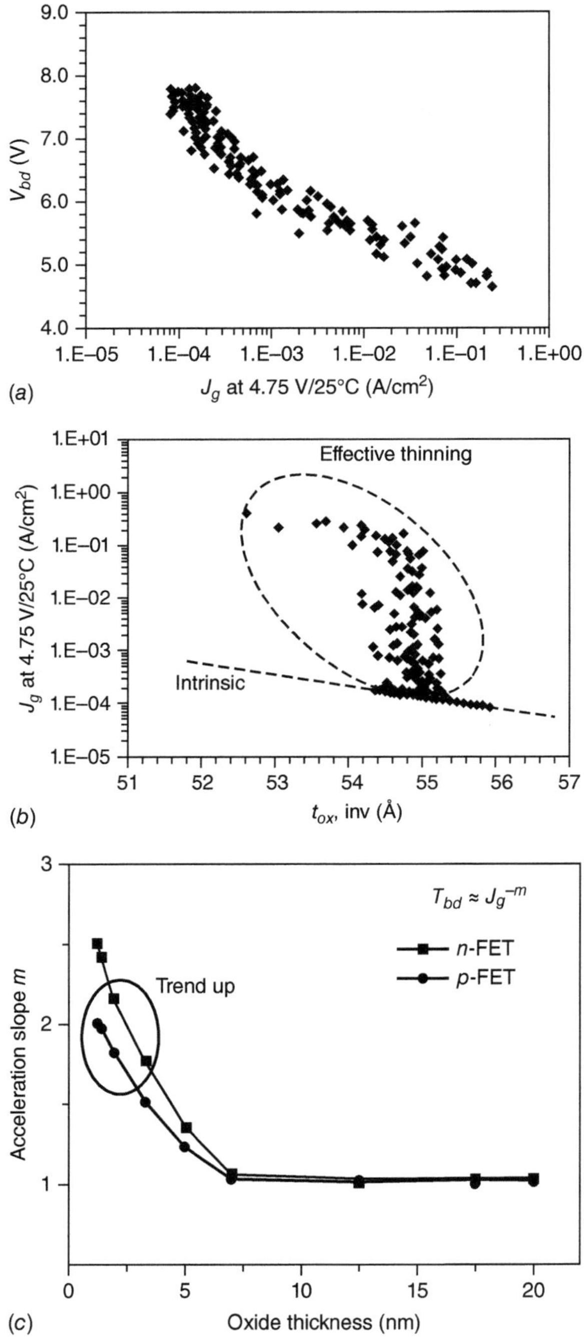

FIGURE 6.12 (*a*) NMOS breakdown voltage V_{bd} versus gate leakage; (*b*) NMOS gate leakage versus inversion t_{ox}. (*Used with permission from IEEE.*[13]) (*c*) The acceleration exponent *m*, log (T_{bd}) versus log (J_g) versus oxide thickness under inversion mode. (*Used with permission from IEEE.*[15])

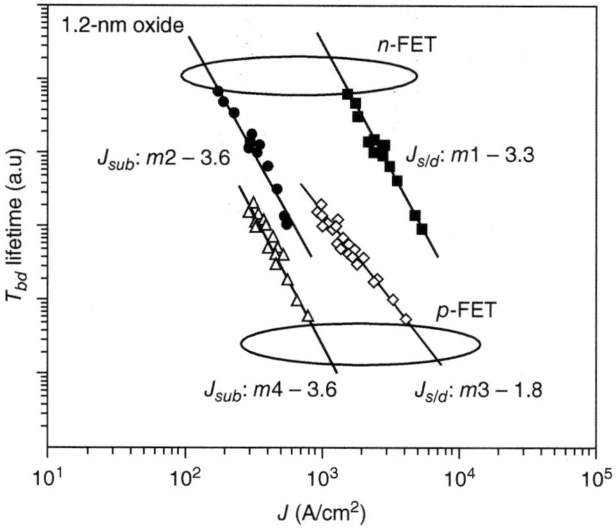

FIGURE 6.13 The logarithm of T_{bd} is plotted as a function of the logarithm of current density J in n-FETs and p-FETs under inversion mode, respectively. (*Used with permission from IEEE.*[15])

constant J_{sub} and J_{sub} is the current density of the substrate region at a constant $J_{s/d}$. It is interesting to note that the acceleration exponents $m1$, $m2$, $m3$, and $m4$ are universal regardless of physical oxide thickness and nitridation process. Besides, the slope $m3$ (from $J_{s/d}$ of the PMOS) is about half that of others, which is attributed to the fact that the *cold-hole injection from the inversion layer of the PMOS is less sensitive on oxide breakdown than on electron injection.*[16]

The strong correlation between the breakdown voltage V_{bd} (and T_{bd}) and the gate leakage can be explained qualitatively by Eq. (6.8), where the Q_{bd} is approximated by a product of T_{bd} and initial stress current J_0 owing to the lack of trapping for ultrathin oxides. From Eq. (6.8), T_{bd} is inversely proportional to gate leakage.

6.2 Gate-Oxide Dielectric Breakdown Mechanism

6.2.1 Extrinsic versus Intrinsic Breakdown

One type of characterization method for GOI is wafer-level voltage-ramp dielectric breakdown (V-ramp) detailed in Joint Electron Device Engineering Council (JEDEC) Standard JESD35.[17] This is sometimes simply referred as the *GOI test*. In the V-ramp test, a voltage that increases linearly with time (in practice, it is stepped, as shown in Fig. 6.14*a*) is applied until the oxide breaks down. The ramp rate is held constant (e.g., at a value of 1 MV/cm-s). Breakdown is defined as the voltage value V_{bd}

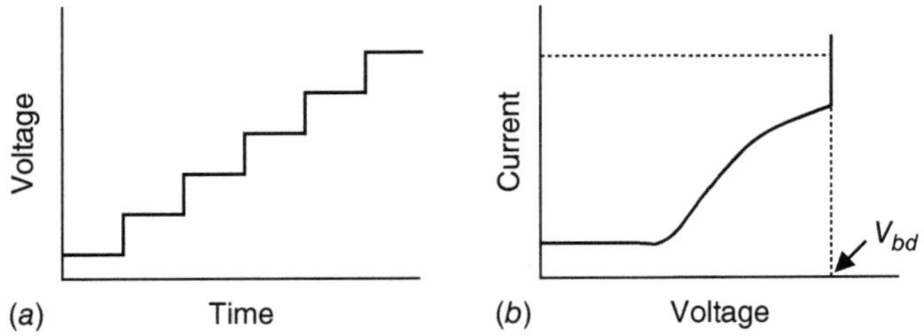

FIGURE 6.14 Schematic for (*a*) the voltage ramp-up with time and (*b*) the corresponding current-voltage (*I-V*) curve recorded for breakdown-voltage V_{bd} characterization.

at which a sudden jump in current is observed (Fig. 6.14*b*). Then statistical distributions of dielectric strength are plotted. The primary purpose of the GOI V-ramp test is to characterize the extrinsic behavior of the oxide being qualified.

The breakdown data are classified into three modes: mode A, mode B, and mode C. An example of the three-mode distributions of gate-oxide breakdown is given in Fig. 6.15. Both mode A and mode B failures are referred as *extrinsic breakdown* and are defect-induced, whereas mode C failure is referred as an *intrinsic breakdown*. Such oxides are defect-free and exhibit wear-out processes during long constant-voltage (or current) stressing as detailed in later sections.

- Mode A is the population that fails less than V_{use} (or V_{dd}). These are early breakdowns or initial failures. The root cause could be a physical short owing to pinholes or particles, very high density of permanent electron/hole traps, or

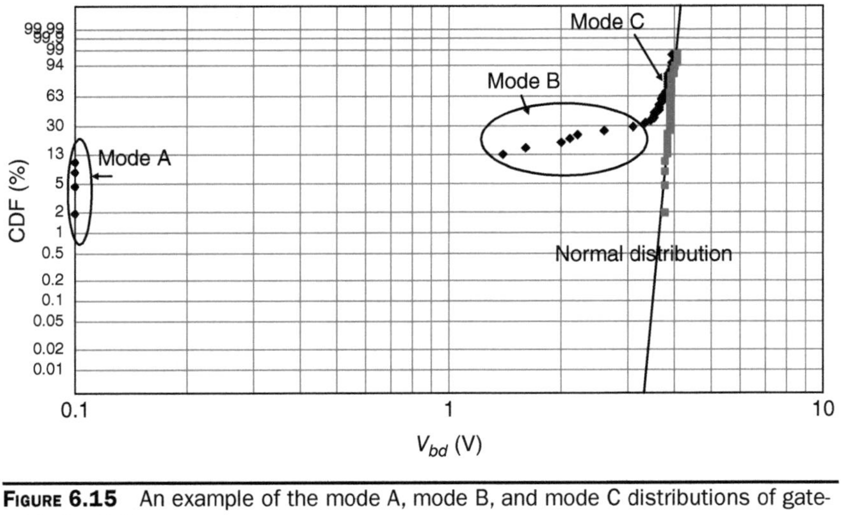

FIGURE 6.15 An example of the mode A, mode B, and mode C distributions of gate-oxide breakdown.

severe heavy metal contamination. Thus, improvements to get a clean oxidation environment and optimization of the oxidation recipe at the ultrathin oxide regime are necessary.

- Mode B is the population that fails at a voltage greater than V_{dd} but less than $2 * V_{dd}$, which represents a potential reliability hazard over the life of the product. The root cause could be the week spot in the oxide, local thinning, a crystal defect and oxygen precipitation in the silicon substrate, electron/hole traps, metal contamination, silicon surface roughness, electrical field concentration at the LOCOS/STI edge, and so on.
- Mode C is the population beyond $2 * V_{dd}$. In this case, the devices can withstand the highest electrical field (>8 to 12 MV/cm) before oxide breakdown. To increase the mode C breakdown field, the oxidation process itself should be examined. One possible method to enhance the inherent dielectric strength is to use a nitrided oxide, which has larger bond strength.

6.2.2 Time-Dependent Dielectric Breakdown

The thin dielectric can breakdown, resulting in gates or capacitor failure in circuits under a constant voltage stress and a given temperature for a certain period. The TDDB test is intended to evaluate the intrinsic oxide breakdown behavior, excluding process-related defects (which are characterized by the GOI V-ramp test discussed earlier). Note that TDDB test is time-consuming and not suitable for a fast process monitor.

Constant-Voltage Stress Test

In this test, a constant voltage is applied to the gate, and variation in the gate leakage I_g is monitored until there is a sudden jump in the gate leakage, an indication of oxide breakdown, and the time is recorded as T_{bd} (time to breakdown). A charge-to-breakdown Q_{bd} also can be calculated from

$$Q_{bd} = \int_0^{T_{bd}} J_g dt \quad \text{(unit: C/cm}^2\text{)} \tag{6.9}$$

where J_g is the constant-current stress density in amperes per square centimeter (A/cm^2), defined by I_g/A, with A as the oxide area. The cumulative distributions of T_{bd} and Q_{bd} are then plotted to evaluate oxide quality.

Constant-Current Stress Test

In this test, a constant current is applied to the gate, and the variation in the gate voltage V_g is monitored and T_{bd} is measured.[18] Q_{bd} is calculated from

$$Q_{bd} = J_g \cdot T_{bd} \quad \text{(unit: C/cm}^2\text{)} \tag{6.10}$$

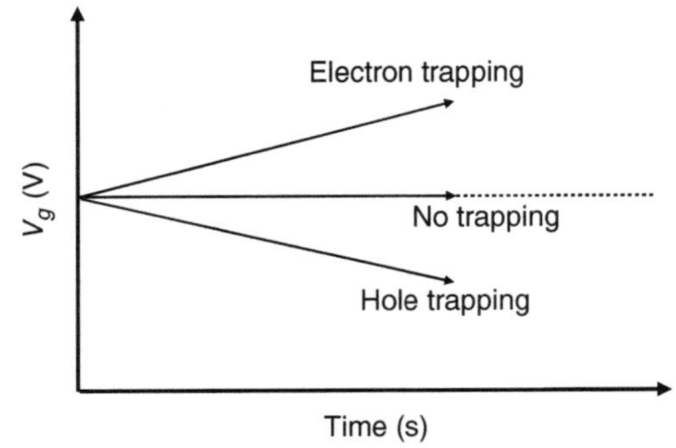

FIGURE 6.16 Schematic for the gate voltage V_g in terms of the CCS time.

The stress-current density is usually 50 to 100 mA/cm^2 when the gate oxide thickness is in the range of 50 to 100 Å.

Figure 6.16 is a schematic for the gate voltage V_g in terms of the constant-current stress (CCS) time, which provides information for either electron or hole trapping during CCS stressing. The slope represents the trapping rate, and usually the thinner the oxide is, the smaller is the trapping rate.

Two-Stage Gate-Oxide Breakdown Process

The gate-oxide breakdown process is related to the generation and accumulation of bulk defects, which can be divided into two stages:

- *Buildup stage:* During this stage, localized high-field and high-current-density regions are formed as a result of charge trapping (defect generation) owing to current flow in the oxide. This is explained by either the *anode-hole-injection (AHI) model,*[19] the *anode-hydrogen-release (AHR) model,*[20] or the *thermo-mechanical model.*[21] Some studies suggested that the AHI rate may be too small to cause the required damage in the time scale of breakdown.[22] On the other hand, the AHR model has been presented only in rather qualitative terms. Therefore, refinements to the AHI model were reported.[23] A quantitative hydrogen-based model for ultrathin oxide breakdown also was reported.[24] The defect-generation and accumulation process continues until the local current density or field or trap (defect) density reaches a critical value. Note that the *thermo-chemical model is field-driven, whereas both the AHI and AHR models are current-driven.*

- *Runaway stage:* At this stage, the local current density or field or defect density has reached a critical value in some parts (or weak spots) of the oxide. Electrical and thermal runaway occurs, resulting in destruction of the oxide in a very short time. This is explained by the *percolation model.*[25]

The preceding defect-generation models will be elaborated in Sec. 6.2.4.

Microscopic Model of Oxide Damage

The silicon dioxide (SiO_2) network consists of silicon atoms mostly in six-member tetrahedral rings with the silicon-oxygen-silicon bridging bond angle at 144°. However, owing to deviations in the amorphous structure, there are also higher- and lower-order rings. The bond angle then will range from 120° for four-member rings to 180° for eight-member rings. The deviation of bond angle leads to the strained bonds (Fig. 6.17).[21] Note that the density of 120° bonds increases toward the oxide-silicon interface. Cleavage of strained bonds in the SiO_2 network, through either electrical stressing or radiation effects, will lead to the formation of a *trivalent silicon (hole trap: ≡ Si•) and a nonbridging oxygen defect (neutral electron trap: ≡ Si−O*•**)** as indicated in Eq. (6.11).

$$\equiv Si - O - Si \equiv \rightarrow \equiv Si - O* + \equiv Si* \tag{6.11}$$

6.2.3 Correlation Between V_{bd} and T_{bd}

There are some efforts to base the V_{bd} dependence on gate-oxide thickness (t_{ox} in angstrom), temperature (T in kelvin), and V-ramp testing ramp rate (R in V/s). An empirical model for gate oxides (20 A < t_{ox} < 32 A) based on the GOI V-ramp test carried out on NMOS and PMOS capacitors under the accumulation mode is given as follows[26]:

$$V_{bd} = V_0 + a \cdot t_{ox} + \frac{E_a}{k_B T} + b \cdot \ln(R) \tag{6.12}$$

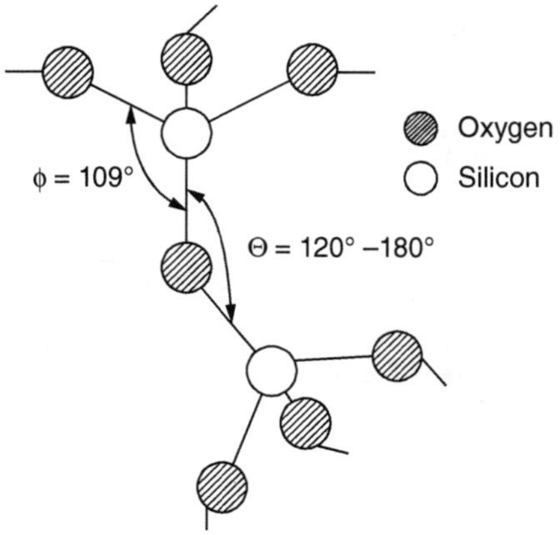

FIGURE 6.17 Schematic SiO_2 network. (*Used with permission from the American Institute of Physics.*[21])

where a is the oxide-thickness acceleration factor (a typical value is 15.5 ± 0.4 MV/cm), E_a is the activation energy (a typical value is 46 ± 4 meV), b is the ramp-rate acceleration factor (a typical value is 82 ± 14 mV), and V_0 is the intercept.

Similarly, the time to breakdown T_{bd} is modeled empirically as follows[26]:

$$T_{bd} = T_0 \cdot \exp\left[\frac{1}{b} \cdot \left(a \cdot t_{ox} + \frac{E_a}{k_B T} - V_g\right)\right] \tag{6.13}$$

Consequently, there is an exponential dependence between V_{bd} and T_{bd} based on Eqs. (6.12) and (6.13):

$$T_{bd} = T_0 \cdot \exp\left[\frac{1}{b} \cdot (V_{bd} - V_0 - V_g) - \ln(R)\right] \tag{6.14}$$

V_{bd} can be obtained by a fast wafer-level V-ramp test, whereas T_{bd} is characterized by extended stressing. Therefore, the correlation between V_{bd} and T_{bd} provides a qualitative approach to quickly accessing the gate-oxide T_{bd}. This is also pretty convenient for production in-line monitoring. Theoretical analysis for the correlation between V_{bd} and T_{bd} is given in Sec. 12.6 of Chap. 12.

6.2.4 Models for Defect Generation

Hole-Trapping Model (Hole-Induced Breakdown Model)

Figure 6.18 provides a schematic illustration for the *hole-trapping model,* which is *the base for the 1/E model* for TDDB lifetime projection for thick oxide used in early times.[27] In this model, electron injection from the cathode results in the generation of hole-electron pairs in the oxide owing to *impact ionization,* which is an important mechanism for *thick oxides.* The impact ionization takes place when the energy of

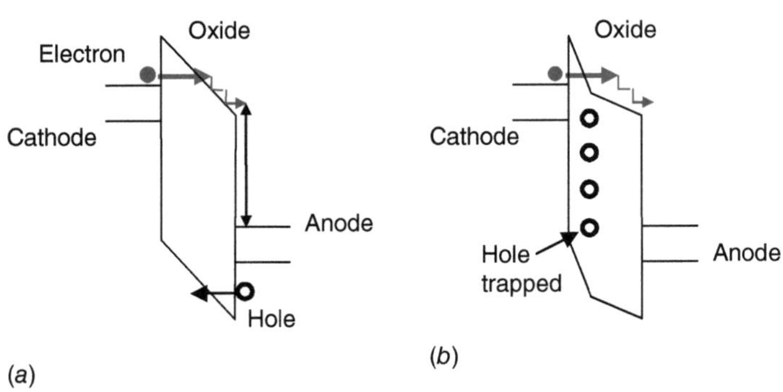

FIGURE 6.18 Schematic illustration of the hole-trapping model.

the tunneling electrons is greater than 9 eV. The generated holes, driven to the cathode because of the applied electrical field (Fig. 6.18*a*), are then trapped and/or recombine with electrons to produce defect states. Hole trapping in localized areas near the cathode increases the cathode electrical field (Fig. 6.18*b*). As a result, the electrons in the cathode will have a thinner barrier to direct tunneling to the gate oxide, which further causes positive feedback and eventual oxide breakdown. The model was demonstrated for gate oxides with thickness in the range of 7.9 to 32 nm.

Anode-Hole-Injection Model

The *AHI model*[19] is an extension of the hole-induced-breakdown model to the case of thin oxides, valid for a thickness range between 2.5 and 13 nm. *In thick oxides, holes are generated primarily by impact ionization within the oxide. On the other hand, in thin oxides, holes originate from injection from the anode.* Electrons tunnel from the cathode to the anode via FN tunneling. At the anode, tunneled electrons give up energy to a deep valence-band electron. A hot hole results and is injected into the oxide, where it may be trapped.

Anode-Hydrogen-Release Model

For pure oxide, hole trapping is common because oxygen vacancy exists in the structure transition region at the SiO_2-Si interface. However, some studies pointed out that the trapping of hot holes is not responsible for the breakdown of silicon dioxide.[28] *The more likely mechanism for silicon dioxide breakdown is that of trap creation by hydrogen release—the AHR model.*[20,29,30]

In the AHR model (Fig. 6.19),[20] tunneling electrons with energy greater than 2 eV transfer their energy and release hydrogen from defect sites near the anode. The mobile hydrogen moves to regions near the cathode, where it can react with, for example, the oxygen vacancies (O_3≡Si–Si≡O_3) and thus produce defect states. *Effective interface softening* by the generated defect states leads to oxide wearout

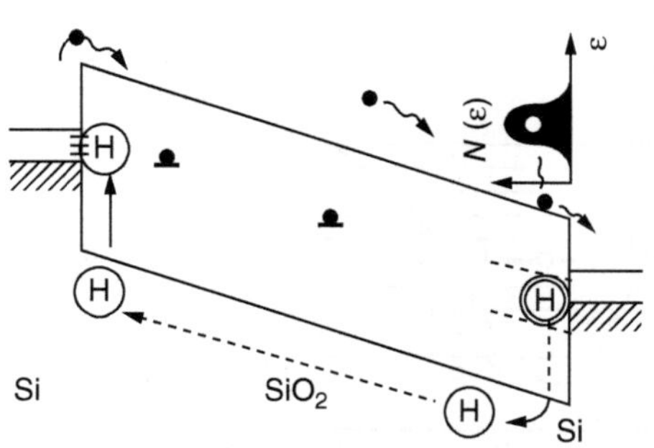

FIGURE 6.19 Schematic illustration of the anode-hole release (AHR) model. (*Used with permission from the American Institute of Physics.*[20])

and breakdown. The *AHR model is the base for the power-law model, which is currently the widely acceptable lifetime extrapolation model for ultrathin gate oxide in advanced technology* (Sec. 6.5.3).

Further study through STM experiments has demonstrated that at a high gate bias (6.5 to 8 V), the single-electron excitation (EE) causes hydrogen release (HR), whereas at very low energies (<2.5 eV) HR is controlled by vibrational excitation (VE), and a transition from single- to multiple-electron mechanisms is reported.[14,31] Figure 6.20 schematically shows the multiple excitation mechanism for HR that are dominant for voltages less than 2.5 V and still existing for voltages up to 6.5 to 8 V.[14]

In both the AHI and AHR models, it is assumed that it is the positive species generated at the anode that is responsible for the oxide degradation that leads to breakdown. The generation rate of the positive species thus is determined by the energy of the electrons at the anode.

However, for the thickest oxide, with a thickness of more than 50 Å, electron trapping is possible (Fig. 6.20). In this oxide thickness regime, in-elastic transport occurs. When V_{ox} is larger than 3.2 eV, the FN tunneling electrons are subjected to scattering in the gate oxide and lose their kinetic energy. After losing their kinetic energy, some electrons may be trapped, and some of them will be accelerated again by the applied electrical field. For the accelerated electrons, another in-elastic scattering will take place. This process will repeat until the

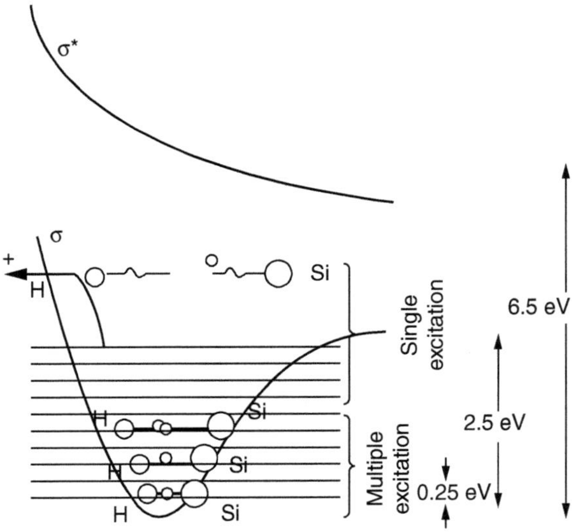

FIGURE 6.20 Hydrogen release through the multiple-excitation mechanism is expected to be dominant for voltages of less than 2.5 V and to exist for voltages of up to 6.5 to 8 V. (*Used with permission from Elsevier.*[14])

electrons arrive at the anode. *The electron energy at the anode and consequently the oxide degradation are determined by the oxide field.*

Thermochemical-Breakdown Model

Oxygen vacancy resulting in a weak Si–Si bond (Fig. 6.21*a*) replacing the normal Si–O–Si bonding (Fig. 6.17*a*) is an important intrinsic defect thought to dominate the low-field time-dependent dielectric breakdown process. As indicated in Fig. 6.21*a*, the permanent dipole moment in the lower half of the $O_3\equiv Si–Si\equiv O_3$ molecule is oriented antiparallel to the local electrical field. This dipolar energy can be used to lower the activation energy for bond breakage by collapsing to the planar *sp*2 configuration. This dipole-field coupling serves to lower the activation energy required for thermal bond breakage and accelerates the dielectric degradation process. Therefore, in this model, the applied electrical field (and not current or tunneling electrons) will cause a reduction in the activation energy required for thermal bond breakage (Fig. 6.21*b*), leading to oxide breakdown under low-field conditions (<10 MV/cm) for gate oxides thicker than 4 nm.[21] It is thus an exponential dependence of reaction rate for failure as follows:

$$\ln(T_{BD}) \propto \frac{\Delta E_a}{k_B T} - \gamma E \tag{6.15}$$

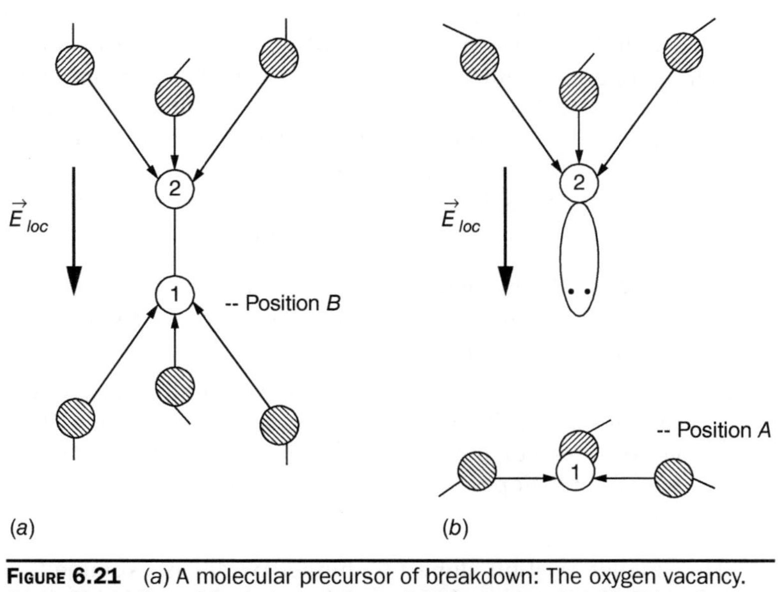

FIGURE 6.21 (*a*) A molecular precursor of breakdown: The oxygen vacancy. (*b*) An illustration of the hole trap that might form when the Si–Si bond collapses to the *sp*2 hybridization state. (*Used with permission from the American Institute of Physics.*[21])

where ΔE_a is the enthalpy of activation (or activation energy), γ is the field acceleration parameter with inverse temperature dependence, E is the electrical field, T is temperature, and k_B is Boltzmann's constant. *The thermochemical model is the base for the E model.*

It is found that *strain in the oxide can reduce the activation energy,* which is expressed as follows[32]:

$$\Delta E_a = \Delta E_a(\text{thermochemical}) - G \cdot \Delta E(\text{strain}) \tag{6.16}$$

$$\Delta E(\text{Strain}) = \frac{\iint E(\theta) f(\theta) d\theta \, dt_{ox}}{t_{ox}} \tag{6.17}$$

where ΔE_a(thermochemical) is the activation energy in Eq. (6.15), ΔE(strain) is the integration of the structural energy $E(\theta)$ for the bond angle (90 to 180°), and t_{ox} is the oxide thickness. G is an enhancement factor. The built-in compressive strain in SiO_2 changes the statistical distribution of the Si−O−Si angle, reducing the activation energy and leading to a decrease in T_{bd} and spread of the distribution.

Percolation Model

The *percolation model* was proposed for oxide breakdown in the 1990s[25] and then was elaborated later.[33,34] The percolation model is illustrated schematically in Fig. 6.22. It is assumed that electron traps are generated inside the oxide at random positions in space. Around these traps, a sphere is defined with a fixed radius r, which is the only parameter of the model (Fig. 6.22*a*). If the spheres of two neighboring traps overlap, conduction between the traps becomes possible. The two interfaces are modeled as an infinite set of traps (Fig. 6.22*b*). This mechanism of trap generation continues until a conducting path is created from one interface to the other, which defines the breakdown condition (Fig. 6.22*c*). Such a conducting path forms when a critical defect density is reached as a result of defect generation during oxide stress.

The percolation model for oxide breakdown has successfully explained two important thickness-dependent observations:

- As the oxide thickness decreases, the density of oxide traps needed to trigger breakdown decreases. In other words, the critical defect density N_{bd} decreases with thinner oxides.
- As the oxide thickness decreases, the Weibull slope β of the time-to-failure distribution decreases; that is, there is a larger spread on the *TTF* values (Fig. 6.23). To the first order the Weibull slopes are essentially independent of injection polarities.[8]

The smaller Weibull slope for thinner oxides is explained as follows: In the thinnest oxides, only a few traps are needed to form the conductive breakdown path, and consequently, there is a large statistical spread on the average density to form such a short path.

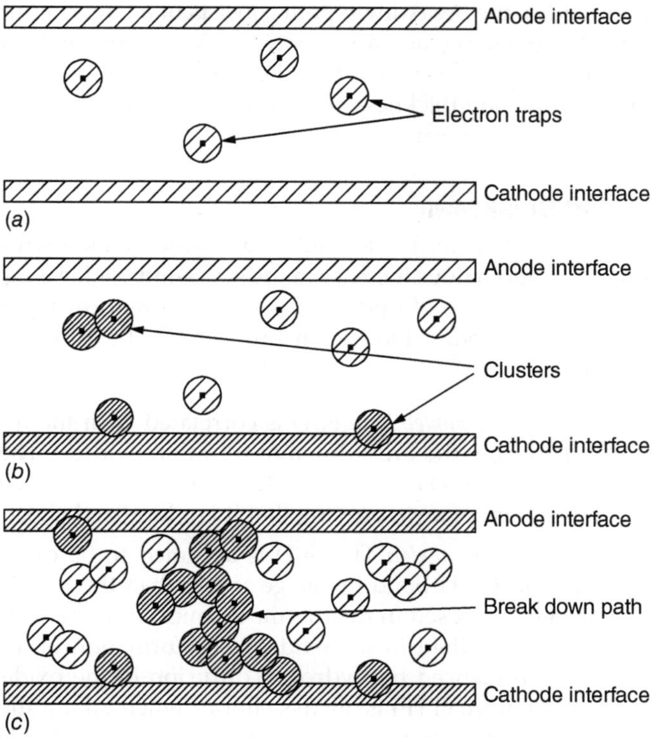

Figure 6.22 The percolation model for oxide breakdown explained step by step. As the density of the neutral electron traps increases, conductive clusters of traps are formed, ultimately leading to the creation of a conductive breakdown path from anode to cathode. (*Used with permission from Elsevier.*[33])

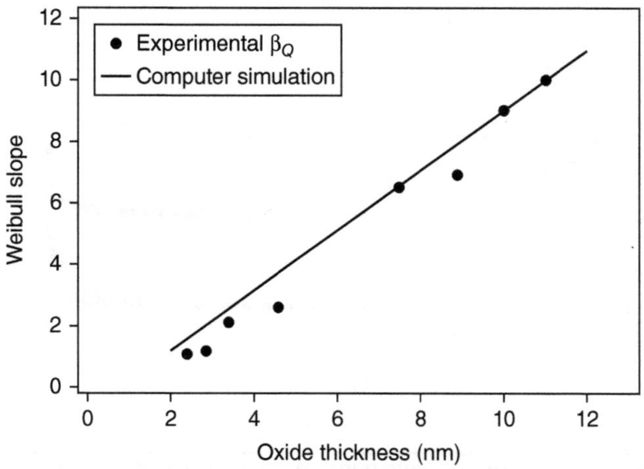

Figure 6.23 The simulated and measured Weibull slope of the Q_{BD} distribution as a function of oxide thickness. (*Used with permission from Elsevier.*[33])

In thicker oxides, the breakdown path consists of a larger number of traps, and the spread on the trap density necessary to generate such a large path is smaller. This means that the change in the Weibull slope as a function of oxide thickness is an intrinsic statistical property of the breakdown mechanism.

6.2.5 Soft Breakdown

Soft breakdown (SBD) can be defined as an oxide breakdown corresponding to an anomalous increase in the stress-induced leakage current and the occurrence of fluctuations in the current during TDDB stressing. Several models have been proposed in literature for the SBD mechanism:

- *Thermal-effect model:* The SBD is correlated with the thermal effects during the breakdown of ultrathin oxides.[35] The gate current after SBD is a unique function of the gate voltage and independent of area, proving that it is a localized effect.[36]
- *Si−H breakage model:* The SBD is assumed to be caused by Si−H and O−H bond breakage rather than the Si−Si bond breakage proposed in the thermochemical *E* model owing to the electrical field in the oxide.[37] The formation of an SBD path is influenced by hydrogen diffusion in the oxide. The time-to-failure (TTF) is always shorter when taking the SBD into account as shown in Fig. 6.24, which has resulted in gate oxide reliability being quite challenging in terms of meeting the qualification requirements.

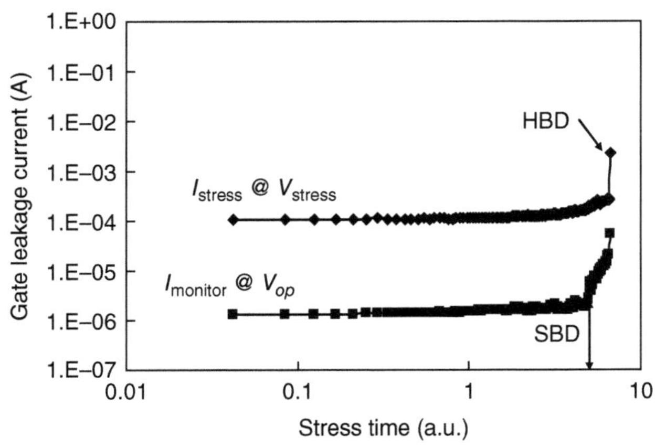

FIGURE 6.24 Measured currents during a constant-voltage stress. I_{stress} is the gate leakage measured at V_{stress}, whereas $I_{monitor}$ is the gate leakage measured at the operating voltage V_{op}. $I_{monitor}$ shows SBD, whereas I_{stress} shows hard breakdown (HBD).

- *Interface-damage model:* When the oxide is thin enough and/or the applied field is low, the traveling distance of electrons in the oxide conduction band after FN tunneling is shorter than electron mean free path[38] (Fig. 6.25*a*). In this case, the electrons travel the oxide conduction band elastically and thereby release most of their energy at the anode region only. Continuous accumulation of energy at the anode could lead to physical damage near the Si-SiO_2 interface in the form of broken bonds. When SBD happens, the total leakage current is a superposition of the tunneling through the undamaged regions (which have a higher tunneling current density) and that through the damaged regions (Fig. 6.25*b*).
- *Percolation model:* During high-field stressing, electrons or hole traps are randomly generated throughout the oxide layer. A local conducting path is created if the traps form a chain between two interfaces (see Fig. 6.22). Then SBD happens. This model can easily explain the random noise signal after SBD and electrical recovery of SBD by opposite-polarity stress.

Since SBD is a localized process at the anode interface, the occurrence of soft (quasi) breakdown will be suppressed by using a low-defect substrate and/or anode interface hardening. It was reported that the stress-induced leakage current (SILC) is higher for oxides on normal wafers (pure oxide or N_2O oxide) than that for oxides on H_2-annealed wafers.[38]

SBD probability increases as t_{ox} decreases owing to the reduction of stored energy in MOS capacitors per unit area. The SBD phenomenon is not observed for oxides thicker than 5 nm,[35] as

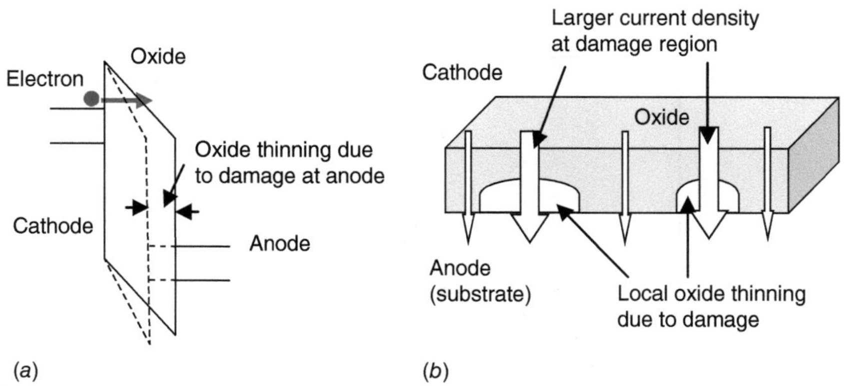

FIGURE 6.25 (*a*) Schematic drawing of electron transport in ultrathin gate oxide under high field stress. (*b*) Schematic drawing of the current path in the oxide after SBD.

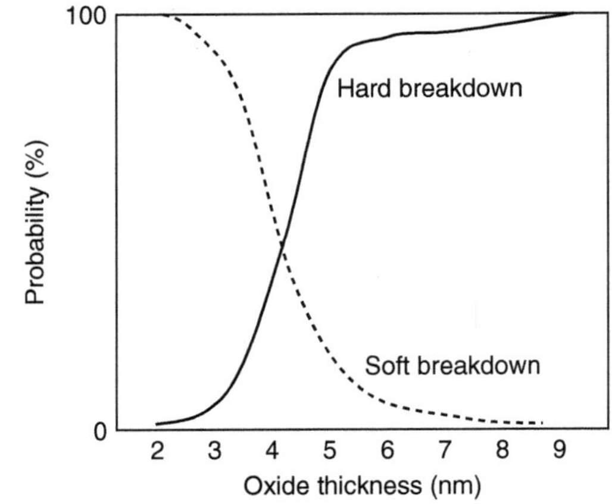

FIGURE 6.26 Breakdown types and probabilities.

shown schematically in Fig. 6.26. For thick gate oxides of higher bias, higher power is applied, thus inducing large thermal effects with hard breakdown (HBD).

6.3 Stress-Induced Leakage Current

SILC, first reported in 1988,[39] is a phenomenon in which a higher gate leakage at low field is observed on a high-field prestressed sample than on a fresh sample.[40,41] Figure 6.27 provides a schematic of SILC. Ricco et al.[41] reported that SILC is not observed for 3.2-nm-thick oxide owing to large DT current. SILC becomes more and more obvious when the oxide thickness increases to 5.0 nm. However, SILC in a 6.8-nm-thick oxide seems less pronounced. In other words, SILC

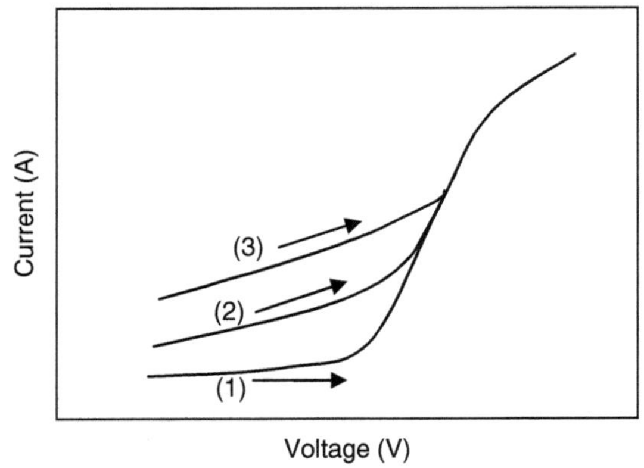

FIGURE 6.27 Schematic of stress-induced leakage current (SILC).

could be a reliability concern for gate oxides around 5 to 6 nm in thickness, which corresponds to the 2.5-V I/O oxide currently used in the industry. A similar trend is observed in a report from Moazzami and Hu,[40] where SILC is reduced when the gate oxide is thicker than 6.5 nm.

Various conduction models for SILC have been proposed in the literature. Figure 6.28 schematically shows the band diagrams for four conduction models. Model A assumes that the enhancement in the local electrical field owing to hole trapping contributes to the SILC (the same as in Fig. 6.18).[39] SILC may not be directly correlated with the defects responsible for oxide breakdown. However, the trapped holes can have an impact on the defect-creation rate and, in turn, on breakdown generation.[42] Model B attributes the SILC to local thinning in oxides owing to local damage at the anode, proposed to explain the soft (quasi) breakdown in oxides thinner than 5 nm (the same as in Fig. 6.25).[38] A thinner oxide will lead to a larger electron

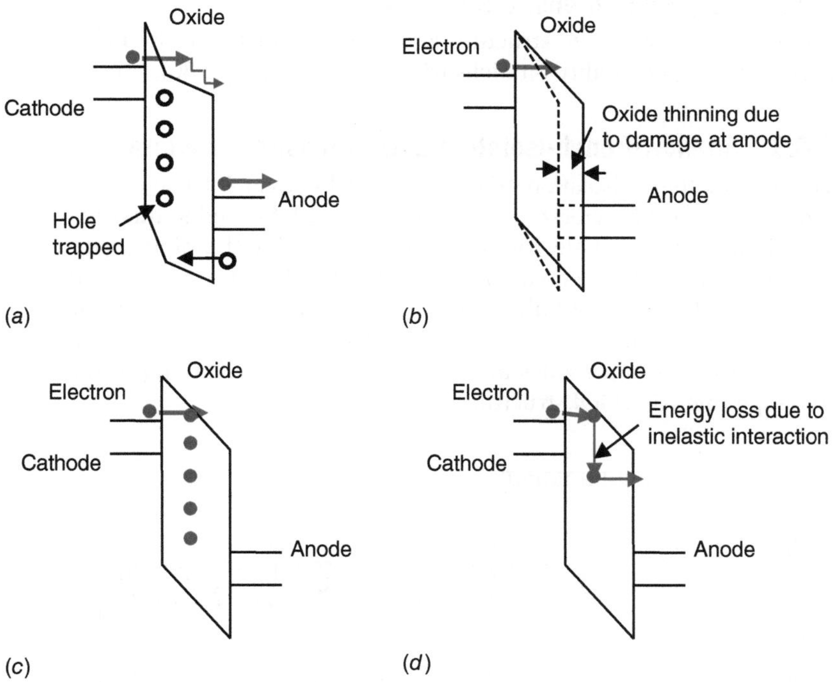

FIGURE 6.28 Schematic band diagrams illustrating the conduction models for SILC. (*a*) Model A: The conduction mechanism is the local field entrancement owing to trapped positive charges (Fig. 6.18). (*b*) Model B: The conduction mechanism is the local thinning of the gate oxide owing to local damage at the anode (see Fig. 6.25). (*c*) Model C: The trap-assisted elastic tunneling is the conduction mechanism for SILC. (*d*) Model D: SILC is attributed to a trap-assisted inelastic tunneling accompanied by a large energy loss/relaxation.

tunneling current. In model C, it is the trap-assisted elastic tunneling that causes the SILC.[40] On the other hand, model D proposes that electrons in SILC are dominated by an inelastic tunneling mechanism accompanied by a large energy loss of around 1.5 eV.[43]

6.4 Gate-Oxide-Integrity Test Structures and Failure Analysis

6.4.1 Bulk Structure

Two types of bulk (or area-plate) capacitors are widely used.[33] One is a small-area capacitor used to measure the intrinsic breakdown properties of the oxide layer. *Small* typically means 10^{-6} to 10^{-4} cm^2 range. The other is a large-area capacitor with area ~10^{-2} cm^2 used to measure the extrinsic breakdown properties of the oxide layer and determine the process-induced defect density (Fig. 6.29).

6.4.2 Poly-Edge-Intensive Structure

The poly-edge-intensive structure is used to characterize the influence of gate-edge quality on dielectric breakdown (Fig. 6.30).

6.4.3 Shallow-Trench-Isolation-Edge-Intensive Structure

The shallow-trench-isolation (STI)–edge (or field-edge)–intensive structure is used to characterize the influence of isolation-edge quality/morphology on the dielectric (Fig. 6.31). Gate-oxide thinning occurring around the active-area corner is common for an unoptimized process.[44,45] This type of failure can be easily characterized using the active-area-edge-intensive GOI structure. The V_{bd} Weibull distribution shown in Fig. 6.15 provides an example of the severe tailing of the STI-edge-intensive GOI structure.

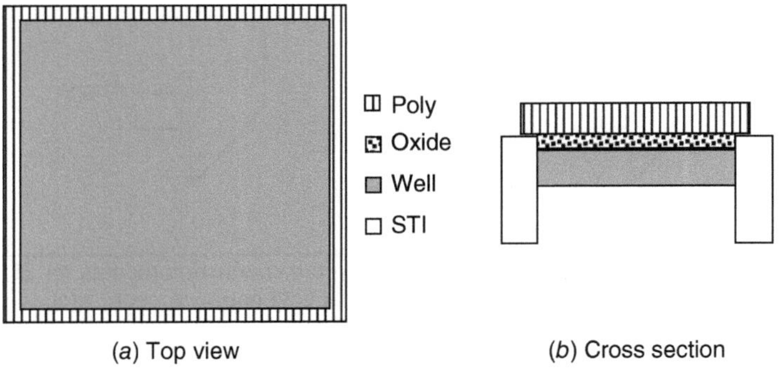

FIGURE 6.29 Schematic of the bulk gate-oxide integrity (GOI) test structure.

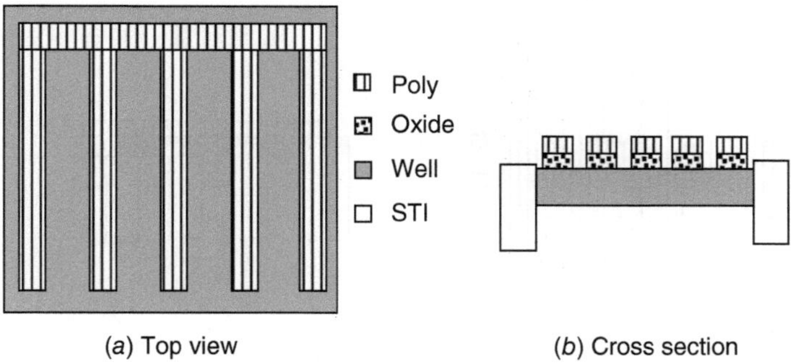

FIGURE 6.30 Schematic of poly-edge-intensive gate-oxide-integrity(GOI) test structure.

6.4.4 Shallow-Trench-Isolation-Corner-Intensive Structure

In the STI-corner (or field-corner)–intensive structure, the active-area small blocks are imbedded in the STI (Fig. 6.32). This structure is more sensitive to isolation-edge processes than the STI-edge-intensive structure (see Fig. 6.31).

Figure 6.33 shows typical Weibull distributions of GOI V_{bd} for 2.5-V I/O bulk, poly-edge-intensive, STI-edge-intensive, and field-corner-intensive test structures.[45]

6.4.5 Failure Analysis for Gate-Oxide Integrity

A common way to observe GOI defects is to use emission microscopy (EMMi),[33] which provides a method to visualize the current flow through an oxide during the oxide degradation and after its breakdown. During the oxide-degradation process, light is generated when the injected electrons reach the anode and their energy is released by the emission of photons, where a uniform light emission from the capacitor is observed. After oxide breakdown, a clear hot spot is

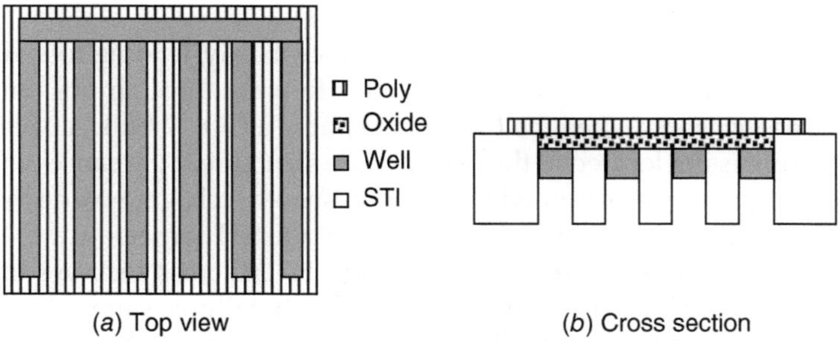

FIGURE 6.31 Schematic of the STI-edge (or field-edge)–intensive GOI test structure.

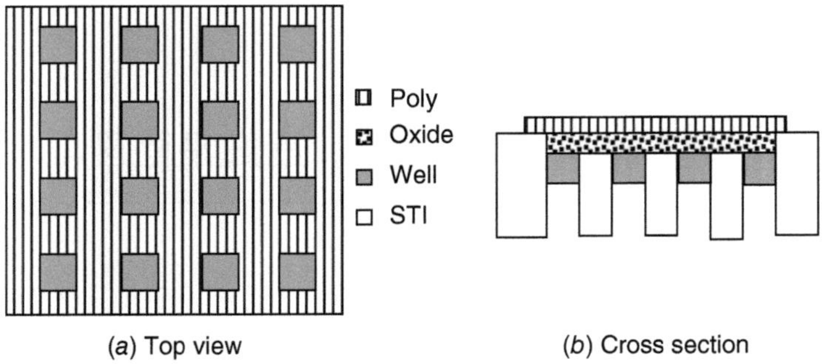

FIGURE 6.32 Schematic of the STI-corner (or field-corner)–intensive GOI test structure.

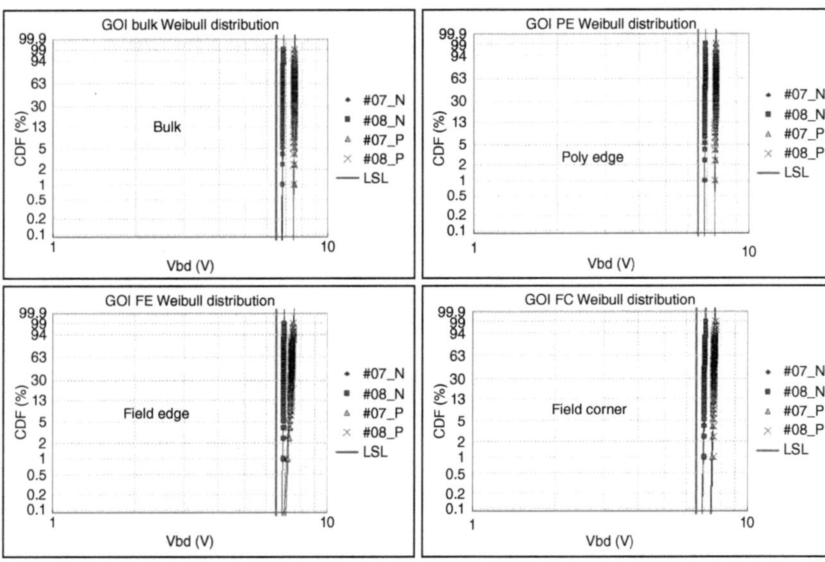

FIGURE 6.33 Typical Weibull distributions of the GOI V_{bd} for different test structures. (*Reproduced with permission from the Electrochemical Society.*[45])

observed where the breakdown takes place locally, leading to a very large FN current in the entire capacitor flowing through this small breakdown spot. When the EMMi hot spot of a GOI *N*-bulk capacitor structure is not located at the center of the structure,[46] the cause of the GOI failure is not attributed to the poly/spacer etch, gate-oxide thinning at the edge, or gate-oxide thinning at the STI corner.

A transmission electron microscopic (TEM) cross section sometimes is necessary to characterize the GOI failure, especially for edge-intensive structures. Figure 6.41 provides the TEM images of STI-corner profiles of the STI-edge-intensive structure with unoptimized processes.

6.5 Gate-Oxide Time-Dependent Dielectric-Breakdown Models, Lifetime Extrapolation Methodology

6.5.1 Weibull Distribution

The TTF of gate-oxide TDDB is commonly assumed to be distributed according to Weibull statistics,[33] which typically are applicable to the weakest-link process. The cumulative distribution function (CDF) of Weibull is expressed as follows:

$$F(t) = 1 - \exp\left[-\left(\frac{t}{\eta}\right)^{\beta}\right] \tag{6.18}$$

where t is the TTF, β is the shape factor of the distribution (or sometimes referred as the *Weibull slope*), and η is the scale factor or 63.2 percent value of the distribution. Equation (6.18) can be rewritten as follows:

$$\ln\{-\ln[1 - F(t)]\} = \beta\ln(t) - \beta\ln(\eta) \tag{6.19}$$

Based on this equation, a plot of ln[–ln(1 – F)] versus ln(t) yields a straight line with a slope of β.

6.5.2 Activation Energy

The gate-oxide TDDB activation energy is found to be linear dependent on the oxide field, for example, for a 3.9-nm-thick oxide considering HBD.[47] E_a decreases with oxide field, and an empirical expression is given as follows:

$$E_a\,(\text{eV}) = -0.1197 \cdot E_{ox} + 1.9535 \tag{6.20}$$

where E_{ox} is oxide field in MV/cm. Kimura[48] gave a similar dependence for a 9.9-nm-thick oxide with HBD as follows:

$$E_a\,(\text{eV}) = -0.07 \cdot E_{ox} + 1.15 \tag{6.21}$$

Similarly, Pompl et al.[37] reported the oxide field–dependent E_a for a 3.4-nm-thick oxide with soft breakdown (SBD) as follows:

$$E_a(eV) = -0.07 \cdot E_{ox} + 0.98 \tag{6.22}$$

In other words, although the slope is the same, SBD has a lower E_a than HBD when the same oxide field is applied. Pompl et al.[37] indicated that it is Si–H bond breakage rather than the Si–Si bond

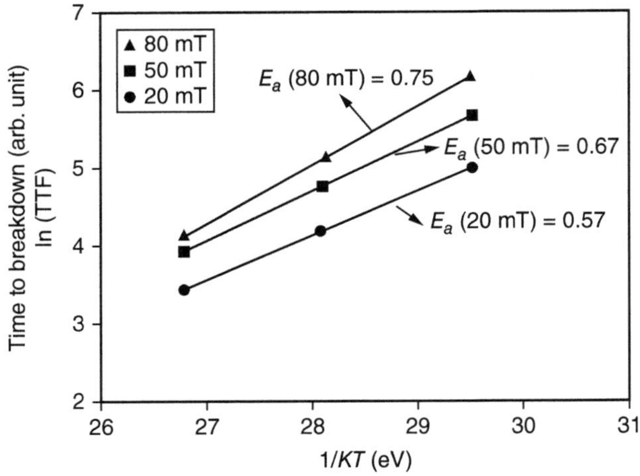

FIGURE 6.34 Arrhenius plot of time-to-63.2-percent-failure data (TTF) for different DPN pressures. (*Used with permission from IEEE.*[49])

breakage proposed in the thermomechanical *E* model to contribute to HBD that is involved in the process leading to SBD. The activation energy for SBD at low oxide fields is determined by hydrogen diffusion (H_2 or H).

Nitrogen concentration also has an impact on the activation energy. The higher decoupled plasma nitridation (DPN) pressure (with lower nitrogen incorporation to the gate oxide) causes a larger activation energy E_a and has longer time-to-breakdown at three different temperatures [49] (Fig. 6.34).

6.5.3 1/*E* Model, *E* Model, *V* Model, and Power-Law Model

There are basically four field-dependent TDDB models that have been well discussed for gate oxides up to the present, that is, the 1/*E* model, the *E* model, the *V* model, and the power-law model. Since integrated-circuit (IC) operating voltage is much less than the stress voltage, the semiconductor industry is interested in the lifetime extrapolated to the operating conditions. However, the lifetimes based on the four TDDB models diverge by several orders of magnitude.

The 1/*E* model is derived from the assumption that the breakdown process is a *current driven process* that depends on FN conduction [Eq. (6.1)].[50,51,19] The 1/*E* model is expressed as

$$\mathrm{TTF} = C\exp\left(\frac{G}{E}\right)\exp\left(\frac{E_a}{k_B T}\right) \tag{6.23}$$

where *C* is a fitted constant (1×10^{-11} s),[52] and *G* is the field acceleration factor (a typical value is 350 MV/cm), which increases with thinner oxides.[19] Based on the $1/E$ model, TDDB data can be collected at very high fields, where there is very high FN conduction, and then can be extrapolated to low fields where there is no FN conduction without any change in the physics of failure of the gate oxide. The TDDB of ultrathin oxides follows the $1/E$ model when the field is larger than 7.2 MV/cm.[47]

The E model has a thermochemical basis,[53] which assumes that the breakdown process is a *field-driven process*, and is expressed as

$$\mathrm{TTF} = C\exp(-\gamma E)\exp\left(\frac{E_a}{k_B T}\right) \tag{6.24}$$

where TTF is time-to-failure, C is a fitted constant, γ is field-acceleration parameter, E is the oxide field, E_a is the thermal activation energy, k_B is Boltzmann's constant, and T is absolute temperature. TDDB of ultrathin oxides follows the E model down to 4.6 MV/cm.[47] Note that the field-acceleration parameter γ in cm/MV was found to be linearly dependent on the inverse of ambient temperature (T in kelvins), as described by the following equation:

$$\gamma = 1{,}549/T - 0.486$$

The V model was proposed for gate oxide thickness values of less than 4 nm with an exponential dependence of voltage rather than electrical field.[23,54] Thus, the TTF is expressed as

$$\mathrm{TTF} = C\exp(-\gamma_V \cdot V)\exp\left(\frac{E_a}{k_B T}\right) \tag{6.25}$$

where γ_V is the voltage-acceleration factor, and V is the applied voltage.

The power-law model is now widely accepted in the low-voltage region with gate oxide thicknesses of less than 2 nm based on a *current-driven breakdown*[55,56]:

$$\mathrm{TTF} = AV^{-n}\exp\left(\frac{E_a}{k_B T}\right) \tag{6.26}$$

The power-law model was proposed originally based on experimental work on ultrathin oxides from 5 nm down to 1 nm.[55] The power-law model preserves important breakdown characteristics, including the universally accepted Poisson random statistics and weakest-link property. The STM experiments have demonstrated that the hydrogen release rate gives a power-law-dependent tunneling current for low gate bias (i.e., < 2.5 V) through multiple local excitations (see Fig. 6.20).[14] The power-law model can be extended to describe a large range of oxide thicknesses up to 10 nm and at stress voltages as large as 12 V.[31] This model is also consistent with the current-driven breakdown property, as demonstrated by independent experimental evidence using substrate-carrier-injection technique.[57]

Hiraiwa et al.[56] compared the median lifetime (MTTF) of gate oxides as a function of gate voltage (absolute value) for different types of MOS transistors and polarities of gate voltages. The TDDB lifetimes are well approximated by power functions in the low-voltage region irrespective of the type of MOS transistor and the polarity of the gate voltage applied. Figure 6.35 shows the exponents produced by the method as a function of the maximum voltage (absolute value) below which the exponents were calculated. When the power law applies, the exponent is independent of the maximum voltage. Figure 6.35 clearly supports the power-law model for low voltages. Hiraiwa et al.[56] found that the lifetimes follow the $1/E$ model for higher voltages. The transition from the $1/E$ model to the power-law model occurs from 3.5 to 5 V in NMOS devices under inversion and from 4 to 5.5 V (absolute value) in the other three cases. The transition voltages approximately correspond to where the dominant component of the gate current changes from DT to the FN tunneling. Hiraiwa et al.[56] thus proposed that the *breakdown is caused by the current flow.*

6.5.4 Area Scaling

For the scale factors η_1 and η_2 of two Weibull distributions (either T_{bd} or Q_{bd} distributions) of capacitors with identical oxide thicknesses, but different areas A_1 and A_2, respectively, it has been shown that the following relationship holds[18]:

$$\frac{\eta_1}{\eta_2} = \left(\frac{A_2}{A_1}\right)^{1/\beta} \tag{6.27}$$

or

$$\eta \approx A^{-1/\beta} \tag{6.28}$$

As β decreases, the area dependence becomes stronger. Therefore, according to Fig. 6.23, there is a stronger area dependence of the T_{bd} or Q_{bd} for thinner oxides.

6.6 Process Effects on Gate-Oxide Integrity and Time-Dependent Dielectric Breakdown Improvement

The quality of a gate-oxide dielectric depends on the quality of the wafer surface prior to oxide growth, the oxide growth cycle, and the processing the oxide is exposed to after it is grown.[58] Since many process parameters can have an impact on gate-oxide quality, the Taguchi method derivated from the design of experiment (DOE) is an effective method to statistically isolate the critical parameters for GOI

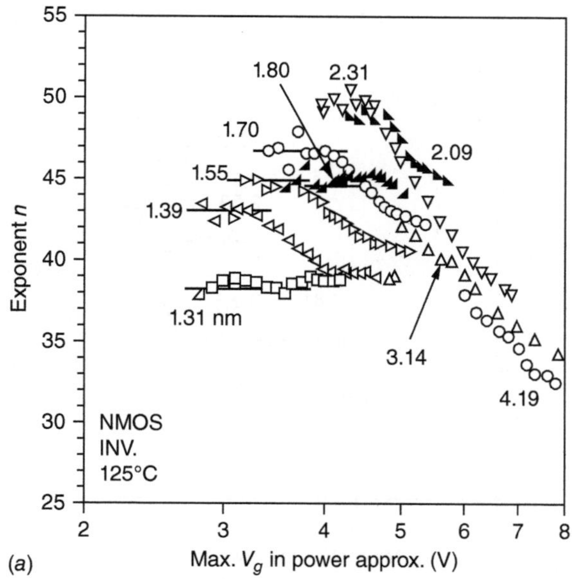

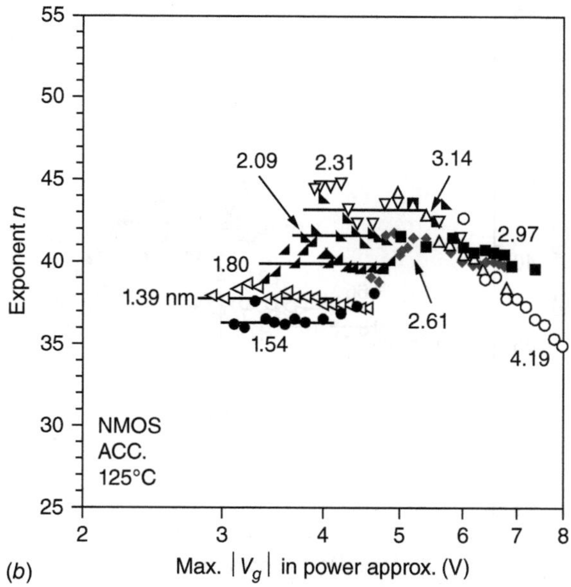

Figure 6.35 Exponent versus maximum gate voltage (absolute value) in power approximation based on the median lifetime (MTTF) of gate oxides as a function of gate voltage. Closed and open symbols stand for the results of oxide and NO oxide, respectively. The numbers outside and inside the parentheses are equivalent oxide thicknesses (EOTs) and exponents, respectively. (*a*) NMOS inversion; (*b*) NMOS accumulation; (*c*) PMOS inversion; (*d*) PMOS accumulation.[56] (*Used with permission from IEEE.*)

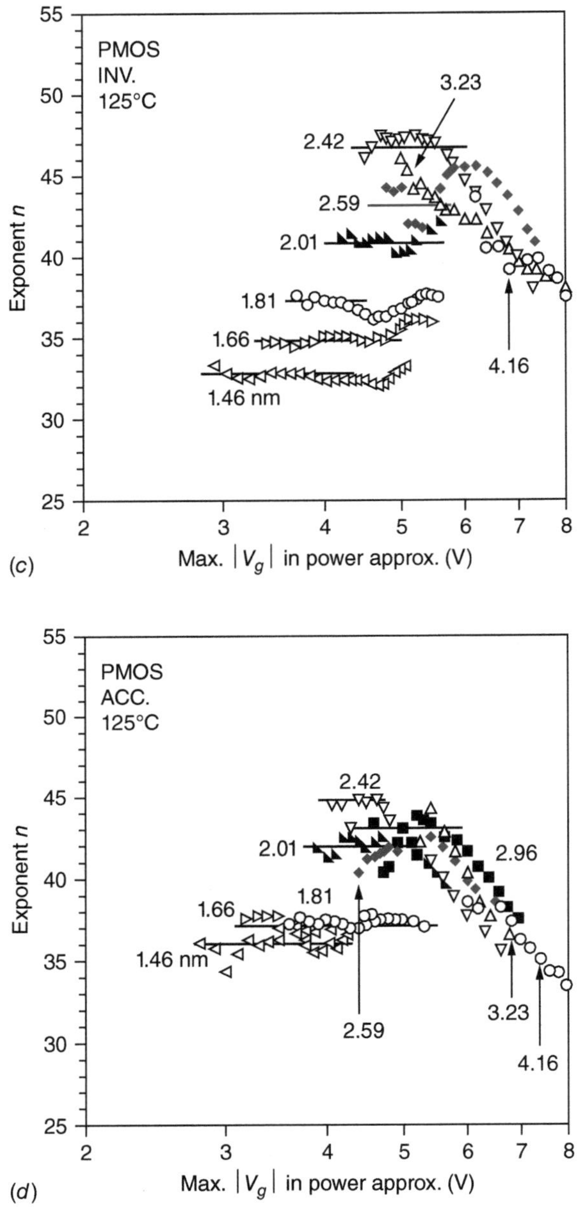

Figure 6.35 *(Continued)*

improvement when sample size is limited.[59] This section shows how the process parameters affect the gate-oxide quality and its TDDB.

6.6.1 Effect of Oxide Thickness

Equations (6.12) and (6.13) show that V_{bd} and T_{bd} depend on oxide thickness linearly and exponentially, respectively. Figure 6.23 shows that a thinner oxide will give a larger spread of the TTF distribution, that is, a smaller Weibull slope (β). From a statistical point of view, oxide thickness variations have been demonstrated to cause significant variation in TDDB lifetimes.[60] Lifetime reduction can be as high as 50 percent when the wafer t_{ox} is reduced by -3σ, mainly through variation of the voltage acceleration factor (γ) (~6 percent). On the other hand, the corresponding shift of the shape factor β in response to wafer t_{ox} variations is much less (<0.4 percent). Therefore, for an oxide with a low lifetime (e.g., ~20 years compared with the 10-year specification), the t_{ox} variation should be well controlled. Otherwise, TDDB could be a serious issue.

6.6.2 Effect of Nitridation

Nitridation to the gate dielectrics has become an effective method to reduce boron penetration in PMOS devices and leakage current in NMOS devices. For a given equivalent oxide thickness (EOT), gate oxides with nitridation will have higher dielectric constants, which allows a greater physical thickness while maintaining the same gate capacitance. Gate leakage-current reduction thus is proportional to the degree of nitridation. It is noted that gate leakage-current reduction is smaller for PMOS devices than for NMOS devices owing to a higher tunneling barrier to holes from the valence band in nitrided oxides. The probability of electron tunneling from the conduction band is still much more significant and, thus, gate leakage currents in NMOS devices are larger than those in PMOS devices. One side effect of nitridation is carrier mobility degradation (i.e., I_{dsat} degradation) caused by additional oxide fixed charges and/or donor-like interface-state generation. The effects of nitrogen incorporation to the oxide and the interfaces are detailed in this section.

Nitrogen at the Si-SiO_2 Interface

An early study in 1999 on wet pyrogenic oxide (thickness range from 40 to 100 Å) annealed in an N_2O ambient[61] showed that for a certain annealing time and temperature, the N concentration (at%) increases with decreasing wet oxide thickness, and the location of the peak of *N is observed near the Si-SiO_2 interface.* Constant-current stressing showed that high-field-charge trapping will lead to fewer hole traps in the thinner gate oxide. A longer TDDB T_{bd} also was observed for oxides annealed in N_2O.[32] Devices with nitrided gate dielectrics exhibit enhanced interface resistance against stress degradation owing to the incorporation of interfacial nitrogen. Note that nitrogen incorporation introduces Si−N bonds (Si−N 4.6 eV)[62] in the interface and relaxes interfacial strained Si−O bonds.[63] It also was suggested that *nitrogen can be either substitutional (replaces Si−H, which is good) or interstitial (sits close to Si−H, which is bad).*[13]

N_2O-ISSG [in-situ steam generation (ISSG)] oxynitride [grown by rapid thermal processing (RTP) in a 99%N_2O/1%H_2 ambient at 1,000°C] exhibited better interface quality, higher mobility, and excellent reliability when compared with rapid thermal nitrided oxide (RTNO) dry oxide (grown by RTP in an NO ambient at 750°C and then in an O_2 ambient at 1,000°C) and N_2O dry oxide (grown by RTP in an N_2O ambient at 1,000°C).[64] A lower nitrogen concentration presents at the Si-SiO_2 interface, which results in a worse hot-carrier injection (HCI) but better NBTI performance.

Among the methods for nitridation, plasma nitridation has attracted the most attention because of its ability to allow engineering of the nitridation profile in the gate oxide. However, Chen et al.[65] observed oxide regrowth at the Si-SiO_2 interface for oxide by plasma nitridation, and this is seen in cross-sectional TEM images, which indicate that the physical thickness of the oxide by remote plasma nitridation (RPN) is 24 Å, increased from 15 Å over the base oxide. It was postulated that the energetic nitrogen radical from the plasma may penetrate into the Si-SiO_2 interface and then may break the Si−O bonds. The oxygen released will move toward the Si substrate, resulting in suboxide growth, which sometimes is referred as *radical-induced reoxidation.*[65]

Nitrogen at the Poly-SiO_2 Interface

The plasma nitridation process, especially for heavy plasma nitridation, can lead to excessive electron traps at the poly-SiO_2 interface because nitrogen is closer in this interface produced by plasma nitridation.[65] Nitrogen radical–induced broken bonds are more severe at the poly-SiO_2 interface than at the Si-SiO_2 interface. Such a phenomenon becomes even more significant under substrate injection polarity-inversion mode for NMOS devices. Therefore, the plasma nitridation process should be carefully optimized because the nitrogen located at the poly-SiO_2 interface can seriously degrade the TDDB performance of NMOS devices under inversion mode. This adverse impact is also demonstrated in Fig. 6.10, where the NMOS device under inversion mode (substrate injection) gives a much smaller power-law exponent than the NMOS device under accumulation mode (gate injection). Similarly, a PMOS device under accumulation mode (substrate injection) gives a much smaller power-law exponent than a PMOS device under inversion mode (gate injection).

In addition, *the poly-Si-SiO_2 interface is rougher than the Si-SiO_2 interface,*[66] which may have some influence on conduction because a rougher interface means a lower injection barrier or an equivalent reduction in oxide thickness.

However, there is another story. For PMOS devices, *high concentrations of nitrogen atoms piled up near the polysilicon-dielectric interface in the SiNO film have been found to effectively prevent boron diffusion and thus improve Q_{bd} characteristics.*[12] It was found that the Q_{bd} value of the SiNO film (a high-density nitrogen at the polysilicon-dielectric interface) is four times larger than that of the SiON film (a high-density nitrogen at the dielectric-substrate interface) even though they have the same total nitrogen content. An SiON film with a higher nitrogen concentration near the Si-SiO_2 interface can prevent boron penetration into the channel but cannot effectively stop its diffusion into the dielectric film. Thus it is not conclusive whether the nitrogen located at the polysilicon-oxide interface is beneficial to breakdown characteristics or not. This should be tackled on a case-by-case basis.

Nitrogen Implantation into the Channel

Nitrogen implantation into the channel before gate oxidation was used to form the NMOS low-threshold thin-oxide device, which can result in different thicknesses of gate oxides.[67] The nitrogen implant improves the oxide breakdown field,[68] suppresses the SiO_2 defect density, and reduces the gate leakage current but induces dislocation loops if the dose is large than 5×10^{14} cm^{-2}.

6.6.3 Effect of Hydrogen/D_2

The ISSG process is a wet-oxide process in which steam is generated in close proximity to the wafer surface. The reliability of the ISSG oxide is considerably improved from the viewpoint of SILC and charge-to-breakdown (Q_{bd}) as the H_2 percentage increases. The presence of hydrogen during the oxidation can reduce defects in the SiO_2 network within the STL, such as Si dangling bonds and weak Si−Si and strained Si−O bonds, by creating highly reactive oxygen atoms that are hypothesized to be dissociated from the molecular oxygen owing to the presence of hydrogen[69] (Fig. 6.36).

Forming-gas annealing (FGA) using hydrogen is a common practice to passivate the dangling bonds in the gate oxide caused by plasma damage. It is found that FGA can affect the TDDB performance, especially for small devices, attributed to the interaction of a larger mechanical stress and excess hydrogen, if any.[70] With the continuous shrinkage of static random-access memory (SRAM) and logic circuits, the size of active region area becomes smaller and smaller. Therefore, FGA turns to be more and more critical for reliability control.

Morifuji et al.[70] simulated hydrostatic stress in a narrow active area. It was shown that compressive stress extends from the STI edge

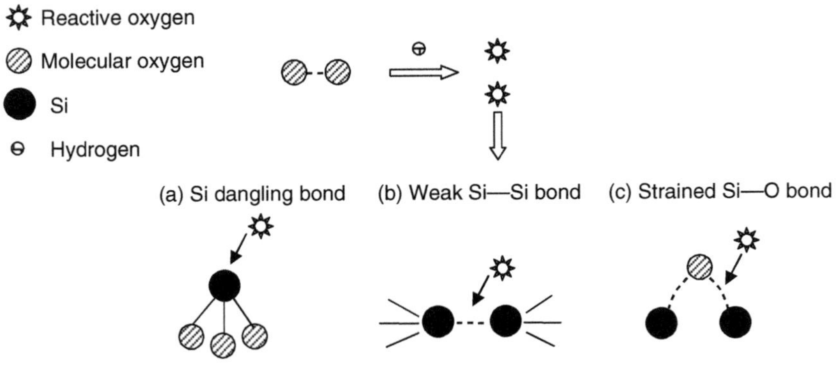

FIGURE 6.36 Hypothesized interactions of hydrogen and oxygen in the SiO interfacial region. The presence of hydrogen accelerates the dissociation of molecular oxygen into reactive atomic oxygens, which reduce intrinsic oxide defects by repairing Si dangling bonds, Si−Si bonds, and strained Si−O bonds in the structural transition layer.

into the channel ~2 μm. In other words, the center of the channel is subject to considerable compressive stress in narrow-active-area devices. These authors further reported that TDDB degrades in small NMOS devices, attributed to the mechanical stress caused by STI degrading TDDB significantly. The TDDB test under the smaller thermal budget of FGA (370°C/60 minutes) is less sensitive to device size than FGA at 400°C for 80 minutes. This observation that excess FGA accelerates TDDB degradation in small gate capacitors was attributed to the interaction of mechanical stress and hydrogen. The amount of H introduced into systems and the thermal budget during the process steps can affect stress, defect formation, and defect agglomeration.

As discussed in Chap. 5 (HCI), the replacement of hydrogen annealing with deuterium annealing to passivate interface states at the Si-SiO_2 interface can improve HCI performance. However, the influence of deuterium incorporation into CMOS processing on GOI is quite arguable in terms of Q_{bd} and SILC.[71] Kim and Hwang[72] reported an improvement in SILC and Q_{bd} characteristics when deuterium was incorporated into MOS capacitors by either annealing dry oxide in a D_2 ambient or growing the gate oxide using D_2O. Mitani et al.[73,74] confirmed the improvement in SILC characteristics when using pyrogenic oxidation with D_2 and O_2 to grow the gate oxide. However, Wu et al.[75] reported that CMOS devices underwent an H_2 or D_2 anneal after the FGA did not show improved SILC or Q_{bd} characteristics. Esseni et al.[76] again used either hydrogen or deuterium during the FGA, but they did not showed an improvement if deuterium was incorporated. Hof et al.[71] argued that for deuterium to be beneficial for bulk-oxide quality, it has to be incorporated an early stage of processing, preferably during or directly after growth of the gate oxide.

6.6.4 Metal Contamination

Metal contamination especially from transition metals, could be one of the factors leading to fatal GOI issues. Examples include iron (Fe),[77,78] copper (Cu),[77,79–81] calcium (Ca),[77] nickel (Ni),[79] molybdenum (Mo),[46] hafnium (Hf),[82] zirconium (Zr),[82] and tungsten (W),[83] the common materials for process requirements or in process equipment. These materials diffuse fast in silicon.[84] Metal contamination in the gate oxide can result in a high leakage current, reduced minority carrier lifetime, and even gate-oxide breakdown. It is generally accepted that thicker oxides (i.e., oxides 5 nm thick or thicker) are more sensitive to metal contamination. This section summaries the contribution of metal contamination to gate-oxide integrity.

Burte and Aderhold[77] reported that the iron, copper, and calcium contamination can influence dielectric breakdown field strength and charge-to-breakdown values if the surface contamination level is raised beyond 10^9 to 10^{10} atoms/cm^2. Choi and Schroder[78] studied the effect of iron contamination on gate-oxide integrity with oxide thickness ranges of from 3 to 5 nm and iron density ranges of from 4×10^{10} to 1.4×10^{12} cm^{-3}. They found that 5-nm oxide degraded the most, whereas the thinner oxides still exhibited some degradation, even for the 3-nm oxide.

Cu contamination (as low as 10 ppm) can lead to a much higher leakage current, a lower breakdown effective field, a poorer charge-to-breakdown, and a worse stress-induced leakage current in an ultrathin 3-nm-thick oxide.[80] The strong degradation of the ultrathin gate oxide integrity can be explained by the lowering of the tunneling barrier and the increased interface-trap tunneling owing to the presence of Cu in the oxide and at the oxide-Si interface. Cu contamination also can degrade Al_2O_3 dielectric on Si.[85]

It also has been reported that low levels of copper contamination present on preoxidation silicon surfaces significantly affect not only the area but also the field-overlap defect density of the gate oxide.[81] A detailed study of the effects of Cu and Ni contamination on 5- to 17-nm oxides showed that thicker oxides ($t_{ox} \geq 7.5$ nm) were more seriously degraded than thinner oxides ($t_{ox} < 7.5$ nm).[79] Cu contamination had a higher impact on GOI than Ni. GOI also was influenced by the type of substrate, that is, whether the substrates were bulk Si wafers or p/p^+ or p/p^{++} epitaxial wafers. This substrate effect is related to the gettering ability of the substrates.

Mo contamination was observed[46] in the gate oxide of NMOS devices and was introduced in the process of germanium (Ge) preamorphization implantation (PAI) mainly in the form of $(^{98}Mo^{12}C^{19}F_2)^{++}$, which has the same nominal mass-to-charge ratio as $^{74}Ge^+$.[46] During the Ge PAI process, Ge the source is GeF_4, and the source arc chamber is made of Mo. The undesirable introduction of Mo into the gate oxide during the PAI process is related to the arc current. A higher arc current results in more F free radicals, which enhances the erosion of Mo arc chamber. Therefore, reducing the arc current is one of the solutions for the Mo contamination issue. Other solution paths include using a W chamber instead of a Mo chamber, changing the implant species from ^{74}Ge to ^{72}Ge to avoid the mass interference of Mo-contained clusters, and changing the source from the gas source of GeF_4 to solid Ge source to avoid the F impact.[46]

However, it is interesting that PMOS gate oxides have not suffered from Mo contamination, which is attributed to the different properties of the poly-Si in PMOS and NMOS devices. In NMOS devices, the Ge PAI is performed before *N*-type lightly doped drain (NLDD) annealing, whereas it is done after *P*-type lightly doped drain (PLDD) annealing in PMOS devices. It is known that annealing can lead to grain-size growth of poly-Si. Therefore, prior to Ge PAI, the poly-Si grain size in NMOS devices is smaller than that in PMOS devices. It is believed that the unannealed poly-Si in NMOS devices is more favorable for Mo diffusion during subsequent thermal processes and the Mo segregates at the gate oxide because the grain boundary provides a fast diffusion path. Furthermore, the boron getting effect in *P*-poly is helpful in blocking Mo from segregating to gate- oxide region.

6.6.5 Effect of Poly Grain Structure

Poly grain structure has a large impact on the interface roughness between poly and the gate oxide and thus on GOI.[86] The grain size and structure of poly can be adjusted based on the deposition conditions, including temperature and poly thickness, as well as the source-drain (S/D) rapid-thermal-anneal (RTA) conditions. A lower poly deposition temperature results in a denser poly with a smaller grain size, which will minimize hydrofluoric acid (HF) slip through the poly and give a better GOI. Similarly, a thicker poly can reduce the possibility for HF penetration. As a result, experiments have demonstrated that gate-dielectric defectivity is greatly reduced when the poly deposition temperature is reduced from 550 to 530°C and when the amorphous poly thickness is increased from 600 to 650 Å.[58]

Kamgar et al.[86] reported the effect of N^+ S/D RTA temperature on TDDB for various gate oxide thicknesses. It appears that the RTA temperature effect on TDDB mean time to failure (MTTF) depends on gate-oxide thickness. For thick oxides (50 and 26 Å), when RTA temperature is increased from 950 to 1,050°C, an approximate 25 percent TDDB MTTF gain is obtained, which is attributed to reduction of electron traps in the oxide.[87] However, with gate oxides be coming thinner and thinner, higher RTA temperature degrades the oxide TDDB performance. For 15-Å-thick oxides, the TDDB MTTF is halved when RTA temperature is increased from 1,000 to 1,100°C. This is explained by the poly protrusions into the gate oxide leading to a rougher interface between the oxide and the overlaying gate poly.[86] The extent of the roughness becomes a larger fraction of the oxide thickness for thin oxides and hence would give rise to a larger impact on gate-oxide breakdown. Note that poly grains continue to grow during the various thermal cycles in device processing. At the S/D activation step, however, when the temperature can go above the viscoelastic point of SiO_2, which is between 950 and 1,000°C depending on the oxide composition,[88] the oxide becomes viscous, and protrusion of the poly grains into the gate oxide is possible.

6.6.6 Effect of Poly Profile (Poly Footing)

Unoptimized poly profile with footing near the source/drain edge showed degraded oxide TDDB performance, especially for the poly-edge structure.[89] Figure 6.37 shows cross sections of poly with different profiles: poly with foot, straight-wall poly, and poly with notch, respectively. Figure 6.38*a* clearly shows that a narrower poly bottom can significantly reduced the source/drain ovelap capacitance C_{ov}—poly footing gives the largest C_{ov}, whereas poly notch has the smallest C_{ov}. Figure 6.38*b*,*c* compare, NMOS TDDB of area (30 × 15 μm) and S/D edge (2,000 μm × 60 nm) structures as a function of the poly profile. It is clear that there is no difference in the TTF distribution of area structure among the three splits, whereas edge structure shows ~2.5× reduction in MTTF for the poly-foot split than for the notch and straight-wall splits. The poly notch has a slightly higher edge-structure MTTF than the straight-wall split. Poly-footing profile allows high-energy implants to penetrate through the edge of the gate oxide and also corresponds to a higher *E*-field stress as well, which causes oxide damage and the worst TDDB performance and hence failures during product burn-in. The problem is more pronounced in NMOS devices owing to their stronger sensitivity to voltage stress and the specific implant species compared to PMOS devices.

6.6.7 Effect of Gate-Oxide Preclean and Etch

In a typical dual gate oxide (DGO) process flow, the thick gate oxide for I/O devices is grown first. Then the region corresponding to the core thin oxide is etched away by either a heavy or a diluted HF solution, followed by growth of the thin oxide in this region. The etching step using a heavy HF solution has been demonstrated to benefit core device reliability such as GOI and NBTI because better oxide-channel interface quality is achieved.[90] A higher F incorporation at the interface by using a heavy HF etching solution is believed to be the reason for the reliability improvement.

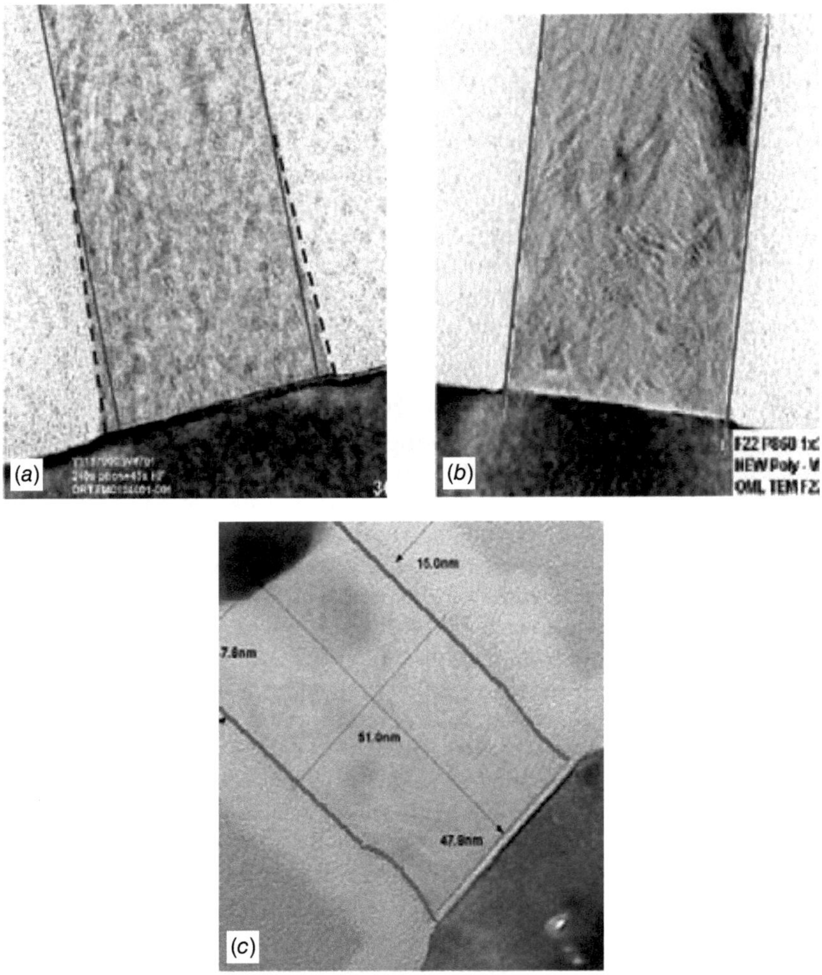

FIGURE 6.37 Cross sections of poly show the difference in profile at the poly bottom near the gate oxide and Si interface: (*a*) poly with foot; (*b*) straight-wall poly; (*c*) poly with notch. The critical dimension of the poly (width in cross section) is around 60 nm.[89] (*Used with permission from IEEE.*)

6.6.8 Effect of Annealing Environment after Sacrificial Oxidation

A high-temperature annealing after sacrificial oxidation is necessary to achieve rounded top corners in STI.[91] The annealing gas could be N_2 or Ar. A sharp corner in the Si is seen in cross-sectional TEM images of the STI edge when no annealing is performed after the sacrificial oxidation. This is basically attributed to the different oxidation rates in the different directions of the crystal plane because of the

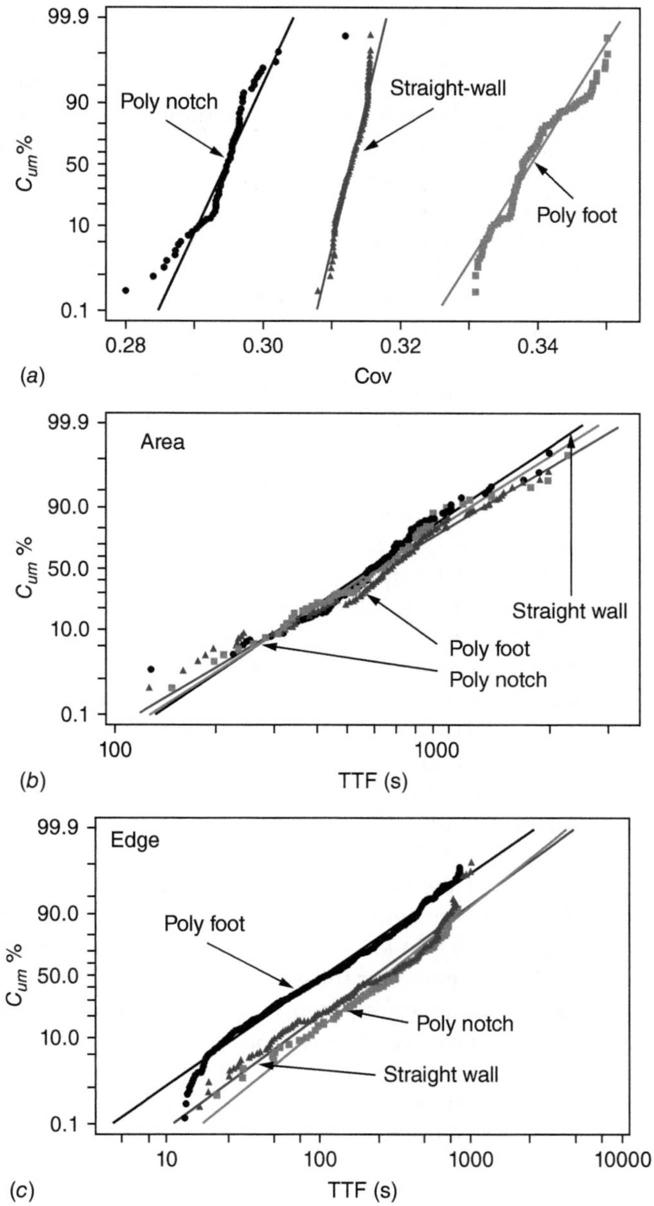

FIGURE 6.38 (*a*) The variation in NMOS source/drain overlap capacitance C_{ov} as a function of the poly profile. Comparison of NMOS TDDB of (*b*) area (30 × 15 μm²) and (*c*) S/D edge (2,000 μm * 60 nm) structures as a function of the poly profile. (*Used with permission from IEEE.*[89])

(100) and (110) crystal orientations in the flat region of MOSFET and in the STI sidewall region, respectively. As mentioned in Chap. 3, a lining oxide after trench etching is used for STI corner rounding. However, sometimes the lining oxide alone may not be sufficient to round the STI edges. In contrast, N_2 or Ar annealing after sacrificial oxidation is an effective way to round the corners. In other words, the Si atoms can migrate into the STI corner region during annealing, even though sacrificial oxide covers the Si surface. It is believed that a rounded STI corner will provide a platform for a more uniform oxidation rate. Thus the gate oxide at the STI corners is thinner in the sample without annealing than in those annealed in an N_2 or Ar ambient. In short, high-temperature annealing in an inert-gas ambient after sacrificial oxidation is effective in rounding the STI corners as well as in achieving a more uniform gate oxide thickness at the corners.

However, contrary to intuition, when compared with a device without annealing, the N_2 annealing gives an increased subthrehold current (hump) in the I_d-V_g curve and a degraded TDDB lifetime. Normally, an increased hump in the I_d-V_g curve is thought to be due to a lower threshold voltage arising from gate-oxide thinning at the sharp top corners of the trench. In other words, the increased hump here is not related to gate-oxide thickness. On the other hand, it is more likely related to the quality change of the gate oxide that occurred at local spots in the STI edge region with N_2 annealing. Nitrogen that accumulated at the interface between the substrate and the sacrificial oxide could not be removed completely at the STI edges even after removal of sacrificial oxide.[91] The local remaining nitrogen suppressed the gate oxidation (something like the Kooi effect[92]) and formed SiON, which lowered the barrier height for leakage current. Under conditions of constant-current stress, large current flows selectively at local spots that have thin low-barrier-height gate dielectrics, leading to a short TDDB lifetime for the case with N_2 annealing.

In contrast, with Ar annealing, there is no I_d-V_g hump, and the TDDB lifetime also is improved over that without annealing. It is presumed that the Ar does not accumulate at the sacrificial oxide–substrate interface. Therefore, Ar annealing after gap filling is promising in improving both the performance and the reliability of transistors with STI.

6.6.9 Effect of Sacrificial-Oxide-Free

A sacrificial-oxide (SAC-OX)-free scheme is employed in a 65-nm STI process integration to reduce STI divots at the STI–active-area (AA) corner and thus to improve device performance.[45] The baseline (with SAC OX) and the SAC-OX-free process flows are compared schematically in Fig. 6.39. Figure 6.40 shows that divots can be reduced from 105 to 115 Å of the baseline process to 20 to 40 Å of

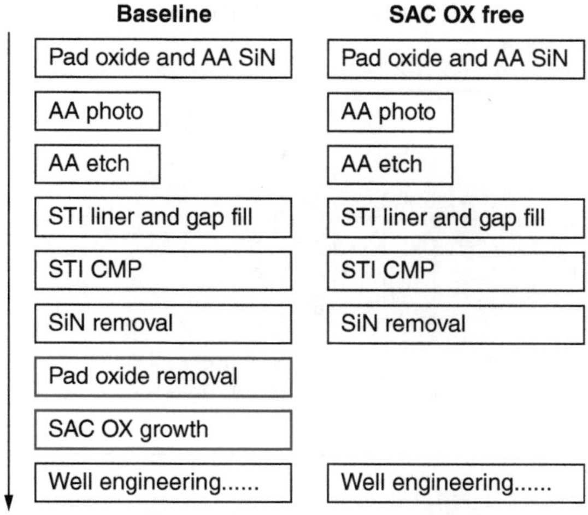

FIGURE 6.39 Baseline (with SAC OX) and SAC-OX-free process flow comparison. (*Reproduced with permission from the Electrochemical Society.*[45])

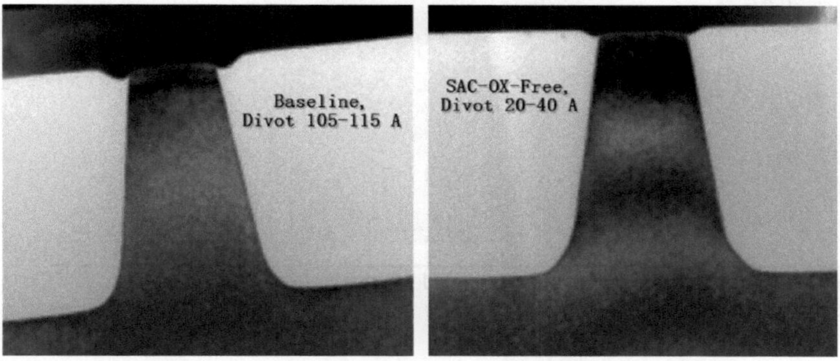

FIGURE 6.40 Baseline (with SAC OX) and SAC-OX-free process divot comparison. (*Reproduced with permission from the Electrochemical Society.*[45])

the SAC-OX-free process. The SAC-OX-free scheme also will make STI process integration less complex and more cost-effective, as shown in Fig. 6.39.

However, the unoptimized SAC-OX-free process flow shows degraded GOI performance (Fig. 6.41), where I/O devices suffer and located at the wafer center. Failure analysis indicates that the AA corners of the failed devices are damaged and become abnormally thinner. The gate-oxide thickness on the damaged AA surfaces was ~37 Å, whereas it was ~57 Å on undamaged AA surfaces.

The AA damage in the SAC-OX-free scheme is attributed to *AA clipping* in the STI HDP gap-fill process,[93] which is usually found on

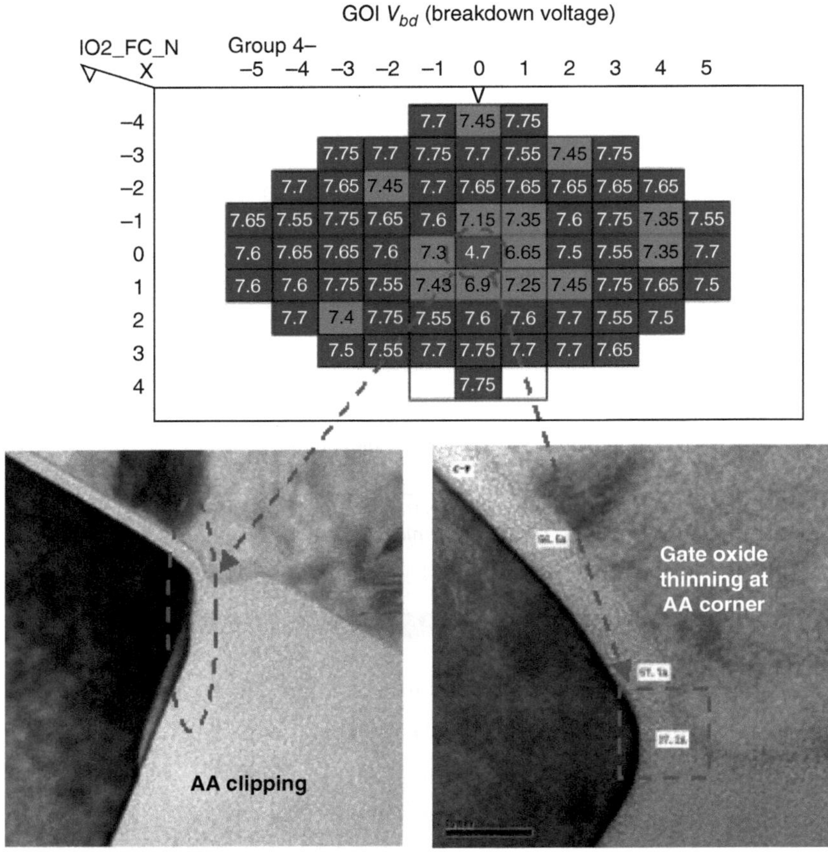

Figure 6.41 AA clipping and AA corner gate-oxide thinning. (*Reproduced with permission from the Electrochemical Society.*[45])

the wafer center owing to the plasma distribution characteristics of the HDP process. With a series of process-parameter optimizations (including deposition-etch ratio, deposition rate, deposition-etch cycles, deposition and etch rate within wafer uniformity, etc.) and hardware configuration optimizations implemented,[94] the AA clipping is effectively reduced and GOI performance improved (see Fig. 6.33).

The I/O-device (but not core-device) GOI degradation in SAC-OX-free flow is explained as follows: In the SAC-OX-free scheme, the exposed AA Si corners that are damaged during the HDP process are not repaired, whereas they can be repaired in the scheme with SAC OX, where a high-temperature oxidation process takes place. This model is supported by the observation that the core device does not

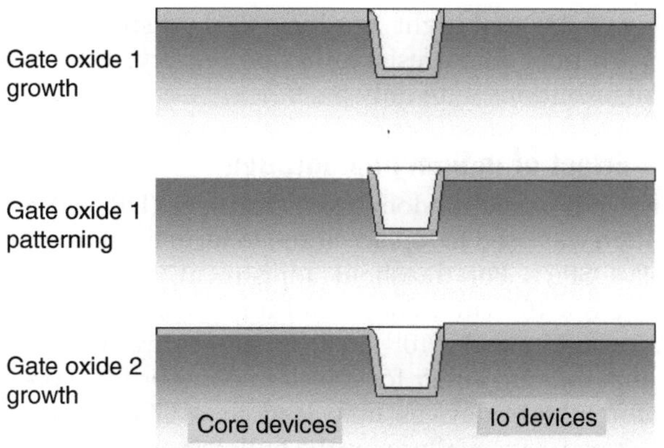

FIGURE 6.42 Schematic of dual-gate-oxide process flow. (*Reproduced with permission from the Electrochemical Society.*[45])

suffer this issue. As shown in Fig. 6.42, the exposed AA Si corners in the I/O area that were damaged during the HDP process did not get consumed and repaired, whereas those in core areas did.[45]

6.6.10 Effect of the Adhesion of Photoresist

The adhesion of the photoresist, controlled by the time interval t_{int} between the oxidation and the photoresist application [hexamethyldisilazane (HMDS) + UV46], plays an important role in GOI.[13] The shorter t_{int} is, the poorer is the adhesion of photoresist to the oxide surface, which is moisture-enhanced. The mechanism is explained as follows. After oxidation, the wafer has a very dry oxide surface for both the STI and active regions. Thus there's very little water to crack the HMDS. The longer the interval t_{int} is, the more water adsorbs and the more efficient the HMDS cracking will be, and the better the adhesion and undercut become. For example, $t_{int} < 1$ hour gives a poor adhesion, whereas $t_{int} > 10$ hours gives a good adhesion. This is applicable to both NMOS and PMOS devices, indicating that the phenomenon is indeed independent of doping species. Figure 6.12*a* is for the poor adhesion case, giving a wide V_{bd}-J_g spread corresponding to a variation of ~20 Å in oxide thickness, much larger than the corresponding measured value (~2.5 Å). What's more, Fig. 6.12*b* indicates that there is effective thinning, although physically there is no obvious global and corner thinning found on TEM.[13] On the other hand, the gate leakage could be used to as a fast check for gate-oxide breakdown and TDDB performance because the effect of the localized effective thinning of the gate oxide is also included (Fig. 6.12*a*). Liu et al.[13] reported that devices with good photoresist adhesion showed a reasonable β of ~ 2.5 for 50-Å-thick oxide (Fig. 6.23) and tight random V_{bd}-I_g distribution, whereas devices with poor photoresist adhesion (empty) exhibited spread Weibull distributions with only $\beta \approx 0.7$.

6.6.11 Effect of Indium Implantation

In advanced dynamic random-access memory (DRAM) technology, indium often was used to replace boron to form the retrograde profile of the NMOSFET. For threshold adjustment, the dose range was around 3 to 8×10^{12} cm^{-2}. For pocket implants, the indium implant dose was around 1 to 3×10^{13} cm^{-2}.[67] High-energy and/or high-dose indium implantation either for a well implant or for a pocket (halo) implant for NMOS devices may lead to a GOI issue.[67] For well implants, it may cause failure for the bulk test structure, whereas for pocket implants, the poly-edge-intensive test structure is a likely to be damaged. When there is a indium and nitrogen co-implant, the GOI degradation will be even worse.[67]

6.6.12 Process Factors on the Power-Law Model Exponent

The power-law model-exponent [n in Eq.(6.26)] value is not necessarily a constant; rather, it is affected by certain factors. Some work suggested that the n value is an indicator of oxide-interface quality.[14,95] The exponent is found to decrease when the gate-oxide thickness is reduced.[56,96] In addition, it was seen that PMOS gate insulators under inversion had a smaller exponent than others.[49,97] Plasma nitridation changes the exponent.[49] The details are given in this section.

Gate-Oxide Thickness Dependence

Figure 6.43 shows the exponents as a function of EOT of gate dielectrics (ranging from 1.3 to 3.0 nm) with or without NO annealing.[56] The exponent of all the cases first increased and then saturated when the insulator increased to a threshold thickness of around 2 nm. Liao et al.[15] show a similar trend for PMOS devices, where n = 31 for 1.6-nm-thick oxide and n increases to 35 when oxide thickness is 5.0 nm.

However, Ohgata et al.[97] reported that the exponents are *independent of oxide thickness* (ranging from 1.6 to 3.0 nm), which is annealed in an NO gas ambient or nitrided by N_2 plasma. They plotted the normalized V_g dependence of T_{bd} on a log-log scale in NMOS and PMOS devices under inversion and accumulation mode. The normalization to a reference area size and thickness was performed by using Weibull slope and thickness coefficients. The power-law exponent n of 45 NMOS under inversion, 40 NMOS under accumulation, and 44 PMOS under accumulation are independent of thickness. The voltage acceleration under inversion mode in a PMOS

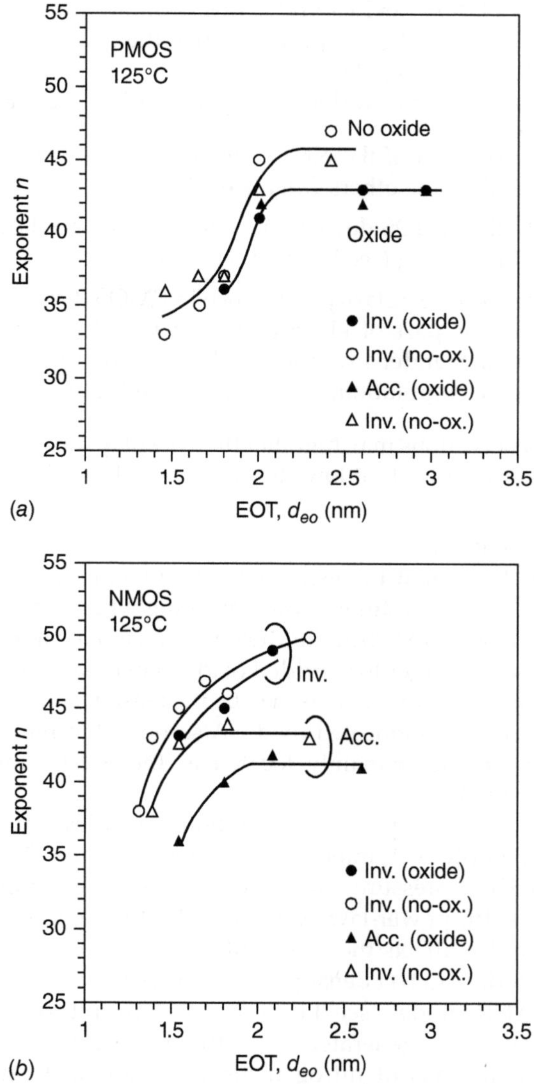

FIGURE 6.43 Exponent *n* versus EOT. The lines are added for an eye's guide. (*Used with permission from IEEE.*[56])

device depends on the gate voltage regardless of thickness ($n \approx 45$ for $|V_g| > 3.8$ V and $n = 33$ for $|V_g| < 3.8$ V).

MOS Type and Polarity Dependence

Ohgata et al.[97] showed that there was little difference between the exponents of PMOS devices of different polarities. However, the exponents of NMOS devices under inversion are much larger than those under accumulation, especially when the oxide is thicker than the threshold thickness (~2 nm).

There are a few observations that can be linked together:

- The exponent of the NMOS device under inversion is much larger than the others Ohgata et al.[97]
- Only the NMOS device under inversion lacks hole-valence-band tunneling (see Fig. 6.4).
- There is little polarity difference in PMOS devices, although the major part of electric current is fed by valence-band tunneling under inversion and conduction-band tunneling under accumulation, respectively (see Fig. 6.4).

These observations may hint that the hole plays an important role in the breakdown events at low stress voltages ($|V_g| < 5$ V).

Nitrogen Dependence

Figure 6.43 shows that the exponents of NO-annealed oxides were larger than those of oxides without annealing when compared at the same EOT for both NMOS and PMOS devices. Note that the physical oxide thickness is larger for nitrided oxide to maintain the same EOT than for nonnitrided oxide. This may be the reason leading to a larger exponent in the NO-annealed oxide. Also note that the nitrogen in NO-annealed oxide is mainly located at the Si-SiO_2 interface, as shown in Chap. 3.

In the case with plasma nitridation, the nitrogen is located mainly at the poly-SiO_2 interface, as shown in Chap. 3. It is found that higher DPN pressure and lower incorporated nitrogen in the oxide increase the power-law exponent in PMOS devices (Fig. 6.44).[49] DPN power also affects the exponent, which is correlated with the resulting interface defect density.[95] This is explained by the nitrogen-enhanced AHR model (see Fig. 6.19). For oxynitride, donor-like interface states are generated by nitrogen incorporation.[98] Since larger concentrations of nitrogen incorporation will cause excess interface states, those energy levels may reduce the proton tunneling barrier in AHR behavior, thus further enhancing the AHR effect to degrade oxide breakdown strength.[95] The model is based on two steps related to chemical reactions involving protons[24,49]: (1) Electrons dissipate energy at the anode interface and release protons (H^+) from the interface suboxide bonds, and (2) the released protons react with oxygen vacancies (Si–Si). Normally, the degradation rate is limited by the slowest process (i.e., step 1). With lower DPN pressure, more nitrogen is incorporated into the oxide, and the height of the total barrier for proton release decreases. Thus the exponent and time-to-breakdown reduce.

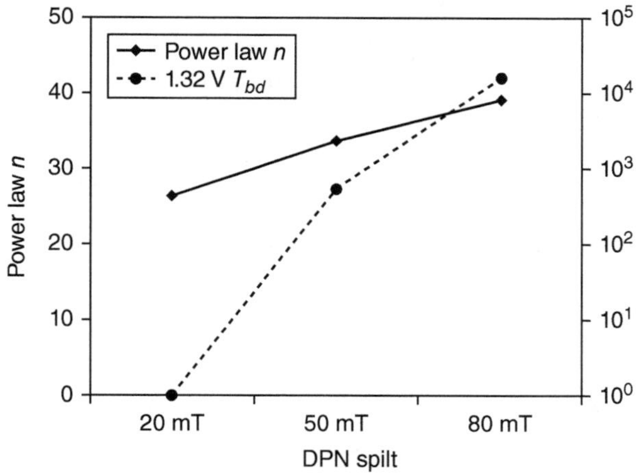

FIGURE 6.44 Higher DPN pressure will cause a steeper power-law *n* value with a longer predicted TDDB lifetime for ultrathin oxide PMOS devices in 1.32-V operation. (*Used with permission from IEEE.*[49])

Boron-Penetration Effect

Boron penetration can degrade TDDB and lead to a smaller power-law exponent *n*. Figure 6.45*a* shows a positive V_t voltage shift and a smaller voltage-acceleration slope for a heavier boron dosage.

Figure 6.45*b* shows the T_{bd} of nitrided oxides with various nitrogen concentrations, but the same 1.2-nm EOT. The *n* value of NMOS stays at ~41. However, the *n* value of PMOS increases from 33 to 40 with the increase in nitrogen concentration. This is contradictory to the universality of power-law voltage dependence regardless of the nitridation process shown by Ohgata et al.[97] and the higher *n* value for a lower nitrogen concentration shown by Kang et al.[49] (see Fig. 6.44). The higher *n* value for a heavier nitrogen concentration sample for PMOS devices is attributed to the larger oxide thickness to maintain the same EOT. Heavier nitrogen concentration is more effective in reducing the boron penetration, leading to reduced holes and electron tunneling currents. As a result, a larger slope is expected. Therefore, when studying the power-law-model exponent, all these process factors should be considered as a whole.

6.7 Process Qualification Practice

A typical GOI V-ramp test procedure can be found at JEDEC STD 35.[17] One common oxide lifetime measurement is the TDDB method,[99] which applies bias to the gate and measures breakdown time where the gate current meets a critical limit. Both the GOI

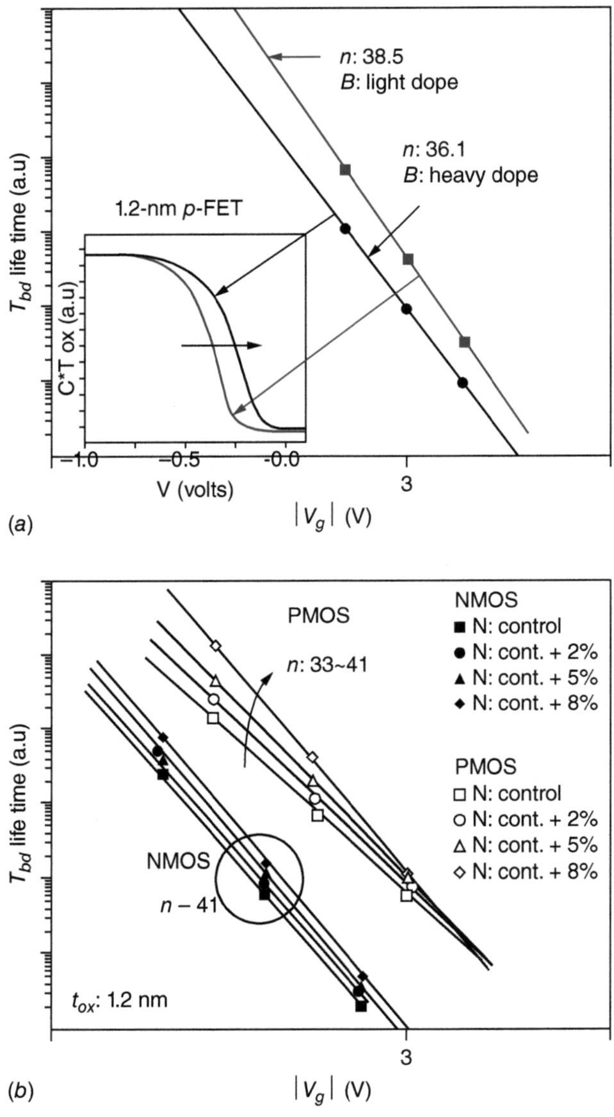

FIGURE 6.45 (*a*) The logarithm of T_{bd} is plotted as a function of the logarithm of V_g under inversion mode with varied boron dope. The inset is the corresponding *C-V*. (*b*) The logarithm of T_{bd} is plotted as a function of the logarithm of V_g under inversion mode with varied nitrogen concentrations, all with an EOT of 1.2 nm. (*Used with permission from IEEE.*[15])

V-ramp and the TDDB methodology for foundry process qualification are summarized at JEDEC JP-001.[100] Table 6.3 provides a summary of the typical process-qualification practices for GOI V-ramp and TDDB methodologies.

	V-ramp	TDDB
Failure criteria	Use one or more of the following criteria for indicating the failure point: 10× increase in current, slope change, low-voltage current increase, increase in noise level.	• For thicker oxides (>4 nm), a 2× to 10× increase in current is typical. • For thinner oxides, the first quasi-breakdown (i.e., soft breakdown) is considered as the failure point or depends on the particular application.
Test structures	N/PMOS large-area capacitor (block, AA finger, poly finger)	N/P MOS transistor array with various areas (for area-scaling purposes)
Sample size	For each oxide type: • Minimum of three lots • Both NMOS and PMOS capacitor test structures • Minimum 10 cm^2 total area	≥15 sites per stress condition: • Minimum of three lots • Both NMOS and PMOS transistor array
Stress conditions	1. Linear voltage ramp from V_{use} to breakdown 2. Ramp rate: 1 MV/cm·s; step duration: 0.1 s 3. Under accumulation mode 4. Temperature at 25°C	1. Constant-voltage stress (CVS) at various stress voltages (typically three) (for voltage acceleration) applied at gate and temperatures (typically three) (e.g., 85 to 150°C) (for activation energy evaluation) 2. Stress under inversion mode as a minimum; stress under accumulation mode as a plus. 3. Stress until at least 63% of the samples have failed
Wafer/ package level	Wafer level	Wafer level or package level
Acceptance criteria	1. Mode A ($V_{bd} < V_{use}$): $D_0 < 5/cm^2$ 2. Mode B ($V_{use} \leq V_{bd} \leq 2 * V_{use}$): $D_0 < 1/cm^2$	TTF (0.1%) > 10 years at 125°C/V_{use} and extrapolated to a large area (e.g., 1 or 0.1 cm^2)
Model	Weibull distribution	Power-law model: TTF = CV^{-n} TTF follows the Weibull distribution
Parameters	V_{bd}	TTF (at V_{use}), activation energy E_a, power-law exponent n, area scaling β

Table 6.3 Typical Process-Qualification Practice for GOI V-Ramp and TDDB Methodologies

6.8 References

1. B. P. Linder and N. W. Cheung, "Calculating plasma damage as a function of gate oxide thickness," in *Proceedings of the 3rd International Symposium on Plasma Process-Induced Damage*, 1998, pp. 42–45.
2. M. Lezlinger and E. H. Snow, "Fowler-Nordheim tunneling into thermally grown SiO_2," *Journal of Applied Physics*, Vol. 40, No. 1, 1969, pp. 278–283.
3. J. C. Ranuarez, M. J. Deen, and C. H. Chen, "A review of gate tunneling current in MOS devices," *Microelectronics Reliability*, Vol. 46, 2006, pp. 1939–1956.
4. G. Chakraborty, S. Chattopadhyay, C. K. Sarkar, and C. Pramanik "Tunneling current at the interface of silicon and silicon dioxide partly embedded with silicon nanocrystals in metal oxide semiconductor structures," *Journal of Applied Physics*, Vol. 101, No. 1, 2007, pp. 024315-1–024315-6.
5. S. Park, J. Kang, B. So, and D. Baek, "Gate oxide integrity by initial gate current," *IEEE International Integrated Reliability Workshop (IRW) Final Report*, 2009, pp. 113–116.
6. Y. Shi, T. P. Ma, and S. Prasad, "Polarity dependent gate tunneling currents in dual-gate CMOSFETs," *IEEE Transactions on Electron Devices*, Vol. 45, No. 11, 1998, pp. 2355–2360.
7. W.-C. Lee, T.-J. King, and C. Hu, "Evidence of hole direct tunneling through ultrathin gate oxide using P poly-SiGe gate," *IEEE Electron Device Letters*, Vol. 20, No. 6, 1999, pp. 268–270.
8. E. Wu, W. Lai, M. Khare, J. Sune, L.-K. Han, J. McKenna, R. Bolam, D. Haimon, and A. Strong, "Polarity-dependent oxide breakdown of NFET devices for ultra-thin gate oxide," in *Proceedings of the 40th IEEE International Annual Reliability Physics Symposium*, 2002, pp. 60–72.
9. L. K. Han, M. Bhat, D. Wristers, J. Fulford, and D. L. Kwong, "Polarity dependence of dielectric breakdown in scaled SiO_2," in *Proceedings of the International Electron Devices Meeting (IEDM)*, 1994, pp. 617–620.
10. C. C. Chen, H. C. Lin, C. Y. Chang, M. S. Liang, C. H. Chien, S. K. Hsien, and T. Y. Huang, "Improved immunity to plasma damage in ultrathin nitrided oxides," *IEEE Electron Device Letters*, Vol. 21, No. 1, 2000, pp. 15–17.
11. Y. Shi, T. P. Ma, S. Prasad, and S. Dhanda, "Polarity-dependent tunneling current and oxide breakdown in dual-gate CMOSFETs," *IEEE Electron Device Letters*, Vol. 19, No. 10, 1998, pp. 391–393.
12. E. Hasegawa, M. Kawata, K. Ando, M. Makabe, M. Kitakata, A. Ishitani, L. Manchanda, M. L. Green, K. S. Krisch, and L. C. Feldman, "The impact of nitrogen profile engineering on ultra-thin nitrided oxide films for dual-gate CMOS ULSI," in *Proceedings of the International Electron Devices Meeting (IEDM)*, 1995, pp. 327–330.
13. N. Liu, A. Haggag, J. Peschke, M. Moosa, C. Weintraub, H. Lazar, G. Campbell, A. Srivastava, J. Liu, J. Porter, K. Picone, J. Parrish, and J. Jiang, "Impacts of process induced interfacial defects on gate oxide integrity," in *Proceedings of the 46th IEEE International Annual Reliability Physics Symposium*, 2008, pp. 725–726.
14. A. Haggag, N. Liu, D. Menke, and M. Moosa, "Physical model for the power-law voltage and current acceleration of TDDB," *Microelectronics Reliability*, Vol. 45, 2005, pp. 1855–1860.
15. P. J. Liao, C. L. Chen, C. J. Wang, and K. Wu, "A new multi-step power-law TDDB lifetime model and boron penetration effect on TDDB of ultra thin oxide," in *Proceedings of the 45th IEEE International Annual Reliability Physics Symposium*, 2007, pp. 574–575.
16. K. Deguchi, S. Uno, A. Ishida, T. Hirose, Y. Kamakura, and K. Taniguchi, "Degradation of ultrathin gate oxides accompanied by hole direct tunneling: Can we keep long-term reliability of p-MOSFETs?" in *Proceedings of the IEEE International Electron Devices Meeting (IEDM)*, 2000, pp. 327–330.
17. JESD35-A, *Procedure for the Wafer-Level Testing of Thin Dielectric*, Electronics Industries Association, Washington, D C, 2001.
18. T. Nigam, R. Degraeve, G. Groeseneken, M. M. Heyns, and H. E. Maes, "Constant current charge-to-breakdown: Still a valid tool to study the

reliability of MOS structures?" in *Proceedings of the 36th IEEE International Annual Reliability Physics Symposium*, 1998, pp. 62–69.

19. K. F. Schuegraf and C. Hu, "Hole injection SiO_2 breakdown model for very low voltage lifetime extrapolation," *IEEE Transactions on Electron Devices*, Vol. 41, No. 5, 1994, pp. 761–767.
20. D. J. DiMaria, E. Cartier, and D. Arnold, "Impact ionization, trap creation, degradation and breakdown in silicon dioxide films on silicon," *Journal of Applied Physics*, Vol. 73, 1993, pp. 3367–3384.
21. J. W. Mcpherson and H. Mogul, "Underlying physics of the thermochemical *E* model in describing low-field time-dependent dielectric breakdown in SiO_2 thin films," *Journal of Applied Physics*, Vol. 84, No. 3, 1998, pp. 1513–1523.
22. D. J. DiMaria, "Defect generation in ultrathin silicon dioxide films produced by anode hole injection," *Applied Physics Letters*, Vol. 77, No. 17, 2000, pp. 2716–2718.
23. M. A. Alam, J. Bude, and A. Ghetti, "Field acceleration for oxide breakdown: Can an accurate anode hole injection model resolve the *E* vs. 1/*E* controversy?" in *Proceedings of the 38th IEEE International Annual Reliability Physics Symposium*, 2000, pp. 21–26.
24. J. Sune and E. Wu, "A new quantitative hydrogen-based model for ultra-thin oxide breakdown," in *Symposium on VLSI Technology Digest of Technical Papers*, 2001, pp. 97–98.
25. H. Z. Massoud and R. Deaton, "Percolation model for the extreme-value statistics of dielectric breakdown in rapid thermal oxides," in *Proceedings of the Third Symposium on Silicon Nitride and Silicon Dioxide Thin Insulating Films*, San Francisco, California, ECS Spring Meeting, 1994, pp. 583–589.
26. F. Monsieur, E. Vincent, D. Roy, S. Bruyere, G. Pananakakis, and G. Ghibaudo, "Time to breakdown and voltage to breakdown modeling for ultra-thin oxides (T_{ox} < 32 A)," in *IEEE International Integrated Reliability Workshop (IRW) Final Report*, 2001, pp. 20–25.
27. I. C. Chen, S. E. Holland, and C. Hu, "Electrical breakdown in thin gate and tunneling oxides," *IEEE Transactions on Electron Devices*, Vol. 32, No. 2, 1985, pp. 413–422.
28. E. M. Vogel, D. W. Heh, J. B. Bernstein, and J. S. Suehle, "Impact of the trapping of anode hot holes on silicon dioxide breakdown," *IEEE Electron Device Letters*, Vol. 23, No. 11, 2002, pp. 667–669.
29. D. A. Buchanan, A. D. Marwick, D. J. DiMaria, and L. Dori, "Hot-electron-induced hydrogen redistribution and defect generation in metal-oxide-semiconductor capacitors," *Journal of Applied Physics*, Vol. 76, 1994, pp. 3595–3608.
30. P. E. Blochl and J. H. Stathis, "Hydrogen electrochemistry and stressinduced leakage current in silica," *Physical Review Letters*, Vol. 83, 1999, pp. 372–375.
31. J. Suñé and E. Wu, "Mechanisms of hydrogen release in the breakdown of SiO_2-based gate oxides," in *Proceedings of the IEEE International Electron Devices Meeting*, 2005, pp. 388–391.
32. Y. Harada, K. Eriguchi, M. Niwa, T. Watanabe, and I. Ohdomari, "Impacts of strained SiO_2 on TDDB lifetime projection," in *Symposium on VLSl Technology Digest of Technical Papers*, 2000, pp. 216–217.
33. R. Degraeve, B. Kaczer, and G. Groeseneken, "Degradation and breakdown in thin oxide layers: Mechanisms, models and reliability prediction," *Microelectronics Reliability*, Vol. 39, 1999, pp. 1445–1460.
34. R. Degraeve, G. Groeseneken, R. Bellens, M. Depas, and H. E. Maes, "A consistent model for the thickness dependence of intrinsic breakdown in ultrathin oxides," in *Proceedings of the International Electron Devices Meeting (IEDM)*, 1995, pp. 863–866.
35. M. Depas, T. Nigam, and M. Heyns, "Soft breakdown of ultra-thin gate oxide layers," *IEEE Transactions on Electron Devices*, Vol. 43, No. 9, 1996, pp. 1499–1504.
36. F. Crupi, R. Degraeve, G. Groeseneken, T. Nigam, and H. E. Maes, "On the properties of the gate and substrate current after soft breakdown in ultrathin oxide layers," *IEEE Transactions on Electron Devices*, Vol. 45, No. 11, 1998, pp. 2329–2334.

37. T. Pompl, H. Wurzer, M. Kerber, and I. Eisele, "Investigation of ultrathin gate oxide reliability behavior by separate characterization of soft breakdown and hard breakdown," in *Proceedings of the 38th IEEE International Annual Reliability Physics Symposium*, 2000, pp. 40–47.
38. S. H. Lee, B. J. Cho, J. C. Kim, and S. H. Choi, "Quasi-breakdown of ultrathin gate oxide under high field stress," in *Proceedings of the International Electron Devices Meeting (IEDM)*, 1994, pp. 605–608.
39. P. Olivo, T. Nguyen, and B. Ricco, "High field induced degradation in ultra-thin SiO_2 films," *IEEE Transactions on Electron Devices*, Vol. 35, No. 12, 1988, pp. 2259–2267.
40. R. Moazzami and C. Hu, "Stress-induced current in thin silicon dioxide films," in *Proceedings of the International Electron Devices Meeting (IEDM)*, 1992, pp. 139–142.
41. B. Ricco, G. Gozzi, and M. Lanzoni, "Modeling and simulation of stress-induced leakage current in ultrathin SiO_2 films," *IEEE Transactions on Electron Devices*, Vol. 45, No. 7, 1998, pp. 1554–1560.
42. P. Riess, G. Ghibaudo, and G. Pananakakis, "Stress-induced leakage current generation kinetics based on anode hole injection and hole dispersive transport," *Journal of Applied Physics*, Vol. 87, No. 9, 2000, pp. 4626–4628.
43. S. Takagi, N. Yasuda, and A. Toriumi, "Experimental evidence of inelastic tunneling and new *I-V* model for stress-induced leakage current," in *Proceedings of the International Electron Devices Meeting (IEDM)*, 1996, pp. 323–326.
44. Y. M. Chan, C. B. Moey, and H. P. Kuan, "Localized TDDB failures related to STI corner profile in advanced embedded high voltage CMOS technologies for power management units," in *Proceedings of the International Symposium on Semiconductor Manufacturing (ISSM)*, 2007, pp. 1–4.
45. Y. P. Deng, H. R. Ren, Q. Fan, H. M. Ho, Q. Xu, Z. Y. Zhang, K. Zheng, Y. J. Wu, Z. H. Gan, and X. J. Nin, "GOI improvement in 65nm sacrificial oxide free process integration," *Electrochemical Society Transactions*, Vol. 27, No. 1, 2010, pp. 383–387.
46. D. Gui, Y. H. Huang, G. B. Ang, Z. X. Xing, Z. Q. Mo, Y. N. Hua, and J. Teong, "Gate oxide integrity failure caused by molybdenum contamination introduced in the ion implantation," in *Proceedings of the IEEE International Symposium on the Physical and Failure Analysis of Integrated Circuits (IFPA)*, 2008, pp. 1–4.
47. A. Yassine, H. E. Nariman, and K. Olasupo, "Field and temperature dependence of TDDB of ultrathin gate oxide," *IEEE Electron Device Letters*, Vol. 20, No. 8, 1999, pp. 390–392.
48. M. Kimura, "Field and temperature acceleration model for time-dependent dielectric breakdown," *IEEE Transactions on Electron Devices*, Vol. 46, No. 1, 1999, pp. 220–229.
49. T. K. Kang, J. Shieh, O. Lo, J. P. Chen, C. L. Lin and K. C. Su, "A comprehensive solution for ultrathin oxide reliability issue including a novel explanation of power-law exponent variations," in *Proceedings of the 43rd IEEE International Annual Reliability Physics Symposium*, 2005, pp. 596–597.
50. I. C. Chen, S. Holland, and C. Hu, "A quantitave physical model for time-dependent breakdown in SiO_2," in *Proceedings of the 23rd IEEE International Annual Reliability Physics Symposium*, 1985, pp. 24–31.
51. J. Lee, I. C. Chen, and C. Hu, "Statistical modeling of silicon dioxide reliability," in *Proceedings of the 26th IEEE International Annual Reliability Physics Symposium*, 1988, pp. 131–138.
52. R. Moazzami, J. Lee, I. C. Chen, and C. Hu, "Projecting the minimum acceptable oxide thickness for time-dependent dielectric breakdown," in *Proceedings of the International Electron Devices Meeting (IEDM)*, 1988, pp. 710–713.
53. J. W. McPherson and D. A. Baglee, "Acceleration parameters for thin gate oxide stressing," in *Proceedings of the 23rd IEEE International Annual Reliability Physics Symposium*, 1985, pp. 1–5.
54. P. E. Nicollian, W. R. Hunter, and J. C . Hu, "Experimental evidence for voltage driven breakdown models in ultrathin gate oxide," in *Proceedings of the 38th IEEE International Reliability Physics Symposium*, 2000, pp. 7–15.

55. E. Y. Wu, A. Vayshenker, E. Nowak, J. Suñé, R.-P. Vollertsen, W. Lai, and D. Harmon, "Experimental evidence of TBD power-law for voltage dependence of oxide breakdown in ultrathin gate oxides," *IEEE Transactions on Electron Devices*, Vol. 49, No. 12, 2002, pp. 2244–2253.
56. A. Hiraiwa and D. Ishikawa, "Comprehensive thickness-dependent power-law of breakdown in CMOS gate oxides," in *Proceedings of the 44th Annual International Reliability Physics Symposium*, San Jose, 2006, pp. 617–618.
57. E. M. Vogel, J. S. Suehle, M. D. Edelstein, B. Wang, Y. Chen, and J. B. Bernstein, "Reliability of ultrathin silicon dioxide under combined substrate hot-electron and constant voltage tunneling stress," *IEEE Transactions on Electron Devices*, Vol. 47, 2000, pp. 1183–1191.
58. T. K. Ng, A. Yap, K. F. Lo, and P. C. Ang, "Gate oxide integrity improvement by optimizing poly deposition process," in *IEEE International Integrated Reliability Workshop (IRW) Final Report*, 2004, pp. 148–150.
59. T. Cahyadi, P. Y. Tan, M. T. Ng, T. Yeo, J. J. Boh, and B. Fun, "The use of Taguchi method for process design of experiment to resolve gate oxide integrity issue," in *IEEE International Integrated Reliability Workshop (IRW) Final Report*, 2009, pp. 128–132.
60. A. A. Keshavarz and L. F. Dion, "Effects of statistical thin-oxide thickness variations on the time-dependent dielectric breakdown (TDDB) parameters for wafer level reliability," in *IEEE International Integrated Reliability Workshop (IRW) Final Report*, 2009, pp. 155–158.
61. M. K. Mazumder, A. Teramoto, J. Komori, M. Sekine, S. Kawazu, and Y. Mashiko, "Effects of N distribution on charge trapping and TDDB characteristics of N_2O annealed wet oxide," *IEEE Transactions on Electron Devices*, Vol. 46, No. 6, June 1999, pp. 1121–1126.
62. T. Hori, H. Iwasaki, and K. Tsuji, "Electrical and physical properties of ultrathin reoxidized nitrided oxides prepared by rapid thermal Proceedings," *IEEE Transactions on Electron Devices*, Vol. 36, 1989, pp. 340–350.
63. R. P. Vasquez and A. Madhukar, "Strain-dependent defect formation kinetics and a correlation between flatband voltage and nitrogen distribution in thermally nitrided SiO_xN_y/Si structures," *Applied Physics Letters*, Vol. 47, 1985, pp. 998–1000.
64. T. M. Pan, H. S. Lin, M. G. Chen, C. H. Liu and Y. J. Chang, "Comparison of electrical and reliability characteristics of different 14-A oxynitride gate dielectrics," *IEEE Electron Device Letters*, Vol. 23, No. 7, 2002, pp. 416–418.
65. C. C. Chen, M. C. Yu, J. Y. , Cheng, M. F. Wang, T. L. Lee, S. C. Chen, C. H. Yu, M. S. Liang, C. H. Chen, C. W. Yang, and Y. K. Fang, "Characterization of plasma damage in plasma nitrided gate dielectrics for advanced CMOS dual gate oxide process," in *Proceedings of the 7th International Symposium on Plasma and Process-Induced Damage*, 2002, pp. 41–44.
66. S. S. Gong, M. E. Bumhani, N. D. Theodore, and D. K. Schroder, "Evaluation of QBD for electron tunneling from the Si/SiO_2 interface compared to electron tunneling from the Poly-Si/SiO_2 interface," *IEEE Transactions on Electron Devices*, Vol. 40, No. 7, 1993, pp. 1251–1257.
67. J. S. Wang, N. C. Wu, T. Wang, J. Hsieh, J. Chen, H. K. Hsu, D. Chen, and T. Fong, "Characterization of indium and nitrogen co-implant of NMOSFET for advanced DRAM technologies with dual-gate oxide," in *Proceedings of the IEEE International Conference on Semiconductor Electronics (ICSE)*, 2004, pp. 48–51.
68. I. H. Nam, J. S. Sim, S. I. Hong, B. G. Park, J. D. Lee, S. W. Lee, M. S. Kang, Y. W. Kim, K. P. Suh, and W. S. Lee, "Ultrathin gate oxide grown on nitrogen-implanted silicon for deep submicron CMOS transistors," *IEEE Transactions on Electron Devices*, Vol. 48, No. 10, 2001, pp. 2310–2316.
69. T. Y. Luo, M. Laughery, G. A. Brown, H. N. Al-Shareef, V. H. C. Watt, A. Karamcheti, M. D. Jackson, and H. R. Huff, "Effect of H2 content on reliability of ultrathin in situ steam generated (ISSG) SiO_2," *IEEE Electron Device Letters*, Vol. 21, 2000, pp. 430–432.

70. E. Morifuji, T. Kumamori, M. Muta, K. Suzuki, M. S. Krishnan, T. Brozek, X. Li, W. Asano, M. Nishigori, N. Yanagiya, S. Yamada, K. Miyamoto, T. Noguchi, and M. Kakumu, "New guideline for hydrogen treatment in advanced system LSI," in *Symposium on VLSI Technology Digest of Technical Papers*, 2002, pp. 218–219.
71. A. J. Hof, E. Hoekstra, A. Y. Kovalgin, R. van Schaijk, W. M. Baks, and J. Schmitz, "The impact of deuterated CMOS Proceedings on gate oxide reliability," *IEEE Transactions on Electron Devices*, Vol. 52, No. 9, 2005, pp. 2111–2115.
72. H. Kim and H. Hwang, "High-quality ultrathin gate oxide prepared by oxidation in D_2O," *Applied Physics Letters*, Vol. 74, 1999, pp. 709–710.
73. Y. Mitani, H. Satake, H. Itoh, and A. Toriumi, "Highly reliable gate oxide under Fowler-Nordheim electron injection by deuterium pyrogenic oxidation and deuterated poly-Si deposition," in *Proceedings of the International Electron Devices Meeting (IEDM)*, 2000, pp. 343–346.
74. Y. Mitani, H. Satake, H. Itoh, and A. Toriumi, "Suppression of stress-induced leakage current after Fowler-Nordheim stressing by deuterium pyrogenic oxidation and deuterated poly-Si deposition," *IEEE Transactions on Electron Devices*, Vol. 49, No. 7, 2002, pp. 1192–1197.
75. J. Wu, E. Rosenbaum, B. MacDonald, E. Li, J. Tao, B. Tracy, and P. Fang, "Anode hole injection versus hydrogen release: the mechanism for gate oxide breakdown," in *Proceedings of the 38th IEEE International Annual Reliability Physics Symposium*, 2000, pp. 27–32.
76. D. Esseni, J. D. Bude, and L. Selmi, "Deuterium effect on interface states and SILC generation in CHE stress conditions: A comparative study," in *Proceedings of the International Electron Devices Meeting (IEDM)*, 2000, pp. 339–342.
77. E. P. Burte and W. Aderhold, "The impact of iron, copper, and calcium contamination of silicon surfaces on the yield of a MOS DRAM test process," *Solid-State Electronics*, Vol. 41, No. 7, 1997, pp. 1021–1025.
78. B. D. Choi and D. K. Schroder, "Degradation of ultrathin oxides by iron contamination," *Applied Physics Letters*, Vol. 79, No. 16, 2001, pp. 2645–2647.
79. R. Holzl, A. Huber, L. Fabry, K.-J. Range, and M. Blietz, "Integrity of ultrathin gate oxides with different oxide thickness, substrate wafers and metallic contaminations," *Applied Physics A Materials Science and Proceedings*, Vol. 72, No. 3, 2001, pp. 351–356.
80. Y. H. Lin, Y. C. Chen, K. T. Chan, F. M. Pan, I. J. Hsieh, and A. Chin, "The strong degradation of 30 Å gate oxide integrity contaminated by copper," *Journal of the Electrochemical Society*, Vol. 148, No. 4, 2001, pp. F73–F76.
81. B. Vermeire and H. G. Parks, "The influence of copper contamination on gate oxide integrity," IEEE/SEMI Advanced Semiconductor Manufacturing Conference and Workshop, 1997, pp. 30–32.
82. B. Vermeire, K. Delbridge, V. Pandit, H. G. Parks, S. Raghavan, K. Ramkumar, S. Geha, and J. Jeon, "The effect of hafnium or zirconium contamination on MOS processes," in *Proceedings of the IEEE/SEMI Advanced Semiconductor Manufacturing Conference and Workshop*, 2002, pp. 299–303.
83. D. Gui, Z. X. Xing, Z. Q. Mo, Y. N. Hua, and S. P. Zhao, "SIMS analysis of gate oxide breakdown due to tungsten contamination," in *Proceedings of the IEEE International Conference on Semiconductor Electronics (ICSE)*, 2006, pp. 477– 480.
84. D. L. Kwong, "Si device Proceedings," in R. W. Cahn (ed.) *Materials Scinece and Technology: A Comprehensive Treatment*, Vol. 16, *Proceedings of Semiconductors*: Vol. ed. : K. A. Jackson, VCH Verlagsgesellschaft, Weinheim, 1996.
85. C. C. Liao, C. F. Cheng, D. S. Yu, and A. Chin, "The copper contamination effect of Al_2O_3 gate dielectric on Si," *Journal of the Electrochemical Society*, Vol. 151, No. 10, 2004, pp. G693–G696.
86. A. Kamgar, H. M. Vaidya, F. H. Baumann, and S. Nakahara, "Impact of gate-poly grain structure on the gate-oxide reliability (CMOS)," *IEEE Electron Device Letters*, Vol. 23, No. 1, 2002, pp. 22–24.
87. L. Fonseca and F. Campabadal, "Electrical characteristics of postoxidation annealed very thin SiO_2 films," *Journal of the Electrochemical Society*, Vol. 143, 1996, pp. 356–360.

88. E. P. EerNisse, "Stress in thermal SiO_2 during growth," *Applied Physics Letters*, Vol. 35, No. 1, 1979, pp. 8–10.
89. Y.-H. Lee, R. Nachman, S. Hu, N. Mielke, and J. Liu, "Implant damage and gate-oxide-edge effects on product reliability," in *Proceedings of the International Electron Devices Meeting (IEDM)*, 2004, pp. 481–484.
90. D. Y. Lee, H. C. Lin, C. L. Chen, T. Y. Huang, T. Wang, T. L. Lee, S. C. Chen, and M. S. Liang, "Impacts of HF etching on ultra-thin core gate oxide integrity in dual gate oxide CMOS technology," in *Proceedings of the 8th International Symposium on Plasma- and Process-Induced Damage*, 2003, pp. 77–80.
91. T. Ohashi, T. Kubota, and A. Nakajima, "Ar annealing for suppression of gate oxide thinning at shallow trench isolation edge," *IEEE Electron Device Letters*, Vol. 28, No. 7, 2007, pp. 562–564.
92. S. Wolf, Deep-Submicron Process Technology, Vol. 4: *Silicon Proceedings for the VLSI Era*, Lattice Press, Sunset Beach, CA, 2002, p. 454.
93. S. Chen, C. Y. Fu, S. M. Jang, C. H. Yu, and M. S. Liang, "Plasma damage reduction for high density plasma CVD phosphosilicate glass process," in *Proceedings of the 7th International Symposium on Plasma- and Process-Induced Damage*, 2002, pp. 76–79.
94. Y. Chen, H. Yuan, Z. Zhang, N. Li, D. Chan, J. C. Chen, X. Li, and G. Zhao, "Advanced HDP STI gap-fill development in 65-nm logic device," *Electrochemical Society Transactions*, Vol. 27, No. 1, 2010, pp. 679–683.
95. J. Y. C. Yang, C.-L. Lin, C.-Y. Hu, J.-P. Chen, C.-J. N. Kao, and K. C. Su, "The correlation of interface defect density and power-law exponent factor on ultra-thin gate dielectric reliability," in *IEEE International Integrated Reliability Workshop (IRW) Final Report*, 2006, pp. 179–181.
96. A. Hiraiwa, and D. Ishikawa, "Thickness-dependent power-law of dielectric breakdown in ultrathin NMOS gate oxides," *Microelectronics Engineering*, Vol. 80, 2005, pp. 374–377.
97. K. Ohgata, M. Ogasawara, K. Shiga, S. Tsujikawa, E. Murakami, H. Kato, H. Umeda, and K. Kubota, "Universality of power-law voltage dependence for TDDB lifetime in thin gate oxide PMOSFETs," in *Proceedings of the IEEE 43th Annual International Reliability Physics Symposium*, San Jose, 2005, pp. 372–376.
98. T. Ohguro, T. Nagano, M. Fujiwara, M. Takayanagi, T. Shimizu, H. S. Momose, S. Nakamura and Y. Toyoshima,"A study of analog characteristics of CMOS with heavily nitrided NO oxynitrides," in *Symposium on VLSI Technology Digest of Technical Papers*, 2001, pp. 91–92.
99. JEP122E, *Failure Mechanisms and Models for Semiconductor Devices*, JEDEC Publication, March 2009, pp. 7–12.
100. JP-001, *Gate Oxide Integrity (GOI), Foundry Process Qualification Guidelines* (Wafer Fabrication Manufacturing Sites), JEDEC/FSA Joint Publication, September 2002, pp. 16–21.

CHAPTER 7

Negative-Bias Temperature Instability

Dual-poly gate [i.e., *N*+ poly for N-channel metal-oxide semiconductor (NMOS) devices and *P*+ poly for P-channel metal-oxide semiconductor (PMOS) devices] is used widely in today's advanced technologies. In PMOS devices, boron in the *P*+ poly gate can penetrate into the gate oxide during high-temperature annealing, which will generate a positive oxide charge and lead to oxide reliability degradation and device parameter shift.[1] To overcome the boron penetration, the oxide generally is nitrided.[2–5] An oxynitride film with two nitrogen-rich layers was suggested in 1990s, with one layer at the poly-SiO_2 interface to suppress boron diffusion and the other layer at the SiO_2–Si-substrate interface to improve NMOS hot-carrier immunity.[6,7] However, the nitrogen-rich layer near the SiO_2–Si-substrate interface has been shown to lower channel mobility and transconductance,[8] as well as degrade PMOS negative-bias temperature instability (NBTI) performance very much.[9] NBTI-induced degradation in a single PMOS will further impact to circuit's aging performance, for example, maximum operating frequency f_{max} and total standby leakage current I_{DDQ} of digital complementary metal-oxide semiconductor (CMOS) circuits,[10] radio-frequency (rf) performance[11] (such as flicker noise[12] and linearity[13]), static random-access memory (SRAM) high-temperature operating life (HTOL),[14] stability of the field programmable gate arrays (FPGAs) owing to the constant stressing of the SRAM cells storing the configuration bits,[15] and so on. As a current common practice, the nitrogen is kept away from the SiO_2–Si-substrate interface, which will be elaborated later in this chapter.

Presently NBTI of PMOS devices is an important reliability concern for the following reasons: (1) introduction of the dual-poly process with surface-channel PMOS devices, (2) increased effective field across the oxide, (3) thinner oxides making the poly-silicon gate closer to the Si-SiO_2 interface, (4) the introduction of nitrogen to reduce gate leakage and inhibit boron penetration, and (5) increased operating temperature of metal-oxide-semicondutor field-effect transistors (MOSFETs). NBTI results in power-law time-dependent degradation of device parameters, such as threshold voltage V_t, saturation drain current I_{dsat}, and so on. The degradation generally is believed to be driven by the oxide electrical field E_{ox}, but not by the gate voltage V_g.[16–18] The degradation aggravated at higher temperature T. Although the device parameter shift is small over short time periods, the progressive shift over several years can lead to device/circuit failure.

NBTI has become one of the hot topics of semiconductor process reliability in recent years. Figure 7.1 shows the number of papers published on NBTI each year. The papers were counted by the IEEE/IEE electronic library (IEL), the American Institute of Physics (AIP), the Institution of Engineering and Technology (IET), the American Vacuum Society (AVS), and the *IBM Journal*. It can be seen that the number of publications has increased abruptly since year 2001–2002, and there has been a continuous increase in recent years. It is interesting to see that both academia and industry have put great effort into understanding the underlying mechanisms of NBTI and NBTI immunity improvement as well. There are some well-written summary reports on NBTI that provide a comprehensive understanding of the NBTI phenomenon.[18–23]

In common practice, device parameter degradation (e.g., ΔI_{dsat}, ΔV_t) induced by NBTI depends on the gate voltage V_g, the temperature T, and the stress time t, as described by the *exponential model* [Eq. (7.1)][24,25]

$$\Delta P\left(\text{or } \frac{\Delta P}{P_0}\cdot 100\%\right) = A\cdot \exp(\gamma_V \cdot V_g)\cdot \exp\left(-\frac{E_a}{k_B T}\right) t^n \tag{7.1}$$

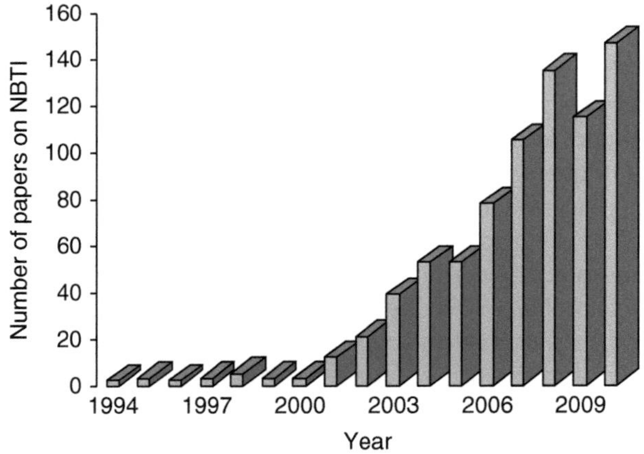

FIGURE 7.1 The number of papers published on NBTI in each year between 1994 and 2010. The papers were counted by the IEEE/IEE electronic library (IEL), AIP, IET, AVS, and the *IBM Journal*. It is seen that the number of publications has increased abruptly since 2001–2002, and there has been a continuous increase in recent years.

or the *power-law model* [Eq. (7.2)][24,25]

$$\Delta P\left(\text{or } \frac{\Delta P}{P_0}\cdot 100\%\right) = A\cdot (V_g)^m \cdot \exp\left(-\frac{E_a}{k_B T}\right) t^n \tag{7.2}$$

where A is a constant, γ_V is the gate-voltage acceleration factor (units are the reciprocal of voltage), m is the gate-voltage exponent (measured values range between 3 to 4), k_B is Bolzmann's constant, E_a is the activation energy (experimentally measured values range between –0.01 to +0.15 eV), and n is the degradation-time exponent (which is typically equal to 0.25 and ranges from 0.2 to 0.3 for most published experimental data).

In some cases, the gate-voltage V_g is replaced by the oxide electrical field E_{ox}, and we get the *exponential model* [Eq. (7.3)][26]

$$\Delta P\left(\text{or } \frac{\Delta P}{P_0}\cdot 100\%\right) = A\cdot \exp(\gamma_E \cdot E_{ox})\cdot \exp\left(-\frac{E_a}{k_B T}\right) t^n \tag{7.3}$$

and the *power-law model* [Eq. (7.4)][27,26]

$$\Delta P\left(\text{or } \frac{\Delta P}{P_0}\cdot 100\%\right) = A\cdot E_{ox}^m \cdot \exp\left(-\frac{E_a}{k_B T}\right) t^n \tag{7.4}$$

where E_{ox} is the oxide electrical field, γ_E is the oxide field-acceleration factor (units are cm/MV), and m is the oxide field exponent.

The *mixed model* is proposed as well as follows[18,28,29]

$$\Delta P\left(\text{or } \frac{\Delta P}{P_0}\cdot 100\%\right) = A\cdot E_{ox} \cdot \exp(\gamma_E \cdot E_{ox})\cdot \exp\left(-\frac{E_a}{k_B T}\right) t^n \tag{7.5}$$

This chapter will give a comprehensive summary for the NBTI phenomenon that is observed more clearly in thinner oxides. NBTI mechanisms and the corresponding models for lifetime extrapolation then are introduced. For NBTI characterization, a specific observation is the so-called recovery effect after removing the stressing voltage, which is discussed in detail in terms of various measurement methods taking this effect into account. Again, the structure effect on NBTI performance is summarized, although there are quite controversial reports in the literature. Important approaches to improve NBTI through process optimization are also detailed. Finally, the NBTI process-qualification requirements are listed as a reference.

7.1 Negative-Bias Temperature Instability Degradation Mechanism

NBTI is linked to the layer of low-energy holes in the channel, which leads to the Si–H bond broken at the Si-SiO_2 interface.[17] In contrast, the gate-oxide damage during Fowler-Nordheim (FN) current injection (as detailed in Chap. 6) generally modeled by anode hole injection (AHI),[30] anode hydrogen release (AHR),[31] or thermochemical breakdown.[32]

It is generally believed that NBTI generates traps at the SiON–Si-substrate interface and/or fixed charges in the SiON bulk. The former defects are predominantly P_b-type centers, whereas the latter are mostly related to hydrogenous species, E' centers, or both in SiON. The NBTI-induced V_t shift is expressed as

$$\Delta V_t = -\frac{\Delta Q_{ot}}{C_{ox}} - \frac{\Delta Q_{it}}{C_{ox}} = -\frac{q\Delta N_{ot}}{C_{ox}} - \frac{q\Delta N_{it}}{C_{ox}} = \Delta V_{ot} + \Delta V_{it} \tag{7.6}$$

where Q_{ot} and Q_{it} are oxide charge and interface trap, respectively; ΔN_{ot} and ΔN_{it} are oxide charge and interface trap, respectively; C_{ox} is the oxide capacitance; and ΔV_{ot} and ΔV_{it} are oxide-trap-component- and interface-trap-component-induced V_t shift, respectively.

7.1.1 Reaction-Diffusion Model

For SiO_2 gate dielectric films, NBTI is believed mainly to be due to generation of interface traps (N_{it}) at the Si-SiO_2 interface. Several phenomenologic models have been proposed to explain the formation of interface traps associated with NBTI degradation, and to list a few, they include the models of Jeppson et al.[16] (diffusion-controlled mechanism), Blat et al.[33] (diffusion mechanism based on release of a neutral hydroxyl species), Ogawa et al.[34] (mechanism based on diffusing species H^0, H^+), and Chakravarthi et al.[35] [quantitative model based on the reaction-diffusion framework (R-D model) to predict interface state generation during NBTI stress].

Among these various models explaining the NBTI phenomenon,[16,33–35] the *R-D model*[35] *has been one of the most promising physical models*, and it will be detailed in this section. In short, according to the R-D model, ≡Si–H bonds at the Si-SiO_2 interface are broken by inversion-layer holes (*reaction*), and the H species released subsequently move away from the interface (*diffusion*) and create N_{it} (≡Si−). Figure 7.2 schematically shows the two-dimensional Si-SiO_2 interface, including (*a*) the ≡SiH defect, (*b*) how this defect may be electrically activated during NBTI to form an interface trap, a fixed oxide charge, and a hydroxyl group, and (*c*) the OH diffusion through the oxide.

In the proposed R-D system,[35] a passivated $Si_3{\equiv}Si{-}H$ (sometimes expressed as SiX) bond at the interface dissociates to form a dangling silicon bond $Si_3{\equiv}Si\bullet$ (sometimes expressed as Si^+) and an unidentified mobile species (X) that is possibly a hydrogen atom (H), a hydrogen molecule (H_2), or a positively charged hydrogen ion (H^+). The R-D model is expressed as follows[16]:

- Reaction (case A):

$$Si_3{\equiv}Si{-}H + h^+ \Leftrightarrow Si_3{\equiv}Si\bullet + X_{\text{interface}} \tag{7.7}$$

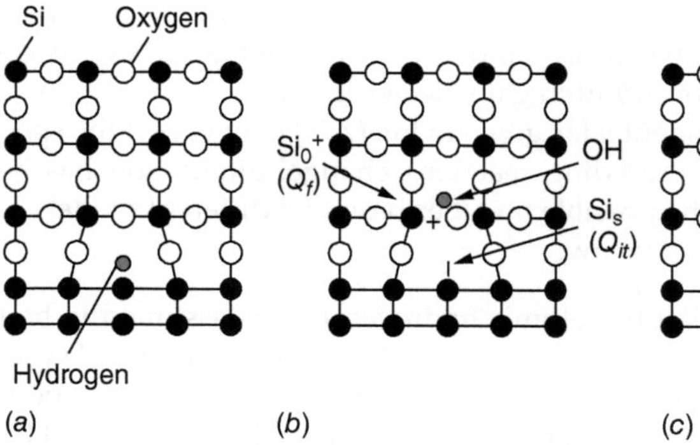

FIGURE 7.2 Schematic two-dimensional representation of the Si-SiO_2 interface showing (*a*) the ≡SiH defect, (*b*) how this defect may be electrically activated during NBTI to form an interface trap, a fixed oxide charge, and a hydroxyl group, and (*c*) the OH diffusion through the oxide. (*Used with permission from the American Institute of Physics.*[22])

- Reaction (case B):

$$O_3{\equiv}Si{-}H + h^+ \Leftrightarrow O_3{\equiv}Si{\bullet} + X_{interface} \tag{7.8}$$

- Diffusion:

$$X_{interface} \Leftrightarrow X^+_{bulk} \tag{7.9}$$

Reactions in Eqs. (7.7) and (7.8) generate the positively charged defect (Si^+) at the interface, referred as the *donor-interface-trap state* ΔN_{it}.[36] The donor interface trap is positive for a PMOS device during stress (biased to inversion). $Si_3{\equiv}Si{\bullet}$ and $O_3{\equiv}Si{\bullet}$ are referred as *fast-state* and *slow-state interface trap*, respectively.[37] According to Eq. (7.9), the species X diffuses into the gate dielectric, and a positive fixed charge is generated (ΔN_{ot}). Equation (7.9) can be further expressed as follows in terms of SiO_2 [Eq. (7.10)] and oxynitride [Eq. (7.11)]:

$$H^+ + O{-}Si{\equiv} \Leftrightarrow HO^+{-}Si{\equiv} \tag{7.10}$$

$$H^+ + N{-}Si{\equiv} \Leftrightarrow HN^+{-}Si{\equiv} \tag{7.11}$$

Equation (7.11) in oxynitride is more likely to take place than Eq. (7.10) to generate the positive oxide traps because of its lower activation energy.[38]

Interestingly, Blat et al.[33] reported that the numbers of interface traps (ΔN_{it}) and positive fixed charges (ΔN_{ot}) generated by a negative-bias temperature (NBT) stress are almost equal for a gate dielectric of SiO_2 with a thickness of 56 nm. Similar results were observed by Tan et al.[39] for ultrathin SiO_2 and SiON gate dielectrics, demonstrating a one-to-one correspondence between the numbers of interface traps and positive fixed charges generated by the NBT stress. However, this observation is not necessarily always true. For example, there is no correlation between ΔN_{ot} and ΔN_{it}[18,40] or negligible ΔN_{ot} as compared with ΔN_{it} for ultrathin gate dielectrics.[36]

The time exponent n [as given in Eq. (7.1)] may vary if the species involving in the preceding reactions change. Simulation has been performed, and the possible n values based on different models[35] are briefly summarized below:

- *Model I:* Neutral atomic hydrogen (H^0) is assumed to be the unknown species X in Eq. (7.7). For a diffusion-limited system, the neutral atomic hydrogen model exhibits a time exponent of 0.25. The corresponding reaction is $SiH + h^+ \Leftrightarrow Si^+ + H^0$
- *Model II:* The species X is molecular hydrogen (H_2), giving a time exponent of 0.165. The corresponding reaction is $SiH + h^+ \Leftrightarrow Si^+ + 0.5H_2^0$
- *Model III:* This is a combination of model I and model II where both atomic hydrogen and molecular hydrogen are assumed to take effect. Therefore, the final slope of the reaction according to the superposition effect has a range from 0.165 to 0.25.
- *Model IV:* The species X is assumed to be the proton hydrogen (H^+). Owing to the additional drift term, the H^+ system shows a slope greater than 0.25. The corresponding reaction is $SiH + 2h^+ \Leftrightarrow Si^+ + H^+$. When there is a combination of H^+ and a H^0 species, the time exponent will range from 0.25 to 0.5.
- *Model V:* A combination of models I, II, and IV will give the following two consecutive reactions: $SiH + 2h^+ \Leftrightarrow Si^+ + H^+$ and $SiH + H^+ \Leftrightarrow Si^+ + H_2^0$. This is so because H^+ reacts with SiH further to form molecular hydrogen if molecular hydrogen is a more stable defect than atomic H^+. In this case, the time exponent will range from 0.165 to 0.5.
- *Model VI:* This case is possible if an empty trap (T) forms an immobile hydrogen-trap complex (TH) with the capture of a mobile hydrogen (H^0). The corresponding time exponent will be larger than 0.25. One example is the presence of hydrogen gettering sites (traps) that significantly alters NBTI degradation. Another example is the increase in NBTI slope observed after plasma damage,[41] that can be explained by the increase in unsaturated bonds in polysilicon or oxide (empty traps) that can trap the diffusing hydrogen atoms.
- *Model VII:* Contrary to model VI, the presence of a hydrogen source or a higher background hydrogen concentration reduces the NBTI slope (<0.25) owing to the release of H.
- *Models VIII and IX:* These are analogous to models VI and VII for trapping and release of molecular hydrogen, respectively.

For trapping molecular hydrogen, the time exponent is larger than 0.165, whereas it is less than 0.165 for releasing molecular hydrogen.

In reality, a system will be most likely mixed by all species (atomic H^0, H^+, and H_0^2) and hydrogen traps; thus a large range of time exponents (0.165 to 0.5) is expected. However, most literature data show that the time exponent ranges between 0.2 and 0.3, indicating that the NBTI degradation reaction is dominated by a neutral hydrogen-related species (e.g., H^0) with minor contributions from other species (e.g., molecular hydrogen, atomic positive hydrogen, or hydrogen traps).

Although there is much consensus on NBTI-induced interface-trap generation by the R-D model, there is more argument on oxide-trap generation in the literatures.

7.1.2 Recovery

Recovery is a well-known phenomenon in NBTI degradation, especially for nitrided oxide, when the stress voltage is removed. The recovery makes an accurate measurement of the degradation difficult. The recovery (or passivation) rate depends on the electrical field; that is, a larger positive bias applied during the recovery stage will lead to a more rapid recovery.[19,42] Figure 7.3 shows ΔI_{dlin} under a

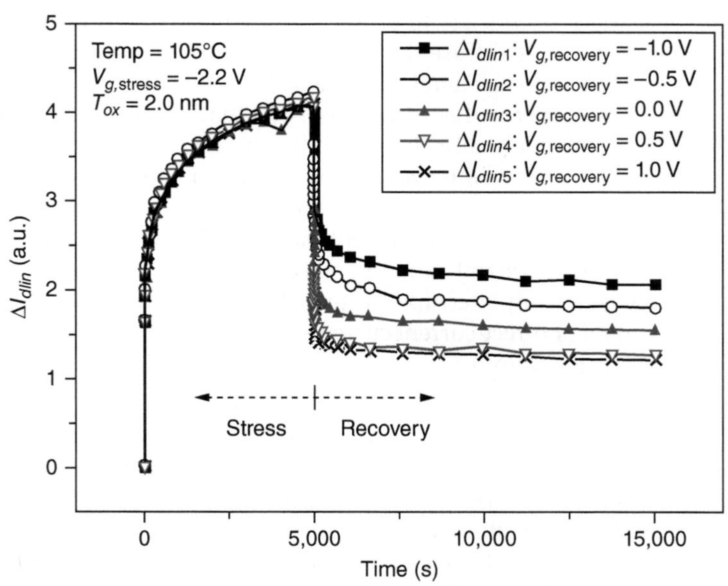

Figure 7.3 Variation of ΔI_{dlin} under stress-recovery-stress. The recovery rate increases with increasing recovery of gate voltage. (*Courtesy of Dr. Lei Jin.*)

stress–recovery (passivation)–stress sequence where the stress conditions remain the same while the passivation gate voltage is changed. The data indicate that a higher electrical field favors the passivation process. The recovery effect could be clearly observed even under room temperature,[42] indicating that the passivation (recovery) is not due to a thermal-activation effect.

Basically, two different recovery mechanisms are proposed in the literature, as detailed below:

- The first recovery mechanism is recovery of the interface traps (ΔN_{it}) attributed to the repassivation of the Si-SiO_2 interface by a reverse reaction between dangling Si bonds and the released hydrogen species.[42–44] This is based on the R–D model discussed in Sec. 7.1.1. It is believed that hydrogen species diffuse to poly-oxide interface during stress, which can be reflected from poly-oxide interface for recovery. If hydrogen is totally reflected from the poly-oxide interface, the recovery rate should be closer to the degradation rate owing to the reverse diffusion of hydrogen (diffusion limited).[35,45] Tsujikawa et al.[45] indeed showed the same time-dependence exponent for both the stressing and recovery stages of NBTI degradation. However, this observation is not always true. One study found that when no saturation phenomenon is seen on degradation curves, the slope of recovery is much smaller than the slope of degradation (e.g., the slope changes from 0.221 to 0.157),[25] implying that hydrogen species partially diffuse through the oxide to the poly gate electrode but do not totally diffuse back.

  However, it has been observed that the interface traps [e.g., as widely directly probed by charge-pumping (CP) current I_{cp}], once created, are always present , no matter how, degradation is monitored.[19] It is the interface-trap creation that contributes to the permanent part of NBTI degradation. For example, no significant ΔN_{it} relaxation can be induced during measurement interruptions (e.g., I_d-V_g sweep for V_t characterization or direct-current current-voltage (DCIV) and CP for N_{it} characterization).[46]

- The second mechanism considers that it is the trapped-holes component (ΔN_{ot}) that is responsible for transient effects with recovery.[47,48] The recovery of the oxide traps (both trapped holes and/or positively charged slow states) is correlated with their neutralization processes.[19] There is a larger recovery effect in nitrided oxide than in pure oxide, as explained below. The hole-trapping centers are the Si_2O bridges in pure oxide and Si_3N local environments in nitrided oxide.[47] Hole trapping is enhanced in nitrided oxides because hole trapping on nitrogen-related centers has a lower reaction energy than on oxygen-related centers. The oxide traps then are linked to the nitrogen-related defects and mostly preexist the stress.[19] Besides, once a hole is trapped in the oxide, the complex formed by the nitrogen-related center plus the trapped hole acts as a hydrogen-trap center for recovery. The neutralized site can return to the positive state when the stress is applied again.[49]

7.1.3 Degradation Saturation Mechanism

As mentioned earlier, NBTI degradations usually are assumed to have a power-law dependence on stress time [Eq. (7.1)]. However, this assumption is not necessarily always true.[35,26] Figure 7.4*a* shows the I_{dsat} shifts versus stress time in PMOS devices with ~20-Å-thick

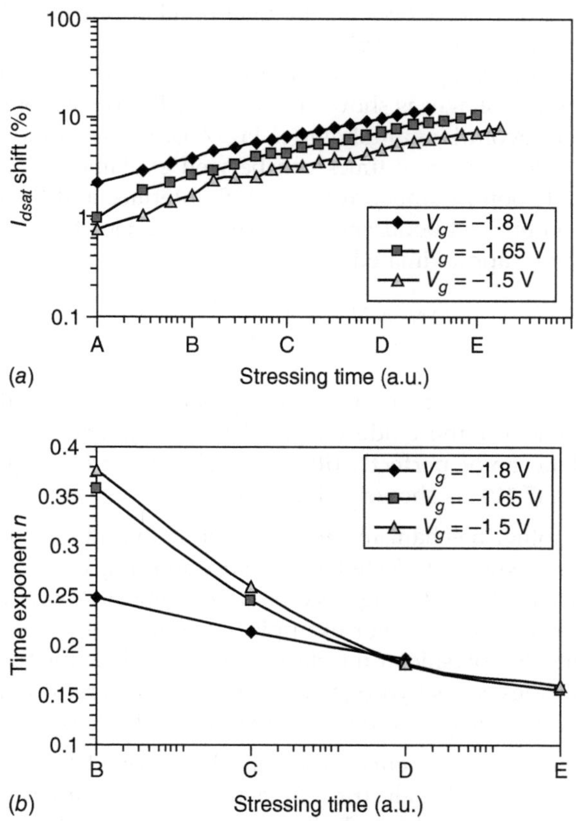

FIGURE 7.4 (*a*) I_{dsat}-shift versus stress time for various stresses V_g at 125°C. Degradation saturates as stress time increases. (*b*) Degradation time-exponent *n*-value power law versus stress time. The *n* value decreases with increasing stress time.

gate oxides for various stress gate voltages at 125°C. I_{dsat} shift has a tendency to saturate with increasing stress time. Under higher stress voltage $|V_g|$, the saturation value becomes higher. As shown in Fig. 7.4*b*, the degradation-time-exponent n decreases from 0.25 to 0.4 with increasing stress time and reaches ~0.16 finally. I_{dsat} shift versus stress V_g at various stress times is well fitted by the exponential dependence on V_g [Eq. (7.1)]. It is noted that the similar saturation phenomenon of NBTI degradation is observed regardless of gate-oxide thickness, indicating that degradation saturation is universal, for which a few explanations have been given in literature as follows:

- One mechanism attributed the degradation saturation to the reaction limitation arising from the *maximum available free SiH sites* [Eq. (7.7)][35] based on the R-D model. The generation of Si^+ decreases as the number of available Si–H bonds decreases with time. Based on this model, the N_{it} generated at saturation converges because the total number of available sites is independent of voltage.

  An example of the evolution of ΔV_t over stress time at different voltages is shown in Fig. 7.5*a*. The corresponding ΔN_{it} time evolutions are shown in Fig. 7.5*b*. The saturation trend at the longer stress times can be well explained by the finite Si–H bond number at the SiO_2-Si interface and the distribution of Si–H bond energies, where interface trap generation can be approximated as[47]

  $$\Delta N_{it} = N_{it,\max}\left(1 - \frac{1}{1+\left(t/\tau\right)^{\alpha}}\right) \tag{7.12}$$

  where $N_{it,max}$, τ, and α are fitting parameters, and all depend on V_g (or the oxide electrical field E_{ox}). The experimental data can be perfectly fitted in the whole stress region using Eq. (7.12), as shown in Fig. 7.5*b*.

- Another mechanism for saturation is that the diffusing species may be reflected when encountering a new interface where the diffusing species is energetically unfavorable to transfer across.[35] For example, such an interface could be a nitride-polysilicon interface. Owing to their higher density, nitrides act as hydrogen barriers[50] and thus reduce the transfer of hydrogen into the nitride from polysilicon.

- The saturation effect also may be ascribed to the gradual decrease in the electrical field at the oxide-Si interface owing to a positive charge buildup in the oxide and to a gradual increase in the favorability of the reverse reaction [Eq. (7.7)] owing to increasing ΔN_{it}.[51] Increasing $|V_g|$ can increase the consumption rate of the Si–H bonds so that the ΔN_{it} saturation appears at an earlier stress time.

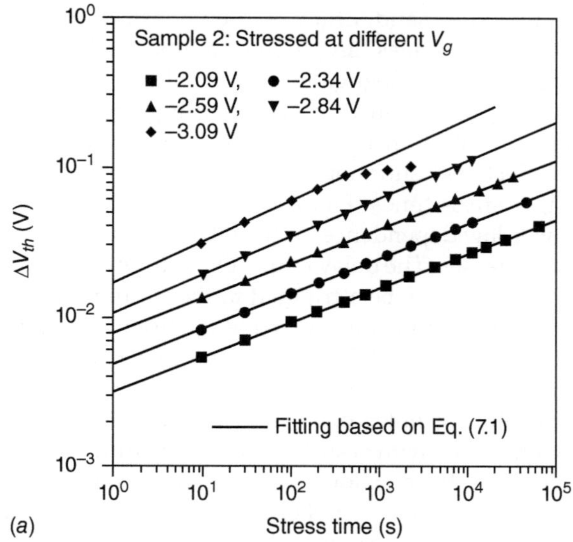

(*a*)

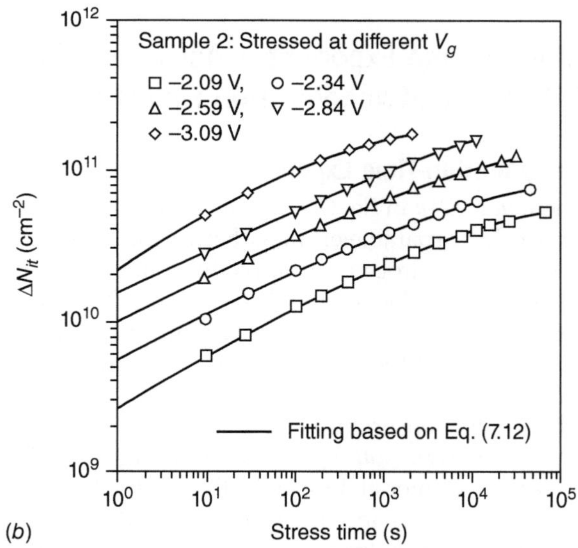

(*b*)

FIGURE 7.5 (*a*) Stress-time evolution of threshold voltage shift ΔV_{th} for a sample stressed at 125°C at different voltages and linear fitting based on assumption of the power-law dependence [Eq. (7.1)]. (*b*) Stress-time evolution of interface trap generation ΔN_{it} for the preceding sample. Fitting lines are based on Eq. (7.12). (*Used with permission from the American Institute of Physics.*[37])

- On the other hand, Zhu et al.[37] emphasized that in the early NBTI stress stages, ΔN_{it} follows the power-law dependence, whereas ΔN_{ot} is negligible, and not all the released species X seem to form positive charges during reactions in Eqs. (7.7) and (7.8). With increasing stress time, the generation rate of ΔN_{it} declines continuously, showing a saturation trend at the longer stress times (Fig. 7.5*b*), whereas ΔN_{ot} still follows the power-law dependence, so the relative contribution of ΔN_{ot} to ΔV_t increases. The delay of the ΔN_{ot} saturation compared with that of ΔN_{it} may be attributed to the continuous dissociation of $O_3\equiv$Si–H bonds in the oxide far from the interface, even after the interfacial Si–H bonds have been substantially consumed. Because these $O_3\equiv$Si• defects can be positively charged by capturing holes or H^+ and the trapped charges cannot communicate with Si surface carriers by tunneling, the defects behave as fixed oxide charges, not as slow-state interface traps. When the $O_3\equiv$Si–H bonds in the SiO_2 layer are mostly consumed, ΔN_{ot} becomes saturated.

7.2 Degradation-Time Exponent *n*, Activation Energy E_a, and Voltage/Field Acceleration Factor γ

7.2.1 Degradation-Time Exponent *n*

The exponent values [*n* in Eq. (7.1)] of the power-law time-dependent degradation of device parameters range between 0.165 and 0.5 based on the R-D model and hydrogen release model, as described in Sec. 7.2.1. Experimentally, the *n* value is found to be between 0.15 and 0.35.[52–55] As a matter of fact, some factors such as gate length, oxide thickness, nitridation, fluorination, and testing methodology are found to affect the *n* value. A summary is given below.

One noticeable observation is that the *n value continues to increase with gate-length scaling*, as shown in Fig. 7.6.[57] The geometry-dependent R-D model indicates that a relatively high time exponent ($n \approx 0.35$) probably arises from two-dimensional hydrogen release and diffusion from drain and source ends, where Si–H bonds are densely populated.

The *degradation-time slope n is found depending on the testing methodology*[62]: An uninterrupted stress (UIS) method (Sec. 7.4.3) indicates no change in *n* value with process change, whereas there is a clearly process-dependent *n* value when it is measured by the conventional stress-measure-stress (SMS) technique, attributed to the recovery after stress removal.

- For UIS by which the recovery effect is negligible, the time slope *n* of the I_{dlin} degradation is ~0.16 to 0.20, which does not show any significant dependence on nitrogen, fluorine, stress

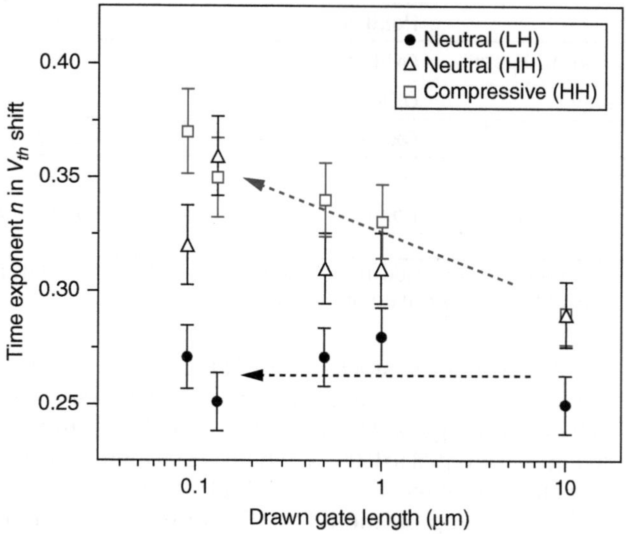

FIGURE 7.6 Gate-length-dependent fitted time exponent n Based on the V_{th} shift under NBTI stress. "Neutral LH" and "Neutral HH" mean low stress (~200 MPa) of the contact etch stop layer (CESL) with low and high hydrogen concentrations, respectively. "Compressive HH" means high-compressive nitride film (>1 GPa) containing high hydrogen concentrations. (*Used with permission from IEEE.*[57])

voltage, or temperature. The similarity of n suggests that the degradation mechanism is the same.

- For SMS, by which periodic stress interruptions are performed for sensing degradation, the recovery is unavoidably taking place. The slope n is process/EOT-dependent.[62]
  - The time slope n is lower for plasma nitrided oxide (PNO) (0.15 to 0.25) than for SiO_2 (0.25 to 0.3).
  - The n value decreases with decreasing oxide thickness (for both SiO_2 and nitrided oxide). The n values at equivalent oxide thickness (EOT) 1, 3, and 6.5 nm are about 0.16, 0.27, and 0.30, respectively.
  - Incorporation of fluorine increases n (~0.30) as well as lifetime.
  - The trend for time slope n for different process is n(nitrided) < $n(SiO_2)$ < n(fluorided).

7.2.2 Activation Energy (E_a)

NBTI becomes more serious at higher temperatures. The activation energy shows no channel-length dependence[56] but does depend on the interface/oxide traps, nitridation, and gate voltage, as detailed below.

- *E_a depends on the oxide traps and interface traps induced by the NBTI stressing.* V_t shift measured at different recovery stages will give quite different activation energies, as detailed below.

	Point *A*	**Point *B***
Main contributors to V_t shift	Oxide trap + interface trap	Interface trap
Activation energy	0.06 eV	0.154 eV
	Oxide Trap	**Interface Traps**
Activation energy	0.02 eV	0.154 eV
Contribution to V_t shift	Point *A*	Points *A* and *B*

Table 7.1 The Contributions of Oxide Traps and Interface Traps to V_t Shift and the Corresponding Activation Energy Based on Ref. 19

In other words, *the various E_a listed in the literature may originate from controversial measurement methods.* Table 7.1 shows that the activation energy of ΔV_t at the end of the NBTI stress (point *A* in Fig. 7.7) is equal to 0.06 eV, whereas that after recovery (point *B* in Fig. 7.7) is equal to 0.154 eV, similar to the activation energy of interface-trap creation.[19] Similar results were reported by Zhu et al.[37] At point *A*, the V_t shift is contributed to by both interface traps, and oxide traps, and the corresponding V_t shift is larger than that at point *B*. However, at point *B* after recovery, the oxide traps are recovered, whereas the interface traps are still there. Therefore, at point *B*, the activation energy of the V_t shift is decided by the interface traps.

- *Oxide traps have a much lower activation energy than interface traps.* The activation energy for the oxide traps extracted from Fig. 7.7*b* is $E_{(A\text{-}B)} = 0.02$ eV, indicating that the V_t shift induced by oxide traps has almost no temperature dependence. It has been shown that *defect generation correlated with hole trapping is hardly temperature dependent,*[19] which is the case here for the oxide traps induced by NBTI stressing. The small

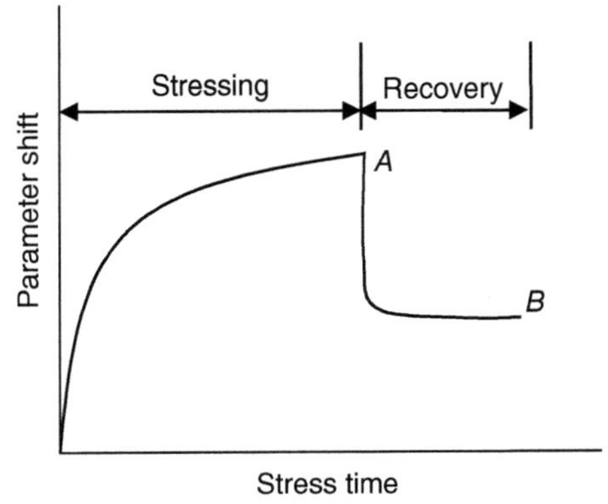

Figure 7.7 Schematic indicating the stressing period and recovery period. Points *A* and *B* denote the end of stressing and recovery periods, respectively.

temperature dependence observed experimentally is explained only by a variation in the effective hole-capture cross section with temperature.

- *Nitrided oxide has a lower activation energy E_a than SiO_2.* E_a is generally between 0.1 and 0.2 eV,[18,45,54] with 0.1 eV for SiON and 0.2 eV for SiO_2 dielectrics.[9,52] Kimizuka et al.[38] reported that the activation energy of NBTI was much smaller for a heavily nitrided NO-oxynitride than for a pure SiO_2. It is well known that nitridation will lead to more oxide traps in the gate oxide and worse NBTI performance. More oxide traps (hole trapping) induced by nitridation thus is correlated with the lower activation energy of the nitrided oxide. Similarly, Tsujikawa et al.[45] reported that the E_a for a nonoptimized oxynitride (0.12 eV) with enhanced NBTI is lower than that of an optimized counterpart (0.14 eV), which can be attributed to the higher number of oxide traps in the nonoptimized oxynitride.
- On the other hand, it looks like the *activation energy for V_t shift is not affected by the hydrogen concentration and stress in the contact-etch stop layer* (*CESL*) *SiN*.[57] All the E_a values are ~0.1 eV for the three cases with different hydrogen concentrations and stresses in the CESL. Therefore, it could be inferred that hydrogen diffusivity from the interface is identical among different types of nitride CESLs.
- *E_a decreases with increasing the stress V_g.* Aono et al.[26] reported that the E_a decreases from 0.25 to 0.16 eV when the stressing voltage increases from 4 to 6.5 V of a PMOS device with a 6.7-nm-thick gate oxide.

7.2.3 Voltage/Field Acceleration Factor γ

The voltage acceleration factor [γ_V in Eq. (7.1)] depends on gate-oxide thickness. γ_V remains proportional to 1/EOT—a thicker oxide corresponds to a smaller voltage acceleration factor.[58] γ_V values can be fitted empirically and correlated with the EOT by the following equation:

$$\gamma_V = 84.6 * \frac{1}{\text{EOT}} \tag{7.13}$$

On the other hand, it is interesting that the oxide field acceleration factor [γ_E in Eq. (7.5)] remains almost constant (~0.6 ± 0.05 cm/MV) when EOT changes from 10 to 50 A.

The voltage acceleration factor is also found to depend on stress voltage. The acceleration factor based on exponential models [Eqs. (7.1) and (7.3)] increases with decreasing stress voltage.[59] Lower stress voltages are desirable for lifetime prediction of operating voltage if the exponential model is used. This is also the reason that the power-law models [Eqs. (7.2) and(7.4)] are proposed to better fit the experimental data.

7.3 Characterization Methodology

NBTI stress is applied to the gate electrode with source, drain, and substrate grounded at elevated temperature (a typical temperature is 125°C to mimic operation). Accelerated testing is usually used to evaluate the NBTI lifetime of a PMOS device. A typical NBTI measurement consists of an iteration of NBTI stress and I_d-V_g measurement. High gate voltages V_g are applied to devices to measure lifetime, and some models of the dependence of lifetime on V_g (E_{ox}) [Eqs. (7.1) to (7.4)] are used to project the lifetime at operating voltage. Device parameter degradation (e.g., ΔI_{dsat} and ΔV_t) induced by NBTI generally can be described by a power-law relationship in terms of stressing time.[16,34,56]

For technology qualifications, typically three or more different acceleration gate voltages are used for a lifetime extrapolation to operating voltage. Figure 7.8 describes the method schematically. The stress time to reach a given parameter-drift criterion is measured for three or more different gate stress voltage (Fig. 7.8*a*). Several transistors/dies are measured in order to evaluate the statistical lifetime. Every device stress test results in a pair of values, that is, gate stress voltage versus time-to-failure for a given criterion (a typical value is a 10 percent I_{dsat} shift). These pairs, plotted in Fig. 7.8*b*, determine the function for lifetime extrapolations. The slope of the extrapolation line defines the voltage acceleration. In this manner, the lifetime under operating conditions (1.1*V_{dd} for a 10 percent margin) can be calculated.

Note that the example shown in Fig. 7.8 for lifetime extrapolation is the conventional *E* model, as described in Eq. (7.1). Other kinds of models [e.g., the power-law model—Eq. (7.2)] are possible. Depending on different models of E_{ox} acceleration, the predicted lifetime can change by several orders of magnitude. This uncertainty of predicted

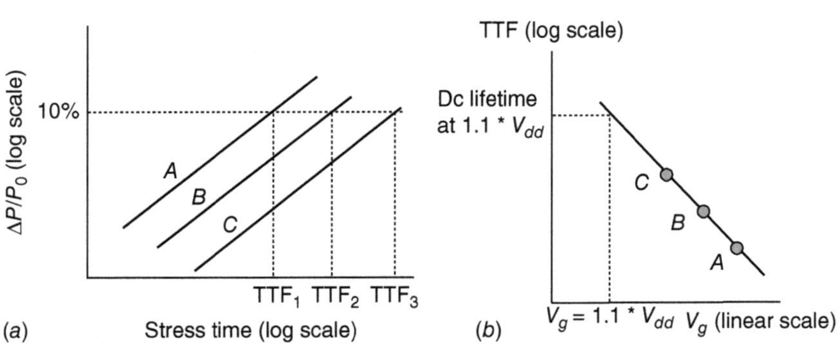

FIGURE 7.8 Schematic for (*a*) degradation of device parameters (e.g., I_{dsat}) with stressing time under three different stress conditions *A*, *B*, and *C*; (*b*) lifetime extrapolation to operating voltage (1.1 * V_{dd}) with the V_g model.

lifetime becomes remarkable as the difference in E_{ox} between operation and accelerated testing is increased. Therefore, the lifetime versus V_g (or E_{ox}) model is the key to correctly evaluating long device lifetime under operating conditions by a short period of stressing under high stressing voltage (or field).

7.3.1 Delay-Time (Recovery) Effect on Characterization

As already discussed in Sec. 7.1.2, NBTI degradation recovers to a large extent after stress is removed. Therefore, it looks promising that the device lifetime will increase when circuits operate under switching (ac) operation.[28,43,60–62] It has been found that owing to the recovery, the magnitude of the NBTI-induced degradation decreases after removal of stress under the conventional stress-measure-stress (SMS) measurement sequence, whereas the power-law time exponent n is higher,[28,43,61–64] as illustrated in Sec. 7.2.1. Therefore, the recovery does affect the NBTI characterization and its lifetime extrapolation.

It has been demonstrated that the recovery will take place less than 1 μs after stress removal.[20,60] Up to 50 percent parameter degradation could be seen in less than 1 second of delay time between the end of stress and the characterization.[67] Rapid recovery arises as a result of rapid capture of hydrogen atoms within the capture cross section of the silicon dangling bond (reaction-limited). After all the atoms residing in this region are captured, further recovery takes place as a result of hydrogen atoms diffusing to the interface again (diffusion-limited).[35]

Such a short a recovery time during NBTI characterization makes it hard (or impractical) to measure the transistor after stress without any recovery, especially when more than one device parameter has to be measured (e.g., because of designer demands). However, in some cases, the transistors within product circuits may not have opportunity to recover under continuous working conditions (e. g., with dc stressing). Therefore, it is easy to postulate that an exaction of the device parameters based on the already recovered state will lead to an overestimation of product lifetime. In other words, the NBTI recovery phenomenon makes lifetime projection unreliable when it is based on already recovered measurement values. Therefore, the need for device reliability improvements leads to a strong demand for accurate reliability assessment methodologies. Some fast measurement techniques (Sec. 7.3.3) have been proposed to try to overcome the recovery effect, for example, on-the-fly technique or use of a special operational amplifier for fast V_t measurement.[67]

However, fortunately, an investigation of measurement delay–induced recovery on NBTI lifetime prediction showed that, contrary to expectations, an undesirable strong overestimation does not occur.[64] If long stress times were used, even delays of 60 seconds resulted in a negligible overestimation. In contrast, too short stress-time

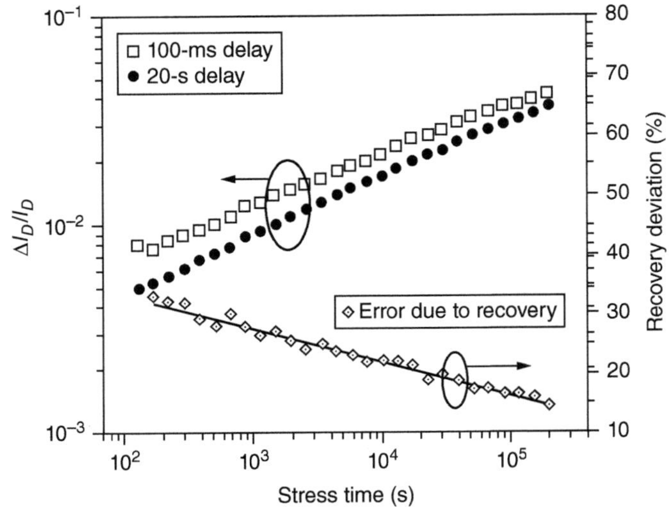

FIGURE 7.9 Degradation of I_D as a function of stress time for 100-ms and 20-s measurement delays for the 65-nm technology. (*Used with permission from IEEE.*[64])

experiments can lead to an underestimation of lifetime. *Both short delays and long stress times the effective ways to improve the accuracy of NBTI lifetime prediction.*

Figure 7.9 shows the degradation of I_D as a function of stress time for 100-ms and 20-s measurement delays for the 65-nm technology. It is seen that the relative deviation between the I_D shift after the varied delays becomes smaller with longer stress times, meaning that with longer and longer stress times, the discrepancies in parameter degradation caused by the recovery become smaller and smaller.

Figure 7.10 shows the time-to-failure (TTF) predicted based on a failure criterion of 10 percent ΔI_{dsat} (the methodology is described in Fig. 7.8) for two different experimental stress groups (short and long), each measured with 100-ms and 60-s delays at the five different stress voltages (V_G = −1.9, −2.0, −2.2, −2.4, and −2.6 V). Lifetime extrapolation to operating voltage for four different measurement groups is also given in Fig. 7.10. Several salient points are highlighted as follows[64]:

- In either the long or short stress groups, the longer delay (60 s) leads to higher recovery and therefore to a higher TTF at the different single-stress voltages than the 0.1-s delay.
- *Regardless of the delay time (slow or fast), the lifetime extrapolated to the operating voltage is almost the same for the long-stress-time group.* This trend is also applicable to the short-stress-time group.

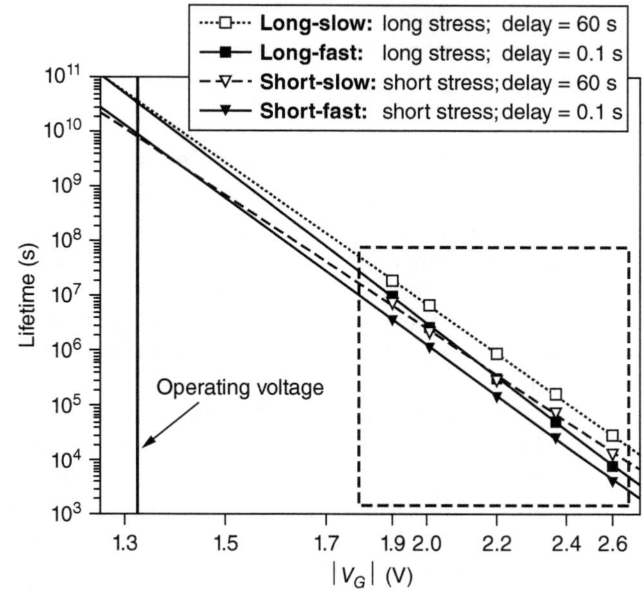

FIGURE 7.10 Lifetime extrapolation to operating voltage of four different measurement groups for the 65-nm technology. Stress time and measurement delay are varied. (*Used with permission from IEEE.*[64])

- The lifetime extrapolated to the operating voltage for the long-stress-time groups is larger than that for the short-stress-time groups, meaning that the extrapolations based on short stress times (groups short-fast and short-slow) even underestimate transistor lifetime. The root cause is that in the short-stress-time groups the degradation curve for each single stress test did not reach the regime with degradation tending to saturate. As a result, the degradation slopes are higher, and therefore, the TTF predicted/extrapolated to 10 percent I_{dsat} shift of each single stress test is underestimated (as shown in Fig. 7.10, the TTF values marked with triangles are lower than those marked with squares).
- Besides the higher TTF values at the different stress voltages, the recovery phenomenon also leads to a change in the voltage acceleration. Figure 7.10 shows that the TTF difference between 60- and 0.1-s delay is smaller at lower stress voltages, resulting in a smaller voltage-acceleration slope in 60-s delay. As a result, the extrapolated lifetimes get close together at the operating voltage.

In short, more recovery happens for longer delays but with a smaller voltage-acceleration slope. *The extrapolated lifetimes under operating voltage based on measurements with different delays should get closer together if the total stress time is the same. Longer stress times giving a better projected lifetime under operating voltage are more desirable because they are closer to reality.*

7.3.2 Stress-Voltage and Stress-Time Effects

For a 1.2-nm-thick ultrathin oxide NBTI lifetime extrapolation, it was found that the *E* model is a better choice for stressing in low fields (<10 MV/cm), whereas it is better to use the power-law model for high-field regions (>10 MV/cm).[25] On the other hand, it has been argued that the power-law model is applicable when there is degradation saturation regardless of gate-oxide thickness and stress voltage.[26]

Figure 7.11*a* shows the 1.2-nm NBTI I_d degradation characteristics, where degradation saturation is not observed.[25] It is seen that the degradation curves show power-law dependence with stress time. Figure 7.11*b* shows the accelerated-degradation slope n versus stress voltage. The n value almost keeps constant at about 0.16 to 0.17 in low-field regions (1.2 to 1.9 V), but it increases from 0.17 to 0.24 with increasing stress voltage in high-field regions (1.9 to 2.7 V). This implies that molecular hydrogen diffusion dominates the degradation in low-field regions ($t^{0.165}$), and additional atomic hydrogen diffusion enhances the degradation slope in high-field regions ($t^{0.25}$) given by the R-D model (Sec. 7.1.1).

Figure 7.12*a* shows the 1.2-nm NBTI degradation characteristics for long-term stress (>1,000 hours) in the stress range from 1.2 to 1.95 V. Again, saturated characteristics are not found on the degradation curves. The degradation slope versus stress time is plotted in Fig. 7.12*b*. It can be seen that the degradation slope does not decrease with increasing stress time and keeps constant at about 0.16 to 0.17. Therefore, the preferred model for NBTI lifetime should be the *E* model rather than the power-law model in the stress range of 1.2 to 1.95 V to predict real application at 1.0 V. In order to further identify the *E* model or the power-law model for lifetime, the dependence of degradation on stress voltage at various stress times is plotted in Fig. 7.13. Chen et al.[25] thus proposed that the degradations should be separated into two regions: *E* field dependence (*E* model) [Eq. (7.1)] with stress voltage in low-field regions (1.2 to 1.9 V) shown in Fig. 7.13*a*, and power-law dependence [Eq. (7.2)] in high-field regions (1.9 to 2.7 V) shown in Fig. 7.13*b*. More specifically, the *E* model is suitable for 1.2-nm nitrided oxide with a stress range of less than 10 MV/cm.

A plausible explanation is given below.[25] Generally speaking, the reaction energy of atomic hydrogen diffusion is higher than that of molecular hydrogen diffusion.[35] In low-field regions, it is easy to form

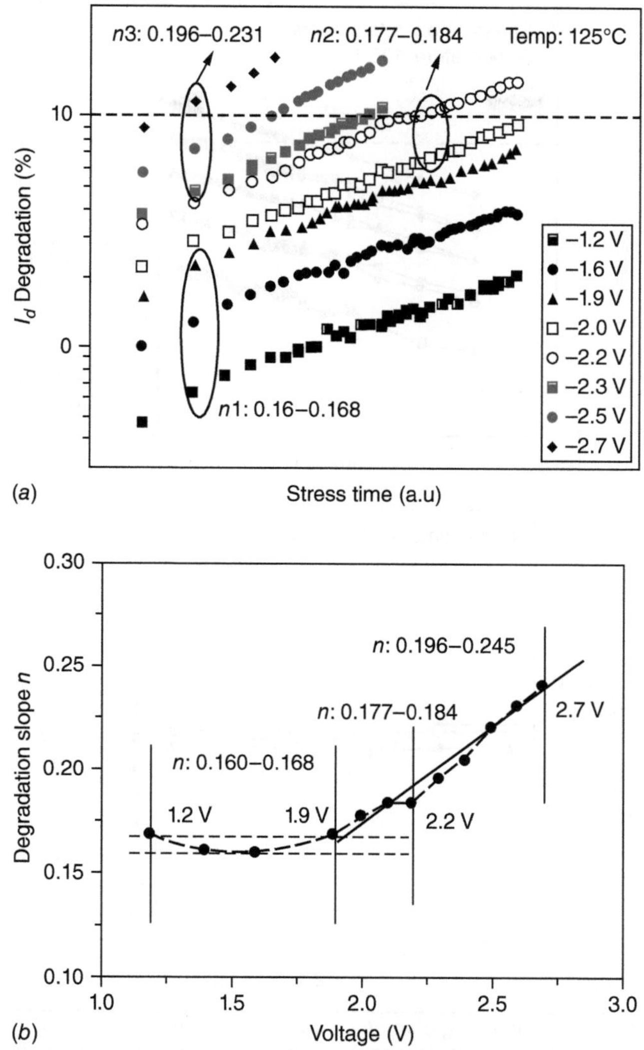

FIGURE 7.11 (*a*) The logarithm of I_d degradation is plotted versus the logarithm of stress time. Degradations show linear dependence with stress time. (*b*) The degradation slope *n* versus stress voltage. The *n* stays around 0.16 to 0.17 from 1.2 to 1.9 V but increases with increasing stress voltage from 1.9 to 2.7 V. (*Used with permission from IEEE.*[25])

the more stable molecular hydrogen owing to the lower energy provided. When the stress voltage is increased, it provides higher energy to form atomic hydrogen. The percentage of atomic hydrogen diffusion increases with increases in stress voltage, which results in power-law characteristics in high-field regions.

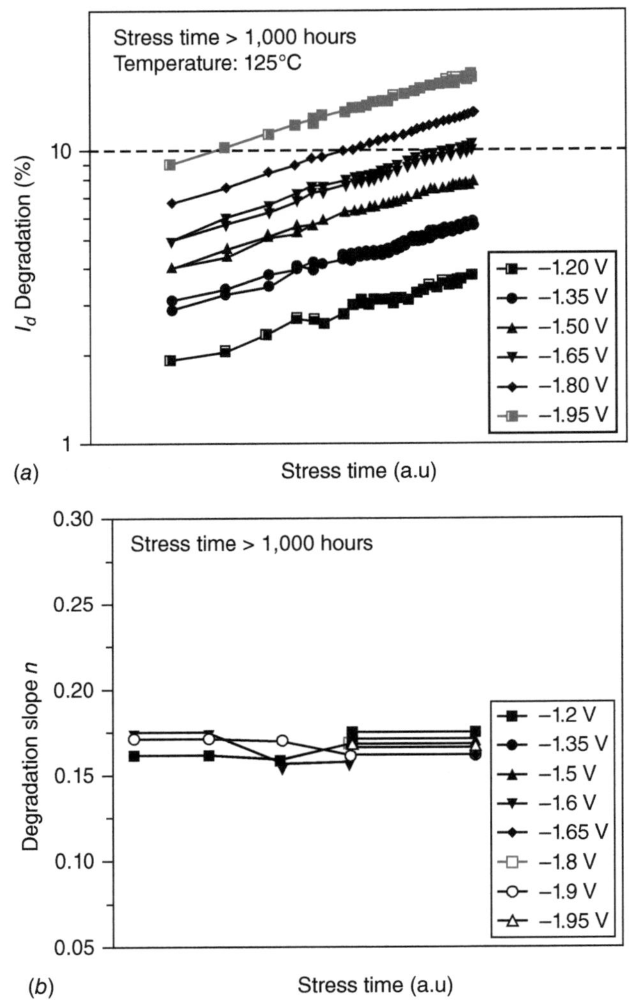

FIGURE 7.12 (*a*) The I_d degradation characteristics. The logarithm of degradations shows linear dependence with the logarithm of stress time, even at stressing up to 1,000 hours. (*b*) The degradation slope *n* versus stress time. The *n* does not decrease with increasing stress time. (*Used with permission from IEEE.*[25])

7.3.3 Uninterrupted-Stress Methods

Owing to the recovery effect, there are many efforts to measure NBTI degradation using no-delay on-the-fly (OTF) I_{dlin}[28,60,65] or minimal-delay fast *I-V* techniques[66,67] for accurate determination of device degradation. These techniques are summarized in this section.

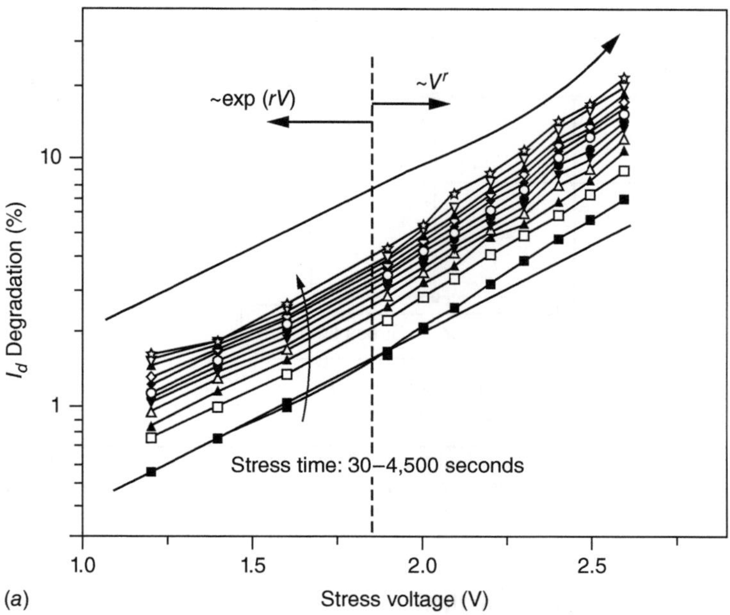

(a)

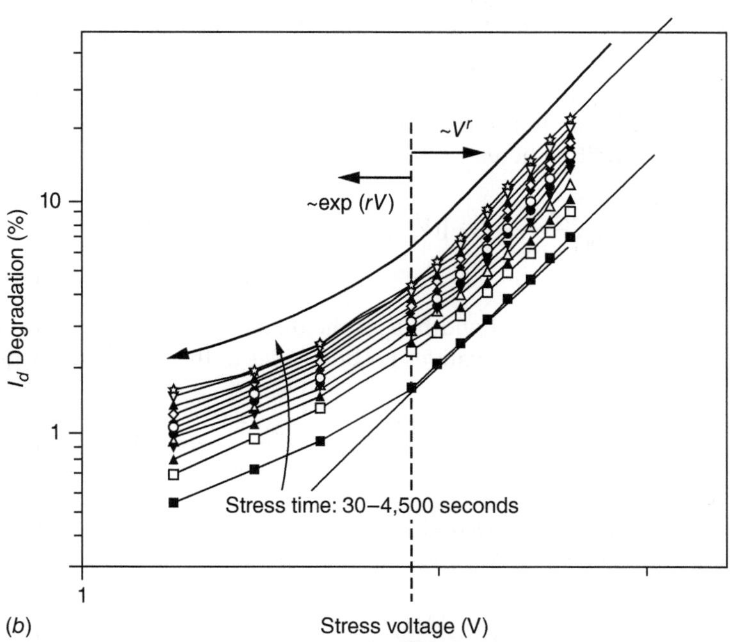

(b)

FIGURE 7.13 (*a*) Degradations versus stress voltage for various stress times. Degradations show *E*-field dependence with stress voltage in low-field regions (1.2 to 1.9 V). (*b*) Degradations versus stress voltage for various stress times. Degradations show power-law dependence with stress voltage in high-field regions (1.9 to 2.7 V). (*Used with permission from IEEE.*[25])

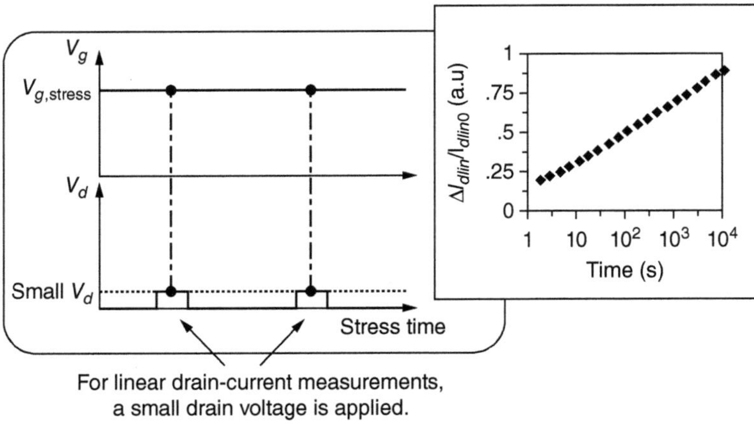

FIGURE 7.14 Schematic of first-level OTF measurement methodology. Linear drain-current shifts can be monitored, with typical evolution, as shown in the inset. (*Used with permission from Elsevier.*[19])

First-Level On-the-Fly Method

In the first-level OTF method (Fig. 7.14), the linear drain current I_{dlin} at preset intervals under $V_g = V_{g,\text{stress}}$ and V_d at a low bias (e.g., a typical $V_d = 50$ mV) is monitored. The corresponding V_t shift can be extracted from the I_{dlin} shift tested at $V_{g,\text{stress}}$ as explained below. The starting point is the linear drain current at a specific V_g, as described in Eq. (2.45), which is rewritten here:

$$I_{dlin} \approx \frac{Z}{L} \cdot \frac{\mu_0 C_{ox}}{1+\theta(V_{g,\text{stress}} - V_t)} \cdot [(V_{g,\text{stress}} - V_t)V_d], \qquad V_d << (V_{g,\text{stress}} - V_t) \tag{7.14}$$

where Z is the channel width, μ_0 is the bulk mobility, and θ is a constant. When taking derivatives for both sides of the equation, we get:

$$\frac{\Delta I_{dlin}}{I_{dlin}} = \frac{-\Delta V_t}{V_{g,\text{stress}} - V_t} + \frac{\Delta \mu_0}{\mu_0} + \frac{V_{g,\text{stress}} - V_t}{[1+\theta(V_{g,\text{stress}} - V_t)]} \cdot \Delta\theta \tag{7.15}$$

When neglecting the second and third terms in Eq. (7.15), a linear relationship between ΔV_t and ΔI_{dlin} is obtained:

$$\Delta V_t = \left|\frac{\Delta I_{dlin}}{I_{dlin}}\right| \cdot (V_{g,\text{stress}} - V_t) \tag{7.16}$$

Second-Level On-the-Fly Method

In the second-level OTF method (Fig. 7.15), the linear drain current I_{dlin} is monitored not only at preset intervals under $V_g = V_{g,\text{stress}}$ and V_d

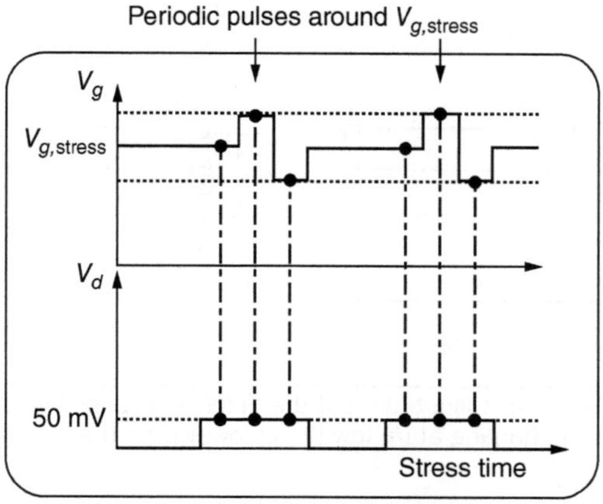

Figure 7.15 Schematic of OTF measurement method. (*Used with permission from Elsevier.*[19])

at a low bias but also at $V_{g,\text{stress}} + \Delta V$ and at $V_{g,\text{stress}} - \Delta V$, where ΔV is a small amount perturbation to $V_{g,\text{stress}}$. The G_m at a time point i can be estimated as follows:

$$G_m(i) = \frac{\partial I_{dlin}}{\partial V_g}\Big|_i = \frac{\Delta I_{dlin}}{\Delta V_g}\Big|_i \approx \frac{I_{dlin}(V_{g,\text{stress}} + \Delta V) - I_{dlin}(V_{g,\text{stress}} - \Delta V)}{2 \cdot \Delta V}\Big|_i \tag{7.17}$$

In general, it is the V_t shift rather than the I_{dlin} shift that is used for design purposes. V_t shift under a constant gate stress voltage at a time point n is approximated/derived as follows:

$$\Delta V_t(n) = \sum_{i=1}^{n} \Delta V_t(i) = -\sum_{i=1}^{n} \frac{I_{dlin}(i) - I_{dlin}(i-1)}{[G_m(i) + G_m(i-1)]/2} \tag{7.18}$$

It was shown that the pulse height does not have a clear impact on the V_t shift observed for ΔV varying from 50 to 250 mV.[19]

The OTF method has attracted much attention for NBTI characterization to exclude the recovery effect. This method is easy to implement using customary parameter analyzers. Some major equipment manufacturers have provided OTF measurement capability.[68,69] However, there are still a few shortcomings to this method that limit its practical application.

- The OTF method determines the electrical parameters at or very close to the stress level $V_{g,\text{stress}}$, which is much larger than the sensing V_g in the conventional method. In the conventional method, I_{dlin} is measured with $V_g = V_{op}$ (operating voltage), and V_t is measured by extrapolating I_d-V_g with the sensing V_g close to V_t. The sensing V_g has little effect on the relative recovery.[70] However, *a larger sensing* $|V_{g,\text{stress}}|$ *will lead to a*

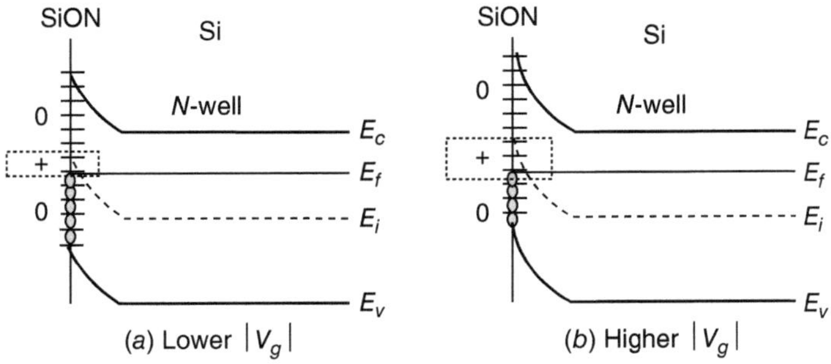

FIGURE 7.16 Schematic illustration of the defects responsible for the sensing V_g effect. Positive charging at (a) low $|V_g|$ is lower than that at (b) high $|V_g|$.

larger measured $|\Delta I_{dlin}|$[71] *and* $|\Delta V_t|$,[70] which can be explained by the higher-charge trap Q sensed at higher $|V_g|$ [Eq. (7.6)], as illustrated schematically in Fig. 7.16.

- In the high-V_g regime (due to $V_{g,\text{stress}}$), the MOSFET is rather insensitive to parameter changes. As an example for stress gate voltage around 2.5 V, a 1-mV gate-voltage change causes a $\Delta I_{dlin}/I_{dlin}$ of only 1×10^{-4}. In other words, this means that an uncertainty in the measured I_{dlin} of 1 percent corresponds to an error in ΔV_t of 100 mV.[71]
- Another problem is that owing to the relatively slow measurement time of commercial quasi-dc parameter analyzers (typically 20 to 150 ms),[70] the first measurement of I_{dlin}—supposed to be unstressed—is done at the stress level (i.e., $V_{g,\text{stress}}$) after a stress of 20 to 150 ms. Therefore, the measured G_m [Eq. (7.17)] and ΔV_t [Eq. (7.18)], also will be underestimated.

Fast-V_t Method

The *fast-V_t method* interrupts stress and then does a very fast measurement and recording of V_t versus recovery time (at a given drain current and voltage). A detailed measurement setup for the fast-V_t method is given by Shen et al.,[66] where V_t is measured using the falling or rising edge of a pulse. In this method, the I_d-V_g is measured during the pulse edge; thus the edge time is the measurement time in the microsecond range. Measurement delays of 0.5 μs have been achieved.[71] Shen et al.[66] observed that as measurement time is increased above 100 μs, the measured V_t-shift of NMOS devices with HfO_2 gate dielectrics after PBTI stressing starts to drop owing to the significant detrapping during the measurement. That is, the recovery effect takes effect at about 100 μs. When the measurement time was increased to several seconds for the conventional quasi-dc technique, ΔV_t was reduced by a factor ~5. However, the shortcoming of this method is that it cannot be performed with customary equipment; specialized measuring equipment is needed.[72]

7.3.4 Influence of Bulk Bias on Negative-Bias Temperature Instability

In the preceding discussion on NBTI stressing, the bulk is grounded. However, it has been found that a bulk bias V_b during NBTI stressing does affect the degradation. However, the V_b effects on NBTI degradation reported in the literature are controversial[74]:

- Mitani et al.[52] and Hurad et al.[47] found that NBTI under an identical V_g stress is independent of the positive bulk bias up to 3 or 4 V.
- Kimizuka et al.[38] observed that the NBTI lifetime begins to decrease significantly at $V_b > {\sim}1.5$ V.
- Mahapatra et al.[54] reported that NBTI at high V_b (≥4 V) shows an identical degradation rate for short stress times, whereas the degradation rate increases dramatically at longer stress times.
- Zhu et al.[74] found that NBTI can be enhanced significantly by applying a positive bulk bias, which depends on both V_g and V_b; that is, a higher $|V_g|$ requires a lower V_b to trigger the enhancement (Fig. 7.17). At low $|V_g|$ (i. e., 1.5 V), the device degradation is almost independent of V_b (up to 2 V). At high $|V_g|$ (≥2.0 V), the degradation is enhanced significantly and has a large degradation rate.

The V_b effects on NBTI degradation generally are attributed to V_b-induced modification of the channel-hole (*so-called cold-hole*) population under inversion,[38,47,52,54,73,74] which is further associated with the dissociation of the Si−H bonds at the SiON-Si interface under NBTI stress.[18,33,34] Note that even though the V_b is varied, the oxide electrical field (E_{ox}) is kept the same (given that V_g is the same) because the inversion channel has the same potential as the grounded source/drain electrodes during NBTI stressing. *However, the cold-hole population in the inversion channel is proportional to* $|V_g - V_t|$ [see Eq. (2.43) in Chap. 2]:

$$|Q_n| = C_{ox}\,|V_g - V_t| \tag{7.19}$$

where $|Q_n|$ is the carrier population in the inversion channel, and C_{ox} is the oxide capacitance. By applying a positive bulk bias V_b, the initial threshold voltage is modified owing to the body effect [see Eq. (2.62) in Chap. 2]. For example, a device with $W/L = 10\ \mu m/0.32\ \mu m$ has a V_t of −0.365, −0.597, and −0.726 V at V_b = 0, 1, and 2 V, respectively.[74] Therefore, *the cold-hole population in the channel decreases with increasing*

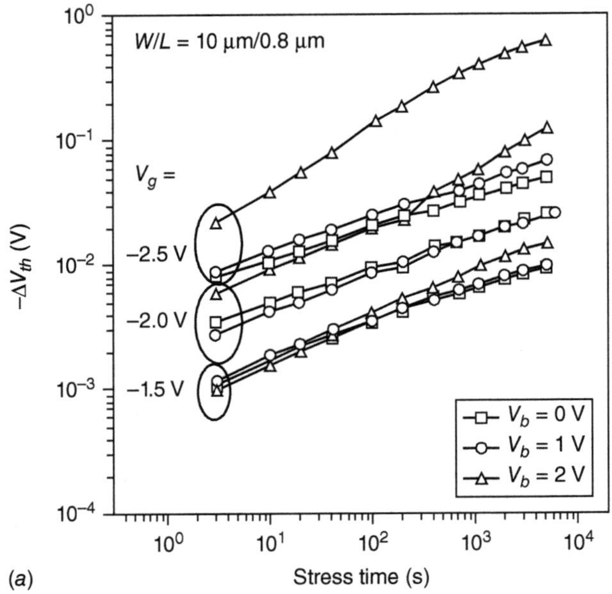

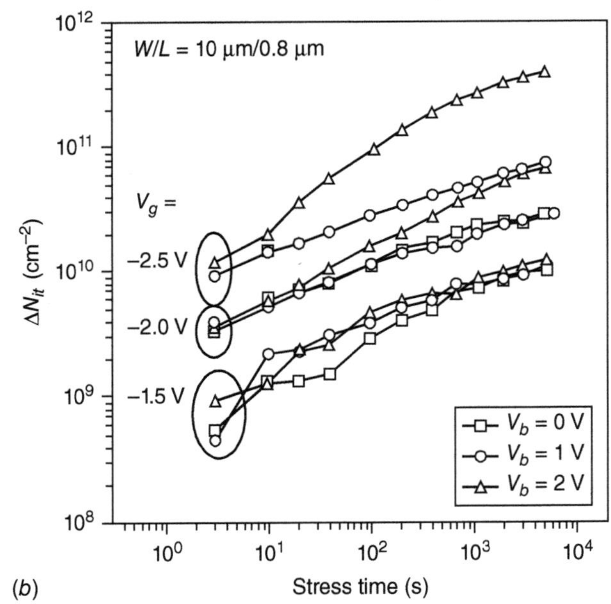

Figure 7.17 (*a*) The threshold voltage shift ΔV_t extracted from the I_d-V_g curves and (*b*) the corresponding interface trap generation (ΔN_{it}) deduced from the modified DCIV curves as a function of stress time under static stress with various V_g and V_b values. PMOS devices with $W/L = 10$ μm/0.8 μm were stressed and measured at 125°C. (*Used with permission from the American Institute of Physics.*[74])

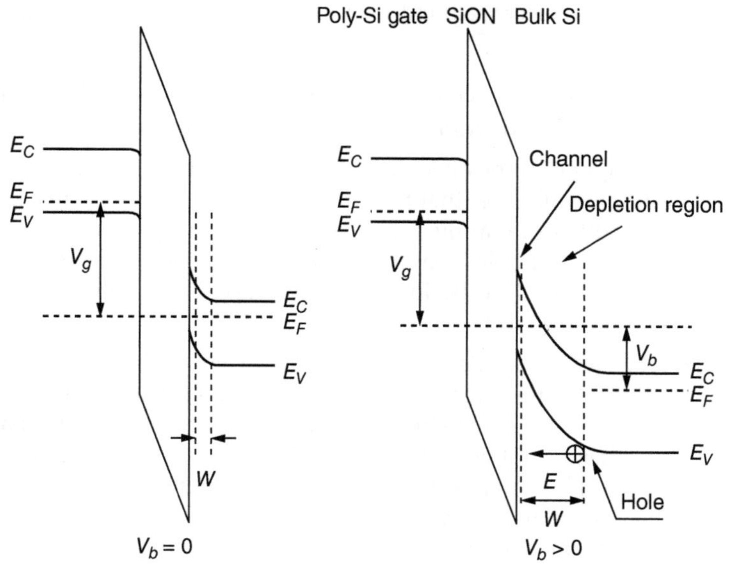

FIGURE 7.18 Energy-band diagrams for a p+ poly-Si gate PMOS under NBT stress with $V_b = 0$ (left) and $V_b > 0$ (*right*). Both the depletion width W and the built-in electrical field E are below the channel-layer increase with increasing V_b. (*Used with permission from the American Institute of Physics.*[74])

V_b at the identical V_g stress. This results in the expection that *a larger V_b will lead to less NBTI degradation if only such cold holes are considered.*

On the other hand, *the V_b-induced NBTI degradation enhancement also has been attributed to the substrate hot-hole injection.*[54,73,74] The energy-band diagrams for a PMOS device under NBT stress with $V_b = 0$ and $V_b > 0$ are schematically shown in Fig. 7.18. The width W of the induced depletion region between the channel and the bulk and the maximum electrical field E_{max} in this region are both proportional to $\sqrt{V_b}$ approximately.[75] Holes from the bulk to the channel can be accelerated when this electrical field is "hot." Thus *the number of hot holes in the channel, as well as their kinetic energy, increases with increasing V_b although the population of the cold inversion holes is reduced.* It is reasonable to assume that only the holes that can tunnel through the interface layer of 1 to 2 Å to be captured by the Si−H bonds can contribute to dissociation of the Si−H bonds.[18] In this way, hot holes are believed to be more effective in inducing degradation than the normal cold-channel holes because hot holes have much higher kinetic momentum normal to the gate oxide.

In short, *with increasing V_b, the degradation induced by the normal cold holes decreases owing to the decrease in channel hole density, whereas additional degradation induced by hot-hole injection increases.* This is consistent with the very small activation energy E_a (0.009 eV) for $V_b = 2$ V compared with $E_a = 0.197$ eV when $V_b = 0$ V because the hot holes energized by the electrical field lead to a tunneling effect depending weakly on temperature.[74]

- At low V_b, the two opposite effects from cold and hot holes make device degradation appear almost independent of V_b.
- At higher V_b, the effect from hot holes increases more quickly and prevails over that from the cold holes, sresulting in a significant degradation enhancement at high V_b. Moreover, the injected hot holes may break the Si–O bonds at or near the SiO_2-Si interface besides the Si–H bonds, as suggested by Varghese et al.[73]
- The field-dependent tunneling coefficient of holes from the channel to the interface layer depends almost exponentially on the local electrical field at the interface. In other words, the barrier height of bond dissociation decreases with increasing E_{ox} (i.e., V_g).[18] Therefore, the V_b effect is more significant at high $|V_g|$ than that at low $|V_g|$, and the minimum V_b to trigger the enhancement of degradation decreases with increasing $|V_g|$.

7.4 Why Is PMOS under Inversion the Worst?

NBTI is observed in both PMOS and NMOS devices, but the effect on PMOS devices is much worse, as shown in Fig. 7.19. Several models explain this observation, one of which is easily understood and is detailed below.[23] Under strong inversion, the V_t shift depends on oxide traps and interface traps as follows:

$$\Delta V_t = -\frac{\Delta Q_{ot}}{C_{ox}} - \frac{\Delta Q_{it}(\psi_s)}{C_{ox}} \tag{7.20}$$

where

$$\Delta Q_{it}(\psi_s) = \pm q\Delta D_{it}(\psi_s - \psi_b) \tag{7.21}$$

Under strong inversion, that is, $\psi_s = 2\psi_b$, we get:

$$\Delta Q_{it} = \pm q\Delta D_{it}\psi_b \tag{7.22}$$

For oxide charges ΔQ_{ox}, $Q_{ox} > 0$ for both NMOS and PMOS devices. However, there are different situations for interface charges (ΔQ_{it}) in NMOS and PMOS devices under strong inversion, according to Fig. 2.22 in Chap. 2: $\Delta Q_{it} < 0$ for NMOS devices, and $\Delta Q_{it} > 0$ for PMOS devices. Therefore, at inversion,

- PMOS: $|\Delta Q_{ox}| + |\Delta Q_{it}|$
- NMOS: $|\Delta Q_{ox}| - |\Delta Q_{it}|$

It is clear that PMOS devices suffer from a higher number of BTI-induced traps because the oxide traps and interface traps have the same sign.

The experimental interface traps (ΔD_{it}) and oxide traps (ΔN_{ot}) in terms of stress time for NMOS NBTI/PBTI and PMOS NBTI/PBTI

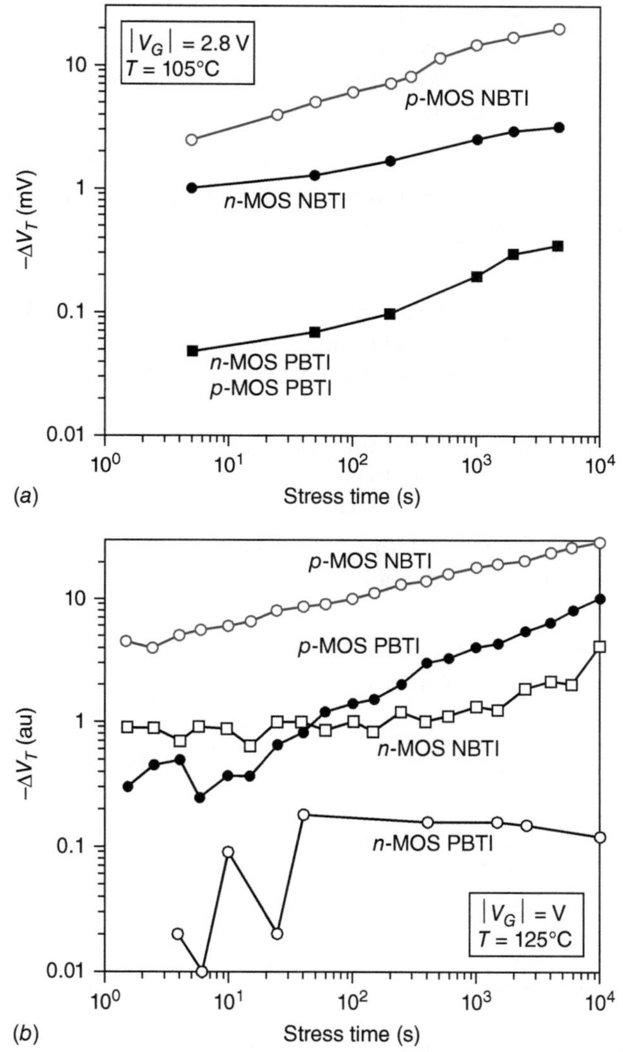

FIGURE 7.19 BTI induced V_t shift from two different research groups when PMOS and NMOS devices are biased with $\pm V_g$. Clearly, PMOS devices under negative gate bias are the most severely affected. (*Used with permission from Elsevier.*[23])

were reported by Denais et al.[76] The observations can be summarized as follows:

- Interface traps:
  - Amount almost the same for PMOS NBTI, NMOS NBTI, and PMOS PBTI, but much less for NMOS PBTI.
  - Correlated with the interaction of the holes and Si−H bonds at the Si-SiO_2 interface. Note that they are positive and negative for PMOS and NMOS devices, respectively.

- Oxide traps (correlated with hole trapping):
  - Positive for PMOS NBTI (the highest trap amount), NMOS NBTI, and PMOS PBTI, but negative for NMOS PBTI (also the smallest trap amount).

These observations can be explained qualitatively by the hole injection from the channel using Fig. 7.20, which gives energy-band diagrams for NMOS and PMOS devices under inversion and accumulation modes:

- For PMOS NBTI (inversion mode; Fig. 7.20*d*) and NMOS NBTI (accumulation mode; Fig. 7.20*a*), holes from valence band (HVB) injected from channel to gate, involving in the interface trap generation (i.e., the Si−H bond dissociation).
- For PMOS PBTI (accumulation mode; Fig. 7.20*c*), HVB holes injected from gate to channel, so the interface trap is something lower than P/NMOS NBTI.
- For NMOS PBTI (inversion mode; Fig. 7.20*b*), no hole is involved. So both interface traps and oxide traps are the lowest.

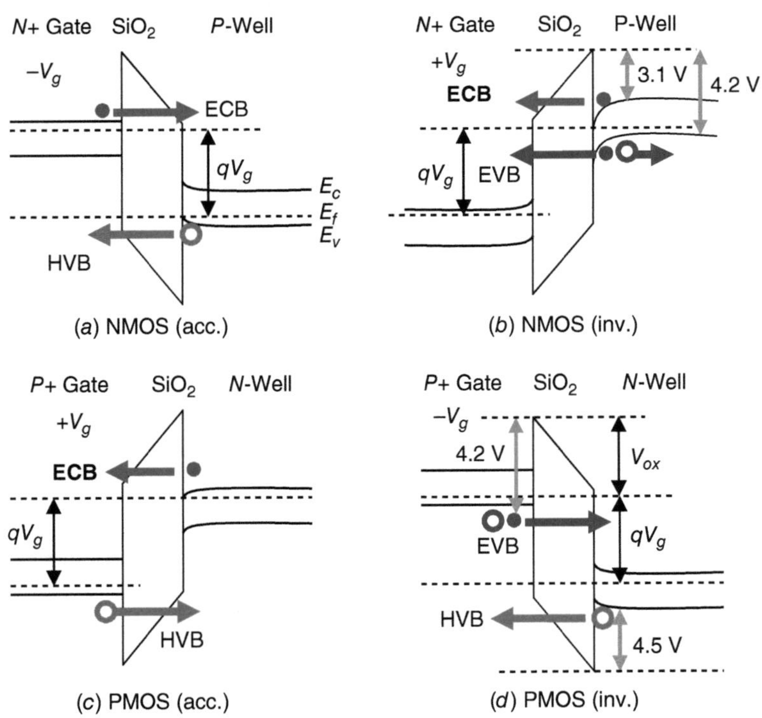

FIGURE 7.20 Energy-band diagrams of NMOS and PMOS devices under inversion and accumulation modes (a replot of Fig. 6.4).

7.5 Effect of Structure on Negative-Bias Temperature Instability

Every VLSI product has some distributions of MOSFET design with respect to channel lengths and widths. For reliability evaluation, the most vulnerable design is to be chosen. Therefore, there is a need to investigate the layout dependence of NBTI for PMOS devices. However, the trends for the length and width dependence reported in the literatures are quite controversially, indicating that such a structure dependence of NBTI is both layout- and process-dependent. A uniform trend cannot be achieved if the manufacturing and layout environments are not the same, which should be dealt with very carefully in the modeling purpose.

7.5.1 Channel Length Dependence

NBTI degradation has a weak dependence on channel length.[77] However, some studies show that very short-channel devices are more sensitive to NBTI.[56,78–80] NBTI originates from both positive charge trapping in the channel and interface traps in the drain/source overlap region,[81] which is the reason that NBTI is sensitive to channel length. A summary of the controversial channel-length-dependent NBTI is given below.

- Terai et al.[105] reported that *for the case without F implantation, the V_t shift decreased when the gate length was increased from 0.2 to 1 μm. However, for the case with F implantation* (dose: 1E15 cm^{-2}), *the V_t shift does not depend much on gate length,* as shown schematically in Fig. 7.21. These results indicate that the fluorine atoms that are implanted into the source/drain regions (see

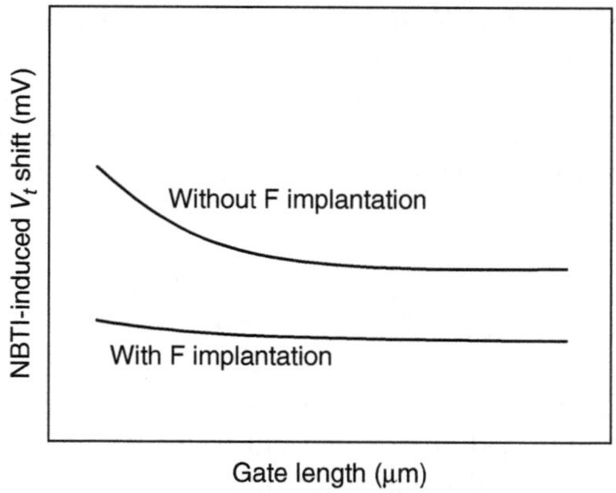

FIGURE 7.21 Schematic of gate-length dependence of ΔV_t for samples with and without fluorine implantation.

Fig. 7.26) diffuse sufficiently toward the center of the long gate electrode (e.g., 1 μm). In addition, it is shown that the V_t shift of the fluorine-implanted sample is smaller (at least 50 percent less) than that of the nonimplanted sample.

The linear V_t shift dependence on gate length without F implantation in Fig. 7.21 can be explained by a model considering two contributions, including interface states (N_{it}) in the drain/source extensions under the gate and positive charge trapping (N_{ot}) in the channel region.[56] N_{it} depends on $2 * L_{S/D}$, where $L_{S/D}$ is the size of the source/drain extension; and N_{ot} depends on $(L - 2 * L_{S/D})$. $L_{S/D}$ should not depend on L because all devices feature exactly the same drain and source engineering. Therefore, a linear-dependent model is proposed as follows[56]:

$$\Delta V_t = \alpha(L - 2L_{S/D}) + \beta(2L_{S/D}) \tag{7.23}$$

where α and β are constants that depend on stress conditions, device technology, etc.

- Math et al.[82] also reported *length independent V_t shift* of NBTI for a 2.3-nm-thick gate oxide, but they did not mention whether the device was implanted with F or not. The four devices under study had the same channel width of 10 μm, but a series of channel lengths: 0.09, 0.12, 0.15, and 10 μm. However, *longer transistors degrade faster in terms of I_{dsat} shift,* meaning that the lifetime based on I_{dsat} shift will decrease with longer gate length. The authors related this observation to the part of defects that are active in saturation. Only defects located in the inverted zone of the channel have an impact on I_{dsat} during measurement. The authors argued that longer transistors present a more prominent inverted zone in saturation and so have more active defects and consequently degrade faster.[82] A more straightforward explanation is given as follows: It is well known that a short-channel device has a smaller V_t owing to short-channel effect (see Fig. 2.15). According to Eq. (2.50) in Chap. 2, which is rewritten here, a smaller V_t for a short-channel-device will lead to a smaller I_{dsat} shift (i.e., $\Delta I_{dsat}/I_{dsat}$):

$$\frac{\Delta I_{dsat}}{I_{Idsat}} = \left[1 + \frac{1}{1 + \theta(V_g - V_t)}\right] \frac{\Delta V_t}{V_g - V_t} \tag{7.24}$$

- Similar to Math et al.,[82] Mooraka et al.[83] also showed that the lifetime (based on a 10 percent I_{dsat} shift) increases when the gate length decreases for low-voltage 65-nm-technology PMOS devices. The compared devices all have the same channel width of 10 μm. However, the authors reported an abrupt increase in lifetime when the channel length was less than 0.1 μm. Such a severe channel-length dependence was attributed to the fact that the greater interface charge needed for short-channel devices to make the same shift in V_t compared with longer-channel devices occurs because of a dominant halo implant. It also was assumed that fluorine passivation to the interface traps has a larger impact on reduced-channel-length devices.

- Cellere et al.[56] reported that V_t *shift decreases when channel length is increased from 0.16 to about 0.5 μm. However, when channel length becomes larger than 0.5μm, NBTI degradation increases again,* which stimulates an empirical double-exponential relationship as follows:

$$\Delta V_t = \alpha \cdot \exp[\xi(L - 2L_{S/D})] + \beta \cdot \exp[\zeta(2L_{S/D})] \qquad (7.25)$$

However, the physical nature of the double-exponential relationship is not well understood.

Another independent work shows the similar length-dependent NBTI degradation: large degradation for the smallest and largest transistors and small degradation for median-sized transistor.[84] The authors reported that the largest transistor with $W/L = 9/9$ μm is most severely degraded. For the series of transistors with the same channel length of 0.4 μm and different channel widths, the least degradation is observed for $W/L = 4/0.4$-μm devices, followed by $W/L = 9/0.4$-μm devices and then the larger degradation for the smallest device of $W/L = 0.42/0.4$-μm.

- *The channel-length-dependent NBTI performance is linked to the hydrogen species,* which is based on the negative effect of H on interface quality *(i.e., high H content at the interface during manufacturing is not desirable).*[116] As shown in Fig. 7.22, Rhee et al.[57] reported that V_t shift strongly depends on gate-length scaling when using SiN CESL with high-hydrogen (HH) contents, and further degradation is observed for highly scaled gate length with compressive HH CESL.

- On the other hand, Ho et al.[85] observed *a significant* $|\Delta V_t|$ *increase for heavily nitrided gate PMOS devices when* $L_g < 100$ *nm. The* $|\Delta V_t|$ *"roll-up" is reduced in lightly nitrided gate devices and is absent in* SiO_2 *gate devices, indicating that it is linked to the presence of nitrogen in the gate dielectric,* as shown schematically in Fig. 7.23. In addition, the authors showed that the V_t shift of the high-N-content sample is larger than (at least 50 percent higher) that of the low-N-content sample. Electron energy-loss spectroscopy (EELS) of the gate dielectric reveals a stronger nitrogen K-edge signal (~401 eV) at the source and drain

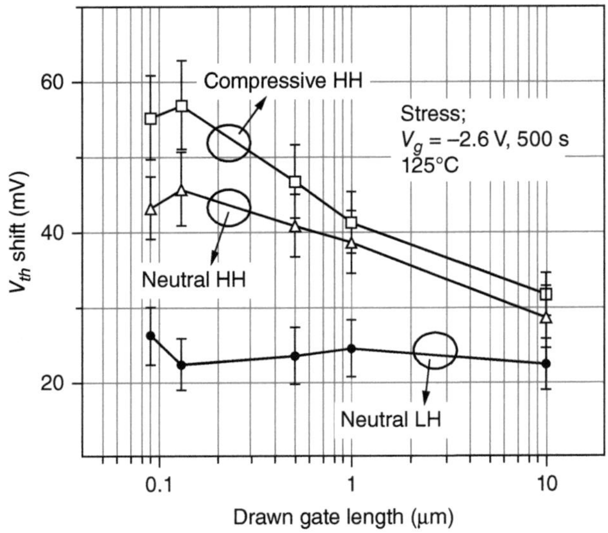

FIGURE 7.22 Gate-length-dependent V_{th} shift for various CESLs under NBTI stress. "Neutral LH" and "Neutral HH" mean low stress (~200 MPa) of the CESL with low and high hydrogen concentrations, respectively. "Compressive HH" means high compressive nitride film (> 1 GPa) containing a high hydrogen concentration. (*Used with permission from IEEE.*[57])

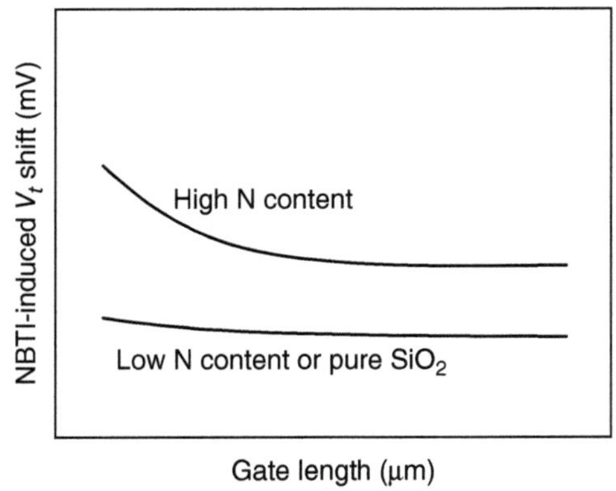

FIGURE 7.23 Schematic of gate-length dependence of ΔV_t for samples with high and low N content.

extensions (SDEs) compared with that in the middle of the channel. The EELS data imply that $|\Delta V_t|$ roll-up in scaled nitrided-gate PMOS devices is a consequence of *locally higher nitrogen concentration near the SDEs.* When the gate length is reduced, these regions of locally high nitrogen concentration overlap, thereby increasing the overall nitrogen concentration in the gate oxide.

7.5.2 Channel Width Dependence

Contradictory channel width dependent–NBTI behaviors are also reported in literatures, as detailed below:

- Mooraka et al.[83] showed that PMOS NBTI lifetime is almost invariant with respect to changes in channel width ranging from 0.2 to 10 μm, where channel length is kept at 10 μm.
- Cellere et al.[56] reported that devices with narrower channels are less sensitive to NBTI (width ranges from 1 to 30 μm), which could be attributed to the different trapping properties for electrons and holes along the channel. It was demonstrated that after channel hot-electron stress, along the channel-width direction, electron trapping in the gate oxide prevails near the STI edge, whereas hole trapping prevails in the middle of the channel.[86] Cellere et al.[56] thus linked the NBTI to holes, not to electron trapping, which explains why wider devices are more prone to NBTI.
- On the contrary, numerical simulations based on the R-D framework considering the geometric dependence of transistor structure and scaling showed that *NBTI-induced interface-trap density increases when the width is scaled down.*[87] The reason for this is that the narrow channel width enhances the diffusion of hydrogen released from the Si−H bonds at the Si-oxide interface. The NBTI degradation time exponent n can range from ¼ to ½ for a narrow-width MOSFET, which is larger than the traditional NBTI exponent of ¼. This trend of *worse NBTI performance for narrower devices* is supported by Math et al.,[82] who reported a lifetime versus channel-width relationship of:

$$TTF = 40{,}650 * \ln(W) + 113{,}768 \tag{7.26}$$

  where TTF is the lifetime in seconds at a V_t shift = 100 mV, W and is the channel width in μm. This relation was derived based on experimental data for channel-width ranges from 0.2 to 10 μm. The narrow device is believed to have higher stress at the active-area perimeter, thus leading to larger NBTI degradation. This observation also may be explained by the much stronger electrical field at the STI corner, as illustrated in Fig. 2.18 (in Chap. 2).

Nitride layers with high amount of hydrogen have strong gate-width dependency (Fig. 7.24*a*) (width ranges from 0.1 to 10 μm),[57] but unlike gate-length dependency, as shown in Fig. 7.22, we cannot see any difference between neutral and compressive mechanical stress. In the active width direction, the stress effect by CESL can be ignored

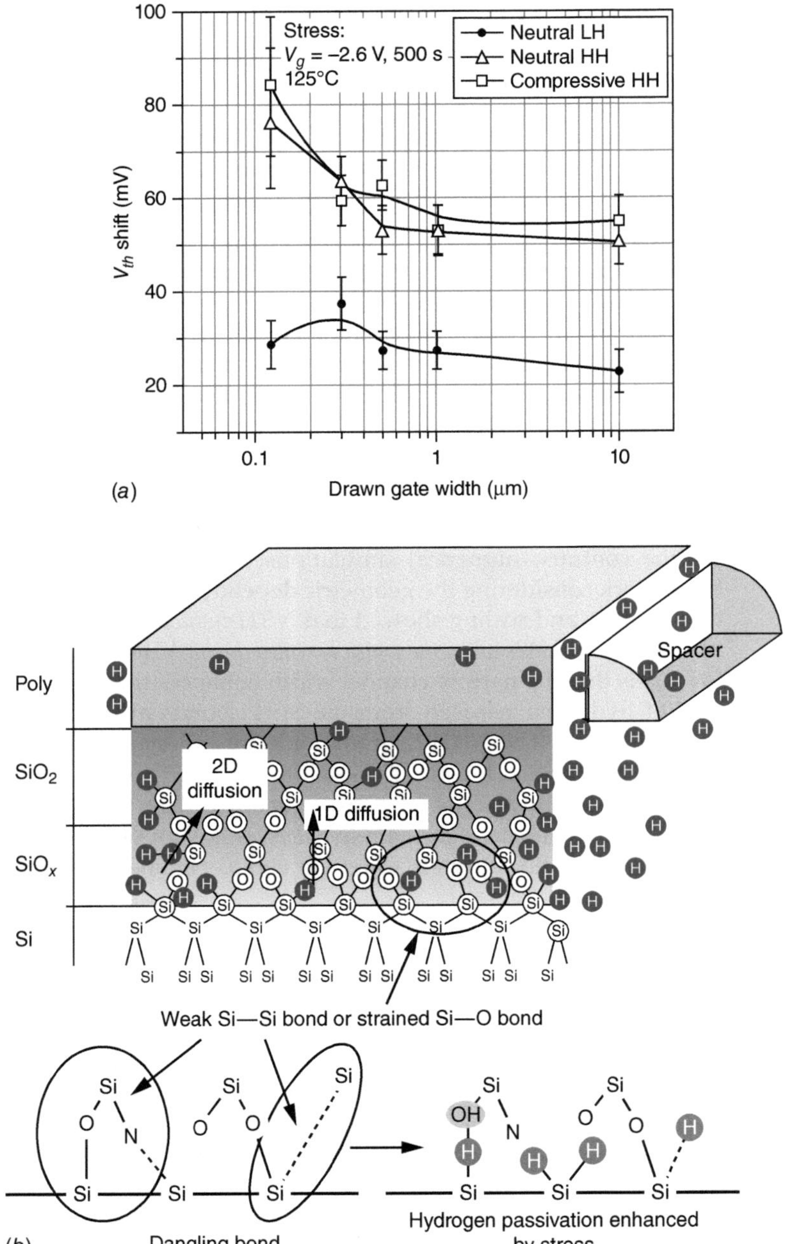

FIGURE 7.24 (a) Gate-width-dependent V_{th} shift for various CESLs under NBTI stress. "Neutral LH" and "neutral HH" mean low stress (~200 MPa) of the CESL with low and high hydrogen concentration, respectively. "Compressive HH" means high compressive nitride film (>1 GPa) containing a high hydrogen concentration. (b) Schematics of geometry effect on NBTI of mechanically stressed CESL with high hydrogen content. (*Used with permission from IEEE.*[57])

because the gate electrode screens the channel region. This geometry dependency of NBTI seems to indicate that increased interface-state passivation (more Si−H bonds) near the edge region of the gate oxide is responsible for enhanced N_{it} generation under given bias condition, resulting in a significant V_t shift in transistors with small active areas, as illustrated in Fig. 7.24*b*. The edge effect at gate spacer and interlayer dielectrics accelerates the deterioration of gate-oxide integrity.[88]

- The edge effect leads to more Si−H passivation bonds, which result in more interface generation during subsequent NBTI stressing.
- The higher is the H concentration in the CESL and the more Si−H passivation bonds are located at the interface, then the higher sensitivity there is for gate-width dependency roll-up at narrower gate width.
- It is the increased mechanical stress[57] and two-dimensional hydrogen diffusion[55] that degrade NBTI reliability more for scaled PMOS devices.

7.5.3 Gate-Oxide Thickness Dependence

When V_t shift is normalized by multiplying it by the capacitance of the gate dielectrics ($C_{ox} = \varepsilon_{ox}/\mathrm{EOT}$) and is plotted against the electrical field of the stress ($E_{\mathrm{stress}} = V_{\mathrm{stress}}/T_{inv}$),[45] it seems to follow a universal line regardless of the gate-dielectric thickness. The normalized ΔV_t represents the summation of positive fixed charges and the total charge of interface traps at a gate voltage of V_t under the assumption that all the charges exist at the bottom interface of the gate dielectric. Moreover, the universal line suggests the relationship between normalized ΔV_t and E_{stress} of Eq. (7.27):

$$\Delta V_t(\text{normalized}) = c \cdot E_{\mathrm{stress}}^m = c \cdot E_{ox}^m \tag{7.27}$$

This is the format of Eq. (7.4). Increasing gate-oxide thickness simply will reduce the stress electrical field E_{ox} and thus reduce the V_t shift. According to Eq. (7.27), by increasing the gate-oxide thickness of ΔT_{inv}, the NBTI-induced V_t shift reduces by a factor ~$[1 + (\Delta T_{inv}/T_{inv})]^m$.[27] This is one of the ways to improve NBTI. However, the side effect of increasing the gate-oxide thickness is the reduction in drive current in the device.

7.6 Effects of Process on Negative-Bias Temperature Instability

NBTI is strongly dependent on process parameters. Thus, NBTI improvement can be achieved through process optimization, as detailed in this section.

7.6.1 Nitrogen and Its Profile

In previous sections it was shown that the introduction of a high concentration of nitrogen atoms into the gate-oxide SiO_2 will increase the NBTI degradation time exponent (Sec. 7.2.1), reduce the NBTI activation energy (Sec. 7.2.2), provide more reaction sites,[39] and thus increase NBTI degradation.[38,89] NBTI stress-induced ΔN_{it} and ΔN_{ot} were reported to increase significantly with increasing nitrogen concentration.[90,91] NBTI depends much on the nitrogen profile in SiON, particularly on the nitrogen concentrations at the SiON-Si substrate interface.[92–97] The NBTI lifetime could be independent of the nitrogen concentration if the peak nitrogen concentration is around the upper half of the dielectric so that the difference in the interfacial nitrogen concentration at the SiON-Si interface is quite small even though the nitrogen concentration is varied.[37] Both ΔV_{ot} and ΔV_{it} were shown to be reduced by controlling the nitrogen profile in SiON.[105] Tables 3.1 and 3.2 in Chap. 3 give a summary of the nitrogen effect on NBTI.

For ultrathin gate dielectrics, rapid thermally nitrided oxides (RTNO) prepared in an NO ambient cannot meet the requirement for suppression of boron penetration and NBTI.[90,98,99] It is also difficult to increase nitrogen concentration in ultrathin base oxide by NO gas annealing because NO gas nitridation will increase the final physical oxide thickness itself. *Higher nitrogen concentration can suppress the boron penetration more effectively but much worsen NBTI performance.* This has led to the application of plasma nitridation, in which nitrogen plasma pressure, plasma immersion time, or plasma generation power are the parameters to be optimized to achieve enough immunity both for boron penetration and for NBTI.[98]

Both N profile and the amount of N implanted into the Si channel were found to depend on the ion kinetic energy.[100] Use of pulsed (turning the source power on and off at kilohertz frequencies) nitrogen plasmas is better to form an abrupt N profile and hence improved NMOS and PMOS device performance and reliability than continuous-wave (nonpulsed) plasmas.[91] An ion striking the surface of the wafer has kinetic energy proportional to the electron temperature (k_BT_e), which describes the Maxwellian distribution of electron energies in the plasma.[101] A reduction in k_BT_e will cause a reduction in the mean energy of the nitrogen ions striking the wafer surface. Pulsed plasmas have been shown to have reduced time-averaged k_gT_e relative to continuous-wave plasmas.[102] It has been found that reducing the density of high-energy electrons in the plasma results in 5 percent improved electron and hole low-field mobilities and 100 percent improved NBTI reliability.[100]

A possible origin of NBTI enhancement by incorporating nitrogen into gate dielectrics is the attraction of H_2O or OH attributed to the existence of positive fixed charges induced by undesirable nitrogen, as demonstrated by thermal desorption spectroscopy (TDS) measurement.[45] H_2O molecules and OH ions promote NBTI.[103,104] Tsujikawa et al.[45] further showed that the heavily nitrided NO-oxynitride and the nonoptimized OI-SiN (silicon nitride gate dielectric with oxygen-enriched interface) have many positive fixed charges even after forming gas anneal (FGA), and serve as sources attracting H_2O or OH and thus worsen NBTI. On the other hand, pure SiO_2 also has many positive fixed charges before FGA, however, that can be annealed away by the FGA, thus giving a better NBTI performance than oxynitride.

Remote Plasma Nitridation

One of the techniques to incorporate nitrogen into the gate oxide is by remote plasma nitridation (RPN).[105] A comparison is given for nitridation by using a radiofrequency (rf) plasma source and an electron-cyclotron-resonance (ECR) plasma source. The study showed the following:

- Nitridation by an ECR plasma source is superior to that by rf plasma source in terms of NBTI lifetime. The SiON dielectric by rf plasma source is thicker with a lower nitrogen concentration than that by ECR plasma source. However, rf plasma nitridation gives a much lower NBTI lifetime. The authors attributed the worse performance by rf plasma source to plasma-induced damage, which correlated with the higher virgin interface-trap density in the rf-nitride sample. The larger plasma damage by the rf plasma source also leads to a higher ratio of oxide traps (ΔV_{ot}) to interface traps (ΔV_{it}) when compared with that by the ECR plasma source.
- Lower N concentration at the SiON–Si-substrate interface gives the lowest NBTI degradation, indicating that it is essential to control the N concentration at the SiON–Si-substrate interface in order to improve NBTI performance of the SiON gate dielectrics. A higher N concentration at the SiON–Si-substrate interface will lead to a larger interface-trap during NBTI stressing.

Terai et al.[105] showed that the peak nitrogen concentration in bulk SiON film can be altered after radical reoxidation. The reoxidation reduced the nitrogen concentration from 17.5 to 9.5 and 4 percent when using middle and high reoxidation conditions, respectively, because nitrogen atoms near the bulk SiON surface were replaced mainly by oxygen owing to reoxidation. However, no clear change in nitrogen concentration at the SiON–Si-substrate interface is attributed to the relatively low temperature for reoxidation using O radicals. As a result, there is no clear difference in the NBTI lifetime for the samples with or without reoxidation, even though the N concentration among the samples varied a few times. Surely, the V_t of the as-prepared samples will increase with less nitrogen incorporated by reoxidation, and it is well known that V_t for the p^+ poly-Si/SiON gate stack becomes more negative with increasing nitrogen concentration.[106] This indicates that the nitrogen atoms near the top interface cause positive charges in the as-fabricated samples, but they do not cause NBTI.

Decoupled Plasma Nitridation

Presently it is a common practice to introduce the nitrogen into the gate oxide by decoupled plasma nitridation (DPN). The plasma could be sustained by continuous-wave (CW) mode or pulsed mode.[91] Figure 7.25*a* shows the I_{dsat} shift comparison after 10 ks of NBTI stressing at $V_g = -2$ V and 125°C between the CW and pulsed DPN. It is clear that the pulsed DPN gives a smaller I_{dsat} shift than the CW DPN. In general, this is believed to relate to the nitrogen peak concentration in the gate oxide. The principle for keeping the injected nitrogen located at the SiON surface by using the pulse plasma can be explained as follows[107]: In a CW plasma nitridation process, the nitrogen is injected continuously into the SiO_2 film and diffuses simultaneously into the gate oxide because of surface activation (an increase in surface temperature) caused by the energetic plasma-beam bombardment. In contrast, in the DPN nitridation process, ionized nitrogen can be injected into the SiO_2 film only during the pulse-on time owing to its higher energy. The subsequent diffusion of injected N_2 into the SiO_2 film can be suppressed during the pulse-off time because at that time the film surface is not activated. A periodic insertion of a few tens of microseconds of pulse-off time can effectively block the diffusion, thus achieving a desirable nitrogen profile in the gate oxide away from the SiO_2-Si interface. Figure 7.25*b* provides a schematic comparison of the nitrogen profile within the SiON oxide between the CW and pulsed DPN. A shallower nitrogen profile in the SiON film closer to the poly-SiON interface is required to improve NBTI lifetime.

Silicon Nitride Gate Dielectrics with Oxygen-Enriched Interface (OI-SiN)

The OI-SiN gate dielectric was developed as a technique to suppress both gate leakage current and boron penetration while keeping carrier mobility high.[108] This high mobility was obtained by eliminating positive fixed charges. The procedure to form the OI-SiN gate dielectrics proposed by Tsujikawa et al.[108] is described as follows: (1) forming an SiN layer on a Si substrate and (2) incorporating oxygen into the SiN layer by N_2O annealing. This direction of reducing the positive fixed charges was found also to be effective to achieve NBTI immunity.[45] Although this method is not employed in the industry,

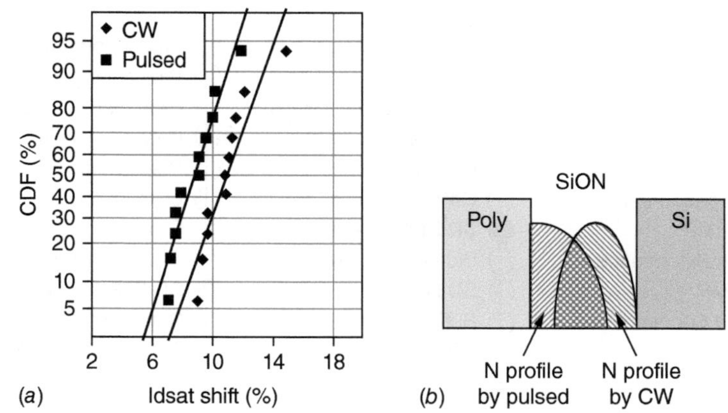

FIGURE 7. 25 (*a*) I_{dsat} shift comparison after 10 ks of NBTI stressing at V_g = −2 V and 125°C between the CW and pulsed DPN. (*b*) A schematic comparison of the nitrogen profile within the SiON oxide between the CW and pulsed DPN.

it clearly indicates that lower nitrogen and higher oxygen concentrations at the interface will benefit NBTI performance.

Nitrogen Implantation the Source/Drain

Lee et al.[109] reported that nitrogen implantation in the channel or source/drain extension can result in serious NBTI degradation. In their study, the nitrogen-implantation energy is 10 keV with different doses (5×10^{13} versus. 5×10^{14} cm^{-2}). A higher nitrogen-implantation dose enhances NBTI degradation. This can be attributed to the locally enhanced degradation reactions between holes and oxide near the S/D, and the nitrogen may diffuse to the interface after the subsequent rapid thermal anneal (RTA) process or lateral scattering of ion implantation.

7.6.2 Incorporation of Fluorine

Fluorine incorporation in the oxides reduces NBTI.[27,52,110,111] Fluorine can be introduced by using BF_2 or F_2 as the PMOS poly predoping, source/drain implant or P-type lightly doped drain (PLDD) implant.[27,91,112] Fluorine implantation with at higher energy but lower dose may give a comparable NBTI lifetime to that case with lower energy but higher dose. The effectiveness of F implantation on NBTI improvement can be attributed to the stronger binding energy of Si–F bonds (540 kJ/mol) formed at the SiON–Si-substrate interface than that of conventional Si–H bonds (298 kJ/mol).[113] *Higher energy combined with lower dose is effective in improving NBTI performance because F atoms with higher energy will have a better chance to penetrate into the poly gate and arrive at the gate oxide.* With further activation annealing, the F atoms thus will have a higher probability to get closer to the oxide-Si interface to passivate the interface with the stronger Si–F bonds.

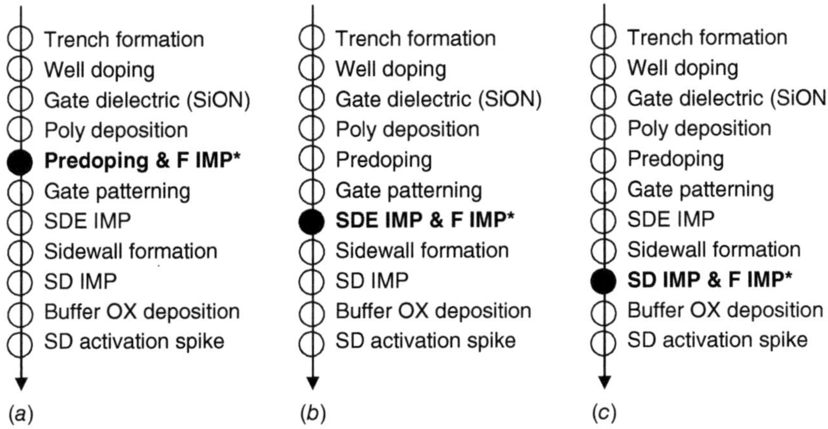

FIGURE 7.26 Three process flows for fluorine incorporation. (*Reproduced with permission from the Electrochemical Society.*[112])

Figure 7.26 schematically shows three process flows for fluorine incorporation,[112] including fluorine implantation after PMOS predoping (Fig. 7.26*a*), after SDE (Fig. 7.26*b*), and after source/drain (S/D) processes (Fig. 7.26*c*). In sequence (*a*), fluorine atoms are distributed within the poly film after implantation Most of the fluorine then will diffuse into the gate oxide and the bulk silicon interface after activation annealing. Sequence (*a*) then would have the largest thermal budget for fluorine diffusion among the three sequences. For sequence (*b*), fluorine atoms are introduced into the active area directly, and hence the dopants are the closest to the gate-oxide interface in the channel. Fluorine atoms in sequence (*c*) also are incorporated directly into the active area similar as in sequence (*b*). However, the F atoms are introduced a bit further away from the channel because of the separation of the sidewall spacer. Table 7.2 shows an NBTI lifetime comparison for different fluorine implant conditions and the control case without F implantation. The results show that for sample 4 with higher energy and lower dose of F implantation, the NBTI lifetime is

Items	Sample 1	Sample 2	Sample 3	Sample 4
F implantation energy	without	Low	Low	High
F implantation dose	without	Low	High	Low
NBTI lifetime	2.7 years	3 years	21 years	20 years

Source: Reproduce with permission from the Electrochemical Society.[112]

TABLE 7.2 NBTI Lifetime Comparison with Fluorine Implantation versus the Control Case (Without F Implantation)

comparable with that of sample 3 with lower energy and higher dose. This indicates that the NBTI lifetime is more sensitive to the F implantation energy than to the F dose. It is understood that F atoms with higher energy will have a better chance to penetrate the poly gate and arrive at the gate oxide. Higher energy implantation also will give the F atoms a higher probability to get closer to the gate oxide through diffusion during the activation annealing. The combination of higher energy and lower dose is preferred owing to the shorter process time.[112]

Fluorine incorporation is proved to decrease the NBTI of plasma-nitrided SiON. Based on the ΔV_t of the rf-nitrided SiON with fluorine implantation by Terai et al.,[105] the F implantation of 5×10^{14} cm^{-2} and 1×10^{15} cm^{-2} decreased ΔV_t by 22 and 36 percent, respectively. ΔV_t was separated into ΔV_{it} and ΔV_{ot} components. Compared with the sample without fluorine implantation, the ΔV_{it} component is reduced by half by introducing fluorine atoms. However, the ΔV_{ot} component is not reduced by fluorine incorporation, indicating that fluorine atoms clearly do not affect the preexisting hole traps within the nitrided oxide but effectively reduce the interface traps.

7.6.3 Gate-Oxide and Si-SiO_2-Interface Quality

In the conventional plasma nitridation oxide (PNO) process, the oxide is produced by base-oxide growth, plasma nitridation, and postanneal and poly-deposition steps in separate tools. It is likely that nitrogen outdiffusion and oxide regrowth will happen when moisture is involved between these steps, which further adversely impacts the gate oxide and interface quality and EOT scaling. Chang et al.[114] developed a cluster PNO process that integrates base oxide, plasma nitridation, postanneal, and poly deposition together (Fig. 7.27). It was found

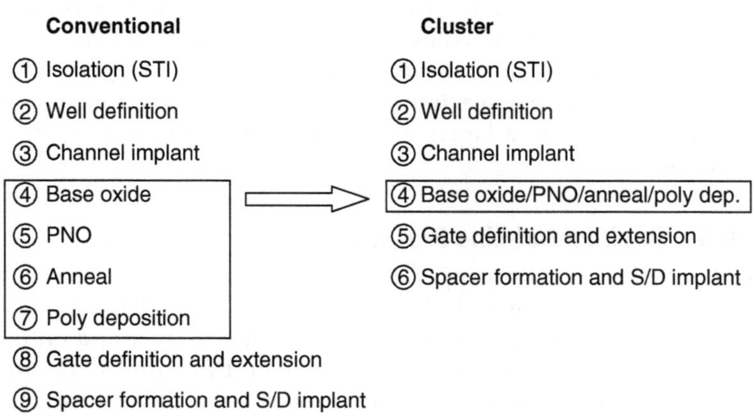

FIGURE 7.27 Key process steps of device fabrication comparing the cluster and conventional techniques. (*Used with permission from the Japan Society of Applied Physics.*[114])

that compared with the conventional PNO process, the cluster PNO process not only reduces the oxide thickness (t_{ox}) but also provides 2× improvement in NBTI immunity, although there is a 3.5× N concentration at the Si-SiO_2 interface produced by the cluster PNO process. The authors attributed the NBTI improvement to the sharper nitrogen profile in the oxide and better Si-SiO_2-interface quality.

Another example of better NBTI immunity caused by improved interface quality is given below. In a typical dual gate oxide (DGO) process flow, the thick gate oxide for I/O devices is grown first. Then the region corresponding to the thin core oxide is etched away by either a heavy or diluted HF solution, followed by growth of the thin oxide at the desired thickness in this region. Lee et al.[115] reported that the etching step using a heavy HF solution will lead to a higher F incorporation at the interface and thus improve core device reliability such as GOI and NBTI because better oxide–channel interface quality is achieved.

7.6.4 H_2/D_2 Anneal

FGA using hydrogen is a common practice to passivate the dangling bonds at the Si-SiO_2 interface (e. g., caused by plasma damage).[111] However, *high H content at the interface (or high density of hydrogen-terminated silicon bonds at the interface) during manufacturing is not desirable.*[38,84] An optimized FGA process can improve NBTI performance. NBTI degradation can be further suppressed if deuterium (D_2) is used in place of hydrogen.[38] Morifuji et al.[116] reported that in both the thin oxide region and the thick I/O gate-oxide region, NBTI lifetime improves significantly by reducing the FGA thermal budget. This observation indicates that hydrogen introduced by FGA plays a key role in the chemical reaction during NBTI degradation. As discussed in previous sections, holes attack Si–H bonds and releases atomic hydrogen. This reaction is reversible because as atomic H can repassivate those broken Si^+ bonds. However, excess H may lead to the formation of molecular H_2, which can slow down the reverse reaction for repassivation. Therefore, excess H (corresponding to a larger FGA thermal budget) worsens NBTI performance.

7.6.5 Back-End-of-Line Process

NBTI of devices with copper–low-*k* interconnects is worse than that with aluminum-SiO_2 interconnects.[117] The NBTI strongly depends on the cap layer over the copper–low-*k* layer and on the intermetal dielectric (IMD) film. A cap-layer–IMD film containing a higher volume of water gives a worse NBTI performance because water penetration and adsorption can worsen NBTI.[33,103] Once the water penetrates into the gate-oxynitride film, it creates an origin of positive fixed charges in the oxynitride. An SiC cap layer is better than an SiN cap layer in terms of NBTI resistance.[117] NBTI performance also depends on the barrier-metal film, which is correlated with hydrogen storage in the barrier metal. Hydrogen can diffuse into the copper during cap-layer (e. g., SiN) deposition and can be absorbed in the barrier (such as Ta) and stored there. Such stored hydrogen can be further desorbed during the subsequent process at high temperature (such as 300°C and above), for example, during the FGA (see Sec. 7.6.4).

7.6.6 Effect of Plasma-Induced Damage

Plasma-damaged devices have worse NBTI performance,[56,41] which is elaborated in Chap. 8. The plasma-induced damage/defects are superimposed on those generated by NBTI stress or efficiently act as precursor for NBTI. In other words, the plasma-induced damage/defects do not change the nature of the V_t degradation phenomenon during NBTI stress. The device with premetal dielectric (PMD) deposited by plasma-enhanced chemical-vapor deposition (PECVD) is much worse than the device with premetal dielectric deposited by subatmospheric CVD—a plasma-free process.[27] On the other hand, NBTI stress can be used as an effective and reliable monitor for latent plasma-induced damage as an alternative to complex solutions,[118] especially for the ultrathin gate oxides.

7.6.7 Boron Penetration

When boron penetration happens at the p^+ poly–gate-oxide interface, it does not alter fresh device characteristics, but it induces severe reliability degradation for the gate oxide,[1] such as worse NBTI performance.[78] Boron penetration–induced degradation becomes even worse with a reduction in gate length. Such degradation enhancement is attributed to local degradation of the gate oxide near the gate edge caused by the electrochemical reaction between holes and oxide defects, which is enhanced by boron penetration through the gate oxide from the p^+ gate. Incorporation of F or BF_2 or both is the choice for reducing the negative impact of boron penetration, as detailed in Sec. 7.6.2.

7.6.8 Effect of Contact-Etch Stop Layer

A liner silicon nitride (SiN) generally is used for a CESL between the PMD film and the gate electrode (see Chap. 3). The SiN deposition process involves a high-temperature anneal in an H atmosphere for up to 1 to 2 hours. The importance of CESL on NBTI is its hydrogen content, which can diffuse into the Si-SiO_2 interface during the deposition step[27] through poly and/or the spacer, acting the same as the hydrogen atoms at the Si-SiO_2 interface and those in the gate oxide. In other words, *a scaled PMOS requires SiN with low hydrogen content to improve NBTI performance.*[57,119] Indeed, a large

NBTI improvement occurs when the PMOS has no contact-etch nitride layer; however, such a layer is not manufacturable in mass production but has only experimental indications. On the other hand, introducing another nitride layer at the back end to replace the oxide liner above the metal makes NBTI even worse, which further confirms the need to optimize the CESL process to achieve an acceptable NBTI performance.[27]

Studies have shown that there is a strong correlation between the number of Si–H bonds and NBTI lifetime whereas the number of N−H bonds has little effect on NBTI.[119] These results suggest that N−H bond is more stable than the Si–H bond and that hydrogen release from N−H bonds under NBT stress is much less than that from Si–H bonds. It is possible to reduce the number of Si–H bonds by post CVD annealing: Through the TDS analysis, *500 and 800°C anneals are necessary for effective Si–H and N–H bond reduction*, respectively. However, as annealing temperature of 500°C is too high to apply for NiSi-formed devices. A process temperature below 450°C is desirable for NiSi integration.

Catalytic CVD SiN thus was proposed as the liner film and sidewall spacer to achieve a low hydrogen concentration in the SiN film with a much better NBTI lifetime.[119] Catalytic CVD is based on the cracking of SiH_4 and NH_3 by a catalytic reaction with tungsten wires heated to approximately 1,700 to 2,000°C. The numbers of radicals of SiH_3, NH, and NH_2 supplied in the catalytic CVD process are more than those in plasma CVD,[120] which makes low-temperature deposition possible. By using a T_{cat} of 2,000°C, the number of Si–H bonds was reduced enough ($<10^{21}$ cm^{-3}) without postdeposition annealing and 10 years lifetime were achieved. The improvement in NBTI lifetime by catalytic CVD SiN application to the sidewall can be much larger than that to the liner film because the diffusion length of hydrogen from the sidewall to the channel is shorter than that from the liner film.

A thicker CESL can lead to a larger V_t shift under NBTI stressing, as reported by Rhee et al.,[57] who studied NBTI-induced V_t shift with V_g stress of −2.6 V for device W/L = 10 μm/0.09 μm. Devices with 100-nm-thick CESLs degrade more than those with 50-nm-thick CESLS.

On the other hand, the SiN film with a certain thickness and/or coverage on the gate electrode was thought to be able to effectively prevent the diffusion of water and thus improve NBTI performance.[103] Three types of SiN films were studied for their effects on NBTI in the channel-length range of 0.4 to 20 μm. In type A, SiN films were deposited all over the wafer. In type B, SiN films were deposited and patterned to cover only the active area. In type C, SiN films covered only the gate area. The results showed the V_t shift in the following sequence after NBTI stressing: Without SiN film > type C SiN film > type B and C SiN films. This trend is observed more clearly at short gate lengths.

At channel lengths of 0.4 μm, the V_t shift without SiN is more than one order higher than that of type B and C SiN films, whereas the V_t shift is almost comparable for channel lengths of 20 μm. The large V_t shift in type C was attributed to water diffusion from the SiN pattern edge to the gate oxide, which leads to the positive charges generated inside the gate oxide of the gate edge region. Type A and type B SiNs have good enough coverage on the gate electrode and thus effectively can prevent water diffusion into gate oxide. However, the SiN thickness should not be less than 6 nm.

7.6.9 Effect of Si Substrate Orientation

Chapter 3 mentioned that the commonly used starting material/wafer is *P*-substrate silicon with a (100) surface orientation because the (100) orientation provides much better Si-SiO_2 interface than other orientations. There is an option to use (110) substrate to boost PMOS performance for advanced technology because a doubled mobility can be achieved.[121] However, NBTI for PMOS devices on (110) substrates was found to be worse than that on (100) substrate because of the increased density of interface states on (110) substrates owing to worse SiO_2 film or interface qualities.[122] An improvement in SiO_2 film quality on (110) substrates is necessary.

7.7 Dynamic Negative-Bias Temperature Instability

Circuits are seldom stressed continually on dc. The conventional NBTI based on dc stressing data overestimates the degradation of the PMOS device in digital CMOS circuits by overlooking the electric passivation effect during normal operation of digital circuits because removal of the applied stress can lead to recovery of NBTI degradation.[16,42,45,123] During operation of a PMOS device in a CMOS inverter, the applied gate bias (input signal) is switched between high and low voltages, whereas the drain bias (output signal) is alternated between low and high voltages correspondingly. During the low (low-output) phase of an inverter, the recovery effect effectively reduces the interface traps generated during the high (high-output) phase and, consequently, recovers the degradations of device parameters to a certain degree.[42] Hence it is important to understand NBTI response to ac stress. Dynamic stress (ac) results in significantly less NBTI degradation (about half) relative to static stress (dc).[42,124]

For ac NBTI, experimental works have indicated no or weak frequency dependence.[42,124] As an example, Fernández et al.[124] studied ac NBTI in the 1-Hz to 2-GHz range based on dedicated on-chip CMOS circuits, showing that ac NBTI is independent of frequency, as detailed below. This observation is applicable to both conventional stress-measure-stress (SMS) and OTF measurements. The circuit for on-chip studies of dc and ac NBTI consisted of a 41-stage ring oscillator, a frequency divider, a buffer, a pass-gate-based multiplexer, and the device under test (DUT)—a single PMOS. When $V_{select} = 0$ V, V_g is connected to the DUT, allowing for direct stressing and characterization. When $V_{select} = V_{cc}$, the DUT is connected to the oscillator, resulting in very high-frequency ac stress with a duty cycle of 50 percent. The V_t shifts owing to ac NBTI stress in a single PMOS device were seen to be independent of frequency in the entire frequency range of 1 Hz to 2 GHz. The result of the corresponding dc NBTI stress obtained on identical devices is ~2× higher than its ac counterpart. The effect of duty cycle at $f = 10$ kHz showed that ΔV_t is increasing monotonically, with an apparent plateau at around 50 percent duty cycle.

Simulation of ac NBTI frequency independence also was performed and derived from the R-D model[16] for lower frequencies.[35,43] There are some key features of the ac response that can be observed from simulations[35]:

- AC degradation is less than dc degradation for any given operation time.
- For intermediate frequencies, the degradation is independent of frequency. For example, 1 Hz and 10 Hz result in similar N_{it} values.
- As frequency is increased further, the degradation decreases with increasing frequency above the kilohertz[35] and the megahetz[43] ranges. This conclusion is drawn based on the characteristic times of the interfacial reactions exceeding that of the ac perturbation. This point looks contradictory to the experimental observations of Fernández et al.,[124] which showed no frequency dependency at all.

7.8 Process Qualification Practice

According to Joint Electron Device Engineering Council (JEDEC) JP-001,[77] the NBTI test is intended to determine the degradation in threshold voltage as a result of gate bias at high temperatures. This degradation is a particular concern in PMOS devices in a dual-poly process owing to boron diffusion. For NBTI qualification, JEDEC JP-001[77] provides good guidance for practioners. Again, we generally obtain device-level NBTI under dc bias. However, the bias is ac under real applications. The recovery effect under ac conditions should be taken into account for determining real lifetime. Therefore, circuit-level ac stress testing and/or simulation of NBTI is highly recommended. Also, it is quite challenging to establish the lifetime correlation between the device-level dc lifetime and the product-level ac lifetime so as to ensure product reliability at the early stage of process development. A brief look at the NBTI qualification practice is given in Table 7.3.

	PMOS NBTI
Failure criteria	• A device is considered to fail when a chosen device parameter changes by more than the specified failure criterion at worst-use condition. • A common practice is to use 10 percent I_{dsat} degradation as the failure criterion. • The failure criteria for the monitored parameters will depend on the specific requirements of the circuit application. Other device parameters such as G_m, I_{dlin}, and V_{tlin}, can be used to monitor NBTI degradation. For example, in some cases, we may use 50- or 100-mV V_t degradation as the criterion.
Test structures	• Devices with the norminal geometry for each transistor type (core, I/O, low/standard/high-V_t devices, etc.) are advised to be evaluated first because they are used mostly in the product design. • A matrix of device lengths/widths as defined in the design rules may be necessary to cover all the application cases. • A safe-operation-area (SOA) matrix of devices in some cases should be provided by the foundry for IC designer reference.
Sample size	For each type of device, test devices from 3 lots, 3 wafers per lot, minimum of 5 sites (typically 8 or 12 sites for advanced-technology nodes) per bias condition.
Stress conditions	• Stress temperature: At the maximum operating temperature (typically 125°C). Higher and lower stress temperatures may be used if the activation energy is measured. • Gate voltage at stress: Source/drain/bulk grounded. • Readouts: Log time scale with at least three points per decade. Times between stress cycles and intermediate measurements must be kept to a minimum.
Wafer/ package level	Wafer level or package level
Acceptance criteria	TTF ≥ 10 years considering 50 percent duty at worst temperature/$V_g = 1.1V_{dd}$
Models	V_g model: Eq. (7.1) or power-law model: Eq. (7.2)
Parameters	TTF (at 1.1 V_{dd}), n, E_a, and γ or m

TABLE 7.3 Typical Process-Qualification Practice for DC Device-Level NBTI

7.9 References

1. M. Y. Hao, D. Nayak, and R. Rakkhit, "Impact of boron penetration at P^+-poly/gate oxide interface on deep-submicron device reliability for dual-gate CMOS technologies," *IEEE Electron Device Letters*, Vol. 18, No. 5, 1997, pp. 215–217.
2. A. Uchiyama, H. Fukuda, T. Hayashi, T. Iwabuchi, and S. Ohno, "High-performance dual-gate sub-halfmicron CMOSFETs with 6-nm-thick nitrided Si% films in an N_2O ambient," in *Proceedings of the International Electron Devices Meeting (IEDM)*, 1990, pp. 425–428.
3. Y. Okada, P. J. Tobin, K. G. Reid, R. I. Hegde, B. Maiti, and S. A. Ajuria, "Gate oxynitride grown in nitric oxide (NO)," in *Symposium on VLSI Technology Digest of Technical Papers*, 1994, pp. 105–106.
4. C. T. Liu, E. J. Lloyd, Y. Ma, M. Du, R. L. Opila, and S. J. Hillenius, "High performance 0.2-μm CMOS with 25-Å gate oxide grown on nitrogen implanted Si substrates," in *Proceedings of the International Electron Devices Meeting (IEDM)*, 1996, pp. 499–503.
5. T. Morimoto, H. S. Mornose, Y. Ozawa, K. Yamabe, and H. Iwai, "Effects of boron penetration and resultant limitations in ultra thin pure-oxide and nitrided-oxide gate-films," in *Proceedings of the International Electron Devices Meeting (IEDM)*, 1990, pp. 429–432.
6. M. L. Green, D. Brasen, K. W. Evans-Lutterodt, L. C. Feldman, K. Krisch, W. Lennard, H. T. Tang, L. Manchanda, and M. T. Tang, "Rapid thermal oxidation of silicon in N_2O between 800 and 1200°C: Incorporated nitrogen and interfacial roughness," *Applied Physics Letters*, Vol. 65, No. 7, 1994, pp. 848–850.
7. E. P. Gusev, H. C. Lu, E. Garfunkel, T. Fustafsson, M. L. Green, D. Brasen, and W. N. Lennard, "Nitrogen engineering of ultrathin oxynitrides by a thermal NO/O_2/NO process," *Journal of Applied Physics*, Vol. 84, No. 5, 1998, pp. 2980–2982.
8. T. Hori, "Inversion layer mobility under high normal field in nitrided-oxide MOSFETs," *IEEE Transactions on Electron Devices*, Vol. 37, No. 9, 1990, pp. 2058–2069.
9. N. Kimizuka, T. Yamamoto, T. Mogami, K. Yamaguchi, K. Imai, and T. Horiuchi, "The impact of bias temperature instability for direct-tunneling ultrathin gate oxide on MOSFET scaling," in *Symposium on VLSI Technology Digest of Technical Papers*, 1999, pp. 73–74.
10. K. Kang, M. A. Alam, and K. Roy, "Estimation of NBTI degradation using IDDQ measurement," in *Proceedings of the 45th Annual International Reliability Physics Symposium*, 2007, pp. 10–16.
11. M. Ruberto, T. Maimon, Y. Shemesh, A. B. Desomeaux, W. Zhang, and C. -S. Yeh, "Consideration of age degradation in the rf performance of CMOS radio chips for high volume manufacturing," in *Proceedings of the IEEE Radio Frequency Integrated Circuits Symposium*, 2005, pp. 549–552.
12. Z. Luo and J. P. Walko, "Physical mechanism of NBTI relaxation by rf and noise performance of rf CMOS devices," in *Proceedings of the 43rd IEEE International Annual Reliability Physics Symposium*, 2005, pp. 712–713.
13. C. Yu, Y. Liu, A. Sadat, and J. S. Yuan, "Impact of temperature-accelerated voltage stress on PMOS rf performance," *IEEE Transactions on Device and Materials Reliability*, Vol. 4, No. 4, 2004, pp. 664–669.
14. A. Haggag, G. Anderson, S. Parihar, D. Burnett, G. Abeln, J. Higman, and M. Moosa, "Understanding SRAM high-temperature-operating-life NBTI: Statistics and permanent vs. recoverable damage," in *Proceedings of the 45th IEEE International Annual Reliability Physics Symposium*, 2007, pp. 452–456.
15. K. Ramakrishnan, S. Suresh, N. Vijaykrishnan, M. J. Irwin, and V. Degalahal, "Impact of NBTI on FPGAs," in *Proceedings of the 20th International Conference on VLSI Design (VLSID'07)*, 2007, pp. 717–722.
16. K. O. Jeppson and C. M. Svensson, "Negative-bias stress of MOS devices at high electric fields and degradation of MOS devices," *Journal of Applied Physics*, Vol. 48, No. 5, 1977, pp. 2004–2014.

17. S. Mahapatra and M. A. Alam, "A predictive reliability model for pMOS bias temperature degradation," in *Proceedings of the International Electron Devices Meeting (IEDM)*, 2002, pp. 505–509.
18. M. Alam and S. Mahapatra, "A comprehensive model of PMOS NBTI degradation," *Microelectronics and Reliability*, Vol. 45, 2005, pp. 71–81.
19. V. Huard, M. Denais, and C. Parthasarathy, "NBTI degradation: From physical mechanism to modeling," *Microelectronics and Reliability*, Vol. 46, 2006, pp. 1–23.
20. J. H. Stathis and S. Zafar, "The negative-bias temperature instability in MOS devices: A review," *Microelectronics and Reliability*, Vol. 46, 2005, pp. 270–286.
21. M. A. Alam, H. Kufluoglu, D. Varghese, and S. Mahapatra, "A comprehensive model for PMOS NBTI degradation: Recent progress," *Microelectronics Reliability*, Vol. 47, 2007, pp. 853–862.
22. D. K. Schroder and J. A. Babcock, "Negative bias temperature instability: Road to cross in deep submicron silicon semiconductor manufacturing," *Journal of Applied Physics*, Vol. 94, No. 1, 2003, pp. 1–18.
23. D. K. Schroder, "Negative bias temperature instability: What do we understand?" *Microelectronics Reliability*, Vol. 47, 2007, pp. 841–852.
24. JEP122E, *Failure Mechanisms and Models for Semiconductor Devices*, JEDEC Solid State Technology Association, Arlington, VA, March 2009.
25. C. L. Chen, Y. M. Lin, C. J. Wang, and K. Wu, "A new finding on NBTI lifetime model and an investigation on NBTI degradation characteristic for 1.2-nm ultrathin oxide," in *Proceedings of the 43rd IEEE International Annual Reliability Physics Symposium*, 2005, pp. 704–705.
26. H. Aono, E. Murakami, K. Okuyama, A. Nishida, M. Minami, Y. Ooji, and K. Kubota, "Modeling of NBTI degradation and its impact on electric field dependence of the lifetime," in *Proceedings of the 42nd IEEE International Annual Reliability Physics Symposium*, 2004, pp. 23–27.
27. A. Scarpa, D. Ward, J. Dubois, L. van Marwijk, S. Gausepohl, R. Campos, K. Y. Sim, A. Cacciato, R. Kho, and M. Bolt, "Negative-bias temperature instability cure by process optimization," *IEEE Transactions on Electron Devices*, Vol. 53, No. 6, 2006, pp. 1331–1339.
28. D. Varghese, D. Saha, S. Mahapatra, K. Ahmed, F. Nouri, and M. Alam, "On the dispersive versus Arrhenius temperature activation of NBTI time evolution in plasma nitrided gate oxides: Measurements, theory, and implications," in *Proceedings of the International Electron Devices Meeting (IEDM)*, 2005, pp. 701–704.
29. A. E. Islam, G. Gupta, S. Mahapatra, A. T. Krishnan, K. Ahmed, F. Nouri, A. Oates, and M. A. Alam, "Gate leakage vs. NBTI in plasma nitrided oxides: Characterization, physical principles, and optimization," in *Proceedings of the International Electron Devices Meeting (IEDM)*, 2006, p. 12.4.1.
30. M. A. Alam, J. Bude, and A. Ghetti, "Field acceleration for oxide breakdown: Can an accurate anode hole injection model resolve the E vs. $1/E$ controversy?" in *Proceedings of the 38th IEEE International Annual Reliability Physics Symposium*, 2000, pp. 21–26.
31. D. J. DiMaria, E. Cartier, and D. Arnold, "Impact ionization, trap creation, degradation and breakdown in silicon dioxide films on silicon," *Journal of Applied Physics*, Vol. 73, 1993, pp. 3367–3384.
32. J. W. Mcpherson and H. Mogul, "Underlying physics of the thermochemical E model in describing low-field time-dependent dielectric breakdown in SiO_2 thin films," *Journal of Applied Physics*, Vol. 84, No. 3, 1998, pp. 1513–1523.
33. C. E. Blat, E. H. Nicollian, and E. H. Poindexter, "Mechansim of negative-bias-temperature instability," *Journal of Applied Physics*, Vol. 69, No. 3, 1991, pp. 1712–1720.
34. S. Ogawa and N. Shiono, "Generalized diffusion-reaction model for the low-field-charge-buildup instability at the Si-SiO_2 interface," *Physical Review B*, Vol. 51, No. 7, 1995, pp. 4218–4230.

35. S. Chakravarthi, A. T. Krishnan, V. Reddy, C. F. Machala, and S. Krishnan, "A comprehensive framework for predictive modeling of negative bias temperature instability," in *Proceedings of the 42nd IEEE International Annual Reliability Physics Symposium*, 2004, pp. 273–282.
36. V. Reddy, A. T. Krishnan, A. Marshall, J. Rodriguez, S. Natarajan, T. Rost, and S. Krishnan, "Impact of negative bias temperature instability on digital circuit reliability," in *Proceedings of the 40th IEEE International Annual Reliability Physics Symposium*, 2002, pp. 248–254.
37. S. Zhu, A. Nakajima, T. Ohashi, and H. Miyake, "Interface trap and oxide charge generation under negative bias temperature instability of *p*-channel metal-oxide-semiconductor field-effect transistors with ultrathin plasma-nitrided SiON gate dielectrics," *Journal of Applied Physics*, Vol. 98, 2005, p. 114504.
38. N. Kimizuka, K. Yamaguchi, K. Imai, T. Iizuka, C. T. Liu, R. C. Keller, and T. Horiuchi, "NBTI enhancement by nitrogen incorporation into ultrathin gate oxide for 0.10-μm gate CMOS generation," in *Symposium on VLSI Technology Digest of Technical Papers*, 2000, pp. 92–93.
39. S. S. Tan, T. P. Chen, C. H. Ang, and L. Chan, "Mechanism of nitrogen-enhanced negative bias temperature instability in pMOSFET," *Microelectronics Reliability*, Vol. 45, No. 1, 2005, pp. 19–30.
40. S. Tsujikawa and J. Yugami, "Positive charge generation due to species of hydrogen during NBTI phenomenon in pMOSFETs with ultrathin SiON gate dielectrics," *Microelectronics Reliability*, Vol. 45, 2005, pp. 65.
41. A. T. Krishnan, V. Reddy, and S. Krishnan, "Impact of charging damage on negative bias temperature instability," in *Proceedings of the International Electron Devices Meeting (IEDM)*, 2001, pp. 865–868.
42. G. Chen, K. Y. Chuah, M. F. Li, D. S. H. Chan, C. H. Ang, J. Z. Zheng, Y. Jin, and D. L. Kwong, "Dynamic NBTI of PMOS transistors and its impact on device lifetime," in *Proceedings of the 41st IEEE International Annual Reliability Physics Symposium*, 2003, pp. 196–202.
43. M. Alam, "A critical examination of the mechanics of dynamic NBTI for *P*-MOSFETs," in *Proceedings of the International Electron Devices Meeting (IEDM)*, 2003, pp. 345–348.
44. T. Yang, C. Shen, M. F. Li, C. H. Ang, C. X. Zhu, Y. -C. Yeo, G. Samudra, and D. -L. Kwong, "Interface trap passivation effect in NBTI measurement for *p*-MOSFET with SiON gate dielectric," *IEEE Electron Device Letters*, Vol. 26, No. 10, 2005, pp. 758–760.
45. S. Tsujikawa, T. Mine, K. Watanabe, Y. Shimamoto, R. Tsucbiya, K. Ohnishi, T. Onai, J. Yugami, and S. Kimura, "Negative bias temperature instability of pMOSFETs with ultrathin SiON gate dielectrics," in *Proceedings of the 41st IEEE International Annual Reliability Physics Symposium*, 2003, pp. 183–188.
46. D. S. Ang and S. Wang, "Insight into the suppressed recovery of NBTI-stressed ultrathin oxynitride gate pMOSFET," *IEEE Electron Device Letters*, Vol. 27, No. 9, 2006, pp. 755–757.
47. V. Huard, M. Denais, F. Perrier, N. Revil, C. Parthasarathy, A. Bravaix, and E. Vincent, "A thorough investigation of MOSFETs NBTI degradation," *Microelectronics Reliability*, Vol. 45, 2005, pp. 83–98.
48. D. S. Ang, "Observation of suppressed interface state relaxation under positive gate biasing of the ultrathin oxynitride gate *p*-MOSFET subjected to negative-bias temperature stressing," *IEEE Electron Device Letters*, Vol. 27, No. 5, 2006, pp. 412–415.
49. S. Tsujikawa and J. Yugami, "Evidence for bulk trap generation during NBTI phenomenon in pMOSFETs with ultrathin SiON gate dielectrics," *IEEE Transactions on Electron Devices*, Vol. 53, No. 1, 2006, pp. 51–55.
50. S. Chakravarthi, P. Kohli, P. R. Chidambaram, H. Bu, A. Jain, and C. F. Machala, "Modeling the effect of source/drain sidewall spacer process on boron ultrashallow junctions," in *Proceedings of the International Conference on Simulation of Semiconductor Processes and Devices*, 2003, pp. 159–162.

51. C. H. Liu, M. T. Lee, C. -Y. Lin, J. Chen, Y. T. Loh, F. -T. Liou, K. Schruefer1, A. A. Katsetos, Z. Yang, N. Rovedo, T. B. Hook, C. Wann, and T. -C. Chen, "Mechanism of threshold voltage shift (ΔV_{th}) caused by negative bias temperature instability (NBTI) in deep submicron pMOSFETs," *Japanese Journal of Applied Physics*, Vol. 41, 2002, pp. 2423–2425.
52. Y. Mitani, M. Nagamine, H. Satake, and A. Toriumi, "NBTI mechanism in ultra-thin gate dielectric—Nitrogen-originated mechanism in SiON," in *Proceedings of the International Electron Devices Meeting (IEDM)*, 2002, pp. 509–512.
53. Y. Mitani, "Influence of nitrogen in ultrathin SiON on negative bias temperature instability under ac stress," in *Proceedings of the International Electron Devices Meeting (IEDM)*, 2004, pp. 117–120.
54. S. Mahapatra, P. B. Kumar, and M. A. Alam, "Investigation and modeling of interface and bulk trap generation during negative bias temperature instability of *p*-MOSFETs," *IEEE Transactions on Electron Devices*, Vol. 51, No. 9, 2004, pp. 1371–1379.
55. H. Kufluoglu and M. A. Alam, "A geometrical unification of the theories of NBTI and HCI time-exponents and its implications for ultra-scaled planar and surround-gate MOSFETs," in *Proceedings of the International Electron Devices Meeting (IEDM)*, 2004, pp. 113–116.
56. G. Cellere, M. G. Valentini, and A. Paccagnella, "Effect of channel width, length, and latent damage on NBTI," in *Proceedings of the IEEE International Conference on Integrated Circuit Design and Technology*, 2004, pp. 303–306.
57. H. S. Rhee, and H. Lee, T. Ueno, D. S. Shin, S. H. Lee, Y. Kim, A. Samoilov, P. O. Hansson, M. Kim, H. S. Kim, and N. I. Lee, "Negative bias temperature instability of carrier-transport enhanced *p*MOSFET with performance boosters," in *Proceedings of the International Electron Devices Meeting (IEDM)*, 2005, pp. 692–695.
58. A. E. Islam, H. Kufluoglo, D. Varghese, S. Mahapatra, and M. A. Alam, "Recent issues in negative-bias temperature instability: Initial degradation, field dependence of interface trap generation, hole trapping effects, and relaxation," *IEEE Transactions on Electron Devices*, Vol. 54, No. 9, 2007, pp. 2143–2154.
59. S. J. Wen, L. Hinh, and H. Puchner, "Voltage acceleration NBTI study for a 90-nm CMOS technology," in *IEEE International Integrated Reliability Workshop Final Report*, 2003, pp. 147–148.
60. S. Rangan, N. Mielke, and E. C. C. Yeh, "Universal recovery behavior of negative bias temperature instability,"in *Proceedings of the International Electron Devices Meeting (IEDM)*, 2003, pp. 341–344.
61. M. Ershov, S. Saxena, H. Karbasi, S. Winters, S. Minehane, J. Babcock, R. Lindley, P. Clifton, M. Redford, and A. Shibkov, "Dynamic recovery of negative bias temperature instability in *p*-type metal-oxide-semiconductor field-effect transistors," *Applied Physics Letters*, Vol. 83, No. 8, 2003, pp. 1647–1649.
62. A. T. Krishnan, C. Chancellor, S. Chakravarthi, P. E. Nicollian, V. Reddy, A. Varghese, R. B. Khamankar, and S. Krishnan, "Material dependence of hydrogen diffusion: Implications for NBTI degradation," in *Proceedings of the International Electron Devices Meeting (IEDM)*, 2005, pp. 688–691.
63. T. Yang, M. F. Li, C. Shen, C. H. Ang, C. Zhu, Y. -C. Yeo, G. Samudra, S. C. Rustagi, M. B. Yu, and D. L. Kwong, "Fast and slow dynamic NBTI components in *p*-MOSFET with SiON dielectric and their impact on device lifetime and circuit application,"in *Symposium on VLSI Technology Digest of Technical Papers*, 2005, pp. 92–93.
64. W. Heinrigs, H. Reisinger, W. Gustin, and C. Schlünder, "Consideration of recovery effects during NBTI measurements for accurate lifetime predictions of state-of-the-art pMOSFETs," in *Proceedings of the 45th IEEE International Annual Reliability Physics Symposium*, 2007, pp. 288–292.
65. M. Denais, C. Parthasarathy, G. Ribes, Y. Rey-Tauriac, N. Revil, A. Bravaix, V. Huard, and F. Perrier, "On-the-fly characterization of NBTI in ultrathin gate oxide *p*-MOSFET's,"in *Proceedings of the International Electron Devices Meeting (IEDM)*, 2004, pp. 109–112.

66. C. Shen, M. -F. Li, X. P. Wang, Y. -C. Yeo, and D. -L. Kwong, "A fast measurement technique of MOSFET *Id-Vg* characteristics," *IEEE Electron Device Letters*, Vol. 27, No. 1, 2006, pp. 55–57.
67. H. Reisinger, O. Blank, W. Heinrigs, A. Muhlhoff, W. Gustin, and C. Schlunder, "Analysis of NBTI degradation and recovery behavior based on ultrafast V_T measurements," in *Proceedings of the 44th IEEE International Annual Reliability Physics Symposium*, 2006, pp. 448–453.
68. Multi-channel parallel timing-on-the-fly NBTI characterization using Agilent B1500A. Agilent application note B1500-6 (online). Available: http://www . home.agilent.com/agilent/redirector. j.
69. On-the-fly VTH measurement for bias temperature instability characterization. Keithley application note 2814. (Online). Available: http://www. keithley. com.
70. Z. Ji, J. F. Zhang, M. H. Chang, B. Kaczer, and G. Groeseneken, "An analysis of the NBTI-induced threshold voltage shift evaluated by different techniques," *IEEE Transactions on Electron Devices*, Vol. 56, No. 5, 2009, pp. 1086–1093.
71. H. Reisinger, O. Blank, W. Heinrigs, W. Gustin, and C. Schlünder, "A comparison of very fast to very slow components in degradation and recovery due to NBTI and bulk hole trapping to existing physical models," *IEEE Transactions on Device and Materials Reliability*, Vol. 7, No. 1, 2007, pp. 119–129.
72. H. Reisinger, U. Brunner, W. Heinrigs, W. Gustin, and C. Schlünder, "A comparison of fast methods for measuring NBTI degradation," *IEEE Transactions on Device and Materials Reliability*, Vol. 7, No. 4, 2007, pp. 531–539.
73. D. Varghese, S. Mahapatra, and M. A. Alam, "Hole energy dependent interface trap generation in MOSFET Si/SiO_2 interface," *IEEE Electron Device Letters*, Vol. 26, No. 8, 2005, pp. 572–574.
74. S. Zhu, A. Nakajima, T. Ohashi, and H. Miyake, "Influence of bulk bias on negative bias temperature instability of *p*-channel metal-oxide-semiconductor field-effect transistors with ultrathin SiON gate dielectrics," *Journal of Applied Physics*, Vol. 99, 2006, p. 064510.
75. S. M. Sze, *Physics of Semiconductor Devices*, 2nd ed., Wiley, New York, 1981, Chap. 7.
76. M. Denais, V. Huard, C. Parthasarathy, G. Ribes, F. Perrier, N. Red, and A. Bravaix, "Interface traps and oxide traps creation under NBTI and PBTI in advanced CMOS technology with a 2-nm gate oxide," in *IEEE International Integrated Reliability Workshop Final Report*, 2003, pp. 1–6.
77. JP001.01, *"Foundry Process Qualification Quidelines (Wafer Fabrication Manufacturing sites)*, JEDEC Solid State Technology Association and Fabless Semiconductor Association, May 2004.
78. T. Yamamoto, K. Uwasawa, and T. Mogami, "Bias temperature instability in scaled *p*+ polysilicon gate *p*-MOSFETs," *IEEE Transactions on Electron Devices*, Vol. 46, No. 5, 1999, pp. 921–926.
79. C. Schlünder, R. Brederlow, B. Ankele, A. Lill, K. Goser, and R. Thewes, "On the degradation of *P*-MOSFETS in analog and rf circuits under inhomogeneous negative bias temperature stress," in *Proceedings of the 41st IEEE International Annual Reliability Physics Symposium*, 2003, pp. 5–10.
80. C. -Y. Lu, H. -C. Lin, Y. -F. Chang, and T. -Y. Huang, "Dc and ac NBTI stresses in PMOSFETs with PE-SiN capping," in *Proceedings of the 44th IEEE International Annual Reliability Physics Symposium*, 2006, pp. 727–728.
81. S. S. Chung, D. K. Lo, J. J. Yang, and T. C. Lin, "Localization of NBTI-induced oxide damage in direct tunneling regime gate oxide *p*MOSFET using a novel low gate leakage gated diode (L^2-GD) method," in *Proceedings of the International Electron Devices Meeting (IEDM)*, 2002, pp. 513–516.
82. G. Math, C. Benard, J. L. Ogier, and D. Goguenheim, "Geometry effects on the NBTI degradation of PMOS transistors,"in *Proceedings of the IEEE International Integrated Reliability Workshop Final Report*, 2008, pp. 60–63.
83. R. Mooraka, K. N. ManjulaRani, D. Ho, N. Mathur, and H. Puchner, "A simulation methodology for the operational life time study of deep submicron technology integrated circuits using channel length dependent NBTI model," in *Proceedings of the International Workshop on Physics of Semiconductor Devices (IWPSD)*, 2007, pp. 146–149.

84. V. Kol'dyaev, "Layout design dependence of NBTI for I/O P-MOSFET," in *Proceedings of the 42nd IEEE International Annual Reliability Physics Symposium*, 2004, pp. 663–664.
85. T. J. J. Ho, D. S. Ang, L. J. Tang, T. W. H. Phua, and C. M. Ng, "Role of nitrogen on the gate length dependence of NBTI," *IEEE Electron Device Letters*, Vol. 30, No. 7, 2009, pp. 772–774.
86. Y. H. Lee, K. Wu, T. Linton, N. Miekle, S. Hu, and B. Wallace, "Channel width dependent hot-carrier degradation of thin-gate pMOSFETs," in *Proceedings of the 38th IEEE International Annual Reliability Physics Symposium*, 2000, pp. 77–82.
87. H. Küflüoglu and M. A. Alam, "Theory of interface-trap-induced NBTI degradation for reduced cross section MOSFETs," *IEEE Transactions on Electron Devices*, Vol. 53, No. 5, 2006, pp. 1120–1130.
88. T. Yamashita, K. Ota, K. Shiga, T. Hayashi, H. Umeda, H. Oda, T. Eimori, M. Inuishi, Y. Ohji, K. Eriguchi , Nakanishi, H. Nakaoka, T. Yamada, M. Nakamura, I. Miyams, A. Kajiya, M. Kubota, and M. Ogura, "Impact of boron penetration from S/D extension on gate-oxide reliability for 65-nm node CMOS and beyond," in *Symposium on VLSI Technology Digest of Technical Papers*, 2004, pp. 136–137.
89. L. Duong, E. Li, V. Gopinath, S. Prasad, J. Lin, D. Pachura, and V. Homback, "Effect of nitrogen incorporation on PMOS negative bias temperature instability in ultrathin oxy-nitrides," in *IEEE International Integrated Reliability Workshop Final Report*, 2003, pp. 128–130.
90. S. S. Tan, T. P. Chen, J. M. Soon, K. P. Loh, C. H. Ang, W. Y. Teo, and L. Chan, "Linear relationship between H^+ trapping reaction energy and defect generation: Insight into nitrogen-enhanced negative bias temperature instability," *Applied Physics Letters*, Vol. 83, No. 3, 2003, pp. 530–532.
91. C. C. Liao, Z. H. Gan, Y. J. Wu , K. Zheng, R. Guo, J. H. Ju, J. Ning, A. He, S. Ye, E. Liu, and W. Wong, "Factors for negative bias temperature instability improvement in deep submicron CMOS technology," in *Proceedings of the 9th International Conference on Solid-State and Integrated-Circuit Technology (ICSICT)*, 2008, pp. 612–615.
92. S. S. Tan, T. P. Chen, C. H. Ang, Y. L. Tan, and L. Chan, "Influence of nitrogen proximity from the Si/SiO_2 interface on negative bias temperature instability," *Japanese Journal of Applied Physics*, Vol. 41, No. 10A, 2002, pp. L1031–L1033.
93. T. Sasaki, K. Kuwazawa, K. Tanaka, J. Kato, and D. -L. Kwong, "Engineering of nitrogen profile in an ultrathin gate insulator to improve transistor performance and NBTI," *IEEE Electron Device Letters*, Vol. 24, No. 3, 2003, pp. 150–152.
94. E. Morifuii, M. Kanda, N. Yanagiya, S. Matsuda, S. Inaba, K. Okano, K. Takahashi, M. Nishigori, H. Tsuno, T. Yamamoto, K. Hiyama, M. Takayanagi, H. Oyamatsu, S. Yamada, T. Noguchi, and M. Kakumu, "High-performance 30-nm bulk CMOS for 65-nm technology node (CMOS5),"in *Proceedings of the International Electron Devices Meeting (IEDM)* 2002, pp. 655–658.
95. D. Ishikawa, S. Sakai, K. Katsuyama, and A. Hiraiwa, "Nitride-sandwiched-oxide gate insulator for low-power CMOS,"in *Proceedings of the International Electron Devices Meeting (IEDM)* 2002, pp. 869–872.
96. N. Yanagiya, S. Matsuda, S. Inaba, M. Takayanagi, I. Mizushima, K. Ohuchi, K. Okano, K. Takahasi, E. Morifuji, M. Kanda, Y. Matsubara, M. Habu, M. Nishigoori, K. Honda, H. Tsuno, K. Yasumoto, T. Yamamoto, K. Hiyama, K. Kokubun, T. Suzuki, J. Yoshikawa, T. Sakurai, T. Ishizuka, Y. Shoda, M. Moriuchi, M. Kishida, H. Matsumori, H. Harakawa, H. Oyamatsu, N. Nagashima, S. Yamada, T. Noguchi, H. Okamoto, and M. Kakumu, 65-nm CMOS technology (CMOS5) with high density embedded memories for broadband microprocessor applications,"in *Proceedings of the International Electron Devices Meeting (IEDM)*, 2002, pp. 57–60.
97. D. Varghese, G. Gupta, L. M. Lakkimsetti, D. Saha, K. Ahmed, F. Nouri, and S. Mahapatra, "Physical mechanism and gate insulator material dependence of generation and recovery of negative-bias temperature instability in *p*-MOSFETs," *IEEE Transactions on Electron Devices*, Vol. 54, No. 7, 2007, pp. 1672–1680.

98. S. S. Tan, C. H. Ang, C. M. Lek, T. P. Chen, B. J. Cho, A. See, and L. Chan, "Characterization of ultrathin plasma nitrided gate dielectrics in *p*MOSFET for 0.18-μm technology and beyond,"in *Proceedings of the 9th International Symposium on the Physical and Failure Analysis of Integrated Circuits (IPFA)*, 2002, pp. 254–258.
99. S. Inaba, T. Shimizu, S. Mori, K. Sekine, K. Saki, H. Suto, H. Fukui, M. Nagamine, M. Fujiwara, T. Yamamoto, M. Takayanagi, I. Mizushima, K. Okano, S. Matsuda, H. Oyamatsu, Y. Tsunashima, S. Yamada, Y. Toyoshima and H. Ishiuchi, "Device performance of sub-50-nm CMOS with ultrathin plasma nitrided gate dielectrics,"in *Proceedings of the International Electron Devices Meeting (IEDM)*, 2002, pp. 651–654.
100. P. A. Kraus, K. Ahmed, T. C. Chua, M. Ershov, H. Karbasi, C. S. Olsen, F. Nouri, J. Holland, R. Zhao, G. Miner, and A. Lepert, "Low-energy nitrogen plasmas for 65-nm node oxynitride gate dielectrics: a correlation of plasma characteristics and device parameters," in *Symposium on VLSI Technology Digest of Technical Papers*, 2003, pp. 143–144.
101. M. A. Liberman and A. J. Lichtenberg, *Principles of Plasma Discharges and Materials Processing* John Wiley and Sons, Hoboken, NJ, 2005, pp. 306–307.
102. D. P. Lymberopoulos, V. I. Kolobov, and D. J. Economou, "Fluid simulation of a pulsed-power inductively coupled argon plasma," *Journal of Vacuum Science and Technology A*, Vol. 16, No. 2, 1998, pp. 564–571.
103. K. Sasada, M. Arimoto, H. Nagasawa, A. Nishida, H. Aoe, T. Dan, S. Fujiwara, Y. Matsushita, and K. Yodoshi, "The influence of SiN films on negative bias temperature instability and characteristics in MOSFETs," in *Proceedings of the IEEE International Conference on Microelectronic Test Structures (ICMTS)*, Vol. 11, 1998, pp. 207–210.
104. J. Ushio, T. Maruizumi, and K. K. Abdelghafar, "Interface structures generated by negative-bias temperature instability in Si/SiO_2 and Si/SiO_xNy interfaces", *Applied Physics Letters*, Vol. 81, No. 10, 2002, pp. 1818–1820.
105. M. Terai, K. Watanabe, and S. Fujieda, "Effect of nitrogen profile and fluorine incorporation on negative-bias temperature instability of ultrathin plasma-nitrided SiON MOSFETs," *IEEE Transactions on Electron Devices*, Vol. 54, No. 7, 2007, pp. 1658–1665.
106. Z. Wang, C. G. Parker, D. W. Hodge, R. T. Croswell, M. Yang, V. Misra, and J. R. Hauser, "Effect of polysilicon gate type on the flatband voltage shift for ultrathin oxide-nitride gate stacks," *IEEE Electron Device Letters*, Vol. 21, No. 4, 2000, pp. 170–172.
107. T. Kawae, Y. Minemura, S. Fukuda, T. Hirano, Y. Suzuki, M. Saito, S. Kadomura, and S. Samukawa, "Drastically improved NBTI lifetime by periodic plasma nitridation for 90-nm mobile applications at low-voltage operation,"in *Proceedings of the International Workshop on Gate Insulator*, 2003, pp. 146–149.
108. S. Tsujikawa, T. Mine, Y. Shimamoto, O. Tonomura, R. Tsuchiya, K. Ohnishi, H. Hamamura, K. Torii, T. Onai, and J. Yugami, "An ultrathin silicon nitride gate dielectrics with oxygen enriched Interface (OI-SiN) for CMOS with EOT of 0.9 nm and beyond," *Symposium on VLSI Technology Digest of Technical Papers*, 2002, pp. 202–203.
109. Y. J. Lee, Y. C. Tang, M. H. Wu, T. S. Chao, P. T. Ho, D. Lai, W. L. Yang, and T. Y. Huang, "NBTI effects of PMOSFETs with different nitrogen dose implantation," in *Proceedings of the 42nd IEEE International Annual Reliability Physics Symposium*, 2004, pp. 681–682.
110. T. B. Hook, E. Adler, F. Guarin, J. Lukaitis, N. Rovedo, and K. Schruefer, "The effect of fluorine on parametrics and reliability in a 0.18-μm 3. 5/6.8-nm dual-gate-oxide CMOS technology," *IEEE Transactions on Electron Devices*, Vol. 48, No. 7, 2001, pp. 1346–1353.
111. D. -Y. Lee, H. -C. Lin, W. -J. Chiang, W. -T. Lu, G. -W. Huang, T. -Y. Huang, and T. Wang, "Process and doping dependence of negative-bias-temeprature instability for *P*-channel MOSFETs," in *Proceedings of the 7th International Symposium on Plasma- and Process-Induced Damage*, 2002, pp. 150–153.

112. A. Zhou, J. Liu, K. Zheng, J. Ju, J. Zhang, H. Gan, H. Ho, and J. Ning, "Fluorine IMP effect on negative bias temperature instability and process induced defects," *Electrochemical Society Transactions*, Vol. 27, No. 1, 2010, pp. 67–72.
113. R. C. Weast and M. J. Astle, *CRC Handbook of Chemistry and Physics*, 62nd ed., Boca Raton, FL, CRC Press, 1981.
114. S. J. Chang, S. Y. Wu, C. L. Chen, T. L. Lee, Y. M. Lin, Y. S. Tsai, H. D. Su, S. B. Chen, Y. S. Chen, M. S. Liang, Y. C. See, and Y. C. Sun, "An integrated gate stack process for sub-90-nm CMOS technology," in *Proceedings of the International Conference on Solid State Devices and Materials*, Tokyo, 2003, pp. 460–461.
115. D. -Y. Lee, H. -C. Lin, C. -L. Chen, T. -Y. Huang, T. Wang, T. -L. Lee, S. -C. Chen, and M. -S. Liang, "Impacts of HF etching on ultra-thin core gate oxide integrity in dual gate oxide CMOS technology," in *Proceedings of the 8th International Symposium on Plasma- and Process-Induced Damage*, 2003, pp. 77–80.
116. E. Morifuji, T. Kumamori, M. Muta, K. Suzuki, M. S. Krishnan, T. Brozek, X. Li, W. Asano, M. Nishigori, N. Yanagiya, S. Yamada, K. Miyamoto, T. Noguchi, and M. Kakumu, "New guideline for hydrogen treatment in advanced system LSI,"in *Symposium on VLSI Technology Digest of Technical Papers*, 2002, pp. 218–219.
117. A. Suzuki, K. Tabuchi, H. Kimura, T. Hasegawa, S. Kadomura, K. Kakamu, H. Kudo, M. Kawano, A Tsukune, and M. Yamada, "A strategy using a copper/low-*k* BEOL process to prevent negative-bias temperature instability (NBTI) in *p*-MOSFETs with ultrathin gate oxide," in *Symposium on VLSI Technology Digest of Technical Papers*, 2002, pp. 216–217.
118. G. Cellere, M. G. Valentini, L. Larcher, and A. Paccagnella, "Plasma induced microbreakdown in small-area MOSFETs," *IEEE Transaction on Electron Devices*, Vol. 49, No. 10, 2002, pp. 1768–1774.
119. M. Yamamura, T. Matsuki, T. Robata, T. Watanabe, S. Inumiya, K. Torii, T. Saitou, H. Amai, Y. Nara, M. Kitazoe, Y. Yuba, and Y. Akasaka, "Improvement in NBTI by catalytic-CVD silicon nitride for hp-65-nm technology," in *Symposium on VLSI Technology Digest of Technical Papers*, 2005, pp. 88–89.
120. H. Matsumura, "Silicon nitride produced by catalytic chemical vapor deposition method," *Journal of Applied Physics*, Vol. 66, No. 8, 1989, pp. 3612–3617.
121. S. S. Chung, "Reliability issues for high-performance nanoscale CMOS technologies with channel mobility enhancing schemes,"in *Proceedings of the International Workshop on Nano CMOS*, 2006, pp. 128–131.
122. H. S. Momose, T. Ohguro, K. Kojima, S. Nakamura, and Y. Toyoshima, "1. 5–nm gate oxide CMOS on (110) surface-oriented Si substrate," *IEEE Transactions on Electron Devices*, Vol. 50, No. 4, 2003, pp. 1001–1008.
123. A. T. Krishnan, V. Reddy, S. Chakravarthi, J. Rodriguez, S. John, and S. Krishnan, "NBTI impact on transistor and circuit: Models, mechanisms and scaling effects," in *Proceedings of the International Electron Devices Meeting (IEDM)*, 2003, pp. 349–352.
124. R. Fernández, B. Kaczer, A. Nackaerts, S. Demuynck, R. Rodríguez, M. Nafría, and G. Groeseneken, "Ac NBTI studied in the 1 Hz–2 GHz range on dedicated on-chip CMOS circuits," in *Proceedings of the International Electron Devices Meeting (IEDM)*, 2006, pp. 1–4.

CHAPTER 8
Plasma-Induced Damage

8.1 Introduction

Plasma processes are used widely in the manufacturing of ultra-large-scale-integration (ULSI) devices, for example, for plasma implantation, etching, and deposition. With the continue scaling of integrated circuits (ICS), the requirement for better control of the thermal budget has led to more plasma steps involved in the integrated circuit (IC) fabrication, and the need for higher directionality of etch and deposition has resulted in the use of higher plasma current densities. However, the side effect for almost all processing steps involving the use of plasma is undesirable charging damage to the devices.[1] The plasma-induced damage (PID), from a macroscopic viewpoint, includes radiation by photons, physical bombardment by ions, contamination, and plasma charging by ions and electrons. Plasma charging is the buildup of charges on the wafer surface during plasma processing. Charging-induced oxide damage during plasma processing has become one of the critical reliability concerns as the gate-oxide thickness is rapidly scaled (especially when the gate-oxide thickness is around 3 nm) and the number of dry-etching and other plasma processing steps increase. The extent of charging damage suffered by a device is determined by the electrical stress (current or voltage) experienced by the gate dielectric during the plasma process, which causes trap creation in the oxide and at the oxide-channel interface,[2] increased stress-induced leakage current (SILC),[3] and early breakdown,[4] thus leading to the adverse impact of charging damage on yield loss,[5] gate-oxide integrity (GOI),[3] hot-carrier injection (HCI) lifetime,[6] negative-bias temperature instability (NBTI) lifetime,[7] and so on. When charging stress is milder, interface traps are likely generated,[8] resulting in large metal-oxide-semiconductor (MOS) device characteristic variations (either die-to-die, wafer-to-wafer, or lot-to-lot variation).[9,10]

Plasma damage is design-related. When a gate electrode is attached to a conductive layer (such as poly, contact, metal interconnect, and via),

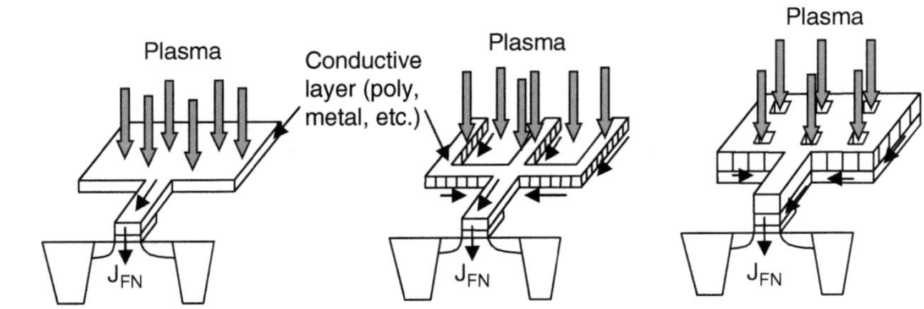

FIGURE 8.1 Schematic illustration of the three types of antennas.

plasma charging to the gate likely happens through some collection paths of the conductive layer, which is in general referred as an *antenna*. On the one hand, the antenna area determines the amount of charge collected by a gate dielectric exposed to plasma. On the other hand, the gate area determines the current density passing through the gate dielectric. As a result, the term *antenna ratio* (*AR* = antenna area/gate area) has been used extensively as an index to determine susceptibility to charging damage in a specific process. Basically, the larger the antenna ratio, the larger is the amount of charge buildup, and thus the greater plasma damage to the gate oxide will be induced. The antenna can collect the charge through the surface of the conductive layer (area-intensive), through the edge of the conductive layer (edge-intensive), and/or through a contact/via. Figure 8.1 schematically shows the three types of antennas. When the charges from the plasma accumulate on the gate electrode, they will lead to the Fowler-Nordheim (FN) tunneling current through the gate dielectric and thus damage it. Table 8.1 lists the typical processes

	Area-Intensive Type	**Edge-Intensive Type**	**Contact/Via Type**
Antenna	Metal, poly, pad (parasitic antenna)[12]	Metal, poly	Contact, via
Typical processes involved in the damage	Implantation, chemical-vapor deposition (CVD) deposition, ashing, and LDD spacer etching	Metal/poly etching, CVD with electron shading effect	Contact/via etching and ashing
Antenna ratio	Antenna-area ratio	Antenna sidewall area	Contact/via area and number

TABLE 8.1 List of Typical Processes Leading to Plasma Damage and Antenna Ratio for the Three Types of Antennas

involved in plasma damage and the main parameters in antenna structures for the three types of antennas. *Design/antenna rules are usually set to limit the antenna area or perimeter connected to the gate dielectric of metal-oxide-semiconductor field-effect transistors (MOSFETs), in order to limit the gate-film degradation caused by charging damage.*[11]

Plasma damage is also process-related. A Cu damascene process differs from the conventional Al process in that the metal lines are formed by depositing Cu in trenches formed in the dielectric and then polishing away the excess Cu. Because the damascene process does not have the metal etch steps that conventional Al processes have, Cu damascene wafers have been demonstrated to have less process-induced damage.[13] However, damascene processes still expose the wafer to plasma environments during dielectric etching and plasma preclean/deposition steps (such as the barrier/seed preclean/deposition and dielectric deposition/preclean), and thus controlling plasma process–induced damage is still of significance.[14,15] Advanced-technology nodes have called for five to six or even higher levels of metallization, resulting in multiple wafer exposures to the plasma process. In addition, with the integration of ultralow-*k* (ULK) materials in back-end-of-line (BEOL) interconnects for 45 nm and beyond, film degradation caused by plasma etch/ash[16–18] has become an unavoidable challenge.[19]

In this chapter the details of the PID mechanisms and the concept of antenna ratio are given first. This is followed by introduction of characterization methods used for PID assessment. The characteristics of plasma itself then are presented because such knowledge is very critical to an understanding of the damage caused during plasma processing. Then the three key ways to achieve a PID-free process are provided in detail: (1) use of protection devices (such as diodes), (2) optimization of the design of the antenna layout, and (3) process optimization, including optimization of the plasma itself (uniformity, intensity, etc.) and plasma-involved processes such as chemical-vapor deposition (CVD), etch, and so on. The PID effects on GOI, NBTI, and HCI then are briefly introduced. Finally, the process-qualification requirements for PID are listed as a reference.

8.2 Mechanisms for Plasma-Induced Damage

Several mechanisms have been identified as the root cause for PID. The most important factors are listed below and detailed in sequence:

- Plasma density
- Plasma nonuniformities across the wafer
- Electron-shading effect (ESE)
- Reverse electron-shading effect (RESE)
- Ultraviolet (UV) radiation

8.2.1 Plasma Density

From the charging and then damage prospective, high-density plasmas have the potential for more severe damage if the needed voltage across the gate oxide is satisfied because plasma tends to act as a current-limited voltage source, and high density means more current.[1] Krishnan et al.[20] demonstrated that with a decrease in the chamber height from 8 to 5 cm, a fivefold increase in the azimuthal component of the inductively coupled electrical field (associated with a source radio frequency) close to the wafer surface is observed. Thus serious plasma charging stress from an inductively coupled plasma (ICP) metal etch is imposed to the 60 Å devices with the 5-cm-high chamber. In contrast, the plasma damage is negligible when the chamber height is 8 cm, resulting from the decreased ion/electron density as well as the decreased electrical field near the wafer surface.

8.2.2 Plasma Nonuniformity Across the Wafer

Plasma nonuniformity across the wafer is a well-known factor leading to plasma damage.[21,22] Figure 8.2 schematically illustrates the plasma-induced current path in gate oxide in a uniform and a nonuniform plasma. In a uniform plasma, the oxide voltage V_{ox} could be small owing to the fact that local time-average current $J_e = J_i$. However, in a nonuniform plasma, there is another current path (current path II in Fig. 8.2*b*) from the high-potential part to the lower-potential part of the wafer, leading to much larger damage to the gate oxide. The severity of the stress/failure depends on the magnitude of the current flux and the wafer temperature, as detailed later.

Plasma source and reactor design are the main factors that affect plasma uniformity, including magnetic field, gas-injection mode, chemistry and pressure, electrode edge effect,[23] source/bias power, and so on.

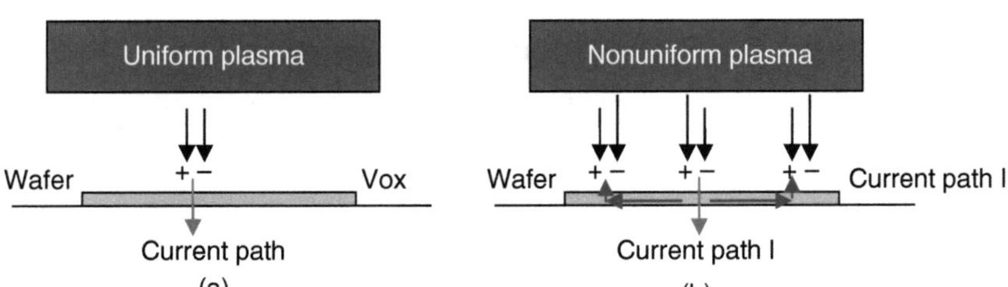

FIGURE 8.2 Schematic to illustrate the plasma-induced current path in gate oxide. (*a*) In a uniform plasma, there is one current path passing through the substrate. (*b*) In a nonuniform plasma, there is another current path (current path II) from the high-potential part to the lower-potential part of the wafer.

8.2.3 Electron-Shading Effect

Plasma consists of electrons, ions, and neutral radicals. The electrons are of more anisotropic characteristic than the ions. In other words, the electron incident angle during a plasma process has a much larger spread compared with that of the ions. The difference in incident angles and the shading of electrons by photoresist result in a net *positive potential stress* applied to the device during a plasma process, which is referred as the *electron-shading effect* (ESE).[24] Figure 8.3 schematically shows the reactive plasma clean (RPC) process using H_2 plasma to clean the metal surface prior to barrier/seed deposition. It is likely that the H^+ ions (i.e., positive charges) will accumulate on the metal surface, which will further electrically stress the gate oxide. ESE has been shown to play a critical role in antenna charging, especially in advanced-technology nodes with increasing aspect-ratio features, electron temperature, and ion current density.[25]

When electron shading is the key mechanism of charging, wafers patterned with thicker resist will tend to have greater plasma damage.

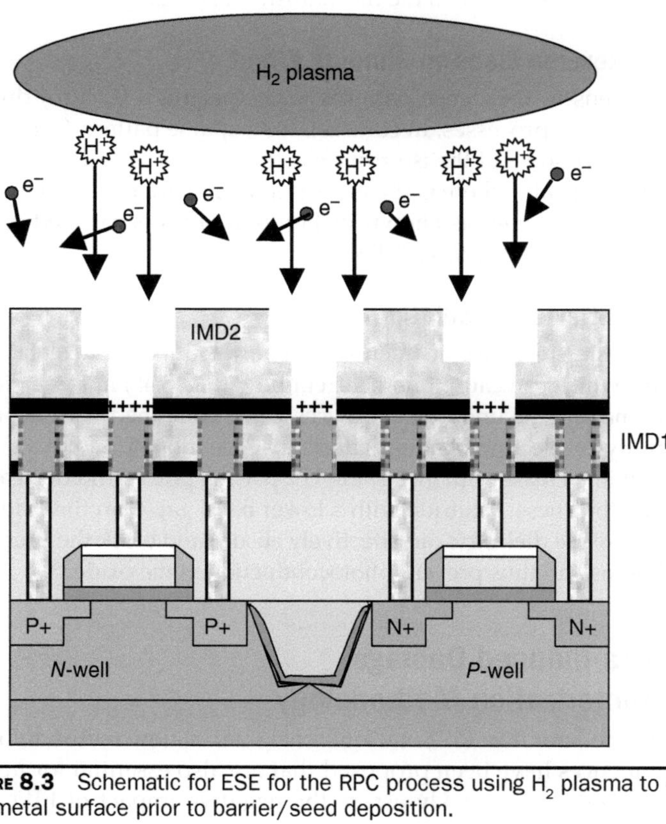

Figure 8.3 Schematic for ESE for the RPC process using H_2 plasma to clean the metal surface prior to barrier/seed deposition.

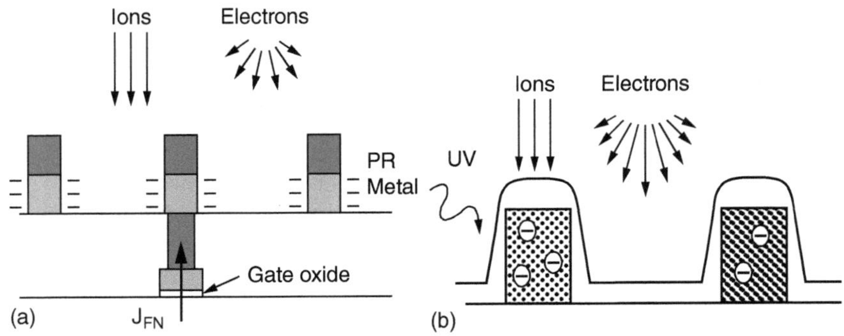

FIGURE 8.4 (*a*) Schematic for reverse electron-shading effect (RESE). (*b*) Schematic for the effect of UV radiation. (*Used with permission from Elsevier.*[25])

Krishnan et al.[20] reported that in contrast to a 1.3-μm-thick resist, resists of 1.6- or 2.0-μm thicknesses suffer catastrophic increases in I_g for both N-channel metal-oxide semiconductor (NMOS) and P-channel metal-oxide semiconductor (PMOS) devices. This type of aspect-ratio-dependent charging and the invariance of charging with overetch prove that ESE is the mechanism of charging.

8.2.4 Reverse Electron-Shading Effect

ESE happens to the dense patterns (e.g., spacing = 0.5 μm) during plasma-related processes. In contrast, for a sparse pattern (e.g., spacing = 2.0 μm), some of the isotropic electrons are collected by the pattern-line sidewall, whereas the anisotropic ions are not, leading to a net *negative charging,* as shown in Fig. 8.4*a*. This is referred as the *reverse electron-shading effect* (RESE).[25,26]

8.2.5 Ultraviolet Radiation

The processing plasma is an intense source of vacuum ultraviolet (VUV) photons, which can cause photocurrent flow (Fig. 8.4*b*) in these dielectrics, damaging the underlying devices when combined with the nonuniformity or electron-shading effects just mentioned.[27] Photocurrent density will increase with increasing UV plasma power and decreasing dielectric thickness. A nitride with a lower band gap than the oxide on top of the oxide dielectric can effectively absorb and block the energetic UV photons and thus prevent photoconduction in the oxide.[28]

8.3 Plasma-Induced Damage Characterization Methodology

In order to monitor PID, various characterization methodologies and sequences have been proposed that are also essential for in-line characterization to quickly assess the negative impact of PID on a minimum set of test structures. However, until now no general standard has been established for PID assessment. Currently, a Joint Electron Device Engineering Council (JEDEC) standard is under preparation to fill this gap.[12]

In the early days, comparison was done directly for such device parameters as I_{dsat}, V_t, transconductance G_m, interface state D_{it}, and gate-oxide leakage I_g between the devices connected with an antenna and the reference device without an antenna.[29–32]

For monitoring gate leakage, the gate is tested under inversion or accumulation at the operating voltage. A high initial leakage current is a strong indicator for PID. A common practice to assess the PID performance is to compute the PID failure rate based on the percentage of failure events where the initial I_g exceeds the preset failure I_g criterion (e.g., 10 pA for thick gates and 10 nA for thin gates).[33] This method of evaluating gate leakage is suitable for monitoring heavy damage, which has become the most widely used parameter for today's advanced-technology nodes. Some details of PID characterization can be found in Ref. 34.

Latent damage also can be generated by PID, and the so-called revealing or diagnostic stress methodology is applied.[35] This approach is also suggested by a JEDEC standard.[36] In this methodology, parameter measurements are performed both before and after an electrical stress. The purpose of such stress is to reveal any potential plasma damage that had been annealed by subsequent processing steps at high temperatures.[31,37]

- One method to reveal the latent damage is to apply a constant gate-current density (e.g., 0.1 A/cm^2) to the transistors for a few seconds (e.g, 4 seconds)[38] so as to simulate the FN wafer-charging stress situation. Others refer to this as constant-current stress (CCS).[32,35,37] Through a simple calculation, 4 seconds of CCS with 0.1 A/cm^2 current density will lead to a 0.4 C/cm^2 charge injected through the gate oxide. It is noted that such a revealing stress is not intended to cause breakdown, and the quantity of injected charge is required for correct interpretation of the stress data.[12,39] In terms of device parameters, the shifts before and after FN stress of the threshold voltage V_t, G_m, subthreshold swing (SS), and interface state D_{it} of a MOSFET (either *N*-channel or *P*-channel or both) with a large antenna ratio is compared with the reference counterparts without an antenna. In most cases, a negative V_t shift is observed,[40] indicating that plasma treatment/deposition gave rise to positive charge traps in the gate oxide. The larger the V_t shift, the greater is the PID. The interface state D_{it} generally is characterized by charge pumping or direct-current current-voltage (DCIV), as elaborated in Chap. 4. However, the D_{it} method is limited by leakage because $t_{ox} < 3$ nm.[8,41]

- The other types of measurements to reveal latent damage include hot-carrier stress[31] and NBTI,[7] *both of which are effective methods to check the latent PID.* Device HCI degradation owing to PID can follow a different lifetime extrapolation model from device degradation without PID.[42] The disadvantage of a hot-carrier stress is that the gate dielectric is not degraded homogeneously along the channel region. In this case, the drain side of the MOS transistor experiences the largest stress, whereas the PID usually is homogeneously distributed across the oxide layer along the channel. As a result, only part of the gate-dielectric damage can be checked by a hot-carrier stress. Additionally, hot-carrier stress and NBTI have the disadvantage of long stress times, which are not suitable for fast process monitoring during mass production.

A short revealing CCS has been proposed by Martin et al.,[35] and this has been demonstrated to be effective and as sensitive as a 100-s hot-carrier stress, providing a fast, productive wafer-level test to reveal latent damage through short stress times, but with high efficiency.[43] Cumulative V_t-shift distributions after a revealing 100-s hot-carrier stress of NMOS transistors are shown in Fig. 8.5 for a reference device (without an antenna) and devices with antennas (antenna I and antenna II). The reference device has no clear V_t shift after the revealing stress. However, devices with either antenna I or antenna II have a clear V_t shift, and the antenna II device is much worse.

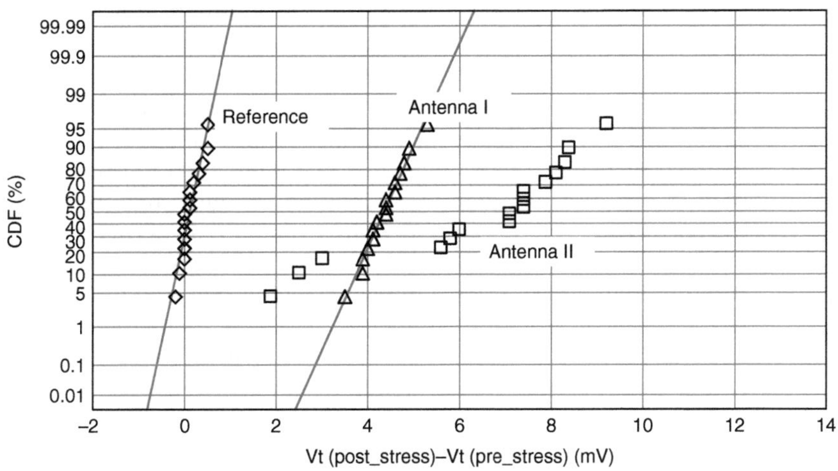

FIGURE 8.5 Cumulative V_t-shift distributions after a revealing 100-s hot-carrier stress of NMOS transistors: reference device (without an antenna) and devices with antennas (antenna I and antenna II). It is clear that the antenna II devices are much worse.

Q_{bd} or V_{bd} is another parameter that is used for the PID characterization. However, they are destructive measurements. A fast exponentially ramped current breakdown stress including soft breakdown detection (in inversion and accumulation) was proposed.[44] Q_{bd} or V_{bd} also may sometimes not be able to detect the plasma damage as well as the initial I_g and V_t measurements. Results show that the gate-oxide-breakdown voltages do not indicate any type of PID because all test keys show the expected intrinsic characteristics between 9 and 10 V, meaning that this method is not as effective as CCS stressing, as shown in Fig. 8.5.

Initial electron-trapping rate (IETR) or slope (IETS) was proposed[45] *suitable for evaluating PID in thick oxides.* In this method, the sample is first annealed (e.g., at 850°C for 60 minutes in an Ar atmosphere) after plasma processing; then a voltage-versus-time curve (*V-t* curve) is measured when sustaining a constant current injection (e.g., 10 mA/cm^2). This voltage variation relates directly to the net charges trapped under the current stress. The slope of the *V-t* curve reflects the electron-trapping rate (ETR) corresponding to the true PID.

Noise has also been used for PID characterization, and the test is nondestructive and sensitive even when the gate oxide is thinner than 4 nm.[46] There is a strong dependence of MOSFET low-frequency noise on the polarity of plasma-charging damage. When the plasma-charging damage is of the gate-injection type, the MOSFET low-frequency noise is insensitive to damage, whereas when the polarity is of the substrate-injection type, the MOSFET low-frequency noise is very sensitive to damage.

8.4 Plasma Characteristics

8.4.1 Plasma Characterization Methodology

Stanford plasma on-wafer real-time (SPORT) testing[23,47–50] is a real-time monitor with metal probes fabricated on the wafer that is suitable for equipment development. The basic concept for this probe is to simulate the antenna conductor located on the top of a dielectric insulating layer and to measure the voltage difference across the dielectric through real-time direct-voltage measurement. This method may be good only for low-bias conditions owing to high impedance between the line and the chamber wall during measurement.

CHARging Monitor-2 (CHARM-2) monitor wafers[51–53] are Electrically Erasable Programmable Read-Only Memory (EEPROM)–type sensors that provide users with wafer maps of wafer surface potential (positive or negative) and charge flux during plasma-involved processes. The principle is described briefly as follows: When measuring surface potential of a given polarity (either positive or negative), a charge-collection electrode (CCE) is placed on the surface of the wafer, connected to a high-input impedance voltmeter (i.e., EEPROM transistor here) shunted by a diode of opposite polarity. On the other hand, for measuring charge flux, the diode is replaced by resistors so that the voltage drop across the resistors is measured. To perform the desired measurements using CHARM-2 monitors, the threshold voltages V_t of the EEPROM transistors are preprogrammed, and then the CHARM-2 wafers are placed in process equipment. During the processing, the resulted V_t values on a parametric tester are measured. Because the EEPROM transistors are reprogrammable, this cycle is repeated many times using the same wafer. The conversion of the postprocess EEPROM V_t values to surface potentials is done using the V_t-V_g calibration curve. This method is suitable for monitoring specific tools (e.g., ion implanters).

Semiconductor process-induced damage effect revealer and remover (SPIDER^2)[54] is a test-structure chip designed to monitor process-induced charge damage. For example, Aum et al.[54] provided a 20 × 20 mm chip-size SPIDER^2 with more than 800 systematically arranged test structures to simulate the advanced-device IC interconnection layout design. Within this chip, charge-collection antennas are connected not only to the MOS gate but also to the transistor source, drain, and substrate with mutually different directions, distances, sizes, spacings, layers, and shapes to detect the interactions of these antennas across the wafer. In addition, reference and guard-ring devices, which protect the devices from charge damage, are built in for the purposes of reference and quantitative comparison.

Surface photo voltage (SPV) measurement[55–58] is a noncontact direct-measurement technology for plasma charges at the dielectric surface and those trapped in the dielectric as well. A typical commercial example is Quantox, provided by KLA-Tencor.[59] The principle of SPV measurement by Quantox is shown schematically in Fig. 8.6 and explained below.

- The first step in the measurement is to supply negative or positive charges (i.e., Q_1 in Fig. 8.6*a*) to the dielectric film surface on the substrate (such as SiO_2 or low-*k* film) by corona discharge. This is similar to applying an external voltage V_g to the gate, as shown in Fig. 2.10. Q_2 in Fig. 8.6*a* consists of the interface traps at the substrate-dielectric interface induced, for example, by the plasma process. There also could be fixed charges in the dielectric film, which are ignored here to simplify the illustration.
- The second step (Fig. 8.6*b*) is to measure the surface voltage V_{surf} using a kelvin probe. The measured V_{surf} is equivalent to the V_g in Fig. 2.10. Note that the surface voltage V_{surf} here results from all the charge components (Q_1 and Q_2).

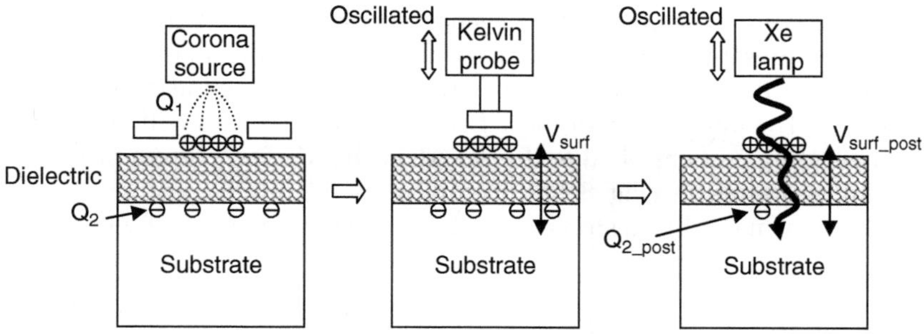

FIGURE 8.6 Schematic of the SPV measurement principle by Quantox.

- In the third step (Fig. 8.6*c*), a Xe lamp is initiated to produce photons to penetrate into the dielectric film (SiO_2), which generates electron-hole pairs in Si. For example, if a negative charge is generated at the Si-SiO_2 interface by the plasma process, the hole from the electron-hole pair may recombine with the negative charge, and the charge Q_2 would be changed to Q_{2_post}. Therefore, surface voltage that results from the Q_1 and Q_{2_post} charges is measured again by the kelvin probe (referred as V_{surf_post}).
- In the fourth step, the SPV is defined as the difference between V_{surf} and V_{surf_post} [Eq. (8.1)]:

$$\text{SPV} = |V_{surf} - V_{surf\_post}| \tag{8.1}$$

The SPV is the voltage resulting from the difference between Q_2 and Q_{2_post}. In other words, *SPV is the voltage resulting from the number of charge traps near the interface* and becomes an index showing charging damage. Equation (2.64) [which is repeated here as Eq. (8.2)] clearly shows that oxide traps near the Si-SiO_2 interface will lead to larger flat-band changes and thus a larger V_t shift.

$$\Delta V_{fb} = -\frac{\int_{x=0}^{x=t_{ox}} x \cdot \rho(x)\,dx}{t_{ox} \cdot C_{ox}} \tag{8.2}$$

- In some cases, sweeping the charge Q_1 from positive to negative or vice versa is necessary by repeating the preceding steps. A full SPV response then is obtained to measure the interface states.

8.4.2 Plasma *I-V* Characteristics and Plasma Parameter Effects on Plasma *I-V* and Damage

In general, to determine the amount of damage experienced by the gate oxide during a plasma process, the first step is to characterize the electrical stress created by the plasma process through a load-line analysis—presented by a plasma *I-V* curve. Then the plasma *I-V* curve is plotted in the same plot with the oxide tunneling-current load line. The intersection of these two types of *I-V* curves will yield the operating point applied to the gate dielectric during plasma processing, as detailed in Fig. 8.7.

The plasma-induced stress current during charging in Fig. 8.7 can be modeled as follows[62]:

$$I_{\text{plasma}} = A_{ant} J_i \{1 - \exp[-(V_{oc} - V)/k_B T_e]\} \tag{8.3}$$

where A_{ant} is the antenna area, J_i is the ion current density collected by the antenna, V_{oc} is the voltage without tunneling (i.e., open-circuit voltage) in gate dielectric, V is the voltage with tunneling, and T_e is the electron temperature.

According to Eq. (8.3), the factors affecting the plasma-induced stress current are detailed below:

- Open-circuit voltage V_{oc}:
  - Minimizing V_{oc} is the prime approach to reducing plasma damage.[60]
  - V_{oc} can be affected by plasma nonuniformities as follows:

    $$V_{oc} \propto T_e \ln\left(\frac{n_{\text{max}}}{n_{\text{min}}}\right) \tag{8.4}$$

    where n_{max} and n_{min} are the maximum and minimum ion densities, respectively. Therefore, plasma nonuniformity should be reduced.
  - V_{oc} also can be affected by electron shading as follows:

    $$V_{oc} \propto T_e \ln\left(\frac{k_i}{k_e}\right) \tag{8.5}$$

    where k_i and k_e are the shadowing factors for ions and electrons.[61]
- Electron temperature T_e:
  - A lower T_e produces less damage. This is easy to understand because the lower T_e will proportionally reduce the voltage applied to the gate dielectric according to Eqs. (8.4) and (8.5), thus exponentially reducing the tunneling current and plasma damage. The plasma *I-V* curve will shift leftward as the electron temperature decreases (Fig. 8.7).
- Ion current density collected by the antenna J_i:
  - J_i is directly proportional to plasma density, and higher J_i values lead to greater degradation. Higher ion densities increase the electron and ion-saturation currents, leading to an upward shift in the plasma *I-V* curve (Fig. 8.7), thus revealing a higher potential for damage.

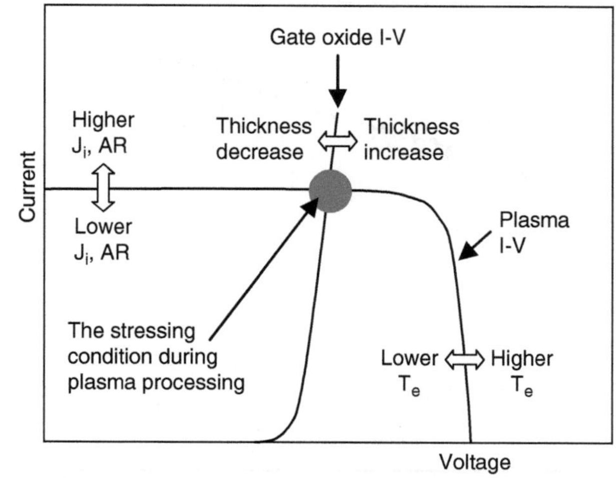

FIGURE 8.7 Schematic shows the stressing condition applied on the gate oxide during plasma processing. The trends for the effects of gate-oxide thickness, electron density, antenna area, and electron temperature on the *I-V* characteristics are also shown.

  - Higher ion densities delay the transition from constant-voltage stress to constant-current stress with oxide scaling (because the higher-density plasma can supply more current), shifting the damage peak to thinner oxides.
- Antenna area A_{ant} or antenna ratio (*AR*):
  - From the plasma collection point of view, a larger antenna can be analogous to a higher ion density, so the same results apply (Fig. 8.7).
  - The difference between *AR* and the ion density is that in practices, *AR* can be scaled by a very large amount such as five orders, whereas ion density has a range of only about three orders. *AR* is currently the preferred method for amplifying plasma damage for detection purposes, especially during the process-development stage.

8.5 Effect of Substrate on Plasma-Induced Damage

8.5.1 Why PMOS Is Worse Than NMOS

A number of works have reported that NMOS devices are less damaged by plasma processes than PMOS devices.[38,62] Figure 8.8 gives an example. Some literature attributes this phenomenon to boron penetration into the gate oxide in PMOS devices.[34,63] On the other hand, this phenomenon is explained qualitatively by the interaction between the plasma *I-V* characteristic and the oxide-tunneling-current loading curve (Fig. 8.9). The plasma can be regarded

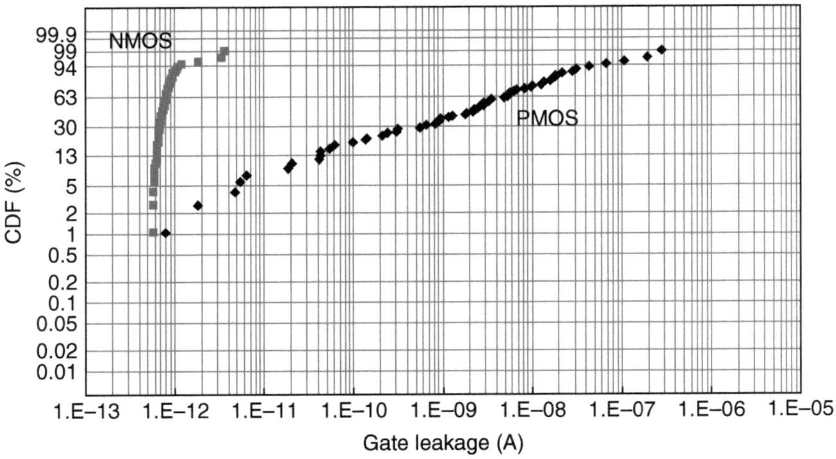

FIGURE 8.8 Cumulative probability versus gate current, showing that NMOS devices degrad less than PMOS devices for the same antenna ratio.

as a voltage-dependent current source. As discussed earlier, the stress condition (i.e., stress voltage and stress current) during a plasma process can be characterized as the intersection point between the *I-V* characteristics of the plasma and those of the oxide tunneling. In Fig. 8.9, *I-V* values for both NMOS and PMOS devices are provided and are overlaid with the plasma *I-V* curve. It is seen that the NMOS device conducts more current than the PMOS device at a specific voltage, which can be attributed to the larger band bending in the NMOS device at a specific positive voltage. It is quite obvious that it is the positive charge that imposes on the gate oxide during the plasma process, for example, attributed to ESE (see Fig. 8.3). When

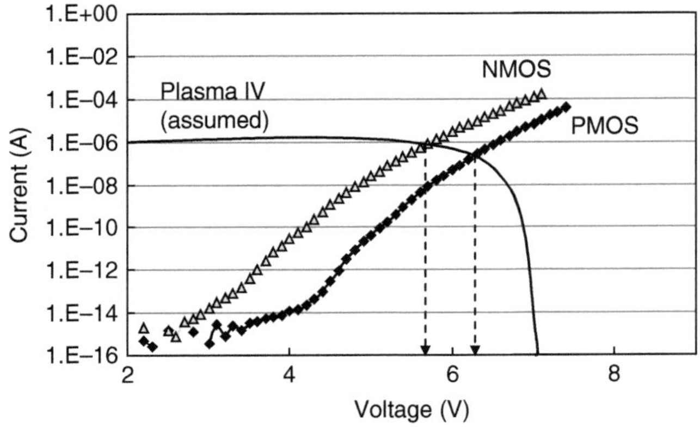

FIGURE 8.9 Current-versus-voltage characteristic curve of plasma overlaid with that of oxide showing the dependence of stress voltage on MOS type. Stress voltage in NMOS devices is lower than that in PMOS devices.

considering positive charges on the gate, NMOS and PMOS devices are under inversion and accumulation, respectively. According to Eq. (6.7),

$$|V_{ox}|_{inv} - |V_{ox}|_{acc} = E_g/q - |V_{poly}| > 0 \tag{8.6}$$

Therefore, the voltage applied to the gate oxide V_{ox} is higher in NMOS devices than that in PMOS devices.

The overlaid *I-V* curves shown in Fig. 8.9 indicate that for a specific antenna area, the PMOS device would suffer from a higher stress voltage. As described by the anode-hole-injection (AHI) model[64] detailed in Chap. 6, a higher stress voltage in PMOS devices means that such devices could fail with a higher probability than NMOS devices.

Krishnan et al.[20] reported that NMOS devices exhibited much less V_t shift (~70-mV shift) with an antenna compared with PMOS devices (~140-mV shift) after being subjected to ICP metal etch. However, this does *not* mean that NMOS devices are free of any stress from the plasma. As described in Chap. 4, the charge-pumping current I_{cp} is a linear function of the average interface-state density. Charge-pumping measurements by Krishnan et al.[20] indicated a distinct degradation of NMOS devices that increases with antenna size. The I_{cp} was about 60 pA for the control test structure without an antenna. When the antenna perimeter was 17,000 μm, the I_{cp} was about 80 pA (i.e., 33 percent increased). What's more, the I_{cp} doubled when antenna perimeter increased to 59,000 μm. It also was seen that the trend in the NMOS I_{cp} across the wafer was similar to the V_t shift in PMOS devices (i.e., wafer center was subjected to plasma damage more seriously), suggesting that the NMOS device is subjected to a comparable severity of plasma stressing. In addition, the NMOS device with a 59,000-μm antenna perimeter showed 5 percent degradation in linear drive current I_{dlin} compared with the one with the 17,000-μm antenna perimeter, and this could not be recovered even after a forming-gas anneal at 400°C.

8.5.2 Effect of Protection Devices

In 1989, Shone et al.[65] proposed protection diodes (PDs) as a means of avoiding plasma charging. The diodes are connected in parallel to the gate, providing an alternate leakage path for the plasma-induced current during processing. In general, the diode protection structures use an *N*+/*P*-well diode for the *N*-channel transistors (Fig. 8.10*a*) and a *P*+/*N*-well diode for the *P*-channel transistors (Fig. 8.10*b*). Such an arrangement ensures that the diode does not affect the functionality of the circuit during normal operation. There are several explanations for the protection mechanism provided by the PDs under reverse bias, as detailed below:

Model I: The junction breakdown voltage is lower than the gate-oxide breakdown voltage. Therefore, even under reverse bias during

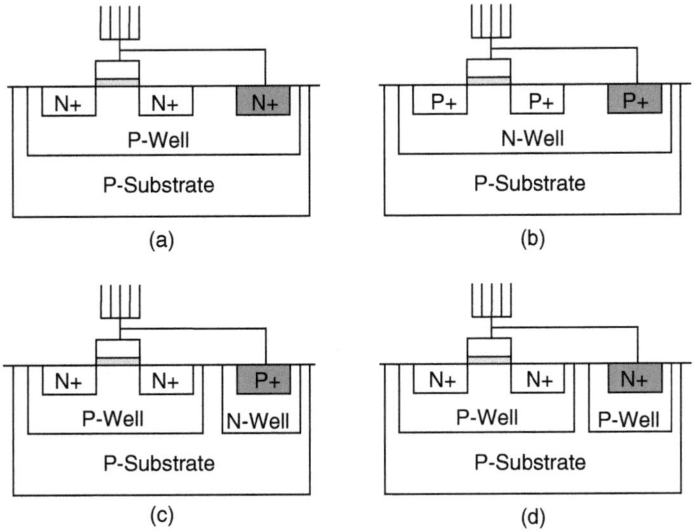

FIGURE 8.10 Schematics of different protection schemes: (*a*) *N*+/*P*-well single diode suitable for NMOS devices; (*b*) *P*+/*N*-well single diode suitable PMOS devices; (*c*) *P*+/*N*-well/*P*-substrate with floating *N*-well double-sided diode suitable for both NMOS and PMOS devices; (*d*) *N*+/*P*-well diode with a separate *P*-well.

plasma processing, the protection diode effectively could protect the gate oxide from damage. The reverse-bias leakage in the diode can provide protection and can be adequate if the diode has a large enough area.[66]

Model II: The presence of light during plasma processing can enhance the diode reverse leakage current through photon-generated current resulting from the light emission of the plasma.[66] However, this may not be able to explain the effectiveness of a PD for a metal-etching process because the diode is beneath or covered by the metal, which may block any plasma illumination from reaching the diode.

Model III: It is the high ambient (or substrate) temperature during the plasma processes that leads to the large diode reverse leakage current.[67] The high temperature ensures low impedance of the diode during processing so the charge can easily flow to the substrate. During plasma etching, the wafers can be heated to about 70°C, and during the plasma deposition processes, the ambient temperature is 400°C or even higher. Basically, the current flowing through the diode doubles for about every 10°C increase in the diode's temperature.

In contrast, a so-called double-sided diode (Fig. 8.10*c*), which is a floating *N*-well base *PNP* diode, can be used for both PMOS and

NMOS devices.[67] The reason is that the leakage of this structure is very large when the deposition/etch temperature is high. A simulation by SILVACO[68] showed that if the temperature is increased by 100°C, the leakage current of this *N*-well floating diode is about three orders of magnitude higher. On the other hand, its leakage current is negligible at room temperature. As a comparison, considering the working potential of the *N*+/*P*-well (Fig. 8.10*a*) and *P*+/*N*-well (Fig. 8.10*b*) MOSFET. A so-called single-diode can only be used for NMOS and PMOS devices, respectively. Experiments have demonstrated that an *N*+/*P*-well diode inside the same *P*-well for an NMOS device can be located at a distance of up to 100 μm away from the MOS without degrading MOS gate-leakage performance.[69] Therefore, the diode does not need to be placed too close to the gate but still achieves an excellent protection capability.

More interestingly, simulation[67] of the protection diode at 400°C showed that for positive charges, the double-sided diode can provide better efficiency than an *N*+/*P*-well diode does because the leakage current of the double-sided diode is about 10 μA, which is two orders of magnitude higher than that of the reverse-biased *N*+/*P*-well diode (i.e., 0.1 μA). Under negative charging, the leakage current of the double-sided diode is still about 10 μA, whereas the *N*+/*P*-well diode has a higher current when the voltage is larger than 0.2 V because it is forward-biased. What's more, the leakage current of an *N*+/*P*-well diode under negative charging will increase continuously with increases in voltage, and the leakage current can reach about 1 mA when the voltage is about 1 V. The simulation results indicate that the double-sided diode can be more effective because it can provide a leakage-current path during plasma processing for both positive and negative charges. It was verified experimentally that the antenna structures protected by double-sided diodes fail less than those protected by *N*+/*P*-well diodes for both NMOS and PMOS devices, indicating that the double-sided diode is more effective.[67]

However, one should be careful when using double-sided diodes for protection when the process is not subjected to a high-temperature plasma treatment. Zwingman et al.[69] reported that for a metal antenna ratio of 2000, the double-sided diode (Fig. 8.10*c*) and the *N*+/*P*-well diode with a separate well (Fig. 8.10*d*) did not provide effective protection from plasma damage (i.e., the gate leakage was larger than 1 μA/μm in many samples). On the other hand, the *N*+/*P*-well diode with a shared well (Fig. 8.10*a*) did provide much better protection from plasma damage (i.e., the gate leakage was around 0.1 nA/μm).

Along with the effectiveness of a diode for its robustness in preventing a product from charging, a major disadvantage of the application of protection diodes is that they may degrade wafer productivity and device performance. Depending on the layout, the use of additional protection diodes resulted in an area increase of 0 to 10 percent.[70]

In addition, a diode connected to a gate leads to extra input capacitance, which definitely will result in lower speed and higher power consumption. For high-performance devices, such a solution therefore may be unacceptable.

8.5.3 Effect of Gate-Oxide Thickness on Plasma-Induced Damage

It has been found that transistors with different gate-oxide thicknesses have different sensitivities to PID. The analysis starts from calculation of the plasma *I-V*, which is then compared with the oxide tunneling-current load lines, as depicted in Fig. 8.7, which show the voltage and current operating points (i.e., the intersection between the plasma load line and the oxide *I-V* curves) during the plasma process with oxide-thickness effect.

Figure 8.11 shows schematically the plasma-induced damage as a function of gate-oxide thickness. The plasma-induced damage shows a bell-shaped dependence on gate-oxide thickness. That is, the damage peaks at an intermediate gate-oxide thickness, which is generally around 3-4 nm.[60,71] The appearance of a peak for plasma-induced damage at certain gate-oxide thickness is resulted from the competing effects of gate-oxide thickness and that of plasma-induced stress voltage. The phenomenon is explained as follows.

- *Effect of gate-oxide thickness:* PID decreases with thickness since thicker gate oxide leads to lower electrical field applied.

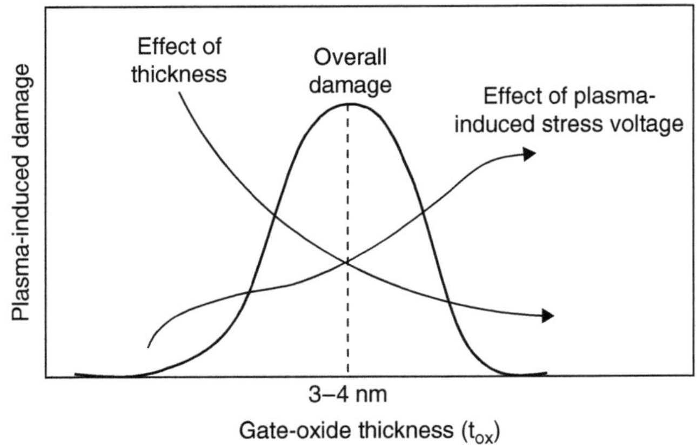

FIGURE 8.11 Schematic shows that there is a peak for plasma-induced damage at certain gate-oxide thickness, resulted from the competing effects of gate-oxide thickness (PID decreases with thickness) and that of plasma-induced stress voltage (PID increases with higher plasma-induced stress voltage). The gate-oxide thickness for the peak plasma-induced damage is generally around 3–4 nm.

The electrical field follows a $1/t_{ox}$ dependence, since $E_{ox} = V_{ox}/t_{ox}$.

- *Effect of plasma-induced stress voltage:* PID increases with higher plasma-induced stress voltage. The stress voltage during plasma processing increases with oxide thickness. The stress voltage can be extracted based on the intersection points as depicted in Fig. 8.7, which shows that a thicker oxide will result in higher stress voltage. However, for a thicker oxide, the stress voltage increases slower than that for an intermediate gate-oxide thickness owing to the dropping down of the plasma *I-V* (Fig. 8.7). On the other hand, for a thinner oxide, it operates in the direct tunneling regime, and the stress voltage drops faster with thickness decrease. Indeed, numerous studies reported that thin oxide is less damaged by plasma attributed to the large intrinsic direct tunneling current inherent in thin gate oxides.[72]

In short, with oxide thickness increase, the stress voltage increases, but the electrical field decreases. From the AHI model,[64] *a lower voltage or a lower electrical field will lead to less damage.* As a result, the bell-shaped dependence of plasma damage on oxide thickness is observed. What's more, with increases in antenna ratio, plasma nonuniformity, ion density, ESE, and electron temperature, the plasma *I-V* curve will shift upward, which will lead to the peak plasma damage shift to thinner oxides as shown in Fig. 8.12.

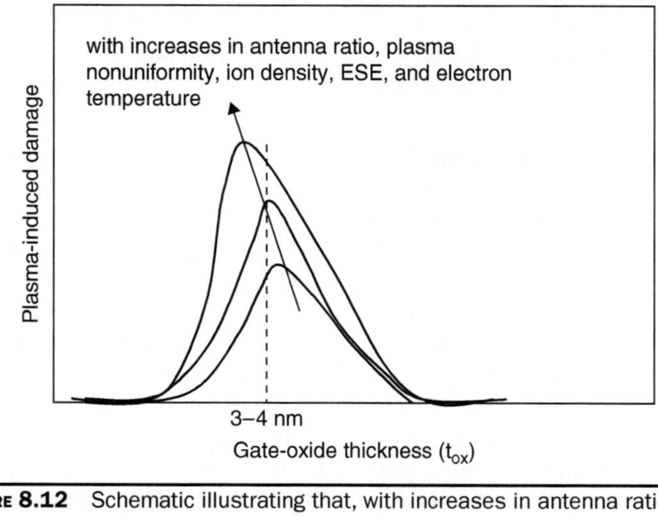

FIGURE 8.12 Schematic illustrating that, with increases in antenna ratio, plasma nonuniformity, ion density, ESE, and electron temperature, the plasma I-V curve will shift upward, which will lead to the peak plasma damage shift to thinner oxides.

8.5.4 Effect on Silicon-on-Insulator Devices

Silicon-on-insulator (SOI) devices could be more robust with respect to plasma-process charging damage than bulk MOSFET devices.[73–75] The buried oxide in an SOI device could prevent the electron flow from the gate oxide to the substrate so that the gate oxide is damaged much less. Reimbold and Poiroux[25] reported that the gate-oxide leakage-current distribution of bulk MOSFET devices is seriously tailed, whereas that for SOI devices is well distributed. However, one should be cautious because damage still can be induced by the current flow between the gate oxide and the source/drain, which will be elaborated in Sec. 8.5.5. Charging currents then can flow from one device to another depending on design.[25]

8.5.5 Antenna Attached to the Source/Drain and the Substrate

The preceding discussion focused on antennas connected to the gate electrode only, and is has been shown that a larger antenna and/or plasma nonuniformity will lead to more plasma damage. However, for transistors with large antennas attached to the gate/source/drain/substrate, damage may not happen as long as the antennas have the same shape, similar sizes, and occur at the similar locations even with nonuniform plasma across the wafer.[54] This is due to the fact that the current across the gate oxide is determined by the potential difference between the gate antenna and the local substrate under the gate. Antennas with the same shape, similar sizes, and occuring at similar locations will not give a potential difference across the wafer even though the plasma is nonuniform. This effect is called the *source / drain / substrate antenna subtraction effect* and accounts for the substrate potential adjustment via the source/drain and substrate.

However, the situation becomes more complicated for some cases as detailed below

- *Case I:* In this case, the antenna ratio is the same for the antennas connected to gate and source/drain, but the spatial separation for the two antennas is varied (i.e., antennas location is not similar). It was found that a large spatial separation between the antenna may induce plasma damage to the transistor even though the antenna ratio is exactly the same. This observation may be explained by a potential difference generated between the large-spaced antennas by a lateral nonuniformity in the plasma. Lai et al.[74] reported identical antennas (10,000 μm^2) attached to both the gate and the diffusion (i.e., source/drain) nodes, but with different spatial separations. Charging damage was observed for all gate to diffusion antenna spacings ranging from 1200 to 6200 μm. However, when both antennas are in close proximity (<100 μm), there is no PID to the device. The severity of damage was observed to decrease with decreasing spatial separation between gate and diffusion antennas as indicated by the decrease in the number of high initial gate-leakage outliers.

- *Case II:* In this case, the spatial separation between the two antennas connected to gate and source/drain is kept the same, but there is a difference in the antenna ratio between the gate antenna and source/drain antenna. Plasma damage also likely does take place for this case. Lai et al.[74] studied a series of test structures, among which the spatial separation between the two antennas was kept as 1200 μm, the gate-node antenna area was kept to 10,000 μm^2, but the source/drain-node antenna area was varied from 200 to 10,000 μm^2. It is interesting to see that the charging damage to the gate dielectric of a device is reduced when the source/drain-node antenna area is smaller. That is, the device with the ~200-μm^2 source/drain-node antenna area showed much less damage than that with 10,000-μm^2 antenna. This observation is not well explained.

8.5.6 Effect of Well Configuration

The plasma during manufacturing may change the local substrate potential, which should be taken into consideration for plasma damage analysis. This is indeed the case for PMOS devices with *N*-wells placed in *P*-wells, as demonstrated by Krishnan et al.[71] They studied the impact of substrate potential on PID by using PMOS devices attached with a 250,000-μm metal antenna perimeter placed in three different types of well configurations, that is, a small, isolated *N*-well, a small *N*-well tied to *P*-substrate, and a large, isolated *N*-well (Fig. 8.13). The small *N*-well case (Fig. 8.13*a*) shows much less device damage than that with the *N*-well tied to the *P*-substrate (Fig. 8.13*b*). This is due to fact that there is a substantial potential drop across the *N*-well–*P*-well junction so that the voltage drop across the gate oxide during manufacturing is reduced compared with the case of *N*-well/*P*-substrate. In addition, when comparing the small *N*-well case (Fig. 8.13*a*) with the large *N*-well case (Fig. 8.13*c*), it is clear that large *N*-well device suffered more seriously from plasma damage because the PMOS device in a large *N*-well cancels the floating effect attributed to its much larger well capacitance as well as photoconduction. In short, for the purpose of catching PID, different structures may have quite different sensitivities. The worst-case (e.g., Fig. 8.13*c*) structure should be employed to effectively detect any potential damage, especially during the process-development stage.

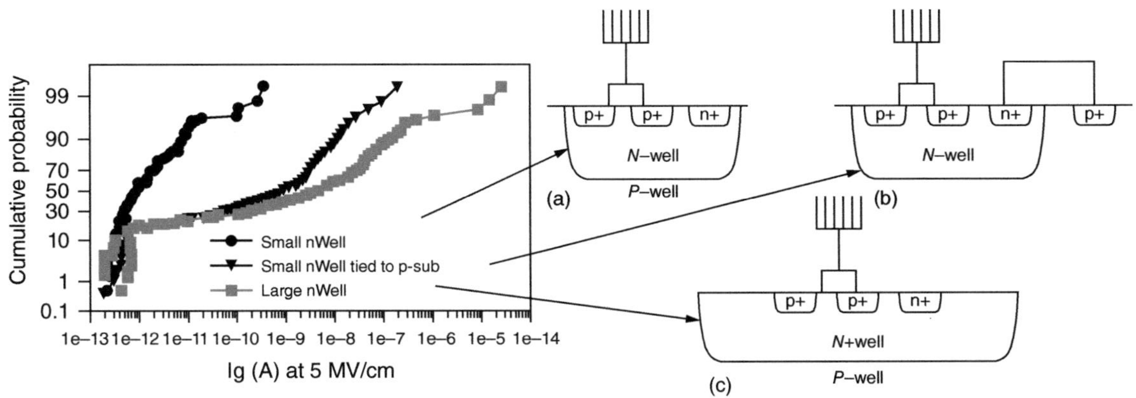

FIGURE 8.13 Local substrate potential effect: Shown here is the antenna damage during metal etch for a PMOS device placed in an *N*-well. Notice that the exent of the damage depends on the well configuration. A transistor placed in a small floating *N*-well degrades much less than the one with the well tied down to substrate. This arises from part of the plasma stress voltage appearing across the well junction. (*Used with permission from IEEE.*[71])

8.6 Effect of Structures on Plasma-Induced Damage

PID is very sensitive to the test structures used. Therefore, for each technology node, the effects of PID should be investigated in various structures to simulate practical situations. National Semiconductor has proposed a template for automatic structure generation, placement, routing, and test-plan development.[69] Eighteen parameters are considered in their designs for antenna effect, including MOS type (i.e., N/PMOS), gate length, gate width, antenna ratio, antenna layer (e.g., contact, poly, metal, or via), antenna type (area and edge intensive), finger contact, contact ratio, finger separation, finger width, metal-jump layer, protection diode area, protection diode type, protection diode on/off, diode offset, diode placement (in well and outside well), with or without pad protection diode, and nominal bias. The advantages of this template for automatic generation are time and cost savings. It also holds the potential for easier migration to new technology. In this section, the effects of various structure designs on PID are summarized.

8.6.1 Effect of Antenna Finger Density

In general, dense finger antennas collect positive charging, and the damage increases with a decrease in the interline spacing owing to the ESE (in Fig. 8.3), whereas the sparse (or isolated) finger antennas collect negative charging, and the damage increases with an increase in the interline spacing owing to the reverse electron shading effect (RESE; in Fig. 8.4*a*).[76,77] Figure 8.14 shows that there is a turnover

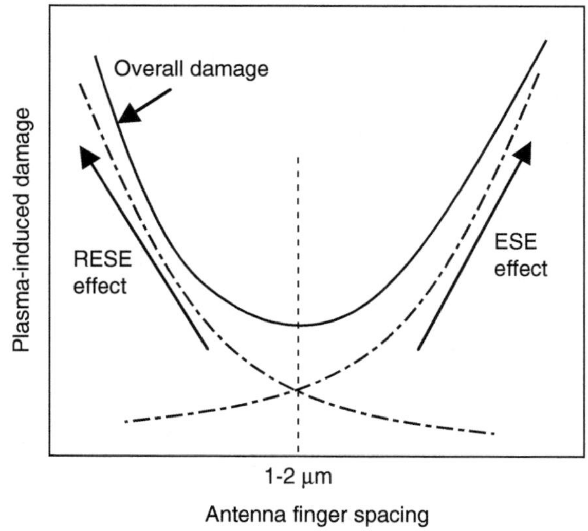

FIGURE 8.14 Dependence of PID on the spacing between antenna fingers.

point where the interline spacing is around 1 to 2 μm attributed to the superposition effects from ESE and RESE. The electron shading dominates when spacing is less than 2 μm and a positive charge is collected, whereas the reverse electron shading dominates when the spacing is larger than 2 μm and a negative charge is collected.

8.6.2 Avoiding Plasma-Induced Damage Through a Bridging Design

The plasma charging issue could be resolved or much reduced by constructing the routing in such a way that all plasma-related metal processing at a certain time is not connected to a gate yet.[70] In Fig. 8.15*a*, since the long M_2 part of the interconnect during processing is already connected the gate, it can lead to large plasma damage. In contrast, as shown in Fig. 8.15*b*, by using a bridge at M_3 or above (here is M_5 as an example), the long M_2 part of the wire is no longer connected to the gate during M_2 processing. In this manner, one of the upper metal layers is always used to complete the connection from the gate to the other connection point, for example, the output (drain area) of another transistor. However, such additional stacks of vias and extra wire routing to each gate likely will lead to many routing obstructions, and thus more space is needed for all connections. The result is a larger chip, which may not be desirable.

8.6.3 Latent Antenna Effect

Another important layout dependency is the latent antenna effect caused by reactive ion etching (RIE) that takes place for the Al interconnection with the etching process.[70] The underlying meaning of the latent antenna effect is that the real antenna ratio changes in time during the etching process. The latent antenna effect is illustrated qualitatively in Fig. 8.16, which gives a schematic plot of the antenna ratio seen by the gate in terms of the four stages of the metal etch,[78] including (*a*) the Al deposition stage, (*b*) the main metal etch stage, (*c*) the latent antenna

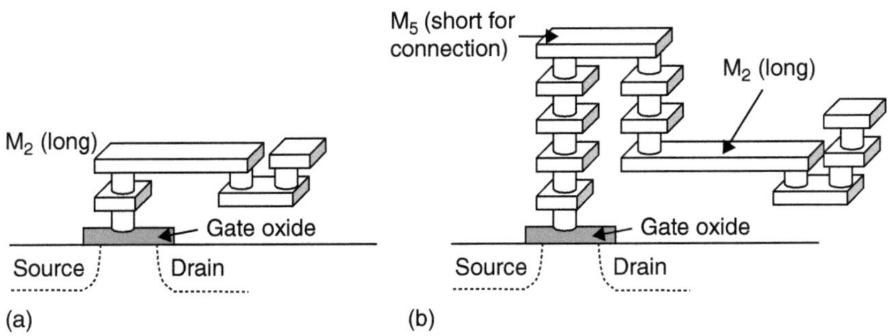

FIGURE 8.15 A "bridging" design scheme to avoid PID to the gate oxide.

stage, and (*d*) the overetch stage. The four stages are elaborated as follows:

- *Stage I:* For the Al interconnect process, the full wafer is first deposited with metal, followed by the main plasma etch after lithography. At that moment, no transistors can be damaged because the metal is connected to the drain (diffusion) areas, which functions as a protection diode and provides a leakage path for plasma-induced charges. In short, the exact antenna ratio is zero at this point, and no damage occurs owing to blanket-metal coverage.
- *Stage II:* When the etch continues, the etching rate in between the metal lines with larger spacing is faster than in those with smaller spacing. At this moment, although there is positive charging in the narrower spacing part of the structure owing to ESE, the positive charging can be balanced (or canceled out) by electrons from the substrate coming from a drain of a transistor that is still connected. Again, the exact antenna ratio is zero.
- *Stage III:* With more metal etched away, it is most likely that somewhere with the least pattern density will be isolated, and islands of metal will be formed. When such an island is connected to a gate but not to a drain area, there will be no supply of the electrons to balance the positive charge, and the antenna potential will rise. Consequently, the antenna potential rises continuously, which will lead to further electrons tunneling through the gate oxide and eventually may damage the gate oxide. At this point it is likely that there is a large transient antenna (referred as a *latent antenna effect*) that may be much larger than the antenna in a traditional sense owing to the undesired connection of other metal pattern during etch.
- *Stage IV:* With etching continuing, the metal with narrow spacing is cleared, and charging damage also can occur, but it is much less likely. It is possible to get a 10-fold increase in stress levels during the latent antenna stage compared with the overetch stage[79] (Fig. 8.16).

Therefore, for the Al interconnection process, in order to design a robust circuit for immunity from PID, the latent antenna effect should be taken into consideration. In other words, the layout configuration, not just a fixed antenna ratio number, also should be considered at the design stage to avoid any potential damage owing to the latent antenna effect because the latent antenna effect strongly depends on the neighborhood of the layout/structure that is connected to the gate.

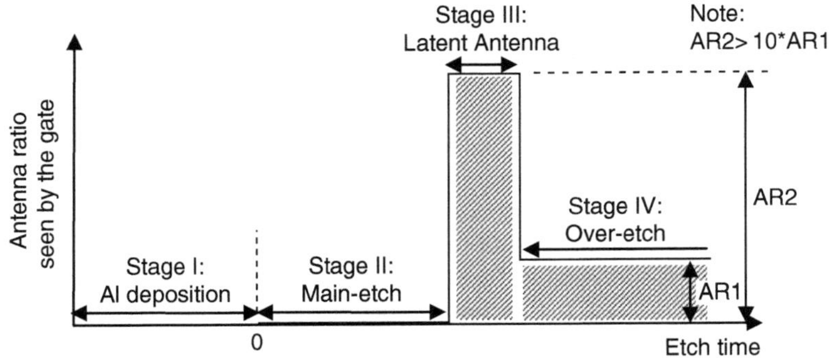

FIGURE 8.16 Schematic plot of the antenna ratio seen by the gate during metal deposition and etch processes in the Al interconnection process.

An example of the impact of the latent antenna effect is given in Fig. 8.17, which shows schematics of three types of antenna layout. In Fig. 8.17*a*, all fingers are connected to the gate, whereas in Fig. 8.17*b* and *c* fewer fingers are connected. Figure 8.17*a* is the conventional type, with the other two being the latent type antennas. However, the PID caused V_t shift distribution could be the same for these three configurations for the Al etch process, as reported by Krishnan et al.,[78] owing to the latent antenna effect—the metal clearing last in small spaces between the lines where the damage occurs primarily. Thus, the perimeter antenna effects of Fig. 8.17*b* or *c* are effectively enhanced over that of Fig. 8.17*a*. Therefore, this should be seriously considered for antenna design rules when designing the interconnect layout for etch-related process.

8.6.4 Spreading Antenna Effect

It has been found that, for the Cu damascene interconnect process, the deposition processes for Cu diffusion barrier films and

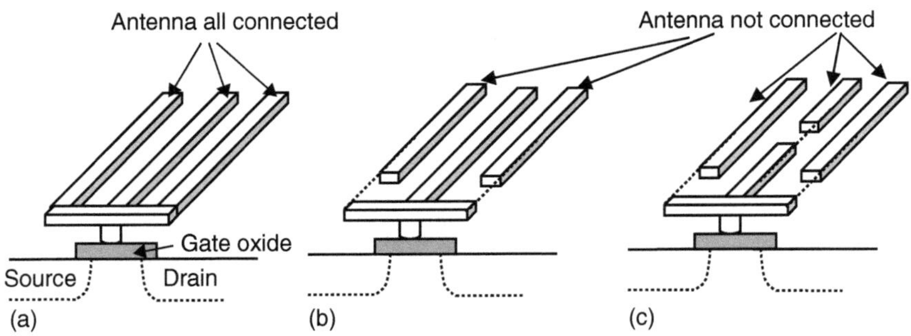

FIGURE 8.17 Schematics of the three types of antenna layout. In (*a*), all fingers are connected to the gate, whereas in (*b*) and (*c*) fewer fingers are connected.

intermetallic dielectrics (IMDs) can induce serious PID.[15] A detailed study on both comb-type (edge-intensive) and square-type (area-intensive) metal antennas showed that the dependence of PID on antenna area in the damascene interconnect process is not a simple relation to the top surface area of wiring. For the edge-intensive situation, the top surface area is defined as the wiring width times the unit length times the finger number. However, owing to the so-called spreading-antenna effect, the effective antenna area that can collect the charges should be much larger than the top surface area. The effective antenna area should be defined as the wiring width plus space times the unit length times the finger number, that is, the total area including both the metal surface and the spacing area, as demonstrated by Matsunaga et al.[15] The antenna area defined by the top metal surface area only (rather than the total area including both the metal surface and the spacing area) for comb-type antennas will underestimate the PID. A similar spreading-antenna effect was observed for the barrier deposition process. In short, the spreading-antenna effect should be well assessed during process development and be taken into consideration when calculating the antenna ratio during design.

The spreading-antenna effect is caused by the photoconductivity[80] of the IMD material during plasma deposition. The simulated electrical field by surface charge in the IMD film deposition process spreads over the wiring-space portion, indicating that the 1.0-μm wiring space is able to work as an antenna in the IMD deposition process.

8.6.5 Capacitor versus Transistor as Detector

The test structures used for PID characterization should be sensitive to any questionable plasma step and should be able to be monitored through easily accessible electrical testing. The common MOS transistor with source/drain is used widely to meet such requirements, as discussed earlier. On the other hand, MOS capacitor structures provide cheaper and faster fabrication than transistors and also can be pretested to ensure their calibration.[81] Ma et al.[81] compared several capacitor tests used for evaluating PID, for example, ramp breakdown measurements (which are quick and easy to automate but lack sensitivity), charge-to-breakdown measurements (which offer better sensitivity but with long measurement times), *V-t* measurements using the slope dV/dt after initial charging (which is sensitive to charge damage but involves tradeoffs between the antenna ratio, testing current, and testing time), leakage measurements (which offer short measurement times and high sensitivity but are limited by the noise level of the measurement system), and *C–V* measurement (which is sensitive for plasma damage in many cases,[21,82] but may not be preferred because the antenna capacitance signal could be much large when thin oxides and larger testing areas are needed).

MOS transistor structures are currently preferred for qualification of PID.[36] By using these structures, device parameters such as I_{dsat}, V_t, G_m, interface state D_{it} and gate-oxide leakage I_g between the device connected with an antenna and the reference device without an antenna are compared to detect PID.

8.7 Process Effect on Plasma-Induced Damage

PID also can be minimized through process optimization, which will be covered in detail in this section.

8.7.1 Effect of Anneal on Process-Induced Damage to the Gate Oxide

One way to repair (part of) the gate-oxide damage is to use a forming-gas annealing (FGA) step at the very end of the process—after the passivation anneal with H_2 or D_2 (mixed with N_2). In fact, the underlying mechanism is that the molecular hydrogen can react with certain kinds of electrically active defects at the Si-SiO_2 interface,[83] thus passivating them. Therefore, hydrogen chemistry in thin SiO_2 layers is a key aspect as the reliability of CMOS technologies.[84] This improvement is demonstrated through the charge pumping measurements of before and after the anneal.[8] As illustrated in Chap. 4, charge-pumping current I_{cp} is directly related to the average interface state density. I_{cp} for NMOS antenna devices is reduced from ~90 pA before H_2 anneal to ~10 pA after H_2 anneal. D_2 anneal shows a further reduction of I_{cp} to ~5 pA. This improvement in interface states is also reflected in transistor performance such as transconductance G_m. It has been speculated that the improved interface might be due to closer matching of the Si–D vibrational mode frequency to that of the bulk phonons of the Si lattice.[85] In the presence of numerous additional states generated by plasma damage, the Si–D bond may be favored because of the match in the vibrational frequency.

King and Hu[86] simulated the plasma charging effect by carrying out dc FN stress at either constant gate voltage or constant gate current. Then the wafers were subjected to anneal for 20 minutes at 400°C in forming gas (N_2/H_2) or at 800 or 900°C in an N_2 environment. Three types of traps induced by such FN stress were identified as follows:

- *Interface traps* (characterized as the midbandgap interface state density D_{it}), which disappeared after the anneal when the devices was subjected to relatively lower-level stress (e.g., injection charge less than 20 percent of charge to breakdown Q_{bd}).
- *Oxide charge traps,* which also can be removed effectively by the anneal.

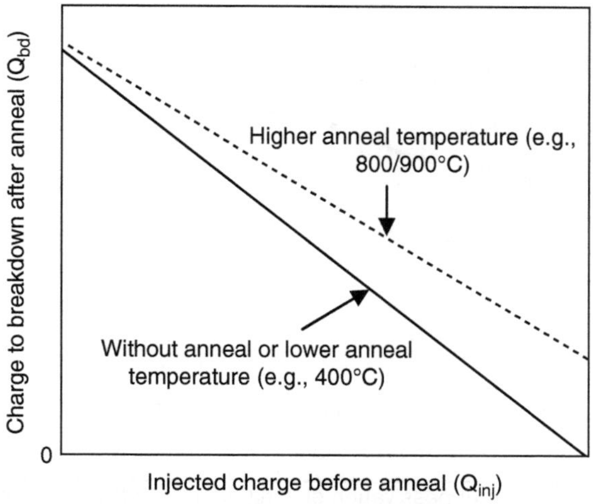

FIGURE 8.18 Schematic illustration of charge to breakdown as a function of a previously injected charge for samples annealed at different temperatures.

- *Neutral electron traps,* which cannot be removed effectively by the 400°C anneal. An 800°C or 900°C anneal can partly remove such neutral electron traps or the damage responsible for gate-oxide breakdown. As such, a 400°C anneal in forming gas after the prestressing showed no breakdown improvement.

Another finding of this work[86] is that the *effect of process-induced damage (stress) on gate-oxide breakdown is cumulative,* even with subsequent high-temperature anneals. As shown in Fig. 8.18, for oxides with consecutive stresses without annealing, the straight line indicates that the total injected charge ($Q_{inj} + Q_{bd}$) is a constant. The line for the lower anneal (e.g., 400°C) in between the two stresses is overlapped with the one without annealing, showing no improvement. However, a higher anneal temperature (800 or 900°C) provides some hints for breakdown improvement because its charge to breakdown Q_{bd} is higher even though the injected charge before annealing Q_{inj} is the same as the one without anneal.

8.7.2 Effect of Passivation Etch

Experiments with various magnetic field conditions were carried out on passivation RIE that opens a window to expose the Al pad embedded in the passivation layer.[92] Results are shown in Fig. 8.19. The split condition consists of recipes with high-intensity B-fields, low-intensity B-fields, and without B-fields. The applied B-field here is to get a higher etching rate and better etching uniformity. The results showed that without B-field, devices are PID-free with perfect I_g distributions.

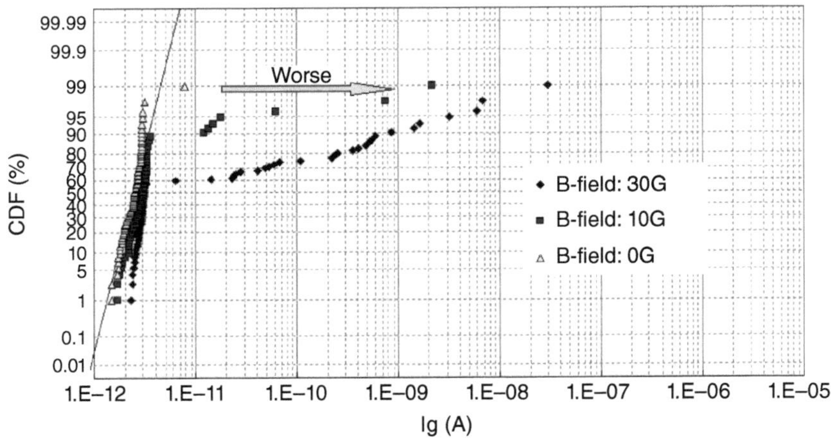

Figure 8.19 Magnetic field of passivation etching effect on PID. It is seen that the RIE recipe without B-fields has the best PID performance.

8.7.3 Effect of SiN Cap NH_3 Plasma Pretreatment Process on Plasma-Induced Damage

SiN film has been used widely as the capping material for Cu interconnects. The in situ NH_3 plasma preclean process has been employed prior to SiN cap deposition to promote adhesion by removing residual Cu oxide after Cu CMP.[87] There are a few reports on the charging damage originating from the NH_3 pretreatment process.[33,88] Preclean with He$^+$-based plasma with an RF generator also may result in plasma damage.[40]

Ang et al.[33] compared thick-gate PMOS PID performance for three SiN cap-deposition splits. The first two processes were employed in tool A with the same NH_3 pretreatment time of 15 seconds, but differed in the RF plasma power (150 versus 210 W), whereas the third process was performed in tool B with 7.5 seconds of NH_3 pretreatment and RF plasma power of 500 W. In all the splits, the same SiN cap process was used repetitively from M_1 up to M_7. The authors found that no PID occurred in the poly area and contact antenna test structures. This is justifiable because the NH_3 pretreatment did nothing to the poly and contact. In contrast, tool A with an NH_3 RF power of 150 W suffered serious PID on M_1 antenna test structures. It is frustrating that contrary to intuition, the other two SiN cap processes with a higher NH_3 RF power (210 and 500 W) produced much less PID on M_1 antenna test structures. Ang et al.[33] attributed this observation to the RF-induced heating. The RF heating may cause a temperature spike in the Cu interconnects during the NH_3 pretreatment process. A higher RF power will result in a higher-temperature spike, then greater increase in resistance in the Cu lines, and hence a lower charging current to the gate oxide for a certain plasma voltage.

8.7.4 Effect of Plasma Parameters on Plasma-Induced Damage

Plasma parameters such as pressure, source power, and bias power have a large impact to PID, but luckily they are easy to modify so as to minimize the PID.[89] Leakage current can be greatly reduced at higher pressures by reducing the electron temperature T_e and at lower bias by reducing the plasma density, as detailed in Sec. 8.4.2.

There have been attempts to use CO_2 to replace O_2 for plasma ashing of ultra-low-k organosilicate glass (ULK OSG) films in advanced-technology devices. The extent of CO_2 plasma damage was found to depend on the plasma density, plasma energy, and low-*k* chemistry. Compared with O_2 plasma, CO_2 plasma may induce less damage probably owing to the compensation effect of the carbon in the CO_2 plasma.[18] While the capacitive-coupled RIE CO_2 plasma induced mostly surface damage having less dielectric loss and plasma damage, the inductive-coupled plasma (ICP) generated more damage in the bulk of the low-*k* material. CO/O_2 and CO_2/N_2 plasmas were further proposed to replace O_2 plasma for the ashing process to reduce plasma ashing damage to porous low-*k* patterned structures.[90] With increasing CO in CO/O_2 plasma or N_2 in CO_2/N_2 plasma, the extent of PID decreases. The reduction in PID was attributed to a decrease in oxygen-radical density and the formation of a C- and N-rich passivation layer on the sidewall and pore surfaces of the patterned low-*k* structure. Higher carbon concentration or smaller pore radius in OSG low-*k* materials can improve material resistance to plasma damage, as indicated by a thinner carbon depletion layer.[91]

8.7.5 Premetal Dielectric Deposition

High-Density Plasma versus Subatmospheric Chemical-Vapor Deposition

When the premetal dielectric (PMD) is deposited using a high-density plasma (HDP) process with high RF power and bias power, the charges are likely to accumulate quickly on the poly gate and thus damage the gate oxide. It has been demonstrated that replacing the HDP process with subatmospheric CVD with a plasma-free process is a choice that reduces PID.[34] The side effect caused by such a process change may be the incomplete gap filling that needs efforts to improve, which is challenging but achievable.

Ultraviolet Absorption Layer

An ultraviolet (UV) light–blocking layer (silicon nitride) was applied to the ULSI device below the interconnection layer to dramatically reduce PID.[28] Another practical solution is to deposit a plasma-free buffer layer (liner oxide) prior to the HDP dielectric.[92] It was demonstrated that the liner oxide can absorb the UV photons from the HDP plasma and, thus, reduce the interface states in the gate oxide generated by the UV radiation from the plasma. Figure 8.20*a* shows that the liner oxide at 200- to 300-Å thickness can decrease the plasma damage significantly when compared with a 100-Å-thickness liner oxide or without a liner oxide. Figure 8.20*b* also indicates a decrease in 10th-percentile gate leakage with liner oxide thickness increases. However, the gate leakage does not show such a trend but varies with the liner oxide thickness. This phenomenon is related to the fact that the I_g distribution has a large tailing for all cases, even for the case with a 300-Å-thick liner, which is attributed to the other plasma damage sources.

Similarly, a conductive top film (CTF); a thin undoped Si layer was deposited in between the contact etch-stop layer (CESL) and the HDP PMD layer to significantly reduce the plasma damage from HDP.[93] The so-called CTF provides a conduction path during the initial stage of the HDP process, thus making the plasma more

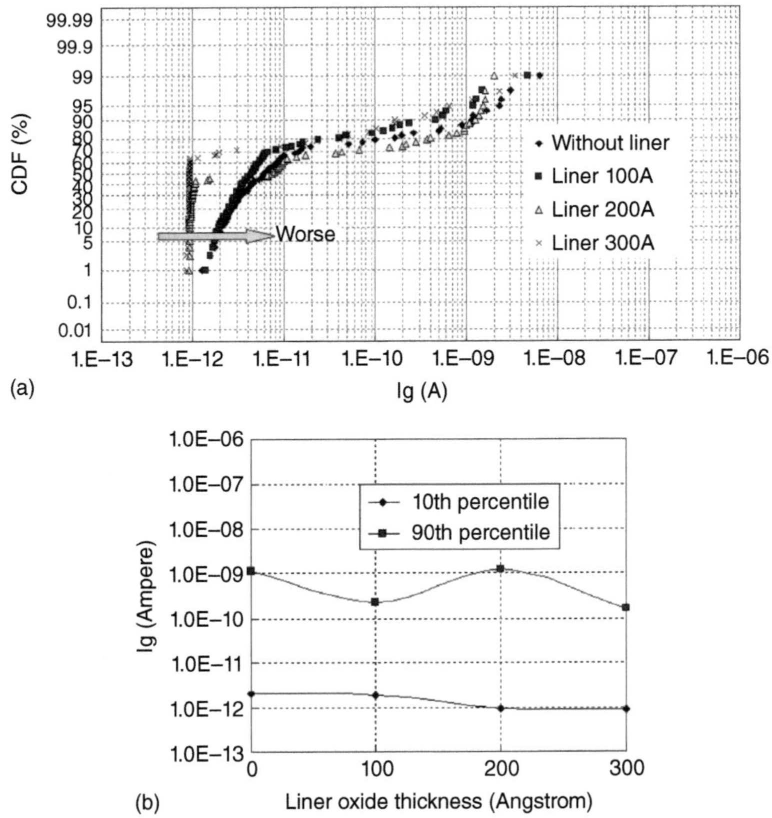

FIGURE 8.20 Gate-leakage decreases with increasing PMD liner oxide thickness. (*a*) I_g cumulative distribution; (*b*) I_g distribution of 10th-, 90th-percentile versus liner oxide thickness.

uniform, which can help to reduce PID. Once PMD deposition starts, the conduction path disappears. Furthermore, UV radiation generated by the PMD process can be significantly blocked by the CTF. The PMD process generates photons with energies ranging from 4.96 to 1.65 eV (the corresponding wavelength ranges between 250 and 750 nm). In layered structure with Si CTF/SiN (CESL)/poly gate/SiO_2/Si substrates in sequence, four interaction events can take place as follows: In event I, photons with energy greater than 2.1 eV can excite electrons at the Si (CTF)–SiN (CESL) interface to inject electrons into the SiN layer, where some electrons will recombine with holes and some will flow into the gate poly. In event II, photons with energy greater than the Si bandgap (i.e., 1.1 eV) can generate electron-hole pairs in the poly Si. In event III, some of generated electrons are excited to overcome the Si-SiO_2 barrier by producing photons with energies greater than 3.2 eV. In event IV, the injected electrons lose energy on arriving at the Si substrate, causing impact ionization and trapping the generated holes in the SiO_2, or tunnelling into the poly Si. Note that events III and IV can be reduced significantly by the presence of Si CTF, which absorbs over 40 percent of the photons with energies greater than 3.2 eV.

8.7.6 Effect of Intermetallic Etch and Via Etch

Plasma damage can be induced by overetching during the barrier open and the preclean/sputter during the barrier deposition process.[30] Krishnan et al.[20] reported that plasma damage to PMOS devices with 35-Å-thick gate oxides induced by metal etch is sensitive to the condition of the chamber with the least damage after a wet clean of the chamber walls.

Tabara[94] reported that introducing extralight ions (e.g., HCl chemistry in metal etcher) can help to neutralize the electrons trapped in the resist and reduce electron shading and notching effects so as to improve immunity to plasma damage.

Kim et al.[95] reported that the lower magnetic field and longer main etch time during the via etching process can reduce the plasma-charging damage to the gate oxide dramatically with low via contact resistance. Their work was based on investigation on N/PMOS capacitors.

The RPC before Cu barrier deposition may have a negative impact on plasma damage.[92] RPC uses ionized hydrogen as a reactant to remove the copper oxide on the exposed Cu surface. The hydrogen ion has the smallest charge-to-mass ratio among various ions and hence the highest mobility under a specific voltage bias. In other words, the via bottom Cu surface is exposed to the RPC plasma during the cleaning. It is likely that such plasma charges can be transferred to the oxide gate through the Cu interconnection and lead to the gate-oxide damage. Thus, this process should be optimized to reduce the PID. Figure 8.21 compares the I_g distributions as a function

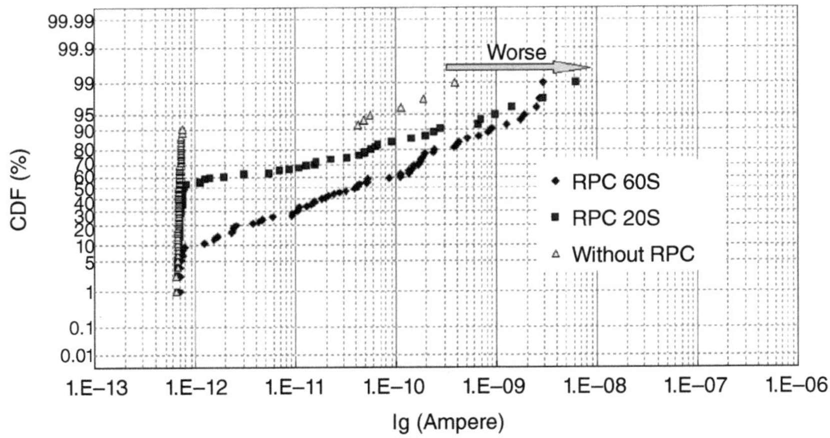

FIGURE 8.21 I_g cumulative distribution showing that gate leakage decreases with decreasing RPC process time.

of RPC time. It is clear that I_g distribution becomes much less tailed by reducing the preclean time. Besides, RPC power and pressure can be optimized to minimize the PID impact.

8.7.7 Effect of Process Temperature

The probability of PID increases significantly with increases in wafer temperature. Therefore, plasma during CVD deposition (in general, wafer temperatures of ~250 to 480°C) will induce much greater damage than etching plasma does (in general, wafer temperatures ≤ 100°C). The relatively high temperature during IMD deposition can exacerbate the plasma damage to the gate oxide as a result of very small tunneling currents.[96] The gate-oxide damage caused by plasma during processing can be analogized to Q_{bd} measurement–induced damage (without plasma), which demonstrated that higher Q_{bd} measurement temperatures can seriously worsen the damage.[97] The underlying mechanism is that at a higher temperature, the lattice has higher-level vibration, and thus the trap generation in the gate oxide will increase, leading to an effective increase in the collision cross section of carriers. In addition, at higher temperatures, molecular bonds are weakened and easier to be break, and thus hydrogen will be more likely to be released, resulting in increased interface trap formation. This is the reason that a D_2 anneal reduces PID because Si—D bonds are stronger.[8]

8.7.8 Reducing Plasma Charge Damage by Equipment Modification

To reduce PID, one can work on equipment improvements.[98] One of the efforts in this area involves ensuring plasma homogeneity based on the principle shown in Fig. 8.2, for example, through coil position optimization.[99]

The electron temperature T_e can be decreased (accompanied by reduced plasma density) by increasing the chamber gap and thus reducing PID.[20,100]

Pulsed plasma is promising in decreasing charging because the electron temperature T_e can be reduced drastically during the off period of the cycle, leading to a reduced or reversed sheath. Electrons then are able to enter narrow pattern spaces to reduce the electron shading effect and thus to prevent charge accumulation. Pulsed plasma also can create low-energy positive ions in the first part of the off pulse and also negative ion flux in the case of electronegative plasmas (such as Cl_2-containing plasmas). Low-energy positive ions can be deflected by the negative charge of the resist and neutralize the upper-mask trapped electrons. Negative ion flux can help to neutralize the bottom-surface positive-ion charging.[25]

8.7.9 Progressive Deterioration Characteristics of Plasma-Induced Damage

PID to the gate oxide generally is believed/assumed to be cumulative, that is, of a progressive deterioration characteristic,[101] even with subsequent high-temperature anneals.[86] This means that the damage induced in each process step will contribute to the eventual failure of the oxide. To check such cumulative effects, a delicate experiment was designed by testing the wafers at eight different steps in the backend process.[112] V_t fluence tests were performed at each step to evaluate gate-oxide quality. In this test, a constant gate current (6.7 mA/cm^2) is forced through the gate oxide in the FN tunneling regime for 200 seconds. The threshold voltage of the transistor with an antenna ratio of 960:1 (no protection diode connected) is monitored periodically during the stress. Electron injection stress was used, which corresponds to a positive V_t shift. Results indeed showed that the V_t shift increases as the wafers are processed through the backend process step by step.

8.7.10 Gate-Oxide Robustness

One way of alleviating PID is to make the gate dielectrics more immune or robust to charging damage, for example, through the incorporation of nitrogen into the gate dielectric film. The methodologies are elaborated below.

Improve the Quality of Gate Oxides

Optimization of the wafer cleaning process and the oxide growth environment, for example, using cluster tools, is some employed in processes to improve the quality of the gate oxides and thus to minimize PID. In situ steam-generation (ISSG) gate oxides grown within an RTP chamber have been demonstrated to be more tolerant to PID than conventional thermal gate oxides generated by atmospheric furnace.[102] More details on gate-oxide quality improvement were provided in Chap. 6.

Nitrided Oxide

Nitrided oxides have been demonstrated to be more immune to plasma-induced charging damage.[38,103–105] A few examples are given below:

- Chen et al.[103] reported that SILC caused by plasma damage can be significantly suppressed in N_2O-nitrided oxides compared with pure oxides, especially for PMOS devices. The nitrided oxides are also more robust when subjected to high-temperature stressing.
- Cheung et al.[105] demonstrated that the PID immunity of gate oxides grown on very low-dose ($2 \times 10^{13}/cm^2$) N^+-implanted silicon can be improved comparing with regular gate oxides of similar thickness. Both hole trapping and electron trapping are suppressed by the incorporation of nitrogen into the gate oxide.
- Chong et al.[38] showed that there is no significant difference in immunity to charging stresses between the oxide nitrided by rapid thermal nitridation (RTNO) and that nitrided by decoupled plasma nitridation (DPNO) for NMOS devices. However, RTNO is more immune to charging damage than DPNO for PMOS devices, as indicated by the larger Q_{bd} and lesser transconductance shift in RTNO. Direct-current current-voltage (DCIV; principle is given in Chap. 4) measurements showed that the interface traps induced by FN stress are fewer in RTNO. Thus it is seen that the SiO_2-Si interface of the RTNO is more resistant to charging damage than DPNO. On the other hand, there is no clear difference in oxide traps between RTNO and DPNO, as indicated by the *C-V* flat-band shift measurement.

Fluorinated Oxide

Chen et al.[106] observed that medium-dose fluorine implantation (e.g. 40 keV, 1×10^{14} cm^{-2}) into the Si substrate prior to oxidation can significantly reduce charging damage, as characterized by gate-leakage measurements, on antenna devices. The reduced plasma charging–induced gate leakage current owing to F incorporation is attributed to the stronger Si−F bonds (5.73 eV) that replace the weak Si−H (3.18 eV) bonds because it is generally believed that the trap-creation mechanism responsible for SILC is hydrogen-related. The incorporation of fluorine also can substantially improve the GOI through structural relaxation of the strained SiO_2 by formation of strong Si−F bonds that replace weak Si−H bonds. However, excess fluorine is also known to deteriorate GOI by creating trap charges in the bulk oxide as well as at the interface. Therefore, the implantation dose should be optimized to maximize the benefit from F incorporation.

8.7.11 Effect of Intermetallic Dielectric Deposition

IMD deposition is one of the important concerns for PID.[107,108] At least two mechanisms are considered to lead to gate-oxide damage during this process step, that is, nonconformal oxide coverage of exposed metal lines[109] and the photoemission owing to UV photons generated inside the plasma[96,110] (see Fig. 8.4*b*). In particular, the gate oxide can be damaged even by very small tunneling currents because it is weakened by the relatively high temperature used during IMD deposition[96] (Sec. 8.7.7).

8.7.12 Effect of Barrier/Seed Deposition

It is easy to understand that the barrier/seed deposition process may lead to PID because it is direct deposition on an exposed wiring metal. When PID of the barrier film deposited using a PE-CVD tool whose conditions are not optimized occurs, a large V_t shift will be observed compared with the reference device with a protection diode.[15]

8.7.13 Insulating Layer at the Backside of the Wafer

In an SOI structure, the buried oxide (BOX) layer can prevent electron flow from the gate to the substrate, as illustrated in Sec. 8.5.4. Similarly, an insulating layer left at the backside of the wafer (such as oxide or nitride) also may serve as a blocking layer for electron flow so as to make the substrate float from the chuck during plasma-related processes. This, in turn, can minimize PID induced in both NMOS and PMOS devices.[111] The underlying mechanism is that when the substrate is prevented from discharging through the backside, the charge is eventually dissipated through the substrate and the gates simultaneously from the front contacts and, therefore, no potential develops across the gate oxide.

8.8 Other Reliability Issues Related to Plasma-Induced Damage

It has been observed that PID has an adverse impact on HCI/NBTI/GOI reliability, which will be elaborated in this section.

8.8.1 Hot-Carrier Injection

Mechanisms of HCI reliability are detailed in Chap. 5. Rakkhit et al.[112] showed clearly that under the same HCI stressing conditions, a much higher I_{dsat} shift in devices suffering from PID was observed, especially in PMOS devices. Devices connected with a protection diode were immune from plasma damage and thus showed better HCI performance than those without a protection diode. However, it seems that there is not much difference between NMOS devices with large metal antennas (and no protection diode connected) and those without antennas. This is so, probably because PMOS devices suffered more plasma damage than NMOS devices, as elaborated in Sec. 8.5.1.

Krishnan et al.[71] reported that with suitable sintering, there is no negative impact from PID on device HCI performance. The

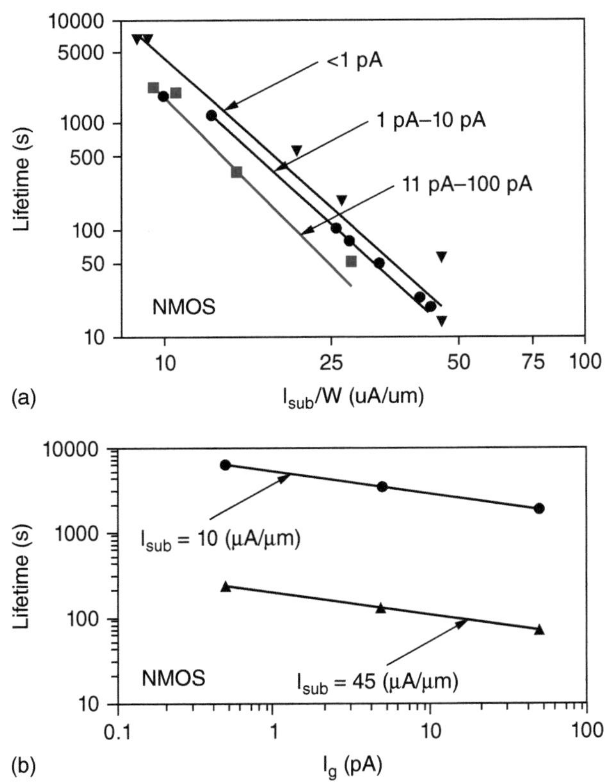

FIGURE 8.22 (*a*) Performing the HCI stress prior to sinter shows large variation and scatter in antenna-damaged device lifetime based on 10 percent I_{dlin} degradation. This variation correlates with the gate leakage-current increase owing to the antenna effect. (*b*) The lifetime of an NMOS device decreases by a factor of 2 for a 10× increase in I_g.(*Used with permission from IEEE.*[71])

hot-carrier-induced degradation did not vary with antenna ratio. However, for devices without sintering, the 10 percent I_{dlin} degradation lifetime as a function of current scattered significantly (Fig. 8.22*a*). These authors established a correlation between HCI lifetime variation and the increase in gate leakage as a result of plasma damage. Figure 8.22 shows the device lifetime characteristics for low, medium, and high gate leakage. It is seen that a 10× increase in I_g is correlated with a 2× reduction in NMOS lifetime. Similar results were obtained for PMOS devices (plots are not shown here): A 10× increase in I_g results in a greater than 10× reduction in lifetime. *This relationship between post–plasma process gate leakage and HCI lifetime is very useful because post–plasma stress I_g is easily measured and can be used as a routine monitor, and the device lifetime impact thus can be assessed faster and more easily using this relationship.*

An effort was under taken to derive the relation between HCI lifetime and process-related damage.[6] The derivation starts from estimation/extrapolation of HCI lifetime for a device using the following empirical relation[113]:

$$\text{TTF} = A \cdot (I_{sub})^{-m} \tag{8.7}$$

where time-to-failure (TTF) is the device lifetime, I_{sub} is the substrate current, and m and A are constants. As illustrated in Chap. 5, the exponent m (~3) is the ratio of the barrier height (between the oxide and the semiconductor conduction bands) and the energy required for impact ionization,[114] whereas A depends on the technology processes. Hence, increasing the value of A is an effective way to improve device lifetime.

Rangan et al.[6] thus correlated the PID impact on gate leakage current (I_g) as given in Fig. 8.22 with the change in the constant A in Eq. (8.7) and thus with HCI lifetime as follows:

$$A = B \cdot (I_g)^{-p} \tag{8.8}$$

Then the HCI TTF in Eq. (8.7) is further expressed as

$$\text{TTF} = B \cdot (I_g)^{-p} \cdot (I_{sub})^{-m} \tag{8.9}$$

where the component $(I_g)^{-p}$ is correlated with the process and plasma with antenna-ratio dependence, and the component $(I_{sub})^{-m}$ is the intrinsic HCI result.

The AR as defined in Sec. 8.1 is the design parameter critical in determining the extent of PID. Since I_g is an index for PID, a relation between gate leakage (e.g., by using the I_g value at CDF 90 or 95 percent) and antenna ratio can be obtained as follows:

$$I_g = C \cdot (AR)^a \tag{8.10}$$

where C and a are constants. C is a process constant that strongly depends on process-related parameters, such as chamber configuration and process conditions. From Eq. (8.10), the constant C corresponds to the case without antenna attached (i.e., $AR = 1$). A process with a larger C value would result in devices having an overall higher magnitude of I_g. The value of a depends on the type of charging, for example, electron-shading effect or plasma nonuniformity. The constant a will go to zero if the net amount of charge collected by the antenna is zero; that is, the number of ions and electrons collected is equal at all times. Empirical values of C and a are 10^{-15} Å and 1.72, respectively, for an inductively coupled plasma (ICP) metal etch step.[6]

Substituting Eq. (8.10) into Eq. (8.9) and normalizing TTF, we get

$$\frac{\text{TTF}}{(I_{sub})^{-m}} = B \cdot C^{-p} \cdot (AR)^{-pa} = B \cdot C^{-p} \cdot (AR)^{-r} \tag{8.11}$$

where $r = p*a$. Equation (8.11) provides the lifetime dependence as a function of the antenna ratio (through AR) and the processing parameters (through C). Rangan et al.[6] reported experimentally that the HCI $\text{TTF}/(I_{sub})^{-m}$ has a power-law dependence on the AR for both NMOS and PMOS devices: A 10× increase in AR causes a 7× reduction in NMOS TTF, whereas a 10× increase in AR causes a 10× reduction in PMOS TTF.

8.8.2 Negative-Bias Temperature Instability

Mechanisms of NBTI are detailed in Chap. 7. The plasma damage can make NBTI in PMOS devices worse, but it, can be effectively suppressed by the CTF process (see Sec. 8.7.5).[93] Weng et al.[115] reported the PMOS NBTI lifetime and breakdown voltage V_{bd} as a function of gate leakage current resulting from plasma damage from various antenna ratios. When the antenna ratio increases from 3× to 500×, NBTI lifetime decreased to about one-fifth, although the gate leakage and the gate-oxide breakdown voltage V_{bd} did not change much. With a further increase in AR to 5000×, all parameters (NBTI lifetime, V_{bd}, and I_g) degraded with a much larger (die-to-die) variation. In other words, for intermediate ARs (<500 for the poly antenna case), I_g and V_{bd} may not be sensitive to plasma damage, whereas NBTI lifetime is still worsen—meaning that NBTI stressing is more promising in revealing latent PID to the device.

8.8.3 Gate-Oxide Integrity

Mechanisms of GOI and SILCs are detailed in Chap. 6. PID to the gate oxide will lead to SILC. Figure 8.23 clearly shows some abnormal I-V

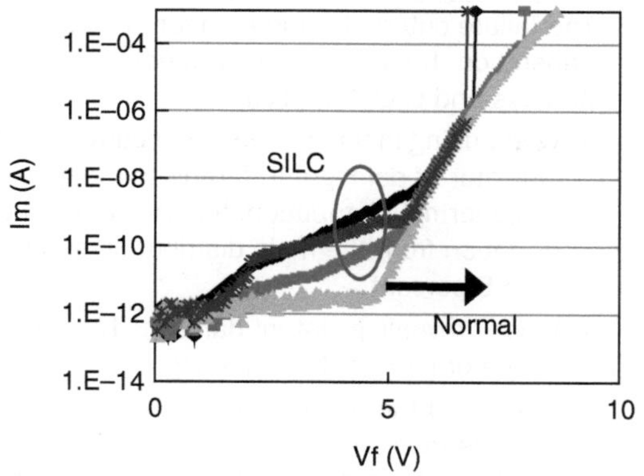

Figure 8.23 PID to the gate oxide will lead to SILC.

curves (SILC) tested on conventional GOI test structures. The larger leakage current below 6 V is caused by the plasma prestress, which will lead to a lower breakdown voltage. Similar SILC induced by PID also has reported by other parties.[106,116]

8.9 Process-Qualification Practice

No general JEDEC standard has been established yet in the industry for PID process qualification, and a standard is under preparation to fill this gap.[12] JEDEC JP-001[36] provides some guidelines as part of the qualification or as an engineering study using MOS transistor test structures. Two typical characterization methodologies for the detection of PID are summarized in Sec. 8.3, including

- Initial gate leakage current I_g with the gate at low voltage (e.g., operating voltage V_{op}) and bulk/source/drain grounding comparing devices connected with an antenna and the reference one without an antenna
- "Revealing" latent damage such as time-dependent dielectric breakdown (TDDB), HCI, and NBTI comparing the distributions before and after stress of one or more of the following parameters: threshold voltage V_t, transconductance G_m, drain saturation current I_{dsat}, and gate leakage I_g at low voltage

Table 8.2 provides some guidelines or practices in detail for the PID qualification.

Failure criteria	The failure criteria for the monitored parameters will depend on the specific requirements of the specific process and circuit application.
	• When using initial gate leakage current as an indicator of damage: Referring to the gate current in lognormal distribution, the tailing of 20 percent deviation from the main distribution should be less than 5 percent.
	• When "revealing" latent damage: The failure criterion is one or more of the following:
	  • Value of the monitored parameter is not within the predefined range;
	  • Difference in the monitored parameter between reference and antenna devices is larger than the predefined limit;
	  • After "revealing" stressing, shift of the monitored parameter in the antenna device is larger than the predefined limit;
	  • Either before or after "revealing" stressing, the distribution of the monitored parameter (e.g., 3σ) in the antenna device is larger than the predefined limit.

Test structures	• Nominal geometry NMOS and PMOS devices for each gate oxide thickness.
	• The following structures of each type are to be tested. Comparison between reference and antenna devices should be performed on the same identical geometry and same site.
	  • Reference device: Free from plasma damage by protection scheme (e.g., protection diode) and without antenna attached.
	  • Antenna devices with antenna ratios defined in design rules for the following typical layers: poly, contact, metal, via, etc.

Sample size	• For I_g monitor: Wafer mapping for three wafers per lot and three lots • For revealing latent damage: Suggested 20 sites per wafer, 3 wafers per lot, and 3 lots. The sampling should be uniformly allocated in the wafer.

Table 8.2 Typical Process Qualification Practices for PID

Stress/test conditions	• For I_g monitor: Measuring gate leakage with gate at low voltage (e.g., operating voltage V_{op}) and bulk/source/drain grounded comparing the reference and antenna devices • For "revealing" latent damage: Identical steps should be applied to both the reference and antenna devices as follows: 1. Measure prestress parameter values. Parameters to be measured include one or more of the following: V_t, G_m, I_{dsat} and I_g. 2. Apply stress to reveal latent damage that is thought to be necessary by using one or more of the following. (*a*) PID-TDDB stress: Optimized stress level to be determined for the specific process. An example of the stress level is 1.4 to 1.8 × V_{op} for 10 seconds. (*b*) PID-HCI stress: Stress at the worst temperature (e.g., room temperature) for I/O device and maximum temperature of operation, say, 125°C for core device. (*c*) PID-NBTI stress: Stress at the worst temperature (e.g., at maximum temperature of operation, say, 125°C). 3. Remeasure parameter values after the "revealing" stress.
Wafer/package level	Wafer level
Model	Not applicable
Parameters	Transistor device parameters: • CDF of predefined parameter variation for within-wafer, wafer-to-wafer, and lot-to-lot • CDF of predefined parameter difference between reference and antenna structures • CDF of predefined parameter shift for before and after stress

Table 8.2 (*Continued*)

8.10 References

1. K. P. Cheung, *Plasma Charging Damage*. London: Springer-Verlag, 2001.
2. D. Dumin, "Characterizing wearout, breakdown and trap generation in thin silicon oxide," *Microelectronics Reliability*, Vol. 37, No. 7, 1997, pp. 1029–1038.
3. K. P. Cheung, N. A. Ciampa, C. T. Liu, C. P. Chang, J. I. Colonell,W. Y. C. Lai, R. Liu, J. F. Miner, H. Vaidya, C. S. Pai, and J. T. Clemens, "Relationship between plasma damage, SILC and gate-oxide reliability," in *Proceedings of the 4th International Symposium on Plasma Process-Induced Damage*, 1999, pp. 137–140.
4. H. C. Lin, C. C. Chen, C. H. Chien, S. K. Hsein, M. F. Wang, T. S. Chao, T. Y. Huang, and C. Y. Chang, "Evaluation of plasma charging in ultrathin gate oxides," *IEEE Electron Device Letters*, Vol. 19, 1998, pp. 68–72.
5. P. W. Mason, D. DeBusk, J. K. McDaniel, A. S. Oates, and K. P. Cheung, "Relationship between yield and reliability impact of plasma damage to gate oxide," in *Proceedings of the 5th International Symposium on Plasma Process-Induced Damage*, 2000, pp. 96–99.
6. S. Rangan, S. Krishnan, A. Amerasckara, S. Aur, and S. Ashok, "A model for channel hot carrier reliability degradation due to plasma damage in MOS Devices," in *Proceedings of the 37th IEEE International Annual Reliability Physics Symposium*, 1999, pp. 370–374.
7. A. T. Krishnan, V. Reddy, and S. Krishnan, "Impact of charging damage on negative bias temperature instability," in *Proceedings of the International Electron Devices Meeting (IEDM)*, 2001, pp. 865–868.
8. S. Krishnan, S. Rangan, S. Hattangady, G. Xing, K. Brennan, M. Rodder, and S. Ashok, "Assessment of charge-induced damage to ultra-thin gate MOSFETs," in *Proceedings of the International Electron Devices Meeting (IEDM)*, 1997, pp. 445–448.
9. H. Shin, C. C. King, T. Horiuchi, and C. Hu, "Thin oxide charging current during plasma etching of aluminum," *IEEE Electron Device Letters*, Vol. 12, No. 8, 1991, pp. 404–406.
10. A. B. Joshi, L. Chung, B. W. Min, and D. L. Kwong, "Gate oxide thickness dependence of RIE-induced damages on *N*-channel MOSFET reliability," in *Proceedings of the 34th IEEE International Annual Reliability Physics Symposium*, 1996, pp. 300–304.
11. A. Scarpa, D. Ward, J. Dubois, L. van Marwijk, S. Gausepohl, R. Campos, K. Y. Sim, A. Cacciato, R. Kho, and M. Bolt, "Negative-bias temperature instability cure by process optimization," *IEEE Tractions on Electron Devices*, Vol. 53, No. 6, 2006, pp. 1331–1339.
12. A. Martin, C. Siol, and C. Schlunder, "Fast productive WLR characterisation methods of plasma induced damage of thin and thick MOS gate oxides," in *IEEE International Integrated Reliability Workshop Final Report*, 2006, pp. 98–104.
13. D. S. Bang, M. Y. Hao, S. Chen, Q. Xiang, G. Yeap, and M. R. Lin, "Effect of Cu damascene metallization on gate SiO_2 plasma damage," in *Proceedings of the 3rd International Symposium on Plasma Process-Induced Damage*, 1998, pp. 64–67.
14. N. Matsunaga, H. Yoshinari, N. Yamada, and H. Shibata, "Plasma-induced damage reduction by spin-on low-*k* dielectrics process," in *ULSI Process Integration II, Electrochemical Society Proceedingss*, Vol. 2001–2, pp. 380–385.
15. N. Matsunaga, H. Yamaguchi, and H. Shibata, "Spreading antenna effect of plasma induced charging damage in damascene interconnect process," in *Proceedingss of the IEEE International Interconnect Technology Conference*, 2003, pp. 198–200.
16. N. Nakamura, T. Yoshizawa, T. Watanabe, H. Miyajima, S. Nakao, N. Yamada, K. Fujita, N. Matsunaga, and H. Shibata, "A plasma damage-resistant ultralow-*k* hybrid dielectric structure for 45-nm node copper dual-damascene interconnects," in *Proceedingss of the IEEE International Interconnect Technology Conference*, 2004, pp. 228–230.
17. L. Broussous, W. Puyrenier, D. Rebiscoul, V. Rouessac, and A. Ayral, "Post-etch cleaning for porous low k integration: impact of HF wet etch

on 'pore-sealing' and '*k* recovery,'" in *Proceedingss of the IEEE International Interconnect Technology Conference*, 2008, pp. 87–89.
18. H. Shi, J. Bao, H. Huang, B. Chao, S. Smith, Y. Sun, P. S. Ho, A. Li, M. Armacost, and D. Kyser, "Mechanistic study of CO_2 plasma damage to OSG low-*k* dielectrics," in *Proceedingss of the IEEE International Interconnect Technology Conference*, 2008, pp. 31–33.
19. H. Liu, J. Widodo, S. L. Liew, Z. H. Wang, Y. H. Wang, B. F. Lin, L. Z. Wu, C. S. Seet, W. Lu, C. H. Low, W. P. Liu, M. S. Zhou, and L. C. Hsia, "Challenges of ultralow-*k* integration in BEOL interconnect for 45-nm and beyond," in *Proceedingss of the IEEE International Interconnect Technology Conference*, 2009, pp. 258–260.
20. S. Krishnan, W. W. Dostalik, K. Brennan, S. Aur, S. Rangan, and S. Ashok, "Inductively coupled plasma (ICP) metal etch damage to 35–60Å gate oxide," in *Proceedings of the International Electron Devices Meeting (IEDM)*, 1996, pp. 731–734.
21. S. Fang and J. McVittie, "Thin oxide damage from gate charging during plasma processing," *IEEE Electron Device Letters*, Vol. 13, No. 5, 1992, pp. 288–290.
22. K. P. Cheung and C. P. Chang, "Plasma-charging damage: A physical model," *Journal of Applied Physics*, Vol. 75, 1994, pp. 4415–4426.
23. S. Ma and J. P. McVittie, "Real time measurement of transients and electrode edge effects for plasma charging induced damage," in *Proceedings of the International Electron Devices Meeting (IEDM)*, 1994, pp. 463–466.
24. K. Hashimoto, "Charging damage caused by electron shading effect," *Japanese Journal of Applied Physics*, Part I, Vol. 33, 1994, pp. 6013–6018.
25. G. Reimbold and T. Poiroux, "Plasma charging damage mechanisms and impact on new technologies," *Microelectronics Reliability*, Vol. 41, 2001, pp. 959–965.
26. T. Poiroux, J. L. Pelloie, K. Rödde, G. Turban, and G. Reimbold, "Plasma process-induced damage in SOI devices," in *Proceedings of the International Electron Devices Meeting (IEDM)*, 1999, pp. 97–100.
27. M. Joshi, J. P. McVittie, and K. Saraswat, "Direct experimental determination and modeling of VUV induced bulk conduction in dielectrics during plasma processing," in *Proceedings of the 5th International Symposium on Plasma Process-Induced Damage*, 2000, pp. 157–160.
28. S. Shuto, I. Kunishima, and S. Tanaka, "UV-blocking technology to reduce plasma-induced transistor damage in ferroelectric devices with low hydrogen resistance," in *Proceedings of the 37th IEEE International Annual Reliability Physics Symposium*, 1999, pp. 356–361.
29. J. R. Shih, A. Huang, Y. H. Chiou, C. C. Chiu, Y. Peng, and J. T. Yue, "Diagnostics of plasma process induced failure on analog device's mismatch characteristics," in *Proceedings of the 7th International Symposium on Plasma Process-Induced Damage*, 2002, pp. 64–67.
30. E. Li, D. Pachura, L. Duong, S. Prasad, and D. Vijay, "Plasma-induced charge damage and its effect on reliability in 0. 115-pm technology," in *Proceedings of the 8th International Symposium on Plasma Process-Induced Damage*, 2003, pp. 69–72.
31. T. B. Hook, D. Harmon, and C. Lin, "Detection of thin oxide (3. 5 nm) dielectric degradation due to charging damage by rapid-ramp breakdown," in *Proceedings of the 38th IEEE International Annual Reliability Physics Symposium*, 2000, pp. 377–388.
32. C. T. Gabriel and J. L. Educato, "Application of damage measurement techniques to a study of antenna structure charging", in *Proceedings of the 2nd International Symposium on Plasma Process-Induced Damage*, 1997, pp. 91–94.
33. C. H. Ang, W. H. Lu, Andrew K. L. Yap, L. C. Goh, Luona N. L. Goh, Y. K. Lim, C. S. Chua, L. H. Ko, Tracy H. S. Tan, S. L. Tob, and L. C. Hsia, "A study of SiN cap NH_3 plasma pre-treatment process on the PID, EM, GOI performance and BEOL defectivity in Cu dual damascene technology," in *Proceedings of the IEEE International Conference on Integrated Circuit Design and Technology*, 2004, pp. 119–122.
34. W. Y. Teo, Y. T. Hou, M. F. Li, P. Chen, L. H. KO, X. Zeng, Y. Lin, F. H. Gn, and L. H. Chan, "Investigation on dual gate oxide charging damage in 0.13-μm copper damascene technology," in *Proceedings of the 7th International Symposium on Plasma Process-Induced Damage*, 2002, pp. 14–17.

35. A. Martin, C. Bukethal, K. H. Ryden, S. Baier, and M. Schwerd, "Quantitative reliability assessment of plasma induced damage on product wafers with fast WLR measurements," in *IEEE International Integrated Reliability Workshop Final Report*, 2008, pp. 81–85.
36. JP-001, "Plasma process induced damage (P2ID)," in *Foundry Process Qualification Guidelines.* Wafer Fabrication Manufacturing Sites, JEDEC/FSA Joint Publication, September 2002, pp. 21–23.
37. D. Smeets, A. Martin, and A. Scarpa, "A general concept for monitoring plasma induced charging damage," in *Proceedings of the 8th International Symposium on Plasma Process-Induced Damage*, 2003, pp. 36–39.
38. D. Chong, W. J. Yoo, and C. M. Lek, "Plasma charging damage immunities of rapid thermal nitrided oxide and decoupled plasma nitrided oxide," in *Proceedings of the 10th International Symposium on the Physical and Failure Analysis of Integrated Circuits*, 2003, pp. 141–145.
39. A. Martin, "Investigation of initial charge trapping and oxide breakdown under Fowler-Nordheim injection," in *IEEE International Integrated Reliability Workshop Final Report*, 1998, pp. 99–104.
40. R. C. J. Wang, D. S. Su, C. T. Yang, D. H. Chen, Y. Y. Doong, J. R. Shih, S. Y. Lee, C. C. Chiu, Y. K. Peng, and J. T. Yue, "Investigation of electromigration properties and plasma charging damages for plasma treatment process in Cu interconnects," in *Proceedings of the 7th International Symposium on Plasma Process-Induced Damage*, 2002, pp. 166–168.
41. H. Guan, Y. Zhang, B. B. Jie, Y. D. He, M. -F. Li, Z. Dong, J. Xie, J. L. F. Wang, A. C. Yen, G. T. T. Sheng, and W. Li, "Nondestructive DCIV method to evaluate plasma charging damage in ultrathin gate oxides," *IEEE Electron Device Letters*, Vol. 20, No. 5, 1999, pp. 238–240.
42. A. Martin, "New approach for the assessment of the effect of plasma induced damage on MOS devices and subsequent design manual rules," in *IEEE International Integrated Reliability Workshop Final Report*, 2007, pp. 57–62.
43. A. Martin, and R. P. Vollertsen, "An introduction to fast wafer level reliability monitoring for integrated circuit mass production," *Microelectronics Reliability*, Vol. 44, 2004, pp. 1209–1231.
44. A. Martin, J. von Hagen, and G. B. Alers, "Ramped current stress for fast and reliable wafer level reliability monitoring of thin gate oxide reliability," *Microelectronics Reliability*, Vol. 43, 2003, pp. 1215–1220.
45. K. P. Cheung, "An efficient method for plasma-charging damage measurement," *IEEE Electron Device Letters*, Vol. 15, No. 11, 1994, pp. 460–462.
46. K. P. Cheung, S. Martin, D. Misra, K. Steiner, J. I. Colonell, C. P. Chang, W. Y. C. Lai, C. T. Liu, R. Liu and C. S. Pai, "Impact of plasma-charging damage polarity on MOSFET noise," in *Proceedings of the International Electron Devices Meeting (IEDM)*, 1997, pp. 437–440.
47. G. A. Roche and J. P. McVittie, "Application of plasma charging probe to production HDP CVD tool," in *Proceedings of the 1st International Symposium on Plasma Process-Induced Damage*, 1996, pp. 71–74.
48. S. Ma and J. P. McVittie, "Prediction of plasma charging induced gate oxide tunneling current and antenna dependence by plasma charging probe," in *Proceedings of the 1st International Symposium on Plasma Process-Induced Damage*, 1996, pp. 20–23.
49. R. Patrick and P. Jones, "Techniques for studying charging in plasma processing," in *Proceedings of the 1st International Symposium on Plasma Process-Induced Damage*, 1996, pp. 91–93.
50. S. Ma, J. P. McVittie, and K. C. Saraswat, "Prediction of plasma charging induced gate oxide damage by plasma charging probe," *IEEE Electron Device Letters*, Vol. 18, 1997, No. 10, 1997, pp. 468–470.
51. A. Cowley, "Characterisation of a Gasnoics A1000 microwave photoresist ashing system using the CHARM-2 plasma charging monitor," in *Proceedings of the 1st International Symposium on Plasma Process-Induced Damage*, 1996, pp. 98–101.

52. W. Lukaszek and A. H. Birrell, "Quantifying wafer charging during via etch," in *Proceedings of the 1st International Symposium on Plasma Process-Induced Damage*, 1996, pp. 30–33.
53. W. Lukaszek, W. Dixon, M. Vella, C. Messick, S. Reno, and J. Shideler, "Characterization of wafer charging mechanisms and oxide survival prediction methodology," in *Proceedings of the 32nd IEEE International Annual Reliability Physics Symposium*, 1994, pp. 334–338.
54. P. K. Aum, R. Brandshaft, D. Brandshaft, and T. B. Dao, "Controlling plasma charge damage in advanced semiconductor manufacturing—challenge of small feature size device, large chip size, and large wafer size," *IEEE Transactions on Electron Devices*, Vol. 45, No. 3, 1998, pp. 722–730.
55. K. Nauka, J. Theil, J. Lagowski, L. Jastrzebski, and S. Sawtchouk, "A complete approach to the in-line monitoring of materials defects introduced in dielectric and Si by plasma processing," in *Proceedings of the 2nd International Symposium on Plasma Process-Induced Damage*, 1997, pp. 127–130.
56. M. W. Goss and A. Findlay, "Correlation of nitride spacer plasma damage results from conventional gate capacitor electrical tests and a new, non-contact approach," in *Proceedings of the 2nd International Symposium on Plasma Process-Induced Damage*, 1997, pp. 135–138.
57. Y. Karzhavin, K. Lao, W. Wu, and C. Gelatos, "Plasma induced charging evaluation using SCA and PDM tools," in *Proceedings of the 2nd International Symposium on Plasma Process-Induced Damage*, 1997, pp. 143–146.
58. C. Cismaru, J. L. Shohet, K. Nauka, and B. Friedmann, "Relationship between the charging damage of test structures and the deposited charge on unpatterned wafers exposed to an electron cyclotron resonance plasma," in *Proceedings of the 2nd International Symposium on Plasma Process-Induced Damage*, 1997, pp. 131–134.
59. Website: http://www.kla-tencor.com/Access date: December 2010.
60. B. P. Linder and N. W. Cheung, "Calculating plasma damage as a function of gate oxide thickness," in *Proceedings of the 3rd International Symposium on Plasma Process-Induced Damage*, 1998, pp. 42–45.
61. V. Vahedi, N. Benjamin, and A. Perry, "Topographic dependence of plasma charging induced device damage," in *Proceedings of the 2nd International Symposium on Plasma Process-Induced Damage*, 1997, pp. 41–44.
62. A. T. Krishnan, S. Krishnan, and P. Nicollian, "Impact of gate area on plasma charging damage: The "reverse" antenna effect," in *Proceedings of the International Electron Devices Meeting (IEDM)*, 2002, pp. 525–528.
63. K. P. Chueng, C. T. Liu, C. P. Chang, J. I. Colonell, W. Y. C. Lai, C. Pai, H. Vaidya, R. Liu, J. T. Clemens, and E. Hasegawa, "Charging damage in thin gate-oxides – better or worse," in *Proceedings of the 3rd International Symposium on Plasma Process-Induced Damage*, 1998, p. 34.
64. K. F. Schuegraf and C. Hu, "Hole injection SiO_2 breakdown model for very low voltage lifetime extrapolation," *IEEE Transactions on Electron Devices*, Vol. 41, No. 5, 1994, pp. 761–767.
65. F. Shone, K. Wu, J. Shaw, E. Hokelet, S. Mittal, and A. Haranahalli, "Gate oxide charging and its elimination for metal antenna capacitor and transistor in VLSI CMOS double layer metal technology," in *Symposium on VLSI Technology Digest of Technical Papers*, 1989, pp. 73–74.
66. D. Park and C. Hu, "The prospect of process induced charging damage in future thin gate oxides," *Microelectronics Reliability*, Vol. 39, 1999, pp. 567–577.
67. Z. Wang, A. Scarpa, S. Smits, C. Salm, and F. Kuper, "Temperature effects on antenna protection strategy for plasma process induced charging damage," in *Proceedings of the 7th International Symposium on Plasma Process-Induced Damage*, 2002, pp. 134–137.
68. Website: www. silvaco.com.
69. T. Zwingman, A. Gabrys, and A. J. West, "Automated test structure generation for characterizing plasma induced damage in MOSFET devices," in *Proceedings of the IEEE International Conference on Microelectronic Test Structures (ICMTS)*, 2009, pp. 102–105.

70. P. Simon, J. M. Luchies, and W. Maly, "Antenna ratio definition for VLSI circuits," in *Proceedings of the 4th International Symposium on Plasma Process-Induced Damage*, 1999, pp. 16–20.
71. S. Krishnan, A. Amerasekera, S. Rangan, and S. Aur, "Antenna device reliability for ULSI processing," in *Proceedings of the International Electron Devices Meeting (IEDM)*, 1998, pp. 601–604.
72. K. P. Cheung, "Plasma charging damage to gate dielectric: Past, present and future," in *Proceedings of the 9th International Symposium on the Physical and Failure Analysis of Integrated Circuits*, 2002, pp. 237–241.
73. T. B. Hook et al. , *44th IEEE International Annual Reliability Physics Symposium*, 2006, Tutorial 110.
74. W. Lai, D. Harmon, T. Hook, V. Ontalus, and J. Gambino, "Ultrathin gate dielectric plasma charging damage in SOI technology," in *Proceedings of the 44th IEEE International Annual Reliability Physics Symposium*, 2006, pp. 370–373.
75. R. Bolam, G. Shahidi, F. Assaderaghi, M. Khare, A. Mocuta, T. Hook, E. Wu, E. Leobandung, S. Voldman, and D. Badami, "Reliability issues for silicon-on-insulator," in *Proceedings of the International Electron Devices Meeting (IEDM)*, 2000, pp. 131–134.
76. J. P. Carrere, J. C. Oberlin, and M. Haond, "Topographical dependence of charging and new phenomenon during inductively coupled plasma CVD process," in *Proceedings of the 5th International Symposium on Plasma Process-Induced Damage*, 2000, pp. 164–167.
77. A. Hasegawa, F. Shimpuku, M. Aoyama, K. Hashimoto and M. Nakamura, "Direction of topography dependent damage current during plasma etching," in *Proceedings of the 3rd International Symposium on Plasma Process-Induced Damage*, 1998, pp. 168–171.
78. S. Krishnan, K. Brennan, and G. Xing, "A transient fuse scheme for plasma etch damage detection," in *Proceedings of the 3rd International Symposium on Plasma Process-Induced Damage*, 1998, pp. 201–204.
79. S. Krishnan, S. Rangan, S. Hattangady, G. Xing, K. Brennan, M. Rodder, and S. Ashok, "Assessment of charge-induced damage to ultra-thin gate MOSFETs," in *Proceedings of the International Electron Devices Meeting (IEDM)*, 1997, pp. 445–448.
80. M. Joshi, J. P. McVittie, and K. Saraswat, "Development of a physical model of UV induced bulk photoconduction in silicon dioxide and application to charging damage," in *Proceedings of the 7th International Symposium on Plasma- and Process-Induced Damage*, 2002, pp. 23–26.
81. S. Ma, W. L. N. Abdel-Ati, and J. P. McVittie, "Sensitivity and limitations of plasma charging damage measurements using MOS capacitors structures," *IEEE Electron Device Letter*, Vol. 18, No. 9, 1997, pp. 420–422.
82. S. Fang and J. P. McVittie, "Model for oxide damage from gate charging during magnetron etching," *Applied Physics Letters*, Vol. 62, No. 13, 1993, pp. 1507–1509.
83. A. Stensman, "Interaction of *P* defects at the (111) Si/SiO interface with molecular hydrogen: Simultaneous action of passivation and dissociation," *Journal of Applied Physics*, Vol. 88, 2000, pp. 489–497.
84. G. Cellere, A. Paccagnella, and M. G. Valentini, "FGA effects on plasma-induced damage: beyond the appearances," *IEEE Transactions on Electron Devices*, Vol. 51, No. 3, 2004, pp. 332–338.
85. R. A. B. Devine, J.-L. Autran, W. L. Warren, K. L. Vanheusdan, and J.-C. Rostaing, "Interfacial hardness enhancement in deuterium annealed 0. 25 μm channel metal oxide semiconductor transistors," *Applied Physics Letters*, Vol. 70, No. 22, 1997, pp. 2999–3001.
86. J. C. King and C. Hu, "Effect of low and high temperature anneal on process induced damage of gate oxide," *IEEE Electron Device Letter*, Vol. 15, No. 11, 1994, pp. 475–476.
87. A. V. Glasow, A. H. Fischer, D. Bunel, G. Friese, A. Hausmann, O. Heitzsch, M. Hommel, I. Kriz, S. Penka, P. Raffin, C. Robin, H. P. Sperlich, F. Ungar, and A. E. Zitelsherger, "The influence of the SiN cap process on the

electromigration and stressvoiding performance of dual damascene Cu interconnect," in *Proceedings of the 41th IEEE International Annual Reliability Physics Symposium*, 2003, pp. 146–150.
88. N. Nakamura, K. Higashi, N. Matsunaga, H. Miyajima, S. Sato, and H. Shibata, "Plasma process-induced wire-to-wire leakage current for low-k SiOC/Cu damascene structure," in *Proceedings of the 7th International Symposium on Plasma Process-Induced Damage*, 2002, pp. 162–165.
89. T. Poiroux, F. Pascal, M. Heitzmann, P. Berruyer, G. Turban, and G. Reimbold, "Study of the influence of process parameters on gate oxide degradation during contact etching in MERE arid HDP reactors," in *Proceedings of the 4th International Symposium on Plasma Process-Induced Damage*, 1999, pp. 12–15.
90. H. Shi, H. Huang, J. Im, P. S. Ho, Y. Zhou, J. T. Pender, M. Armacost, and D. Kyser, "Minimization of plasma ashing damage to OSG low-k dielectrics," in *Proceedings of the International Interconnect Technology Conference (IITC)*, 2010, pp. 1–3.
91. H. Shi, H. Huang, J. Bao, J. Im, P. S. Ho, Y. Zhou, J. T. Pender, M. Armacost, and D. Kyser, "Plasma altered layer model for plasma damage characterization of porous OSG films," in *Proceedings of the International Interconnect Technology Conference (IITC)*, 2009, pp. 78–80.
92. C. C. Liao, Z. H. Gan, Y. J. Wu, K. Zheng, R. Guo, L. F. Zhang, and J. Ning, "A comprehensive study of reliability improvement for 65nm Cu/low-k process," in *Proceedings of the 16th International Symposium on the Physical and Failure Analysis of Integrated Circuits*, 2009, pp. 685–689.
93. S. C. Song, S. Filipiak, A. Perera, M. Tumer, F. Huang, S. G. H. Anderson, L. Kang, B. Min, D. Menke, S. Tukunang, and S. Venkatesan, "Avoiding plasma induced damage to gate oxide with conductive top film (CTF) on PECVD contact etch stop layer," in *Symposium on VLSI Technology Digest of Technical Papers*, 2002, pp. 72–73.
94. S. Tabara, "A method for reducing notching and electron shading damage in a continuous wave ECR metal etcher," in *Proceedings of the 4th International Symposium on Plasma Process-Induced Damage*, 1999, pp. 108–111.
95. N. S. Kim, H. G. Yoon, C. K. Lee, J. Zhao, C. Y. Tuck, Y. S. Cheah, W. W. Yew, P. Southworth, S. H. Han, and K. S. Pey, "Effect of magnetic field on plasma damage during via etching in sub-micron CMOS technology," in *Proceedings of the 42nd IEEE International Annual Reliability Physics Symposium*, 2004, pp. 665–666.
96. K. P. Cheung, "On the mechanism of plasma enhanced deposition charging damage," in *Proceedings of the 5th International Symposium on Plasma Process-Induced Damage*, 2000, pp. 161–164.
97. P. R. Apte, T. Kubota, and K. C. Saraswat, "Constant current stress breakdown in ultrathin SiO_2 films," *Journal of the Electrochemical Society*, Vol. 140, No. 3, 1993, pp. 770–773.
98. M. Okigawa, Y. Ishigawa, S. Kuagai, and S. Samuhawa, "Reduction of ultra violet radiation induced damage and its time-resolved measurement using time-modulated plasma," in *Proceedings of the 7th International Symposium on Plasma Process-Induced Damage*, 2002, pp. 122–125.
99. H. Shan, C. H. Bjorkman, R. A. Lindley, K. Collins, M. Rice, G. Z. Yin, M. D. Welch, R. Ramanathan, J. Werking, and D. Galley, "Minimizing charge-up damage during dielectric etchers' hardware and process development stages," in *Proceedings of the 4th International Symposium on Plasma Process-Induced Damage*, 1999, pp. 3–7.
100. S. Siu, R. Patrick, V. Vahedi, S. Alba, G. Valentini, and P. Colombo, "Effect of plasma density and uniformity, electron temperature, process gas, and chamber on electron shading damage," in *Proceedings of the 4th International Symposium on Plasma Process-Induced Damage*, 1999, pp. 25–28.
101. K. Eriguchi, Y. Uraoka, and S. Odanaka, "A new gate oxide lifetime prediction method using cumulative damage law and its applications to plasma-damaged oxides," in *Proceedings of the International Electron Devices Meeting (IEDM)*, 1995, pp. 323–326.

102. G. Cellere, M. G. Valentini, M. Caminati, M. E. Vitali, A. Morol, and A. Paccagnella, "Plasma damage reduction by using ISSG gate oxides," in *Proceedings of the 8th International Symposium on Plasma Process-Induced Damage*, 2003, pp. 65–68.
103. C. C. Chen, H. C. Lin, C. Y. Chang, M. S. Liang, C. H. Chien, S. K. Hsien, T. Y. Huang, and T. S. Chao, "Plasma induced charging damage in ultrathin (3-nm) gate oxides," *IEEE Transaction on Electron Devices*, Vol. 47, 2000, pp. 1355–1360.
104. L. K. Han, M. Bhat, D. Wristers, H. H. Wang, and D. L. Kwong, "Recent developments in ultra thin oxynitride gate dielectrics," *Microelectronic Engineering*, Vol. 28, 1995, pp. 89–96.
105. K. P. Cheung, D. Misra, I. Colonell, C. T. Liu, Y. Ma, C. P. Chang, W. Y. C. Lai, R. Liu, and C. S. Pai, "Plasma damage immunity of thin gate oxide grown on very lightly *N*+ implanted silicon," *IEEE Electron Device Letters*, Vol. 19, 1998, pp. 231–233.
106. C. -C. Chen, H. -C. Lin, C. -Y Chang, C. -C. Huang, C. -H. Chien, T. -Y Huang, and M. -S. Liang, "Improved plasma charging immunity in ultra-thin gate oxide with fluorine and nitrogen implantation," in *Proceedings of the 5th International Symposium on Plasma Process-Induced Damage*, 2000, pp. 121–124.
107. C. T. Gabriel and R. Y. Kim, "Transient fuse structures: the role of metal etching vs. dielectric deposition," in *Proceedings of the 5th Symposium on Plasma Process-Induced Damage*, 2000, pp. 168–171.
108. A. Scarpa, M. Diekema, C. van der Schaar, H. Valk, A. Harke, and F. G. Kuper, "Process dependent antenna ratio rules for HSQ and FSG back-ends of (embedded flash) 0.18 pm CMOS technology," in *Proceedings of the 7th International Symposium on Plasma Process-Induced Damage*, 2002, pp. 138–141.
109. G. S. Hwang and K. P. Giapis, "Modeling of charging damage during interlevel oxide deposition in high-density plasmas," *Journal of Applied Physics*, Vol. 84, 1998, pp. 154–160.
110. M. Joshi, J. P. McVittie, K. Saraswat, C. Cismaru, and J. L. Shohet, "Measurement of VUV induced surface conduction in dielectrics using synchrotron radiation," in *Proceedings of the 5th International Symposium on Plasma Process-Induced Damage*, 2000, pp. 14–17.
111. T. B. Hook, "Backside films and charging during via etch in LOCOS and STI technologies," in *Proceedings of the 3rd International Symposium on Plasma Process-Induced Damage*, 1998, pp. 11–14.
112. R. Rakkhit, F. P. Heiler, P. Fang, and C. Sander, "Process induced oxide damage and its implications to device reliability of submicron transistors," in *Proceedings of the 31st IEEE International Annual Reliability Physics Symposium*, 1993, pp. 293–296.
113. E. Takeda and N. Suzuki, "An empirical model for device degradation due to hot-carrier injection," *IEEE Electron Device Letter*, Vol. 4, 1983, pp. 111–113.
114. C. Hu, S. C. Tam, F. C Hsu, P. K. Ko, T. Y. Chan, and K. W. Terrill, "Hot electron induced MOSFET degradation: Model, monitor, and improvement," *IEEE Transactions on Electron Device*, Vol. 48, No. 4, 1985, pp. 375–385.
115. W. T. Weng, S. Oates, and T.-Y. Huang, "A comprehensive model for plasma damage enhanced transistor reliability degradation," in *Proceedings of the 45th IEEE International Annual Reliability Physics Symposium*, 2007, pp. 364–369.
116. D. Park, M. Kennard, Y. Melaku, N. Benjamin, T. J. King, and C. Hu, "Stress-induced leakage current due to charging damage: gate oxide thickness and gate poly-Si etching condition dependence," in *Proceedings of the 3rd International Symposium on Plasma Process-Induced Damage*, 1998, pp. 56–59.

CHAPTER 9

Electrostatic Discharge Protection of Integrated Circuits

9.1 Background of Electrostatic Discharge Events

Electrostatic discharge (ESD) is one of the most prevalent threats to electronic components. It is an event in which a finite amount of charge is transferred from one object (e. g., a human body) to another (e. g., a microchip). This process can result in a very high current passing through the object within a very short period of time.[1,2] A common ESD phenomenon is shown in Fig. 9.1, where a person is shocked by an ESD from the metal doorknob to the ground via the human body. When a microchip or electronic system is subject to an ESD event, the huge ESD-induced current can damage the microchip and cause malfunction to the electronic system if the heat generated in the object cannot be dissipated quickly enough. Figure 9.2 shows the damage in a microchip resulting from improper handling by a charged person, a well-known ESD event called the *human-body model*. It is estimated that about 35 percent of all damage to microchips is ESD-related, resulting in a loss of revenue of several hundred million dollars in the global semiconductor industry every year.[3] The continuing scaling of complementary metal-oxide-semiconductor (CMOS) technology makes the ESD-induced failures even more prominent, and one can predict with certainty that the availability of effective and robust ESD protection solutions will become a critical and essential component of the successful advancement and commercialization of next-generation CMOS technology.[4–7]

In this chapter, the fundamentals of ESD, including its mechanisms, standards, and protection principles, will be introduced first.

FIGURE 9.1 A commonly known ESD phenomenon.

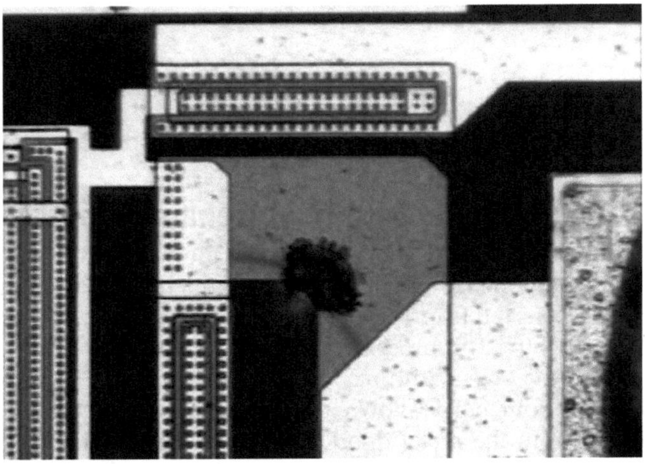

FIGURE 9.2 Damage (middle of photo) in microchip caused by ESD event.

This is followed by a discussions of modeling of ESD protection devices and ESD testing characterization. Finally, ESD protection designs for the input/output (I/O) and power-supply pins will be developed and illustrated.

The following three processes can generate charges on an object: the *triboelectrification process*, where the charges are generated owing to the contacting and rubbing of two different objects; the *inductive process*, where the charges are generated owing to the presence of an electric field; and the *conductive process*, where a previously uncharged object is charged after it is in touch with another charged object. The amount of charge generated increases with decreasing air humidity, and the severity of the charge is described by the voltage. For example, a human body can carry more than 10,000 V of ESD in a dry environment and a few hundred volts of ESD in a wet environment.[8]

Various ESD events taking place in our daily life and in microchip manufacturing environment can be categorized into the following four standards or models: the *human-body model* (HBM), which describes the event when a charged person touches a microchip; the *machine model* (MM), which describes the event when a charged metal object is in contact with a microchip; the *charged-device model* (CDM), which describes the event when a charged device is in contact with a grounded object; and the *international electrotechnical commission model* (IEC), which describes an event when a charged cable/wire is in contact with an electronic component.

Some comments on ESD standards are in order. The human-body model is a mature, well-understood ESD model for simulating charge transfer from a person's finger to an electronic component. However, recent industry data indicate that the HBM rarely simulates real-world ESD failures. Latest-generation package styles such as micro ball-grid arrays (mBGAs), small outline transistors (SOTs), shrink small outline transistor package (SC70s), and chip scale packages (CSPs) with millimeter-range dimensions are often effectively too small for people to handle with their fingers. Even in cases of relatively large components, most high-volume component and board manufacturing uses automated equipment, so humans rarely touch the components. The CDM can more successfully replicate in-house and customer integrated-circuit (IC) failures at the component level. The CDM simulates the damage induced when a metal pin or solder ball on a charged IC package is instantaneously discharged via contact with a metallic object at ground potential.

An effective way to protect an electronics system against ESD events is to incorporate an ESD protection structure onto the microchip (i.e., called *on-chip ESD protection*) to increase the survivability of the core circuit when an ESD strikes, and a schematic of such a structure is shown in Fig. 9.3, where all the input, output, and power-supply pins are connected to the ground bus rail via ESD protection devices, labeled as *uSCR*. These ESD devices must be off during normal system operation (i.e., in the absence of an ESD event), must be turned on quickly when an ESD event takes place so that the current generated by the ESD event can be conducted by the ESD protection devices and discharged to the ground, must not be damaged by the ESD stress themselves, must clamp the pins to a sufficiently low voltage during the ESD event, and must return to the off-state after the ESD event has passed.[9,10]

There are a number of technologies for which the above-mentioned on-chip ESD protection is not feasible, and using an off-chip ESD protection solution becomes necessary. The off-chip ESD protection structure can be incorporated into the electrical cables, connectors, and ceramic carriers or on the circuit boards. Figure 9.4 shows a Global System for Mobile Communications (GSM) wireless module with two off-chip ESD protection devices included.

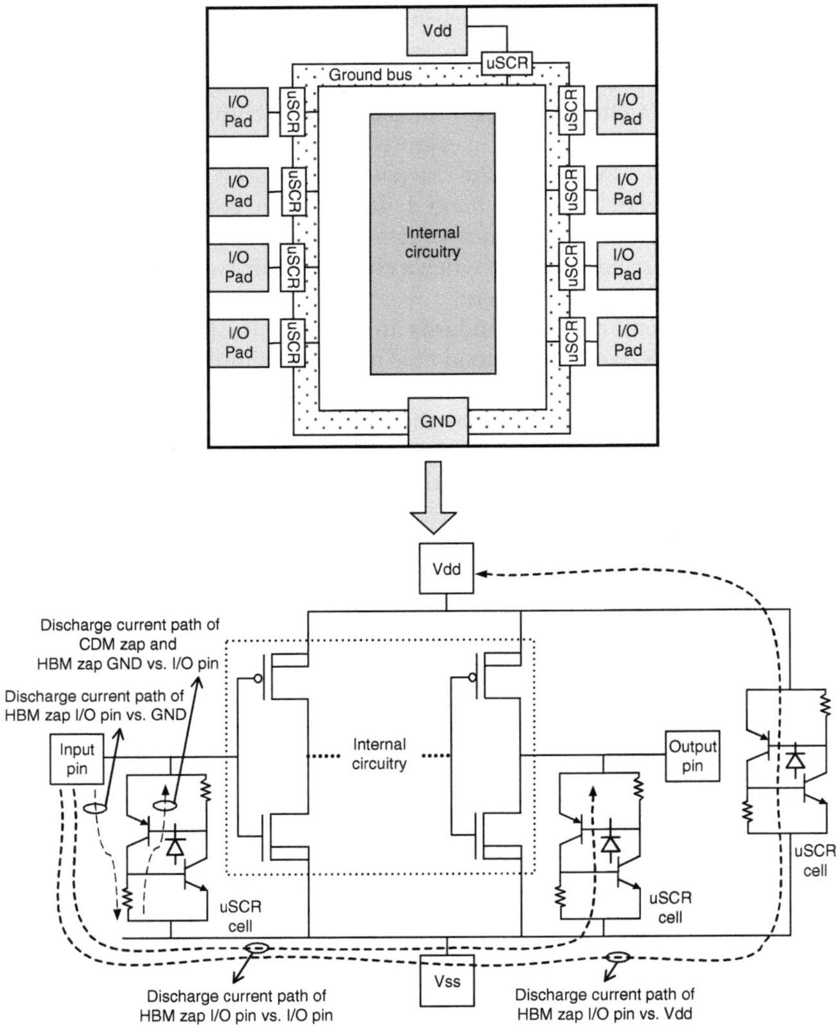

FIGURE 9.3 Schematic of an on-chip ESD protection device (*above*) and its equivalent circuit (*below*).

In the future, designing effective ESD protection solutions will become increasingly difficult and costly. As shown in Fig. 9.5, the cost of designing ESD protection solutions in general increases with CMOS technology advancement (i.e., the technology node reducing from 180 to 45 nm).[11] Furthermore, using custom solutions (ESD protection solutions designed with customized and optimized approaches) has the advantage of cutting the cost over using public solutions (ESD protection solutions designed with generalized and off-the-shelf approaches). The operation of an ESD protection device

Figure 9.4 Off-chip ESD protection (*circled*) for a wireless GSM module.

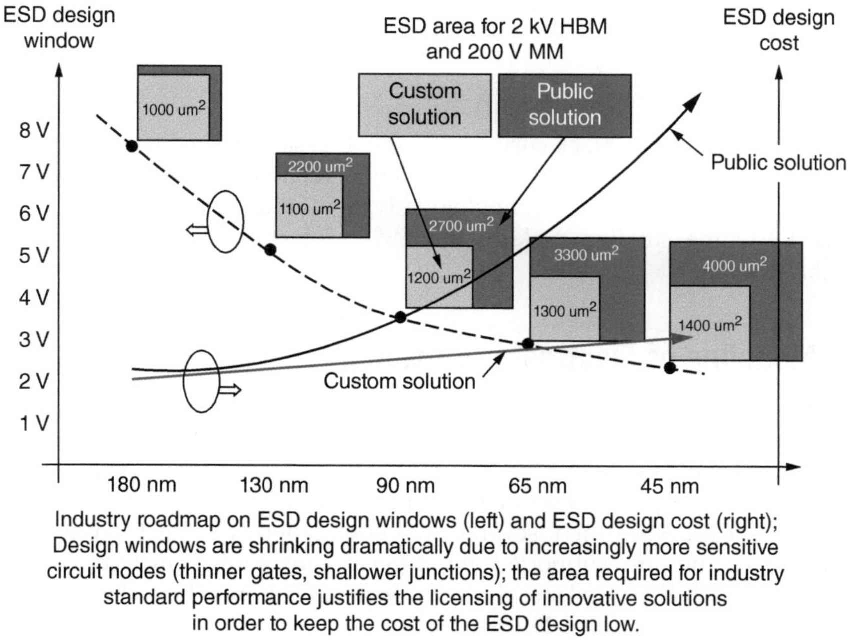

Figure 9.5 ESD design cost as a function of the CMOS technology using the custom solution and the public solution. (*Used with permission from Sofics.*[11])

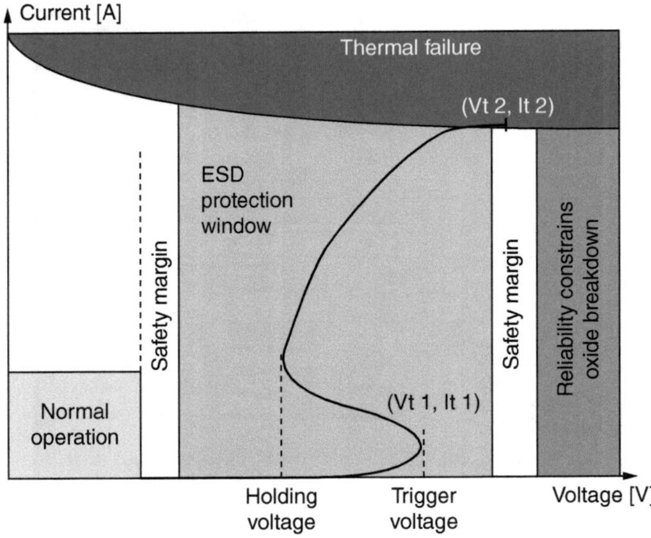

FIGURE 9.6 Illustration of IC operating area (the left side "normal operation" region), ESD design window (the central "ESD protection window" region), I-V curve of ESD protection device (*black line*), and IC reliability constraint (the right side "oxide breakdown" region).

is described by the black curve in Fig. 9.6. It is sandwiched between the IC operating area on the left and the IC reliability constraint on the right. As the CMOS technology advances, this design window will become smaller, making the ESD protection design more challenging.

9.2 Modeling of Electrostatic-Discharge-Protection Device

Simulation Program with Integrated-Circuit Emphasis (SPICE) is the standard software tool for circuit simulation. The backbone of such a tool was developed by a research group at the University of California in the 1970s. Since then, several different versions of SPICE have emerged and are supported by different companies, including Cadence SPICE by Cadence Systems, Smart SPICE by Silvaco International, and PSPICE by Microsim Corp.

To carry out SPICE simulation, a circuit, including active components (i.e., transistors) and passive components (i.e., resistors, capacitors, and inductors), first needs to be specified. A library containing the models of various active components is available in SPICE, and depending on the transistors specified in the circuit, appropriate models then are taken from the library to replace these active components. A numerical algorithm then is executed to perform the simulation of the circuit.

Let us focus on the metal-oxide-semiconductor field-effect transistor (MOSFET) because this device is frequently used to build ESD protection circuits. There are several different levels of MOS models in SPICE, with the lowest-level model being the most compact but least accurate. Unfortunately, all the existing MOS models were developed with the intention of simulating typical MOS operations under relatively low current/voltage conditions, and as such, these models are not applicable to and are highly inaccurate for ESD applications.

Many studies have been conducted in recent years to gain insights into the behavior of MOSFETs subjected to ESD stresses. Krabbenborg et al.[12] and Hong et al.[13] have demonstrated that two-dimensional (2D) electrothermal simulation is superior to one-dimensional (1D) simulation because a number of factors, especially the temperature and current density distributions, are affected by the existence of the second dimension. Later, Voldman et al.[14] described the importance of three-dimensional (3D) simulation in advanced CMOS technology. Pinto-Guedes and Chan[15] broke new ground by including a parasitic bipolar induced breakdown model in short-channel MOS devices for circuit simulation. Electrothermal circuit simulators, especially circuit-level electrothermal simulation tool (iETSIM),[16] also have made significant progress toward the analysis of ESD effects in circuits. Diaz et al.[17] and Ramaswamy et al.[18] reported an electrothermal model that can be used to simulate MOS devices operating in the snapback region and up to the onset of second breakdown. They also presented the implementation of such a model in the circuit-level simulator iETSIM. Compact models that are compatible with the industry-standard SPICE, however, are more desirable owing to their simplicity and easy integration. To this end, Amerasekera et al.[19] developed a nonelectrothermal CMOS model including the parasitic bipolar transistor for SPICE circuit simulation, but their simple approximation for the substrate resistance can lead to considerable errors.

In this section, a physics-based and compact MOS model applicable for SPICE simulation under the ESD event will be developed and discussed. First, the device physics for the *n*-channel MOSFET and its parasitic bipolar transistor are described. Physics-based equations describing MOSFET behavior subject to the ESD event then are derived. This is followed by discussions of implementation of the MOS model developed into Cadence SPICE. Finally, SPICE simulation results are presented.

9.2.1 Physical Behavior of NMOS Devices with Parasitic Bipolar Transistors

As the distance between the drain and source regions in deep-submicron MOS devices shrinks to a value comparable with the base-layer thickness in a typical bipolar junction transistor (BJT), the parasitic lateral BJT in the MOS device becomes prominent and can be easily activated under the ESD event. Once the parasitic BJT is turned on, the behavior of the MOSFET changes dramatically. Let us consider an *n*-channel MOSFET (NMOS). When the drain voltage is sufficiently large, as would be the case with an ESD applied to the drain terminal, most carriers in the channel (i.e., electrons) are able to gain considerable kinetic energy when crossing the drain-substrate reverse-biased depletion region. These electrons, when colliding with the lattice, break covalent bonds and generate electron-hole pairs (EHPs). The generated EHPs then may have enough energy themselves to break more covalent bonds, thus generating even more EHPs. This phenomenon is called *avalanche multiplication*. As shown in Fig. 9.7, the generated electrons are swept into the drain by the electric field and increase the drain current. On the other hand, the holes generated flow into the substrate (i.e., quasi-neutral bulk) and constitute the substrate current. Since the semiconductor substrate has a finite resistance, this hole current gives rise to a potential drop across the source-substrate junction. As the drain voltage increases, the substrate current also increases, and eventually, the voltage drop at the source-substrate junction becomes large enough to turn on the junction and cause electrons to inject from the source (i.e., the emitter of the parasitic BJT) into the substrate (i.e., the base of the parasitic BJT). These electrons flow to the drain area (i.e., the collector of the parasitic BJT) and add to the

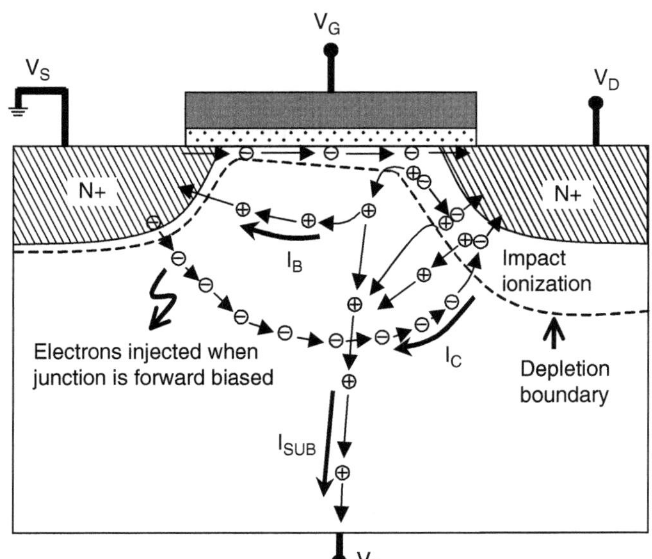

FIGURE 9.7 NMOS transistor with parasitic bipolar transistor.

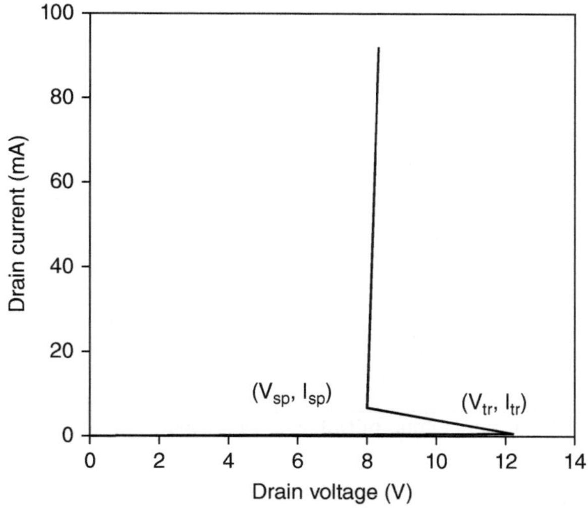

Figure 9.8 *I-V* curve with snapback behavior at $V_g = 0$.

drain current, which, in turn, results in more avalanche-generated holes near the drain junction and thus a positive-feedback action. When this occurs, the drain voltage is reduced quickly and stays at a relatively constant level, a phenomenon called the *snapback*, as illustrated in Fig. 9.8.

A paper[20] has shown that snapback may be associated with the "body effect" of the MOS transistor, but snapback is unquestionably involved in the parasitic bipolar mechanism at high current and high voltage levels.[10, 15, 21] In Fig. 9.8, the trigger voltage V_{tr} is where avalanche multiplication first occurs and turns on the lateral parasitic bipolar transistor. Once the bipolar device turns on, the drain voltage is reduced, and the negative-resistance phenomenon is observed because more electrons from the source drift to the drain junction until a minimum snapback voltage V_{sp} is reached. After that, the resistance generally is positive and small owing to the conductivity modulation of the substrate associated with the increased current in the substrate.

9.2.2 Development of a Compact Model for Electrostatic Discharge

The equivalent circuit of an NMOS device with the parasitic BJT is depicted in Fig. 9.9. The total drain current I_{DT} is comprised of the following three components: (1) the drain current I_D through the channel, (2) the collector current I_C of the parasitic BJT, and (3) the electron current $I_{e,gen}$ owing to impact ionization, which is the same as the impact-generated hole current $I_{h,gen}$. Furthermore, $I_{h,gen}$ consists of the

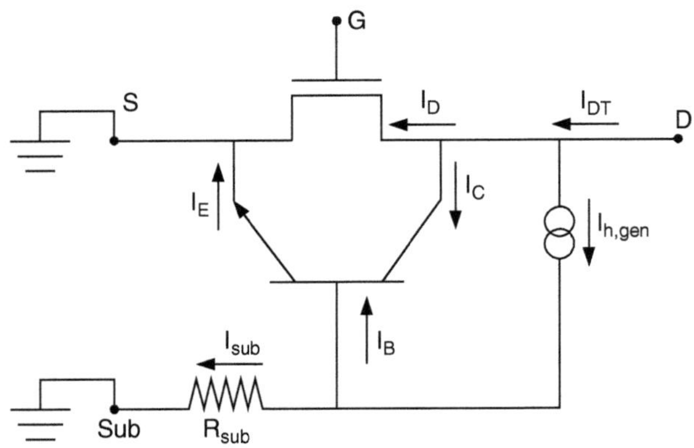

Figure 9.9 Equivalent circuit of NMOS devices, including a parasitic bipolar mechanism.

substrate current I_{sub} passing through the substrate and the base current I_B of the parasitic BJT.

It should be pointed out that the existing NMOS model (i.e., conventional NMOS model) in SPICE does not include the components of $I_{h,gen}$, R_{sub}, and parasitic bipolar transistor. As will be shown later, such a simplification will result in a large error in ESD circuit simulations. In the following subsections, in developing an NMOS model suitable for ESD simulation, these three components will be modeled and incorporated into the conventional NMOS model.

Avalanche Multiplication

As mentioned earlier, electron-hole pairs are produced near the drain junction owing to impact ionization. The holes then are swept into the substrate region and constitute the impact-generated hole current $I_{h,gen}$. For zero gate voltage, $I_{h,gen}$ is associated with thermal generation and minority carrier diffusion.[8,22] When the gate voltage is greater than the threshold voltage, $I_{h,gen}$ becomes a function of I_D. Prior to the turn-on of the parasitic BJT, such a current can be expressed as

$$I_{h,gen} = (M-1)I_D \tag{9.1}$$

where M is the avalanche multiplication factor. Note that I_{sub} is equal to $I_{h,gen}$ here because there is no parasitic bipolar action. After the parasitic bipolar is turned on, in addition to the typical drain current I_D, there is a collector current I_C flowing from the source to the drain, and

$$I_{h,gen} = (M-1)(I_D + I_C) \tag{9.2}$$

$I_{h,gen}$ also can be given by

$$I_{h,gen} = I_{sub} + I_B \tag{9.3}$$

where I_B is the base current of the parasitic bipolar transistor. The total drain current I_{DT} is

$$I_{DT} = I_D + I_C + I_{h,gen} \tag{9.4}$$

Substituting Eq. (9.2) into Eq. (9.4), gives

$$I_{DT} = M(I_D + I_C) \tag{9.5}$$

The multiplication factor M can be modeled as[10,23]

$$M = \frac{1}{1 - I_{\text{ion}}} \tag{9.6}$$

where I_{ion} is the impact ionization integral:

$$I_{ion} = \int_0^{l_d} \alpha_n \exp\left[-\int_0^x (\alpha_n - \alpha_p) dx'\right] dx \tag{9.7}$$

Here α_n is the impact ionization coefficient of electrons, α_p is the impact ionization coefficient of holes, and l_d is the length of the high-field region in the depletion region. For the case of $\alpha_n = \alpha_p = \alpha$, the impact ionization integral can be reduced to

$$I_{ion} = \int_0^{l_d} \alpha dx \tag{9.8}$$

and α is approximated by[24–26]

$$\alpha = A \cdot \exp\left(\frac{-B}{E}\right) \tag{9.9}$$

where E is the electric field, and A and B are constant parameters. The electric field typically is a function of position and has the form of $E = E_{sat} \cosh(x/l_d)$,[24,25] where E_{sat} is the electric field at which electrons possess saturation drift velocity. For simplicity, E can be approximated as a constant at snapback conditions and expressed by

$$E = \frac{V_D - V_{dsat}}{l_d} \tag{9.10}$$

Substituting Eqs. (9.8), (9.9), and (9.10) into Eq. (9.6), we have

$$M = \frac{1}{1 - P_1 \exp[-P_2/(V_D - V_{dsat})]} \tag{9.11}$$

where $P_1 = Al_d$ and $P_2 = Bl_d$ are model parameters that need to be determined from measurements. In order to avoid the pole in the denominator (i.e., when $V_D = V_{dsat}$) in Eq. (9.11), a programming technique can be employed in the SPICE program using C or the Analog Hardware Definition/Description Language (AHDL) language, the details of which will be presented later in Sec. 9.2.3. Lim et al.[26] reported an alternative formulation for M to resolve the discontinuity problem when the denominator approaches zero in Eq. (9.11). However, their new function deviates from physical reality in the high-current region,[27] and other limitations also took place during simulation of a circuit using an initially negatively CDM pulse at the I/O pad.[28]

A simple yet effective method for the extraction of parameters P_1 and P_2 has been reported in Ref. 19, and such a method will be used here. The procedure involves dc sweep measurements of the NMOS device at a gate voltage above threshold voltage using an HP 4156A parameter analyzer. Equation (9.11) can be rewritten as

$$\ln\left(1-\frac{1}{M}\right)=\frac{-P_2}{V_D-V_{dsat}}+\ln(P_1) \tag{9.12}$$

M can be measured experimentally prior to the snapback from the following relationship:

$$M=\frac{I_{DT}}{I_{DT}-I_{sub}} \tag{9.13}$$

By plotting $\ln(1 - 1/M)$ versus $1/(V_D - V_{dsat})$ and recognizing that Eq. (9.12) is in the form of $y = mx + b$, one can determine the parameter P_1 from the y-axis intercept and P_2 from the slope of the curve, as shown in Fig. 9.10.

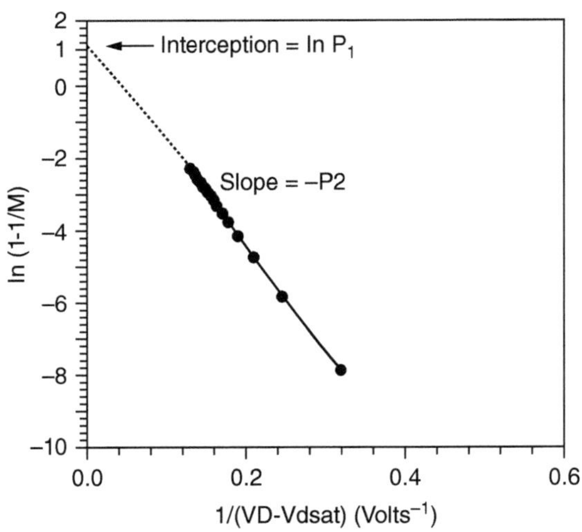

FIGURE 9.10 Extraction of parameters P_1 and P_2 for modeling the multiplication factor.

Substrate Resistance

Following the occurrence of avalanche multiplication, the source-substrate junction (i.e., emitter-base junction of the parasitic bipolar transistor) is turned on when the voltage drop of $I_{sub}R_{sub}$ reaches about 0.8 V. Thus the substrate resistance is a very important parameter for determining the on/off state of the parasitic BJT. In this work, we will neglect the effects of conductivity modulation on R_{sub}, and we will develop an R_{sub} model based on the geometry and resistivity of the substrate.[22,29] A two-part section is applied to modeling the substrate resistance. Figure 9.11 shows the schematic of the two sections in the substrate region. Assuming that the hole current flows uniformly, the substrate resistance R_{top} in the top section with the rectangular shape can be expressed as

$$R_{top} = \rho \cdot [Y'/(WL')]$$

and

$$Y' = X_j + \frac{X_d}{2} \tag{9.14}$$

where ρ is the substrate resistivity, X_j is the depth of source/drain junction, X_d is the width of the depletion region at the drain junction, and L' is the effective length of the rectangle.

Consider the case in which the hole current spreads at an angle of 45° in the bottom section of the substrate; the electric field E in this region can be written as

$$E = \frac{\rho \cdot I_{sub}}{A} \tag{9.15}$$

with $A = (L+2y)(W+2y)$

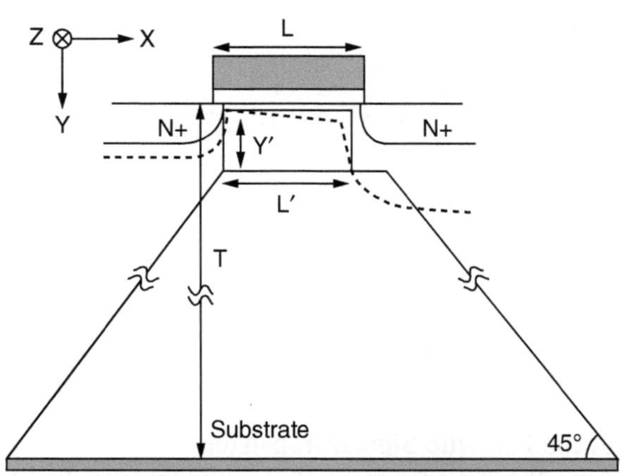

FIGURE 9.11 Illustration of the substrate resistance consisting of the top and bottom regions.

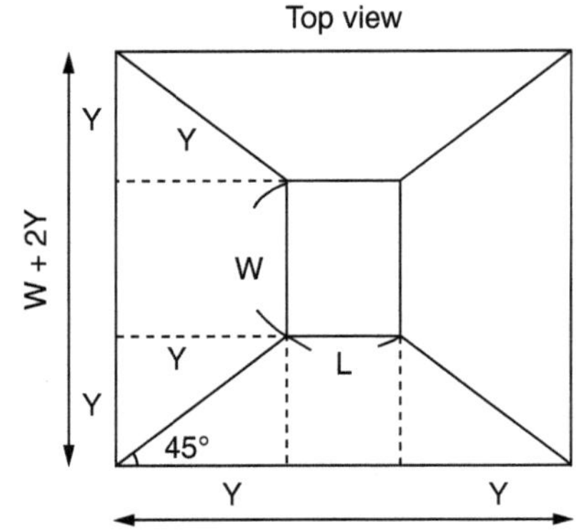

FIGURE 9.12 Top view of the bottom substrate region with a spreading angle of 45°.

where A is the spreading area of the bottom section of the substrate (Fig. 9.12), L is the channel length, and W is the channel width. Integrating Eq. (9.15) with respect to y from the end of the top-section region Y' to the substrate contact T (see Fig. 9.11), we obtain

$$R_{bott} = \int_{Y'}^{T} Edy/I_{sub} \tag{9.16}$$

Substituting Eq. (9.15) into Eq. (9.16), we have

$$R_{bott} = \frac{1}{2\rho(W-L)}\left[\ln\left(\frac{L+2T}{L+2Y'}\right) - \ln\left(\frac{W+2T}{W+2Y'}\right)\right] \tag{9.17}$$

When $L = W$, Eq. (9.17) can be simplified to

$$R_{bott} = \frac{T-Y'}{\rho[(W+2T)(W+2Y')]} \tag{9.18}$$

Finally, the total substrate resistance is

$$R_{sub} = R_{top} + R_{bott} \tag{9.19}$$

For the MOS device considered with $\rho = 9$ Ω-cm, $L = 1.2$ μm, and $W = 160$ μm, the substrate resistance R_{sub} is approximately 2000 Ω.

Currents of the Parasitic Bipolar Transistor

In this section, the currents associated with the parasitic BJT are analyzed. The parasitic BJT is analogous to a lateral BJT, and some characteristics and features of the parasitic BJT are different from the vertical bipolar counterpart.[30,31] In order to keep the model compact, we do not follow the approach of Gummel-Poon-like models,[32,33] but rather simple and empirical equations are used in this work. The first-order collector current I_C can be expressed by

$$I_C = I_{OC}\left[\exp\left(\frac{V_{BE}}{V_T}\right) - \exp\left(\frac{V_{BC}}{V_T}\right)\right] \tag{9.20}$$

where I_{OC} is the reverse saturation current for electrons diffusing at the collector junction, V_T is the thermal voltage, V_{BE} is the base-emitter voltage drop, and V_{BC} is the base-collector voltage drop. The base current I_B is written by

$$I_B = I_{OB}\left[\exp\left(\frac{V_{BE}}{V_T}\right) - 1\right] \tag{9.21}$$

where I_{OB} is the reverse saturation current for holes diffusing at the emitter junction. Dividing Eq. (9.20) by Eq. (9.21) with the assumption that $V_{BC} < 0$, that is, that the base-collector junction is reverse-biased, the common emitter current gain β is obtained as

$$\beta \approx \frac{I_{OC}}{I_{OB}} \tag{9.22}$$

The operation of the parasitic BJT is related to the holding voltage V_{sp}, which occurs when[10,27]

$$\beta(M-1) \geq 1 \tag{9.23}$$

The reverse saturation currents I_{OC} and I_{OB} are directly proportional to the collector and emitter areas (sidewall for the parasitic BJT), respectively. Thus, β changes with the geometry (size) of the MOS device. Based on Eq. (9.20), we have

$$I_{OC} = \frac{I_C}{\exp(V_{BE}/V_T)} \quad \text{for} \quad V_{BC} < 0 \tag{9.24}$$

From Eq. (9.21), we have

$$I_{OB} = \frac{I_B}{\exp(V_{BE}/V_T)} \quad \text{for} \quad \exp(V_{BE}/V_T) >> 1 \tag{9.25}$$

Introducing Eq. (9.5) into Eq. (9.24), with $V_{BE} = I_{sub}R_{sub}$, yields

$$I_{OC} = \frac{\left(\dfrac{I_{DT}}{M - I_D}\right)}{\exp\left(\dfrac{I_{sub}R_{sub}}{V_T}\right)} \tag{9.26}$$

In addition,

$$I_{DT} = M\left(I_{DT} - I_{h,gen}\right) \tag{9.27}$$

and substituting Eq. (9.3) into Eq. (9.27), the base current I_B is given as

$$I_B = \frac{(M-1)I_{DT}}{M} - I_{sub} \tag{9.28}$$

Therefore, I_{OB} can be rewritten by putting Eq. (9.28) into Eq. (9.25), that is,

$$I_{OB} = \frac{\dfrac{(M-1)\,I_{DT}}{M} - I_{sub}}{\exp\left(\dfrac{I_{sub}R_{sub}}{V_T}\right)} \tag{9.29}$$

Equations (9.26) and (9.29) can be used to extract I_{OC} and I_{OB}, respectively. Let us consider an example. At $V_G = 0$, the measured *I-V* characteristics show a snapback holding voltage $V_{sp} = 8$ V and holding current $I_{DT} = 7$ mA. $P_1 = 5$ and $P_2 = 26$ are extracted using the method described previously, and $M = 1.24$ from Eq. (9.11). This, together with onset voltage for the parasitic BJT of $V_{BE} = I_{sub}R_{sub} = 0.8$ V, yields $I_{OC} = 2.17 \times 10^{-16}$ A and $I_{OB} = 3.68 \times 10^{-17}$ A. Using these values, $\beta = 6$ is determined from Eq. (9.22), which is consistent with the condition given in Eq. (9.23).

9.2.3 Model Implementation in SPICE

The new MOS model can be implemented into Cadence SPICE using macro modeling for the equivalent circuit depicted in Fig. 9.9. Conventionally, macro modeling means creating a model of a circuit "block" using predefined circuit primitives (elements) available in a circuit simulator such as linear/nonlinear controlled sources, resistors, capacitors, etc.[34,35] However, the hole-generation current source (i.e., dependent current source) and the parasitic bipolar transistor cannot be predefined. There are two methods offered in some SPICE-like simulators to solve this problem. One is the AHDL modeling tool, and the other is the C-code modeling tool. The C-code modeling tool is used to change the SPICE intrinsic models written via C-code subroutines or even to develop a completely new C-code subroutine that describes the device model equations. In general, the AHDL modeling tool is more user friendly than its C-code counterpart. Models can be written using the AHDL to define the current-voltage relationships that exist between the electrical terminals of a block, and this language is very similar to the C programming language in syntax.[36] The AHDL is available in Cadence SPICE and

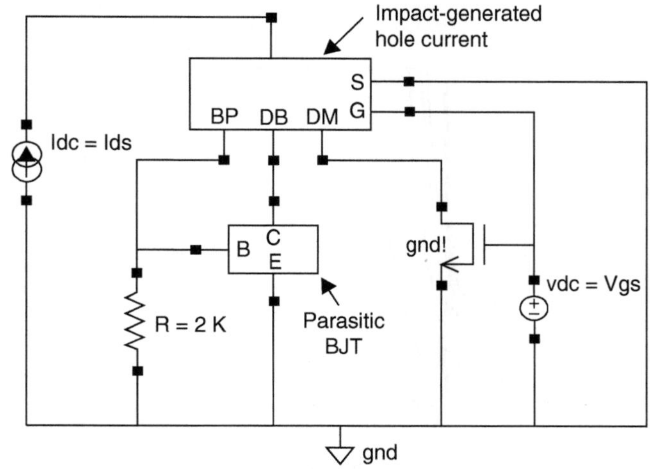

FIGURE 9.13 Schematic of the improved MOS model implemented in Cadence SPICE.

will be used here to create the impact ionization-generated hole current, parasitic BJT, and substrate-resistance blocks, as illustrated in Fig. 9.13.

It is important to note that since the saturation drain voltage V_{dsat} is not a standard SPICE MOS model parameter but an operating-point parameter, the MOS model equations of V_{dsat} must be written into the created block of the hole-generation current source including the multiplication factor $M = f(V_{dsat})$ in Eq. (9.11). Also noted is that the syntax *max* can be used to write max(0.001, $V_D - V_{dsat}$) to avoid the pole in the denominator. The 0.001 is not fixed but depends on the simulation incremental-step set point. Figure 9.14 shows the flowchart that describes the implementation of the improved MOS models into the Cadence SPICE tool.

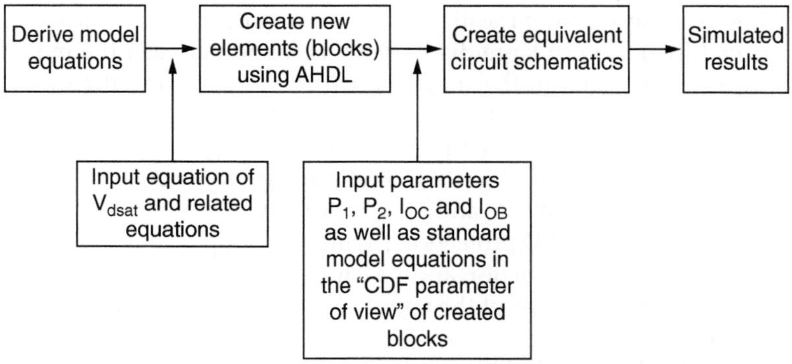

FIGURE 9.14 Flowchart of implementing the improved model into SPICE.

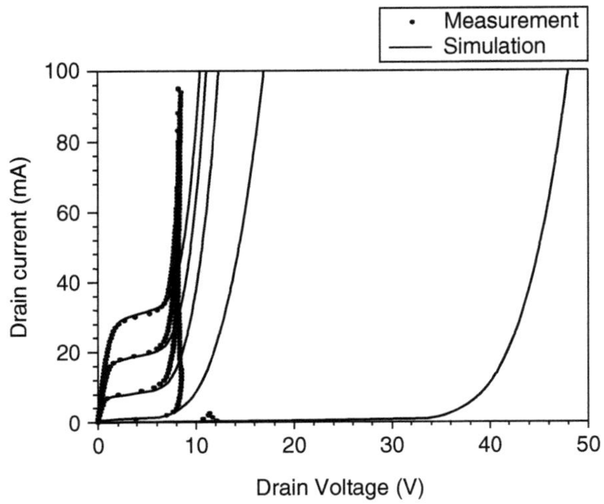

FIGURE 9.15 Comparison of *I-V* characteristics obtained from measurements and SPICE simulation based on the conventional MOS model.

9.2.4 Results and Discussion

The Hewlett-Packard (HP) 4156A parameter analyzer was used to measure the *I-V* family of curves covering the snapback phenomena. A current source is applied to the drain terminal, and the gate is set at various specified voltages. A voltage source cannot be used in this case because it is difficult to observe the sudden decrease in voltage (i.e., snapback) using the voltage sweep. Instead, the current must be swept incrementally to measure the snapback behavior of the drain voltage. NMOS transistors with a channel length 1.2 μm and a channel width of 160 μm are used. Figure 9.15 shows the MOS *I-V* characteristics obtained from SPICE simulation using the conventional MOS model (curves;) i.e., existing MOS model in Cadence SPICE, which does not include the impact-generated current source, substrate resistance, and parasitic bipolar transistor) and from measurements (dots). A very large discrepancy is found between the model and measured data in the snapback region. When using the model developed in this work, the accuracy of SPICE simulation is much improved, as evidenced by the results given in Fig. 9.16. The parameters extracted and used in this simulation are $R_{sub} = 2000\ \Omega$, $P_1 = 5$, $P_2 = 26$, $I_{OC} = 1.23 \times 10^{-16}$ A, and $I_{OB} = 5 \times 10^{-17}$ A.

Next, a simple ESD protection circuit (i.e., supply clamp) consisting of a MOSFET, capacitor, and resistor is considered.[37] Figure 9.17 illustrates the schematic of the supply clamp and the (HBM)[38] implemented in the SPICE simulator. To account for the effect of overshoot frequently observed in HBM tester measurements, parasitic capacitance and inductance have been added in the HBM equivalent

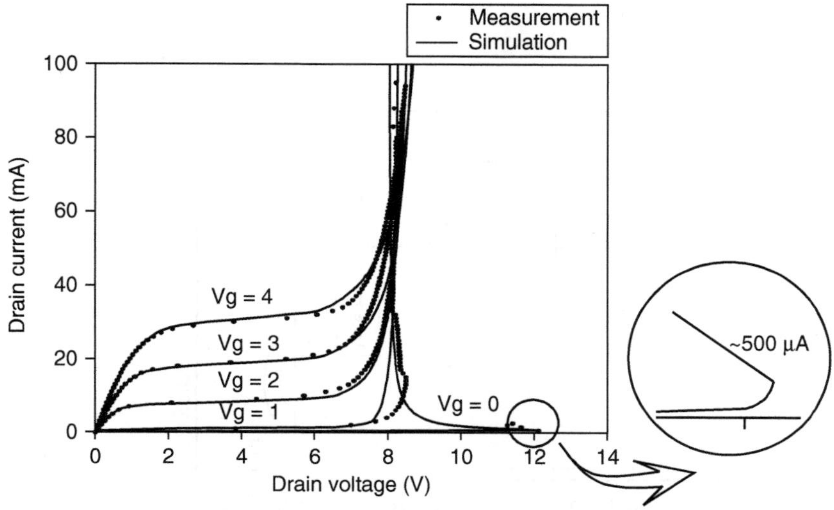

Figure 9.16 Comparison of *I-V* characteristics obtained from measurements and SPICE simulation based on the improved MOS model developed in this work.

circuit.[39,40] Figures 9.18 and 9.19 compare the simulated and measured transient responses of the drain current and voltage, respectively, subjected to an ESD charging voltage of 150 V. Note that such a drain current consists of the drain current of the regular MOSFET, the collector current of the parasitic BJT, and the impact-generated current (see Fig. 9.9). Clearly, Fig. 9.18 shows that the peak drain current obtained from the improved model (105 mA) is in much better agreement with measurements (105 mA) than that obtained from the conventional model (90 mA). Similarly, Fig. 9.19 indicates that the measured clamped drain voltage (8.3 V) is also in better agreement with the improved model (8.5 V) than the conventional model (9.5 V).

9.2.5 Advanced Metal-Oxide-Semiconductor Models

The compact MOS model developed in the preceding section consists of the substrate resistance R_{sub} and the substrate current I_{sub}. These two components determine the turn-on mechanism of the parasitic bipolar transistor (i.e., snapback region) and are critical for accurate ESD modeling. However, the models for R_{sub} and I_{sub} derived previously were oversimplified because some important physical effects were overlooked. As shown in Sec. 9.2.4, these models work well for describing the behavior of an ESD protection circuit subjected to relatively low ESD charging voltages (i.e., a charging voltage of 150 V was used in Fig. 9.19), but they become questionable when a large ESD charging voltage is applied. Improved and advanced R_{sub} and I_{sub} models will be developed in this section.

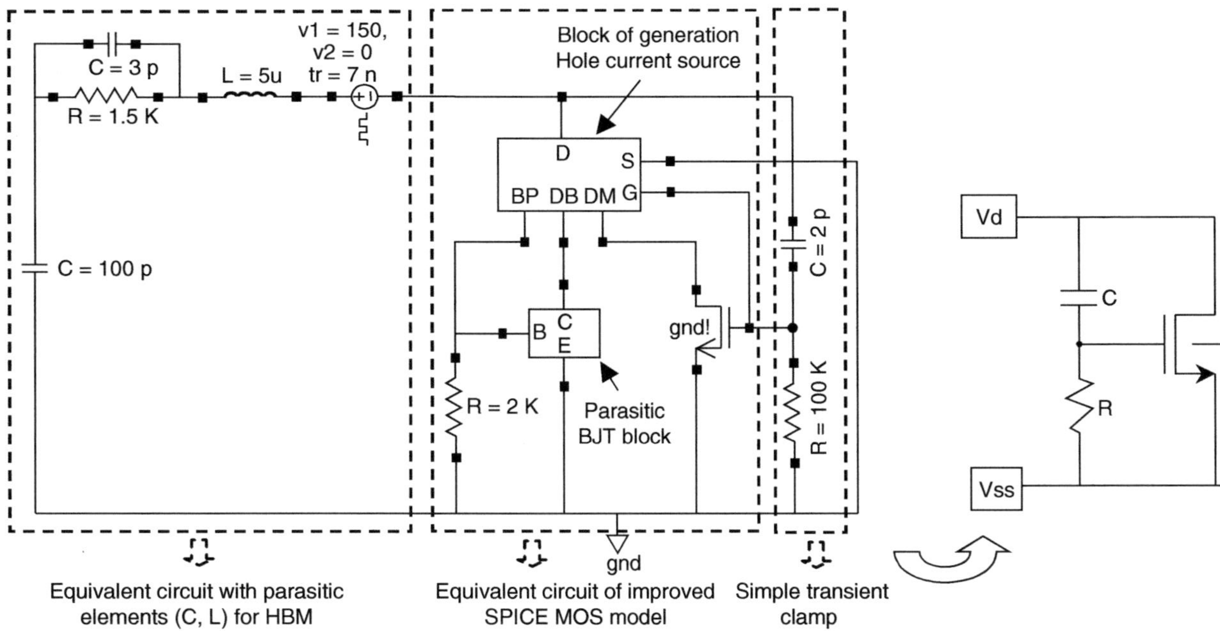

Figure 9.17 Schematic of an ESD protection circuit (*right*) and an HBM source (*left*) implemented in SPICE.

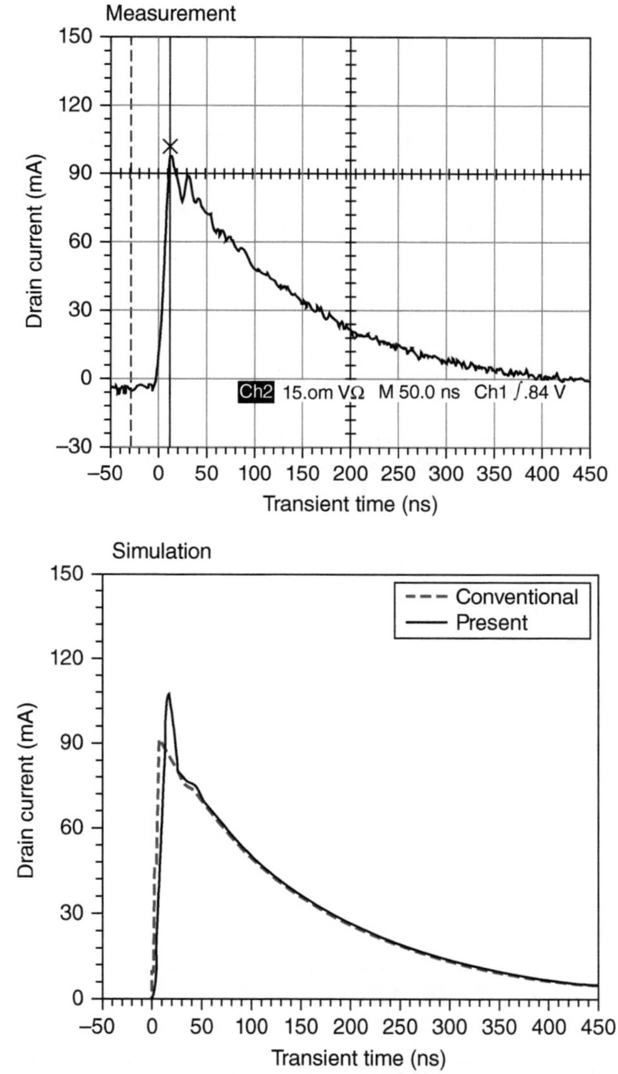

Figure 9.18 Comparison of drain-current transient responses obtained from measurements (*top*) and simulations (*bottom*) based on the conventional and present MOS models subject to an HBM charging voltage of 150 V.

Advanced Substrate-Resistance Model

The snapback characteristics of MOSFETs depend strongly on the substrate resistance R_{sub}.[41] Several studies on modeling such a resistance have been reported in the literature. Hsu et al.[22] developed an R_{sub} model by assuming that the hole current flows uniformly in an effective substrate area with a spreading angle of 45° and that the

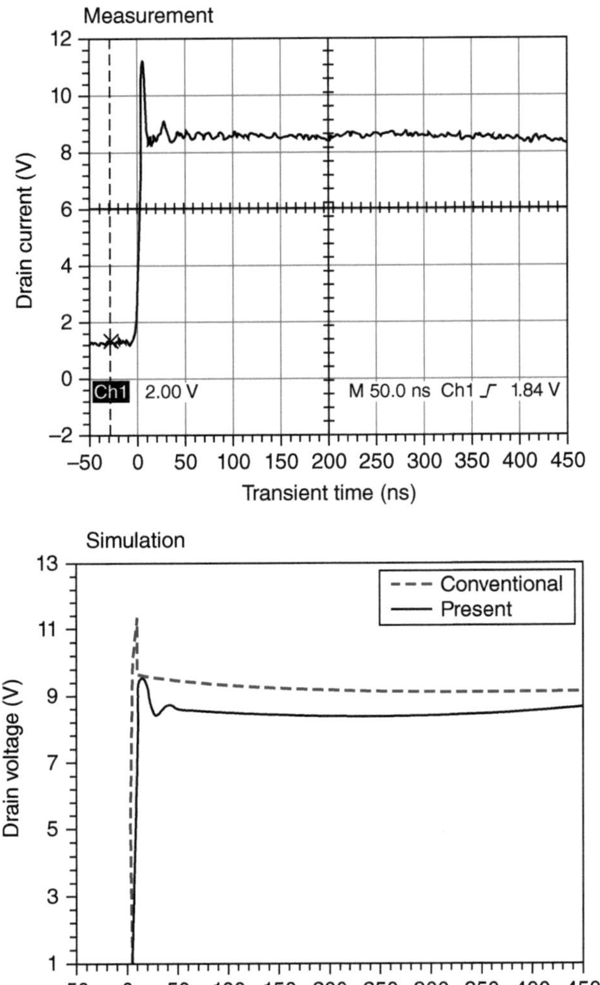

FIGURE 9.19 Comparison of drain-voltage transient responses obtained from measurements (*top*) and simulations (*bottom*) based on the conventional and present MOS models subject to an HBM charging voltage of 150 V.

effect of conductivity modulation associated with the injection of holes is negligible. Another R_{sub} model was derived by empirically fitting to the measured R_{sub} versus effective channel length L_{eff} data for NMOS devices having channel lengths ranging from 1.15 to 9.15 μm.[42] It has the form of

$$R_{sub} = 129(L_{eff})^{-0.33} \tag{9.30}$$

A simple R_{sub} model also was reported in Ref. 19:

$$R_{sub} = \frac{V_{BE}}{I_{sub}} \tag{9.31}$$

where $V_{BE} = 0.8$ V is the emitter-base turn-on voltage of the parasitic bipolar transistor in the MOSFET, and I_{sub} is the substrate current. The model is intended for describing MOSFET behavior after the parasitic bipolar transistor is activated (postsnapback region).

In this subsection, an R_{sub} model that is more comprehensive and robust than the existing ones in the literature will be developed. It is derived based on the transport of free carriers and other relevant physics in the MOSFET rather than the substrate current[19] or empirical fitting.[42] Results obtained from device simulation will be included in support of the model. Comparison with the existing models also will be made to illustrate the improvement of the model developed.

Hole-Injection Mechanism in Substrate

To gain physical insight into the distribution of hole density in the substrate, we first carry out simulation using a 2D device simulator called *Atlas*.[43] Basically, Atlas is a software package that numerically solves the five basic semiconductor equations: the Poisson equation, electron- and hole-continuity equations, and electron- and hole-current equations. The approach presented below applies generally to all MOS devices, and the device considered is a bulk NMOS with a channel length L of 1.2 μm and width W of 160 μm. In addition, the Atlas device file used in this study has been calibrated against experimental data.

Figure 9.20 shows the simulated hole-density contours in a MOSFET biased at $V_{gs} = 1$ V and $V_{ds} = 6.0$ V. In the figure, the origins of the

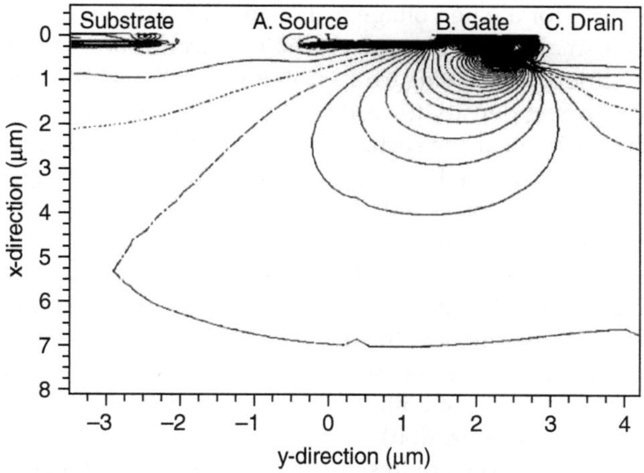

FIGURE 9.20 Simulated hole-density contours at $V_{gs} = 1$ V and $V_{ds} = 6.0$ V.

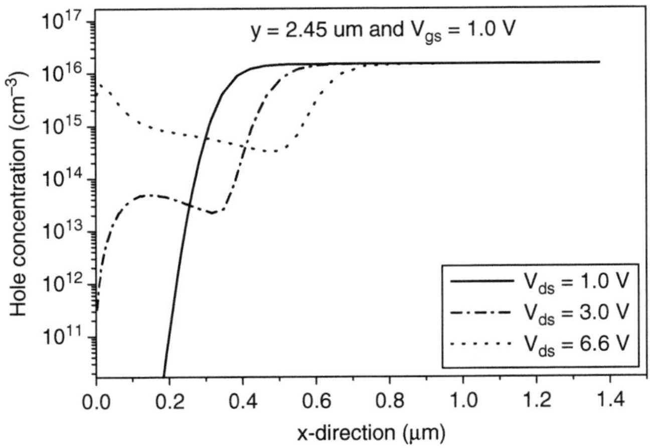

FIGURE 9.21 Simulated hole-density distribution underneath the gate at y = 2.45 μm.

x and y axes are taken at the Si-SiO_2 interface and the outmost edge of the source region, respectively. The results indicate that the majority of hole density is confined to the region underneath the gate and that the maximum hole density occurs near the drain junction. Outside the gate region, the hole density decreases rapidly toward to edges of the MOSFET. This suggests that the resistance associated with the substrate region outside the gate is of secondary importance and can be accounted for by using a fitting parameter k $(0 < k < 1)$ to be developed later.

We now turn our attention to the substrate region underneath the gate. Figure 9.21 shows the simulated hole concentration versus the x direction taken at a y position underneath the gate near the drain ($y = 2.45$ μm) for a gate voltage of 1 V and three different drain voltages. At a small drain voltage, the hole concentration is very small near the surface (i.e., owing to electron inversion) and approaches the substrate doping density toward the bulk. When the drain voltage is large, the region near the surface is filled with holes owing to significant impact ionization. Below the surface is the space-charge region, followed by the region of bulk substrate.

Based on the results given in Figs. 9.20 and 9.21, we can now focus on the region underneath the gate and divide the region into three subregions, as shown in Fig. 9.22. The horizontal-direction (x-direction) boundaries of the device are taken at the outmost edges of the drain and source regions. Subregion 1 covers the region near the surface consisting of both electron inversion and hole generation (owing to impact ionization), subregion 2 denotes the space-charge region, and subregion 3 represents the bulk substrate, where the hole density is approximately uniform. In general, the hole density and

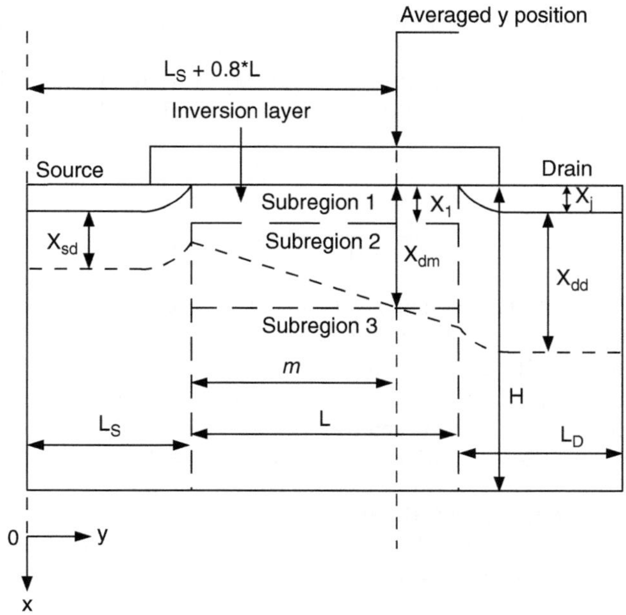

Figure 9.22 Schematic of an MOS structure showing the three subregions.

layer thickness in the three subregions are functions of x and y directions. To simplify the analysis and make the model sufficiently compact, however, we will consider only the x-direction dependence and will carry out the analysis at an averaged y position of $y = L_S + 0.8L$, where L_S and L are the lengths of the source/drain region and channel, respectively (see Fig. 9.22). Such a y position was selected as the position having the averaged hole density at the Si-SiO_2 interface based on Atlas simulation. The errors induced from such a simplified approach should be minimal, as will be verified later by device simulation and measurements, because the dimension of the y direction is much smaller than that of the x direction.

The vertical-direction (x-direction) boundaries of the three subregions, X_1, X_{dm}, and H, are the thicknesses of the inversion layer, space-charge layer, and substrate, respectively. These boundaries will be defined later.

Impact Ionization

In the BSIM3v3 model, the saturation drain current I_{ds} from the subthreshold to the strong-inversion regime can be formulated in a unified expression given by[44]

$$I_{ds} = \frac{I_{dso}(V_{dsat})}{1 + R_{ds} \cdot \frac{I_{dso}(V_{dsat})}{V_{dsat}}} \cdot F_A \tag{9.32}$$

where V_{dsat} is the saturation drain voltage, R_{ds} is the parasitic source/drain resistance, F_A is a compound expression accounting for second-order effects, and I_{ds0} is the saturation current without the second-order effects.[44] The impact generation current I_{gen} is related I_{ds} as[19]

$$I_{gen} = I_{ds} \cdot (M-1) \tag{9.33}$$

$$M = \frac{1}{1 - P_1 \cdot \exp(-P_2/(V_{ds} - V_{dsat}))} \tag{9.34}$$

Here V_{ds} is the drain voltage, M is the avalanche multiplication factor, and parameters P_1 and P_2 in Eq. (9.33) can be extracted from the characteristics of $\ln(1-1/M)$ versus $1/(V_{ds} - V_{dsat})$ at a constant gate voltage. The data in such a plot exhibit a linear relationship, the slope of the linear line is $-P_2$, and the intercept of the linear line on the y axis is $\ln(P_1)$. For the device under study, we found $P_1 = 2.0312$ and $P_2 = 16.37$.

Hole Density and Substrate Resistance

We now proceed to develop models for the hole-density distributions and substrate resistances in the three subregions.

Subregion 1 The hole density p in this region results from impact ionization and is related to the electron density associated with the drain current and the avalanche multiplication factor. Thus,

$$p = (M-1) \cdot n = (M-1) \cdot I_{ds}/(A_1 \cdot q \cdot v_d) \tag{9.35}$$

where n is the electron density associated with the drain current, $A_1 = W \cdot X_1$, and v_d is electron saturation velocity in the pinch-off region. X_1 is the thickness of subregion 1 and is the same as the thickness of the inversion layer. The x-direction electrostatic potential ϕ can be expressed as[45]

$$\phi(x) = 2 \cdot \phi_B \cdot (1 - x/X_{dm})^2 \tag{9.36}$$

where X_{dm} is the space-charge layer thickness and ϕ_B is the bulk potential. At $x = X_1$, $\phi(x)$ will cross over the bulk potential [i.e., $\varphi(X_1) = \varphi_B$], and X_1 can be determined as a function of X_{dm}. For the device under study, $X_1 = 0.293 \cdot X_{dm}$, and X_{dm} is given by

$$X_{dm} = (X_{sd} + X_j) + (X_{dd} - X_{sd}) \cdot (m/L) \tag{9.37}$$

Here m is the lateral dimension shown in Fig. 9.22, and X_{sd} and X_{dd} are the space-charge layer thicknesses at the source and drain junctions, respectively:

$$X_{dd} = \left[2 \cdot \frac{\varepsilon_{si}}{q} \cdot \left(\frac{1}{p_0} + \frac{1}{N_D}\right)(V_{ds} + \zeta)\right]^{0.5} \tag{9.38}$$

$$X_{sd} = X_{dd}\big|_{V_{ds}=0} \tag{9.39}$$

where ζ is the built-in voltage for the drain/source and substrate junctions.

Since both electrons (owing to inversion) and holes (owing to impact ionization) are present in this region, the substrate resistance R_1 in subregion 1, including conductivity modulation, is

$$R_1 = \frac{1}{q \cdot n \cdot \mu_n + q \cdot p \cdot \mu_p} \cdot \frac{X_1}{W \cdot (L - l_d)} \tag{9.40}$$

where μ_n and μ_p are the electron and hole mobilities, and l_d is the length of pinch-off region.[46]

Subregion 2 The hole density in this subregion can be derived from the drift and diffusion current equation and continuity equation. The difficulty in solving these equations is the fact that the electric field $E(x)$ in subregion 2, which is influenced by the injected free carriers and the bias conditions, is not known. To avoid this difficulty, we propose the following expression for $E(x)$:

$$E(x) = f_1(V_{ds}) \cdot f_2(V_{gs}) \cdot E'(x) \tag{9.41}$$

where $E'(x)$ is the electric field based on the depletion approximation, and f_1 and f_2 are empirical parameters accounting for the drain and gate bias dependencies, respectively:

$$E'(x) = \frac{q \cdot p_0}{\varepsilon_{si}}(X_{dm} - x) \tag{9.42}$$

$$f_1(V_{ds}) = 0.7183 \cdot \exp(-0.137 V_{ds}) \tag{9.43}$$

$$f_2(V_{gs}) = 0.8698 + 0.14783 V_{gs} \tag{9.44}$$

Equations (9.43) and (9.44) were obtained by fitting the Atlas simulation results using Microcal Origin 5.0.[47] Putting Eqs. (9.42) through (9.44) into the drift-diffusion and continuity equations, we can solve for $p(x)$. Figure 9.23 compares the hole distributions in subregion 2 obtained from the model and from Atlas simulations.

The resistance R_2 of this region can be modeled as

$$R_2 = k \cdot \int_{X_1}^{X_{dm}} \frac{1}{q \cdot p(x) \cdot \mu_p} \cdot \frac{dx}{A} \tag{9.45}$$

where $A = W \cdot L$, and k is an empirical parameter accounting for the spread of hole density in the substrate regions outside the gate, which is modeled by considering only the effect of the lateral dimension of the substrate:

$$k = \frac{L}{L + 0.5 \cdot L_D + 0.5 \cdot L_S} \tag{9.46}$$

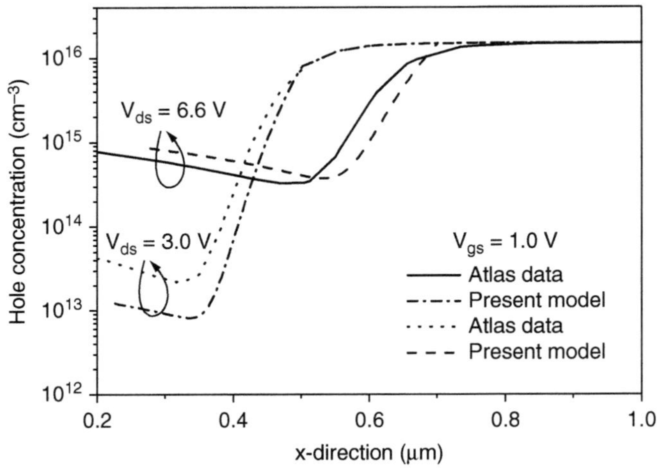

Figure 9.23 Comparison of hole distributions obtained from Atlas simulation.

L_S and L_D are the lengths of the source and drain junctions, respectively (see Fig. 9.16).

Subregion 3 This subregion is the bulk region, and the resistance R_3 is

$$R_3 = k \cdot \int_{X_{dm}}^{H} \frac{1}{q \cdot p \cdot \mu_p} \cdot \frac{dx}{A} \tag{9.47}$$

and the hole density is approximately equal to the substrate doping density ($p \approx N_A$).

Finally, the total substrate resistance R_{sub} can be given by

$$R_{sub} = R_1 + R_2 + R_3 \tag{9.48}$$

Figure 9.24 compares the present R_{sub} model with three other existing models calculated for different drain and gate voltages. Note that the model in Eq. (9.30) and the model by Hsu et al.,[22] which suggested a substrate region with a spreading angle of 45°, are independent of the drain and gate voltages. This is due to the fact that such models neglect conductivity modulation caused by the injection of impact-generated holes in the substrate. The model in Eq. (9.31) is bias-dependent because it is a function of substrate current I_{sub}, but it yields a much larger R_{sub} than the present model for a wide range of V_{ds}. Also included in the figure are data (Atlas simulation) calculated from $R_{sub} = V'/I_{sub}$, where V' is the potential of the initial point of impact-generated hole flow (i.e., at the averaged y position and the

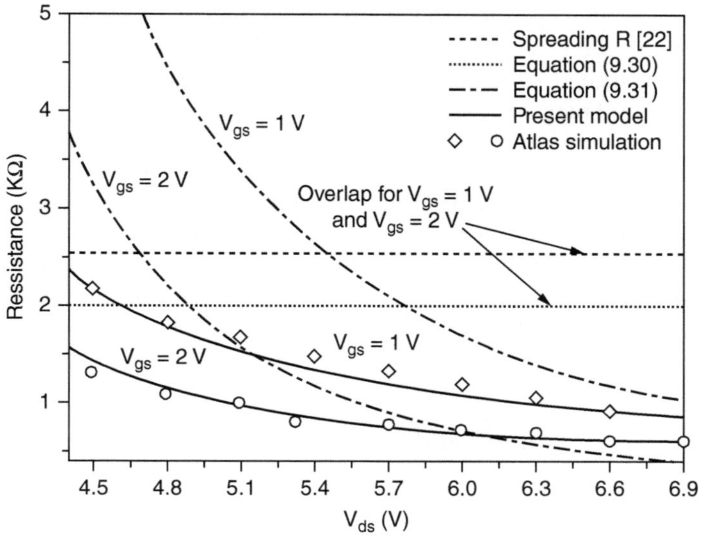

FIGURE 9.24 Comparison of the present and existing substrate-resistance models calculated at different drain and gate voltages. Also included are data calculated from $R_{sub} = V'/I_{sub}$ (denoted "Atlas simulation").

Si-SiO_2 interface). Both V' and I_{sub} are obtained from Atlas simulation. Note that V' is not equal to V_{ds} because the holes do not originate from the drain terminal. Clearly, the present model compares more favorably with Atlas simulation than the existing models. It should be pointed out that the model in Eq. (9.31) is intended only for the modeling of turn-on behavior of the parasitic bipolar transistor (snapback) under ESD stress. As a result, such a model is not applicable for standard MOS operation and thus the large discrepancies observed for relatively small V_{ds}. The present model, on the other hand, may not be valid for the postsnapback regime because it does not account for the additional conduction mechanism in the bulk of the MOSFET owing to activation of the parasitic bipolar transistor.

Advanced Substrate-Current Model

Substrate leakage current caused by the impact ionization of high-energy carriers near the drain is becoming increasingly prominent with the decrease in MOSFET dimensions.[48] Such a current plays an important role in determining the ESD robustness and hot-carrier lifetime of the device. The velocity–saturation-region length and electric field near the drain junction are two of the most important parameters in modeling substrate current. Several analytical substrate current models based on these two parameters have been reported in the literature. For example, Wong and Poon[49] and Kolhatkar and Dutta[50] have studied issues related to the substrate current.

Arora and Sharma[51] also developed an empirical substrate-current model based on the results of MINIMOS device simulation. As will be demonstrated later, however, these existing models have limited accuracy and may be questionable for deep-submicron MOS devices.

In this subsection, an improved and robust substrate-current model will be developed. Comparisons between the present and several existing substrate-current models will be made to illustrate the improvement of this work. Both device simulation results and measurement data will be included in support of model development.

As mentioned earlier, the impact-generated hole current $I_{h,gen}$ is closely related to the substrate current I_{sub}. For an n-channel MOSFET, I_{sub} can be given in the following general expression:

$$I_{sub} = I_D \cdot \int_{L-l_d}^{L} \alpha_n dx = I_D \cdot A_i \cdot \int_{L-l_d}^{L} e^{-B_i/E(x)} dx \tag{9.49}$$

Here, I_D is the drain current, L is the effective channel length, l_d is the length of the high-field region (i.e., velocity-saturation region) near the drain junction, α_n is the impact ionization rate, A_i and B_i are the ionization constants, and $E(x)$ is the electric field in the high-field region. As can be seen in Fig. 9.25, prior to the position of the $L - l_d$, the field is below the critical field E_s ($\simeq 4 \times 10^4$ V/cm) and is a linear function with respect to the position in the channel. Beyond this point, however, the field increases rapidly and reaches a peak value of E_m at position x_m. The impact ionization rate α_n can be expressed as

$$\alpha_n = A_i \cdot \exp[-B_i/E(x)] \tag{9.50}$$

Clearly, in modeling I_{sub}, l_d and $E(x)$ are two critical parameters.

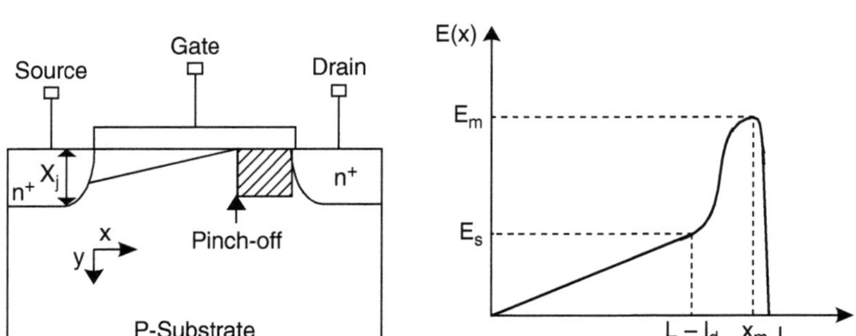

FIGURE 9.25 Schematic of MOS device and the electric field distribution in the channel.

We first focus on l_d. Since the electric field is distributed linearly from zero to E_s between $x = 0$ and $x = L - l_d$, we have

$$\left.\frac{dE(x)}{dx}\right|_{x=L-l_d} = \frac{E_s}{L - l_d} \tag{9.51}$$

Using this boundary condition and the approach given in Ref. 52; the electrostatic potential $V(x)$ in the high-field region (i.e., between $x = L - l_d$ and $x = L$) can be obtained:

$$V(x) = \frac{\lambda^2 \cdot E_s}{L - l_d}\left(\cosh\frac{y - L + l_d}{\lambda} - 1\right) + \lambda \cdot E_s \cdot \sinh\frac{y - L + l_d}{\lambda} + V_{dsat} \tag{9.52}$$

where V_{dsat} is the drain saturation voltage and

$$\lambda = \sqrt{(\varepsilon_{si}/\varepsilon_{ox}) \cdot t_{ox} \cdot x_j} \tag{9.53}$$

t_{ox} is the oxide-layer thickness, x_j is the drain-source junction depth, and ε is the dielectric permittivity. Let $(u = V_{ds} - V_{dsat})/(\lambda \cdot E_s)$, $a = \lambda/(L - l_d)$ using the boundary condition of $E(x = L) = 0$, and solving for u from Eq. (9.52) yields

$$u = a \cdot \left(\cosh\frac{l_d}{\lambda} - 1\right) + \sinh\frac{l_d}{\lambda} \tag{9.54}$$

From Eq. (9.54), l_d can be derived as

$$l_d = \lambda \cdot \ln\left(\frac{a + u + \sqrt{u^2 + 2 \cdot a \cdot u + 1}}{a + 1}\right) \tag{9.55}$$

Equation (9.55) is a transcendental equation and can be solved numerically. However, to solve Eq. (9.55) analytically so that it can be easily implemented into a software tool such as SPICE, knowledge of parameter a is required before l_d can be calculated.

To this end, Wong and Poon suggested $l_d \approx 2 \cdot \lambda$ and thus $a = \lambda/(L - l_d) \approx \lambda/(L - 2 \cdot \lambda)$. This works very well for long-channel devices but can give rise to a large error if L is reduced below 0.5 μm. Kolhatkar and Dutta used $l_d \approx k \cdot \lambda$ and $a \approx \lambda/(L - k \cdot \lambda)$, where k is a variable. Using this method, l_d could be quite accurate, but the procedure may be very complex because the value of k varies and needs to be determined for each channel length.

We propose a more effective method to model the parameter a. Consider a reference for l_d called l_{ref} for a reference device. Thus, a

becomes $a \approx \lambda/(L - l_{ref})$, and l_{ref} can be expressed as a polynomial and a function of $(V_{ds} - V_{dsat})$:

$$l_{ref} = \lambda \cdot [P_0 + P_1(V_{ds} - V_{dsat}) + P_2(V_{ds} - V_{dsat})^2 + P_3(V_{ds} - V_{dsat})^3 + P_4(V_{ds} - V_{dsat})^4] \tag{9.56}$$

Here P_0, P_1, P_2, P_3 and P_4 are fitting parameters, and their values can be determined by fitting Eq. (9.56) to l_d of the reference MOS device obtained from device simulation. The fact that l_{ref} is bias-dependent makes the fitting easier and the parameter a more robust. For the devices under study, with channel lengths ranging from 1.0 to 0.2 μm, we have found that a reference device with a channel length of 0.75 μm is appropriate. This approach works well for a wide range of MOS devices, and once l_{ref} is established, no further fitting is required when a different device is used. Putting Eq. (9.56) into Eq. (9.55), l_d can be calculated analytically.

Another critical parameter for modeling I_{sub} is the electric field. Differentiating Eq. (9.52) leads to the electric field in the high-field region:

$$E(x) = \sqrt{\left(\frac{V(x) - V_{dsat}}{\lambda} + \frac{\lambda \cdot E_s}{L - l_d}\right)^2 + E_s^{\,2} - \frac{\lambda^2 \cdot E_s^{\,2}}{(L - l_d)^2}} \tag{9.57}$$

Putting this into Eq. (9.49), changing the variable of integration from dx to dE, and integrating from E_s to E_m, we have

$$I_{sub} = I_D \cdot A_i \cdot \lambda \cdot \int_{E_s}^{E_m} \frac{e^{-B_i/E(y)}}{\sqrt{E(y)^2 - E_s^2 + \lambda^2 \cdot E_s^2/(L - l_d)^2}}\, dE \tag{9.58}$$

where E_m is the maximum electric field at $x = x_m$. The simplest expression for E_m used conventionally has the form of[53]

$$E_m = \frac{V_{ds} - V_{dsat}}{l_d} \tag{9.59}$$

This was later modified by Arora and Sharma:[51]

$$E_m = \frac{V_{ds} - \eta_0 \cdot V_{dsat}}{l_d} \tag{9.60}$$

where η_0 is an empirical parameter obtained by fitting Eq. (9.60) to the electric field simulated by MINIMOS device simulator. To further improve the accuracy, Kolhatkar and Dutta proposed the following expression for E_m:[50]

$$E_m = \frac{V_{ds} - \eta_1 \cdot V_{dsat}}{\beta \cdot l_d} \tag{9.61}$$

Here η_1 and β are also fitting parameters. We will employ Eq. (9.61) in our model, but our fitting approach is different from that of Kolhatkar and Dutta. In their model, η_1 and β were fitted to the substrate current, whereas the values of ionization constants A_i and B_i were kept constant. In our approach, η_1 and β are first fitted to E_m simulated by the Atlas device simulator[43] for the reference device. Then B_i is kept to 1.92×10^6 V/cm,[43] and A_i is determined by fitting the substrate current of the reference device obtained from the present model with that obtained from measurements or device simulation. As will be shown later, this approach yields more accurate results because A_i is sensitive to E_m.[45] Furthermore, our approach determines the fitting parameters from the reference device, and the same parameters are used for all other devices, whereas the existing models require repeated fitting processes, and different fitting parameters are needed for different devices considered.

The drain saturation voltage V_{dsat} in Eq. (9.61) can be determined by using the BSIM3v3.2.2 model as[44]

$$V_{dsat} = \frac{E_s \cdot L \cdot (V_{gs} - V_{th} + 2 \cdot V_t)}{E_s \cdot \alpha \cdot L + V_{gs} - V_{th} + 2 \cdot V_t} \tag{9.62}$$

where V_{gs} is the gate voltage, V_{th} is the threshold voltage, V_t is the thermal voltage, and α is a fitting parameter.

Using an approach different from that in Eq. (9.49), Sing and Sudlow[54] proposed an alternative and empirical equation for I_{sub}:

$$I_{sub} = a \cdot I_D \cdot (V_{ds} - V_{dsat})^b \tag{9.63}$$

where a and b are fitting parameters.

As mentioned earlier, our modeling approach requires the use of a 0.75-μm NMOS as the reference device. To this end, the device listed in Table 9.1 is considered.

From this reference device, the values of fitting parameters P_0, P_1, P_2, P_3, and P_4 are obtained as 0.11581, 1.37034, −0.33244, 0.04217, and −0.00202, respectively. Figure 9.26 compares the high-field-region lengths obtained from the present, Wong-Poon,[49] and Kolhatkar-Dutta models.[50] Atlas device simulation results are also included in

X_j	0.3 μm
L	0.75 μm
p_{sub}	5.2×10^{16} (/cm^3)
T_{ox}	18 nm

TABLE 9.1 Parameters for the NMOS Device Considered

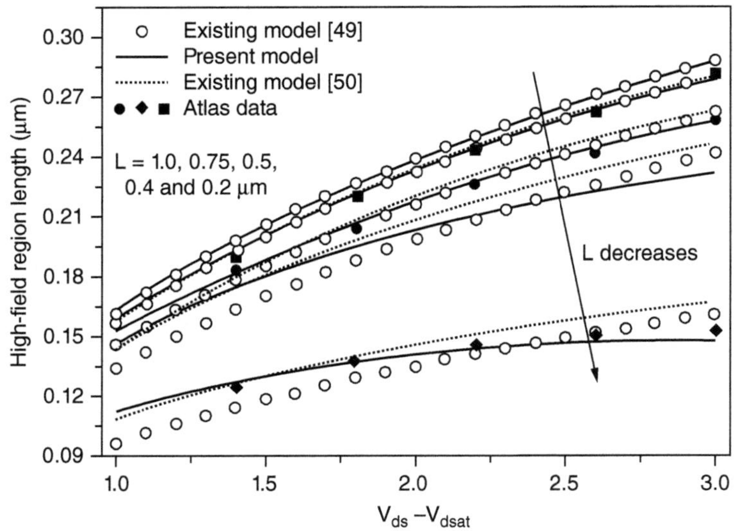

FIGURE 9.26 Characteristics of high-field-region length for MOS devices having several different channel lengths obtained from the present model, the Wong-Poon model,[49] the Kolhatkar-Dutta model,[50] and Atlas device simulation.

the figure. Clearly, the Wong-Poon model works quite well for long-channel MOS devices, but it becomes questionable for devices having a channel length below 0.5 μm. On the other hand, the accuracy of the Kolhatkar-Dutta model is reduced with increasing drain voltage. Good agreement is found between the simulation results and the present model.

Figure 9.27 shows the maximum fields obtained from the present approach, two existing models [Eqs. (9.60) and (9.61)], and Atlas device simulation. The following parameters were used in the various models: $\eta_0 = 0.717$ in Eq. (9.61) and $\eta_1 = 1.803$ and $\beta = 0.39831$ in the present model. Note that η_0 is device-dependent, and a value of 0.717 was suggested for the 0.75-μm device considered in Ref. 51. Hence only the 0.75-μm result from Eq. (9.61) is shown in Fig. 9.27. The results suggest that the existing models considerably underestimate the maximum field, particularly for high drain voltages.

Figure 9.28 compares the substrate currents obtained from the present model, existing models, and measurements for a 0.75-μm MOSFET. All models predict similar peak substrate currents, but the existing models either underestimate or overestimate the substrate current after the current has reached its maximum. The better accuracy of the present model clearly demonstrates the improvement of this work.

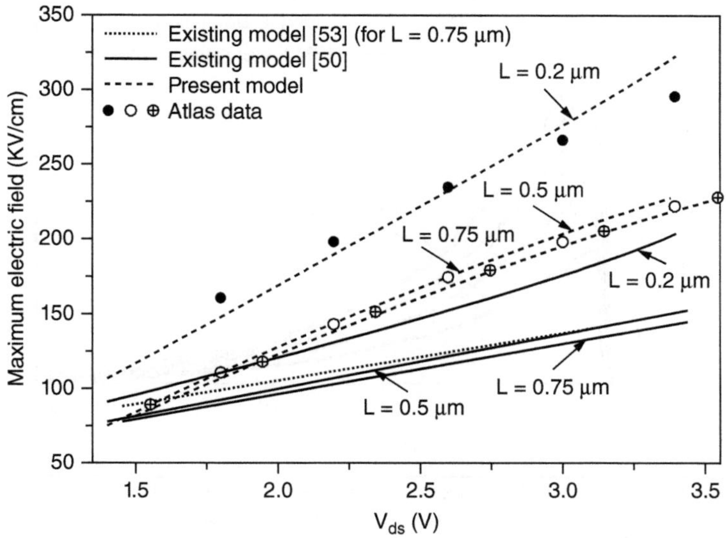

FIGURE 9.27 Maximum electric fields obtained from the present model, the Chung et al. model,[53] the Kolhatkar-Dutta model,[50] and Atlas device simulation. For the Arora and Sharma model, only the result for a 0.75-μm device is shown.

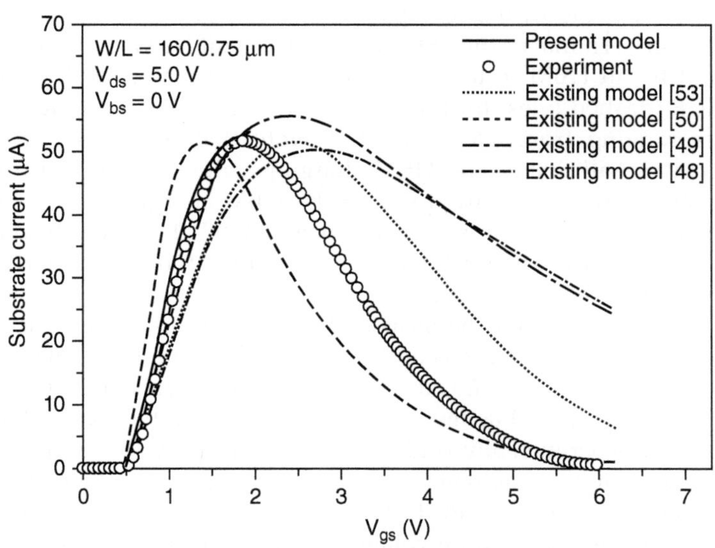

FIGURE 9.28 Substrate currents obtained from the present model, four existing models,[48–50,53] and measurements for a 0.75-μm *N*-channel MOSFET.

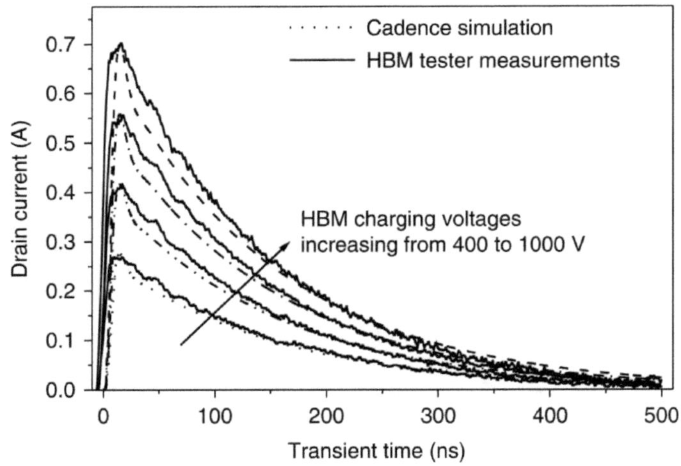

FIGURE 9.29 Simulated and measured transient responses of currents passing through the drain terminal of an ESD protection circuit subjected to four different HBM charging voltages.

Cadence Simulations Based on Improved R_{sub} and I_{sub} Models

Improved R_{sub} and I_{sub} models developed earlier can be implemented into Cadence SPICE in the same manner as that described in Sec. 9.2.3. Figure 9.29 compares the current I_{DT} passing through the drain terminal of the ESD protection circuit (see Fig. 9.11) measured by an HBM tester (solid lines) and simulated by Cadence SPICE (dashed lines) for four relatively large HBM charging voltages ranging from 400 to 1000 V. Note that the peak current increases with increasing HBM charging voltage. Such a current can be approximated by dividing the HBM charging voltage by the sum of the human skin resistance of 1500 Ω and the turn-on resistance of the MOSFET. Good agreement is found between the model and measured data.

Figure 9.30 shows the measured and simulated voltages at the drain terminal of the protection circuit for the same HBM charging voltages. Unlike the drain current, the drain voltage is insensitive to charging voltage. While the oscillatory behavior of the drain voltage observed experimentally could not be described by the model, the averaged clamp voltages measured are in good agreement with those simulated by Cadence SPICE. We suggest that oscillatory behavior seen in the measured drain voltage in Fig. 9.30 is due entirely to the parasitic effects of the probe of the oscilloscope (i.e., probe P5100 was used in Fig. 9.30). To demonstrate this, I have measured the drain voltage using an HBM charging voltage of 1000 V and three different types of voltage probes (P5100, P6106, and P6131), and the results are

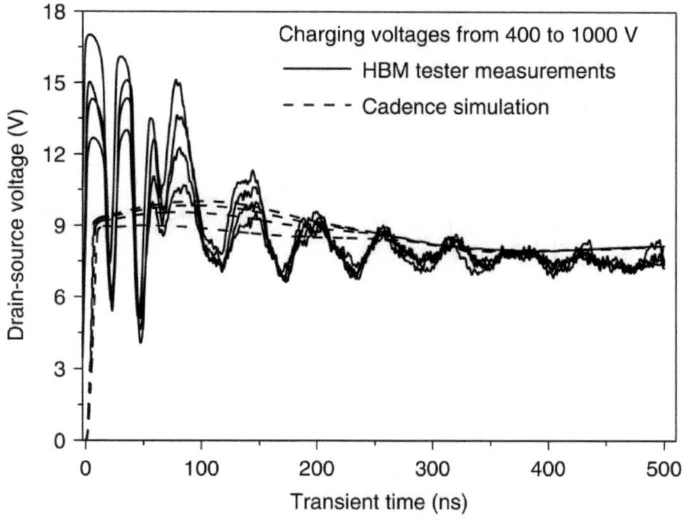

FIGURE 9.30 Simulated and measured transient responses of drain voltages of the ESD protection circuit subjected to four different HBM charging voltages.

given in Fig. 9.31. Clearly, the different probes, and thus different parasitic components, produce different oscillatory behavior, thereby indicating that the oscillation is not generated by the HBM or the MOS device under study.

More insights are needed to understand the observed behavior of the ESD protection circuit. To investigate the importance of the parasitic BJT, I show in Fig. 9.32*a* the voltage V_B at the base terminal of the parasitic BJT simulated for four different HBM charging voltages. This voltage is critical because it determines the turn-on of the parasitic BJT. Clearly, the MOSFET operates in the snapback region ($V_B > 0.7$ V) soon after an HBM pulse is applied, which indicates that the parasitic BJT is turned on during the entire ESD transients. Figure 9.32*b* illustrates the three components of the current I_{DT} passing through the drain terminal, namely, the drain current I_{ds} of the regular MOSFET, impact-generated current $I_{h,gen}$, and collector current I_C of the parasitic BJT. It is shown that I_C is the dominant component for I_{DT}, and the significance of including the turn-on mechanism of the parasitic BJT in ESD modeling is clearly demonstrated.

Finally, Fig. 9.33 compares the drain voltages simulated from the advanced model (present model) and a simple model (model developed in the preceding section) using constant R_{sub} and drain-voltage-dependent M (simple model A) and a simple model using constant R_{sub} and constant M (simple model B).

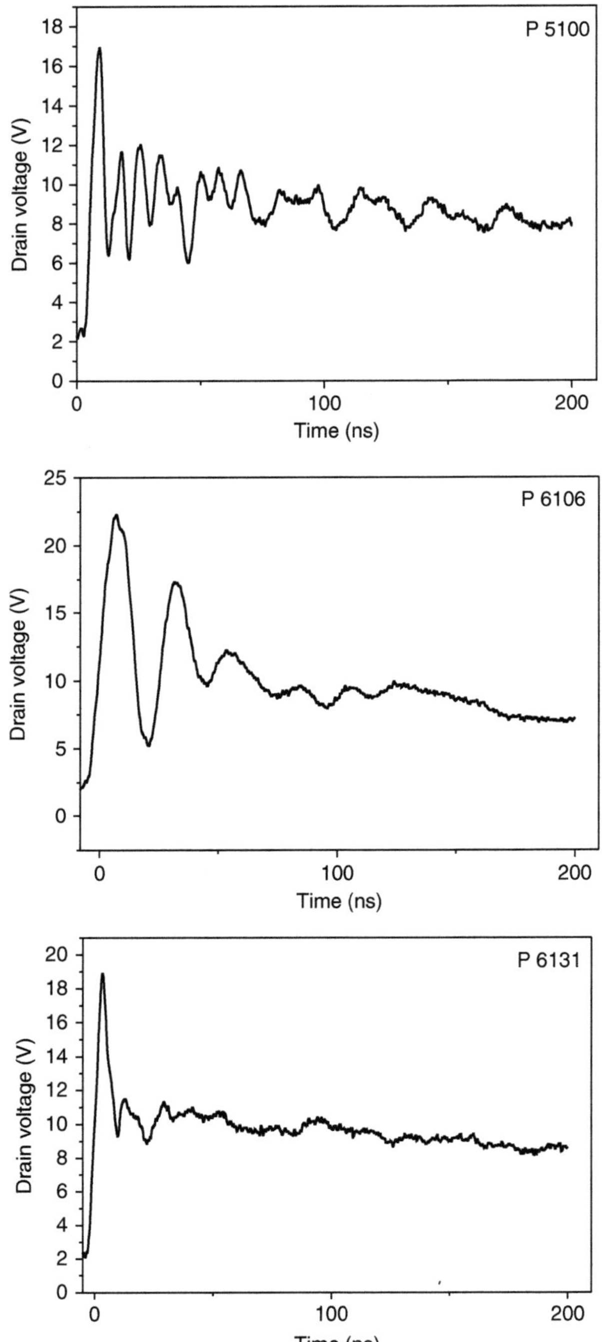

FIGURE 9.31 Voltages at the drain terminal of the ESD protection circuit subjected to an HBM charging voltage of 1000 V measured using three different voltage probes.

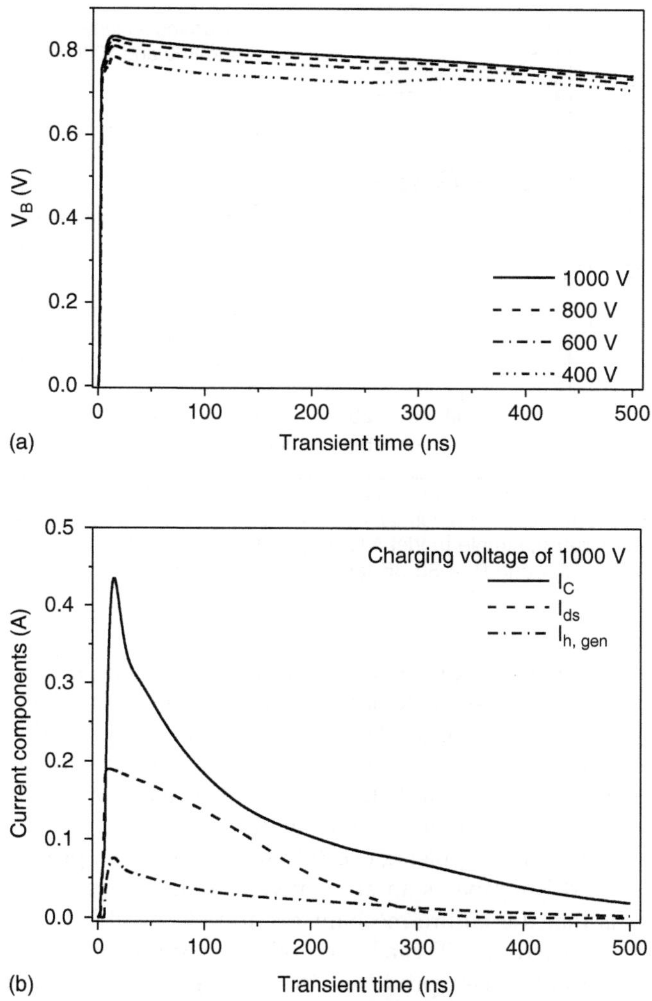

FIGURE 9.32 (*a*) Simulated voltages at the base terminal of the parasitic BJT for four different HBM charging voltages, and (*b*) simulated current components in the MOSFET subject to a 1000-V HBM stress.

9.3 Electrostatic-Discharge Measurements and Testing

Characterization and testing are important parts for the design and verification of ESD protection structures. Because of the rapid transients of ESDs (on the order of nanoseconds), measurements of the ESD response have been proven to be not straightforward.[55] When using the HBM, testing of device and circuit susceptibility normally are carried out using the HBM tester. However, the exponential nature of the HBM ESD waveform, having a very short duration and

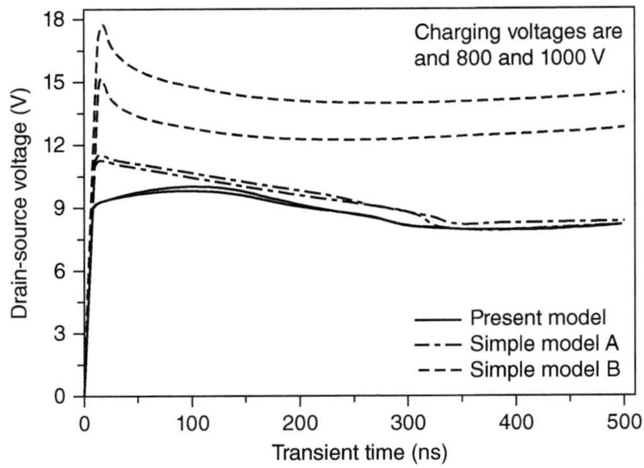

FIGURE 9.33 Comparison of the drain voltages simulated from the improved model (present model with voltage-dependent substrate resistance and substrate current), simple model A (using constant R_{sub} = 2000 Ω and constant P_1 = 8.4 and P_2 = 35.36 for *M*), and simple model B (using constant R_{sub} = 2000 Ω and constant *M* = 1.05).

high peak current (typically tens of nanoseconds and amps of current), makes it difficult to use for device characterization.[56] Figure 9.34 shows the equivalent circuit of the HBM tester, where 100 pF and 1500 Ω represent the capacitance and skin resistance of a typical human body, respectively, and DUT denotes the "device under test," which can be a semiconductor device or a circuit.

Transmission-line pulsing (TLP) technique was first proposed by Maloney and Khurana as an alternative to ESD testing.[57] The basic concept is that the failure mechanisms and electrical behavior of a device influenced by ESD can be reproduced with constant and short voltage/current pulses produced by a charged transmission line. Owing to its versatility and low cost, the TLP technique has gained popularity in recent years to characterize and measure ESD events in

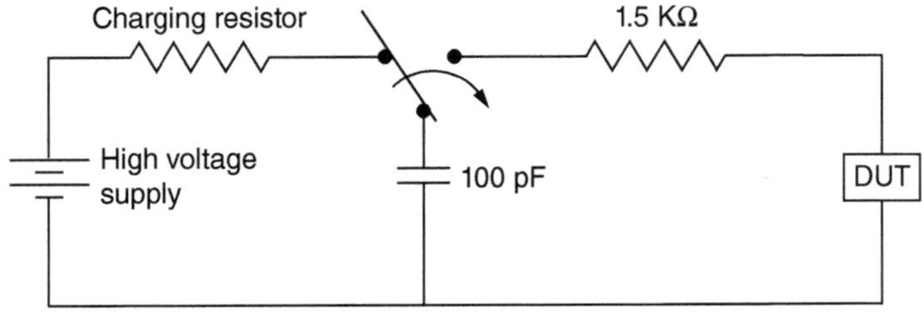

FIGURE 9.34 Equivalent circuits of the HBM tester consisting of a 100-pF capacitor in series with a 1500-Ω resistor connected to the device under test (DUT).

semiconductor devices.[58–62] In addition, the TLP technique is a preferred research method in ESD studies because of the excellent correlation between the pulses generated by the TLP and those associated with the HBM. Using the TLP technique, high-current square pulses are generated by means of discharging a predevice or circuit. A more detailed description of the TLP technique will be given later.

When generating the TLP waveforms, however, engineers and researchers frequently encounter the following two issues. First, the TLP waveforms generated often have considerable distortion and oscillation, which must be reduced or eliminated to obtain a high degree of confidence in using the TLP technique for ESD measurements. Second, the precise TLP pulse width that produces the equivalent effect of the HBM is not known. The magnitude of the TLP pulse is less a problem because it is known to be equal to the HBM charging voltage divided by the human-body resistance of 1500 Ω. On the other hand, the required pulse width is less clear. Over the years, TLP pulse widths ranging from 50 to 200 ns have been used, but a more precise pulse width is needed.

9.3.1 Experimental Setup for Electrostatic-Discharge Measurements Based on the Transmission-Line-Pulsing Technique

The typical TLP setup incorporates a 50-Ω coaxial cable charged by a dc voltage generator, and subsequently, a TLP waveform is generated by discharging the coaxial cable into the load circuit. Figure 9.35*a* shows the equivalent circuit of the TLP technique, and Fig. 9.35*b* shows the simplified TLP equivalent circuit without the capacitive and inductive elements. TLP pulse width is a function of the transmission-line cable length, and the parallel capacitance C_p and the series inductance L_p determine the pulse rise time. MOS devices and circuits will be used as the DUT in this study.

The typical matched load circuit includes the 50-Ω terminator, R_S, and the DUT, as shown in Fig. 9.35*b*. To minimize the waveform reflection, it is desirable to have a load matched to the 50-Ω coaxial cable. To this end, the ideal impedance of the load circuit (i.e., the entire circuit after $Z = 50\ \Omega$ in Fig. 9.35*b*) should be 50 Ω. Since the impedance of the DUT (i.e., a turn-on MOS device) is normally very small during the ESD event, a large R_S is needed to achieve this goal. It has been suggested that R_S must be at least 500 Ω to minimize the pulse distortion owing to reflections.[9] As will be shown later, however, such an approach cannot effectively produce smooth TLP pulses. Moreover, a large R_S can significantly affect the waveform shape and reduce the current supplied to the DUT.

An issue regarding the TLP input and output voltages needs to be clarified. The input voltage (called the *dc precharge voltage*) is the dc voltage charging to the 50-Ω coaxial cable, and the output voltage

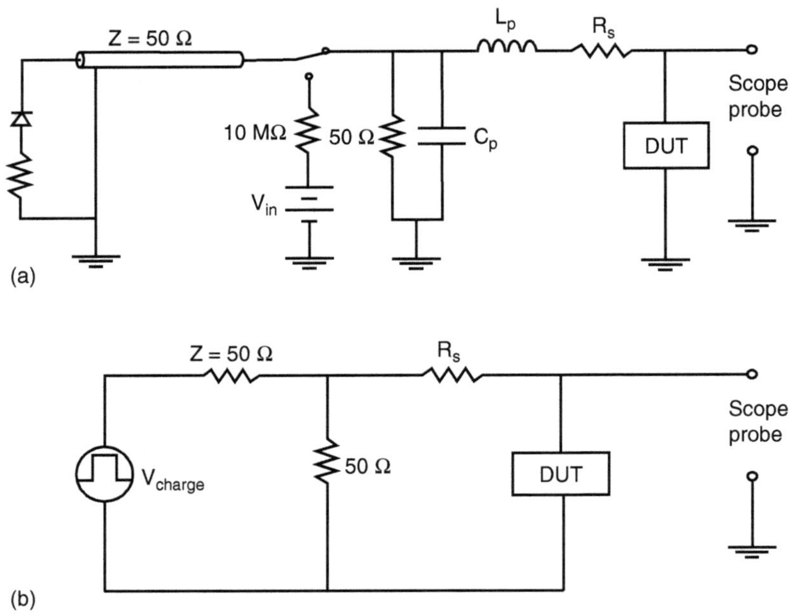

FIGURE 9.35 (*a*) Schematic of the typical TLP technique including the parallel capacitor and series inductor and (*b*) simplified TLP equivalent circuit without the *L-C* elements.

(called the *TLP voltage waveform*) is the TLP voltage waveform generated from discharging the cable and taken at the DUT (i.e., drain terminal of the MOS device). A high input voltage does not always result in a high output-voltage waveform owing to different device behavior, such as snapback and typical MOS operations. As will be shown later, the TLP waveforms would have a higher degree of distortion if a relatively low dc precharge voltage were used. Furthermore, the TLP voltage waveform may decrease when the dc precharge voltage is increased owing to MOS snapback operation.

To demonstrate the TLP waveform distortion and oscillation, we use a TLP setup having a typical matched load circuit with $R_S = 750\ \Omega$ (to show 750 Ω in the figure) (see Fig. 9.35*b*) and a conventional voltage probe with a ground lead. A TLP having an 80-ns width is applied to the drain terminal of a 160-μm/1.2-μm (channel width/channel length) NMOS device. Figure 9.36 shows the resulting 9-V voltage and current waveforms generated using a relatively large dc precharge voltage of 150 V. Note that a large voltage spike at the rising edge of the pulse and subsequent oscillations are found in the voltage waveform, which can be attributed to the nonideal matched load and parasitic inductance associated with the ground lead of the voltage probe.[63] The distortion and oscillation would be even worse if a smaller dc precharge voltage were used to generate the TLP waveforms.

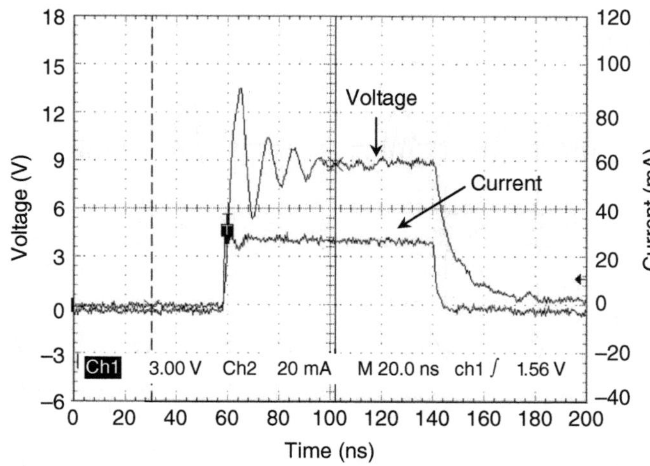

FIGURE 9.36 Voltage and current waveforms generated using a large dc precharge voltage, the conventional load circuit, and conventional voltage probe with ground lead.

9.3.2 Development of a Matched-Load Circuit

To reduce the parasitic inductance associated with the voltage probe and thus to minimize the oscillation of the TLP waveform, the typical matched-load circuit (see Fig. 9.35*b*) with a 50-Ω terminator and R_s must be modified and improved. To this end, a key issue is to develop a matched-load circuit, including the DUT, that would have a load of 50 Ω (i.e., matched to the 50-Ω impedance of the transmission line) independent of the variable impedance associated with the DUT (i.e., the MOS device). The improved matched-load circuit, called the *R-2R circuit,* shown in Fig. 9.37 can accomplish this goal. It consists of the DUT impedance R_3 and an adjustable number of stages of R_1 and R_2, where R_1 (i.e., 50 Ω) denotes the transmission-line impedance, and R_2 is set at $R_2 = 2R_1$.

First, let us assume that $R_3 = R_1$; then the equivalent resistances $R_{eq,x}$ (i.e., subscript x denotes the number of different nodes in the circuit) of the R-2R circuit are given by

$$R_{eq,x} = \frac{1}{\dfrac{1}{R_1 + R_1} + \dfrac{1}{R_2}} = \frac{1}{\dfrac{1}{R_2} + \dfrac{1}{R_2}} = \frac{1}{\dfrac{2}{R_2}} = \frac{R_2}{2} = R_1 \tag{9.64}$$

$$R_{eq,x-1} = \frac{1}{\dfrac{1}{R_{eq,x} + R_1} + \dfrac{1}{R_2}} = \frac{1}{\dfrac{1}{R_1 + R_1} + \dfrac{1}{R_2}} = R_1 \tag{9.65}$$

$$R_{eq,1} = \frac{1}{\dfrac{1}{R_{eq,2} + R_1} + \dfrac{1}{R_2}} = \frac{1}{\dfrac{1}{R_1 + R_1} + \dfrac{1}{R_2}} = R_1 \tag{9.66}$$

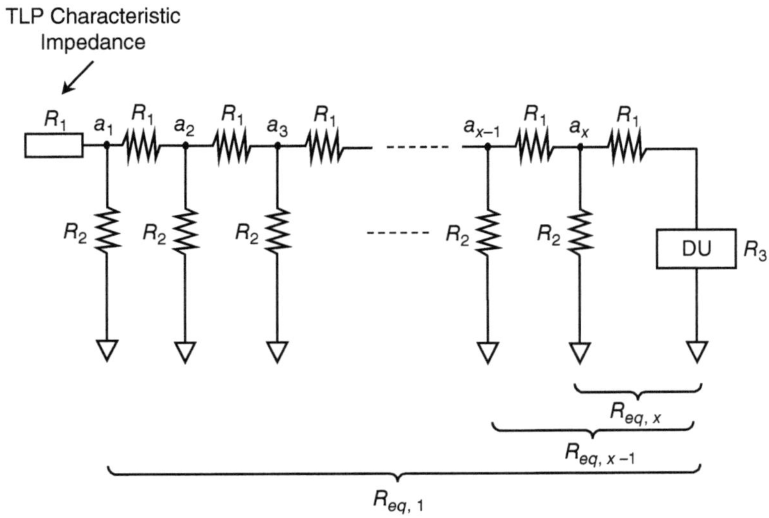

FIGURE 9.37 Equivalent circuit for the improved R-2R matched load.

Thus the same impedance of R_1 at different nodes $a_1, a_2, a_3, \ldots, a_{x-1}, a_x$ can be obtained. Note that $R_{eq,1}$ is the impedance at node a_1 and thus the impedance of the R-2R matched-load circuit. From the TLP theory, the amplitude of the reflected-voltage wave normalized to the amplitude of the incident-voltage wave is called the *voltage reflection coefficient* Γ, which is given by[64]

$$\Gamma = V_o^- / V_o^+ = (Z_L - Z_0)/(Z_L + Z_0) \tag{9.67}$$

where Z_L is the load impedance and Z_0 is the characteristic impedance of the transmission line. Since Z_0 is equal to R_1 and Z_L is equal to the equivalent resistor R_1 at point a_1, Eq. (9.67) becomes

$$\Gamma = V_o^- / V_o^+ \tag{9.68}$$

The zero-reflection coefficient given in Eq. (9.68) suggests that there is no reflected wave.

Let us consider the voltage drop at different nodes of the R-2R circuit. Consider an initial discharge voltage of V_{a1}. According to the current divider rule,

$$V_{a3} = \frac{V_{a2}}{2} = \frac{V_{a1}}{2^2} \tag{9.69}$$

$$\vdots$$

$$V_{ax} = \frac{V_{a(x-1)}}{2} = \frac{V_{a1}}{2^{x-1}} \tag{9.70}$$

Thus, except for the initial stage, the voltage at each node is half that at the previous node.

Number of Stages	$R_{eq,1}$ (Ω)				
	$R_3 = 0$	$R_3 = 100\ \Omega$	$R_3 = 1000\ \Omega$	$R_3 = 10000\ \Omega$	$R_3 = \infty$
	33.3	60.0	91.3	99.01	100.0
2	45.5	52.4	58.6	59.8	60.0
3	48.8	50.6	52.05	52.3	52.4
4	49.7	50.1	50.5	50.6	50.6
5	49.9	50.04	50.1	50.1	50.1

TABLE 9.2. Impedances of the Improved Matched-Load Circuit for Various R-2R Stages and DUT Resistances

The preceding theory was derived based on the assumption that $R_3 = R_1$ (i.e., the impedance of the transmission line is the same as that of the DUT). Let us now consider a generalized case of an arbitrary R_3 (i.e., $R_3 \neq R_1$). For this case, close-form expressions for $R_{eq,x}$ are more complicated and will not provide clear insights. Thus, instead of showing the expressions, numerical values of the impedance $R_{eq,1}$ using different R-2R stages and different R_3 values are summarized in Table 9.2 to illustrate the matched load of the R-2R circuit. The results indicate clearly that $R_{eq,1}$ approaches R_1 (=50 Ω) when the number of stages is increased even if R_3 deviates considerably from R_1.

It should be pointed out that ideal resistors with no parasitic inductance or capacitance have been assumed in the preceding analysis. This R-2R network is designed to be a broadband-matching circuit with an infinite frequency response. The frequency response of an actual network will be limited by the parasitic inductance and capacitance introduced by the type of resistors and the construction techniques employed. The test circuit was constructed to match a 20-ns to 100-ns pulse with 6-ns rise and fall times. The rise and fall times were set primarily by the additional components C_p and L_p in the TLP setup. Since the matched-load circuit was capable of producing the desired 6-ns rise and fall times, its bandwidth should exceeds 167 MHz (i.e., estimated frequency range in which the improved test circuit is effective is from low frequency to more than 167 MHz). This bandwidth is more than adequate because the frequency of the main oscillations observed in the conventional test circuit is about 60 MHz.

Based on the concept just presented, an improved matched-load circuit can be constructed by using precision carbon and chip resistors connected to the TLP test system through a Bayonet Neill-Concelman (BNC) connector. An experiment was performed setting the gate voltage at 1 V and the TLP pulses applied to five-, three-, and one-stage matched-load circuits. The resulting voltage and current waveforms are shown in Fig. 9.38. Comparing the waveforms in Fig. 9.38 with those in Fig. 9.36, it is clear that use of the improved TLP setup can significantly reduce the spike and oscillation in TLP

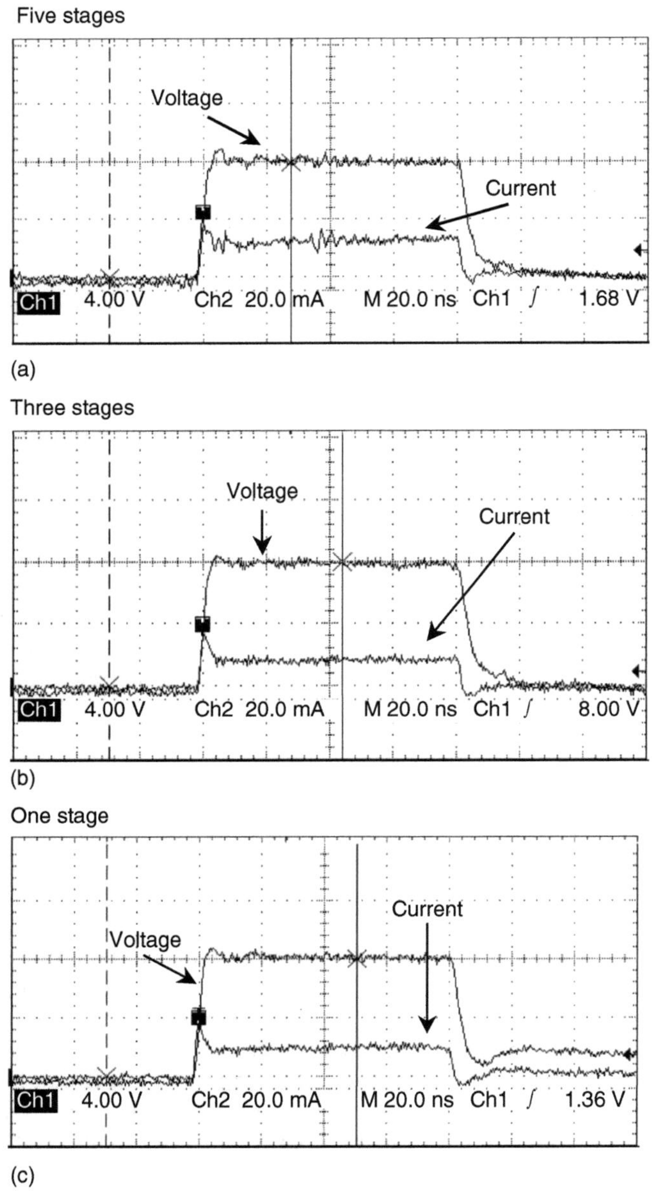

FIGURE 9.38 Voltage and current waveforms generated using the (*a*) five-stage, (*b*) three-stage, and (*c*) one-stage matched-load circuits and *1-V* gate voltage.

waveforms under a wide range of dc precharge voltages. The benefit of using a high number of stages in the matched-load circuit, however, is less clear. In fact, a closer look into the results in Fig. 9.38 reveals that the waveform obtained from the three-stage circuit is actually smoother than those obtained from the one- and five-stage circuits. The reason that the three-stage circuit yields a better waveform than the one-stage circuit is because of the improved matched load (i.e., the matched load is closer to 50 Ω for the three-stage circuit). But it is a little puzzling as to why the waveform deteriorates when the number of stages is increased from three to five. One possible explanation is that while the increasing number of stages improves the transmission-line matching, the total magnitude of parasitic elements is increased with increasing stage count. Therefore, the extra parasitic elements added by the additional stages of chip resistors could negate the enhancement gained by their inclusion. As a result, a TLP setup with the improved voltage probe and a three-stage improved matched-load circuit should render optimal TLP waveforms.

When characterizing semiconductor devices under ESD stress, the magnitude of the TLP output current is of great importance. If the output current is not sufficiently large, snapback or other impending breakdown mechanisms may never be observed. Figure 9.39 illustrates equivalent TLP circuits employing the typical matched-load approach having an R_S = 583.35 Ω and the two-stage improved matched-load circuit, respectively, each containing a 5-Ω DUT. Note that the typical

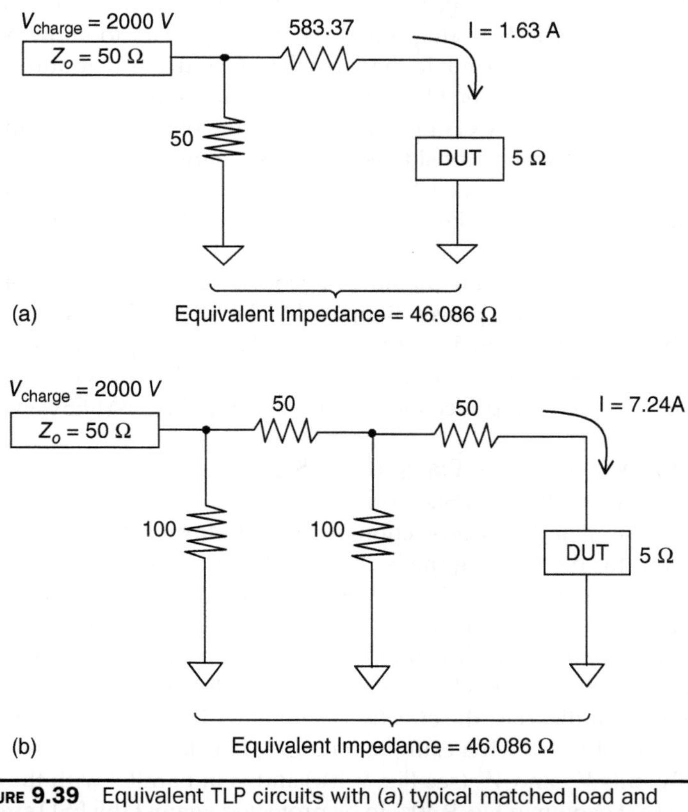

FIGURE 9.39 Equivalent TLP circuits with (*a*) typical matched load and (*b*) improved matched load. Under the same equivalent impedance, the improved matched-load circuit has the advantage of generating a larger output current.

and improved matched-load circuits have the same equivalent impedance of 46.086 Ω. The output current obtained for the typical matched-load circuit I_{typ} is 1.63 A when the dc precharge voltage is set at 2000 V. Correspondingly, the output current I_{imp} for the improved matched-load circuit is 7.24 A with the same precharge voltage. In other words, I_{imp} is several times larger than I_{typ} under the same dc precharge voltage. As a result, when higher TLP output current levels are needed, the improved matched-load circuit again holds a significant advantage over the typical matched-load circuit.

9.3.3 Determination of Transmission-Line Pulse Width Equivalent to the Human-Body Model

ESD produces large currents and high electric fields in semiconductor devices. Energy released during an ESD event generates "self-heating" in the device and causes a rapid temperature increase.[65] The three basic MOS device failures are (1) junction burnout, (2) oxide punch-through, and (3) metallization burnout. Whether the MOSFET failure owing to ESD is decided by the peak lattice temperature or the total energy dissipated in the MOSFET is the subject of this section.

The TLP tester can generate pulses with different pulse widths and magnitudes. For ESD testing purposes, it is important to know what TLP pulse shape will result in the same device failure as that caused by an ESD event such as the HBM. First, a commonly used ESD preventive circuit (i.e., supply-clamp circuit) consisting of a MOSFET, capacitor, and resistor will be considered. The ESD charging voltage causing the MOSFET in this circuit to fail will be measured using the HBM tester. ESD transients of the MOSFET subjected to this charging voltage then will be simulated using a 2D device simulator called Atlas.[43] The simulation results include the time-dependent current, voltage, lattice temperature, and total power dissipated in the MOSFET. This information then will be used to determine what TLP pulse shape is equivalent to the HBM ESD and, therefore, should be used in the TLP technique to characterize the ESD protection device subject to the HBM ESD event.

Electrostatic-Discharge Clamp Under Study

It is well known that an ESD protection structure consisting of voltage-supply clamp and pad-protection devices can form a robust conductive path for discharging the high current associated with ESD flow between any pair of pins.[66,67] In general, an ESD pad-protection network shunts input/output pins to the ground bus under stress events. The ESD protection circuit shown in Fig. 9.40*a* is considered. It consists of a 2-pF capacitor, a 100-kΩ resistor, and two identical MOSFETs connected in parallel (i.e., the effective channel width is twice that of the single MOSFET). Such a clamp circuit is connected between the two supply rails V_{dd} and V_{ss} (see Fig. 9.40*a*) and is in parallel with the circuitry (not shown in Fig. 9.40*a*) to be protected against the ESD stress.

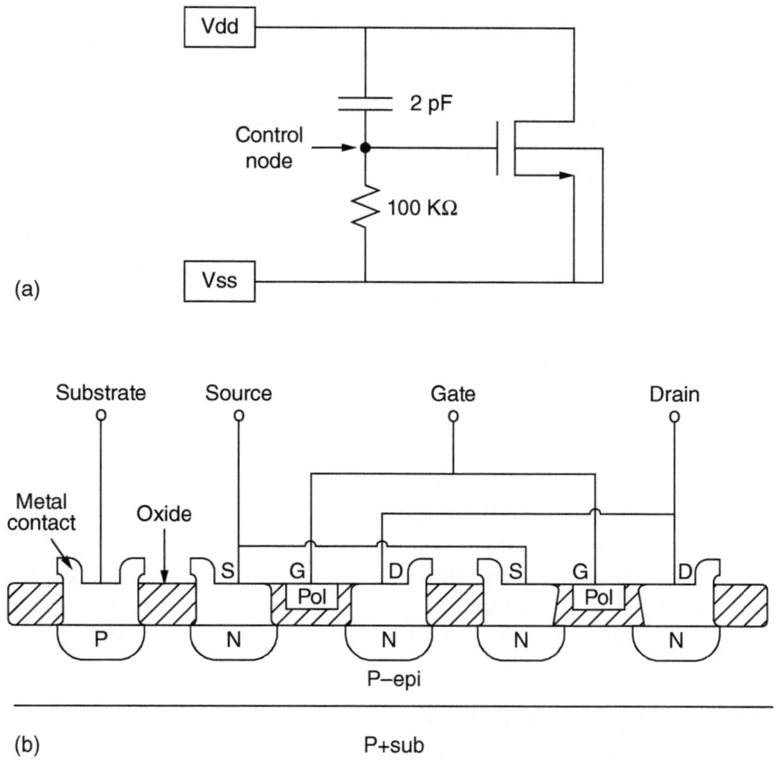

FIGURE 9.40 (*a*) ESD protection circuit (i.e., clamp circuit) considered, which consists of a MOSFET, a capacitor, and a resistor, and (*b*) a cross section of the MOSFET (two identical MOSFETs connected in parallel) used in the clamp circuit.

The cross-sectional layout of the MOSFET is shown in Fig. 9.40*b*. It is an *n*-channel MOS device having a channel length of 1.2 μm, a channel doping density of 1.5×10^{15} cm^{-3}, a peak drain/source doping density of 1×10^{20} cm^{-3}, a gate-oxide thickness of 25 nm, and different channel widths. Furthermore, the MOS channel width, not the channel length, can be changed relatively dramatically to alter the size of the MOS device. As a result, MOS devices with a fixed channel length (i.e., 1.2 μm), but several different channel widths were used to investigate the ESD effects.

When an HBM ESD pulse with a relatively large rise time is introduced to the pad of the clamp circuit (i.e., V_{dd} in Fig. 9.40*a*), the capacitor initially will be a short circuit, which turns on the MOS device quickly. Consequently, a large drain current associated with the HBM ESD will pass through the MOS device during the initial transients. At the same time, the high voltage associated with the ESD pulse applied to the drain can give rise to avalanche multiplication near the reverse-biased drain junction. This results in a flow of avalanche-generated holes to the substrate and a voltage drop across the substrate resistance. As this voltage drop approaches 0.7 V, the parasitic *n/p/n* BJT in the MOSFET is turned on. Thus, during the initial stage of an ESD event, the MOSFET acts like a short circuit with a very high total drain current consisting of the collector current from the parasitic BJT, the drain current from the normal MOS, and the generation current from the avalanche multiplication. In the next sequence, the capacitor will be charged through the resistor, and the charging rate depends on the *RC* time constant of the clamp circuit. As the voltage across the capacitor increases, the gate-to-source voltage V_{GS} of the NMOS decreases, and the capacitor acts as an open circuit when fully charged. The MOS device is turned off, whereas the parasitic BJT is still on, when V_{GS} decays below its threshold voltage. This leads to a quick decrease in the drain current and a relatively constant drain voltage during subsequent stage of the ESD event (i.e., the voltage across the circuit is clamped to a fixed value). Eventually, the parasitic BJT is turned off when the ESD pulse decays to a value below the critical voltage for avalanche. For an ESD pulse with a relatively small rise time, the voltage drop across the resistor connected to the gate in the clamp circuit may not be high enough to turn on the NMOS in the initial stage, and only the parasitic BJT is operational during the ESD event.

Determination of Failure Mechanism

In this subsection, simulations and measurements of MOS behavior during the HBM ESD transients will be carried out, the failure mechanism associated with the HBM ESD will be studied, and the TLP pulse (i.e., pulse magnitude, pulse width, and pulse rise time and fall time) producing the effect equivalent to the HBM will be determined.

Our first step is to determine the HBM charging voltage causing failure of the MOSFET in the clamp circuit shown in Fig. 9.40. The testing criterion will be set at the HBM charging voltage that causes a soft failure in the MOSFET. This *soft failure* is defined as the point where the gate leakage current of the MOSFET increases rapidly after the device is subjected to an ESD stress using an HBM tester. The leakage current of the MOSFET was measured after the clamp circuit was stressed by an HBM charging voltage, and subsequently the MOS device was biased under the condition that the gate, source, and body were connected to the ground, and 5 V was applied to the drain. Such a failure, which is nondestructive, is caused by the temporary junction short, alloy spike, and/or oxide short in the MOS device during the ESD event. At an even higher stress condition, a *hard failure* that is destructive and permanent will take place. Typically, the stress conditions causing the soft and hard failures are not very far apart. As a result, the soft failure is frequently used as a critical criterion for ESD testing and characterization.

Figure 9.41 shows the leakage-current versus HBM charging-voltage characteristics measured using the HBM tester, with the point

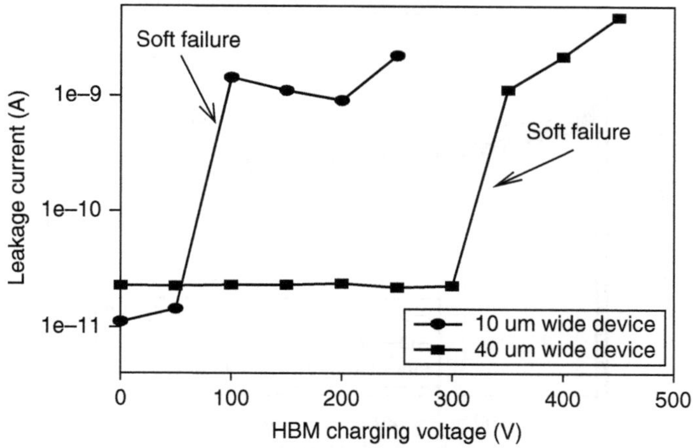

FIGURE 9.41 Measured leakage-current characteristics of the MOS devices after being subjected to the HBM ESD, with the positions of soft failure indicated by arrows.

of soft failure indicated by the arrows (i.e., the point where the leakage current starts to increase very rapidly). It can be seen that the HBM charging voltage causing the soft failure increases with increasing device width. This is due to the fact that a larger device can withstand higher energy without being damaged. The HBM charging voltage that induces the soft failure is called the *HBM failed voltage*, and Table 9.3 lists the measured HBM failed voltages for three different MOSFET widths. As expected, the HBM failed voltage increases with increasing MOSFET width. Note that all three MOSFETs have the same channel length of 1.2 μm and identical device makeups such as oxide thickness, drain-source junction depth, and substrate doping density.

2D device simulations require knowledge of the MOSFET doping profiles and layout. The circuit components (i.e., capacitors and resistors) in the clamp circuit (see Fig. 9.40*a*) and the components associated with the HBM then are incorporated into Atlas using the mix-mode option. Subjecting the device to the HBM failed voltages just mentioned, the time-dependent drain voltage, drain current, and lattice temperature of the MOSFET in the clamp circuit can be simulated, and the results for MOS devices with channel widths of 20, 40, and 80 μm are shown in Fig. 9.42. It should be pointed out that the

Device Size (μm)	Failed Voltage (V)
80	545
40	275
20	150

TABLE 9.3 HBM Failed Voltages Measured for MOS Devices with Three Different Channel Widths

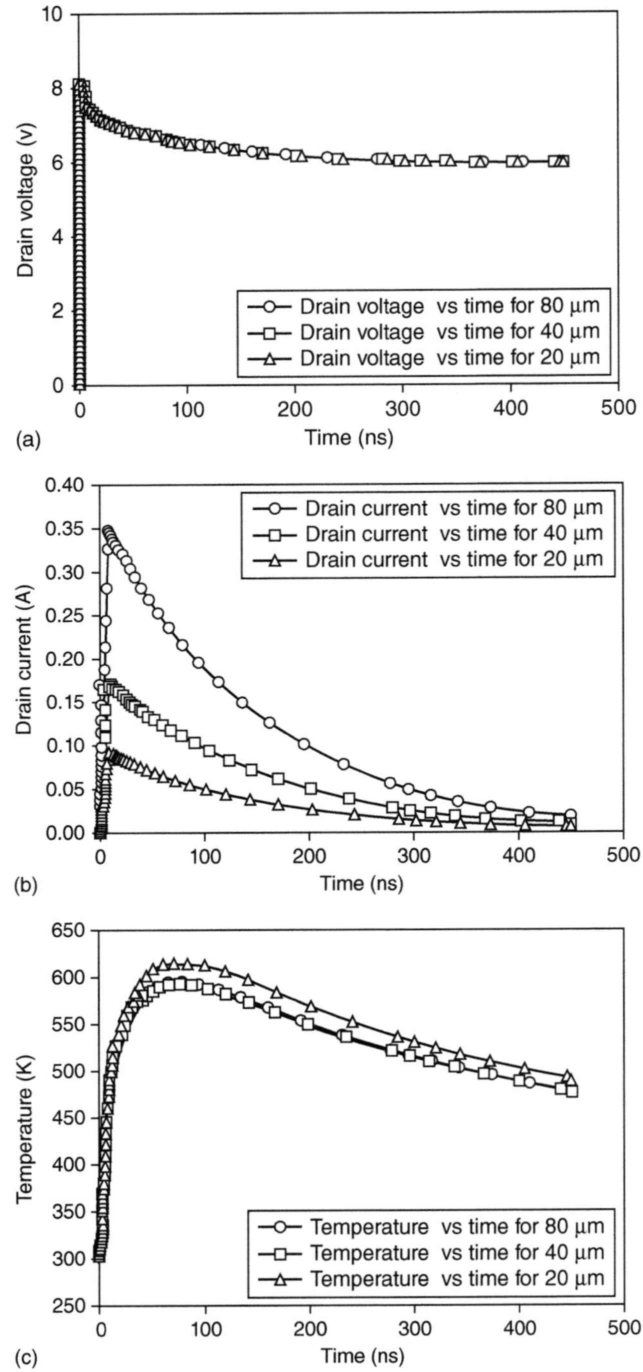

FIGURE 9.42 Simulated time-dependent (*a*) drain voltage, (*b*) drain current, and (*c*) lattice temperature of MOSFETs with three different sizes in the clamp circuit during the HBM ESD transients.

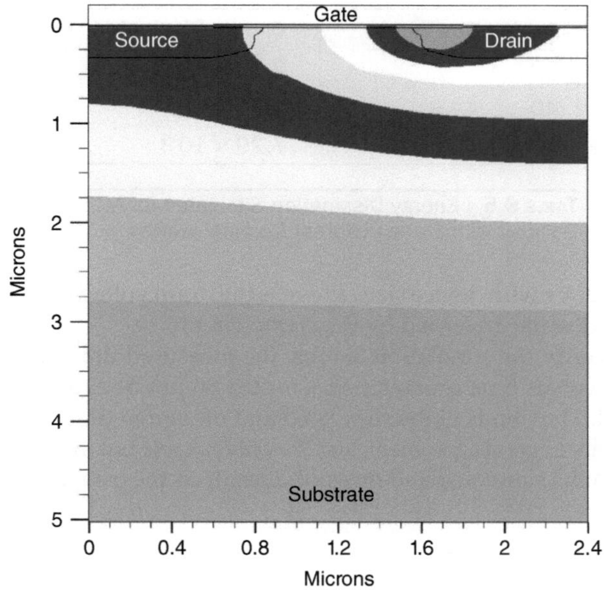

FIGURE 9.43 Simulated lattice-temperature contours in the MOSFETs, which indicate that the hot spot (location of highest lattice temperature) is located at the surface near the drain junction.

lattice temperature is not uniform in the MOS device, and the lattice temperature shown in Fig. 9.42*c* is taken at the location (i.e., hot spot) where the lattice temperature is the highest. Figure 9.43 illustrates the temperature contours in the MOSFET, indicating that the hot spot is located at the surface near the drain junction.

Table 9.4 lists the simulated peak lattice temperatures at the hot spot of MOS devices. Also, from the time-dependent current and voltage characteristics, the total energy dissipated in the MOSFETs during the ESD event can be calculated, and the values are given in Table 9.5. The results in Fig. 9.42 indicate that the drain currents of the three devices changed considerable, but the drain voltages are relatively insensitive to MOS device size. This is so because the drain current is proportional to the magnitude of the TLP current pulse,

Device Size (μm)	Peak Temperature (K)
80	595
40	600
20	616

TABLE 9.4 Peak Lattice Temperature Simulated for MOS Devices with Three Different Channel Widths

Device Size (μm)	Energy Dissipated (J)
80	2.63×10^{-7}
40	1.37×10^{-7}
20	7.29×10^{-8}

TABLE 9.5 Energy Dissipation Simulated for MOS Devices with Three Different Channel Widths

which varies with device size, whereas the drain voltage is "clamped" to a fixed value governed by the clamp circuit.

To verify the simulation results, the measured drain-voltage and current versus time characteristics for the 80-μm device are shown in Fig. 9.44. The trends of the simulated and measured drain voltage and current are in good agreement, and the only notable but minor difference between the simulated and measured results is the overshoot found in

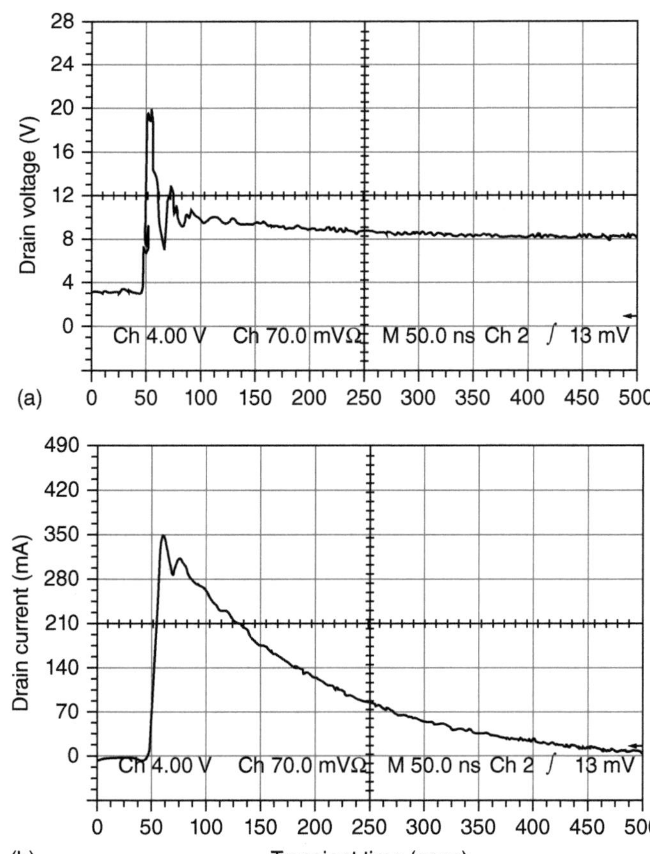

FIGURE 9.44 Measured time-dependent (*a*) drain voltage and (*b*) drain current of the 80-μm-wide MOSFET in the clamp circuit during the HBM ESD transients.

the measured drain-voltage characteristics. Note that both the drain voltage and current reach their peak values in a very short period of time. After that, the drain current decays quickly toward zero, whereas the drain voltage holds a relatively constant value.

Next, the TLP pulse shape equivalent to the HBM based on the criteria of peak temperature and total energy dissipated in the MOSFET is studied and determined. It will be shown later that the total energy dissipated is the main factor governing device failure and deciding the equivalency of the TLP and HBM test techniques. The current level of the TLP pulse (i.e., pulse magnitude) should be the same as the peak drain current during the HBM transients, as shown in Fig. 9.42*b*, which is approximately equal to the HBM failed voltage divided by the skin resistance of human body (i.e., 1500 Ω). The TLP rise and fall times were set to 6 ns using an *LC* circuit to match the rise time of the HBM pulse. As a result, the only unknown and to-be-determine parameter for the TLP pulse shape is its width.

To determine the TLP pulse width, simulation of transient characteristics of the MOSFET in the clamp circuit subjected to the TLP technique is also required, and the MOS device layout and the TLP equivalent circuit components are specified in Atlas using the mix-mode option. The correct TLP pulse width is found when it gives rise to the same peak temperature or energy dissipated in the MOSFET, depending on the criterion, as that simulated from the HBM equivalent circuit.

Figure 9.45 shows the TLP pulse widths versus the MOSFET width simulated based on the criteria of peak temperature and total energy dissipated. TLP pulse widths of about 95 and 25 ns were

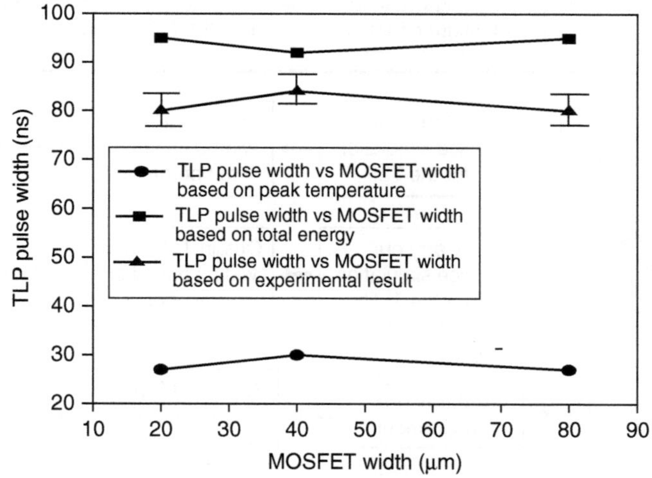

FIGURE 9.45 TLP pulse widths (equivalent to the HBM) determined from simulations based on peak lattice temperature in the MOSFET, simulations based on total energy dissipated in the MOSFET, and measured from the leakage characteristics of the MOSFET after the ESD stress.

obtained based on the energy-dissipated and peak-temperature criteria, respectively. Also shown in Fig. 9.45 are the measured TLP pulse widths that give rise to the identical MOSFET soft failure caused by the HBM failed voltage. The measurement procedure is the same as that described previously using the HBM tester, and the only difference here is that a TLP pulse, instead of an HBM charging voltage, is used to stress the MOS device in the clamp circuit. For every MOS size (i.e., channel width), 10 identical devices were measured, and the upper and lower limits of measured data are indicated by the error bars and the mean values are indicated by the symbols in Fig. 9.45. Clearly, the width of 95 ns simulated based on the energy-dissipation criterion is in better agreement with the width of about 85 ns obtained from measurements than that of 25 ns simulated based on the peak-temperature criterion. Thus, the results in Fig. 9.45 suggest that the total energy dissipated in the MOSFET is the main mechanism deciding ESD failure and that this energy should be used as the criterion for correlating the TLP technique and HBM. Also, while the TLP pulse-current level is a function of MOSFET width (i.e., or function of HBM charging voltage), the TLP pulse width is insensitive to the size of the MOS device.

A flowchart is given in Fig. 9.46 to summarize the procedure for determining the TLP pulse width suitable for HBM testing and characterization.

In the preceding analysis, it has been demonstrated that using the soft failure as the testing criterion, a pulse width of about 80 ns should

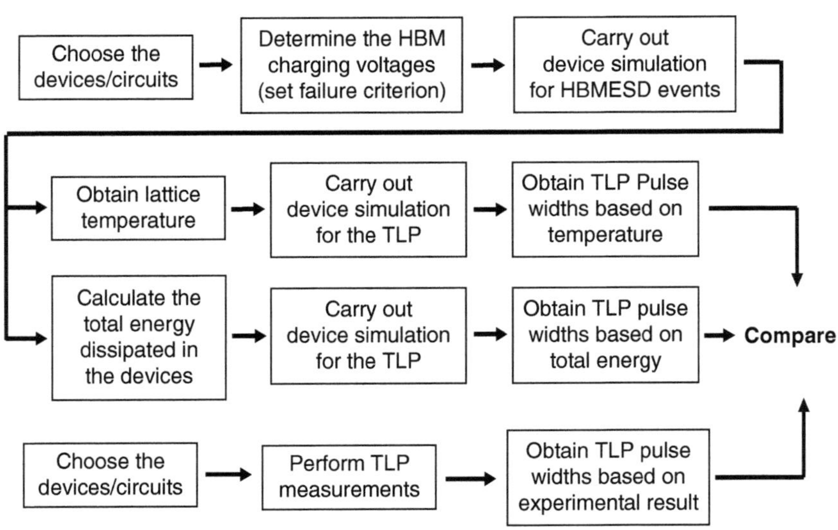

FIGURE 9.46 Flowchart summarizing the procedure to determine the TLP pulse width equivalent to the HBM ESD stress.

be used in the TLP technique for measurements of ESD response. In addition, it was shown that such a pulse width is insensitive to MOS width. We further suggest that the TLP pulse width should not be a function of other MOS device makeup, such as channel length and doping density. This reasoning is derived from the observation that the total power dissipated in the MOS device is the main factor for soft failure, and such a power is different for different device makeups. This power is related to the HBM failed voltage, which then determines the TLP pulse-current level to be used. Consequently, the power variation owing to the different device makeups is accounted for by the change in TLP pulse-current level, and the TLP width should be independent of MOS device makeup. Under certain circumstances, however, this assertion becomes invalid. To gain the relevant physical insight, a simple theoretical analysis will be given in the next subsection.

Theoretical Analysis

Knowing that the total energy dissipated in a semiconductor device is the main factor causing ESD-induced failure, as was demonstrated in the preceding section, we can now proceed to derive a simple theory for estimating the equivalent TLP pulse width. The length of cable needed to generate the correct TLP pulse width for ESD measurements will be calculated as well.

Consider the condition of soft failure. The peak currents passing through the device under test with the HBM tester ($I_{\text{HBM},pk}$) and TLP (I_{TLP}) are the same:

$$I_{\text{HBM},pk} = I_{\text{TLP}} = \frac{V_{\text{HBM},pk}}{R_{\text{HBM}}} \tag{9.71}$$

where $V_{\text{HBM},pk}$ is the HBM voltage causing MOSFET failure, and R_{HBM} (=1500 Ω) is the resistance associated with the HBM. Another condition specified is that the total energy causing MOS device failure is the same for both the HBM and the TLP technique:

$$E_{f,\text{HBM}} = E_{f,\text{TLP}} \tag{9.72}$$

$$= \int_0^\infty i^2 R_{\text{DUT}}\, dt \tag{9.73}$$

where $i(t)$ is the time-dependent current, and R_{DUT} is the resistance of the device under test. Typically, $R_{\text{HBM}} >> R_{\text{DUT}}$ and $C_{\text{HBM}} >> C_{\text{DUT}}$. Thus,

$$i_{\text{HBM}}(t) \approx I_{\text{HBM},pk} \exp\left(\frac{-t}{R_{\text{HBM}} C_{\text{HBM}}}\right) \tag{9.74}$$

where C_{HBM} (=100 pF) is the capacitance associated with the HBM, and C_{DUT} is the capacitance associated with the DUT, including the capacitances in the MOS device and the 2-pF capacitance in the clamp circuit. Substituting Eq. (9.74) into Eq. (9.73) leads to

$$E_{f,\text{HBM}} = 0.5 I^2_{\text{HBM},pk} R_{\text{DUT}} R_{\text{HBM}} C_{\text{HBM}} \tag{9.75}$$

For the TLP technique, I_{TLP} is constant, and if the very small portions of the rise and fall of the pulse are neglected, we have

$$E_{f,\text{TLP}} = I^2_{\text{TLP}} R_{\text{DUT}} t_{\text{TLP}} \tag{9.76}$$

where t_{TLP} is the pulse width. Putting Eqs. (9.76), (9.75), and (9.71) into Eq. (9.72) yields

$$t_{\text{TLP}} = 0.5 R_{\text{HBM}} C_{\text{HBM}} \tag{9.77}$$

Thus, according to this simple theory, the width of the TLP pulse equivalent to the HBM is estimated to be 75 ns. While this is a rough estimation, it agrees reasonably well with the simulated and measured results shown previously. The length L of a 50-Ω coaxial cable needed to generate such a pulse is

$$L = \frac{t_{\text{TLP}} x V}{2} \tag{9.78}$$

where V (=2 × 10^8 m/s) is the propagation velocity. Thus a cable length of about 25 feet is needed to generate the TLP pulse equivalent to the HBM.

It is important to point out that the constant TLP pulse width (i.e., independent of the MOS device makeup) predicted by this simple model is a result of the assumptions, ($R_{HBM} >> R_{DUT}$ and $C_{HBM} >> C_{DUT}$) used in the derivation. If these assumptions become invalid, then the decay of $i_{HBM}(t)$ will be controlled by a time constant related to $R_{DUT}C_{DUT}$ rather than $R_{HBM}C_{HBM}$ given in Eq. (9.74). Consequently, the equivalent TLP pulse width would be a function of R_{DUT} and C_{DUT} and therefore influenced by the MOS device makeup, such as the device size, channel length, and doping density.

Procedure to Determine Transmission-Line Pulse Equivalent to the Human-Body Model

The following procedure can be recommended to determine the TLP pulse equivalent to the HBM ESD in a computer-aided design environment:

1. Choose the semiconductor device and/or clamp circuit (i.e., DUT) for ESD testing.
2. Select and determine the HBM charging voltage for ESD testing.

3. Carry out device simulation of HBM ESD transient behavior based on this HBM charging voltage.
4. Based on the simulation results, calculate the total energy dissipated in the semiconductor device.
5. Carry out device simulation for the TLP technique. Use a TLP current pulse with a magnitude equal to the HBM charging voltage divided by 1500 Ω, a typical rise and fall time of 6 ns, and several different pulse widths.
6. The correct pulse width is the one that gives rise to the same energy dissipated as that obtained from the HBM ESD simulation. This pulse width is relatively insensitive to MOS device makeup unless the RC time constant associated with the HBM is smaller than that associated with the device under test.
7. This TLP pulse produces ESD effects identical to those of the HBM.

9.4 Design of On-Chip Electrostatic-Discharge Protection Solutions

As mentioned at the beginning of this chapter, effective ESD protection solutions are must-haves for modern ICs. The design of ESD protection solutions, however, is challenging and difficult owing to the constraints imposed by the technology, IC operation, and customer demand. This section will focus on issues pertinent to the design of on-chip ESD protection solutions.

A widely used on-chip ESD protection scheme is shown in Fig. 9.47, where two ESD protection devices (i.e., diodes) are connected to each I/O pin, and several ESD protection devices/circuits, called *supply clamps*, are connected between the supply-pin V_{dd} and the ground pin V_{ss}. These devices are used to protect the core circuit against the positive- or negative-polarity ESD stress by creating a conducting loop when a pin is subject to an ESD stress and another pin is grounded. Such a conducting path, consisting of the ESD protection device and metal lines, renders the following two objectives during an ESD event: (1) It prevents the current generated by the ESD event from entering the core circuit, and (2) It clamps the voltages at the pins to an acceptable level. Both these objectives therefore can minimize the likelihood of ESD-induced damage to the core circuit.

While the principle of ESD protection is fairly straightforward, several requirements must be considered and met before a protection solution is designed successfully:[1,8–10]

1. The ESD protection devices must be in the off state during normal operation (ESD event is absent).

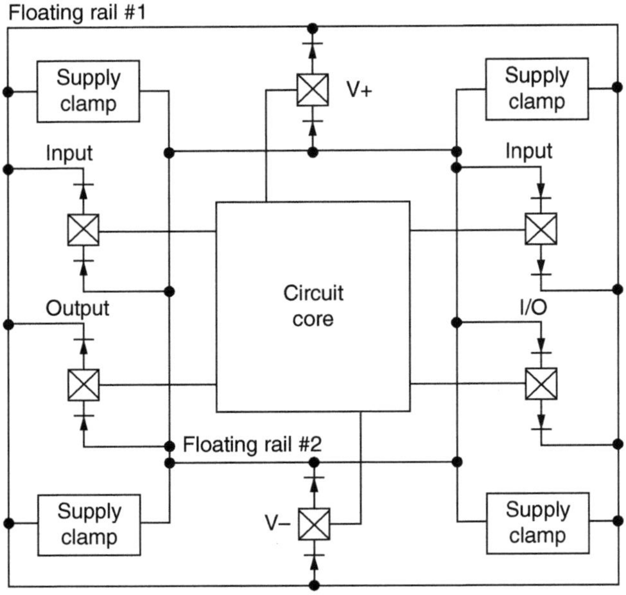

FIGURE 9.47 Typical ESD protection scheme constructed using diodes at the I/O pins and supply clamps between the power supply and the ground pins.

2. The ESD protection devices must turn on sufficient quickly when an ESD event occurs.
3. The voltages at the pins to which the ESD protection devices are connected must be maintained at a sufficiently low value during the ESD event to avoid core circuit failures.
4. The ESD protection devices as well as the core circuit cannot be damaged by the ESD stress.
5. The ESD protection devices must return to the off state after the ESD event has passed, otherwise, the devices will operate in the prohibited latch-up state.

ESD protection devices can be classified into nonsnapback and snapback devices. Figure 9.48 shows the current-voltage characteristics of nonsnapback and snapback ESD protection devices, respectively. Let us first discuss the snapback behavior in Fig. 9.48*b*. There are three important points: the trigger point (point of device turn-on), the holding point (point of device operation), and the failure point (point of device being damaged). The voltage at the trigger point, called the *trigger voltage* V_{t1}, must be located within a window. The lower bound of the window is the operating voltage at the pin to which the ESD protection device is connected, and the higher bound is the maximum voltage the pin can tolerate without damaging the

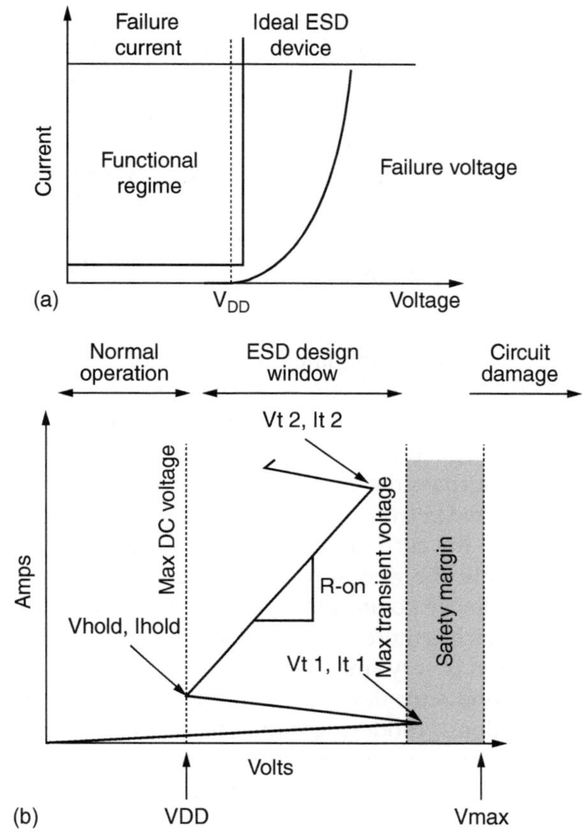

Figure 9.48 *I-V* curve of (*a*) nonsnapback and (*b*) snapback ESD protection devices.

core circuit. The values of these two boundaries depend strongly on the type of core circuit (digital versus analog, low-voltage versus high-voltage, etc.). The snapback reduces the voltage drop and thus the power dissipation on the protection device, hence resulting in an increase in the robustness of the device. The voltage at the holding point, called the *holding voltage* V_h, can influence the failure point. The smaller the holding voltage, the higher is the failure current It2 (current at the failure point). From this perspective, it is advantageous to reduce the V_h of the protection device. However, having a too small V_h can induce latch-up unless V_h is larger than the operating voltage at the pin. As such, an ideal protection device would possess a holding voltage slightly larger than the operating voltage at the pin. Moving upward along the *I-V* curve in Fig. 9.48*b*, the curve slope represents the on-state resistance, which should be minimized to reduce the likelihood of the voltage being larger than the upper-bound voltage discussed earlier, which results in ESD-induced damage to the core circuit. Finally, a high failure current It2 is desirable because it describes robustness or the protection capability of the ESD protection device. As mentioned earlier, It2 is a strong function of V_h. In addition, a higher robustness can be achieved by increasing the device size, but at the expenses of increasing Si consumption and increasing parasitic capacitance associated with the ESD protection device. All the concepts just discussed also apply to the nonsnapback device, except that the trigger voltage is the same as the holding voltage for such devices.

Correlation between the holding point (i.e., holding voltage and holding current) and the latch-up needs to be elaborated. Two approaches can be used to eliminate the latch-up. The first is to make sure that the ESD protection device's holding voltage is larger than the operating voltage at the pin. The second is to make the device's holding current surpass the current available at the pin. For the supply clamp connected between V_{dd} and V_{ss}, the second approach would not work because the current available at V_{dd} is normally very large. Consequently, only the first approach can be implemented. For an ESD protection device connected to an I/O pin, on the other hand, the second approach is more desirable because the current level available at the I/O pin is very small.

Three semiconductor devices are frequently used to realize ESD protection solutions: diodes, grounded-gate *n*-channel MOSFETs (ggNMOSs), and silicon-controlled rectifiers (SCRs). Comparison of the *I-V* curves of SCRs, ggNMOSs, and diodes is given in Fig. 9.49.

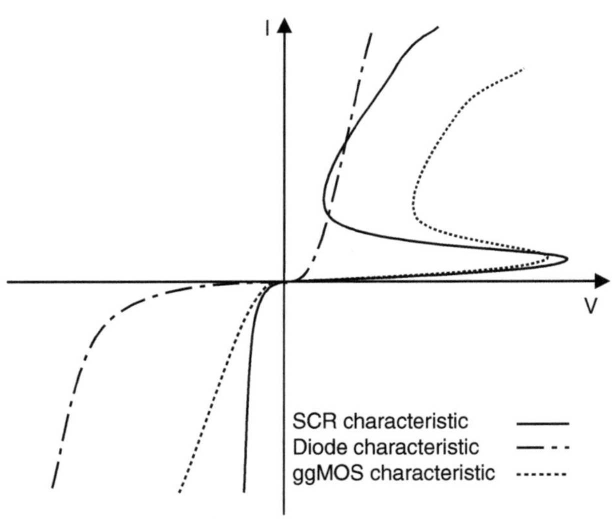

FIGURE 9.49 *I-V* curves of diodes, ggNMOSs, and SCRs.

Clearly, the diode is a nonsnapback-type device, whereas the SCR and ggNMOS are snapback-type devices. All devices can be triggered and conduct current in both the positive and negative directions. For the diode in the forward direction, the trigger and holding voltages are identical, about 0.7 V. This is quite good for the design of ESD protection for low-voltage ICs, because the trigger voltage required for such an application is relatively low. But sometimes a few diodes need to be connected in series to increase the trigger voltage (i.e., trigger voltage is the number of diodes $\times$ 0.7 V). This, however, has the disadvantage of consuming a large die area and increasing the parasitic resistance at the pin. A diode also can trigger and conduct current in the breakdown region in the negative direction. The trigger voltage in operation is relatively large, so it is potentially useful for ESD protection of high-voltage ICs. But since the holding voltage is the same as the trigger voltage (no snapback), the diode in this operation has a very low robustness. As a result, the diode is normally called a *one-directional device* and is suitable for use in the positive ESD operation but not the negative ESD operation. The ggNMOS and SCR, on the other hand, have snapback behavior in the positive direction, and they exhibit a forward-biased diode characteristic and trigger at -0.7 V in the negative direction. Thus these two devices can provide relatively robust bidirectional ESD protection.

Voltage level is commonly used to gauge ESD protection capability. For the HBM, the maximum ESD protection a device can provide, in terms of voltage, is the failure current multiplied by the human-body resistance of 1500 Ω. For other ESD models, the correlation between failure current and voltage level is yet to be clearly established. The *I-V* curve of ESD protection devices needs to be obtained from the TLP tester rather than the conventional curve tracer.

In an effort to determine the suitability of the devices for ESD application, eight different devices (shown in Fig. 9.50) were fabricated in a 0.18-μm CMOS technology.[68] Four of these devices are diodes, one is a ggNMOS, and three are SCRs. The areas, parasitic capacitances, and maximum HBM voltages of these devices are listed in Table 9.6. Also included in the table is the figure of merit (FOM) = HBM voltage/(area $\times$ capacitance). The higher the FOM, the more suitable it is for ESD application. The results suggest that the ggNMOS is the least suited and the SCR is the most attractive. This FOM cannot be considered as the only ESD design indicator because other factors such as the trigger voltage and holding voltage, as well as the ease of processing, also should be taken into consideration.

The following subsections will cover the design of ESD protection solutions based on SCRs and diodes.

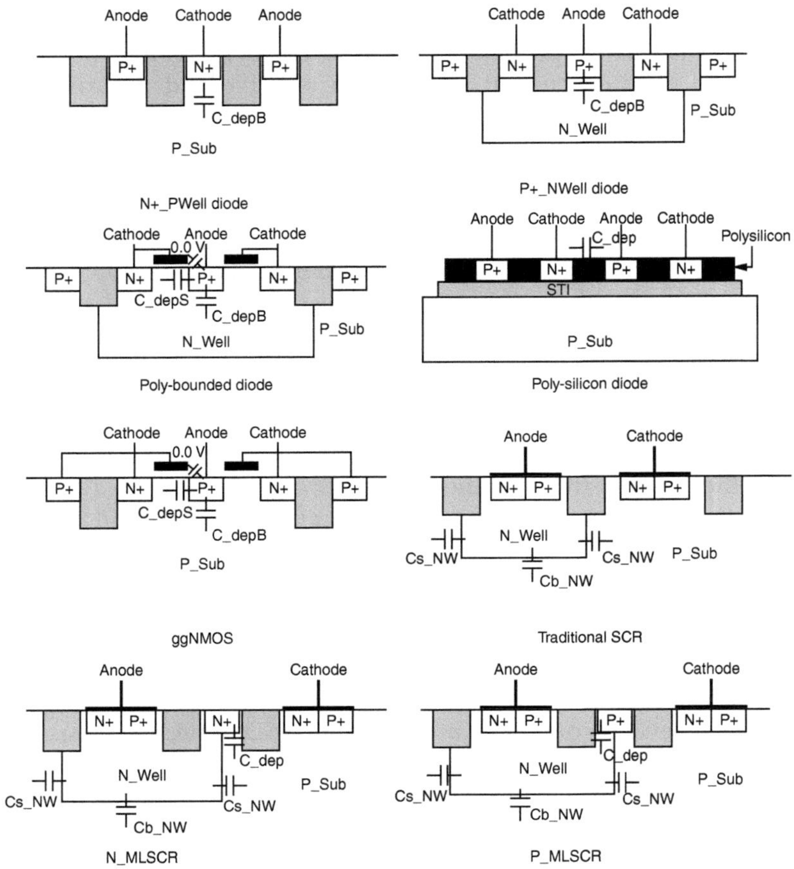

FIGURE 9.50 Schematics of eight different devices for ESD applications.

Device	Area (μm²)	Capacitance C_{ESD} (pF)	V_{HBM} per unit Area (V/μm²)	Figure of Merit V_{HBM}/Area × C_{ESD}
N^+_*P*-well diode	1815.47	0.3794	4.610376	12
P^+_*N*-well diode	1813.424	0.3825	4.532861	12
Poly-bounded diode	1838.545	0.8091	4.617782	6
Poly-silicon diode	1780.695	0.22	2.493409	11
ggNMOS	1341.294	0.7895	1.722218	2
Traditional SCR	1861.5	0.2226	8.058018	36
*N*_MLSCR	2505.63	0.3295	5.986518	18
*P*_MLSCR	2505.63	0.3553	5.986518	17

TABLE 9.6 ESD Performance of the Eight Devices Considered in Fig. 9.50

9.4.1 Electrostatic-Discharge Design Based on Silicon-Controlled Rectifiers

As discussed in the preceding section, the SCR is a highly suitable device for ESD protection applications. Figure 9.51*a* shows a schematic of a typical SCR structure consisting four main regions of p^+, n, p, and n^+. Thus an SCR is a coupling of $n/p/n$ and $p/n/p$ BJTs. Two more regions, n^+ next to p^+ and p^+ next to n^+, are called the *well-tie regions* because they are the contact regions for the *N*-well and *P*-well regions. The SCR equivalent circuit is given in Fig. 9.51*b*. In a typical ESD protection scheme, the anode is connected to the pin, and the cathode is connected to the ground. During normal operation, the voltage at the pin is too low to turn on the two BJTs, and hence the SCR is off. When a positive-polarity ESD event takes place, the very large voltage associated with the ESD can cause the base-collector junction of the $p/n/p$ BJT to break down. This gives rise to a current passing through the $p/n/p$ BJT and the well-tie resistor connected between the $n/p/n$ BJT. When the voltage drop across this resistance reaches 0.7 V, the n/p/n BJT is turned on, and the SCR is triggered. Once the BJTs are turned on, the voltage drop between the anode and the cathode is suddenly reduced, causing the *I-V* curve to snap back. When subjected to a negative-polarity ESD event, the SCR behaves like a forward bias and triggers at -0.7 V. This bidirectional current conduction provides an excellent means for protecting the

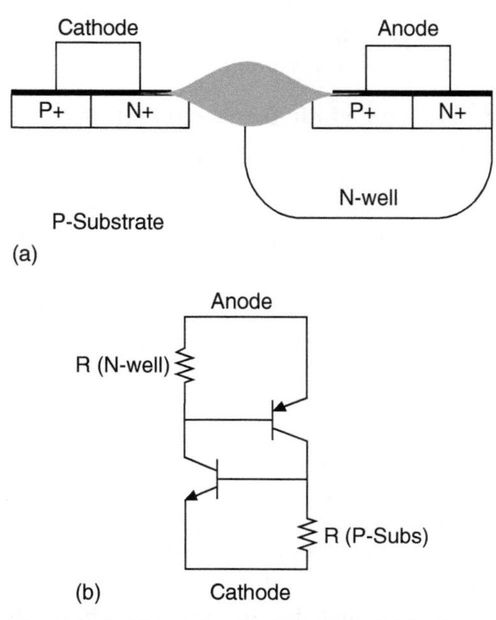

FIGURE 9.51 (*a*) Schematic of an SCR and (*b*) equivalent circuit of an SCR.

core circuit against both that the positive- and negative-polarity ESD events, provided that the positive and negative trigger voltages are designed outside the operating-voltage range at the pin. As such, the operating voltage at the pin cannot go below zero (i.e., negative operating voltages); otherwise, the SCR will be triggered during normal operation.

The trigger voltage in an SCR is determined by the reverse blocking junction.[69] There are three junctions going from the anode to cathode of SCR, as shown in Fig. 9.51*a*, when the device is subject to a positive ESD (anode is positive-biased and cathode is grounded): the forward-biased p^+/n, the reverse-biased n/p, and the p/n^+ forward-biased junctions. The two forward-biased junctions have a relatively small and negligible voltage drop of 0.7 V, and the reversed-biased junction (i.e., reverse blocking junction) can be turned on only when the applied voltage exceeds its breakdown voltage. Thus the SCR trigger voltage is approximately the breakdown voltage of the reverse blocking junction. Since the p/n junction breakdown voltage is directly proportional to the impurity doping densities on both sides of the junction, one can choose a junction configuration available in a particular processing to come up with a desirable trigger voltage. When the SCR is subject to a negative ESD (anode is revised-biased and cathode is grounded), there are several forward-biased diodes in parallel embedded in the SCR, and the trigger voltage under this condition is -0.7 V independent of the junction configuration used. The holding voltage of the SCR is affected largely by the current gain of the two BJTs in the SCR; the smaller the current gain, the larger is the BJT voltage drop, and thus the larger is the SCR holding voltage.

Input/Output-Pin Electrostatic-Discharge Protection

Since the current level available at the I/O pin in most ICs is very low, the design of ESD protection device at the I/O pin becomes relatively easy. This is so because the latch-up essentially is not an issue as long as the holding current of the protection device is higher than the operating current at the pin, which is normally the case for the great majority of protection devices. Thus the main ESD design requirements here are minimizing the holding voltage and ensuring that the trigger voltage is located within the window sandwiched between the operating voltage at the pin (lower bound) and the maximum voltage the pin can tolerate (upper bound). Small footprint, on-state resistance, and parasitic capacitance are also important considerations.

Let us consider a case where the I/O pin operating voltage V_{pin} is in a range of a positive voltage $V(+)$ and a negative voltage $V(-)$, as shown by the dotted lines in Fig. 9.52. Also, the maximum voltages the core circuit can tolerate in the positive and negative directions are denoted by the dashed line in Fig. 9.52. Therefore, the trigger voltages of the I/O ESD protection device in the positive and

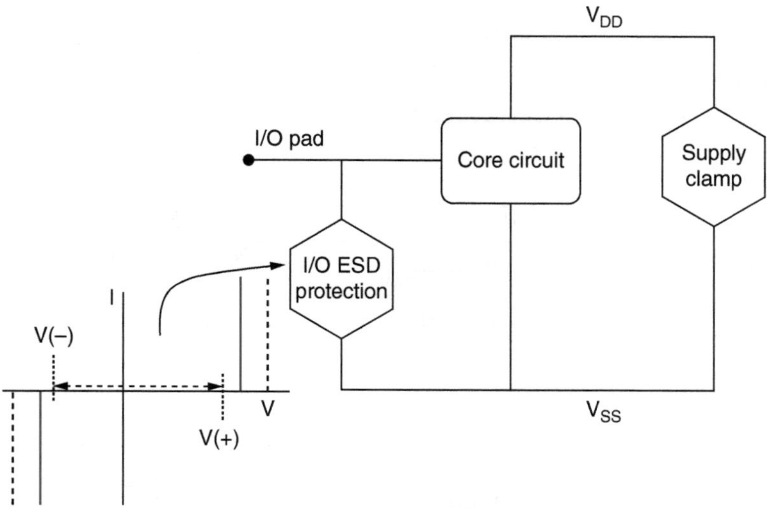

FIGURE 9.52 Trigger voltage (solid lines) required at an I/O pin having a variable operating voltage (dotted lines) and maximum tolerable voltage (dashed lines).

negative directions, denoted by the solid lines in Fig. 9.52, must be located between the dotted and dashed lines.

Figure 9.53 shows the structure of an SCR that can be used for the I/O ESD application with $V(-) = 0$. As mentioned earlier, the SCR trigger voltage in the positive ESD direction is governed by the configuration of the reverse blocking junction. For a typical biCMOS technology, nine different junction configurations by combining the different doping density pair can be obtained, as shown in Fig. 9.54. The corresponding trigger voltages for these junction configurations

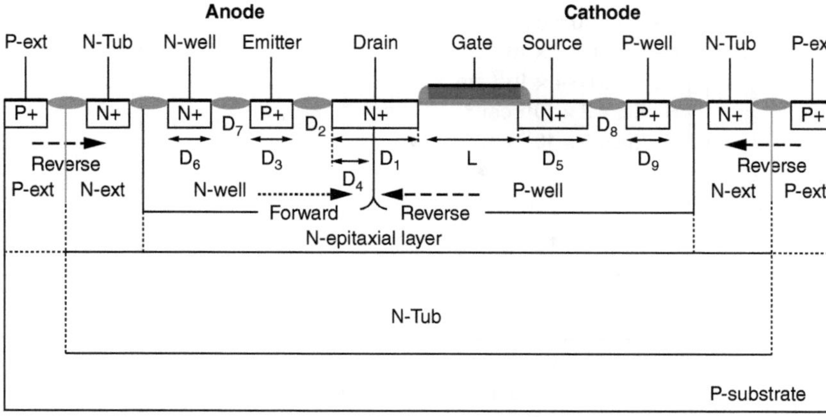

FIGURE 9.53 An SCR structure in the biCMOS technology.

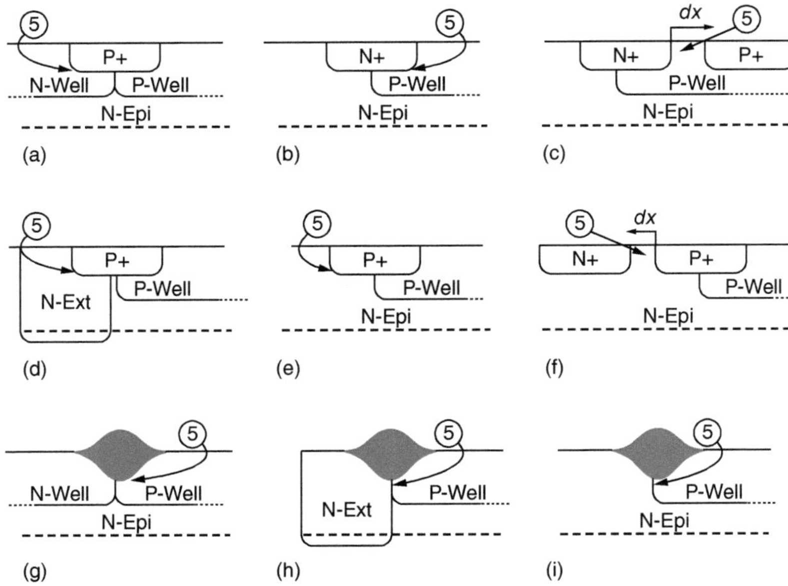

FIGURE 9.54 Nine different reverse-blocking-junction configurations available in the biCMOS technology.

are given in Fig. 9.55. For the negative ESD, the trigger voltage of the SCR is –0.7 V, which is below the operating voltage $V(-) = 0$ regardless of the junction configuration. When $V(-)$ dips below zero, an SCR with a trigger voltage more negative than –0.7 V in the negative direction is needed. This can be done by placing two SCRs in parallel,

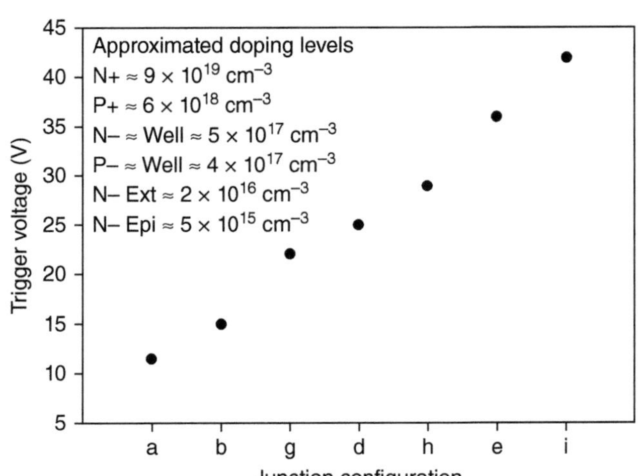

FIGURE 9.55 Trigger voltages of the nine junction configurations given in Fig. 9.54.

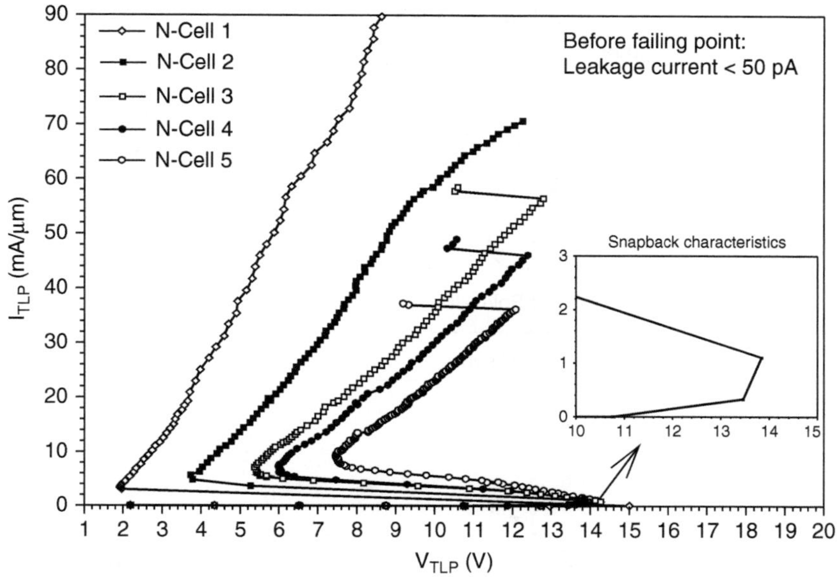

Figure 9.56 TLP *I-V* curves of five SCRs having different holding voltages.

one with anode to the pad and cathode to the ground and the other with the opposite polarity. Such a structure, called a *dual-polarity SCR,* can generate snapback *I-V* curves in the position and negative directions with the trigger voltage satisfying the requirement of exceeding the operating voltages.[70]

The holding voltage is influenced largely by the current gain of the BJTs in the SCR. Increasing the base width (D_1 in Fig. 9.53) decreases the BJT current gain and consequently increases the holding voltage. Figure 9.56 shows the TLP *I-V* curves of five SCRs having five different D_1 values ranging from 1.6 to 8.0 μm (i.e., cells 1 to 5 have D_1 = 1.6, 3.2, 5.1, 7.3, and 8.0 μm, respectively). Obviously, the holding voltage increases with increasing D_1 and the failure current drops when the holding voltage is increased. For the I/O pin protection under consideration, the best SCR to choose is the one with the smallest holding voltage because there is no latch-up issue at the pin, and the smallest holding voltage yields the highest failure current.

It is much more difficult to design an SCR for low-voltage ESD protection that demands a relatively small trigger voltage of, say, less than 5 V. An approach for achieving this is to use the punch-through mechanism rather than junction breakdown, to control the triggering of the SCR. The two depletion regions associated with the two N^+/P-well junctions in the SCR in Fig. 9.57*a* are likely to touch each other by making the length D_6 small, thus resulting in the punch-through, high current conduction, and SCR triggering. Four different SCRs

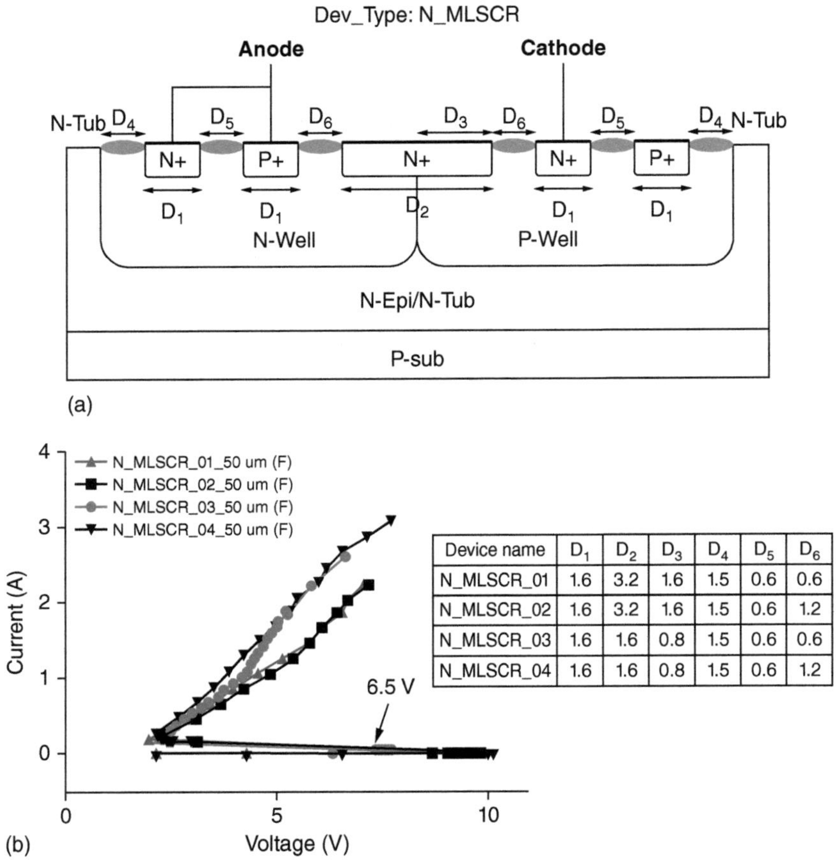

Device name	D_1	D_2	D_3	D_4	D_5	D_6
N_MLSCR_01	1.6	3.2	1.6	1.5	0.6	0.6
N_MLSCR_02	1.6	3.2	1.6	1.5	0.6	1.2
N_MLSCR_03	1.6	1.6	0.8	1.5	0.6	0.6
N_MLSCR_04	1.6	1.6	0.8	1.5	0.6	1.2

FIGURE 9.57 (*a*) Structure of an SCR with a reduced trigger voltage using the punch-through mechanism and (*b*) dimensions and TLP current-voltage curves of four such SCRs.

have been fabricated, and their device dimensions and TLP *I-V* curves are given in Fig. 9.57*b*. Two different D_6 values, 0.6 and 1.2 μm, are considered, and the devices with $D_6 = 0.6$ μm yield a relatively low trigger voltage of about 6.5 V. On the other hand, a larger trigger voltage of about 11 V is found in devices with $D_6 = 1.2$ μm because the trigger mechanism in such devices is governed by the breakdown of the N^+/P-well junction. A main drawback of the punch-through SCR is the substantially large leakage current, which is on the order of microamperes as opposed to nanoamperes for its junction breakdown counterpart.

Another approach to reducing the trigger voltage is to add external trigger circuitry to the SCR, as shown in Fig. 9.58. Either the circuitry G_1 or G_2 can be used to make the SCR trigger at a lower

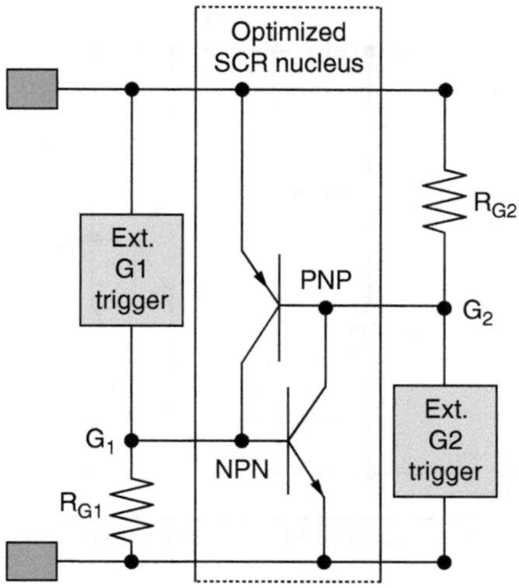

FIGURE 9.58 Schematic of an SCR with external trigger circuitry to reduce the trigger voltage.

trigger voltage.[71] The concept is that G_1 can be triggered before the SCR, and once it is triggered, the current flowing through R_{g1} will turn on the *NPN* BJT and the SCR. If G_2 is used, then G_2 can be triggered at a small voltage, which then turns on the *PNP* BJT and the SCR. The equivalent circuits of several diode-assisted trigger SCRs are given in Fig. 9.59. Each structure has four junction diodes connected to the SCR and a trigger voltage of about 2.8 V. The low trigger voltage results because when the voltage at the top pin exceeds 2.8 V, the four diode conducts, the emitter-base voltage of the *n*/*p*/*n* or *p*/*n*/*p* BJT reaches 0.7 V, and the SCR triggers. These external diodes,

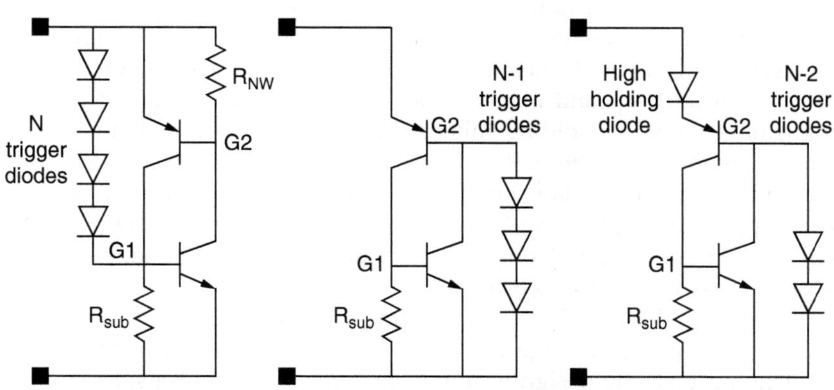

FIGURE 9.59 Three different an SCRs with diode-string external trigger-assisted circuitries.

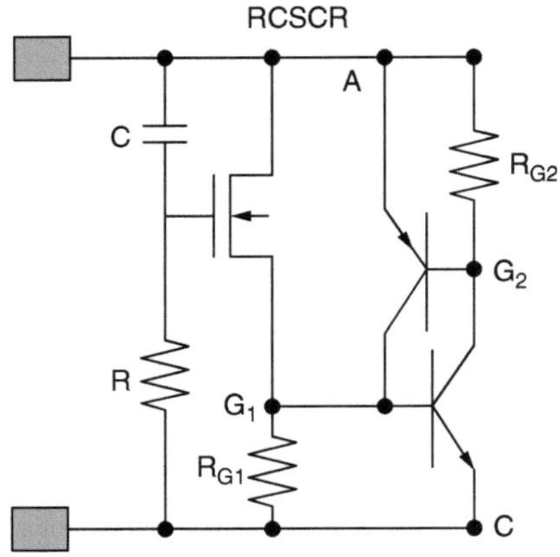

FIGURE 9.60 SCR with *RC*-coupled MOS external trigger-assisted circuitry.

unfortunately, give rises to a relatively large leakage current when the SCR is in the off state. The *RC*-MOS-assisted trigger SCR given in Fig. 9.60 provides a remedy to this problem, but such a design may have the issue of false triggering owing to the voltage fluctuation at the pin.

Power-Supply-Pin Electrostatic-Discharge Protection

Latch-up is a major concern for the supply clamp connected between the V_{dd} and V_{ss} pins because the current available at V_{dd} is extremely high. This, together with the relatively low holding voltage, makes the SCR latch-up-prone when it is used as a power clamp.

Consider a case of $V_{dd} = 5$ V and $V_{ss} = 0$, and the maximum voltage the V_{dd} pin can tolerate is 16 V. The procedures for designing the trigger and holding voltages discussed in Figs. 9.53 through 9.56 still apply here, but the junction configuration and the value of D_3 are different. This time, a junction configuration that yields a trigger voltage between 5 and 16 V in the positive ESD direction needs to be selected. To prevent latch-up and at the same time maintain a satisfactory robustness, the SCR's holding voltage must exceed 5 V, but not much larger than that. Thus the SCR with $D_1 = 5.1$ μm (i.e., cell 3 in Fig. 9.56) would be a good candidate. Figure 9.61 shows the TLP *I-V* results of two supply clamps, *SC*-1 and *SC*-2. Both devices satisfy the trigger-voltage requirement, but only *SC*-2, having a holding voltage of about 6 V, is immune to latch-up.

Achieving the very high SCR holding voltage required for ESD protection of a high-voltage IC has been proven to be much more challenging than the design examples just given. Let us consider a high-voltage IC with two I/O pins and a supply voltage of 40 V, as shown in Fig. 9.62. Here both the trigger and holding voltages must be larger than 40 V, in addition to the requirements of a compact size

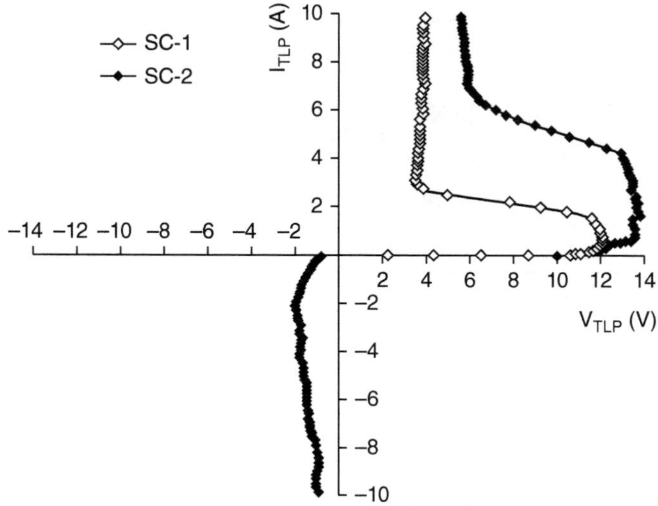

FIGURE 9.61 TLP *I-V* curves of two different supply clamps *SC*-1 and *SC*-2.

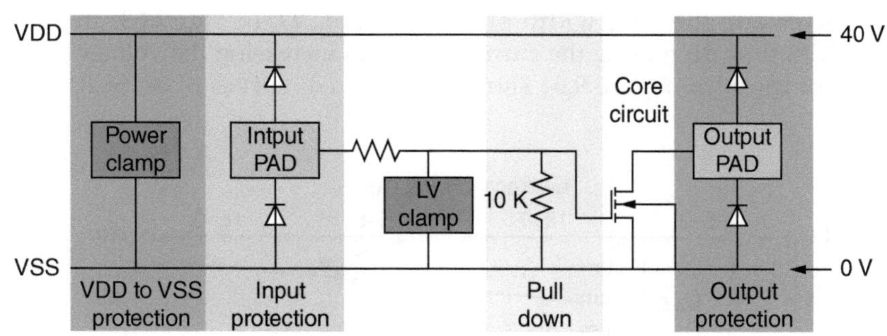

FIGURE 9.62 Schematic of a high-voltage IC with an ESD protection structure.

and good robustness. Obviously, a simple but not optimal scheme would be using a reverse-biased diode (i.e., anode tied to the supply pin and cathode tied to ground), and the trigger voltage (or breakdown voltage, which is the same as the holding voltage) of the diode is designed to be larger than 40 V. But the robustness of the diode would be very poor unless the diode size was very large.

Alternatively, it would be more desirable to use an SCR because it would be more robust than its diode counterpart. A high SCR trigger voltage is not difficult, but an SCR typically possesses a relatively low holding voltage (i.e., less than 5 V) which can induce latch-up when used in high-voltage ESD applications.[72] Figure 9.63 shows an SCR structure aimed at rendering a no-snapback *I-V* characteristic. To this end, a relatively low trigger voltage and a relatively high holding

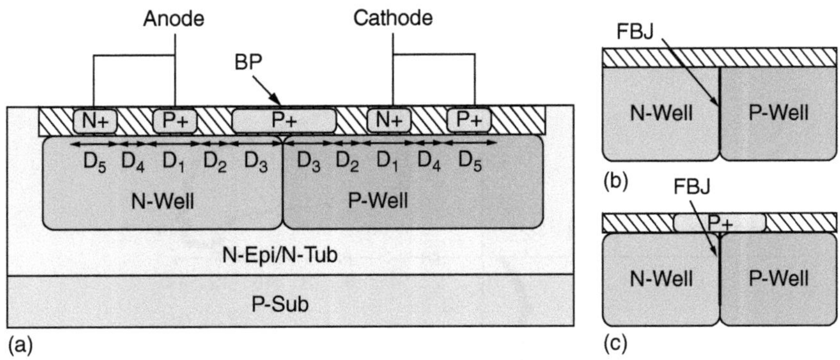

FIGURE 9.63 Single SCR cell with a small snapback behavior.

voltage are needed. The former can be achieved by adding a p^+ region between the *N*-well and *P*-well junction. Thus the SCR's trigger voltage is reduced (to about 12 V) because the reverse blocking junction of the SCR is the p^+/*N*-well junction rather than the *N*-well/*P*-well junction. For the latter, increasing the SCR holding voltage is possible by increasing the base width of the BJTs (i.e., D_3 in Fig. 9.63) in the SCR, thus decreasing the current gain and increasing the voltage drop of the BJTs. Figure 9.64 shows the TLP *I-V* curves of an SCR

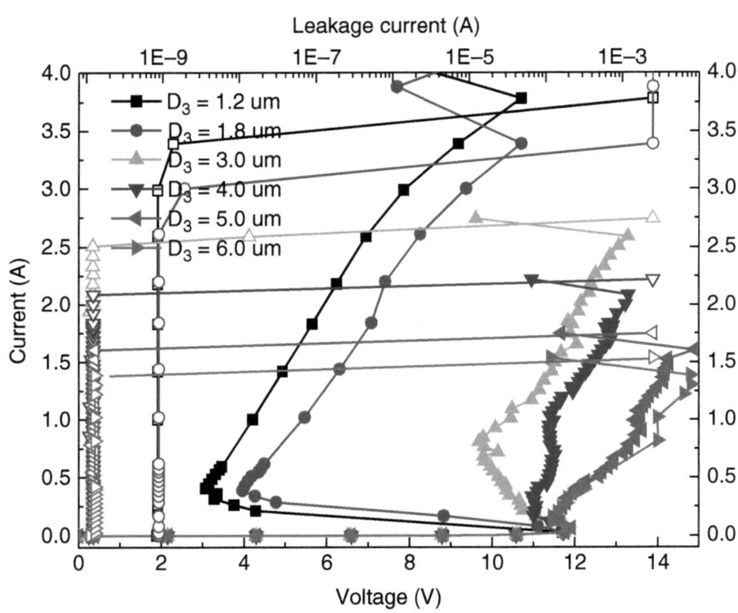

FIGURE 9.64 TLP *I-V* curves of the SCR cell in Fig. 9.63 having different D_3 values.

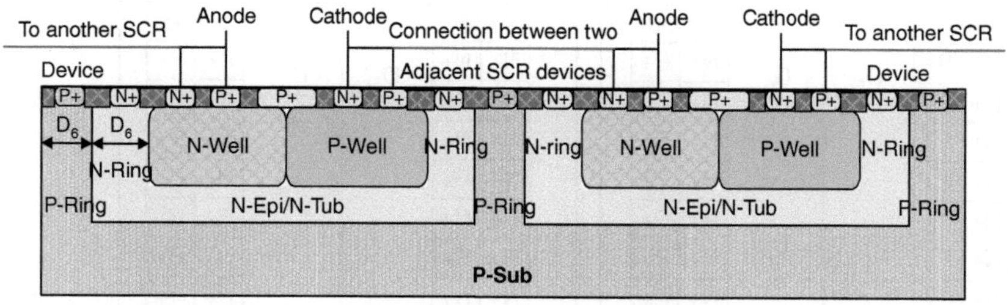

Figure 9.65 Stacking of two SCR cells including the isolation regions using the *n*- and *p*-guard rings.

having a width of 50 μm and different D_3 values. The single SCR cell with $D_3 = 4.0$ μm exhibits a nearly no-snapback behavior with a holding voltage of about 11 V.

Customized trigger/holding voltage therefore can be made possible by stacking a different number of SCR cells. Figure 9.65 shows cross-sectional views of two stacking SCRs with the *n*- and *p*-guard rings used to fully isolate the two cells. The TLP *I-V* curves of several stacking SCR structures are illustrated in Fig. 9.66. Clearly, a desirable trigger/holding voltage can be achieved by stacking a proper number of SCR cells. The stacking structures also exhibit a satisfactory robustness. For instance, the four-cell structure with a relative small size of 4 × 50 μm (stacking number × cell width) possesses a failure

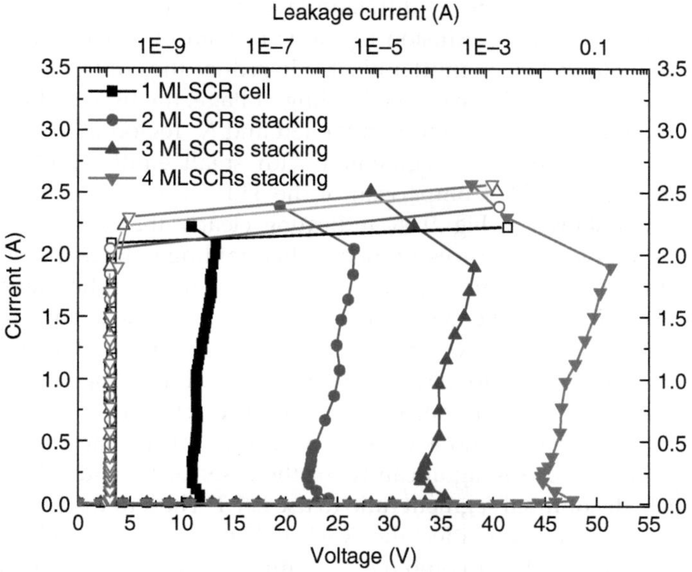

Figure 9.66 TLP *I-V* curves of SCRs having different stacking numbers.

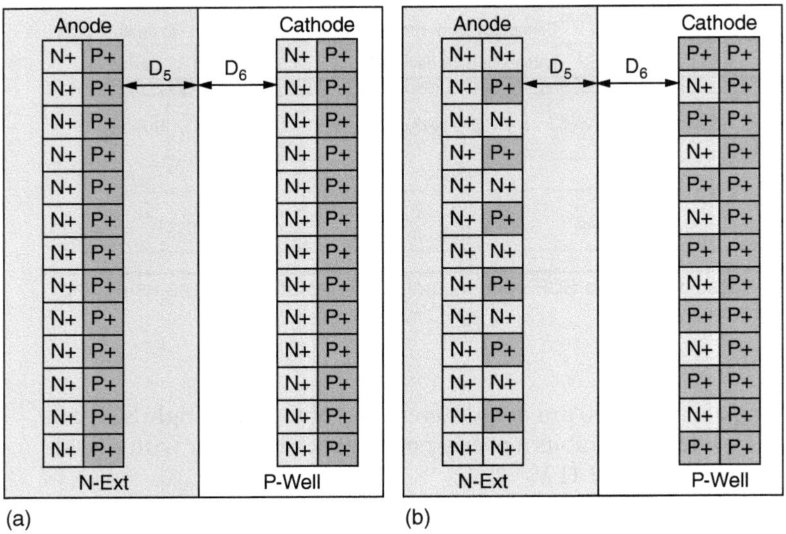

Figure 9.67 Top views of SCRs with (*a*) conventional strip N^+ and P^+ topology (left) and (*b*) novel segmented N^+ and P^+ topology (right).

current of about 2.3 A. As such, the combination of flexibility, ease of design, compactness, and robustness makes such an approach attractive for high-voltage ESD applications.

Sometimes it is possible to realize a high holding voltage with engineering of the SCR's junction layout.[73] Figure 9.67*a* shows the topology of the N^+ and P^+ diffusion layout of a conventional SCR, where each N^+ or P^+ diffusion region is formed in a strip topology (i.e., a single and large emitter block). To reduce the emitter injection efficiency and thus increase the holding voltage, the two emitter diffusion regions (e.g., P^+ region on the left and N^+ region on the right) can be segmented (i.e., segmented with small emitter blocks) as shown in Fig. 9.67*b*. Comparison of the TLP *I-V* curves of the two devices is shown in Fig. 9.68, indicating clearly that the new segmented topology produces a much higher holding voltage (i.e., 40 V) than the conventional strip topology. The segment ratio (ratio of number of P^+ blocks versus number of N^+ blocks on the left and symmetry on the right) plays a role in the holding voltage, as shown in Fig. 9.69, but only up to the ratio of 1:1. The holding voltage increases when the segment ratio decreases from 3:1 to 1:1, but it stays about the same when the ratio is decreased beyond 1:1. Furthermore, the failure current drops significantly for the cases of 1:2 and 1:3 ratios. As such, the optimal segment ratio to use is 1:1. The lengths D_5 and D_6 (see Fig. 9.67) also can affect the SCR's ESD performance, and a relatively large D_5/D_6, in combination with the segmented emitter, is needed to ensure a high holding voltage, as illustrated in Fig. 9.70.

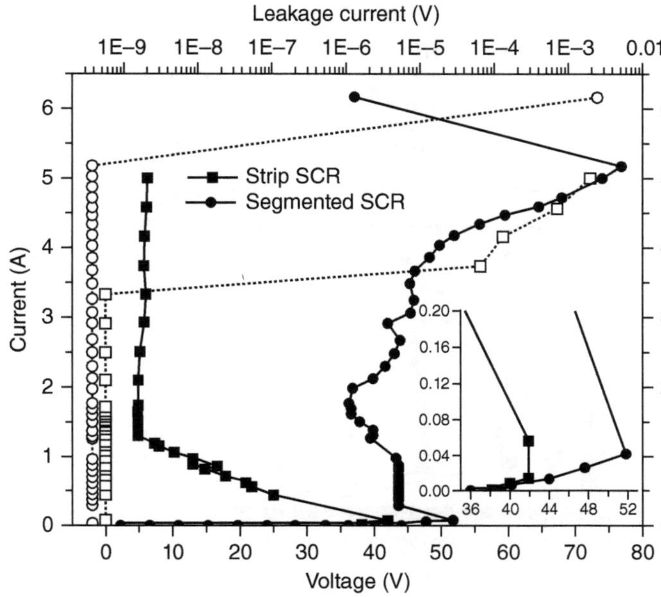

FIGURE 9.68 Comparison of TLP *I-V* curves of the conventional and segmented SCRs.

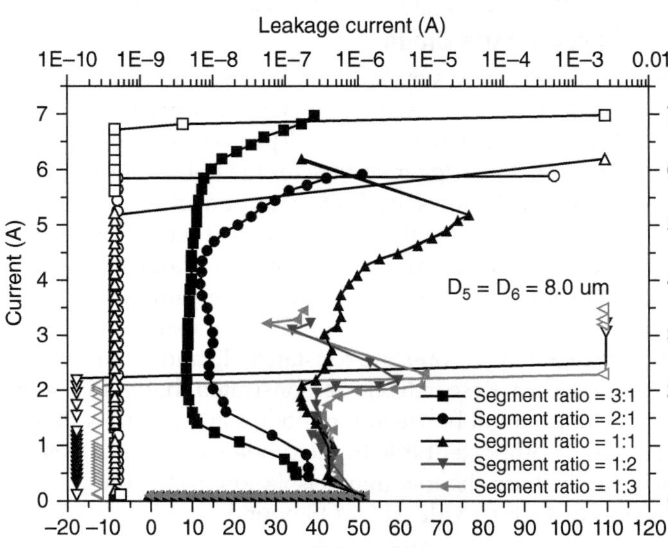

FIGURE 9.69 TLP *I-V* curves of segmented SCRs having five different segment ratios.

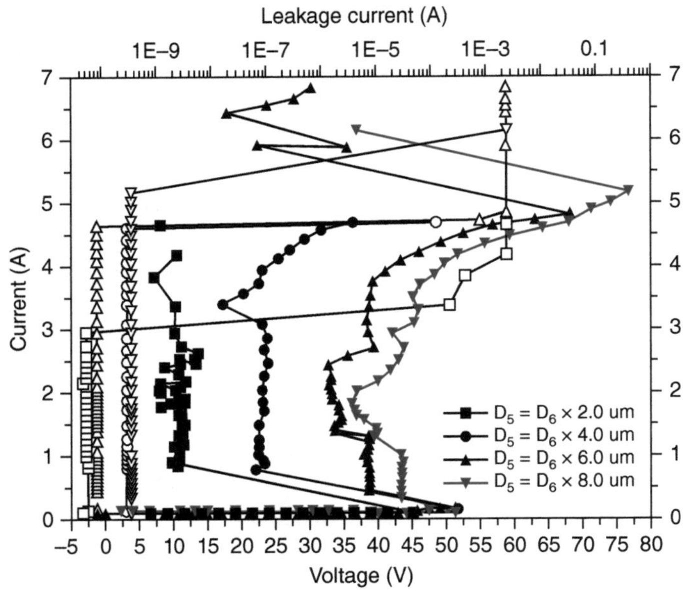

FIGURE 9.70 TLP I-V curves of segmented SCRs having a segment ratio of 1:1 and four different D_5/D_6 ratios.

9.4.2 Electrostatic-Discharge-Protection Design Based on the Diode

Because of its relatively simple structure and good perfomance, the junction diode is widely used in on-chip ESD protection applications.[74,75] This is particularly true for ESD protection of low-voltage ICs where a relatively low trigger voltage for the ESD protection device is required. Figure 9.71 shows the ESD protection scheme built based on diodes. In this protection approach, each I/O pin is connected with two back-to-back diodes, and several supply clamps are connected between the supply rail and the negative (or ground) rail.

Nonetheless, the relatively poor ESD robustness of the diode, commonly gaged by the on-state resistance R_{on} and failure current It2, is of major concern. Research has demonstrated that different technologies for the isolation between the diode's anode and cathode regions can play an important role in the diode's ESD robustness. These technologies include shallow trench isolation (STI), LOCal Oxidation of Silicon (LOCOS) oxide (LOCOS-bound), and polysilicon gate (poly-bound). It has been reported that for ESD-protection purposes, diodes with STI are inferior to those with other isolations.[76,77]

Diode Optimization

In this section we will first study and compare the ESD performances of the LOCOS- and poly-bound diodes fabricated in a BiCMOS

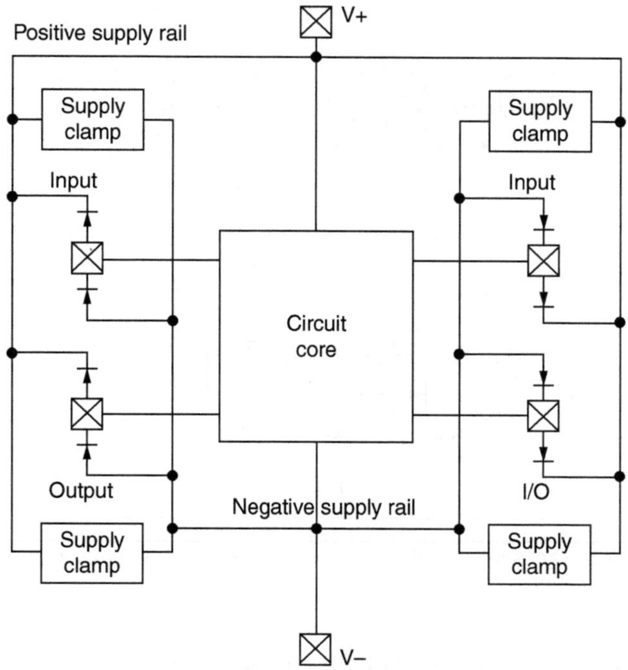

FIGURE 9.71 Schematic of ESD protection scheme constructured using two back-to-back diodes at each I/O pin and supply clamps conneected between the positive and negative supply pins.

technology. The better-performing device then will be investigated in more detail in an effort to identify an optimal diode structure for robust ESD protection applications.

Figure 9.72 shows cross sections of N^+/P_{well} LOCOS- and poly-bound diodes. The critical ESD figures of merit, namely, the failure current It2 and on-state resistance R_{on}, of the diodes depend on various factors, including the diode's dimensions, geometry, junction configuration, terminal connection, and metal pattern.[78,79] For the diode structure considered to have two anode regions and one cathode region (see Fig. 9.72), it has been found that the anode width W_a plays a marginal role in the diode's ESD performance.[80] Moreover, the crossing pattern was found to be the best metal layout for the diode's ESD robustness.[79] As such, the following will not account for the effect of W_a and will consider only diodes using the crossing metal pattern. Furthermore, unless noted otherwise, all measurements will be performed using pulses generated from Barth 4002 transmission-line-pulsing (TLP) tester with a pulse width of 100 ns and a rise time of 10 ns, a stress condition equivalent to the well-known HBM.

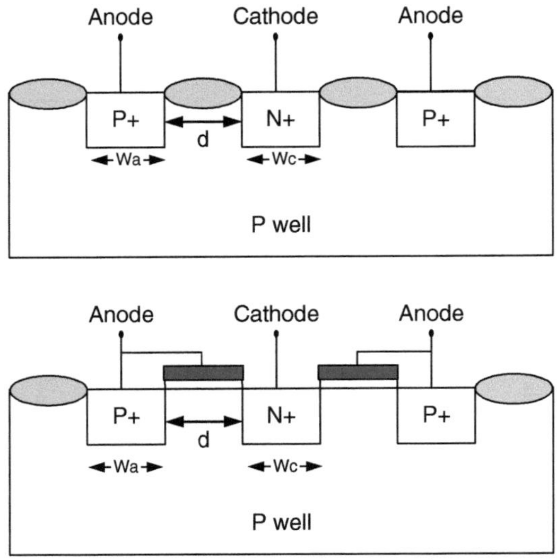

FIGURE 9.72 Cross section of N^+/P_{well} diodes with LOCOS (*top*) and polysilicon (*bottom*) isolations.

Figure 9.73 shows the TLP *I-V* characteristics of N^+/P_{well} and P^+/N_{well} LOCOS- and poly-bound diodes with the same dimensions and metal layout. The anode width W_a and cathode width W_c are both 1.6 μm. The diode length *L*, or the length of anode/cathode, is 40 μm. The anode-to-cathode distance (i.e., isolation width) *d* for both diode types is 2 μm. It can be seen the failure current It2 (i.e., the currents at which the *I-V* curves ended in Fig. 9.73) of the

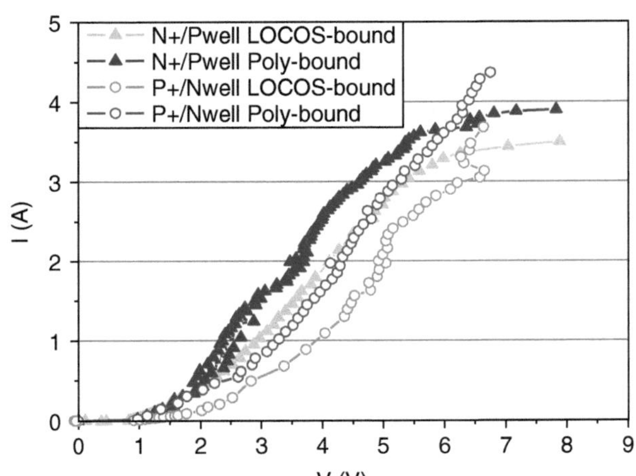

FIGURE 9.73 *I-V* characteristics of LOCOS- and poly-bound diodes.

poly-bound diode is higher than that of the LOCOS-bound diode for both the N^+/P_{well} and P^+/N_{well} junction configurations (i.e., 4.3 versus 3.7 A for a P^+/N_{well} and 3.9 versus 3.4 A for an N^+/P_{well}). Thus, the poly-bound diode possesses a better ESD current-carrying capability than the LOCOS-bound diode. The lower It2 in the LOCOS-bound diode is due mainly to the higher current density taking place in the bird-beak regions.

In addition to It2, the voltage drop or voltage clamping on a diode is also important to ESD applications, and a good voltage-clamping capability (i.e., producing a low voltage drop) is needed to minimize the possibility of ESD-induced core-circuit damage. Such a capability is directly related to the on-state resistance R_{on} of the diode because a lower on-state resistance leads to a smaller voltage drop after the diode is triggered by an ESD event. For the LOCOS-bound diode, the current between the anode and cathode must pass underneath the curved LOCOS oxide. For the poly-bound diode, on the other hand, since the polysilicon gate is flat and does not penetrating into the silicon, the current flows straight between the anode and cathode. As a result, the current path for the poly-bound diode is shorter than that for the LOCOS-bound diode and hence a smaller on-state resistance for the poly-bound diode. This reasoning is consistent with the R_{on} values given in Table 9.7 extracted from the TLP *I-V* curves in Fig. 9.73. Clearly, the poly-bound diode shows a better voltage-clamping capability than the LOCOS-bound diode for both the N^+/P_{well} and P^+/N_{well} junction configurations.

For a fast ESD transient event such as the CDM, the diode's turn-on speed becomes another key consideration. Figure 9.74 shows the voltage-versus-time waveforms of the N^+/P_{well} and P^+/N_{well} LOCOS- and poly-bound diodes subject to a very-fast TLP pulse with 2-ns duration and 15-V amplitude. The device's turn-on speed can be characterized from such a transient as the time it takes from the point where the voltage peaks to the point where the voltage becomes relatively constant. Clearly, for both the N^+/P_{well} and P^+/N_{well} junction configurations, the poly-bound diode possesses a faster turn-on speed than the LOCOS-bound diode. This stems mainly from the fact that the voltage overshoot is less prominent in the poly-bound diode.[7]

The preceding analysis has illustrated clearly that the poly-bound diode is superior to the LOCOS-bound diode for ESD protection

Isolation	N^+/P_{well}	P^+/N_{well}
LOCOS	1.25 Ω	1.4 Ω
Polysilicon	1.02 Ω	1.21 Ω

TABLE 9.7 On-State Resistances of LOCOS- and Poly-Bound Diodes.

applications. In an effort to design an optimal ESD diode, the study below will investigate the factors that can influence the ESD robustness of the poly-bound device. The optimization will focus on the failure current and on-state resistance of the N^+/P_{well} structure, but the approach applies generally to the P^+/N_{well} diode as well. Different device parameters will be varied, and their effects on the poly-bound diode's ESD performance will be studied based on TLP experimental data. In changing the parameters, a typical nominal value will be selected first, and an increment of about 50 percent of the nominal value will be used.

The first consideration is diode length L (i.e., length of anode/cathode), which can considerably influence the poly-bound diode's

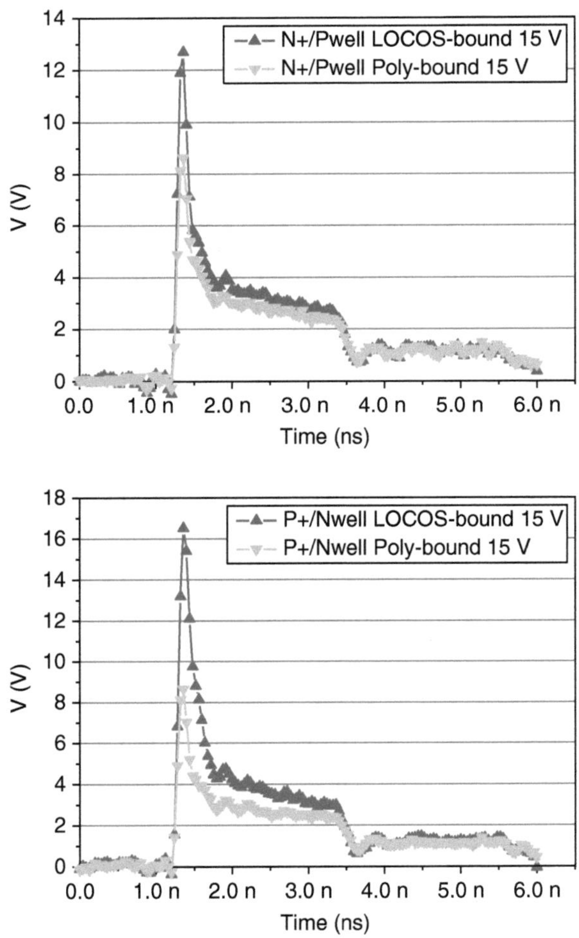

FIGURE 9.74 Transient voltage waveforms for N^+/P_{well} (*top*) and P^+/N_{well} (*bottom*) LOCOS- and poly-bound diodes subject to a very-fast TLP pulse with a 2-ns duration and 15-V amplitude.

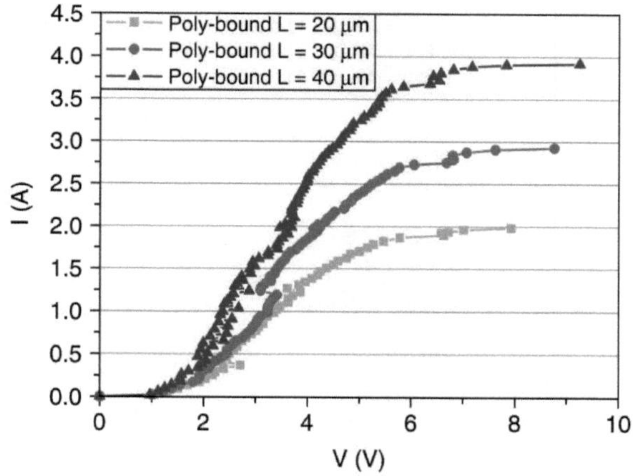

FIGURE 9.75 *I-V* characteristics of N^+/P_{well} poly-bound diodes having different lengths.

failure current It2 and on-state resistance R_{on}. It is expected that increasing L will increase It2 and decrease R_{on} owing to the enlarged device size. This is indeed the case, as evidenced from the trends demonstrated in Fig. 9.75 and Fig. 9.76. Figure 9.75 shows the *I-V* characteristics of N^+/P_{well} poly-bound diodes having an anode width W_a = 1.6 µm, cathode width W_c = 1.6 µm, polysilicon gate width d = 2 µm, and different diode lengths L of 20, 30, and 40 µm (i.e., 20 µm is the nominal value, and the increment is 50 percent of the nominal). It can be seen that It2 increases linearly with increasing diode length (i.e., It2 values are 1.98, 2.93, and 3.92 A, respectively). Figure 9.76 shows that

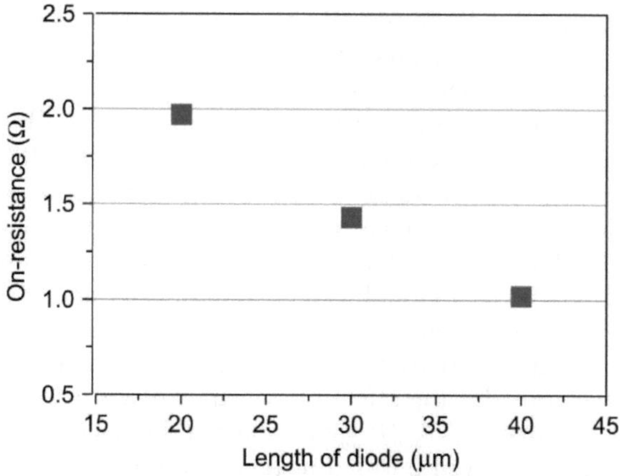

FIGURE 9.76 On-state resistances of N^+/P_{well} poly-bound diodes having different lengths.

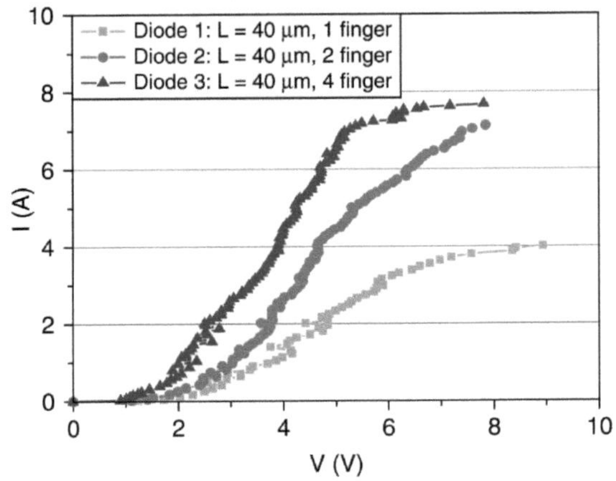

FIGURE 9.77 *I-V* characteristics of N^+/P_{well} poly-bound diodes having different finger numbers.

the on resistance R_{on} decreases with increasing diode length *L*. The drawback of the increased diode's size and parasitic associated with *L* increases will be accounted for later.

The ESD robustness is also a function of the number of anode/cathode fingers. Figure 9.77 shows the *I-V* characteristics of poly-bound diodes having the same diode length of 40 μm, but three different finger numbers. Diodes 3 and 1 have the highest and lowest It2 values and the smallest and largest R_{on} values, respectively, indicating that increasing the finger number offers an appealing means of enhancing the poly-bound diode's ESD robustness. Nonetheless, when the total area is fixed and the finger number is increased (i.e., the diode length is reduced with increasing finger number), it has been found that the diode actually becomes less robust owing to the fact that the maximum metal width allowed on each finger is inversely proportional to the finger number.

Changing the cathode width W_c also can affect the ESD performance of the poly-bound diode. Figure 9.78 shows that It2 increases from 2 to 3 A when W_c increases from 1.6 to 3.8 μm. The on-state resistance R_{on} decreases with increasing W_c, as can be seen in Fig. 9.79. The improved ESD robustness stems from the fact that increasing the cathode width gives rise to a reduced current density and thus a lower temperature in the region. The downside of increasing W_c, of course, is an increased device size and increased parasitic capacitance.

Next, we consider the effect of polysilicon width *d*. Figure 9.80 shows the *I-V* curves of N^+/P_{well} poly-bound diodes with polysilicon width *d* increasing from 2 to 7 μm. It can be seen that It2 is almost unchanged for the different *d* values. On the other hand, the larger polysilicon width gives rise to a longer current path from the anode

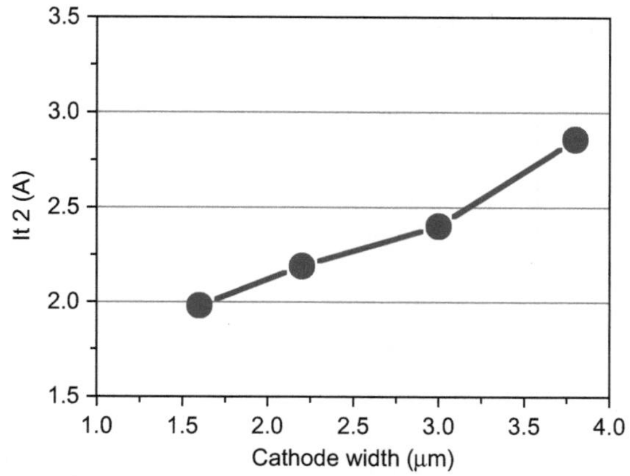

FIGURE 9.78 It2 of $n+/P$-well poly-bound diodes with different cathode widths.

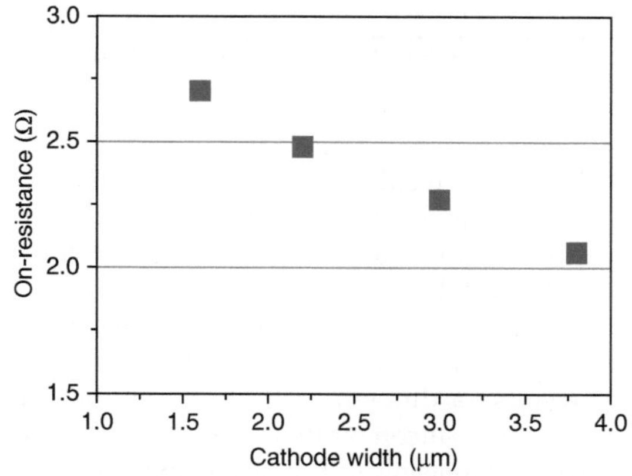

FIGURE 9.79 On-state resistances of N^+/P_{well} poly-bound diodes with different cathode widths

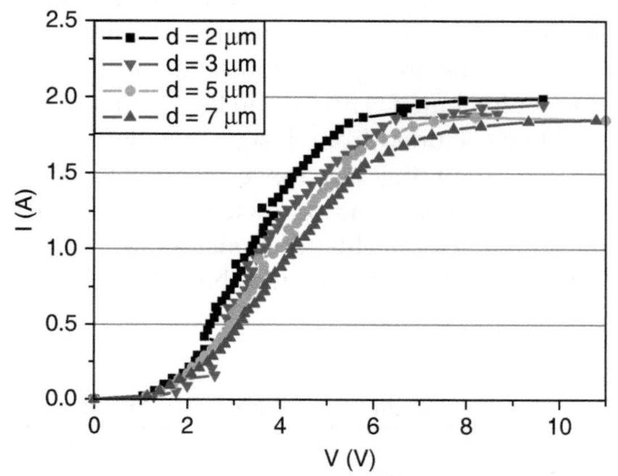

FIGURE 9.80 I-V characteristics of n^+/p_{well} poly-bound diodes with different polysilicon widths.

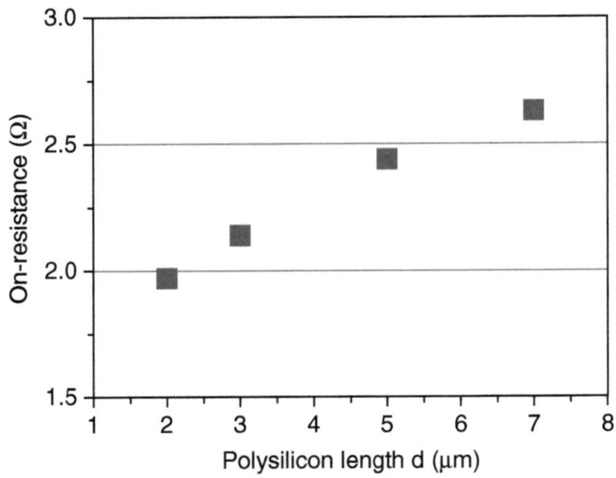

FIGURE 9.81 On-state resistances of N^+/P_{well} poly-bound diodes with different polysilicon widths.

to the cathode and thus results in a higher on-state resistance R_{on}, as evidenced from the data given in Fig. 9.81. Thus, the use of a relatively small *d* seems to be advantagous from the perspective of the diode's voltage-clamping capability.

However, the width of polysilicon cannot be reduced indefinitely and is subjected to a minimal value as explained below. Table 9.8 shows the It2 results for different polysilicon widths ranging from 1 to 7 μm. When the polysilicon widths are higher than 2 μm, It2 increases slightly with decreasing *d*. But when the polysilicon width is reduced further from 2 to 1 μm, It2 is decreased by 30 percent. This It2 reduction stems from metal failure, because the maximum width of the metal layer above the N^+ cathode region is mandated by the layout design rule. Having a very small *d* reduces the metal width, increases the current density in the metal, and consequently decreases It2. Junction punch-through is another factor limiting how small *d* value can be implemented. Thus, it can be concluded that the use of $d < 2$ μm should be excluded from design consideration.

As illustrated, changing the above-mentioned design parameters, that is, diode length *L*, cathode width W_c, polysilicon width *d*, and finger number, can influence considerably the poly-bound diode's It2 and R_{on} values. Moreover, these changes also alter the

D	1 μm	2 μm	3 μm	5 μm	7 μm
It2	1.37 A	1.98 A	1.95 A	1.90 A	1.85 A

TABLE 9.8 It2 of Poly-Bound Diodes with Different Polysilicon Widths

L(μm)	**20**	**30**	**40**	
Capacitance (fF)	38.72	58.08	77.44	
W_c (μm)	**1.6**	**2.2**	**3.0**	**3.8**
Capacitance (fF)	38.72	47.4	58.92	70.22
d(μm)	**2**	**3**	**5**	**7**
Capacitance (fF)	38.72	43	51.54	60.08
Finger numbers	**1**	**2**	**4**	
Capacitance (fF)	38.72	77.8	155.9	

TABLE 9.9 Parasitic Capacitances of Poly-Bound Diodes Having Different Design Parameters

device size and capacitance, two other important criteria related to the ESD device's compactness and parasitics. Table 9.9 shows the capacitances of the poly-bound diodes with different design parameters. It should be pointed out that the capacitance is a function not just of device size but also of junction perimeter length.

To account for the effects of changing these design parameters on ESD device robusteness, device size, and capacitance of the poly-bound diode in a more unified manner, we define the following two FOM for It2 and R_{on} for the purpose of determining quantitatively how much ESD robustness is enhanced versus an increased design parameter:

$$\text{FOM\_It2} = \frac{\Delta\text{It2}}{\Delta\text{device\_size} * \Delta\text{cap}} \tag{9.79}$$

$$\text{FOM\_}R_{on} = \frac{\Delta R_{on}}{\Delta\text{device\_size} * \Delta\text{cap}} \tag{9.80}$$

where ΔIt2, ΔR_{on}, Δdevice_size, and Δcap are the increase in It2, R_{on}, diode size, and parasitic capacitance, respectively, owing to the increase in a specific design parameter. These two FOMs can more effectively reflect the ESD robustness of the diode than It2 and R_{on} alone. The more positive the FOM_It2 and the more negative the FOM_R_{on} (i.e., R_{on} decreases with increasing design parameter), the more suitable is increasing the particular design parameter for ESD applications. Conversely, the more negative the FOM_It2 and the more positive the FOM_R_{on}, the more beneficial is decreasing the particular design parameter.

Fig. 9.82 compares FOM_It2 and FOM_R_{on} values obtained for the four different design parameters considered. The results suggest that increasing the cathode width is the best way to enhance the diode's ESD robustness, followed by increasing the diode length

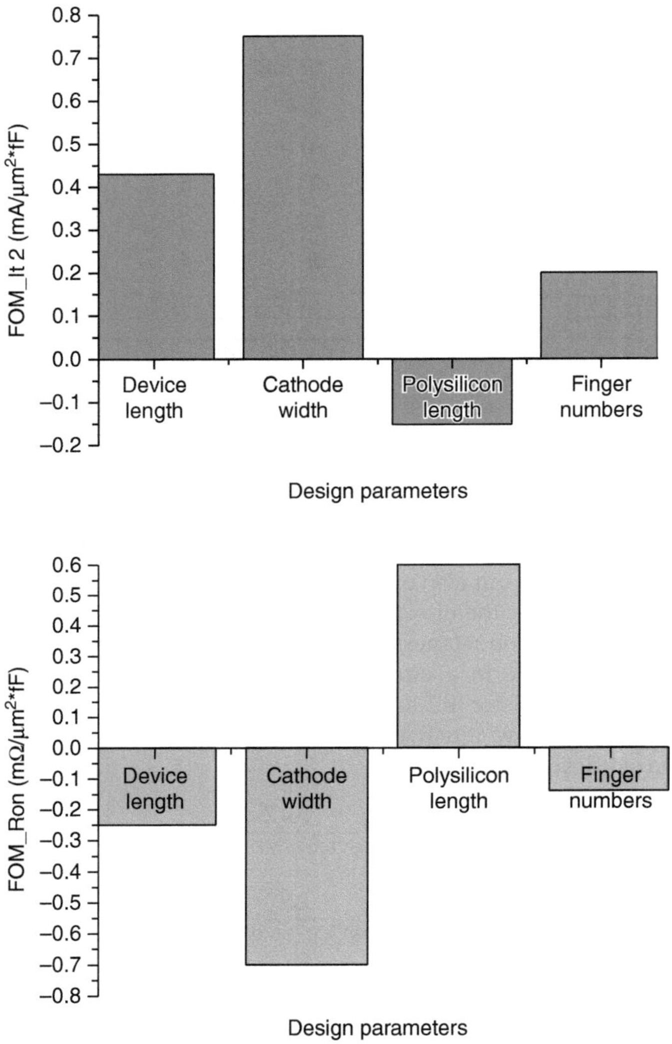

FIGURE 9.82 Figures of merit for It2 (*top*) and R_{on} (*bottom*) obtained for the four different design parameters.

and increasing finger number. On the other hand, decreasing the polysilicon width is advantageous from the viewpoint of ESD diode optimization.

Two more design parameters can affect the diode's ESD performance, but changing these parameters does not vary the device size and alters the capacitance only minimally. As such, these two parameters were not considered in the above-mentioned FOM_It2 and FOM_R_{on}. The first is the different ways in which the polysilicon gate

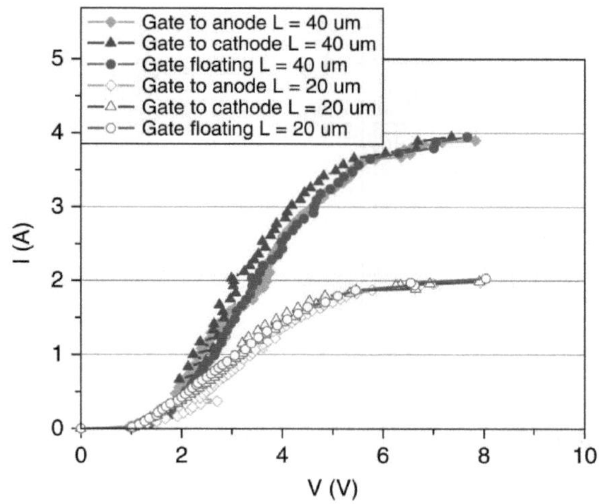

FIGURE 9.83 *I-V* characteristics of poly-bound diodes with gate-to-anode, gate-to-cathode, and floating-gate connections.

can be connected. The three possible ways are gate-to-anode, gate-to-cathode, and gate floating. Figure 9.83 shows the *I-V* characteristics of the poly-bound diode having the three different terminal connections and two different device lengths of 20 and 40 μm. As can be seen in the figure, It2 is insensitive to the types of terminal connections for both the diode lengths considered. However, allowing the polysilicon gate to float will be of concern because of the issue of gate potential uncertainty. When the polysilicon is connected to the cathode, gate-oxide breakdown can occur if a sufficiently high voltage is built between the polysilicon and *P*-well region. Thus, the best connection is to tie the polysilicon gate to the P^+ anode.

Lastly, we examine the effect of metal topology on the diode's ESD performance. Figure 9.84 shows N^+/P_{well} poly-bound diodes having the same area but five different metal topologies using the crossing pattern. In the figure, the grey region denotes the level 1 metal M_1 placed on the diode's diffusion regions, and the shaded region denotes the level 2 metal M_2 connected to the vias. For example, diode_A1C1 has one M_2 line connected to the anode and one M_2 line to the cathode. On the other hand, diode_A4C4 has four M_2 lines connected to the anode and four M_2 line connected to the cathode. The widths of each M_2 line in diode_A1C1 and diode_A4C4 are the widest and narrowest, respectively, among the five cases considered. Table 9.10 compares It2 and on-state resistance R_{on} obtained for the five different metal topologies. Diode_A4C4 has the highest It2 value because its current flow can use the most inward and outward paths. The on-state resistance does not benefit from having a large number of metal lines, however, and diode_A2C2 exhibits the smallest

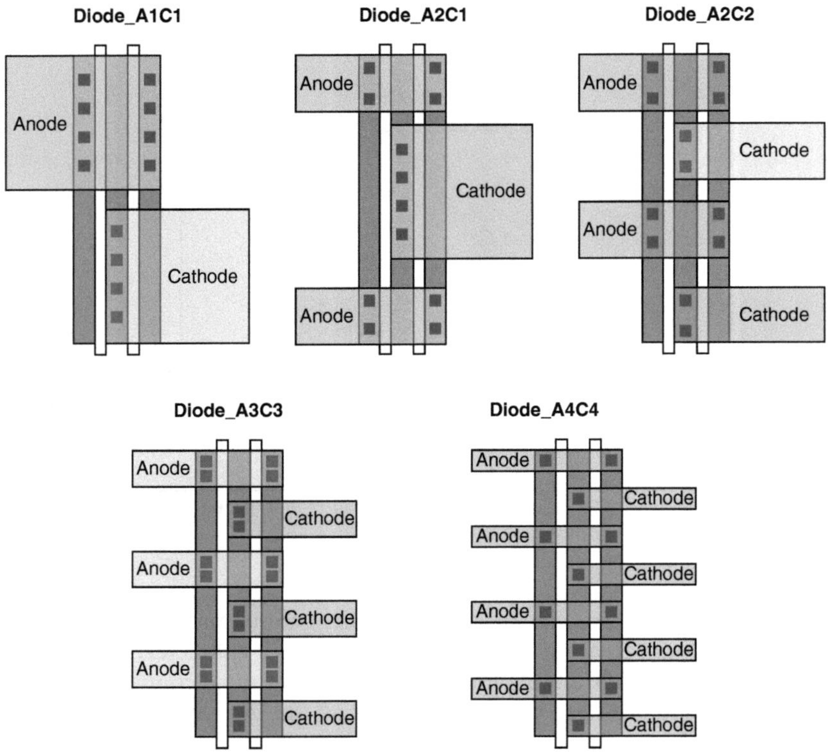

FIGURE 9.84 Poly-bound diodes having the same area but five different crossing metal topologies.

	A1C1	A2C1	A2C2	A3C3	A4C4
It2	3.05 A	3.52 A	3.49 A	3.55 A	3.56 A
R_{on}	1.29 Ω	1.57 Ω	1.16 Ω	1.19 Ω	1.27 Ω

TABLE 9.10 It2 and R_{on} values of Poly-Bound Diodes Having the Different Metal Topologies Shown in Fig. 9.84

R_{on} value among the five different topologies. Overall, topologies A2C2 and A3C3 seem to offer the best ESD performance.

9.4.3 Radio Frequency Optimization

Figure 9.85 shows a typical ESD protection structure consisting of two diodes at each I/O pin and a supply clamp. The purpose of the supply clamp is to limit the maximum differential voltage that can be attained between two power-supply pins (i.e., V_{dd} and V_{ss}). Consider a case in which the pin located on the left-hand side of the core circuit in Fig. 9.85 is subject to a positive-polarity ESD pulse and the

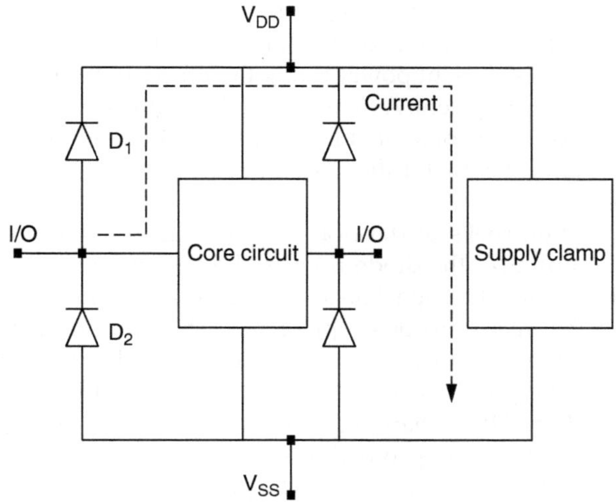

FIGURE 9.85 On-chip ESD protection structure including two diodes and a supply clamp. The dashed lines indicated the discharge path during the ESD event.

V_{SS} terminal is connected to ground. The supply clamp provides a low-impedance discharge path; that is, the turn-on voltage of the supply clamp is smaller than the breakdown voltage of diode D_2. Thus the discharge path (denoted by dashed lines in Fig. 9.85) will be from the pin through the forward biased diode D_1 to the V_{dd} pin and then down to the V_{ss} pin via the supply clamp. It has been demonstrated that with a supply clamp set at the correct voltage level, all the various discharge paths involve only forward-biased diodes and the supply clamp.[81] Diodes that are not operated in the reverse breakdown region do not require a large area to dissipate the heat. Thus, the diodes can be made relatively small and their capacitance can be minimized.[82,83]

To prevent diode D_2 from being operated in the breakdown region, the voltage across this device has to be smaller than its breakdown voltage V_{br}. This implies that during the ESD event, the sum of the maximum voltage drops V_{D1} and V_{clamp} across the forward-biased diode D_1 and across the supply clamp, respectively, must be smaller than V_{br}. Thus,

$$V_{br} > \max(V_{D1} + V_{clamp}) \tag{9.81}$$

For simplicity, the voltage drop on the metal interconnect has been neglected in Eq. (9.81).

The ESD protection structure shown in Fig. 9.85 has been widely used and works well for protection of modern microchips. However, as the operating frequency of the microchip continues to increase, the parasitic capacitance associated with the diodes starts to impose a problem for radio frequency (RF) operations. This cannot be resolved simply by reducing the area of the diodes because a small-size diode may not have a sufficient power rating to withstand the heat dissipation during the ESD event. As a result, diode optimization for minimal parasitic capacitance for a rated ESD peak current handling capability becomes a critical issue for the design of ESD protection structure for RF microchips.

This section seeks to develop a systematic and comprehensive method to optimize the diode used in the ESD protection structure for minimal parasitic capacitance and, thus, for optimal RF performance of the microchip being protected. The optimization will be focused on the dimensions and doping of the diode, taking into account the effects of reverse breakdown, heat dissipation, and different levels of the HBM charging voltage. A case study to implement and optimize the ESD protection structure based on an existing 0.13-μm CMOS technology also will be presented. Silvaco device and mix-mode simulations,[42] calibrated against experimental data, will serve as the basis for development of the optimization method.

Calibration of Device/Mix-Mode Simulator In this section, an Atlas device and mix-mode simulator will be calibrated based on the measured characteristics. In principle, the Atlas device simulator solves the drift-diffusion model, coupled with the heat equation for the semiconductor device specified. In the case of devices having external components, such as resistors and capacitors, mix-mode simulation is required. The following physical mechanisms have been taking into account: Lombardi surface mobility, field-dependent mobility, Shockley-Read-Hall recombination with concentration-dependent lifetime, Auger recombination, bandgap narrowing, and impact ionization.

A typical $P^+/N_{well}/N^+$ diode structure is considered and shown in Fig. 9.86, where the vertical dimension is defined by the p^+-region

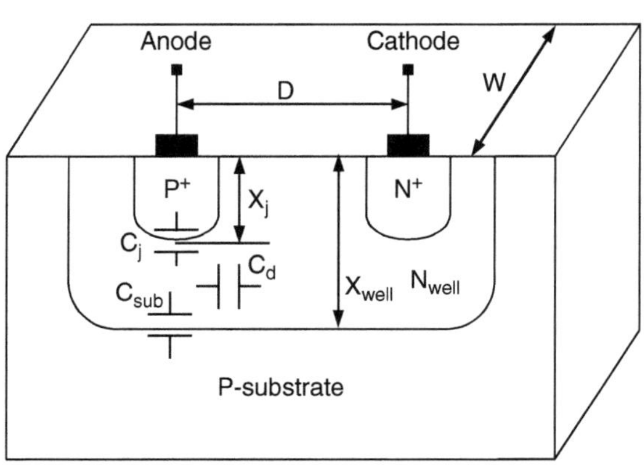

Figure 9.86 Schematic of the diode structure considered.

depth X_j and the N-well depth X_{well}. The lateral and third dimensions are defined by the distance D between the anode and the cathode and the diode width W, respectively. The capacitance associated with the diode consists primarily of the junction capacitance C_j in the p^+/N well space-charge region, the junction capacitance C_{sub} in the N-well/substrate space-charge region, and the diffusion capacitance C_d in the quasi-neutral N-well region (see Fig. 9.86). For the ESD protection structure under study, C_j and C_{sub} are much more important than C_d because the diodes are reverse-biased under normal RF operations.

The Atlas device simulator was calibrated so that good simulation accuracy would be ensured. The simulated and measured dc (current versus voltage), transient (voltage versus time), and ac (capacitance versus voltage) results are shown in Fig. 9.87 for a typical diode of $D = 1$ μm, $W = 50$ μm, peak acceptor doping density $N_A = 10^{20}$ cm^{-3}, peak donor doping density $N_D = 10^{17}$ cm^{-3}, $X_j = 0.15$ μm, and $X_{well} = 2$ μm. Note that the diode considered is an isolated device, so there is no parasitic capacitance associated with the metal interconnect. The dc, transient, and ac measurements were carried out using the parameter analyzer (HP4156B), TLP tester (Barth 4002), and impedance analyzer (E4991A), respectively.

The supply clamp used here, shown in Fig. 9.88, consists of a capacitor of 2 pF, a resistor of 100 kΩ, and an NMOSFET having an oxide thickness of $T_{ox} = 18$ nm, width-to-length ratio of $W/L = 160/1.2$ μm/μm, and substrate doping density of $p_{sub} = 5 \times 10^{16}$ cm^{-3}. Since the supply clamp is disconnected from the I/O pads during regular operation (i.e., the two diodes are reversed-biased by V_{dd} and V_{ss}), the parasitic capacitance associated with the supply clamp is not of main concern, and thus a relatively large MOS device is used to provide a sufficiently high power rating and to withstand the heat dissipation associated with the ESD pulse. The voltage across the supply clamp during the ESD transients is relatively constant, and the clamp voltage is about 8.0 V. Note that the focus here is on determining the minimal diode size for a targeted ESD protection level. The supply clamp also can be optimized to further reduce the diode size, but this will not be considered here. Unless otherwise noted, an HBM charging voltage of 1 kV will be considered to illustrate diode optimization.

Design Approach The design approach is based on the following three steps. The first step is to look for a diode structure that will generate the smallest parasitic capacitance yet satisfy the requirement of reverse breakdown voltage. As discussed earlier, the combined voltage drop across the forward-biased diode D_1 and the supply clamp cannot be larger than the breakdown voltage of the reverse-biased diode D_2 (see Fig. 9.85). This puts an important constraint on the diode design and thus needs to be considered. To this end, the focus will be on the vertical dimension (junction depth X_j and thickness of

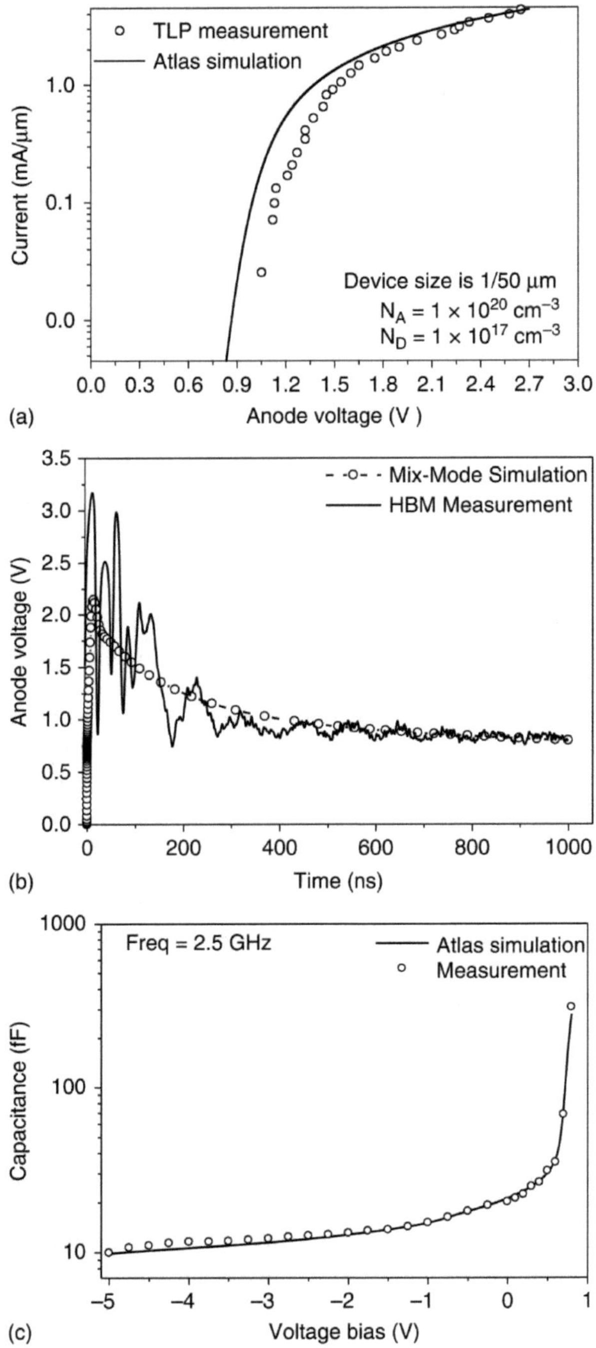

Figure 9.87 Measured and simulated (*a*) current-voltage characteristics, (*b*) anode voltage versus time characteristics, and (*c*) capacitance-voltage characteristics of a typical RF diode.

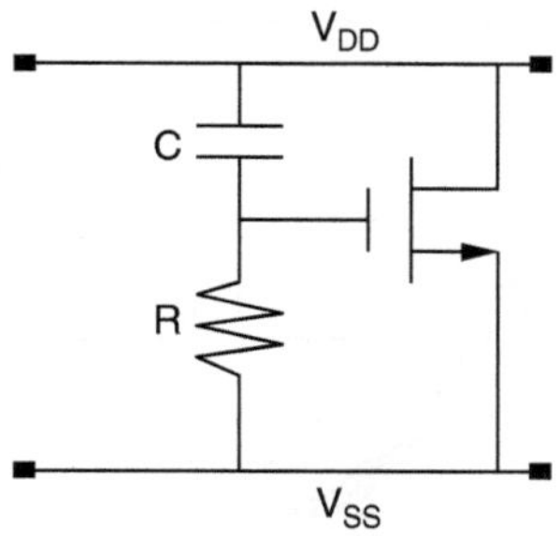

Figure 9.88 ESD supply clamp considered in the ESD optimization.

N-well X_{well}) and the peak doping density N_D in the N-well because these two most strongly influence the breakdown voltage. Since the diodes are normally fabricated concurrently with the MOSFET in the supply clamp, X_j and X_{well} of the diode should be in the vicinity of those for the drain/source regions and the N-well of the MOSFET, respectively. For the 1.2-μm MOSFET considered, the mean values of X_j and X_{well} are 0.3 and 2.0 μm, respectively.[5,6] Thus, when optimizing X_j and X_{well}, we will consider a range centering on these mean values.

Typical doping densities of $N_A = 10^{20}$ cm^{-3} in the p^+-type region and 5×10^{16} cm^{-3} in the p-type substrate will be used. These two parameters are not critical to the requirement of the breakdown voltage or the parasitic capacitance of the diode. The doping profiles in p^+ and N-well regions are assumed Gaussian, and the thickness of the p-type substrate is the wafer thickness.

The second step is to determine the diode width/lateral dimension (i.e., diode width and distance between the anode and the cathode contacts) that will result in minimal parasitic capacitance yet satisfy the requirement of heat dissipation (i.e., power rating).

The last step is global optimization based on the requirements of both the breakdown voltage and the power rating so that an optimal diode structure with minimal parasitic capacitance can be realized.

It is important to point out that the above-mentioned design methodology applies only to the case where there is flexibility in the fabrication process to allow for altering/modifying the doping densities and junction depths in the diode.

Optimization Based on Breakdown Voltage Figure 9.89 shows the simulated diode breakdown voltages versus the N-well doping density N_D for a typical range of three different X_{well}. Clearly, the breakdown voltage depends strongly on N_D but is insensitive to X_{well}. Since the optimal diode structure is not known at this time, we will estimate for now that the voltage drop across the forward-biased diode is 4 V. This voltage was obtained from the current-voltage characteristics of a typical diode, together with knowledge of a typical peak current passing through the supply clamp. Note that the on resistance of the diode, which is a function of the doping density, will affect the

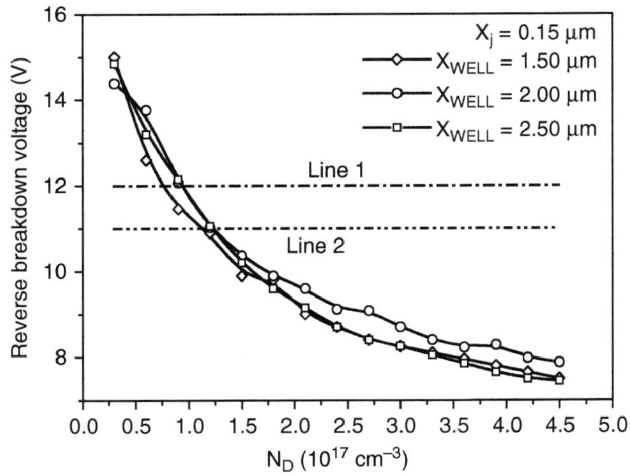

FIGURE 9.89 Reverse breakdown voltage versus the *N*-well doping density simulated for three different *N*-well thicknesses.

voltage across the forward-biased diode. The approximated voltage drop, combined with the maximum clamp voltage in the supply clamp of 8 V, leads to a design criterion that the diode must have a breakdown voltage of at least 12 V (indicated as line 1 in Fig. 9.89). As such, all the N_D and X_{well} values above line 1 satisfy the breakdown-voltage requirement. The farther away the points are above line 1, the smaller is the N_D, and thus the smaller is the parasitic capacitance.[7] However, a smaller N_D will result in a lower critical temperature (details given in the next section) and thereby a lower power rating, a larger diode area needed for heat dissipation, and thus a larger parasitic capacitance. This is an important trade off in the design of diodes for minimal parasitic capacitance. As will be demonstrated later, the effect of heat dissipation outweighs the effect of doping density on parasitic capacitance. Consequently, the points of intercept (i.e., interceptions of the curves and line 1) are the best possible choices for the optimal diode structure. Since X_{well} does not strongly influence the breakdown voltage, nor is it a strong function of the parasitic capacitance, we select a medium X_{well} of 2.0 µm.

Figure 9.90 shows the reverse breakdown voltages for $X_{well} = 2$ µm and a typical range of five different X_j values. The five points of intercept between the curves and line 1 are N_D/X_j (cm^{-3}/µm) = $6 \times 10^{16}/0.10$, $9 \times 10^{16}/0.15$, $1.15 \times 10^{17}/0.30$, $1.45 \times 10^{17}/0.45$, and $1.6 \times 10^{17}/0.60$. As mentioned earlier, a smaller N_D gives rise to a smaller power rating and thus a larger required lateral dimension, which increases the parasitic capacitance. For this reason, the intercept points turn out to be the best possible choices rather than the points above line 1. Since the parasitic capacitances decrease with decreasing N_D,[46] the point of

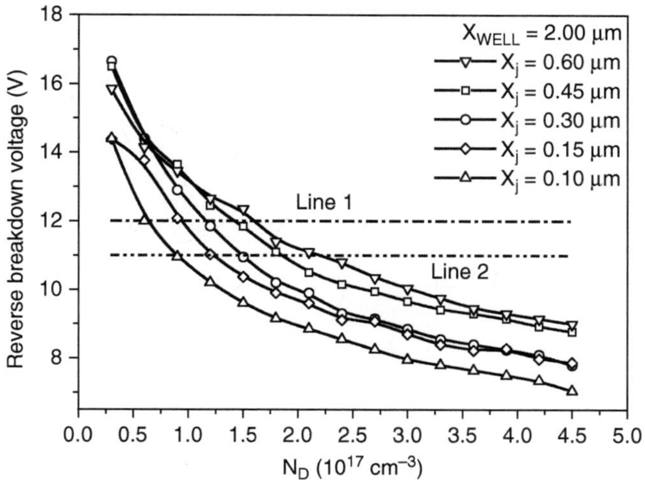

FIGURE 9.90 Reverse breakdown voltage versus the *N*-well doping density simulated for five different p^+-region thicknesses.

$N_D/X_j = 6 \times 10^{16}/0.1$ seems to be the best among the five intercept points. Nevertheless, as will be shown in the next section, this is not the case when the power-rating requirement is taking into account.

Optimization Based on the Power Rating Having identified the best possible vertical dimensions and doping densities in the *N*-well, we can now focus on optimization of width/lateral dimension based on the requirement of heat dissipation in the diode. Such a dimension is defined by $D \times W$, which is called the *diode area A*. Here we will first use the intercept point of $N_D/X_j = 9 \times 10^{16}/0.15$ to demonstrate the requirement of power rating.

It has been suggested that the diode will likely be destroyed when the diode temperature T is beyond a critical temperature T_{crit}.[84] This critical temperature T_{crit} is defined as the temperature for which the intrinsic free-carrier density n_i is equal to the doping density near the hot spot in the semiconductor device. Figure 9.91 shows the 2D temperature contours in the diode simulated by Atlas. It can be seen that the hot spot is located on the surface of the *N*-well near the cathode, and the doping density at the hot spot is N_D. Thus, T_{crit} can be calculated from the following equations:[84]

$$n_i = N_D = n_{i0} \cdot \left(\frac{T_{crit}}{T_0}\right)^{3/2} \cdot \exp\left(\frac{E_g}{2k_B T_{crit}} - \frac{E_{g0}}{2k_B T_0}\right) \tag{9.82}$$

$$E_g = 1.1255 - \frac{4.73 \times 10^{-4} \cdot T_{crit}^{\ 2}}{T_{crit} + 636} \tag{9.83}$$

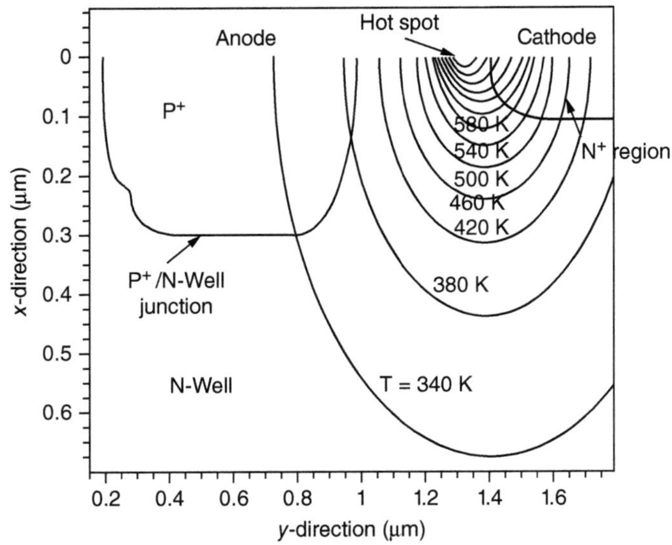

FIGURE 9.91 Simulated 2-D temperature contours in the diode, with the hot spot indicated.

Here $T_o = 300$ K is the ambient temperature, E_g is the energy bandgap, $E_{g0} = E_g(T = T_0)$, $n_{i0} = n_i(T = T_0)$, and k_B is the Boltzmann constant. For $N_D = 9 \times 10^{16}$ cm^{-3} considered, T_{crit} is 795 K.

Figure 9.92*a* compares the temperature distributions in a diode having $N_D = 9 \times 10^{16}$ cm^{-3} and $X_j = 0.15$ µm for four different diode areas. Figure 9.92*b* shows the corresponding peak temperatures versus the diode area with the constraint of $T_{crit} = 795$ K included. The intercept point indicates a required minimum diode area of 120 µm^2. Repeating the same procedure, the minimum diode areas for the other four intercept points of $N_D/X_j = 6 \times 10^{16}/0.10, 1.15 \times 10^{17}/0.30, 1.45 \times 10^{17}/0.45$, and $1.6 \times 10^{17}/0.60$ are found to be 180, 105, 85, and 80 µm^2 respectively.

Global Optimization The preceding analysis has indicated that not only the reverse breakdown but also the heat dissipation needs to be taken into account in designing the diode structure for minimal capacitance, and five best possible diode structures (N_D/X_j/area (cm^{-3}/µm/µm^2) = $6 \times 10^{16}/0.10/180$, $9 \times 10^{16}/0.15/120$, $1.15 \times 10^{17}/0.30/105$, $1.45 \times 10^{17}/0.45/85$, and $1.6 \times 10^{17}/0.60/80$) have been identified. Figure 9.93 compares the capacitances simulated for these five diode structures. Clearly, the diode structure with $X_{well} = 2.0$ µm, $X_j = 0.15$ µm, $N_D = 9 \times 10^{16}$ cm^{-3}, and area = 120 µm^2 is the optimal structure with the smallest parasitic capacitance.

We can now go back and fine-tune the limit of the diode breakdown voltage, which was estimated at 12 V. Mix-mode simulation of the optimized diode and supply clamp yields a combined voltage drop of 11 V across them. Repeating the above-mentioned design

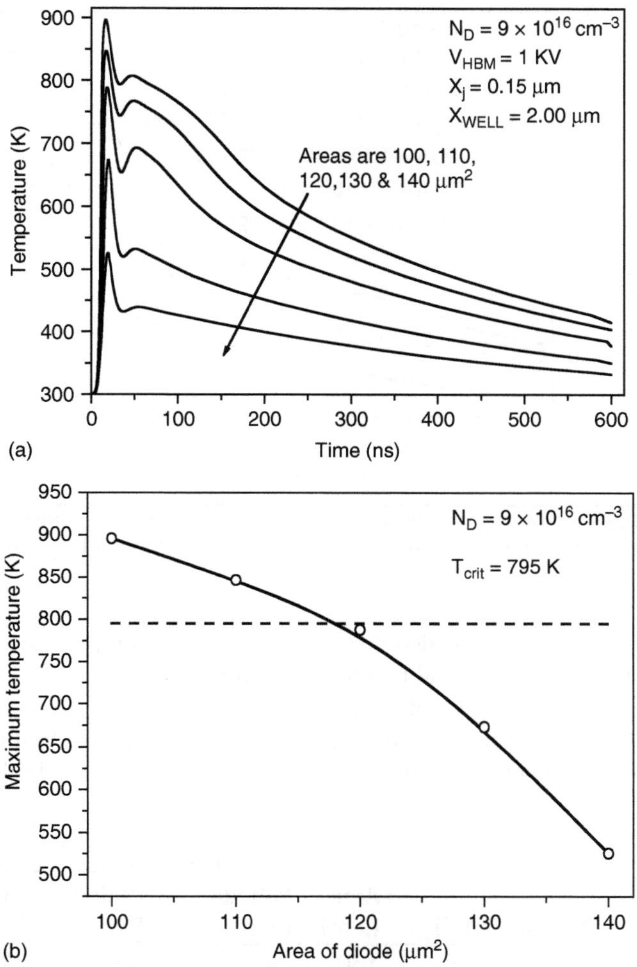

Figure 9.92 (*a*) Temperature versus time in a diode with four different areas, and (*b*) the maximum temperature in the diode plotted as a function of the diode area, with the critical temperature indicated.

procedure based on this new breakdown requirement, the second-round optimization leads to an optimized diode structure of X_j = 0.15 µm, $N_D = 1.2 \times 10^{17}\text{cm}^{-3}$, and area = 105 µm². A third iteration is not necessary because it would render the same solution.

Results and Verification To verify the design, Atlas mix-mode simulations were carried out for the ESD protection structure, which consisted of two optimized diodes at each I/O pin and a supply clamp between V_{dd} and V_{ss}. The results indicate that the combined voltage drop on the forward-biased diode and supply clamp is below the diode breakdown voltage during the ESD transients. The requirement

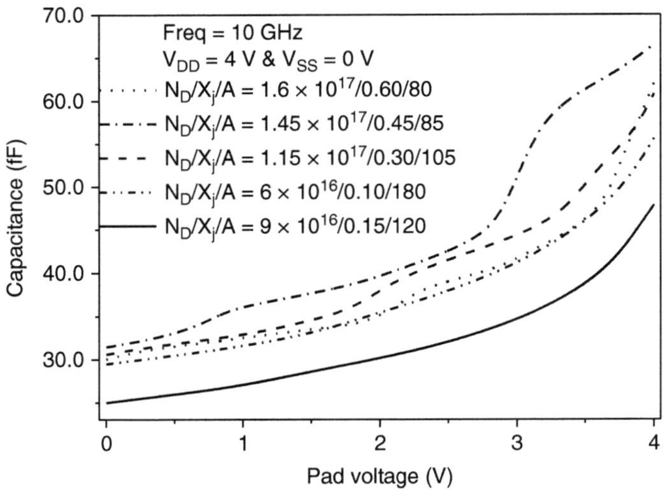

FIGURE 9.93 Parasitic capacitance versus the pad voltage of five best possible diode structures simulated at 10 GHz and supply voltages of V_{dd} = 4 V and V_{ss} = 0.

for the power rating is also met; that is, the peak temperature in the diode is below the critical temperature under ESD stress.

We next verify whether the suggested diode structure is indeed the optimal structure with minimal parasitic capacitance. To this end, we consider seven additional diode structures, as listed in Table 9.11.

Diode	X_J (μm)	N_D (cm^{-3})	Area (μm^2)	Breakdown Voltage Req.	Power Rating Req.	Cap. (fF) (2 V, 10 GHz)
Device 1*	0.10	6×10^{16}	130	Pass	Fail	N/A
Device 2*	0.15	1.5×10^{17}	110	Fail	N/A	N/A
Device 3†	0.15	9×10^{16}	120	Pass	Pass	30.16
Device 4†	0.30	9×10^{16}	180	Pass	Pass	57.90
Device 5†	0.30	1.5×10^{17}	90	Pass	Pass	36.09
Device 6†	0.45	1.45×10^{17}	85	Pass	Pass	39.66
Device 7†	0.60	1.6×10^{17}	80	Pass	Pass	35.13
Optimized	0.15	1.2×10^{17}	105	Pass	Pass	28.70

*Randomly selected structures

†Structures meeting the requirements of breakdown voltage and power rating and having N_D/X_j above the intercept points

TABLE 9.11 Comparison of Nonoptimized (Devices 1–7) and Optimized Diodes for Minimal Parasitic Capacitance for an ESD stress of 1kV HBM

Two of these diodes (devices 1 and 2) are randomly selected, and they may not meet the requirements of breakdown voltage and/or power rating. The other five diodes (devices 3 to 7) are intentionally chosen to have N_D/X_j above the intercept points, and their areas are the minimal areas meeting the requirement of power rating. Note that to satisfy the breakdown-voltage requirement, N_D/X_j cannot be lower than the intercept points. The purpose of considering devices 3 to 7 is to illustrate our earlier suggestion that the intercept points are the best possible points for minimal parasitic capacitance. The results summarized in Table 9.11 suggest that the randomly selected diode structures (devices 1 and 2) fail the requirements of breakdown voltage and/or power rating and that the optimized diode structure yields the smallest parasitic capacitance among all devices considered.

Figure 9.94 compares the parasitic capacitances at the I/O pin having optimized and nonoptimized (device 4) diodes as a function of pad voltage at a frequency of 10 GHz. The two power supplies considered are $V_{dd} = 4$ V and $V_{ss} = 0$. Note that there are two diodes connected to each I/O pin. Clearly, the pin with the optimized diodes possesses a smaller parasitic capacitance than that with the nonoptimized diodes. Furthermore, it is shown that the parasitic capacitance reaches a minimum when the pad voltage is equal to $0.5(V_{dd} - V_{ss})$.

An HBM charging voltage of 1 kV has been used in the preceding diode optimization. Since a larger charging voltage will give rise to elevated requirements of breakdown voltage and heat dissipation, the optimized diode structure is a function of the HBM charging voltage.

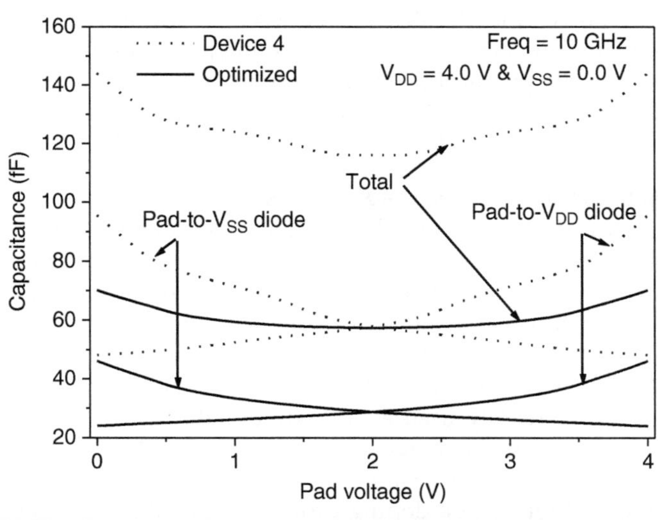

FIGURE 9.94 Parasitic capacitance versus the pad voltage at the I/O pin having the optimized and nonoptimized (device 4) diodes.

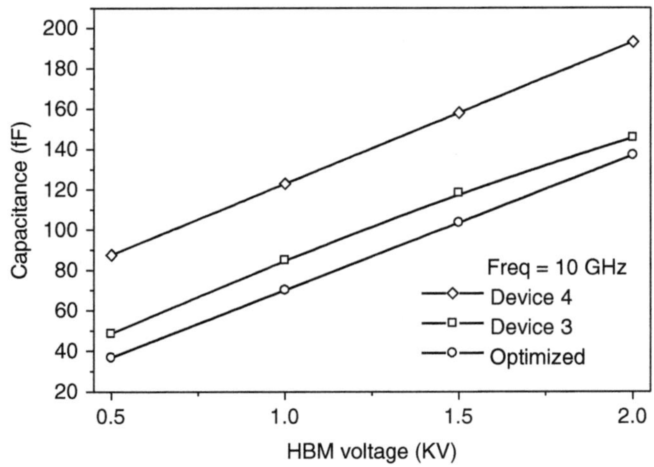

FIGURE 9.95 Parasitic capacitance versus HBM charging voltage at the I/O pin having the optimized and nonoptimized (devices 3 and 4) diodes.

Figure 9.95 shows the I/O pin capacitance versus HBM charging voltage having optimized and nonoptimized diodes. Apparently, the parasitic capacitance is increased if the level of the ESD stress is increased. This is due to the larger diode required as a result of the heat-dissipation constraint.

While the present study focused only on the parasitic capacitance, the parasitic resistance associated with the diode also has been implicitly optimized through consideration of the requirements of the breakdown voltage and power rating. This resistance depends strongly on N_D and the distance between n^+ and p^+ regions in the diode,[83] and the fact that we chose the intercept points in Fig. 9.90 rather than the points above the interception (i.e., these points have smaller N_D values than the intercept points) suggests that the parasitic resistance of the optimized diode is also minimal.

Design Based on an Existing CMOS Process

In most cases, it is not possible to adjust the doping densities and junction depths for the diodes fabricated in a particular CMOS technology using a foundry. In this section we will optimize the ESD protection structure based on an existing 0.13-μm CMOS technology without the possibility of changing the doping profile. A cross-sectional view of a 0.13-μm MOS device is given in Fig. 9.96. Such a device can be readily used as a diode by taking the substrate and drain terminals as the anode and cathode, respectively. The device has a channel length of 0.18 μm, *P*-well, deep *N*-well, threshold-adjust implant, and retrograde doping profile. The width of the MOSFET can be varied and determined using our design approach. Since the

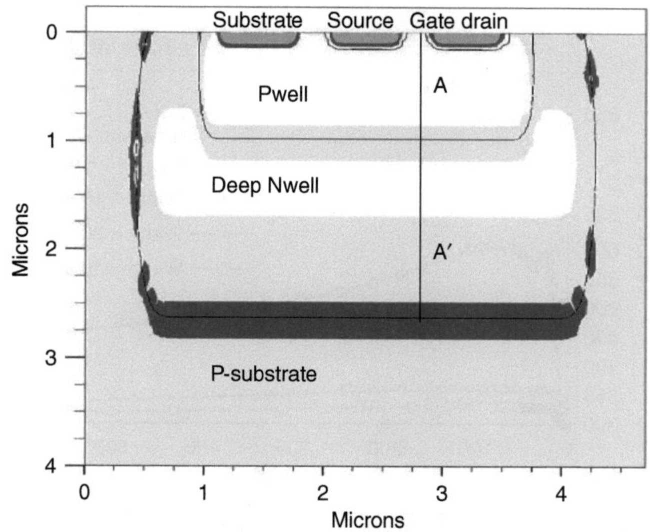

FIGURE 9.96 Cross section of the MOS structure from an existing 0.13-micro-meter CMOS technology.

doping profile for this particular technology is fixed, optimization of the diode breakdown voltage is not possible. For the device in Fig. 9.96, the diode breakdown voltage is 9.3 V. Furthermore, we will follow the typical horizontal dimension in the 0.13-μm CMOS technology (see Fig. 9.96); thus the only dimension to be optimized is the channel width *W*.

Let us now determine the minimum MOSFET width based on the requirement of the critical temperature for an HBM charging voltage of 1 kV. The requirement of the breakdown voltage will be discussed later. Following the procedure discussed earlier, the doping density at the hot spot in the MOSFET (i.e., near the drain region) is determined, and the critical temperature of 860 K is found. Figure 9.97 shows the time-dependent temperature at the hot spot of the MOSFET simulated as a function of channel width. It can be seen that all devices having a width larger than 10 μm can pass the power-rating requirement. The breakdown behavior is given in Fig. 9.98. Since the voltage drop in the forward-biased diode is a function of diode dimensions, different channel widths resulted in different total voltage drops (i.e, $V_{D1} + V_{\text{clamp}}$) on the forward-biased diode and supply clamp. The results suggest that the devices with channel widths of 10 and 20 μm possess total voltage drops that are higher than the diode breakdown voltage of 9.3 V, and thus the minimum width satisfying the breakdown-voltage requirement, as well as the power-rating requirement, is 30 μm.

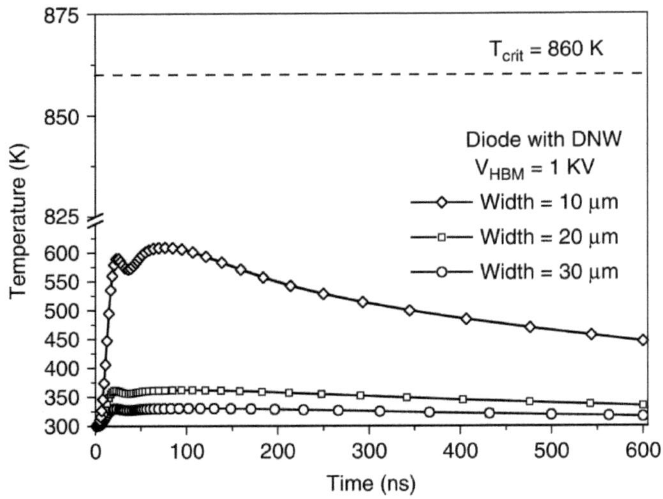

FIGURE 9.97 Temperature versus time simulated for different channel widths. The dashed lines of T_{crit} = 860 K denote the critical temperature of the device.

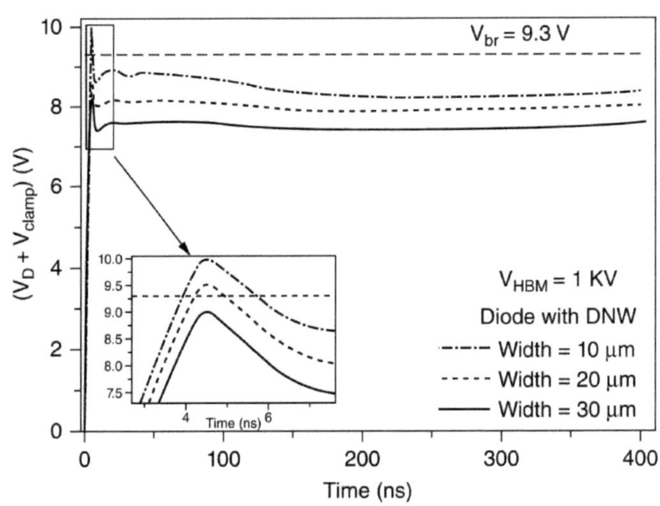

FIGURE 9.98 Simulated total voltage drop on the forward-biased diode and supply clamp for different channel lengths. The dashed lines of V_{br} = 9.3 V denote the diode breakdown voltage.

We have constructed the ESD protection structure of two diodes (each with a width of 30 μm determined earlier) and a supply clamp with an *n*-channel MOSFET (with a relatively large width of 160 μm to prevent the MOS from being damaged), all fabricated from the

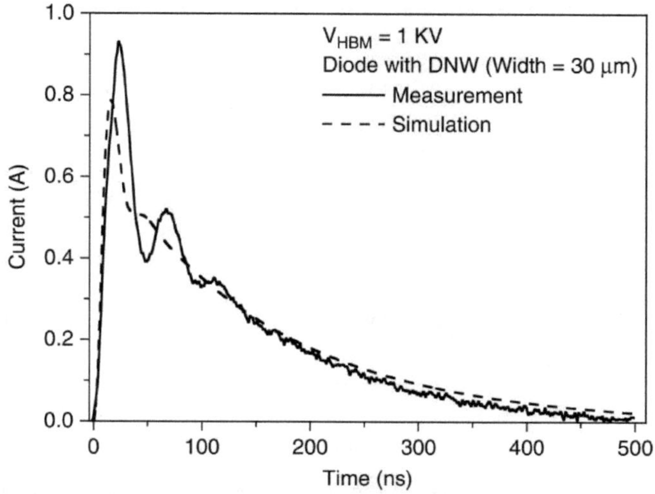

FIGURE 9.99 Comparison of the simulated and measured current passing through the drain terminal of a MOSFET in a supply clamp. The ESD protection structure here is constructed using an optimized diode with a channel width of 30 μm.

0.13-μm CMOS technology mentioned earlier. An HBM test with a stress of 1 kV was conducted, and Fig. 9.99 shows the transient responses of the current passing through the drain terminal in the supply clamp obtained from measurements and Atlas simulation. No damage incurred in the diodes and MOSFET in the supply clamp during the stress test. We also have conducted ESD tests using diodes with a channel width of 20 μm and found that the diode in the reverse path was damaged owing to its operation in the breakdown region. This is consistent with the design criteria presented and used earlier.

The 0.13-μm CMOS technology considered here has the option of not including the deep *N*-well (DNW). For this structure, we found that the breakdown voltage is 8.9 V, the critical temperature is 840 K, and the required width is 40 μm. The simulated capacitance characteristics at the I/O pin using the optimized diodes (forward- and reverse-path diodes) with and without DNW at a frequency of 10 GHz for different HBM charging voltages and different pad voltages are given in Fig. 9.100. Also included are experimental data for MOS devices with DNW. Clearly, the presence of the DNW structure leads to a smaller parasitic capacitance and thus should be used for constructing the ESD protection structure. The parasitic capacitance of the diode with an optimized size and an ESD protection capability of 2 KV HBM is about 120 fF.

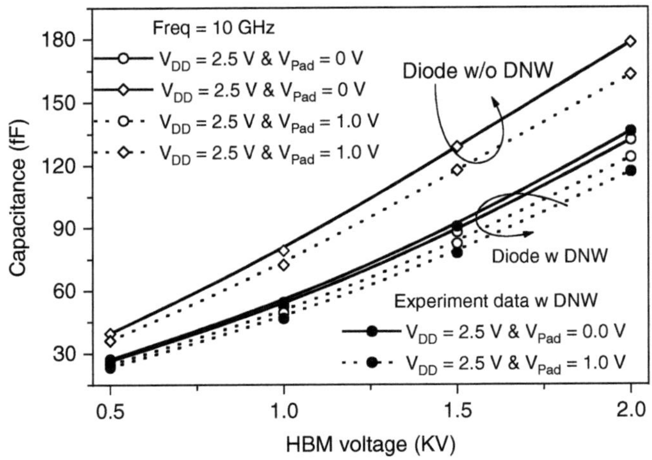

FIGURE 9.100 Capacitances at the I/O pin having two optimized diodes (forward- and reverse-path diodes) with and without the deep *N*-well structure at a frequency of 10 GHz for different HBM charging voltages and different pad voltages.

9.5 References

1. J. Vinson, G. Croft, J. Bernier, and J. J. Liou, *Electrostatic Discharge Analysis and Design Handbook*. Norwell, MA: Kluwer Academic Publishers, 2002.
2. J. Vinson and J. J. Liou, "Electrostatic discharge in semiconductor devices: protection techniques" (Invited Paper), *Proceedings of the IEEE*, Vol. 88, 2000, pp. 1878–1900.
3. M. Brandt and S. Halperin, "What does ESD really cost?" *Circuits Assembly Magazine*, June 1, 2003.
4. Z. Liu, J. Vinson, L. Lou, and J. J. Liou, "An improved bidirectional SCR structure for low-triggering ESD applications," *IEEE Electron Device Letters*, Vol. 29, 2008, pp. 360–362.
5. J. C. Lee, G. D. Croft, J. J. Liou, W. R. Young, and J. Bernier, "Modeling and measurement approaches for electrostatic discharge in semiconductor devices and ICs: An overview," *Microelectronics Reliability*, Vol. 39, 1999, pp. 579–594.
6. W. Liu, J. J. Liou, J. Chung, Y. H. Jeong, W. C. Chen, and H. C. Lin, "Electrostatic discharge (ESD) robustness of Si nanowire field-effect transistors," *IEEE Electron Device Letters*, Vol. 30, 2009, pp. 969–971.
7. Y. Li, J. J. Liou, J. Vinson, and L. Zhang, "Investigation of LOCOS- and polysilicon-bound diodes for robust electrostatic discharge (ESD) applications," *IEEE Transactions on Electron Devices*, Vol. 57, 2010, pp. 814–819.
8. A. Amerasekera and C. Duvvury, *ESD in Silicon Integrated Circuits*. Hoboken, NJ: John Wiley, 2002.
9. S. H. Voldman, *ESD Circuits and Devices*. Hoboken, NJ: John Wiley, 2006.
10. S. H. Voldman, *Latchup*. Hoboken, NJ: John Wiley, 2007.
11. K. Van Mele, "Effective ESD strategies in nano-CMOS IC design," Sarnoff Europe, available: http://www.chipestimate.com/techtalk.php?d=2007-12-04.
12. B. Krabbenborg, R. Beltman, P. Wolbert, and T. Mouthaan, "Physics of Electro-Thermal Effects in ESD Protection Devices," in *EOS/ESD Symposium Proceedings*, 1991, pp. 98–103.
13. S. Hong, J. Kim, K. Yoo, G. Kwon, and T. Won, "Two-dimensional electrothermal simulations and design of electrostatic discharge (ESD) protection circuit," in *EOS/ESD Symposium Proceedings*, 1993, pp. 157–163.

14. S. H. Voldman, S. S. Furkay, and James R. Slinkman, "Three-dimensional transient electrothermal simulation of electrostatic discharge protection circuits," in *EOS/ESD Symposium Proceedings*, 1994, pp. 246–256.
15. M. Pinto-Guedes and P. C. Chan, "A circuit simulation model for bipolar-induced breakdown in MOSFET," *IEEE Transactions on Computer-Aided Design*, Vol. 7, No. 2, February 1988, pp. 289–294.
16. S. Ramaswamy and S.-M. Kang, *iETSIM User's Manual Version 2.0*, 1996.
17. C. H. Diaz, S.-M. Kang, and C. Duvvury, "Circuit-level electrothermal simulation of electrical overstress failures in advanced MOS I/L protection devices," *IEEE Transactions on Computer-Aided Design of Integrated Circuits and Systems*, Vol. 13, No. 4, 1994, pp. 482–493.
18. S. Ramaswamy, E. Li, E. Rosenbaum and S. M. Kang, "Circuit-level simulation of CDM-ESD and EOS in submicron MOS devices," in *EOS/ESD Symposium Proceedings*, 1996, pp. 316–321.
19. A. Amerasekera, S. Ramaswamy, M.-C. Chang and C. Duvvury, "Modeling MOS snapback and parasitic bipolar action for circuit-level ESD and high current simulations,"in *Proceedings of the International Reliability Physics Symposium*, 1996, pp. 318–326.
20. P. Rossel, H. Tranduc, D. Montcoqut, G. Charitat and I. Pages, "Avalanche characteristics of MOS transistors," in *Proceedings of the 21st International Conference on Microelectronics (MIEL'97)*, Vol. 1, 1997, pp. 371–381.
21. T. Toyabe, K. Yamaguchi, S. Asai, and M. S. Mock, "A numerical model of avalanche breakdown in MOSFETs," *IEEE Transactions on Electron Devices*, Vol. ED-25, No. 7, 1978, pp. 825–832.
22. F.-C. Hsu, P.-K. Ko, S. Tam, C. Hu, and R. S. Muller, "An analytical breakdown model for short-channel MOSFETs," *IEEE Transactions on Electron Devices*, Vol. ED-29, No. 11, 1982, pp. 1735–1740.
23. S. E. Laux and F. H. Gaensslen, "A study of channel avalanche breakdown in scaled n-MOSFETs," *IEEE Transactions on Electron Devices*, Vol. ED-34. No. 5, 1987, pp. 1066–1073.
24. Y. A. El-Mansy and A. R. Boothroyd, "A simple two-dimensional model for IGFET operation in the saturation region," *IEEE Transactions on Electron Devices*, Vol. ED-24, No. 3, 1977, pp. 254–262.
25. T. Y. Chan, P. K. Ko, and C. Hu, "Dependence of channel electric field on device scaling," *IEEE Transactions on Electron Devices*, Vol. EDL-6, No. 10, 1985, pp. 551–553.
26. S. L. Lim, X. Y. Zhang, Z. Yu, S. Beebe, and R. W. Dutton, "A computationally stable quasi-empirical compact model for the simulation of MOS breakdown in ESD-protection circuit design," in *Proceedings of the International Conference on Simulation of Semiconductor Processes and Devices*, 1997, pp. 161–164.
27. M. Mergens, W. Wilkening, S. Mettler, H. Wolf, and W. Fichtner, "Modular approach of a high current MOS compact model for circuit-level ESD simulation including transient gate coupling behavior," *IEEE International Reliability Physics Symposium*, 1999, pp. 167–178.
28. S. G. Beebe, "Simulation of complete CMOS I/O circuit response to CDM stress," in *Proceedings of the EOS/ESD Symposium Proceedings*, 1998, pp. 259–270.
29. A. Schutz, S. Selberherr, and H. W. Potzl, "A two-dimensional model of the avalanche effect in MOS transistors," *Solid-State Electronics*, Vol. 25, No. 3, 1982, pp. 177–183.
30. J. Lindmayer and W. Schneider, "Theory of lateral transistors," *Solid-State Electronics*, Vol. 10, 1967, pp. 225–234.
31. S. Verdonckt-Vandebroek and S. S. Wong, "High-gain lateral bipolar action in a MOSFET structure," *IEEE Transactions on Electron Devices*, Vol. 38, No. 11, 1991, pp. 2487–2495.
32. G. Massobrio and P. Antognetti, *Semiconductor Device Modeling with SPICE*, 2nd ed., NY: McGraw-Hill, 1993.
33. M. Reisch, "On bistable behavior and open-base breakdown of bipolar transistors in the avalanche regime: modeling and applications," *IEEE Transactions on Electron Devices*, Vol. 39, No. 6, 1992, pp. 1398–1409.

34. H. Chang, E. Charbon, U. Choudhury, A. Demir, E. Felt, E. Liu, E. Malavasi, A. Sangiovanni-Vincentelli, and L. Vassiliou, *A Top-Down, Constraint-Driven Design Methodology for Analog Integrated Circuits*. Boston: Kluwer Publishers, 1997.
35. A. Maxim and G. Maxim, "A high accuracy power MOSFT SPICE behavioral macromodel including the device self-heating and safe operation area simulation," *IEEE Applied Power Electronics Conference and Exposition (APEC)*, Vol. 1, 1999, pp. 177–183.
36. *SPECTRE Manual*, December 1998.
37. J. C. Lee, G. D. Croft, J. J. Liou, W. R. Young, and J. C. Bernier, "Modeling and measurement approaches for electrostatic discharge in semiconductor devices and ICs: An overview," *Microelectronics Reliability*, Vol. 39, No. 5, May 1999, pp. 579–593.
38. J. C. Bernier, G. D. Croft, and W. R. Young, "A process independent ESD design methodology," in *Proceedings of ISCAS*, May 1999.
39. T. M. Bilodeau, "Theoretical and empirical analyses of the effects of circuit parasitics on the calibration of HBM ESD simulations," in *ESO/ESD Symposium Proceedings*, 1989, pp. 43–49.
40. K. Verhaege, P. J. Roussel, G. Groeseneken, H. E. Maes, H. Gieser, C. Russ, P. Egger, X. Guggenmos, and F. G. Kuper, "Analysis of HBM ESD testers and specifications using a 4th order lumped element model," in *ESO/ESD Symposium Proceedings*, 1993, pp. 129–137.
41. T. Sleotnicki, "New physical model of multiplication induced breakdown in MOSFETs," *Solid-State Electronics*, Vol. 34, 1991, pp. 1297–1307.
42. N. D. Jankovic, "Pre-turn-on source bipolar injection in graded NMOSTs," *IEEE Transactions on Electron Devices*, Vol. 38, No. 11, 1991, pp. 2527–2530.
43. *Atlas Manual*, Silvaco International, 1999.
44. *BSIM3v3 Model Users' Manual*, UC Berkeley, 1999.
45. N. Arora, *MOSFET Models for VLSI Circuit Simulation—Theory and Practice*. NY: Springer-Verlag, 1993.
46. J. J. Liou, *Advanced Semiconductor Device Physics and Modeling*. Boston: Artech House, 1994.
47. *Microcal Origin 5.0 Manual*, 2000.
48. C. Hu, S. C. Tam, F. C. Hsu, P. K. Ko, T. Y. Chan, and K. W. Terrill, "Hot-electron induced MOSFET degradation-model, monitor and improvement," *IEEE Transactions on Electron Devices*, Vol. 32, 1985, pp. 375–385.
49. H. Wong and M. C. Poon, "Approximation of the length of velocity saturation region in MOSFETs," *IEEE Transaction on Electron Devices*, Vol. 44, 1997.
50. J. S. Kolhatkar and A. K. Dutta, "A new substrate current model for submicron MOSFETs," *IEEE Transaction on Electron Devices*, Vol. 47, 2000.
51. N. D. Arora and M. S. Sharma, "MOSFET substrate current model for circuit simulation," *IEEE Transaction on Electron Devices*, Vol. 38, 1991.
52. K. Sonoda, K. Taniguchi, and C. Hamaguchi, "Analytical device model for submicrometer MOSFETs," *IEEE Transaction on Electron Devices*, Vol. 38, 1991.
53. J. Chung, M. C. Jeng, G. May, P. K. Ko, and C. Hu, "Hot-electron currents in deep-submicrometer MOSFETs," in *Proceedings of IEEE, IEDM*, 1988.
54. Y. W. Sing and B. Sudlow, "Modeling and VLSI design constraints of substrate current," in *Proceedings of IEDM*, 1980, p. 732.
55. L. V. Roozendaal, A. Amerasekera, P. Bos, W. Baelde, F. Bontekoe, P. Kersten, E. Korma, P. Rommers, P. Krys, U. Weber, and P. Ashby, "Standard ESD testing of integrated circuits," in *EOS/ESD Symposium Proceedings*, 1990, pp. 119–130.
56. J. Vinson and J. J. Liou, "Electrostatic discharge in semiconductor devices: an overview," *IEEE Proceedings*, Vol. 86, 1998, pp. 399–418.
57. T. Maloney and N. Khurana, "Transmission line pulsing techniques for circuit modeling of ESD phenomena," in *EOS/ESD Symposium Proceedings*, 1985, pp. 49–54.
58. H. Gobner, T. Muller-Lynch, K. Esmark, and M. Stecher, "Wide range control of the sustaining voltage of ESD protection elements realized in a smart power technology," in *EOS/ESD Symposium Proceedings*, 1999, pp. 19–27.

59. H. Wolf, H. Gieser, and W. Wilkening, "Analyzing the switching behavior of ESD-protection transistors by very fast transmission line pulsing," in *EOS/ESD Symposium Proceedings*, 1999, pp. 28–37.
60. J. C. Smith, "An anti-snapback circuit technique for inhibiting parasitic bipolar conduction during EOS/ESD events," in *EOS/ESD Symposium Proceedings*, 1999, pp. 62–69.
61. K. Bock, B. Keppens, V. De Heyn, G. Groeseneken, L. Y. Ching, and A. Naem, "Influence of gate length on ESD-performance for deep submicron CMOS technology," in *EOS/ESD Symposium Proceedings*, 1999, pp. 95–104.
62. C. Y. Chu and E. R. Worley, "Ultra low impedance transmission line tester," in *EOS/ESD Symposium Proceedings*, 1998, pp. 311–319.
63. D. G. Pierce, W. Shiley, B.D. Mulcahy, K. E. Wagner, and M. Wunder, "Electrical overstress testing of a 256K UVEPROM to rectangular and double exponential pulses," in *EOS/ESD Symposium Proceedings*, 1988, pp. 137–146.
64. D. M. Pozar, *Microwave Engineering*. Reading, MA: Addison-Wesley, 1990, pp. 76–77.
65. C. Duvvury and A. Amerasekera, "ESD: A pervasive reliability concern for IC technologies," *IEEE Proceedings*, Vol. 81, May 1993, pp. 690–702.
66. G. D. Croft, "ESD protection using a variable voltage supply clamp," in *EOS/ESD Symposium Proceedings*, 1994, pp. 135–140.
67. C. H. Diaz, S. M. Kang, and C. Duvvury, *Modeling of Electrical Overstress in Integrated Circuits*. Boston: Kluwer Publishers, 1995.
68. X. Du, S. Dong, Y. Han, and J. J. Liou, "Evaluation of RF electrostatic discharge (ESD) protection in 0.18-μm CMOS technology," *Microelectronics Reliability*, Vol. 48, 2008, pp. 1552–1556.
69. J. A. Salcedo, J. J. Liou, and J. C. Bernier, "Design and integration of novel SCR-based devices for ESD protection in CMOS/BiCMOS technologies," *IEEE Transactions on Electron Devices*, Vol. 52, 2005, pp. 2682–2689.
70. J. Salcedo and J. J. Liou, "A novel dual-polarity device with symmetrical/asymmetrical S-type I-V characteristics for ESD protection design," *IEEE Electron Device Letters*, Vol. 27, 2006, pp. 65–67.
71. L. Lou and J. J. Liou, "An unassisted, low-trigger and high holding voltage SCR for on-chip ESD protection applications," *IEEE Electron Device Letters*, Vol. 28, 2007, pp. 1120–1122.
72. Zhiwei Liu, J. J. Liou, Shurong Dong, and Y. Han, "Silicon controlled rectifier stacking structure for high voltage ESD protection applications," *IEEE Electron Device Letters*, Vol. 31, 2010, pp. 845–847.
73. Z. Liu, J. J. Liou, and J. Vinson, "Novel silicon controlled rectifier for high-voltage ESD applications," *IEEE Electron Device Letters*, Vol. 29, 2008, pp. 753–755.
74. R. Gauthier, W. Stadler, K. Esmark, P. Riess, A. Salman, M, Muhammad, and C. Putman, "Evaluation of diode-based and NMOS/Lnpn-based ESD protection strategies in a triple gate oxide thickness 0.13-μm CMOS logic technology," in *EOS/ESD Symposium Proceedings*, 2001, pp. 205–215.
75. C. Richier, P. Salome, G. Mabboux, I. Zara, A. Juge, and P. Mortini, "Investigation on different ESD protection strategies devoted to 3.3 V RF applications (2 GHz) in a 0.18 μm CMOS process," in *EOS/ESD Symposium Proceedings*, 2000, pp. 251–259.
76. S. H. Voldman, S. Geisller, J. Nakos, J. Pekarik, and R. Gauthier, "Semiconductor process and structural optimization of shallow trench isolation-defined and polysilicon-bound source/drain diodes for ESD networks," in *EOS/ESD Symposium Proceedings*, 1998, pp. 151–160.
77. M. D. Ker and C. M. Lee, "Inteference of ESD protection diodes on RF performance in Giga-Hz RF circuits," in *Circuit and System Symposium Proceedings*, 2003, pp. 297–300.
78. K. Chatty, R. Gauthier, C. Putman, M. Muhammad, M. Woo, J. Li, R. Halbach, and C. Seguin, "Study of factors limiting ESD diode performance in 90-nm CMOS technologies and beyond," in *Intational Reliability Physics Symposium Proceedings*, 2005, pp. 98–105.
79. Y. Li, J. J. Liou, and J. Vinson, "Investigation of diode geometry and metal line pattern or robust ESD protection applications," *Microelectronics Reliability*, Vol. 48, No. 10, 2008 pp. 1660–1663.

80. J. R. Manouvrier, J.-R., Manouvrier, P. Fonteneau, C.-A. Legrand, P. Nouet, and F. Azais, "Characterization of the transient behavior of gated/STI diodes and their associated BJT in the CDM time domain," in *EOS/ESD Symposium Proceedings*, 2007, pp. 165–174.
81. N. Maene, J. Vandenbroeck, and L. Bempt, "On chip electrostatic discharge protections for inputs, outputs, and supplied of CMOS circuits," in *EOS/ESD Symposium Proceedings*, 1992, pp. 228–233.
82. Stephen G. Beebe, "*Characterization, modeling and design of ESD protection circuit*," Ph.D. dissertation, Stanford University, 1998.
83. R. Velghe, P. de Vreede, and P. Woerlee, "Diode network used as ESD protection in RF applications," in *Proceedings of EOS/ESD Symposium*, 2001, pp. 337–345.
84. C. Crowell and S. Sze, "Temperature dependence of avalanche multiplication in semiconductors," *Applied Physics Letters*, Vol. 9, September 1966, pp. 242–244.

PART 3

Back-End-of-Line

CHAPTER 10
Electromigration

10.1 Physics of Electromigration

Electromigration (EM) refers to the gradual degradation of interconnects owing to the combined effects of current and temperature. Electromigration is widely accepted as the mass transport of metal ions in a conductor as a result of momentum transfer from electrons traveling under the influence of an applied electrical field.[1] It is basically a diffusion process of bulk and surface atoms under the influence of a driving force. The EM driving force F_{EM} is given by Eq. (10.1):

$$F_{EM} = Z^* \cdot e \cdot \rho \cdot j \tag{10.1}$$

where Z^* is the effective charge number of the interconnect metal, e is the electron charge, ρ is the interconnect resistivity, and j is the current density applied.

In addition to the diffusion of atoms, there is another force in the opposite direction of electron flow that is commonly referred as *backstress force.* This force is related to the stress gradient induced by the the motion of atoms toward the cathode end along the interconnects. As a result, atom flux during electromigration J_a is a combination of both the driving force F_{EM} and the backstress force, which is given by[2]

$$J_a = \frac{DC_a}{k_B T}\left(Z^* \cdot e \cdot \rho \cdot j - \Omega \frac{\partial \sigma}{\partial x}\right) \tag{10.2}$$

where C_a is the atomic concentration, k_B is the Boltzmann constant, Ω is the atomic volume, and $\partial\sigma/\partial x$ is the stress gradient along the line. $\Omega(\partial\sigma/\partial x)$ is the so-called backstress force. D is the effective diffusion coefficient of the conductive material, which is given by Eq. (10.3) as a function of temperature:

$$D = D_0 \exp\left(\frac{E_a}{k_B T}\right) \tag{10.3}$$

where D_0 is the diffusion constant, E_a is the activation energy, and T is the temperature.

EM drift velocity under an applied electrical field v_d when only considering the driving force F_{EM} can be expressed by

$$v_d = \frac{DF_{EM}}{k_B T} = \frac{D_0}{k_B T} \cdot \exp\left(\frac{E_a}{k_B T}\right) \cdot (Z^* \cdot e \cdot \rho \cdot j) \tag{10.4}$$

On the other hand, when also including the effect of backstress force, the EM drift velocity should be revised to be

$$v_d = \frac{D_0}{k_B T} \cdot \exp\left(\frac{E_a}{k_B T}\right) \cdot \left(Z^* \cdot e \cdot \rho \cdot j - \frac{\partial \sigma}{\partial x} \Omega\right) \tag{10.5}$$

As described earlier, EM is a transport/diffusion phenomenon under the influence of an applied electrical field, where the diffusivity is a weighted sum of the atomic diffusivities of all diffusion paths in the interconnect. Because the Cu interconnect process is referred to as *dual damascene,* where the trench and via are processed together (Fig. 10.1), the resulting primary diffusion paths in this interconnect system include bulk (or lattice) diffusion, metal-liner interface diffusion, metal–cap-layer interface diffusion, and grain boundary diffusion. For the Cu interconnect, the Cu-cap interface diffusion (0.9 to 1.2 eV) has a lower activation energy than lattice (2.1 eV) and grain-boundary (1.2 eV) diffusion.[3]

An expression for the product of Z^* and D for a damascene structure can be written using Eq. (10.6) as a weighted sum of all diffusion paths in this interconnects system[4]:

$$Z^* D = Z_L^* n_L D_L + Z_I^* D_I \delta_I (2/w + 1/h) + Z_S^* D_S \delta_S / h + Z_{GB}^* \sum_j^n D_{GB} (\delta_{GB}/d) \tag{10.6}$$

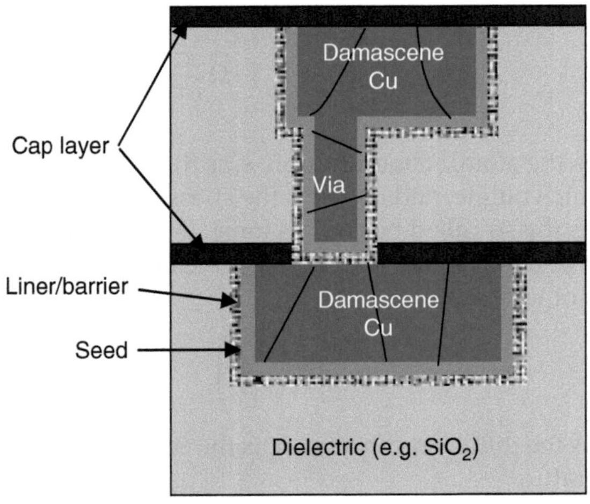

FIGURE 10.1 Schematic cross section of dual damascene interconnect system.

where the subscripts L, I, S, and GB refer to the lattice, metal-liner interface, metal–cap-layer interface, and grain boundary, respectively; δ_I, δ_S, and δ_{GB} are the width of the metal-liner interface, metal–cap-layer interface, and grain boundary, respectively; d is the grain size; w is the line width; h is the line thickness; and n_L, $\delta_I(2/w+1/h)$, δ_S/h, and (δ_{GB}/d) are the fractions of atoms diffusing through the bulk lattice, metal-liner interface, metal–cap-layer interface, and grain boundary of the line, respectively.

In short, EM has been identified as a major failure mode in metallic interconnects of microelectronic integrated circuits ICs for over three decades. Therefore, the problem of EM has been investigated extensively both experimentally and theoretically. This chapter summarizes the EM phenomena, including mechanisms, lifetime prediction methodology, process effect, and test-structure effects. Joule heating–related EM behavior under ac conditions, as defined by the root-mean-square current I_{rms}, is also included. Finally, the process-qualification requirements are listed for reference.

10.2 Electromigration Characterization

10.2.1 Package-Level Reliability versus Wafer-Level Reliability

There are several methods to characterize EM, such as low-frequency noise measurement,[5] ratio of resistance or resistometric test,[6] internal friction method,[7] and temperature–ramp resistance analysis (TRACE),[8,9] as well as waver-level fast tests such as the standard wafer-level EM acceleration test (SWEAT),[10] the breakdown-energy-of-metal (BEM) method,[11] and the wafer-level isothermal joule-heated EM test (WIJET).[12] The drift-velocity method is another important technique that was invented by Blech[13] and is a very elegant method to study EM mass transport.

Today, the resistometric test has been used widely for the study/evaluation of EM. In this test, the change in resistance is monitored as a function of time when the test structure is stressed at high temperature and high current density. Two types of stressing are widely employed as follows:

- *Wafer-level isothermal EM (WLR) test:*[14,15] In this stressing scenario, a very high current density (>10 MA/cm^2) is forced through the test structures while the experiment is carried out in a room-temperature ambient. It is the joule heating that raises the average temperature of the test structure [simulated by the temperature coefficient of resistivity (TCR)] to about 400°C. This technique allowes rapid determination of the possibility of EM damage and reduces some observed effects of postfailure arcing damage.[16] The time-to-failure (TTF) is generally determined under failure criteria of a small resistance increase such as 1 percent, or 3 or 5 percent. However, this method doesn't allow extracting activation energy E_a and current exponent n properly because the joule heating itself cannot separate the temperature stress from the current stress.[17]

- *Package-level reliability (PLR) test:* In this stressing scenario, a long-term stress with a lower stress current density (around several MA/cm^2), but high oven ambient temperature (in excess of 300°C) is applied. In this case, the joule heating from the low-stress current should be able to be ignored. In the testing setup, the dice are attached to ceramic test packages, and Al bond pads are connected to package lead frame using Au or Al wire bonding. The high-resolution Qualitau[18] package-level EM test system is one of the choices to perform the EM characterization. The sample size can be 24 to 30 pieces for each experiment. The drawbacks of the PLR test include limited sample size, long test duration, high package cost, and long packaging time.

10.2.2 Metal-Line Test Structures

Standard Wafer-Level Electromigration Acceleration Test Structure

The first wafer-level EM test structure is the SWEAT structure proposed by Root and Turner.[10,19] This structure consists of a succession of narrow and wide regions linked by tapered zone with a 45° angle (Fig. 10.2). For the narrow segment, width W_n is user-defined, and length L_n is greater than the critical length. For the wide segment, width $W_w = 10W_n$, and length $L_w/W_w \geq 1$. For this structure, it is stressed at high temperature (generated by self-heating) and high current density (>10 MA/cm^2). The disadvantage of this structure is that high-current-density divergences and high temperature gradients are generated in the narrow region. It is also hard to control the peak temperature precisely.

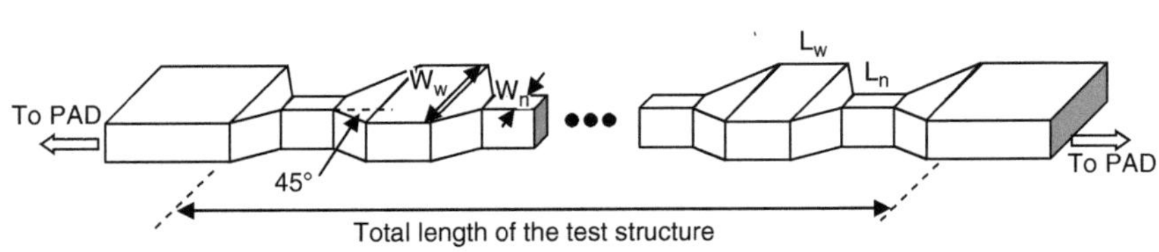

Figure 10.2 SWEAT test structure for EM monitoring.

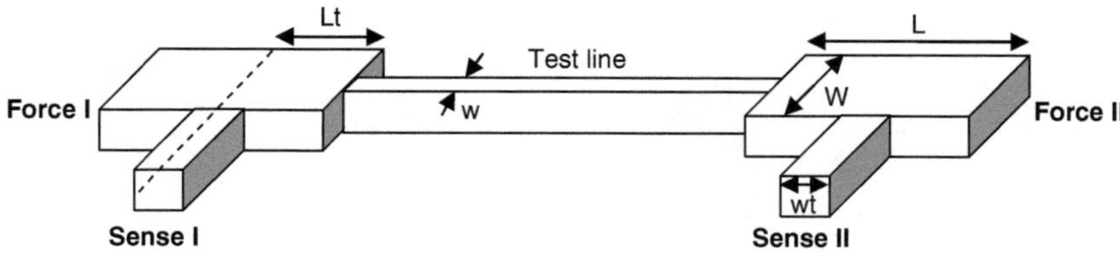

Figure 10.3 Schematic of the NIST EM test structure.

National Institute of Standards and Technology Structure

The National Institute of Standards and Technology (NIST) EM test structure was designed by Schafft et al. in 1987[20,21] and is based on the straight, flat line. It is a one-level structure with a long stripe connected through a transition-line segment to a probe pad. The four-terminal kelvin structure, including two Force and two Sense terminals, is used to reduce the measurement interference (Fig. 10.3). For the design guideline, the test line width is decided by the designer, with one set for the minimum design rule and at least one set much wider. For current Cu interconnects, a length larger than 200 μm is necessary. An example of test line length is 800 μm. For end segment width $W = 2 * w$, and length > 100 μm. For the voltage sense terminal, width $w_t = w$, and length > 100 μm, and position $L_t = 2 * W$. If taper is adopted, the angle will be 45°. The NIST structure is also useful for joule heating studies.

10.2.3 Critical-Length Test Structures

A Blech-Kinsborn edge-displacement test segment was proposed in 1975.[13] In this structure, segments of the conductor material to be characterized are fabricated on long, high-resistance conductor, such as Al over titanium nitride or Cu over tungsten. The current flow is from the higher-resistance conductor to the lower-resistance conductor segments, which undergo EM-induced cathode drift[22] (Fig. 10.4). According to the Blech effect, no edge displacement can occur if the length-current product (jL) is smaller than a threshold value $(jL)_{crit}$, as described by the following equation derived from Eq. (10.2):

$$(jL)_{crit} = \frac{\Omega\sigma_{crit}}{Z^* e\rho} \tag{10.7}$$

where σ_{crit} is the critical back-stress.

As a rule of thumb, the sheet-resistance difference between the high- and low-resistance conductors should be larger than 100. The length of the test structure can range from 1 to 100 μm, which depends on the conductor material. The width of the test structure also can be changed to study the line-width effect.

However, the shortcoming of this structure (see Fig. 10.4) is that a special process is required to manufacture it. To resolve this issue, a

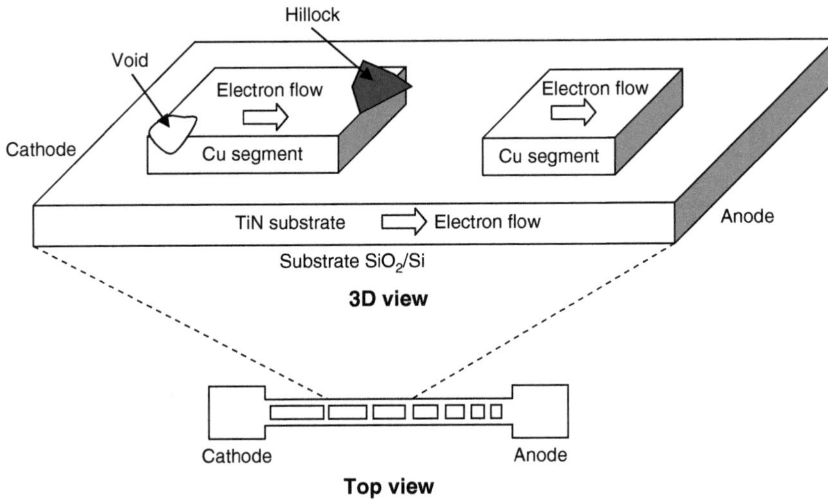

FIGURE 10.4 Schematic for the critical-length test structure with two kinds of resistance conductors.

via chain consisting of multiple segments with different lengths was proposed to characterize the EM critical length.[23] Figure 10.5 is a schematic where the lower-level stripe (M_x) is wider and shorter than the critical length (e.g., 50 μm), which ensures that failure will occur above the M_x level; that is, at either the via or the M_{x+1} trench, to facilitate failure analysis. A series of lengths of the upper-level metal (M_{x+1}) covers the critical length. An example contains 6 repeating sets of 14 interconnects where the M_2 length varies from 10 to 300 μm.[23] The test structure is stressed until failure occurs, and then the shortest failure segment is inspected. Note that the test structure is destroyed once one segment is open. It is expected that the longest segment will fail the earliest, from which the critical length could be estimated. In total, 7 test structures with 7 * 6 = 42 interconnects for a given length (7 * 84 = 588 total interconnects) were examined by Ogawa et al.[24] Within the sampling of 588 interconnects, 99 (16.8 percent) damage sites were identified after stressing at $T = 325°C$ and $j = 1$ MA/cm^2. The results showed that the failure rate increased abruptly for line-lengths longer than 100 μm, which then was characterized as the critical length for EM.

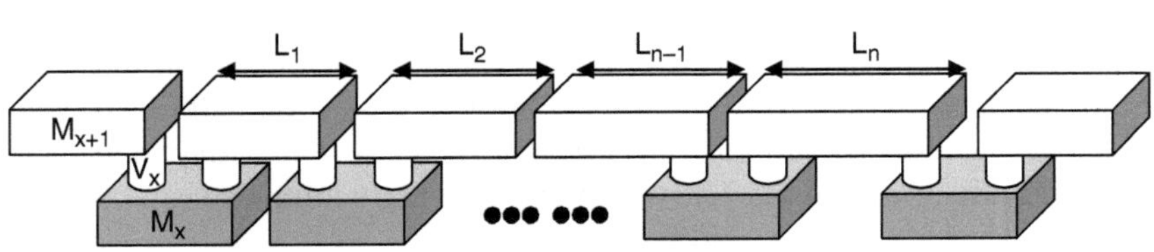

FIGURE 10.5 Schematic of the via chain structure for EM critical-length characterization.

10.2.4 Drift-Velocity Test Structure

To investigate the interconnect (e.g., Al or Cu) ionic drift velocity, a Kawasaki-Hu-type test structure can be used.[25] It is composed of a level of metal line, contact, and a silicided active area layer, as shown in Fig. 10.6*a*. Four contacts (e.g., tungsten) to the silicided AA layer are arranged in series along the length of M_1 at the cathode end. During stressing, voids will form at the cathode end owing to atom migration, as shown schematically in Fig. 10.6*b* and *c*. As a result, step changes in the resistance correspond to the voids in the metal line extending past each contact as expected, which is shown schematically in Fig. 10.6*d*, where the resistance jumps correspond to an open failure of the contacts. t_1 and t_4 are time to first contact fail and last contact fail, respectively. During EM stressing, the W-plug contacts fail one by one as metal (e.g., Al here) depletion proceeds. The displacement that causes each contact failure is known from the test-structure layout, that is, l_1 versus t_1 and l_4 versus t_4 as shown in Fig. 10.6*e*. The *incubation time* t_{inc} is defined as the time during which no Al migrates, so the intersection of the fitted line at zero displacement represents the incubation time. The slope of the line represents the EM drift velocity. This technique is also applicable to Cu interconnects. By determining the displacement from the layout, the Cu ionic drift velocity and effective incubation time of displacement can be estimated.[26–28]

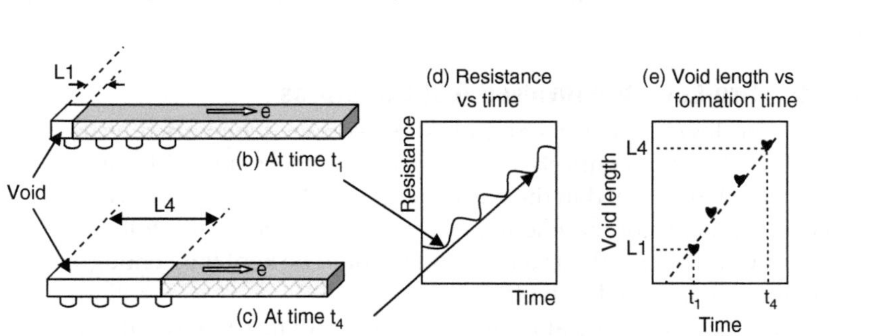

Figure 10.6 (*a*) Kawasaki-Hu-type test structure to investigate the metal interconnect ionic drift velocity. (*b*) Schematic of a void formed for first-contact failure. (*c*) Schematic of the extended void at the cathode end for all four contact failures. (*d*) Schematic of resistance change of the structure during EM stress. (*e*) Plot of void length versus formation time for deciding the EM drift velocity.

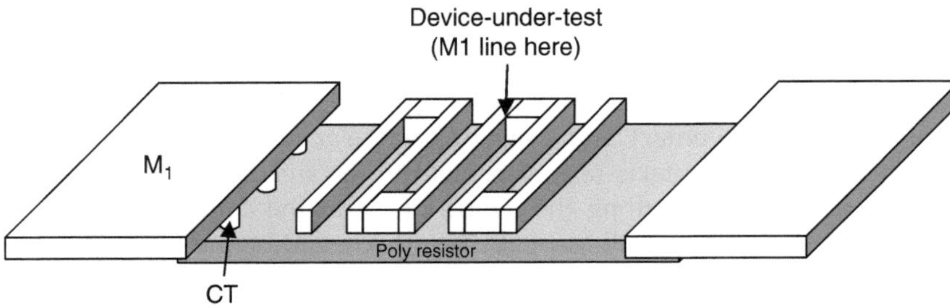

FIGURE 10.7 Schematic diagram of poly-heated via structure.

10.2.5 Heat-Generation Test Structures

To address EM in WLR tests, a polysilicon resistor below the EM metal-line structure is used to heat the structure for WLR EM stressing.[29] Figure 10.7 is a schematic diagram of the test structure for poly-heated via EM test. It consists of a serpentine M_1 on top of a poly resistor. The poly resistor is used to increase the temperature of the via chain. There is a good correlation between the WLR test using polysilicon resistor heating and the package-level test (PLR test) in a high-temperature oven ambient. However, the efficiency could be lower for higher metal levels because of heat dissipation in bulk silicon.

Alternatively, heating-coil structures were proposed to heat a standard mono-level metal line[30] or an EM via-ended metal line.[31] Figure 10.8 provides a schematic. The coils play the role of heater and allow for separating the temperature stress from the current stress. The extracted E_a and n values by this wafer-level technique are similar to the ones from package-level tests. However, one should be careful that thermal gradient on the edge of this structure should be minimized, otherwise, there will be much discrepancy between the wafer-level and package-level tests.[32]

10.2.6 Two-Level Via-Involved Test Structures

Typical two-level EM test structures are shown schematically in Fig. 10.9.[33] The lines connected to the pads are short and wide such that voids will be formed in the narrow and long test lines in between the vias. The width of the line in between the vias generally follows the minimum-design rule, and the length should be kept long enough (>200 μm) to avoid the Blech effect.

As EM proceeds, resistance of the test line will increase smoothly with time before final avalanche failure occurs, which is independent of interconnect length.[34] The increase in resistance is attributed to the presence of many voids located all along the line length.[35] Each of these voids contributes to a resistance increase as low as 0.01 to

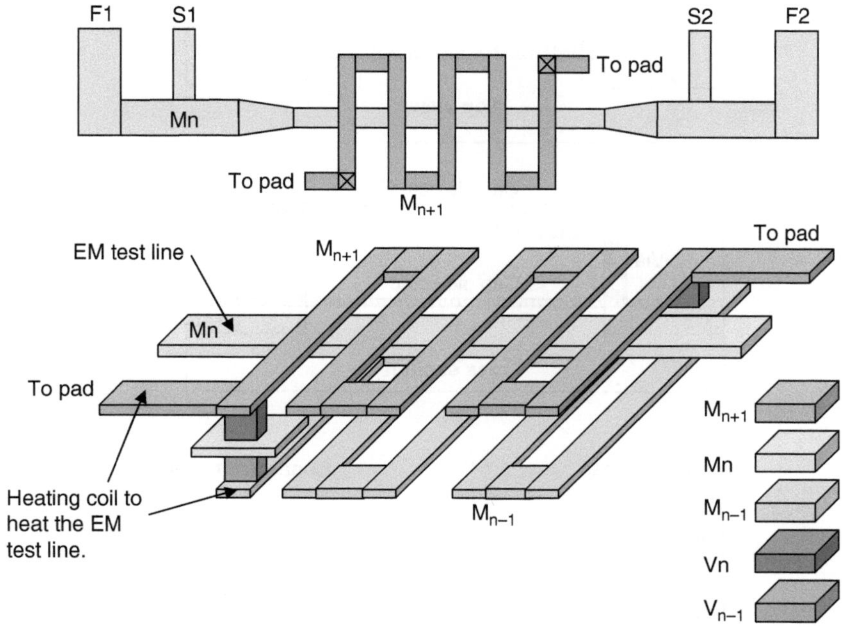

Figure 10.8 Schematic of top view and three-dimensional (3D) view of a heating coil surrounding the metal-line EM structure.

0.05 Ω as soon as its shape does not exceed half the line width or line thickness. For such small void volumes, local joule heating–induced resistance increases can be neglected. Once a void grows to a critical size, the local joule heating will be significant, leading to an abrupt increase in resistance and finally interconnect failure.[36]

Figure 10.10 provides schematics of via-involved EM test structures for dual damascene Cu interconnects. Both upstream and downstream stressing modes can be used with this structure. In the upstream mode (also referred as the *via-depletion mode;* Fig. 10.10*a*), electrons flow from a via below (V_x) to a line above (M_{x+1}) during EM stressing. In the downstream mode (also sometime referred as the *line-depletion mode;* Fig. 10.10*b*), electrons flow from a via above (V_{x+1}) to a line below (M_{x+1}) during EM stressing. For example, the Cu line is 200 μm long with widths ranging from the minimum-design rule to 1 μm.[37] The nominal via size follows design-rule requirements. For via-depletion EM, the early failures are often associated with liner weakness in the via.[38] For line-depletion EM, the configuration of via/line contact is the key factor for failure characteristics. The failure signature can be an abrupt resistance jump or with a gradual increase in resistance, depending on whether the via bottom contacts the line liner below. Both failure mechanisms will be elaborated later in Sec. 10.4.

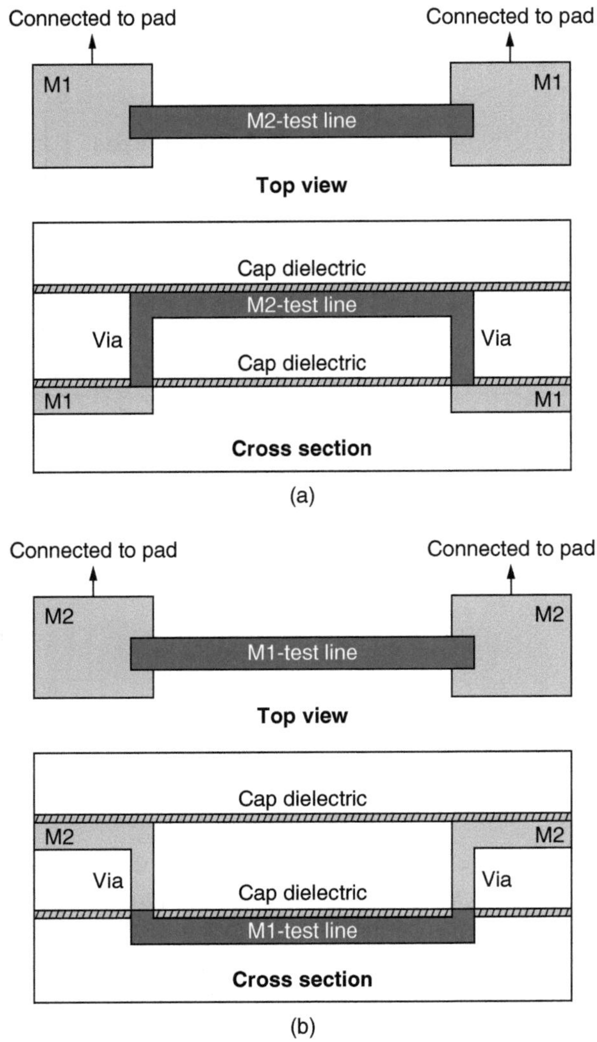

Figure 10.9 Schematic of (*a*) upper M_2 (upstream) and (*b*) lower M_1 test (downstream) structures. (*Reproduce with permission from the Electrochemical Society.*[33])

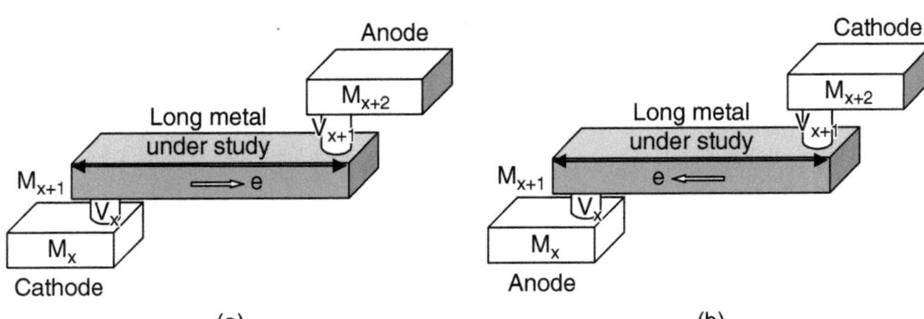

Figure 10.10 Schematic EM test structures of (*a*) via-depletion mode and (*b*) line-depletion mode for dual damascene Cu interconnect.

10.3 Time-to-Failure in Electromigration

10.3.1 Black's Equation (Two-Parameter Lognormal Distribution)

Generally, the lognormal distribution is employed for EM TTF statistics, although a few reports have suggested other distributions.[39] The two-parameter form has parameters including σ = the shape parameter and t_{50} = the median. The lognormal *probability distribution function* (PDF) is defined as[40,41]

$$f(t)=\frac{1}{t\sigma\sqrt{2\pi}}\exp\left[-\frac{(\ln t-\mu)^2}{2\sigma^2}\right] \tag{10.8}$$

TTF has a lognormal distribution, then the (natural) logarithm of TTF has a normal distribution with mean $\mu = \ln t_{50}$ (t_{50} is the median of TTF of a lognormal distribution) and standard deviation σ. The lognormal *cumulative probability* (or *failure*) *distribution function* (CDF), which gives the cumulative probability of failure prior to t, is as follows:

$$F(t)=\int_0^t\frac{1}{t\sigma\sqrt{2\pi}}\exp\left[-\frac{(\ln t-\ln t_{50})^2}{2\sigma^2}\right]\cdot dt=\phi\left(\frac{\ln t-\ln t_{50}}{\sigma}\right) \tag{10.9}$$

where ϕ (z) denotes the standard normal CDF.

The TTF data generally are analyzed using a probability plot, which is a simple visual method for performing goodness-of-fit analysis and also provides estimates of the mean and standard deviation of the data. Equation (10.9) can be rewritten as

$$\ln t=\sigma\phi^{-1}\{F(t)\}+\ln t_{50} \tag{10.10}$$

where ϕ^{-1} denotes the inverse function for the standard normal distribution. If we let $y = t$ and $x = \phi^{-1}\{F(t)\}$, then ln y is linear in x with slope σ and intercept [when $F(t) = 0.5$] of ln $t_{50.}$ If the data are consistent with a lognormal model, the resulting plot will have points that line up roughly on a straight line with slope σ and intercept t_{50} on the y axis.

The electronics industry uses the median time-to-failure (MTF) analysis to predict the lifetime of a device. In 1969, Black provided the following empirical equation to analyze failure in Al interconnects caused by EM[42]:

$$\text{MTF}=A\cdot j^{-n}\cdot\exp\left(\frac{E_a}{k_BT}\right) \tag{10.11}$$

where j is the current density, n is the current density exponent, E_a is the activation energy, k_B is the Boltzmann constant, T is temperature in kelvins, and A is a constant.

In practice, EM reliability assessments are often based on tests conducted at accelerated conditions (i.e., high temperatures and current densities), which are then scaled back to use conditions. These extrapolations are commonly based on the preceding Black's law:

$$\mathrm{TTF}_{\mathrm{use}} = \mathrm{MTF}_{\mathrm{test}} \left(\frac{j_{\mathrm{test}}}{j_{\mathrm{use}}}\right)^{n} \exp\left[\frac{E_a}{k_B T} \cdot \left(\frac{1}{T_{\mathrm{use}}} - \frac{1}{T_{\mathrm{test}}}\right)\right] \exp(-N\sigma) \qquad (10.12)$$

where $\mathrm{TTF}_{\mathrm{use}}$ is the lifetime at use conditions, $\mathrm{MTF}_{\mathrm{test}}$ is the median time-to-failure at accelerated test conditions, j_{test} is the test-current density, j_{use} is the use-current density, n is the current-density exponent, T_{test} is the test temperature, T_{use} is the use temperature, N is a constant that relates MTF with a time-to-failure at a different fail percent, and σ is the deviation in time to failure.

Activation Energy

The activation energy E_a is found independent of the dielectric materials (SiO_2 versus low-k of carbon-based CVD oxide) if the cap layer and the failure mechanism are the same.[43] Table 10.1 lists some activation energies for EM of copper interconnects. More activation energy values will be included in subsequent sections when discussing the cap layer, barrier layer, seed layer, and so on.

Current-Density Exponent *n*

The current-density exponent n in general ranges between 1 and 2. A high value of n approaching 2 could be attributed to overstressing with a high current density, which causes a temperature gradient in the metal line and would not be a real case in properly designed products.[47] Therefore, one should be careful when selecting a suitable stressing current density during EM testing. If the stressing current density is too low, the testing time will be too long. On the other hand, if the stressing current density is too high, the n value obtained will be close to 2, which will give an overestimated lifetime extrapolation. Another reason leading to a high n value could be the onset of the Blech length effect.[48] Further analysis by dividing the EM failure

Test	Activation Energy	Dielectric	Cap	Reference
Cu	0.9 eV	Polymer	SiCN	44
Cu	1.9 eV	Polymer	CoWP	44
Cu	0.87 eV	SiCOH	SiCN	45
Cu/CNT composite	1 to 1.277 eV	Blech test structure		46

Table 10.1 A List of Activation Energies for EM of Copper Interconnects

into nucleation and growth stages[49] shows that a larger n value (with a higher current density) is related to more nucleation contribution ($n \to 2$),[50] and a lower n value (with a lower current density) corresponds to more growth contribution ($n \to 1$).[51] Intermediate values of $n(1 < n < 2)$ indicate that both nucleation and growth are occurring during the EM stress.[52]

The kinetics depend on the initial site for void nucleation.[52] For a dual-damascene downstream structure, if the void initially nucleates directly under the via, failure will occur soon after nucleation, and hence the kinetics are nucleation-limited with n value approaching 2. In contrast, if the void forms initially in the metal line far from the via, considerable void growth (and diffusion to the via) will be required before failure occurs, and the kinetics are limited by void growth with n approaching 1.

The current-density exponent n is found increased when the line length is shorter than 50 μm in the presence of significant backstress for both SiO_2-based and low-k dielectric materials.[43] For line lengths larger than 50 μm, the n value ranges between 1 and 1.5, which is not very length-dependent. However, when the line length becomes shorter than 50 μm, the n value increases abruptly and can reach 3.5 to 4.0. Simulation work showed a similar trend.[53] The increase in n in the presence of significant backstress can be related to higher median time-to-failures.

Standard Deviation σ

According to Eq. (10.12), a tighter distribution, that is, a small value of standard deviation σ, will give a higher extrapolated lifetime because there is a smaller difference between t_{50} (mean TTF) and $t_{0.1}$ (low percentile). Hau-Riege et al.[43] showed that σ is constant ($\sigma \approx 0.2$ to 0.3) for lines longer than 25 μm. At the onset of backstress effects, σ increases for both SiO_2-based and low-k dielectric materials with wider distributions ($\sigma \approx 0.6$ to 1.5), which will lead to lower extrapolated lifetimes. In other words, for short metal lines, although the t_{50} (mean TTF) can be elongated owing to backstress effect, the lifetime at low percentile (e.g., $t_{0.1}$) could not benefit very much owing to the much wider distribution compared with long metal lines. This is especially true in advanced-technology nodes with scaled dimensions, where process variations (e.g., variations in the via barrier integrity, dimension variations, etc.) more likely lead to a wide spread in failure times.

10.3.2 Bimodal Lognormal Distribution

For some via/line EM structures, failure-time distributions have been found to severely deviate from ideal linear behavior, as described by the two-parameter lognormal distribution [Eq. (10.9)]. Two bimodal lognormal distributions were proposed to obtain a satisfactory fit. These models[54] are described briefly below, although they are not widely adopted.

Superposition model

In this model, the overall distribution is the sum of two two-parameter lognormal distributions,

$$F(t) = P(A)^* F_A(t; t_{50,A}, \sigma_A) + [1 - P(A)]^* F_B(t; t_{50,B}, \sigma_B) \qquad (10.13)$$

where F is the overall CDF, F_A and F_B are the CDFs for early and late modes, respectively, and $P(A)$ is the fraction of the early mode.

The resulting cumulative failure distribution appears S-shaped in the probability plot. Two different S-shaped types are observed experimentally (Fig. 10.11):

- *The flat-steep-flat shape:* An example is the TTF distribution measured at highly accelerated conditions on structures with tungsten vias. In this case, $t_{50,A}$ is not much different from $t_{50,B}$. However, σ_B is much smaller than σ_A (i.e., distribution B is much tighter than distribution A). Further, the PDF of the broad distribution may fully cover the tight distribution.
- *The steep-flat-steep shape:* An example is the TTF distribution measured on downstream stressed aluminum vias. In this case, $t_{50,A}$ is much different from $t_{50,B}$. However, σ_B is almost the same as σ_A. In other words, the distribution consists of two parts, where the peaks of the individual PDFs are separated from each other in terms of time. Hence they contribute sequentially to the resulting CDF.

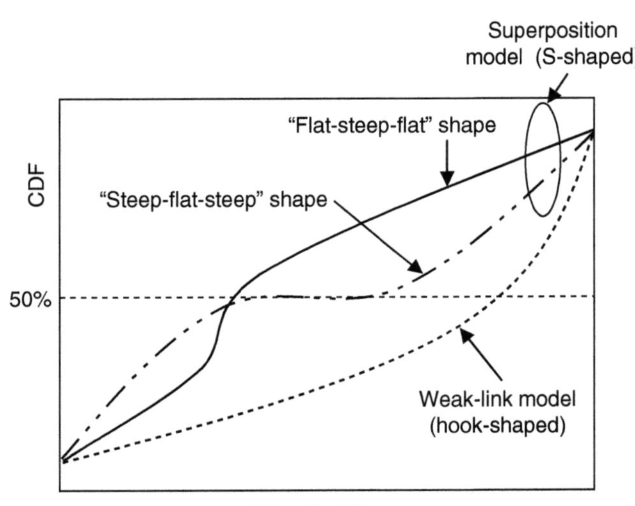

FIGURE 10.11 Schematic illustrating three types of bimodal CDF distributions explained by the superposition model and weakest-link model.

Weakest-Link (or Competing-Risks) Model

In this particular failure scenario, different mechanisms can cause device failure in a serial (chainlike) fashion. [54] This is the case, for example, if both failure mechanisms can act in a single-line or via-line structure. The structure fails if mechanism *A* or mechanism *B* produces a failure. Equivalently, the device does not fail if it fails neither due to mechanism *A* nor due to mechanism *B* until time *t*. The overall CDF can be expressed as follows:

$$F(t) = F_A(t; t_{50,A}, \sigma_A) + F_B(t; t_{50,B}, \sigma_B) - F_A(t; t_{50,A}, \sigma_A)^* F_B(t; t_{50,B}, \sigma_B) \tag{10.14}$$

where F is the overall CDF, and F_A and F_B are the CDFs for mode *A* and mode *B*, respectively.

The weakest-link approach appears as hook-shaped in the probability plot, where the later branch has a steeper slope than the early branch (see Fig. 10.11). An example is the TTF distribution measured on upstream stressed via-line structures with early fails corresponding to void in via and late fails corresponding to void in cathode end (see Fig. 10.13).

10.3.3 Three-Parameter Lognormal Distribution

The two-parameter lognormal distribution discussed in Sec. 10.3.1 has been used widely in both industry and academia to fit EM failure times. However, some studies showed that it encounters a difficulty with EM stresses on interconnects without sufficient liner/via redundancy, in which case early failure is always observed and dominated by slitlike voids under the via as shown in Fig. 10.14. A three-parameter lognormal distribution thus is proposed to fit both the early and the normal portions of the failure-time distributions, thus providing a more accurate lifetime projection for EM. This looks especially true for the line-depletion/void-size-limited EM with early failures.

The three-parameter lognormal distribution is expressed below:[55]

$$f(t; t_{50}, \sigma, X_0) = \frac{1}{(t - X_0)\sigma\sqrt{2\pi}} \exp\left[-\frac{\log\left(\dfrac{t - X_0}{t_{50} - X_0}\right)}{2\sigma^2} \right] \tag{10.15}$$

where X_0 is *the minimum threshold failure time*—the third parameter proposed for the model. The minimum threshold failure time is not the failure time observed from the first failure among the samples in a set of stress but rather is a theoretical limit that the distribution approaches. In theory, the earliest failure time approaches this minimum threshold failure time as sample size is increased, but it will never quite reach it for finite sample sizes. From a data-fitting point of view, X_0 is similar to t_{50} and σ, all of which are fit from a set of experimental data. Both the shape parameter σ and the differences between t_{50} and X_0 have significant impact on the difference between the two distributions, especially for the early portion of the distribution. As a special case, when $X_0 = 0$, the three-parameter lognormal distribution turns out to be the traditional two-parameter distribution.

Figure 10.12*a* shows the fitting lines of the EM TTF data for structure types *A* to *E* with two-parameter lognormal distributions. As the redundancy (via and via/line liner contact) decreases, the failure distribution becomes wider, and the quality of the fitting becomes poorer. For structures having good redundancy, the tight sigma allows good fitting of the data with single two-parameter lognormal distributions (types *A*, *C*, and *E*). However, for types *B* and *D*, a single two-parameter lognormal distribution does not fit the data well. It is clear that the two-parameter lognormal distribution cannot accommodate the bending down of the early portion of these broader distributions. It severely overestimates EM susceptibility.

Figure 10.12*b* shows the fitting curves of the experimental data (the same as those in Fig. 10.12*a*) with the three-parameter lognormal distribution described by Eq. (10.15). It is observed that the three-parameter lognormal distribution produces a much better fit

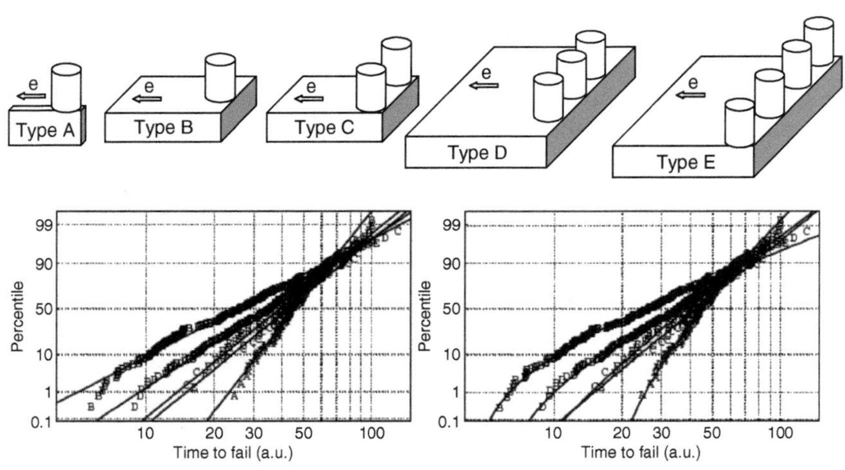

FIGURE 10.12 (*a*) Experimental EM TTF and the corresponding two-parameter lognormal distribution fit; (*b*) Experimental EM TTF and the corresponding three-parameter lognormal distribution fit. (*Used with permission from IEEE.*[55])

than the two-parameter lognormal distribution, especially for the early portion of the broader distributions (*B* and *D*). For EM lifetime projection, the early portion of the distribution is the most important part, and those early failures are the real reliability concerns.

10.4 Electromigration Failure Modes

EM-induced voids have various modes and locations owing to the different factors, such as metal line level, dielectric type, via component and position, and so on. The experimentally observed failure modes of EM are summarized below:

- For the so-called upstream test structure in Fig. 10.9*a* or the via-depletion test structure given in Fig. 10.10*a*, the void will appear inside the via (Fig. 10.13*a*) in the case of EM early failure, suspected to be caused by a weak liner (barrier) region. This type of void usually corresponds to abrupt resistance increase (i.e., either no or very little liner redundancy effect) is observed. For the late-failure case (Fig. 10.13*b*), the void will appear in the Cu line above the via. In this case, a gradual resistance increase with time is usually observed owing to good liner redundancy. Therefore, *for upstream or via-depletion EM, the barrier/seed process and via profile are the two key factors to suppress the early failure mode of the distribution.*
- For the so-called downstream test structure in Fig. 10.9*b* or the line-depletion test structure in Fig. 10.10*b*, void generally is observed in the metal below the via. Figure 10.14 shows four cross sections of poststress samples with failure time increasing from an early failure (image 1) to a late failure (image 4). It is seen that the early and late failures may have the same mechanism, whereas the only difference is the void size and shape that cause failure. The very early failure (image 1) has

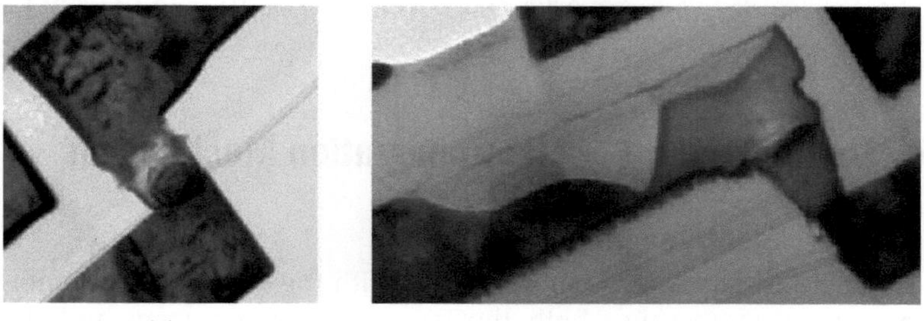

(a) (b)

FIGURE 10.13 SEM cross sections of (*a*) EM early failure with void in via and (*b*) late failure with void in line.

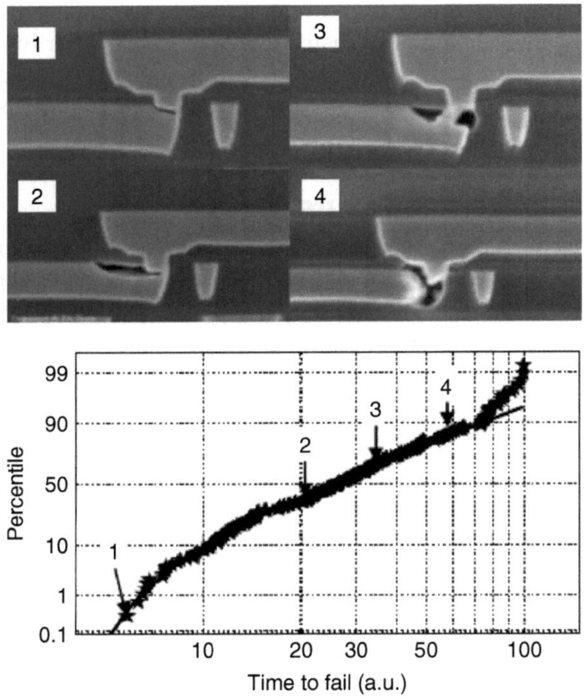

FIGURE 10.14 SEM cross sections of post-EM stress samples with failure times increasing from image 1 to image 4.(*Used with permission from IEEE.*[55])

a very thin, slitlike void right underneath the via. The late failure (image 4) has a larger void not right underneath the via but adjacent to the via, which is the normal EM phenomenon because the void migrates from the anode and eventually adheres to the via bottom to cause the failure. These SEM pictures demonstrate that the failure times are determined by void sizes and shapes. Therefore, *for downstream or line-depletion EM, the via/line liner contact and viagouging are critical to enhancing the minimum failure times of the distribution,* which will be elaborated in Sec. 10.6.[55]

10.5 Understanding of Electromigration Mechanisms

10.5.1 Electromigration at Contact

Contact EM has become more critical with the continuous technology advances because both the increased drive current and the reduced contact size will lead to significant current-density increases in the source/drain contacts of scaled devices.[56] In contrast to the traditional tungsten contact, a Cu contact is being considered for advanced-technology nodes because of advantages related to lower contact resistance and reduced tool cost. However, this raises new challenges for contact EM.[57]

A conventional test structure for contact-EM evaluation is similar to that shown in Fig. 10.9*b*, where M_2 is replaced by M_1, V_1 is replaced by contact, and M_1 is replaced by a poly line (either *N*-type or *P*-type doped poly) or a diffusion line (either *N*-type or *P*-type doped poly). A typical contact-EM test structure consists of two M_1 lines (e.g., 100 μm each) and one sandwich poly or diffusion line (e.g., 100 μm) with two contacts used. It is found that it is critical to reduce thermal stress and improve barrier uniformity through process optimization to achieve a reliable Cu contact with Ta-based barriers as vias in higher metal layers and the traditional W contacts.[57]

A four-terminal kelvin contact structure also was used for contact-EM characterization, as shown in Fig. 10.15. Kauerauf et al.[56] reported that the Cu contact showed voids and diffusion into the Si, whereas W contacts showed no clear damage. However, when looking at the proximity of the stressed contacts, severe damage above the contact in the M_1 was observed, and the distortion extended to the layers above attributed to the local joule heating at the contact interfaces, which was independent of the contact material used.

10.5.2 Electromigration at Al and W Vias

Al and W vias are used widely in relatively old technology nodes. This section gives a brief summary of EM at Al and W vias. Via-interconnect interfaces are sites of potentially large mass divergence owing to EM where Al depletion takes place. A reservoir at line end could be a way to extend EM lifetime because it could supply Al to replace the depleted atoms.

Tao et al. [58] reported EM results of different Al and W via test structures under current density of 35 MA/cm^2 and temperature of 250°C. Figure 10.16 schematically shows the four test structures. Structures *A* and *B* are for Al vias, whereas structures *C* and *D* for W vias. Some results are listed below:

- Al vias:
  - In structure *A*, the Al via has no TiW in its bottom, top, and sidewall. Nine samples were tested, and no failure of any

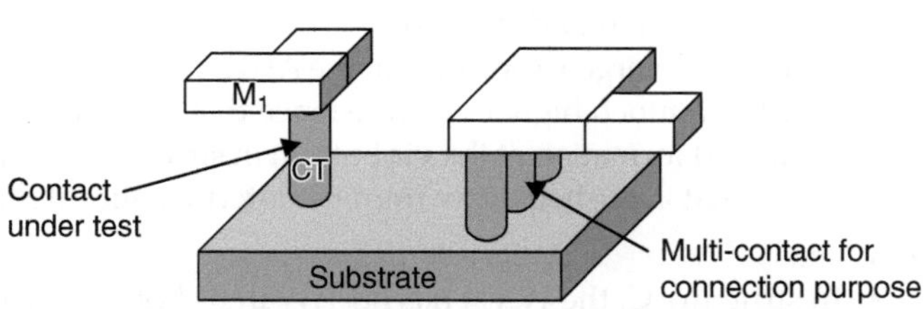

Figure 10.15 Schematic of a kelvin structure used for contact-EM resistance measurements.

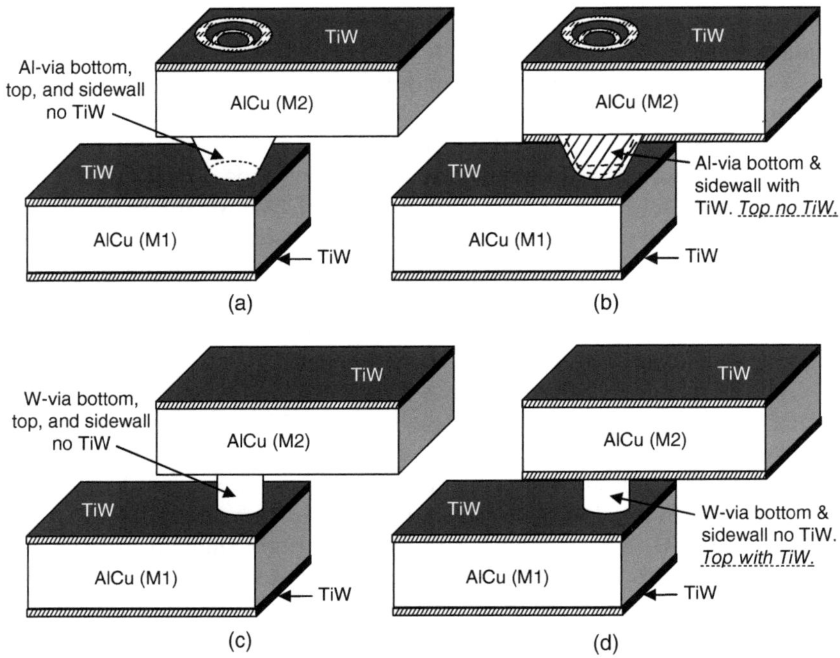

FIGURE 10.16 Schematics of EM test structures: (*a*) Al via with its bottom, top, and sidewall having no TiW; (*b*) Al-via with its bottom and sidewall having TiW but its top having no TiW; (*c*) W-via with its bottom, top, and sidewall having no TiW; (*d*) W-via with its bottom and sidewall having no TiW, but its top having TiW.

Al via was observed before the interconnects became open. The lifetimes of Al vias are not only longer than those of W-plug vias (structure C), but also longer than those of 10-μm-wide interconnects, even though the current density passing through the vias is more than 10× higher than that passing through the interconnects. This phenomenon is attributed to the small via size and thus much smaller grain boundary in the Al via and thus less Al atom flux divergence there.

- In structure B, the Al via has a bottom and sidewall of TiW, but no TiW in its top. Structure *B* is worse than structure *A*, and via failures were observed in both electron flow directions (upstream and downstream). This is attributed to the high flux-divergence location at the Al-TiW interface because TiW has a much higher EM resistance than Al. The increase in Al void formation at the via bottom in structure B is due to the Al that is pushed away from the intermetallic interface.

- W vias:
  - In structure C, the W via has no TiW in its bottom, top, and sidewall. Its EM lifetime is much shorter than the Al via in

structure A. The root cause is that the Al (interconnect)–W (via) structure presents an extreme flux-divergence location at the Al-W interface for Al EM because W has a much higher EM resistance than Al. In other words, the Al tends to migrate off the via, whereas W exhibits negligible diffusion or transport. The shorter lifetime for W-vias can be explained as follows: For this structure, when electrons move from M_1 to M_2, Al will be pushed away from the top surface of the W-vias (i.e., the interface between W and Al-Cu), and voids will develop there. When electrons move from M_2 to M_1, the Al accumulates and decreases at the top and bottom of the W-via area, respectively. In other words, in both cases of electron flow, voids were found at the interface between the W-via and Al-Cu metal line, which is the root cause leading to a large resistance increase and short EM lifetime.

- For structure D, an additional TiW layer is included on top of the W-via. When electrons move from M_1 to M_2, no failure at vias is observed, and the failure location is found in the M_2 lead in the vicinity of the via area. In this case, the bottom TiW layer of M_2 served as a robust redundant layer, which can reduce effective current density. Therefore, open failure did not occur at the via area even after Al had been pushed away over the plug area. On the other hand, in the case with electron flow from M_2 to M_1, which is similar to the situation in structure C, all open failures happened at vias.

10.5.3 Electromigration at Cu Interconnects

There are three basic understandings of EM mechanisms, as detailed below:[59]

- *Tensile-stress mechanism:* In this scenario (Fig. 10.17*a*), high tensile stress develops at the cathode end for both the metal end and the via bottom during EM, where the Ta liner forms a blocking boundary to the diffusing Cu. When the critical tensile stress for void nucleation is reached, a void forms and then grows.
- *Current-crowding mechanism:* In this scenario (Fig. 10.17*b*), at the cathode end, there is a current density gradient between the lower corner and extension in M_2 line above the via, which may induce a flux of vacancies to accumulate at the upper corner. This mechanism was proposed for explaining voids formed in low-current-density regions such as reservoirs. It is the current density gradient–induced vacancy drift[60] that leads to the vacancy accumulation and supersaturation in the upper corner of M_2 line extension.[42] A void will nucleate, grow, and eventually deplete the entire cathode end of

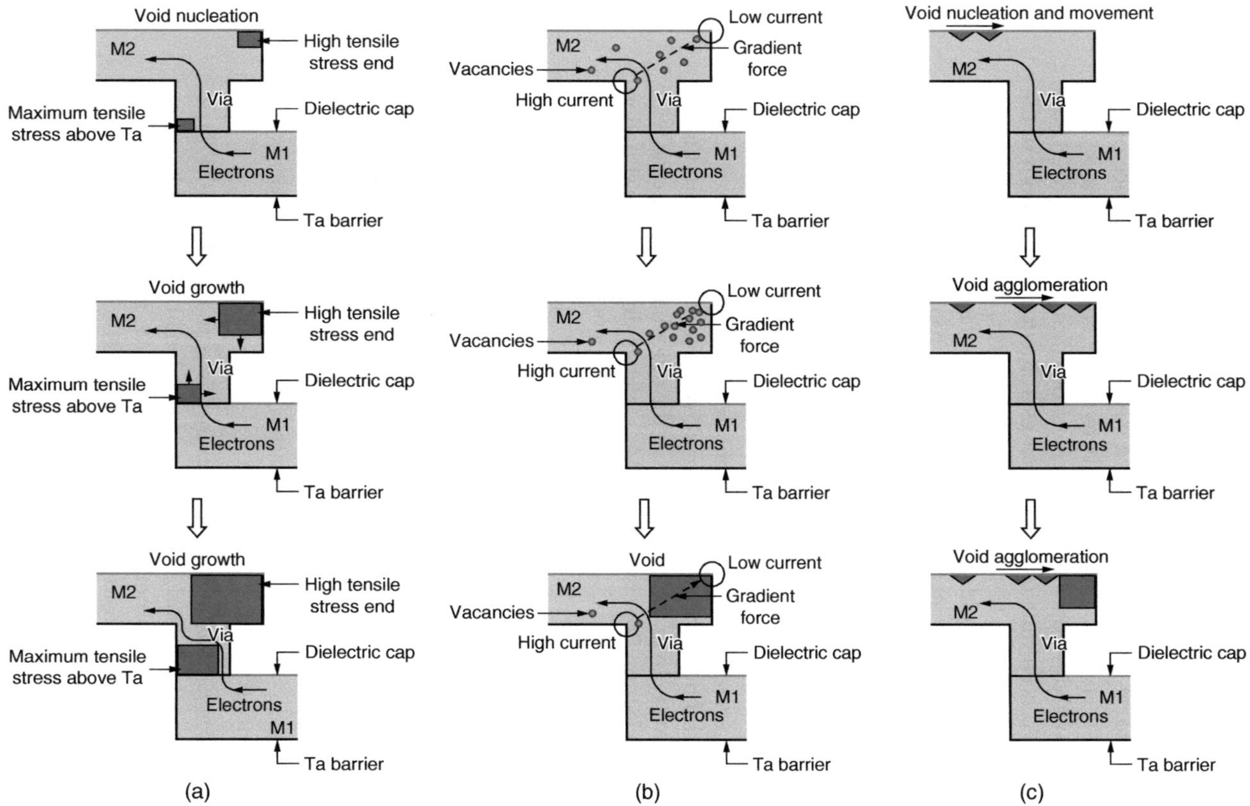

FIGURE 10.17 Current understanding of EM mechanisms: (*a*) illustration of tensile-stress mechanism; (*b*) illustration of current-crowding mechanism; (*c*) illustration of interfacial diffusion mechanism. (*Used with permission from Elsevier.*[59])

the strip. *Both the tensile-stress and current-crowding mechanisms propose that voids should nucleate at the cathode via region and subsequently coalesce and grow.*

- *Interfacial diffusion mechanism:* In this scenario (Fig. 10.17*c*), the nucleation and migration of voids along the Cu–cap-layer interface and finally accumulation at cathode end occurs based on in situ EM observations that voids nucleate homogeneously at the Cu-cap-layer interface and migrate toward the cathode end to cause failures.[61] Some believe that the void nucleation is due to delamination of the copper from the cap layer if a sufficiently high normal stress is achieved.[47] Indeed, the Cu-cap interface provides an easy migration path for ions owing to the high diffusion coefficient associated with the low surface migration activation energy, especially in Cu dual-damascene interconnects.

10.6 Process Effect on Electromigration

10.6.1 Cu/Low-*k* Interaction, Cu/Low-*k* Interface Control

The interfacial diffusion mechanism discussed in Sec. 10.5.3 suggests that the condition of the top interface of the Cu plays a key role on EM reliability. Numerous studies have demonstrated that EM lifetime depends on the adhesion behavior between the cap layer and the Cu surface.[62–64] In order to suppress this fatal interface diffusion in Cu damascene interconnects, great efforts have been undertaken. For example, a two-orders-of-magnitude improvement in Cu EM lifetime was reported when comparing Cu lines coated with a 10-nm-thick electroless CoWP film with lines with a 35-nm-thick SiC_xH_y surface film.[65,66] More details are given in this section.

Effect of an SiN Cap NH_3 Plasma Pretreatment Process on Electromigration

The basic idea is to improve EM reliability by achieving good adhesion in order to suppress Cu migration along the Cu-cap interface. An in situ NH_3 plasma preclean process has been employed prior to SiC_x cap deposition to promote adhesion by removing residual Cu oxide after Cu chemical-mechanical polishing (CMP), providing better performance than in devices treated with hydrogen (H_2) plasma.[67,68] Failure analysis reveals that there is Cu oxide at the interface of SiC_x samples with H_2 plasma treatment, and that accelerates Cu EM diffusion. A more aggressive NH_3 treatment could improve the EM reliability by enhancing the adhesion between Cu and the overlying cap on removal of the copper oxide and hence effectively mitigate the rapid Cu diffusion pathways along the Cu-cap interface.[63] An H_2-plasma-treated sample has a lower E_a

(0.67 eV) than an NH_3-plasma-treated sample (0.72 eV). A cap of SiC_xN_y combined with NH_3 plasma treatment gives an even higher E_a of 0.91 eV.[67]

Similarly, Ang et al.[69] reported that extended NH_3 treatment with an SiN cap could improve the EM reliability considerably by increasing the activation energy E_a from 0.59 eV for the short treatment to 0.85 eV for the long treatment[69] because prolonging the NH_3 treatment could inhibit one of the EM degradation mechanisms—diffusion along the Cu-cap interface.[70] In addition, the lognormal sigma value for the short treatment is 0.69, which is much larger than that for the long treatment (0.36), indicating that the long treatment also can tighten the TTF distribution.

However, Vairagar et al.[33] reported that H_2 treatment with an SiN_x cap provides a better EM performance than NH_3 treatment because there is a much higher number of Cu-Si bonds at the Cu-SiN_x–cap interface in the H_2-treated sample, as detected by x-ray photoemission spectroscopy (XPS) measurements. H_2 plasma treatment can form hydrides on the Cu surface with no change in microstructure.[71] Improvement in EM performance owing to hydrogen also has been attributed to interfacial segregation of hydrogen and the reduction of defects by athermal annealing.[72]

The high-energy bombardment during precleaning of the cap layer promoted good adhesion between the Cu surface and the SiN cap layer, resulting in excellent EM performance. However, it is warned that the high-energy bombardment may seriously damage Cu bulk because of crystal defects and then worsen stress-migration performance,[73] and there also could be potential plasma-induced damage to the devices.[69]

Copper Silicide Formation on the Cu Surface

The formation of Cu_xSi layer on the Cu surface by a plasma enhanced chemical vapor deposition (PECVD) self-aligned barrier (PSAB) process[74] or SiH_4 plasma treatment[33] can greatly improve interface adhesion. This process is selective because the SiH_4 reacts only with the Cu surface to form Cu_xSi, but does not react with the low-k material. Cross-sectional transmission electron microscopy (TEM) before EM tests (Fig. 10.18) clearly indicates copper silicide formation at the Cu-cap dielectric interface for both H_2- and silane-treated specimens. The adhesion between the Cu and the dielectric layer on top thus is much improved, resulting in more than twice EM improvement and a 10 to 1000× increase in time-dependent dielectric breakdown lifetime over the standard PECVD process.

However, the silicide formation process can lead to diffusion of Si into the Cu line through the grain boundary, which causes line resistance to increase.[75] Hayashi et al.[76] proposed a Ti barrier (to replace the traditional Ta barrier) combined with the silicide-cap process to form Ti-silicide at the surface of the Cu line and thus suppress Si in the silicide-cap diffusion into the Cu line and enhance adhesion.

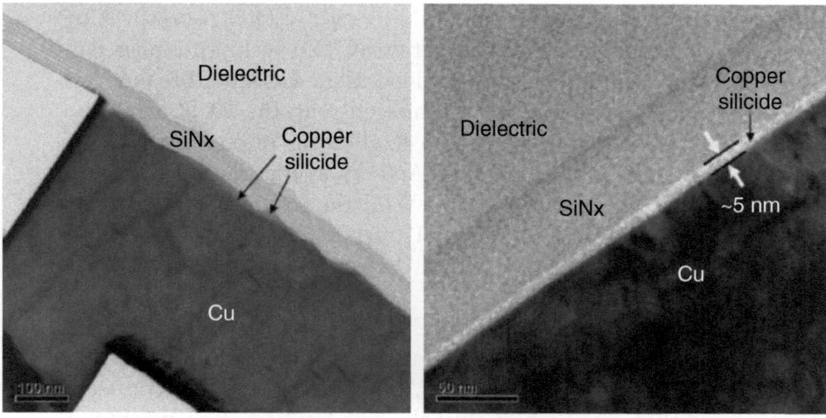

(a) H_2-treated specimen (b) SiH_4 treated specimen

FIGURE 10.18 Cross-sectional TEM images indicating copper silicide formation at the Cu-cap dielectric interface for (*a*) H_2-treated and (*b*) silane-treated specimens. (*Reproduce with permission from the Electrochemical Society.*[33])

A Ti barrier is proposed for the following two reasons: First, Ti in a Ti barrier can diffuse into the Cu line and locate at the Cu-cap interface.[77] Second, Ti-silicide formation energies (TiSi 129.8 kJ/mol and Ti_5Si_3 579.8 kJ/mol) are higher than that of Cu-silicide (Cu_3Si 13.6 kJ/mol),[78] suggesting that Ti-silicide is more stable than Cu-silicide. The results show that the mean-time-to-failure (MTTF) of a device with a Ti barrier (but without a silicide cap) is one order longer than that of a sample with a Ta barrier (also without a silicide cap). What's more impressive is that with both a Ti barrier and a silicide cap, a 100× MTTF is achieved. The corresponding EM activation energy E_a is 1.45 eV.[76] The underlying mechanism is that in the sample with a Ti barrier without a silicide cap, Ti atoms gather at the Cu line surface and reduce Cu atom transport during EM testing only at those parts where Ti appears, and a one order MTTF improvement is gained. On the other hand, in the sample with a Ti barrier with a silicide cap, almost all the Ti is located at the grain boundary of the Cu line surface during EM testing, providing a two order MTTF gain.

Effect of Overlayers (CoWP, Ta/TaN, SiN_x, $SiC_xN_yH_z$)

The top surface of the Cu damascene line is covered with a thin dielectric diffusion barrier layer (often referred as a *cap layer*), for example, SiN_x, SiC_x, $SiC_xN_yH_z$ etc., whereas the bottom surface and two sidewalls are sealed with a metal liner (also often referred as a *barrier*), for example, Ta, TiN, TiSiN-based liner (barrier effect on EM will be covered in Sec. 10.6.3). A thin CoWP or Ta/TaN cap on top of the Cu line surface has been demonstrated to be able to significantly reduce interface diffusion and, thus, improve the EM lifetime when compared with lines capped with SiN_x or $SiC_xN_yH_z$.[79] A cobalt (Co) cap layer by chemical-vapor-deposition (CVD) technique also has proven to improve the temperature-dependent dielectric breakdown (TDDB) and EM performance.[80] Compared with the SiCN cap only, the CoWP-capped samples showed a 40× lifetime improvement with a small lifetime variation ($\sigma = 0.34$).[81] Activation energies for EM were found to be 2.0, 1.4, and 0.85 to 1.1 eV for the Cu lines capped with CoWP, Ta/TaN, and SiN_x or $SiC_xN_yH_z$, respectively.[64] The CoWP cap is promising to improve EM with a dielectric constant $k \leqslant 2.4$ for ultralow-*k* materials.[82]

When compared with a plasma CVD self-aligned barrier (PSAB) (where the barrier is located between the Cu and the SiC cap layer) and the CuAl alloy seed techniques,[83] the CoWP cap interconnects showed not only high reliability but also outstanding efficiency in the suppression of any increased resistance.

A schematic comparison of the SiCN cap and the CoWP cap is given in Fig. 10.19 in terms of resistance shift during EM stressing and the failure analysis (FA) morphology. An SiCN cap (mode I) gives a small initial resistance increase but short TTF, representing void formation at the cathode via corner, as shown in Fig. 10.19*a*. A CoWP cap (mode II) has a large initial resistance jump but longer TTF, representing void formation in the trench line far away from the via, as shown in Fig. 10.19*b*. For the CoWP cap, interfacial diffusion is significantly suppressed by the metal cap. Instead of moving along the interface, microvoids can be readily trapped at the local interface, where either a large grain resides or an interface defect is present. The void will stay there and grow larger until it finally causes interconnect failure. For both modes, the Ta-based barrier remained intact and provided electrical redundancy after void formation.

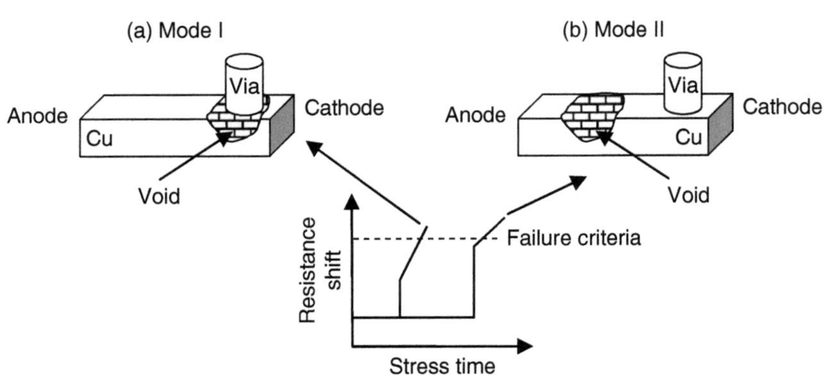

FIGURE 10.19 Schematic of two modes of failure. In mode I, a typical resistance shift has a small initial resistance increase but short time to failure (the case for an SiCN cap). In mode II, the typical resistance shift has a large initial resistance increase but long time to failure (the case for a CoWP cap.)

10.6.2 Control of Cu-Interconnect Microstructure

Interconnect microstructure can affect the nucleation and evolution of EM-induced voids.[84] Microstructure parameters of copper interconnects such as grain size, orientation, and stress should be taken into consideration.[85] Prior to the 65-nm-technology node, EM in Cu conductors with a bamboo-like grain structure differed from that in Al or Al alloys in that the mass-transport mechanism, at typical IC operating conditions, was almost exclusively interfacial diffusion in Cu as opposed to grain boundary diffusion in Al or Al alloys[86–88] as detailed in Sec. 10.6.1. As a result, Cu EM is relatively insensitive to grain sizes compared with Al-based interconnects and is more a function of interfacial quality.

However, with the technology advance to 65-nm node and beyond, small grains agglomerated and mixed with large bamboo grains in Cu damascene lines narrower than 90 nm.[89] This resulted in further degradation of EM lifetime owing to the increased contribution of grain boundary diffusion. The EM activation energy E_a of the Cu grain boundary is found to be 0.79 eV, which is lower than that for Cu along the Cu–amorphous a-$SiC_xN_yH_z$ interface (0.95 eV).[89] Improvement in EM reliability for the 32-nm-node technology and beyond will be a great challenge. The grain size of the Cu wire can be controlled by adjusting the wafer annealing step after Cu electroplating and before Cu CMP,[89] or changing the current profile during electroplating.[81,90]

Zhang et al.[81] reported that the EM lifetime of "large grain" structures (average grain size 215 nm) is twice of that of "small grain" structures (average grain size 123 nm) when a CoWP cap is not used. What's more, when a CoWP cap is applied compared with the case without a CoWP cap, EM lifetime increased about 100× and 24× for "large grain" and "small grain" structures, respectively. If we assume that the interfacial diffusion for the same cap is independent of grain structure, the higher EM lifetime for the "large grain" structures can be attributed to the decrease in mass transport along the grain boundary. When the interfacial diffusion is reduced to a point that it contributes little to the mass transport (e.g., when the CoWP cap is applied here), the grain-boundary diffusion will play a more dominant role in controlling EM lifetime; a small variation in the grain size may induce a large difference in EM performance.[81]

It was found that grain size of the Cu line is proportional to the temperature coefficient of resistance (TCR) of Cu lines based on TEM characterization. The Cu lines with lower TCR values have a higher percentage of fine-grain structure and worse EM performance.[91] Thus, TCR could be a good Cu microstructure index for first-order estimation.

10.6.3 Barrier/Seed-Layer Effect

Barrier-layer (also frequently referred as the *liner*) integrity has been well known as one of the key factors affecting Cu interconnect reliability.[92]

Although the primary functions for the barrier are to provide good adhesion to the surrounding insulator and to prevent Cu from diffusing into the dielectrics to form a leakage path, the liner also has been found to serve as a redundant layer so that the electric current still can pass through in the Cu-voided area. The interface between the liner and the Cu also has been a key from a reliability point of view because good adhesion between the liner, and the Cu can effectively slow down Cu diffusion[93] and help to reduce the probability of void formation within or underneath the vias. In this section, the effects of the barrier/seed layer on EM performance are summarized.

Barrier Coverage in Vias

The weakness of the Ta barrier and Cu seed coverage at via sidewalls can create a reliability concern.[94] It is known that EM early failure is often related to the Ta diffusion barrier coverage in the via.[38,95] It is quite accepted that the integrity of the barrier coverage in the via is one of the keys to reducing or even eliminating EM early failure. High-temperature baking also may lead to oxidation of the Ta barrier, which further leads to easier Cu migration and possible Cu outdiffusion.[96] A TaN barrier will be better for resistance to barrier oxidation[97] and thus a benefit to both EM and stress migration.

The EM lifetime is less bimodally distributed with increasing barrier thickness on the via sidewall, resulting in better EM reliability for the upstream structure[92] (see Fig. 10.9*a*). Failure analysis on the early mode of thin barrier splits (thin at the via wall) shows EM-induced void formation in the middle of the cathode via (Fig. 10.20*a* and *b*). The same via void is also observed in a large-scale via with tetraethylorthosilicate (TEOS) interlevel dielectric (ILD) (Fig. 10.20*c*), indicating that this is not a material-related issue. Figure 10.20*e* shows that there is no (or maybe extremely thin) barrier coverage on the via sidewall toward M_2 before Cu seed deposition. This nonuniform barrier coverage is due to a shadow effect in a physical-vapor-deposition (PVD) barrier process for a high-aspect-ratio via. The angle of PVD barrier beam reaching the via sidewall toward M_2 can be limited by M_2 patterns, resulting in thin and nonconformal barrier coverage in the same region as schematically illustrated in Fig. 10.21. When the barrier is extremely thin, a fast diffusion path can be created at the Cu-barrier interface,[95] leading to via void formation corresponding to early-mode EM.

Another via barrier effect is oxidation of a thin barrier having a temperature dependence on a thin and nonuniform barrier coverage at the via bottom, and a gradual increase in via resistance is expected.[95,98] This also results in extreme bimodal behaviors affecting apparent EM activation energy and lognormal parameters (MTTF and σ). In summary, barrier coverage inside vias has a significant impact on upstream EM reliability.

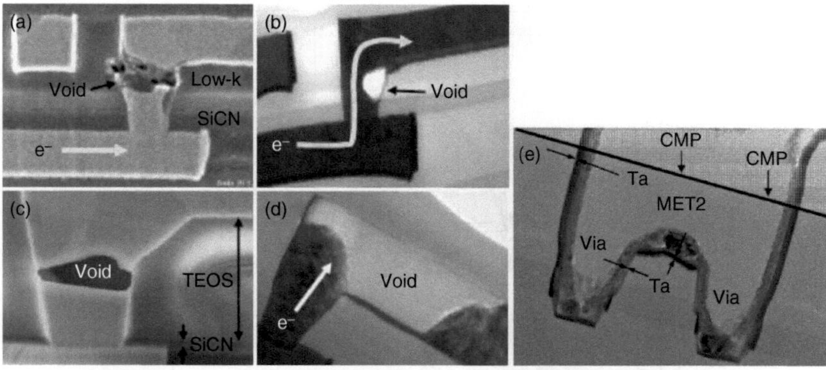

FIGURE 10.20 (*a*) A thin barrier split (thin at the via wall) shows an EM-induced void formation in the middle of via. (*b*) An initial void starts at the thin via sidewall connected to M_2. (*c*) The same via void is observed at the thin barrier area of the large-scale via with TEOS ILD. (*d*) In an optimized barrier split, a normal EM void (for Cu interconnects with an SiCN cap layer and Ta barrier) is fully spanning a metal trench. (*e*) TEM image taken from a dual-damascene (DD) chain (similar to an EM test structure, but with a very short M_2) after PVD barrier deposition shows that there is no (or maybe extremely thin) barrier coverage on the via sidewall toward MET_2.(*Used with permission from IEEE.*[92])

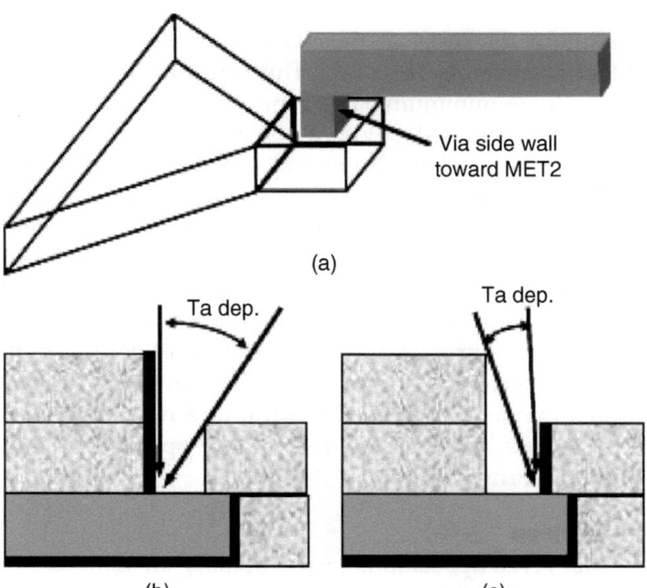

FIGURE 10.21 Schematic diagram of DD Cu interconnects showing that the metal above the via causes a shadow effect for PVD barrier deposition. The via sidewall facing the metal line end has a narrower angle for PVD Ta beams, which will result in thinner barrier deposition in this area.(*Used with permission from IEEE.*[92])

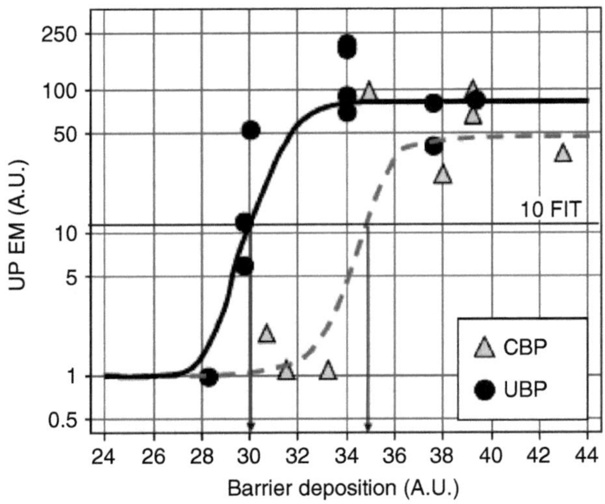

FIGURE 10.22 Upstream EM lifetimes ($t_{0.1\%}$) at a reference condition as a function of barrier deposition. CBP stands for a conventional barrier process. UBP stands for a uniform barrier process. The minimum barrier deposition meeting 10 failures in time (FIT) for CBP and UBP can be determined to be 35 and 30 (a.u.) of barrier deposition. (*Used with permission from IEEE.*[92])

The minimum thickness of the via barrier seems to be more important than the average thickness. This is justifiable because it is the location with the minimum barrier thickness that is the weakest point. Figure 10.22 shows that with a conventional barrier process (CBP), upstream EM improves dramatically above 35 (a.u.) of barrier deposition.[92] This abrupt transition behavior indicates that a critical amount of barrier deposition is needed to achieve a conformal via barrier (or eliminate a weak point in the via barrier). It is interesting to see that with improved uniformity in barrier thickness for the trench and the via through a so-called uniform barrier process (UBP), the minimum barrier deposition meeting 10 failures in time (FIT) can be ~14 percent less than with CBP. Compared with CBP, UBP also has a thinner trench barrier and thus a much lower resistance. Therefore, the barrier process should be optimized for both reliability and performance.

Via Barrier Preclean

Prior to barrier-layer deposition, conventionally an ionized argon (Ar^+) plasma is employed to clean the Cu surfaces (with an oxidation layer) under the via bottoms by physical bombardment in order to reduce via resistance and improve adhesion. The preclean process should be optimized (e.g., through gas flow and bias power of the argon plasma) to achieve reliability requirements because the physical bombardment can induce Cu recessing from the underlying metal layer, and then copper ions are resputtered onto the via bottoms and sidewalls.[73]

Via barrier preclean can result in two different types of via geometries[92]: a cone-shaped via (sometimes referred as *via gauging*[37]); (Fig. 10.23*a*) and a flat via (see Fig. 10.23*b* and *c*). The cone-shaped via is caused by an overetch of the via barrier that also removes the Cu in the lower metal. This kind of deep punch-through of the via into the metal beneath can improve the via bottom interface. On the other hand, the flat via is formed by controlling the via-barrier etch process. In this case, barrier integrity can be improved by limiting Cu resputter on the via sidewall. The via morphology is found to have an impact to the EM behavior, as detailed below:

- *Upstream EM structure.* The cone-shaped via EM is worse than that of the flat via when both have comparable barrier thicknesses. This is explained as follows for an M_1-M_2-M_1 structure. In the cone-shaped via etch, a high-energy Ar^+ sputter process is used in a DD structure. As a result, as shown in Fig. 10.24, clear V_1 punch-through into M_1 and M_2-trench punch-through into the inter-metal dielectric (IMD) (under M_2) are observed. It is likely that the Cu in M_1 and the dielectric in IMD (under M_2) are sputtered and redeposited on the sidewall, which will degrade barrier integrity. A faster diffusion path is likely to form at the interface between the Cu and barrier and reduce upstream EM TTF.
- *Downstream EM structure.* Downstream EM of the flat via is very sensitive to process conditions. For an optimized process, flat-via EM is comparable with cone-shaped via EM. However, for an unoptimized process, flat-via EM could be much worse. Figure 10.23*a* shows that for cone-shaped vias,

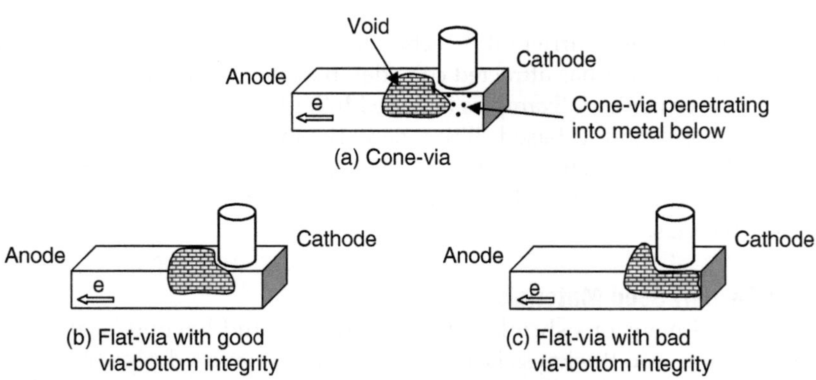

FIGURE 10.23 Schematics of the void location in down stream structure for (*a*) a cone-shaped via penetrating into the metal below; (*b*) a flat via with good via-bottom integrity; and (*c*) a flat via with bad via-bottom integrity.

the EM-induced void has a difficult time migrating to be just beneath the via because the cone-shaped via with barrier can effectively block void growth there. As a result, the cone-shaped via EM TTF is long. On the other hand, for the flat via, two typical void locations were observed, as shown schematically in Fig. 10.23*b* and *c*. When the via-bottom barrier has good integrity, the void will be somewhat blocked near the via corner, as shown in Fig. 10.23*b*, giving a long EM TTF. However, if the via-bottom barrier is not good owing to an unoptimized process, the void will migrate along the via-bottom–metal-below interface so that a thin void grows under the via and causes failure in a very short time, as shown in Fig. 10.23*c*. Therefore, flat vias are sensitive to via processes that affect the via-bottom interface. In contrast, cone-shaped vias are less sensitive to via processes, because the via-bottom interface has been resurfaced by extensive Ar sputtering causing a punch-through during the via barrier etch.

Ultrathin Barrier/Seed Layer

The thickness of the barrier layer should be decreased to reduce the line resistance along with the interconnect scaling. The required thickness is below 3 nm for a 22-nm node.[99] Continuous development is underway to extend the present PVD Ta- or Ti-based barrier technologies. A few efforts are listed below:

- CVD and atomic layer deposition (ALD) have been developed as alternative deposition methods for precise thickness and uniformity control.[100]
- Electroless plating can be a candidate for barrier and metal cap deposition owing to the superior coverage and good selectivity.[101]
- Self-forming barrier (SFB) technique to form an extremely thin barrier layer has attracted considerable attentions as an alternative method. Examples include $MnSi_xO_y$ grown from CuMn seeds[102] and Ti-based SFB, where Ti can react with various dielectrics such as SiOC, SiCN, and SiCO, thus forming a continuous barrier layer along the sidewall of the trench to the vias.[103]

Novel Barrier/Seed Materials

As device dimensions shrink to the 32-nm node and beyond, scaling of the Cu seed thickness is also required to achieve a defect-free Cu gap fill during the subsequent Cu electroplating process, which leads to new challenges to retain both resistance and EM reliability.

However, the constraint of forming a continuous Cu film on Ta/TaN (currently used as barrier layer) sets a fundamental limit on the minimum Cu seed thickness.[104] Therefore, searching solutions to replace Ta/TaN as the barrier is ongoing. Some potential candidates are given below.

Noble metals and their alloys, such as Ru,[104–108] RuTa,[104,109,110] and Co,[110–112] which allow electroplating of Cu directly on them, have been proposed and examined as seed enhancement layers, which improve Cu gap fill by acting as a supplemental wetting layer, thus improving EM performance in advanced-technology nodes.

On the other hand, after being ignored for a number of years, Ti-based barriers recently have regained attention for Cu interconnect applications extendable to the 32-nm node and beyond.[77] Compared with a Ta-based barrier, a Ti barrier has several disadvantages and advantages.[77,113]

10.6.4 Effect of Solute/Doping on Electromigration

Cu doped with impurities (or sometime referred as a *Cu-alloy*), such as Al,[27,28] Ag,[114] Ti,[115,116] Sn,[116] or Mn[117] can effectively improve the EM lifetime.[52,118] The dopants are typically introduced into the Cu seed layer. During subsequent anneals, the impurities segregate at grain boundaries and interfaces, including the critical interface between the Cu and the capping layer. The EM lifetime is improved over 10× by the suppression of Cu diffusion owing to, for example, the piled-up of Al at the Cu-SiCN interface of the Cu interconnects. A higher doping concentration will result in higher EM lifetimes. Iguchi et al.[119] demonstrated that CuAl seed can improve the E_a to 1.03 eV from the pure copper E_a of 0.85 eV. Tonegawa et al.[116] showed that the CuSn seed layer has a higher endurance for EM with tight distributions than a CuTi seed layer, which is also better than pure Cu. The EM lifetime of CuSn was 9× longer than that of pure Cu.

The main issue with this approach is the resistivity increase in Cu owing to the segregation of Al dopant atoms to the interface, leading to increased surface scattering.[27] It is interesting to see that the Cu drift velocity decreases as the Al impurity concentration increases, which is independent of the line width.

A Cu-CNT (carbon nanotube) composite material for via and line interconnect applications was demonstrated to have more than a 5× longer EM lifetime than Cu by using kelvin and Blech test structures promising for application to advanced-technology nodes.[46]

10.6.5 Effect of Dual-Damascene Structure Profile

With technology marching toward the 45nm node and below, EM bimodality behavior with early failures is difficult to be improved on only with barrier/seed process tuning itself, such as TaN/Ta

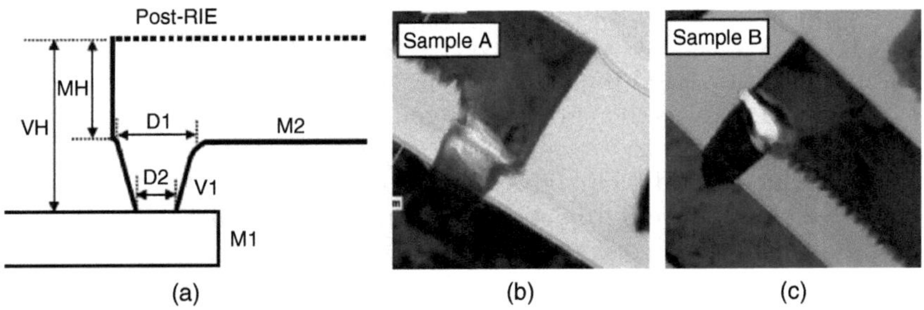

FIGURE 10.24 (*a*) Key factors defined for the DD structure: *MH* is postetch trench depth, *VH* is postetch via height, D_1 is via open CD at the chamfer area, and D_2 is via bottom CD. (*b*) TEM cross-sectional micrograph for via-void mode. (*c*) TEM cross-sectional micrograph for chamfer-void mode. (*Used with permission from IEEE.*[94])

thickness, sputtering energy of deposit species, Cu seed bias power and thickness and so on. Optimization of dual-damascene (DD) structure profile is found necessary to minimize or even eliminate early failures.[94] Figure 10.24*a* defines two key aspect ratios (ARs) as

- *Via AR* = VH/D_2 (postetch via height divided by via bottom CD), and
- *Chamfer AR* = MH/D_1 (postetch trench depth divided by via open CD at chamfer)

These account for the two different early failure modes shown in Fig. 10.24*b* and *c*, respectively. Table 10.2 clearly shows the correlation between the via/chamfer ARs and the via/chamfer void modes. As a result, it is interesting to see that reducing IMD thickness and trench etch depth or increasing via CD can lower the AR and thus effectively minimize early failures in the upstream EM structure.[94] However, it is cautioned that reducing the metal height may increase the *RC* delay, and increasing the via CD may impact TDDB performance. Thus a balance between the EM reliability and *RC* electrical performance and TDDB reliability should be achieved.

	Via AR		Chamfer AR		Early Fail Mode	
Wafer	**Value**	**Level**	**Value**	**Level**	**Via**	**Chamfer**
W1	5.6	High	2.11	High	Yes	Yes
W2	4.84	Medium	2.16	High	Yes	Yes
W3	4.57	Low	1.96	Medium	No	Yes
W4	4.57	Low	1.75	Low	No	No

TABLE 10.2 EM Failure Mode versus AR of Processes with Different Via/Chamfer ARs (*Used with permission from IEEE.*[94])

10.6.6 Effect of Oxygen Content on Cu Interconnects

An oxygen-free interface between the Cu upper surface and the SiCN cap can improve the adhesive strength and thus reduce the EM and stress migration (SM) degradation owing to suppression of vacancy diffusion at the interface.[120,121] Oxygen atoms in the electrochemically deposited (ECD) Cu film also adversely impact EM and SM reliability because oxygen atoms can diffuse to the Ta barrier during crystallization annealing, and Ta is more preferably oxidized than Cu.[96] The oxygen atoms mainly diffuse along the Cu-grain boundary than in the bulk Cu.[122]

To prevent the Ta barrier from oxidizing, a thin Ti layer was deposited on the pure ECD Cu film before crystallization annealing.[115] Since the standard Gibbs-free energy of Ti oxidation is negatively larger than that of Ta as well as Cu, the oxygen atoms from the ECD Cu are absorbed selectively into the *Ti layer as an oxygen absorber (OxA)*. Since Ti is hardly removed by Cu CMP slurries, the Ti OxA was removed by wet etching followed by Cu and Ta/TaN CMPs. The oxidation-free Ta barrier remarkably improved the reliability. Three different methods of Ti addition were proposed; (1) Ti layer insertion under Ta/TaN stacked barrier metal, (2) Ti layer insertion between a Ta/TaN barrier and Cu, and (3) Ti doping from the surface of the ECP Cu film. The first method drastically suppresses the stress-induced voiding (SIV) under the via connected to a wide lower line owing to adhesion improvement by Ti at the via bottom, whereas EM is not improved. In the second method, by contrast, EM is improved, but the SIV resistance is degraded. The Ti doping from the bottom surface of the Cu film restricts the grain growth and increases the tensile stress, enhancing the SIV. The third method improves not only SIV but also EM resistance.

Alternatively, *Al oxygen absorber (Al OxA)* is another option to reduce Ta oxidation because the standard Gibbs-free energy of Al oxidation is smaller than that of Ta and Cu, as shown in Fig. 10.25.[120] Several Cu metallization structures were evaluated, as shown in Fig. 10.26,[120] where the structure III is the so-called low-oxygen-content (LOC) CuAl alloy with which no barrier metal (Ta) oxidation was achieved because the oxygen atoms from the ECD Cu are captured by the Al OxA during crystallization annealing. As a result, structure III with the LOC CuAl alloy has a 140× EM TTF compared with conventional structure I. A brief summary of Fig. 10.26 is given below:

- *Structure I:* This is the conventional ECD Cu with the PVD Cu seed, where the oxygen in the CuO_x on the ECD Cu surface diffuses into the Ta barrier to be oxidized as follows:

$$CuO_x \rightarrow Cu + O \qquad \text{on the ECD Cu surface}$$
$$Ta + O \rightarrow TaO_x \qquad \text{on the Ta barrier}$$

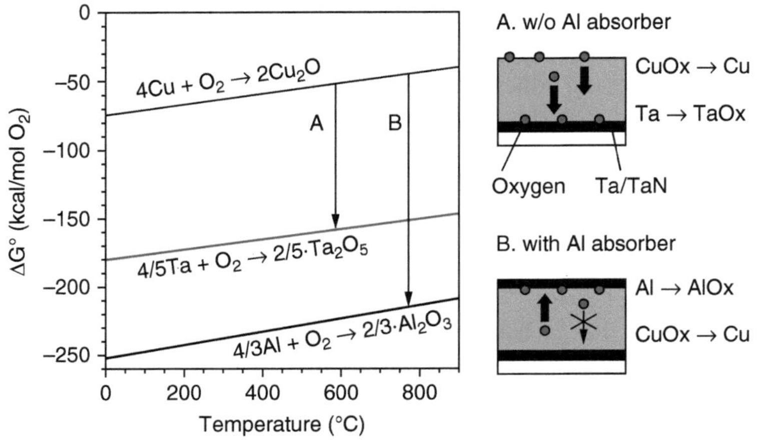

FIGURE 10.25 Standard Gibbs-free energies of oxidation as a function of temperature. Since aluminum (Al) has smallest ΔGO_x, oxygen atoms tend to be captured by Al, not by Ta or Cu, according to metallurgical thermodynamic principles. (*Used with permission from IEEE.*[120])

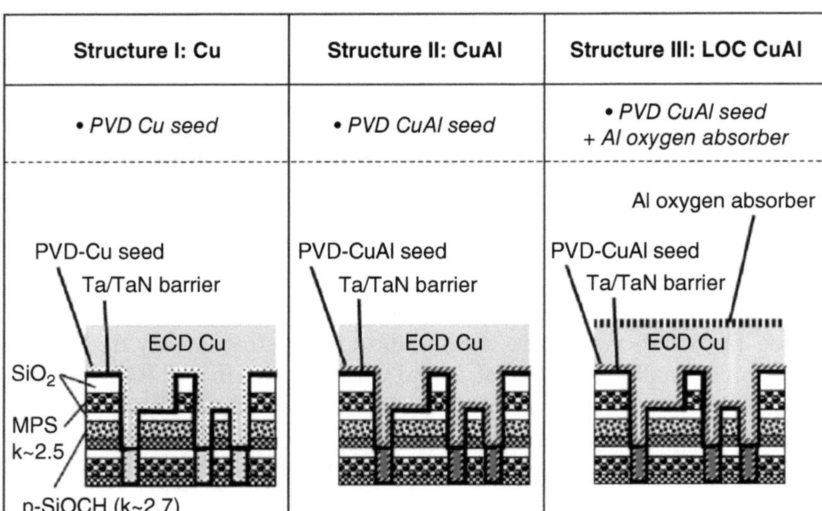

FIGURE 10.26 Cu-metallization structures evaluated. (1) structure I: conventional Cu DDIs using Cu liner; (2) structure II: conventional CuAl alloy DDI using CuAl liner: (3) structure III: LOC CuAl DDI using Al OxA and CuAl liner. (*Used with permission from IEEE.*[120])

Alternatively, if an Al OxA layer is deposited on the ECD Cu with the PVD Cu seed, all the oxygen atoms will be captured in the Al OxA, and no oxygen will diffuse to the Ta barrier, resulting in an oxygen-free interface of the Cu-Ta barrier.

- *Structure II:* When the CuAl alloy seed is used instead of Cu seed, the oxygen atoms on the ECD Cu surface can again diffuse toward the Ta barrier to form TaO_x. In addition, the diluted Al in the CuAl seed captures some of the oxygen atoms before they reach the Ta barrier, resulting in a higher oxygen concentration than that in pure Cu films. However, the diluted Al in the CuAl seed is not strong enough to prevent the Ta from oxidizing.
- *Structure III:* By depositing the Al OxA, all the oxygen atoms on the ECD Cu are captured, and no oxygen diffusion occurs to the Ta barrier, resulting in the LOC CuAl film with an oxygen-free high-quality CuAl-Ta interface. However, there could be an oxidation layer on the surface of the CuAl seed prior to the ECD Cu deposition.

10.6.7 Effect of Preexisting Voids on Electromigration

Voids can form in the interconnect during a series of fabrication processes. Depending on the specific process, the voids can form in any location of the interconnect with wide ranges of sizes and various shapes. Also, these voids can grow, shrink, move, and change their shapes depending on the initial condition of the voids and their surroundings.[85] The shape and location of the preexisting voids have a significant effect on the reliability of the Cu interconnect, especially when the metal lines become narrower and narrower in advanced-technology nodes.[123]

One of the critical processes in Cu interconnect fabrication that leads to void formation is Ta barrier metal and Cu seed layer deposition before electroplating.[123]

- *Side void (i.e., void between Cu and barrier metal)*: This is the case when the Cu seed layer is too thin, with some discontinuous regions exposing the barrier metal. The exposed barrier metal (Ta) easily oxidizes and makes Cu unable to grow during electroplating, which eventually leads to voiding between Cu and the barrier metal. In side-voided samples, no early failure was observed, and the failure shows no bimodal trend.
- *Center void (i.e., void between Cu and capping layer)*: This is the case when the Cu seed layer is too thick, causing overhang at the top of the trench that eventually leads to a void between Cu and the capping layer. *Bimodal TTF distribution* was observed in center-voided samples with abrupt resistance rises during EM testing, especially at high current density. It is believed that it is the higher local current density around the void for the center-void case that leads to the bimodal EM TTF distribution.

There is a range of seed-layer thickness for minimizing both center and side voids, and this range is expected to decrease with decreasing line width. When minimizing both types of voids is unachievable through process optimization and one type of the void must be accepted more than the other, the side void is preferred because it has a less negative impact to EM.[123] However, the side-void impact on other reliability issues such as IMD breakdown also should be assessed to get an overall picture for process optimization.

10.7 Effect of Structures on Electromigration

Interconnect geometry can affect EM performance and lifetime,[24,124–127] and will be detailed in this section. For example, line width, length change, and number of vias affect EM significantly, and so on.

10.7.1 Via/Line Interconnection Configuration

As elaborated in preceding sections, the interface between Cu and its cap layer is the fast Cu diffusion path.[128] However, it is also frequently observed that early EM failures are often associated with vias, either by voiding in the via for the *via-depletion mode* (see Fig. 10.13) or by voiding underneath the via for the *line-depletion mode* (see Fig. 10.14).

- For via-depletion EM, the liner process and via profile are the two key factors to suppress the early failure mode of the distribution.
- For line-depletion EM, the interconnect configuration between the via and the underlying line is critical to the failure characteristics.[55,129] Good integrity between the via and the line liner below can effectively prevent or minimize open-circuit EM failures. Redundant vias can significantly improve EM performance in both the median failure time (t_{50}) and the distribution shape (σ), depending on the arrangement of these vias relative to the line below.

A remarkable work on the effect of via/line interconnection configuration on EM is presented by Li et al.,[129] who studied 12 V_1/M_1 contact configurations at the cathode end of the Cu interconnect EM structures (schematic cross section shown in Fig. 10.10*b*) and categorized them into four groups, each having a few different V_1 arrangements:

- *Group I:* Single-V_1/M_1 connection, with V_1 contacting the M_1 liner (types A and B)

- *Group II:* Single-V_1/M_1 connection, with V_1 away from the M_1 liner (types C and D)
- *Group III:* Multiple-V_1/M_1 connections, with V_1 not contacting the M_1 liner (types E, F, G, and H)
- *Group IV:* Multiple-V_1/M_1 connections, with V_1 contacting the M_1 liner (types I, J, K, and L)

In these configurations, a typical characteristic is use of the liner as a redundant layer to elongate the EM lifetime effectively by carrying current when the Cu line is depleted so as to minimize EM failures with abrupt resistance jump (open circuit).

The EM characteristics of these structures are briefly summarized in this section. Note that in these experiments,[129] the stresses were performed at 300°C and at a current density of 2.5 MA/cm^2 in the Cu line, where the stress current is applied on the basis of the M_1 line width. Therefore, a higher current is applied to a wider line, leading to a larger stress current per via in case that the same number of vias is used.

Group I: Single-V_1/M_1 Connection with V_1 Contacting the M_1 Liner

In group I, two cases were included, as shown schematically in Fig. 10.27. In type A, a single V_1 is connected to an M_1 line below, where the M_1 line is narrower than the via width and the V_1 bottom is always connecting the M_1 sidewall liners. The characteristic is a resistance jump followed by gradual increase with the EM

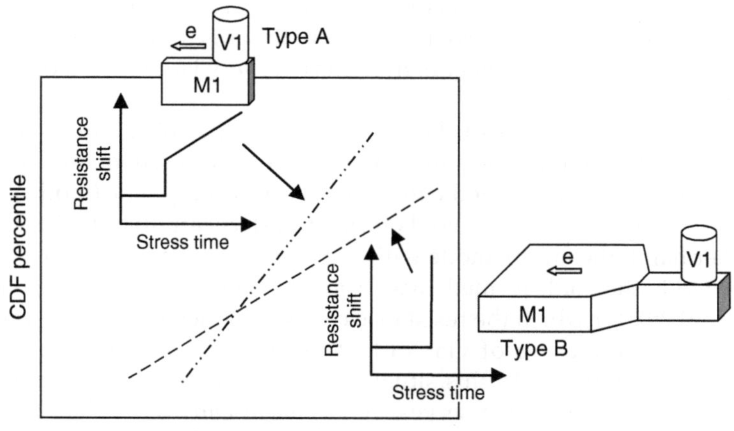

FIGURE 10.27 Schematic of the CDF distribution of the time-to-failure for type A and type B. The insets show the corresponding test structures and the characteristics of resistance shift versus stress time.

stressing time (see the inset in Fig. 10.27) because the via–line-liner connection serves very well as a redundant electric current path after voids form either at the via bottom or in the metal line underneath the via. As a result the failure time distribution in this case has a smaller σ (i.e., tight distribution) and behaves as a single-mode lognormal distribution (type A in Fig. 10.27).

In the type B scenario in Fig. 10.27, the majority part of the line is much wider than the via size, where a tapered line end is designed so that the line width near the end is narrower than the rest portion of the line. The purpose of this kind of design is to let the single via sit on both sides of the line sidewalls below so as to form a redundant current path similar to type A.[130] Compared with type A, the resistance change in type B (see inset in Fig. 10.27) looks typically subject to more abrupt jumping during EM stressing. Sometimes a nonlinear postjump resistance increase is also observed. As a result, a much broader failure distribution (larger σ) is observed in type B. This difference is thought to be caused by the higher stress-current applied in the tapered area (i.e., the narrow necking area) in type B because the untapered area is wider but with the same stress-current density as type A. Therefore, a faster void growth in the tapered section and higher joule heating after void formation under the via are expected to lead to the more abrupt resistance jump.

Group II: Single-V_1/M_1 Connection with V_1 Away from the M_1 Liner

Similar to type B, the metal width is much larger than the via size for types C and D. However, in types C and D, the single via is placed away from the line end and sidewalls and does not contact the liner. The curves of resistance change with stressing time consist of both abrupt jump and gradual increase (Fig. 10.28). The failure time distribution for these via/line-contact configurations is usually bimodal or single mode but with a very high sigma (i.e., very large spread of TTF distribution).

The void locations of the two different types of failure modes (i.e., abrupt jump and gradual increase) are given schematically in Fig. 10.29. The failures of the resistance abrupt-jump type (mode I in Fig. 10.29) are caused by voids formed underneath the via. On the other hand, the failure mode with a gradual resistance increase was related to the voids formed away from the via in the line (mode II in Fig. 10.29), for which the resistance change characteristics look very similar to the cases of via contacting the line liner below (see Fig. 10.27, type A). In this situation, although there is void in the metal line, the line liner serves well as electrically redundant layer requiring a larger void size to cause failure, and then the corresponding lifetime is long.

In short, the resistance abrupt-jump type fails contribute to TTF at small CDF percentages, whereas the gradual-resistance-change type fails contribute to the TTF at large CDF percentages, although the

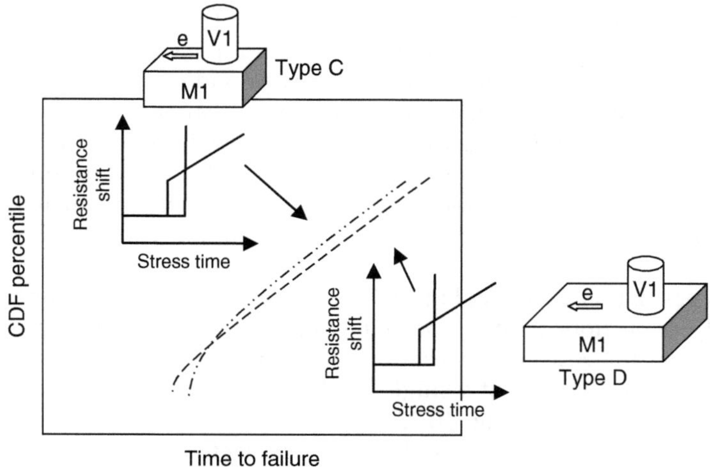

Figure 10.28 Schematic of the CDF distribution of the TTF for types C and D. The insets show the corresponding test structures and the characteristics of resistance shift versus stress time.

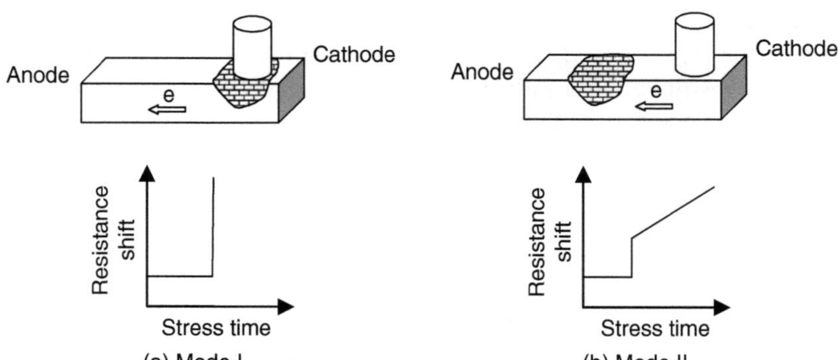

Figure 10.29 Schematics of the two failure modes. In mode I, typical resistance shift has an abrupt jump with the void located near the via. In mode II, typical resistance shift has a small initial increase followed by gradual increase with voids generated away from the void.

final TTF distribution apparently could be single mode, as shown in Fig. 10.28.

Group III: Multiple-V_1/M_1 Connections with V_1 Not Contacting the M_1 Liner

As discussed in Group II, for wide Cu lines connected to a single via, the EM TTF could be very widely distributed, resulting in a very small lifetime at a low percentage (e.g., $CDF_{0.1\%}$), which is not desirable from a production point of view. Multiple via connections have been a common practice because they provide advantages for reliability, yield, and contact-resistance considerations. Again the vias may or may not contact the line liner below. Furthermore, for multiple-via layouts, there are many more possible configurations for the via/line contact, even for the same number of vias, such as along the line length, along the line width, or various combinations thereof.

Types E, F, G, and H in Fig. 10.30 are four examples of multiple V_1/M_1 connections with V_1 not contacting the M_1 liner. The characteristics of the resistance variation with time for these examples are similar to the case of a single V_1/M_1 connection without via–line-liner contact (see Fig. 10.28). Most of the failures are abrupt resistance jumps, owing to the lack of electrical redundancy between V_1 and M_1. Figure 10.30 presents a schematic of the TTF distributions for these four types. It is clear that with redundant vias along the line length (or an extra row of vias) (types E and G), the failure distribution (both failure time and sigma) can be improved significantly comparing with types F and H. This could be attributed to the reservoir effect, as discussed in next section. Compared with the case of a single via without a via–line-liner connection (types C and D), it is expected that the redundant multiple vias result in better failure distributions, even for types F and H with only one row of vias.

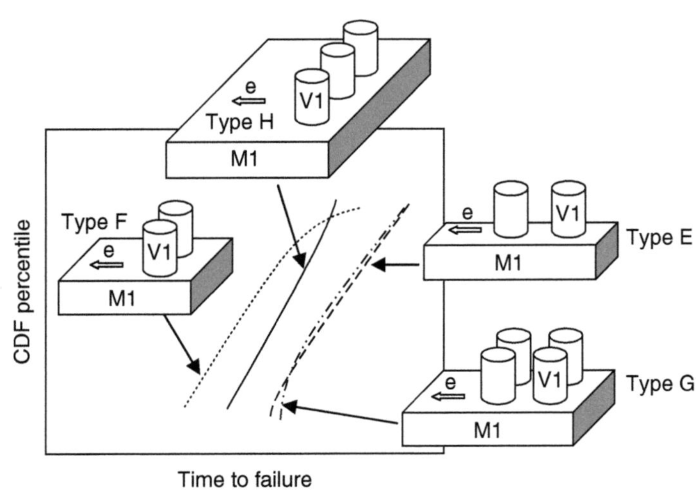

FIGURE 10.30 Schematic of the CDF distributions of the time-to-failure for types E, F, G, and H. The insets show the corresponding test structures.

Group IV: Multiple-V_1/M_1 Connections with V_1 Contacting the M_1 Liner

In group III (types E, F, G, and H), the wide lines are connected through redundant vias without via–line-liner contact, where the EM failures are still dominated by abrupt resistance jump, although the redundant via contacts clearly can improve the EM performance. To achieve electrical redundancy between vias and the line below for multiple vias with even better EM performance, the layout configuration for via–line-liner connection should be optimized. For example, types I, J, K, and L shown in Fig. 10.31 are promising designs for multiple-via/line connections with vias contacting the line liner underneath, and having the same M_1 line width. All these cases showed an increase in linear resistance (via/line electrical redundancy) after an initial resistance jump (as shown in Fig. 10.29*b*). As expected, the multiple-via/line connection with via–line-liner contact can very effectively elongate EM lifetimes by preventing failure from abrupt resistance jump. With the increase of the number of redundant vias (type K versus type J), the EM lifetime increased in terms of not only the better main distribution ($CDF_{50\%}$) but also the samller spread of the distribution (σ). For the same number of redundant vias (type I versus type J), the tapered Cu line end showed shorter TTF and a worse distribution as type B. Furthermore, when checking type K versus type L, it is clear that locating the vias in multiple rows can ensure a much longer EM TTF, even though the total number of vias is the same.

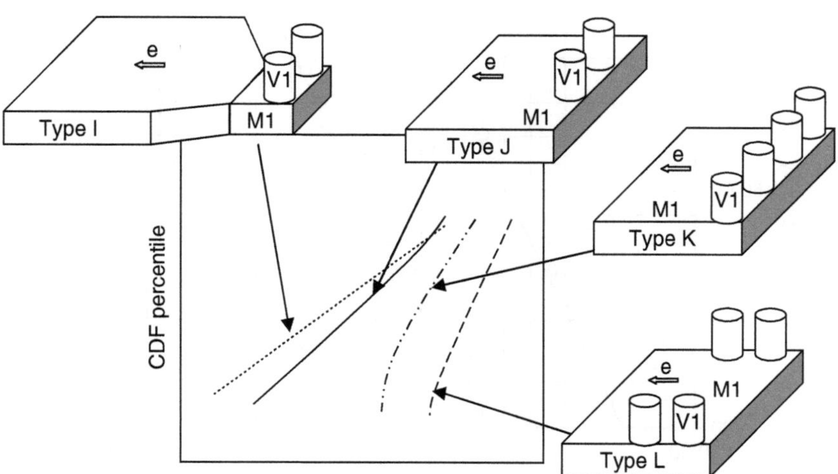

FIGURE 10.31 Schematic of the CDF distribution of the time-to-failure for types I, J, K, and L. The insets show the corresponding test structures.

10.7.2 Reservoir Effect (Line-Extension Effect)

Line extension (sometimes referred as *reservoir*) is effective in improving EM performance and can be easily implemented during design. There are several explanations for the underlying mechanism.

Jeon and Park[131] correlated the reservoir effect on EM with the vacancy concentration distribution in the reservoir region, assuming that no current flow there based on the stress distribution computed by the finite-element method. Gan et al.[132,133] found that in the V_1 upstream structure, as the extension length in M_2 increases from 0 to 60 nm, the median-time-to-failure increased from 50 to 140 hours, representing an ~200 percent improvement in lifetime (Fig. 10.32). However, further increments of the extension length from 60 to 120 nm did not result in any significant improvement in EM lifetime. Through finite-element modeling of the current distribution in the metal-via structure, it thus was proposed that there is a low-current-density zone in the M_2 extension corners above the cathode that retards void migration into the extension. Accordingly, only part of the extension volume could serve as an effective reservoir for void accumulation. Based on this analysis, a 60-nm extension appears to be the critical extension length beyond which increasing extension sizes have no effect on EM lifetime. An analytical model also was established that agrees very well with existing experimental results, indicating that the critical extension length is a function of void size and electrical

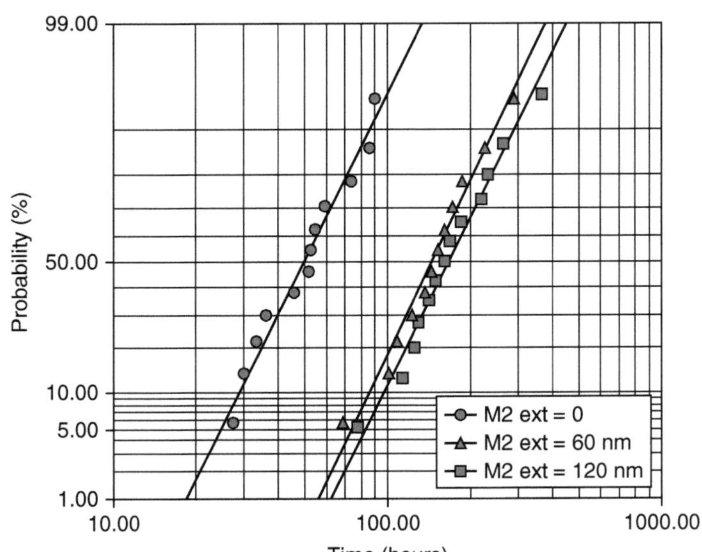

FIGURE 10.32 Cumulative distribution function (CDF) plot of median-time-to-failure for 280-nm-wide M_2 copper lines with three different M_2 extensions. *(Used with permission from Cambridge University Press.*[132]*)*

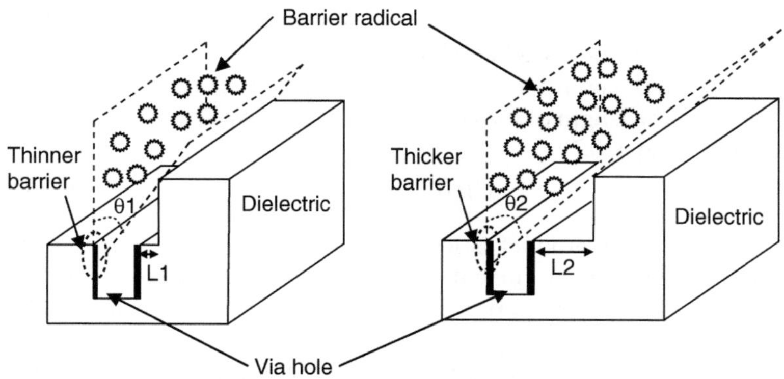

Figure 10.33 A schematic to show the line-extension effect on barrier coverage quality in vias.

field gradient. Some design guidelines for EM-resistant circuits could be generated by the model.

From a process integration point of view, Chen et al.[37] attributed the gain in EM resistance for longer line-end extensions to the barrier quality in the via. Figure 10.33 is a schematic highlighting the interaction between line extension and the PVD barrier-deposition process. With short line extensions (L_1), the real aspect ratio for a via is much higher than that for a longer line extension (L_2). Owing to the shadowing effect of the dielectric at line end, extension L_1 has a smaller entry angle (θ_1) for incoming barrier element species than extension L_2. As a result, good-quality barrier coverage in longer line extensions (L_2) is easier to achieve than in shorter line extensions (L_1).

10.7.3 Metal Critical-Length Effect

The Blech length effect[48] or metal critical-length effect is well known for Al interconnects, and it describes a phenomenon that occurs if the conductor is sufficiently short or the current density is sufficiently low that the mechanical stress gradient (backstress) generated at the line end in response to the EM driving force effectively stops EM behaviour. For Cu interconnects, a similar Blech effect is observed. The TaN/Ta barrier layers at the bottom of the via act as blocking boundaries. Hence, during EM stress, metal atoms will accumulate at anode, resulting in hydrostatic stress. This stress produces a backflux of atoms that is opposite in direction to the flux from EM. For interconnects that are short enough, the backflux of atoms prevents killer voids from forming, and the interconnects are immortal.[134] Therefore, it is possible to avoid EM problems in local interconnects by limiting the length of the wires. However, it is quite common that the backstress in the low-*k* materials may not be able to grow large enough to completely counteract the electron wind force.[43] No immortality for

EM failure was observed even at jL values as low as 375 A/cm for a Cu–low-k (carbon-based CVD SiO_2) system.

The Blech condition is met when the product of the current density j and the length of the conductor L is below a critical value $(jL)_{crit}$ related to Eq. (10.7):

$$j \cdot L \leq (jL)_{crit} \tag{10.16}$$

If the Blech product is less than $(jL)_{crit}$, EM failure will not be observed. For a given current density, the length of the conductor that must be exceeded to permit EM failure is called the *Blech length* or *metal critical length* L_{crit}. L_{crit} is about 20 μm for Al.

Table 10.3 gives a comparison of the critical products of unpassivated and passivated Al and Cu lines. For unpassivated Al and

		Al Line	Cu Line
Unpassivated	$(jL)_{crit}$	500–1260 A/cm at 350°C[48,135]	525 A/cm (300°C)[136] to 900–1600 A/cm (175–225°C)[135]
	Temperature dependence	Decreasing with increasing temperature, attributed to changes in the thermomechanical properties of a given interconnect system[48]	Decreases in the threshold product from about 1200 A/cm to about 400 A/cm in the temperature range 200–280°C[135]
Passivated	$(jL)_{crit}$	2500–4500 A/cm[24,48,137]	jL product is 3000 A/cm at 250°C and 3600 A/cm at 200°C,[34] 2800 A/cm[138] and 3900 A/cm[139] at 300°C, and 1500 A/cm[43] at 350°C
	Temperature dependence	Little temperature dependence[137,140]	Decreases from 3940 down to 2660 A/cm as temperature is decreased from 400 to 295°C[125,139]

TABLE 10.3 A Comparison of the Critical Products of Unpassivated and Passivated Al and Cu Lines

Cu-based interconnects with the same type of test structure, Frankovic and Bernstein[135] showed that the critical product in Cu was as much as 1.8× higher than that in Al. This is presumably a result of the greater amount of surrounding passivation in a damascene interconnect. However, this kind of advantage presumably disappears if both types of interconnects are fully passivated. An interesting report is the much higher products (nearly 40 000 A/cm) in AlCu damascene interconnects, which exhibited little tensile stress.[141]

10.7.4 Metal Thickness/Width Dependence

Metal thickness within a wafer has a range of variations. For example, metals at the wafer edge may be considerably thinner (e.g., 22 percent thickness reduction) than the center depending on the deposition tool.[142] Metal-thickness changes will directly translate into current-density variations when considering EM behavior. Thinner metals will experience higher current densities based on Black's model, EM failure times, and ultimately, the lifetimes will be reduced. It is expected that the EM MTTF is linearly decreased with decreasing line thickness. Cheng et al.[91] showed that the MTTF increased from 90 to 140 units when metal thickness increased from 0.14 to 0.26 μm.

The median lifetime also decreases with decreasing line width, but, with a faster trend compared with thickness dependence. Cheng et al.[91] showed that the MTTF increased from 10 to 78 units when metal width increased from 0.07 to 0.2 μm. The much shorter EM lifetime for the narrower metal line (say, 0.07 μm here) corresponds to a much faster Cu drift velocity, which is explained as follows. It is found that grain size and coefficient-of-resistance (TCR) values of Cu lines decreased with line-width reduction. As described in Sec. 10.6.2, the smaller grain size in the narrow line provides more EM paths through the grain boundary in addition to the Cu–cap-layer interface diffusion for wide lines.[91]

10.8 Electromigration under Alternating-Current Conditions

Presently, high-performance interconnect design applications under ac operating conditions are based on the specified limits for the maximum values of the average (J_{avg}), rms (root-mean-square, J_{rms}), and peak (J_{peak}) current densities.[143] *However, interconnect design rules are generated in a manner that EM and self-heating are not considered simultaneously for generating EM lifetime guidelines for interconnects.* The fact is that with the continuous scaling of devices, the multilevel interconnects suffer from self-heating more severely with the application of Cu and low-k dielectrics.[144] In this section, the quantities of these current densities for ac applications are summarized from a practical point of view.

10.8.1 Definition of Peak, Average, and rms Current Densities

The peak, average, and rms current densities are defined as follows:

1. The *peak current density* J_{peak} is simply the current density corresponding to the peak current I_{peak} of the waveform:

$$J_{peak} = \frac{I_{peak}}{A} \tag{10.17}$$

   where A is the cross-sectional area of the interconnect. Short-duration high peak currents including ESD pulses may lead to thermally induced open-circuit metal failure, which is a potential reliability concern.[145]

2. The *average current density* J_{avg} is defined as

$$J_{avg} = \frac{1}{T}\int_0^T j(t)\,dt \tag{10.18}$$

   where T is the time period of the current waveform $j(t)$. *In common practice, the EM lifetime of interconnects under dc conditions is known to be determined by* J_{avg}.[146]

3. The *rms current density* J_{rms} is defined as

$$J_{rms} = \sqrt{\frac{1}{T}\int_0^T j^2(t)\,dt} \tag{10.19}$$

   Joule-heating of the interconnect is determined by the J_{rms}. The temperature rise of metal interconnects owing to the joule-heating phenomenon should be taken into account because EM capability is highly temperature-dependent [Eq. (10.11)].

As a specific example, Fig. 10.34 illustrates the various current definitions for a unipolar pulsed waveform. The preceding three current densities can be expressed as follows:

$$J_{avg} = \delta \cdot J_{peak} \tag{10.20}$$

$$J_{rms} = \sqrt{\delta} \cdot J_{peak} \tag{10.21}$$

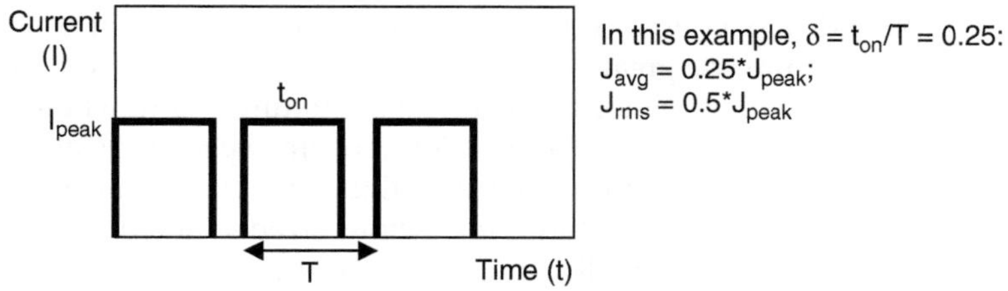

FIGURE 10.34 A unipolar pulsed waveform illustrating the three current densities defined.

where δ is the duty cycle, defined as t_{on}/T with T the period of the pulse in Fig. 10.34.

10.8.2 Characterization of J_{rms}

In addition to the J_{max_dc} limit set for interconnect EM reliability under dc in the design rule, another important current-limit rule for interconnect reliability is the J_{rms} limit for ac lines as discussed earlier, which considers the joule heating resulting from the current passing through. Similar to Eq. (10.19), *joule heating* also can be defined by the rms current I_{rms} as

$$I_{rms} = \sqrt{\frac{1}{T}\int_0^T i(t)dt} \tag{10.22}$$

$$J_{rms} = I_{rms} \cdot A \tag{10.23}$$

Joule heating leads to a temperature increase of the interconnect, which will further limit the EM reliability of the interconnect according to Black's equation of EM [Eq. (10.11)]. A relatively small increase in temperature (ΔT) can seriously degrade EM lifetime. For example, a 5°C temperature increase is sufficient to degrade the EM lifetime by about 30 percent when considering the copper activation energy E_a of 0.95 eV. Therefore, as the system operating temperature increases, the magnitude of the allowed current is decreased in order to maintain the target failure rate.[147] Chen et al.[147] reported that Cu loses 75 percent of its current-carrying capability at an operating temperature of 125°C compared with 100°C and about 90 percent at 150°C. Therefore, the maximum allowed current density in the interconnects decreases with increasing temperature and should be determined by both the joule heating (J_{rms}) and EM (J_{max_dc}) limits. The J_{rms} rule is set to ensure interconnect reliability based on a preselected temperature increase ΔT (e.g., 5 or 10°C).

From a design point of view, ac lines that are capable of generating a temperature rise greater than ΔT must be placed away from the EM-sensitive dc lines. Since the ac line temperature is a function of line dimensions, the required separation (i.e., line spacing) is also a function of these dimensions. Hence, maintaining current densities below both the J_{max_dc} and J_{rms} limits for various-sized ac and dc lines is critical to providing thermally safe designs for very large-scale integration (VLSI) technologies.[147]

The following paragraphs provides the methodology on how to define the J_{rms} rule for a specific process based on measurements. Following the analysis of Schafft,[20] joule heating will generate a steady-state metal-to-metal temperature difference ΔT with a current density J_{rms} given by

$$J_{rms} = \sqrt{\frac{G'}{wd_m\rho_0(TCR + 1/\Delta T)}} \tag{10.24}$$

where w is the metal line width, d_m is the metal line thickness, G' is the total thermal conductance per unit length, ρ_0 is the metal resistivity, and TCR is the thermal coefficient of resistance.

According to the Harmon-Gill (H-G)[148] quasi-analytical heat-conduction model, the total thermal conductance G ($=G'^*L$, where L is the metal line length) for the fully embedded metal line is obtained by the superposition of the contributions from the top, bottom, and sides of the line.

On the other hand, to experimentally determine the thermal conductance G for the metal-line structures, the TCR can be determined at all sites for each structure by measuring line resistance at

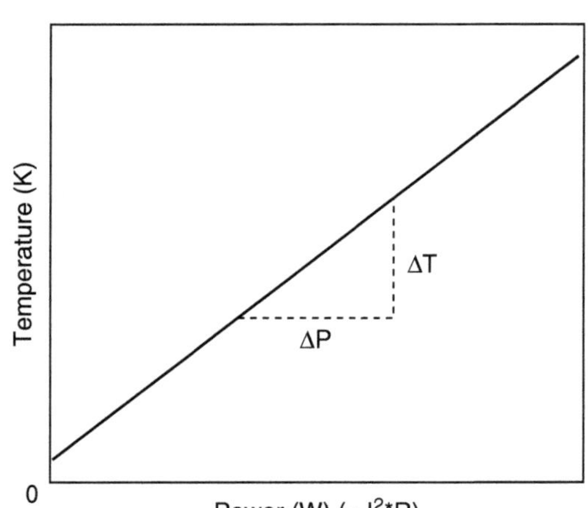

FIGURE 10.35 A schematic illustrating a temperature-versus-power (TVP) curve.

different temperatures, just like the one used for isothermal EM testing. A temperature-versus-power (TVP) relationship is obtained by ramping up the current applied to the metal line. A schematic illustrating a TVP curve is shown in Fig. 10.35. Note that the power on the x axis equals to $I^2 * R$; whereas the temperature on the y axis is determined by the TCR. The thermal conductance G of the lines thus is obtained by taking the inverse slope of a linear fit to these TVP data points according to the following relationship:

$$\Delta T / \Delta P = R_\theta = 1/G \tag{10.25}$$

where R_θ is the thermal impedance of the interconnect line to the substrate.

As discussed earlier, a J_{rms} design rule is defined by limiting the joule heating–induced temperature rise (e.g., $\Delta T = 5°C$). The thermal conductance from Eq. (10.25) then is inserted into Eq. (10.24) to predict the allowable current density versus line width, as shown in Fig. 10.36 as an example.

The J_{rms} is a function of line width, line space, interconnect level, interconnect materials, and dielectric materials, all of which should be taken into consideration when setting the J_{rms} design rule for IC design.

- *Effect of interconnect level.* It is seen in Fig. 10.36 that M_1 has a higher allowable J_{rms} than the other metal levels (M_2 to M_4) because M_1 is closer to the substrate thus has better heat dissipation. In addition, based on Harmon-Gill model,[148] thermal

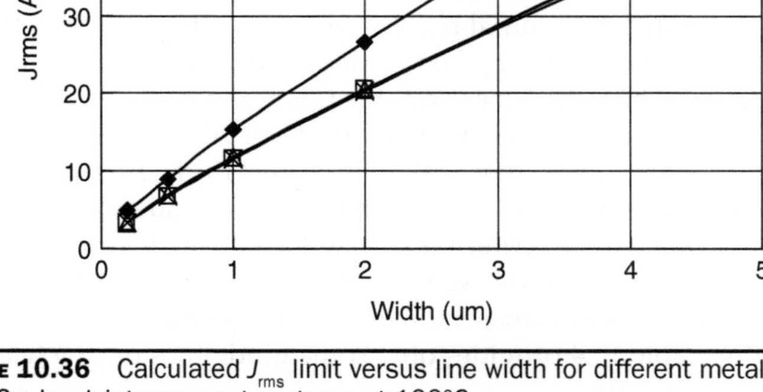

FIGURE 10.36 Calculated J_{rms} limit versus line width for different metal layers with Cu–low-k interconnect systems at 100°C.

conductance G is inversely proportional to d_i, where d_i is the dielectric thickness under the metal line. Thus, according to Eq. (10.24), J_{rms} is approximately proportional to $1/\sqrt{d_i}$, which suggests that the upper-level metal interconnects will be heated to higher temperatures than the lower-level metal wires.

- *Effect of interconnect material.* For the same cross section, current density, and surrounding dielectric, Cu interconnects would experience lower temperature rises than Al interconnect lines owing to their higher thermal conductivity, thermal capacity, and most important, lower resistivity. Eq. (10.24) shows that J_{rms} is proportional to $1/\sqrt{\rho_0}$, indicating that a lower resistivity metal could tolerate higher J_{rms}.
- *Effect of dielectric material.* The application of low-k dielectrics leads to a lower allowable current density according to Eq. (10.24) because the low-k materials have lower thermal conductivities than the conventional SiO_2. In addition, based on the Harmon-Gill model,[148] thermal conductance G is proportional to k, where k is the dielectric constant. Thus, according to Eq. (10.24), J_{rms} is approximately proportional to $\sqrt{k}$. Therefore, a lower k value corresponds to a smaller allowed J_{rms}.

10.9 Process-Qualification Practice

To evaluate an interconnect EM resistance, Joint Electron Device Engineering Council (JEDEC) recommends use of $t_{0.1}$ or J_{max} as a number of merit.[149] The derivation of these merit numbers is based on the stress-failure-time monomodal lognormal distribution (t_{50} and σ). However, one should be aware that for different test structures (e.g., different via/line contact configurations, as shown in Sec. 10.7), mixed stress failure modes (abrupt versus gradual resistance increases) are possible. Furthermore, different failure modes can result in different initial resistance jumps. These failure characteristics can change the t_{50} and σ derivation from the stress data depending on the failure criterion (the fraction of resistance increase: $\Delta R/R_0 * 100\%$) used.[129] $\Delta R/R_0$ equaling 10 or 20 percent may give a much different t_{50} and σ that will finally affect the $t_{0.1}$ lifetime prediction at operating temperature. Therefore, it is prudent to carefully evaluate what failure criterion to use based on the stress-failure characteristics and/or the specific application to ensure that the derived merit numbers reasonably reflect the technologic capability. Therefore, Table 10.4 provides only some guidelines or practices for the EM qualification.

	Package Level	Wafer Level (Isothermal EM)
Failure criteria	$\Delta R \geq 10\%$ or 20% per specific application requirement	$\Delta R \geq 3\%$
Test structures	• Via/line structures to accommodate downstream and upstream current flow • The via size follows the design rule, whereas the metal line has varied metal widths and lengths as defined in the design rules	• Metal line with varied metal widths and lengths • Metal width ranges within design rules
Sample size	≥ 15 devices per condition	≥ 15 devices per condition
Stress conditions	$T = 250$–350°C $J = 1$–5 MA/cm^2	$T = 250$–400°C $J = 10$–50 MA/cm^2
Acceptance criteria	TTF (0.1%) > 10 years @110°C/J_{op}	TTF (0.1%) > 10 years @110°C/J_{op}
Model	$\tau = A * J^{-n} e^{Ea/kT}$	$\tau = A * J^{-n} e^{Ea/kT}$
Distribution	Lognormal	Lognormal
Parameters	TTF (at J_{op}), σ, E_a, n	TTF (at J_{op}), σ, E_a, n

Table 10.4 Typical Process Qualification Practice for EM

10.10 References

1. P. R. Besser, M. C. Madden, and P. A. Flinn, "In situ scanning electron microscopy observation of the dynamic behavior of electromigration voids in passivated aluminum lines," *Journal of Applied Physics*, Vol. 72, 1992, pp. 3792–3797.
2. C. S. Hau-Riege, "An introduction to Cu electromigration," *Microelectronics Reliability*, Vol. 44, 2004, pp. 195–205.
3. T. D. Sullivan, "Reliability consideration for copper metallization in ULSI circuits," in: O. Kraft, E. Arzt, C. A. Volkert, P. S. Ho, H. Okabayashi, ed. *Stress Induced Phenomena in Metallization*, Fifth International Workshop, Stuttgart, Germany, AIP Conference Proceedings, Vol. 491, 1999, pp. 39–50.
4. C. S. Hau-Riege and C. V. Thompson, "Electromigration in Cu interconnects with very different grain structures," *Applied Physics Letters*, Vol. 78, 2001, pp. 3451–3453.
5. P. E. Bagnoli, A. Diligenti, B. Neri, and S. Ciucci, "Noise measurements in thin-film interconnects: A non-destructive technique to characterize electromigration," *Journal of Applied Physics*, Vol. 63, No. 5, 1988, pp. 1448–1451.
6. W. Baerg, K. Wu, P. Davies, G. Dao, and D. Fraser, "The electrical resistance ratio (RR) as a thin film metal monitor," in *Proceedings of the 28th IEEE Annual International Reliability Physics Symposium*, 1990, pp. 119–123.

7. F. Vollkommer, H. G. Bohn, K. Robrock, and W. Schilling, "Internal friction: A fast technique for electromigration failure analysis," in *Proceedings of the 28th IEEE Annual International Reliability Physics Symposium*, 1990, pp. 27–29.
8. R. W. Pasco and J. A. Schwarz, "Temperature-ramp resistance analysis to characterize electromigration," *Solid-State Electronics*, Vol. 26, 1983, pp. 445–452.
9. L. E. Felton, J. A. Schwarz, and J. R. Lloyd, "A comparison of the median time to failure with temperature-ramp resistance analysis to characterize electromigration to determine electromigration kinetic parameters," *Thin solid Films*, Vol. 155, 1987, pp. 209–215.
10. B. J. Root and T. Turner, "Wafer level electromigration tests for production monitoring," in *Proceedings of the 23rd IEEE Annual International Reliability Physics Symposium*, 1985, pp. 100–107.
11. C. C. Hong and D. L. Cook, "Breakdown energy of metal (BEM): A new technique for monitoring reliability at wafer level," in *Proceedings of the 23rd IEEE Annual International Reliability Physics Symposium*, 1985, pp. 108–114.
12. R. E. Jones and L. D. Smith, "A new wafer level isothermal joule heated electromigration test for rapid testing of integrated circuit interconnects," *Journal of Applied Physics*, Vol. 61, 1987, pp. 4670–4678.
13. I. A. Blech and E. Kinsbron, "Electromigration in thin gold films on molybdenum surfaces," *Thin Solid Films*, Vol. 25, 1975, pp. 327–334.
14. R. E. Johes et al., "A new wafer-level isothermal joule heated electromigration test for rapid testing of integrated circuit interconnect," *Journal of Applied Physics*, Vol. 61, No. 9, 1987, p. 4670.
15. S. Y. Lee, J. B. Lai, S. C. Lee, L. H. Chu, Y. S. Huang, R. Y. Shiue, Y. K. Peng, and J. T. Yue, "Real case study for isothermal EM test as a process control methodology," in *Proceedings of the 39th IEEE Annual International Reliability Physics Symposium*, 2001, pp. 184–188.
16. J. R. Lloyd, J. T. Coffin, M. J. Sullivan, E. T. Severn, G. S. Hopper, and J. L. Jozwiak, "Electromigration failures in thin film silicides and polysilicon/silicide (polycide) structures," in *Proceedings of the 21st IEEE Annual International Reliability Physics Symposium*, 1983, pp. 198–202.
17. M. Sakimoto, T. Itoo, T. Fujii, H. Yamaguchi, and K. Eguchi, "Temperature measurement of Al metallization and the study of Black's model in high current density," in *Proceedings of the 33rd IEEE Annual International Reliability Physics Symposium*, 1995, pp. 333–341.
18. www. qualitau.com, Access date: May 2006.
19. C. R. Crowell, C. C. Shih, and V. C. Tyree, "SWEAT structure design and test procedure criteria based upon TEARS characterization and spatial distribution of failures in iterated structures," in *Proceedings of the 29th IEEE Annual International Reliability Physics Symposium*, 1991, pp. 277–286.
20. H. A. Schafft, "Thermal analysis of electromigration test structures," *IEEE Transactions on Electron Devices*, Vol. 34, No. 3, 1987, pp. 664–672.
21. H. A. Schafft, T. C. Staton, J. Mandel, and J. D. Shott, "Reproducibility of electromigration measurements," *IEEE Transactions on Electron Devices*, Vol. 34, No. 3, 1987, pp. 673–681.
22. R. Frankovic and G. H. Bernstein, "Electromigration drift and threshold in Cu thin-film interconnects," *IEEE Transactions on Electron Devices*, Vol. 43, No. 12, 1996, pp. 2233–2239.
23. P. S. Ho, K. -D. Lee, E. T. Ogawa, X. Lu, H. Matsuhashi, V. A. Blaschke and R. Augur, "Electromigration reliability of Cu interconnects and effects of low k dielectrics," in *Proceedings of the International Electron Devices Meeting (IEDM)*, 2002, pp. 741–744.
24. E. T. Ogawa, K. D. Lee, A. V. Blaschke, and Paul S. Ho, "Electromigration reliability issues in dual-damascene Cu interconnects," *IEEE Transactions on Reliability*, Vol. 51, No. 4, 2002, pp. 403–419.
25. H. Kawasaki and C. -K. Hu, "An electromigration failure model of tungsten plug contacts/vias for realistic lifetime prediction," in *Symposium on VLSI Technology Digest of Technical Papers*, 1996, pp. 192–193.

26. S. Yokogawa and H. Tsuchiya, "Effects of Al doping on the electromigration performance of damascene Cu interconnects," *Journal of Applied Physics*, Vol. 101, No. 1, 2007, p. 013 513.
27. S. Yokogawa, H. Tsuchiya, Y. Kakuhara, and K. Kikuta, "Analysis of Al doping effects on resistivity and electromigration of copper interconnects," *IEEE Transactions on Device and Materials Reliability*, Vol. 8, No. 1, 2008, pp. 216–221.
28. M. Tada, M. Abe, N. Furutake, F. Ito, T. Tonegawa, M. Sekine, and Y. Hayashi, "Improving Reliability of copper dual-damascene interconnects by impurity doping and interface strengthening," *IEEE Transactions on Electron Devices*, Vol. 54, No. 8, 2007, pp. 1867–1877.
29. H. -K. Yap, K. -L. Yap, Y. -C. Tan, and K. -F. Lo, "Wafer level electromigration testing on via/line structure with a poly-heated method in comparison to standard package level tests," in *Proceedings of the 10th International Symposium on the Physical and Failure Analysis of Integrated Circuits (IPFA)*, 2003, pp. 75–79.
30. C. Chappaz, P. Waltz, and L. Castellon; "Electromigration: from package to wafer level thanks to a heating coil structure," in *Proceedings of the IEEE International Conference on Microelectronics Test Structures (ICMTS)*, 2008, pp. 43–45.
31. C. Chappaz and P. Nakkala, "Lifetime extrapolation for electromigration tests at wafer level with a dedicated device," in *Proceedings of the 48th IEEE Annual International Reliability Physics Symposium*, 2010, pp. 887–889.
32. X. Federspiel, V. Girault, and D. Ney, "Effect of joule heating on the determination of electromigration parameters," in *IEEE International Integrated Reliability Workshop Final Report*, 2003, pp. 139–142.
33. A. V. Vairagar, Z. H. Gan, W. Shao, S. G. Mhaisalkar, H. Y. Li, K. N. Tu, Z. Chen, E. Zschech, H. J. Engelmann, and S. Zhang, "Improvement of electromigration lifetime of submicrometer dual-damascene Cu interconnects through surface engineering," *Journal of the Electrochemical Society*, Vol. 153, No. 9, 2006, pp. G840–G845.
34. L. Arnaud, "Electromigration threshold length effect in dual damascene copper-oxide interconnects," in *Proceedings of the 40th IEEE Annual International Reliability Physics Symposium*, 2002, pp. 433–434.
35. L. Arnaud, G. Tartavel, T. Gerger, D. Mariolle, Y. Gobil, and I. Touet, Microstructure and electromigration in copper damascene lines, *Microelectronics Reliability*, Vol. 40, 2000, pp. 77–86.
36. J. C. Doan, J. C. Bravman, P. A. Flinn, and T. N. Marieb, "The relationship between resistance changes and void volume changes in passivated aluminum interconnects," in *Proceedings of the 37th IEEE Annual International Reliability Physics Symposium*, 1999, pp. 206–212.
37. F. Chen, B. Li, T. Lee, C. Christiansen, J. Gill, M. Angyal, M. Shinosky, C. Burke, W. Hasting, R. Austin,T. Sullivan, D. Badami, and J. Aitken, "Technology reliability qualification of a 65nm CMOS Cu/low-*k* BEOL interconnect," in *Proceedings of the 13rd International Symposium on the Physical and Failure Analysis of Integrated Circuits (IPFA)*, 2006, pp. 97–105.
38. J. Gill, T. Sullivan, S. Yankee, H. Barth, and A. von Glasow, " Investigation of via-dominated multi-modal EM failure distributions in dual damascene Cu interconnects with a discussion of the statistical implications," in *Proceedings of the 40th IEEE Annual International Reliability Physics Symposium*, 2002, pp. 298–304.
39. J. M. Towner, "Are electromigration failures lognormally distributed?" in *Proceedings of the 28th IEEE Annual International Reliability Physics Symposium*, 1990, pp. 100–105.
40. R. Billinton and R. N. Allan, *Reliability Evaluation of Engineering Systems: Concepts and Techniques*, New York: Plenum Press, 1992.
41. Engineering Statistic handbook; available at: http://www.itl.nist.gov/div898/handbook/index.htm, Access date: December 2010.
42. S. Maitrejean, F. Fusalba, M. Patz, V. Jousseaume, and T. Mourier, "Adhesion studies of thin films on ultralow *k*," in *Proceedings of the International Interconnect Technology Conference (IITC)*, 2002, pp. 206–208.

43. C. S. Hau-Riege, A. P. Marathe, and V. Pham, "The effect of low-*k* ILD on the electromigration reliability of Cu interconnects with different line lengths," in *Proceedings of the 41st IEEE Annual International Reliability Physics Symposium*, 2003, pp. 173–177.
44. C. K. Hu, L. Gignac, R. Rosenberg, E. Liniger, J. Rubino, C. Sambucetti, A. Stamper, A. Domenicucci, and X. Chen, "Reduced Cu interface diffusion by CoWP surface coating," *Microelectronics Reliability*, Vol. 70, 2003, pp. 406–411.
45. B. Li, C. Christiansen, J. Gill, T. Sullivan, E. Yashchin, and R. Filippi, "Threshold electromigration failure time and its statistics for Cu interconnects," *Journal of Applied Physics*, Vol. 100, 2006, p. 114516.
46. Y. Chai and P. Chan, "High electromigration-resistant copper/carbon nanotube composite for interconnect application," in *Proceedings of the International Electron Devices Meeting (IEDM)*, 2008, pp. 1–4.
47. J. R. Lloyd, "New models for interconnect failure in advanced IC technology," in *Proceedings of the 15th International Symposium on the Physical and Failure Analysis of Integrated Circuits (IPFA)*, 2008, pp. 1–7.
48. I. A. Blech, "Electromigration in thin aluminum films on titanium nitride," *Journal of Applied Physics*, Vol. 47, 1976, pp. 1203–1208.
49. J. R. Lloyd, "Black's law revisited: Nucleation and growth in electromigration failure," *Microelectronics Reliability*, Vol. 47, 2007, pp. 1468–1472.
50. R. Kircheim and U. Kaeber, "Atomistic and computer modeling of metallization failure of integrated circuits by electromigration," *Journal of Applied Physics*, Vol. 70, No. 1, 1991, pp. 172–181.
51. A. Korhonen. P. Boergesen, K. N. Tu, and C. -Y. Li, "Stress evolution due to electromigration in confined metal lines," *Journal of Applied Physics*, Vol. 73, 1993, pp. 3790–3799.
52. J. P. Gambino, T. C. Lee, F. Chen, and T. D. Sullivan, "Reliability challenges for advanced copper interconnects: Electromigration and time-dependent dielectric breakdown (TDDB)," in *Proceedings of the 16th International Symposium on the Physical and Failure Analysis of Integrated Circuits (IPFA)*, 2009, pp. 677–684.
53. V. Andleigh, V. T. Srikar, Y. -J. Park, and C. V. Thompson, "Mechanism maps for electromigration-induced failure of metal and alloy interconnects," *Journal of Applied Physics*, Vol. 86, No. 12, 1999, pp. 6737–6745.
54. A. H. Fischer, A. Abel, M. Lepper, A. E. Zitzelsberger, and A. von Glasow "Experimental data and statistical models for bimodal EM failures," in *Proceedings of the 38th IEEE Annual International Reliability Physics Symposium*, 2000, pp. 359–363.
55. B. Li, C. Christiansen, J. Gill, R. Filippi, T. Sullivan, and E. Yashchin, "Minimum void size and 3-parameter lognormal distribution for EM failures in Cu interconnects," in *Proceedings of the 44th IEEE Annual International Reliability Physics Symposium*, 2006, pp. 115–122.
56. T. Kauerauf, G. Butera, K. Croes, S. Demuynck, C. J. Wilson, P. Roussel, C. Drijbooms, H. Bender, M. Lofrano, B. Vandevelde, Z. Tőkei, and G. Groeseneken, "Degradation and failure analysis of copper and tungsten contacts under high fluence stress," in *Proceedings of the 48th IEEE Annual International Reliability Physics Symposium*, 2010, pp. 712–716.
57. K. Wang, C. J. Wilson, A. Cuthbertson, R. Herberholz, H. P. Coulson, A. G. O'Neill, and A. B. Horsfall, "Influence of barriers on the reliability of dual damascene copper contacts," in *Proceedings of the 46th IEEE Annual International Reliability Physics Symposium*, 2008, pp. 677–678.
58. J. Tao, K. K. Young, N. W. Cheung, and C. Hu, "Electromigration reliability of tungsten and aluminum vias and improvements under AC current stress," *IEEE Transactions on Electron Devices*, Vol. 40, No. 8, 1993, pp. 1398–1405.
59. W. Shao, A. V. Vairagar, C. H. Tung, Z. L. Xie, A. Krishnamoorthy, and S. G. Mhaisalkar, "Electromigration in copper damascene interconnects: reservoir effects and failure analysis," *Surface & Coatings Technology*, Vol. 198, 2005, pp. 257–261.
60. C. L. Gan, C. V. Thompson, K. L. Pey, W. K. Choi, H. L. Tay, B. Yu, and M. K. Radhakrishnan, "Effect of current direction on the lifetime of different levels

of Cu dual-damascene metallization," *Applied Physics Letters*, Vol. 79, 2001, pp. 4592–4594.
61. A. V. Vairagar, S. G. Mhaisalkar, Ahila Krishnamoorthy, K. N. Tu, A. M. Gusak, M. A. Meyer, and E. Zschech, "In-situ observation of electromigration induced void migration in dual-damascene Cu interconnect structures," *Applied Physics Letters*, Vol. 85, No. 13, 2004, pp. 2502–2504.
62. R. C. J. Wang, D. S. Su, C. T. Yang, D. H. Chen, Y. Y. Doong, J. R. Shih, S. Y. Lee, C. C. Chiu, Y. K. Peng, and I. T. Yue, "Investigation of electromigration properties and plasma charging damages for plasma treatment process in Cu interconnects," in *Proceedings of the 7th International Symposium on Plasma Process-Induced Damage*, 2002, pp. 166–168.
63. A. V. Glasow, A. H. Fischer, D. Bunel, G. Friese, A. Hausmann, O. Heitzsch, M. Hommel, I. Kriz, S. Penka, P. Raffin, C. Robin, H. P. Sperlich, F. Ungar, and A. E. Zitelsherger, "The influence of the SiN cap process on the electromigration and stress voiding performance of dual damascene Cu interconnects," in *Proceedings of the 41st IEEE Annual International Reliability Physics Symposium*, 2003, pp. 146–150.
64. C-K. Hu, L. Gignac, E. Liniger, B. Herbst, D. Rath, S. T. Chen, S. Kaldor, A. Simon, and W. T. Tseng, "Comparison of Cu electromigration lifetime in Cu interconnects coated various caps," *Applied Physics Letters*, Vol. 83, 2003, pp. 869–871.
65. C. -K. Hu, L. Gignac, R. Rosenberg, E. Liniger, J. Rubino, C. Sambucetti, A. Domenicucci, X. Chen, and A. K. Stamper, „Reduced electromigration of Cu wires by surface coating," *Applied Physics Letters*, Vol. 81, No. 10, 2002, pp. 1782–1784.
66. C. -K. Hu, L. Gignac, R. Rosenberg, B. Herbst, S. Smith, J. Rubino, D. Canaperi, S. T. Chen, S. C. Seo, and D. Restaino, „Atom motion of Cu and Co in Cu damascene lines with a CoWP cap," *Applied Physics Letters*, Vol. 84, No. 24, 2004, pp. 4986–4989.
67. T. Usui, T. Oki, H. Miyajima, K. Tabuchi, K. Watanabe, T. Hasegawa, and H. Shibata, "Identification of electromigration dominant diffusion path for Cu damascene interconnects and effect of plasma treatment and barrier dielectrics on electromigration performance," in *Proceedings of the 42nd IEEE Annual International Reliability Physics Symposium*, 2004, pp. 246–250.
68. T. Kouno, T. Suzuki, S. Otsuka, T. Hosoda, T. Nakamura, Y. Mizushima, M. Shiozu, H. Matsuyama, K. Shono, H. Watatani, Y. Ohkura, M. Sato, S. Fukuyama, and M. Miyajima, "Stress-induced voiding under vias connected to narrow copper lines," in *Proceedings of the International Electron Devices Meeting (IEDM)*, 2005, pp. 187–190.
69. C. H. Ang, W. H. Lu, Andrew K. L. Yap, L. C. Goh, Luona N. L. Goh, Y. K. Lim, C. S. Chua, L. H. Ko, Tracy H. S. Tan, S. L. Tob, and L. C. Hsia, "A study of SiN cap NH_3 plasma pre-treatment process on the PID, EM, GOI performance and BEOL defectivity in Cu dual damascene technology," in *Proceedings of the IEEE International Conference on Integrated Circuit Design and Technology*, 2004, pp. 119–122.
70. C. K. Hu, L. Gignac, E. Linger, R. Rosenberg, and A. Stamper, "Bimodal Electromigration Mechanisms in Dual-Damascene Cu Line/Via on W," *International Interconnect Technology Conference (IITC)*, 2002, pp. 133–135.
71. Y. L. Chan, P. Chuang, and T. J. Chuang, "Vibrational study of CH_2 and CH_3 radicals on the Cu(111) surface by high resolution electron energy loss spectroscopy," *Journal of Vacuum Science and Technology A*, Vol. 16, No. 3, 1998, pp. 1023–1030.
72. K. P. Rodbell, P. J. Ficalora, and R. Koch, "Effect of hydrogen on electromigration and $1/f$ noise in gold films," *Applied Physics Letters*, Vol. 50, No. 20, 1987, pp. 1415–1417.
73. J. -P. Wang, Y. -K. Su, and J. F. Chen, "Effects of surface cleaning on stressvoiding and electromigration of Cu-damascene interconnection," *IEEE Transactions on Device and Materials Reliability*, Vol. 8, No. 1, 2008, pp. 210–215.

74. K. Chattopadhyay, B. van Schravendijk, T. W. Mountsier, G. B. Alers, M. Hornbeck, H. J. Wu, R. Shaviv, G. Harm, D. Vitkavage, E. Apen, Y. Yu, and R. Havemann, "In-situ formation of a copper silicide cap for TDDB and electromigration improvement," in *Proceedings of the 44th IEEE Annual International Reliability Physics Symposium*, 2006, pp. 128–130.
75. S. Yokogawa, K. Kikuta, H. Tsuchiya, T. Takewaki, M. Suzuki, H. Toyoshima, Y. Kakuhara, N. Kawahara, T. Usami, K. Ohto, K. Fujii, Y. Tsuchiya, K. Arita, K. Motoyama, M. Tohara, T. Taiji, T. Kurokawa, and M. Sekine, "A novel resistivity measurement technique for scaled-down Cu interconnects implemented to reliability-focused automobile applications," in *Proceedings of the International Electron Devices Meeting (IEDM)*, 2006, p. S4. 1.
76. Y. Hayashi, N. Matsunaga, M. Wada, S. Nakao, K. Watanabe, A. Sakata, and H. Shibata, "Low resistive and highly reliable copper interconnects in combination of silicide-cap with Ti-barrier for 32-nm-node and beyond," in *Proceedings of the International Interconnect Technology Conference (IITC)*, 2009, pp. 252–254.
77. W. Wu, H. -J. Wu, G. Dixit, R. Shaviv, M. Gao, T. Mountsier, G. Harm, A. Dulkin, N. Uchigami, S. K. Kailasam, E. Klawuhn, and R. H. Havemann, "Ti-based barrier for Cu interconnect applications," in *Proceedings of the International Interconnect Technology Conference (IITC)*, 2008, pp. 202–204.
78. W. F. Gale and T. C. Totemeier, eds., *Smithell's Metals Reference Book*, 8th ed. New York: Elsevier, 2004, pp. 8-23–8-24.
79. C. K. Hu, D. Canaperi, S. T. Chen, L. M. Gignac, B. Herbst, S. Kaldor, M. Krishnan, E. Liniger, D. L. Rath, D. Restaino, R. Rosenberg, J. Rubino, S. -C. Seo, A. Simon, S. Smith, and W. -T. Tseng, "Effect of overlayers on electromigration reliability improvement for Cu/low *K* interconnects," in *Proceedings of the 42nd IEEE Annual International Reliability Physics Symposium*, 2004, pp. 222–227.
80. C. -C. Yang, P. Flaitz, P. -C. Wang, F. Chen, and D. Edelstein, "Characterization of selectively deposited cobalt capping layers: selectivity and electromigration resistance," *IEEE Electron Device Letters*, Vol. 31, No. 7, 2010, pp. 728–730.
81. L. Zhang, J. P. Zhou, J. Im, P. S. Ho, O. Aubel, C. Hennesthal, and E. Zschech, "Effects of cap layer and grain structure on electromigration reliability of Cu/low-k interconnects for 45 nm technology node," in *Proceedings of the 48th IEEE Annual International Reliability Physics Symposium*, 2010, pp. 581–585.
82. J. Bao, N. Lustig, E. Engbrecht, J. Gill, R. Filippi, T. C. Lee, K. Chanda, D. Kioussis, A. Lisi, T. Cheng, S. B. Law, A. Simon, P. Flaitz, J. Choi, W. Tseng, E. Zielinski, S. M. Gates, A. Grill, S. Nguyen, and H. Shobha, "A BEOL multilevel structure with ultra low-k materials ($k \leq 2.4$)," *International Interconnect Technology Conference (IITC)*, 2010, pp. 1–3.
83. S. Yokogawa, K. Kikuta, H. Tsuchiya, T. Takewaki, M. Suzuki, H. Toyoshima, Y. Kakuhara, N. Kawahara, T. Usami, K. Ohto, K. Fujii, Y. Tsuchiya, K. Arita, K. Motoyama, M. Tohara, T. Taiji, T. Kurokawa, and M. Sekine, "Tradeoff characteristics between resistivity and reliability for scaled-down Cu-based interconnects," *IEEE Transactions on Electron Devices*, Vol. 55, No. 1, 2008, pp. 350–357.
84. H. Ceric, R. L. de Orio, J. Cervenka, and S. Selberherr, "Copper microstructure impact on evolution of electromigration induced voids," in *Proceedings of the International Conference on Simulation of Semiconductor Processes and Devices (SISPAD)*, 2009, pp. 1–4.
85. E. Zschech, M. A. Meyer, I. Zienert, E. Langer, H. Geisler, A. Preusse, and P. Huebler, "Electromigration-induced copper interconnect degradation and failure: the role of microstructure," in *Proceedings of the 12th International Symposium on the Physical and Failure Analysis of Integrated Circuits (IPFA)*, 2005, pp. 85–91.
86. R. Lloyd, and J. J. Clement, "Electromigration in Cu conductors," *Thin Solid Films*, Vol. 262, 1995, pp. 135–141.
87. C. -K. Hu, R. Rosenberg, and K. L. Lee, "Electromigration path in Cu thin-film lines," *Applied Physics Letters*, Vol. 74, 1999, pp. 2945–2947.
88. D. Edelstein, C. Uzoh, C. Cabral, P. DeHaven, P. Buchwalter, A. Simon, E. Cooney, S. Malhotra, D. Klaus, H. Rathore, B. Aganvala, and D. Nguyen, "A

high performance liner for copper damascene interconnects," *IEEE International Interconnect Technology Conference (IITC)*, 2001, pp. 9–11.
89. C. K. Hu, L. Gignac, B. Baker, E. Liniger, R. Yu, and P. Flaitz, "Impact of Cu microstructure on electromigration reliability," in *Proceedings of the IEEE International Interconnect Technology Conference (IITC)*, 2007, pp. 93–95.
90. L. Zhang, M. Kraatz, O. Aubel, C. Hennesthal, J. Im, E. Zschech, and P. S. Ho, "Cap layer and grain size effects on electromigration reliability in Cu/low-*k* interconnects," in *Proceedings of the International Interconnect Technology Conference (IITC)*, 2010, pp. 1–3.
91. Y. -L. Cheng, B. -J. Wei, and Y. -L. Wang, "Scaling effect on electromigration in copper interconnects," in *Proceedings of the International Symposium on the Physical and Failure Analysis of Integrated Circuits (IPFA)*, 2009, pp. 698–701.
92. K. D. Lee, Y. J. Park, T. Kim, and W. R. Hunter, "Via processing effects on electromigration in 65nm technology," in *Proceedings of the 44th IEEE Annual International Reliability Physics Symposium*, 2006, pp. 103–106.
93. M. W. Lane, E. G. Liniger, and J. R. Lloyd, "Relationships between interfacial adhesion and electromigration in Cu metallization," *Journal of Applied Physics*, Vol. 93, 2003, pp. 1417–1421.
94. W. Liu, Y. K. Lim, F. Zhang, H. Liu, Y. H. Zhao, A. Y. Du, B. C. Zhang, J. B. Tan, D. K. Sohn, and L. C. Hsia, "Study of upstream electromigration bimodality and its improvement in Cu low-*k* interconnects," in *Proceedings of the 48th IEEE Annual International Reliability Physics Symposium*, 2010, pp. 906–910.
95. E. G. Liniger, C. -K. Hu, L. M. Gignac, and A. Simon, "Effect of liner thickness on electromigration lifetime," *Journal of Applied Physics*, Vol. 93, 2003, p. 9576.
96. W. C. Baek, J. P. Zhou, J. Im, P. S. Ho, J. G. Lee, S. B. Hwang, K. K. Choi, S. K. Park, O. J. Jung, L. Smith, and K. Pfeifer, "Oxidation of the Ta diffusion barrier and its effect on the reliability of Cu interconnects," in *Proceedings of the 44th IEEE Annual International Reliability Physics Symposium*, 2006, pp. 131–135.
97. O. Aubel, W. Yao, M. A. Meyer, H. J. Engelmann, J. Poppe, F. Feustel, and C. Witt, "New failure mechanism during high temperature storage testing and its application on SIV risk evaluation," in *IEEE International Integrated Reliability Workshop Final Report*, 2009, pp. 5–10.
98. J. Tong, D. Martini, N. Magtoto, and J. Kelber, "Ta Metallization of Si-O-C Substrate and Cu Metallization of Ta/Si-O-C Multilayer," *Journal of Vacuum Science and Technology B*, Vol. 21, 2003, pp. 293–300.
99. K. Ueno, "Material and process challenges for interconnects in nanoelectronics era," in *Proceedings of the International Symposium on VLSI Technology Systems and Applications (VLSI-TSA)*, 2010, pp. 64–65.
100. D. -Y. Moon, T. -S. Kwon, B. -W. Kang, W. -S. Kim, B. M. Kim, J. H. Kim, and J. -W. Park, "Copper seed layer using atomic layer deposition for Cu interconnect," in *Proceedings of the 3rd International Nanoelectronics Conference (INEC)*, 2010, pp. 450–451.
101. T. Ishigami, T. Kurokawa, Y. Kakuhara, B. Withers, J. Jacobs, A. Kolics, I. Ivanov, M. Sekine and K. Ueno , "High reliability Cu interconnection utilizing a low contamination CoWP capping layer," in *Proceedings of the International Interconnect Technology Conference (IITC)*, 2004, pp. 75–77.
102. T. Usui, H. Nasu, S. Takahashi, N. Shimizu, T. Nishikawa, M. Yoshimaru, H. Shibata, M. Wada, and J. Koike, "Highly reliable copper dual-damascene interconnects with self-formed MnSixOy barrier layer," *IEEE Transactions on Electron Devices*, Vol. 53, No. 10, 2006, pp. 2492–2499.
103. K. Ohmori, K. Mori, K. Maekawa, K. Kohama, K. Ito, T. Ohnishi, M. Mizuno, M. Fujisawa, M. Murakami, and H. Miyatake, "A key of self-formed barrier technique for reliability improvement of Cu dual damascene interconnects," in *Proceedings of the International Interconnect Technology Conference (IITC)*, 2010, pp. 1–3.
104. C. -C. Yang, S. Cohen, T. Shaw, P. -C. Wang, T. Nogami, and D. Edelstein, "Characterization of "ultrathin-Cu"/Ru(Ta)/TaN liner stack for copper interconnects," *IEEE Transactions on Electron Devices*, Vol. 31, No. 7, 2010, pp. 722–724.

105. M. Damayanti, T. Sritharan, S. G. Mhaisalkar, and Z. H. Gan, "Effects of dissolved nitrogen in improving barrier properties of ruthenium," *Applied Physics Letters*, Vol. 88, 2006, p. 044101.
106. C. -C. Yang, T. Spooner, S. Ponoth, K. Chanda, A. Simon, C. Lavoie M. Lane, C. -K. Hu, E. Liniger, L. Gignac, T. Shaw, S. Cohen, F. McFeely, and D. Edelstein, "Physical, electrical, and reliability characterization of Ru for Cu interconnects," in *Proceedings of the International Interconnect Technology Conference (IITC)*, 2006, pp. 187–189.
107. L. Zhao, Z. Tokei, G. G. Gischia, H. Volders, and G. Beyer, "A new perspective of barrier material evaluation and process optimization," in *Proceedings of the International Interconnect Technology Conference (IITC)*, 2009, pp. 206–208.
108. N. Nakamura, Y. Takigawa, E. Soda, N. Hosoi, Y. Tarumi, H. Aoyama, Y. Tanaka, D. Kawamura, S. Ogawa, N. Oda, S. Kondo, I. Mori, and S. Saito, "Design impact study of wiring size and barrier metal on device performance toward 22 nm-node featuring EUV lithography," *International Interconnect Technology Conference (IITC)*, 2009, pp. 14–16.
109. K. Mori, K. Ohmori, N. Torazawa, S. Hirao, S. Kaneyama, H. Korogi, K. Maekawa, S. Fukui, K. Tomita, M. Inoue, H. Chibahara, Y. Imai, N. Suzumura, K. Asai and M. Kojima, "Effects of Ru-Ta alloy barrier on Cu filling and reliability for Cu interconnects," in *Proceedings of the International Interconnect Technology Conference (IITC)*, 2008, pp. 99–101.
110. L. Carbonell, H. Volders, N. Heylen, K. Kellens, R. Caluwaerts, K. Devriendt, E. A. Sanchez, J. Wouters, V. Gravey, K. Shah, Q. Luo, A. Sundarrajan, J. Lu, J. Aubuchon, P. Ma, M. Narasimhan, A. Cockburn, Z. Tökei, and G. P. Beyer, "Metallization of sub-30-nm interconnects: comparison of different liner/seed combinations," in *Proceedings of the International Interconnect Technology Conference (IITC)*, 2009, pp. 200–202.
111. P. Ma, Q. Luo, A. Sundarrajan, J. Lu, J. Aubuchon, J. Tseng, N. Kumar, M. Okazaki, Y. Wang, Y. Wang, Y. Chen, M. Naik, I. Emesh, and M. Narasimhan, "Optimized integrated copper gap-fill approaches for 2× flash devices," in *Proceedings of the International Interconnect Technology Conference (IITC)*, 2009, pp. 38–40.
112. T. Nogami, J. Maniscalco, A. adan, P. Flaitz, P. DeHaven, C. Parks, L. Tai, B. St. Lawrence, R. Davis, R. Murphy, T. Shaw, S. Cohen, C-K. Hu, C. Cabral, Jr. , S. Hiang, J. Kelly, M. Zaitz, J. Schmatz, S. Choi, K. Tsumura, C. Penny, H.C. Chen, D. Canaperi, T. Vo, F. Ito, O. Straten, A. Simon, S.- H. Rhee, B-Y. Kim, T. Bolom, V. Ryan, P. Ma, J. Ren, J. Aubuchon, J. Fine, P. Kozlowski, T. Spooner, and D. Edelstein, " CVD Co and its application to Cu damascene interconnections," in *Proceedings of the International Interconnect Technology Conference (IITC)*, 2010, pp. 1–3.
113. D. Edelstein, C. Uzoh, C. Cabral, P. DeHaven, P. Buchwalter, A. Simon, E. Cooney, S. Malhotra, D. Klaus, H. Rathore, B. Agarwala, and D. Nguyen, "A high performance liner for copper damascene interconnects," in *Proceedings of the International Interconnect Technology Conference (IITC)* 2001, pp. 9–11.
114. A. Isobayashi, Y. Enomoto, H. Yamada, S. Takahashi, and S. Kadomura, "Thermally robust Cu interconnects with Cu-Ag alloy for sub-45-nm node," in *Proceedings of the International Electron Devices Meeting (IEDM)*, 2004, pp. 953–956.
115. M. Ueki, M. Hiroi, N. Ikarashi, T. Onodera, N. Furutake, N. Inoue, and Y. Hayashi, "Effects of Ti addition on via reliability in Cu dual damascene interconnects," in *Proceedings of the IEEE Transactions on Electron Devices*, Vol. 51, No. 11, 2004, pp. 1883–1891.
116. T. Tonegawa, M. Hiroi, K. Motoyama, K. Fujii, and H. Miyamoto, "Suppression of bimodal stress-induced voiding using high-diffusive dopant from Cu-alloy seed layer," in *Proceedings of the International Interconnect Technology Conference (IITC)*, 2003, pp. 216–218.
117. J. Koike, M. Haneda, J. Iijima, and M. Wada, "Cu alloy metallization for self-forming barrier process," in *Proceedings of the International Interconnect Technology Conference (IITC)*, 2006, pp. 161–163.

118. J. P. Gambino, "Improved reliability of copper interconnects using alloying," in *Proceedings of the International Symposium on the Physical and Failure Analysis of Integrated Circuits (IPFA)*, 2010, pp. 1–7.
119. M. Iguchi, S. Yokogawa, H. Aizawa, Y. Kakuhara, H. Tsuchiya, N. Okada, K. Imai, M. Tohara, K. Fujii, and T. Watanabe, "Optimization of metallization processes for 32-nm-node highly reliable ultralow-k (k = 2. 4)/Cu multilevel interconnects incorporating a bilayer low-k barrier cap (k = 3. 9)," in *Proceedings of the International Electron Devices Meeting (IEDM)*, 2009, pp. 871–874.
120. Y. Hayashi, M. Abe, M. Tada, M. Narihiro, M. Tagami, M. Ueki, N. Inoue, F. Ito, H. Yamamoto, T. Takeuchi, S. Saito, T. Onodera, and N. Furutake, "Robust low oxygen content Cu alloy for scaled-down ULSI interconnects based on metallurgical thermodynamic principles," *IEEE Transactions on Electron Devices*, Vol. 56, No. 8, 2009, pp. 1579–1587.
121. Y. Hayashi, N. Matsunaga, M. Wada, S. Nakao, K. Watanabe, S. Kato, A. Sakata, A. Kajita, and H. Shibata, "Impact of oxygen on Cu surface for highly reliable low-k/Cu interconnects with CuSiN and Ti-based barrier metal," in *Proceedings of the International Interconnect Technology Conference (IITC)*, 2010.
122. K. -M. Yin, L. Chang, F. -R. Chen, J. -J. Kai, C. -C. Chiang, G. Chuang, P. Ding, B. Chin, H. Zhang, and F. Chen, "Oxidation of Ta diffusion barrier layer for Cu metallization in thermal annealing," *Thin Solid Films*, Vol. 388, No. 1/2, 2001, pp. 27–33.
123. Z. Choi, M. Tsukasa, J. M. Lee, G. H. Choi, S. Choi, and J. -T. Moon, "Effect of pre-existing void in sub-30-nm Cu interconnect reliability," in *Proceedings of the 48th IEEE Annual International Reliability Physics Symposium*, 2010, pp. 903–905.
124. E. T. Ogawa, A. J. Bierwag, K. D. Lee, H. Matsuhashi, P. R. Justison, A. N. Ramamurthi, and P. S. Ho, "Direct observation of a critical length effect in dual-damascene Cu/oxide interconnects," *Applied Physics Letters*, Vol. 78, No. 18, 2001, pp. 2652–2654.
125. P. C. Wang, and R. G. Filippi, "Electromigration threshold in copper interconnects," *Applied Physics Letters*, Vol. 78, No. 23, 2001, pp. 3598–3600.
126. K. D. Lee, E. T. Ogawa, H. Matsuhashi, P. R. Justison, K. -S. Ko, and P. S. Ho, "Electromigration critical length effect in Cu/oxide dual-damascene interconnects," *Applied Physics Letters*, Vol. 79, No. 20, 2001, pp. 3236–3239.
127. F. Wei, C. L. Gan, C. V. Thompson, S. P. Hau-Riege, J. J. Clement, H. L. Tay, B. Yu, M. K. Radhakrishnan, K. L. Pey and W. K. Choi. "Length effects on the reliability of dual-damascene Cu interconnects," *Proceedings of Mater Research Society Symposium*, Vol. 716, 2002, p. B133.
128. C-K, Hu, R. Rosenberg, and K. Y. Lee, "Electromigration path in thin-film lines," *Applied Physics Letters*, Vol. 74, No. 20, 1999, pp. 2945–2947.
129. B. Li, J. Gill, C. J. Christiansen, T. D. Sullivan, and P. S. McLaughlin, "Impact of via-line contact on Cu interconnect electromigration performance," in *Proceedings of the 43rd IEEE Annual International Reliability Physics Symposium*, 2005, pp. 24–30.
130. B. Li, T. D. Sullivan, and T. C. Lee, "Line depletion electromigration characteristics of Cu interconnects," in *Proceedings of the 41st IEEE Annual International Reliability Physics Symposium*, 2003, pp. 140–145.
131. I. Jeon and Y. -B. Park, "Analysis of the reservoir effect on electromigration reliability," *Microelectronics Reliability*, Vol. 44, 2004, pp. 917–928.
132. Z. H. Gan, W. Shao, S. G. Mhaisalkar, Z. Chen, H. Li, K. N. Tu, and A. M. Gusak, "Reservoir effect and the role of low current density regions on electromigration lifetimes in copper interconnects," *Journal of Materials Research*, Vol. 21, No. 9, 2006, pp. 2241–2245.
133. Z. H. Gan, A. M. Gusak, W. Shao, Z. Chen, S. G. Mhaisalkar, T. Zaporozhets, and K. N. Tu, "Analytical modeling of reservoir effect on electromigration in Cu interconnects," *Journal of Materials Research*, Vol. 22, No. 1, 2007, pp. 152–156.

134. C. Christiansen, B. Li, and J. Gill, "Blech effect and lifetime projection for Cu/low-*k* interconnects," *International Interconnect Technology Conference (IITC)*, 2008, p. 114–116.
135. R. Frankovic and G. H. Bernstein, "Pulsed-current duty cycle dependence of electromigration-induced stress generation in aluminum conductors," *IEEE Transactions on Electron Devices*, Vol. 17, 1996, pp. 244–246.
136. K. L. Lee, C. K. Hu, and K. N. Tu, "In situ scanning electron microscope comparison studies on electromigration of Cu and Cu(Sn) alloys for advanced chip interconnects," *Journal of Applied Physics*, Vol. 78, 1995, pp. 4428–4437.
137. H. U. Schrieber, "Electromigration threshold in aluminum films," *Solid-State Electronics*, Vol. 28, 1985, pp. 617–626.
138. S. Thrasher, C. Capasso, L. Zhao, R. Hernandez, P. Mulski, S. Rose, T. Nguyen, and H. Kawasaki, "Blech effect in single-inlaid Cu interconnects," in *Proceedings of the International Interconnect Technology Conference (IITC)*, 2001, pp. 177–179.
139. P. -C. Wang, G. S. Cargill, II, I. C. Noyan, and C. -K. Hu, "Electromigration-induced stress in aluminum conductor lines measured by x-ray microdiffraction," *Applied Physics Letters*, Vol. 72, 1998, pp. 1296–1298.
140. R. G. Filippi, G. A. Biery, and R. A. Wachnik, "The electromigration short-length effect in Ti–A1Cu–Ti metallization with tungsten studs," *Journal of Applied Physics*, Vol. 78, 1995, pp. 3756–3768.
141. R. G. Filippi, M. A. Gribelyuk, T. Joseph, T. Kane, and T. D. Sullivan, "Electromigration in AlCu lines: Comparison of dual damascene and metal reactive ion etching," *Thin Solid Films*, Vol. 388, No. 1, 2001, pp. 303–314.
142. A. A. Keshavarz and L. F. Dion, "Effects of metal thickness variations on IC metal lifetime due to electromigration," in *IEEE International Integrated Reliability Workshop Final Report*, 2009, pp. 170–173.
143. N. S. Nagaraj, F. Cano, H. Haznedar, and D. Young, "A practical approach to static signal electromigration analysis," in *Proceedings of the 35th Design Automation Conference*, 1998, pp. 572–577.
144. K. Banerjee and A. Mehrotra, "Global (interconnect) warming," *IEEE Circuits and Devices Magazine*, Vol. 17, No. 5, 2001, pp. 16–32.
145. K. Banerjee, A. Amerasekera, N. Cheung, and C. Hu, "High-current failure model for VLSI interconnects under short-pulse stress conditions," *IEEE Electron Device Letters*, Vol. 18, No. 9, 1997, pp. 405–407.
146. B. K. Liew, N. W. Cheung, and C. Hu, "Projecting interconnect electromigration lifetime for arbitrary current waveforms," *IEEE Transactions on Electron Devices*, Vol. 37, No. 5, 1990, pp. 1343–1351.
147. F. Chen, J. Gill, D. Harmon, T. Sullivan, B. Li, A. Strong, H. Rathore, D. Edelstein, C-C, Yang, A. Cowley, and L. Clevenger, "Measurements of effective thermal conductivity for advanced interconnect structures with various composite low-*k* dielectrics," in *Proceedings of the 42nd IEEE Annual International Reliability Physics Symposium*, 2004, pp. 68–73.
148. D. Harmon, J. Gill, and T. Sullivan, "Thermal conductance of IC interconnects embedded in dielectrics," *IEEE International Reliability Workshop Final Report*, 1998, pp. 54–60.
149. JP-001, "Electromigration," *Foundry Process Qualification Guidelines.* Wafer Fabrication Manufacturing Sites, JEDEC/FSA Joint Publication, September 2002, pp. 7–9.

CHAPTER 11

Stress Migration

11.1 Introduction

The application of complicated Cu/low-*k* structures in advanced-technology nodes has created a complex dynamic of stress concentrations, stress gradients, mass transport, adhesion, and microstructural integrity and has resulted in the recognition of stress migration (SM) or stress-induced voiding (SIV) as a significant failure mechanism that has been studied with great interest.[1–3] In contrast with Al interconnects, whose reliability limitations have been well characterized for 30 years, Cu interconnect reliability recently has been studied extensively. Since Cu has a larger grain-boundary activation energy, which results in a lower mobility, Cu is expected to have more resistance to SIV than Al under the same stress level. However, it seems that sometimes it is overly optimistic to assume that Cu has excellent SIV robustness because Cu doesn't have a natural oxide protection layer, as does Al. In addition, electrodeposited Cu has a much greater initial void concentration than Al.[4] Thus, with decreasing dimensions and the introduction of new materials, Cu has shown vulnerability to stress migration, just as its predecessor, Al, exhibited.

The physics of SM is summarized in Sec. 11.2. The impetus to the numerous studies on SM is the trend of miniaturization of integrated circuits (ICs), where the SM problem is aggravated owing to aggressive demands for higher circuit density and smaller feature size. Furthermore, SM may influence electromigration (EM) when there is significant amount of vacancy accumulation owing to SM in the cathode area, which may accelerate EM lifetime degradation.[5] Ogawa et al.[1] reviewed SM mechanisms such as grain structure, flaws, interfaces, atomic transport mechanisms, and temperature effect. It also has been shown that an *active diffusion volume*, a region where the interconnect geometry, the diffusion-mechanism volume, and the stress-gradient region coexist, is required for voiding to take place.

A number of test structures have been proposed for accessing SM reliability with certain characterization methodologies, which are detailed in Sec. 11.3. The phenomenon of SM in IC metallization was first reported in the mid-1980s[6] and is known to manifest in the temperature range from 423 to 523 K, slightly below EM test temperatures, which are typically above 573 K. Therefore, stress-induced voiding behavior is usually investigated by means of baking tests at elevated temperatures.[7] When stress-induced voids occur, they likely lead to the interconnect failure. Section 11.4 summarizes the different SIV failure modes.

SM is a term that describes the formation and growth of voids owing to the development of large tensile and compressive stresses in a metallization. High tensile stresses evolve in the metal lines during the cooling process from the passivation temperature by coefficient-of-thermal-expansion (CTE) mismatches with the surrounding dielectrics and the Si substrate. These stresses will facilitate void formation within the metal lines for stress relaxation, which eventually can cause metal line opening and thus device failure.[8] The stress level is controlled by the material properties (the metals and their surrounding dielectrics), interconnection layout, processing conditions, and the integration approach used in fabrication of the interconnect system.[1,9,10] Besides thermal stresses, considerable *growth stress*, which is due to grain growth, is known to be developed in Cu interconnects by electroplating.[11] Furthermore, more complicated stress-related problems arise as a result of the adoption of various low-*k* materials for dielectrics. Problems such as voids, cracks, and thermal stress also can lead to device failure or a degraded lifecycle in operation. Therefore, it is important to understand stress distribution and evolution in Cu interconnects through measurements and modeling. Finite-element analysis (either two- or three-dimensional) has been performed in the literature in an attempt to understand the stress conditions of complex multilevel structures, which are elaborated in Sec. 11.5.

Cu SIV is strongly dependent on process, surface modifications, and composition, which are given in Sec. 11.6. From a reliability point of view, any techniques that can improve the interface and grain-boundary quality to slow down copper diffusion and/or optimize the interconnects profile will help to alleviate SM. Careful selection of different cap-layer materials and cap-deposition processes can improve the adhesion to copper. Furthermore, it has been realized that the inclusion of a small amount of a second metal in copper films during electroplating and its subsequent segregation at grain boundaries by thermal treatment suppress the diffusivity at the copper grain boundary. Good liner cover age and void-free copper filling are ways to minimize SM-induced failures in the via.[12] The intermetal dielectric (IMD) affects SM. Especially after low-*k* dielectric material was introduced in Cu damascene interconnects, the SM issue has become more complex. Generally speaking, standard inorganic-based IMDs, such as silica, have relatively low CTE and a high elastic modulus. Thus the hydrostatic stress is concentrated at the via and the top of the lines leading to void nucleation. Whereas in case of organic low-*k* dielectric materials, which have a higher CTE and a very low

Young's modulus, large von Mises stress is maintained at the via, and thus deformation of the via is the anticipated failure mode.[13]

Cu SIV is also strongly structure-dependent, which is the topic of Sec. 11.7. SM is sensitive to interconnect geometry for a given set of materials (metal and insulation). Geometric features such as metal line width, line length, via diameter, via-to-line overlap and design aspects of interconnects are found to have significant impact on SM behavior.[1,10,11] For example, a no-misalignment via profile is necessary to ensure good SIV resistance. The geometric effect on SM should be taken into consideration in a design to ensure the highest possible level of built-in reliability.

Finally, the process-reliability qualification for SM is summarized in Sec. 11.8.

11.2 Physics of Stress Migration

11.2.1 Basic Understanding of the Stress Migration Mechanism

Based on observations from SM failure, it has been generally accepted that tensile stress induced by the thermal treatment in the fabrication process is the driving force for void formation.[14] This stress arises from the CTE mismatch between the metal and its surrounding materials. In the scenario of stress-induced voiding, regions of high-tension stress enable the nucleation of voids, which is the case for interconnect metal fabricated at high temperature and cooled down to low temperature. Note that while thermomechanical stress increases with decreasing temperature, diffusion increases with increasing temperature. Such interplay between driving force and atomic diffusivity means that stress migration typically shows a peak in its voiding rate at an intermediate temperature range (Fig. 11.1), which already has been seen experimentally.[1] At high temperature, stress voiding is limited by metal stress levels that become compressive, and at lower temperatures, the diffusivity for mass transport becomes progressively slower. Therefore, reliability testing should be performed around the temperature corresponding to the maximum drift, that is, the highest acceleration.

Croes et al.[15] reported that the worst temperature is quite structure-dependent: The worst temperature for the kelvin single via (see Fig. 11.6) is 150°C, whereas the worst temperature for the M_1-plane structure is 175°C and that for the M_2-plane structure is 200°C.

In addition to thermomechanical stresses as just explained, microstructural properties play a part in influencing the mechanism of SM. A void is normally nucleated either at the interface between the metal and the dielectric passivation or at the grain boundaries. The grain

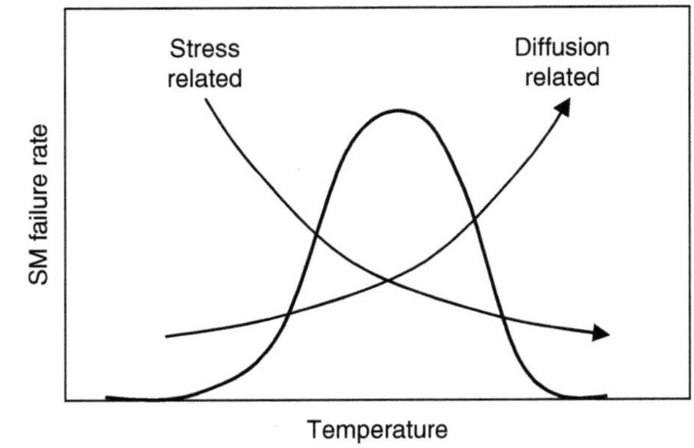

FIGURE 11.1 Schematic showing that there is a peak for SM failure rate at certain temperatures, resulting from the competing effects of tensile stress (decreasing with temperature) and atomic diffusion (increasing with temperature). The temperature for the peak SM failure rate is generally around 150 to 200°C.

boundaries are full of vacancies, and the free volume released by grain growth as a result of grain-boundary elimination creates sizable voids. Void nucleation may be assisted by either interfacial or grain-boundary sliding.[1,16,17] The void grows under the influence of a sufficiently high adjacent stress field that can be due to either dislocation creep or vacancy condensation. The vacancy then diffuses along the grain boundaries and/or interfacial boundaries in a polycrystalline microstructure or through the lattice in a near-bamboo-type structure of the lines. Further stress relaxation is believed to take place by vacancy diffusion accompanied by growth of stress-induced voids.[18]

Figure 11.2 gives a schematic illustrating the micromechanism of stress voiding, where the void formation is the result of the thermal diffusion of vacancies. The driving force of the void formation is the volume contraction of Cu, which leads to a buildup of residual stress. The SM occurs near vias because of the lower stress magnitude there attributed to the complex geometry (structural discontinuity) and multiple material compositions. In addition, the grain boundaries underneath and the surrounding metal-cap interfaces serve as the vacancy sources, as well as the fastest diffusion paths in the copper interconnect. Many small voids can be generated at the grain-boundary area after dielectric barrier deposition even before high-temperature storage (HTS) testing, and some of the voids were grown after HTS, preferentially at the grain-boundary corners.[19] This result implies that unlucky landing of a via over the grain-boundary area would be one of the root causes of SIV under the via.

In situ observation of void evolution in SM test structures with a Cu via diameter of about 100 nm connected to wide (>1-μm) Cu plates was performed by synchrotron-based transmission x-ray

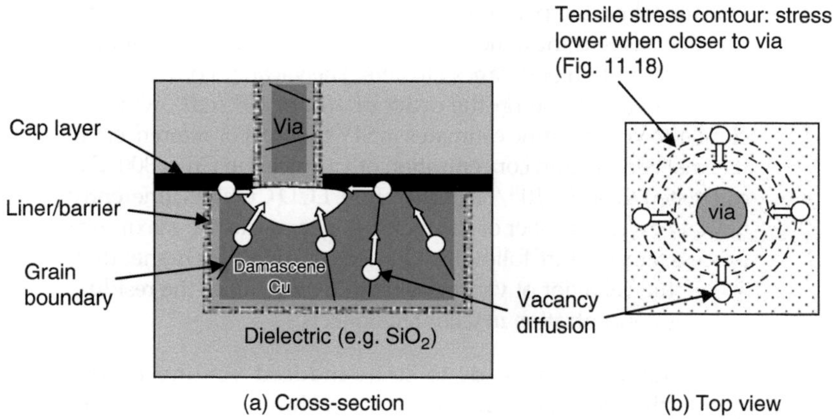

Figure 11.2 Schematic showing the voiding process under vias during high-temperature baking.

microscopy (TXM).[20] It was shown that voids originally formed at the site where eventually the catastrophic failure occurs, that is, directly beneath the via in wide Cu line at the edge of the via bottom. During SM tests, Cu atoms migrated from regions of low tensile (or high compressive) stress to regions of high tensile stress, and simultaneously, vacancies migrated along the stress gradient (within a limited range of some micrometers) in the opposite direction, to the location where vias connect wide Cu lines.

The minimum number of vacancies needed to cause the via to open is fairly small. In electrochemical deposition (ECD) Cu interconnects, there is at least one vacancy for every 500 Cu atoms for a given via connection to a wide line below the via.[21] It is noted that there is one vacancy for every 1.8×10^{15} Cu atoms if thermal equilibrium was achieved.[22] This indicates the defective nature of the Cu being deposited using ECD.

11.2.2 Active Diffusion Volume

Direct observation of vacancy defects in electroplated Cu films with annealing temperature varying from 100 to 700°C were investigated by positron annihilation and high-angle annular dark-field field scanning transmission electron microscopy (HAADF-STEM).[23] For the as-deposited situation, the grain size is small, and most of the vacancies are located at grain boundaries. After annealing, the evolution of vacancy defects can be divided into two regions with respect to the annealing temperature.

- For annealing temperature below 300°C, vacancies agglomerate into small clusters, and the behavior of vacancies is closely related to grain growth.

- For annealing temperature over 300°C, the vacancy concentration decreases as the annealing temperature increases, attributed to the dissolution of vacancy clusters. The vacancy concentration was estimated to be on the order of 10^{19} to $10^{20}/cm^3$, which is similar to void volume estimates in SIV failure. For reference, the thermal equilibrium concentration of vacancies in Cu at 300°C is calculated to be $5 \times 10^{11}/cm^3$. Therefore, ECD Cu films inherently have an excess number of vacancies. In general, as the maximum process temperature following Cu electoplating goes higher than 300°C, the behavior of vacancies mainly determines the residual vacancy concentration in Cu interconnects.

Vacancy migration that leads to stress-induced voiding is controlled by the diffusion mechanisms active within a given material. From the mass-transport point of view, three major volumetric scales that consist of *interconnect volume, diffusion volume,* and *stress gradient* can be thought to define the diffusion problem.[1] The interconnect volume, which is simple to define, is calculated by the product of length, width, and height within the region where the damage formation occurs. The active diffusion volume cannot exist without having these three volumetric scales being present simultaneously at a specific site of the interconnect. As an example, the active diffusion volume and the void volume to cause failure are assumed to be 7.2 and 0.01 μm^3, respectively.[23] Only vacancies that are within the active-diffusion-volume region can contribute to the voiding process.[24]

Diffusion volume states the supply of vacancies that will coalesce to form the void and is specified by the characteristic diffusion path length as shown below[25]:

$$X_D(t,T) = 2\sqrt{K \cdot t} = 2\sqrt{\frac{D \cdot B \cdot \Omega}{k_B \cdot T} \cdot t} \tag{11.1}$$

where D is the atomic diffusivity, B is the effective modulus of the Cu dual-damascene stack, Ω is the atomic volume, k_B is the Boltzmann constant, t is time, and T is the local temperature. For a given time interval and temperature, the supply of vacancies that have the potential to contribute to the voiding is defined by the interconnect volume in a distance $X \leq X_D$ from the voiding location.

With the assumption that the vacancies are uniformly distributed within the interconnect volume, the active volume is approximately equivalent to the diffusion volume. Figure 11.3 schematically illustrates the different diffusion-volume behaviors in different-sized metal plates.[26]

Only one-dimensional (1D) vacancy migration is feasible in narrow lines, which results in a $\sqrt{t}$ dependence of the resulting diffusion volume around the voiding location. However, for the case of wider lines, two-dimensional (2D) vacancy migration from a circular catchment area leads to a linear time-dependence relationship. This explains why wide lines have earlier failure times because they can provide more vacancies per time to the voiding location. When the width of the lines exceeded X_D, the number of captured vacancies within a time t becomes independent of the line width, resulting in a saturation of the mean time to failure (MTF). When the diffusion is the dominant mechanism for stress-induced voiding, the drift of resistance (R shift) is indeed seen to be a linear function of $\sqrt{t}$.[26]

Lin et al.[27] reported that under 200°C baking, the resistance change associated with voiding shows saturation with stress time, which appears following the $\sqrt{t}$ relation given in Fig. 11.3. However, under 150°C baking, a linear time dependence of ΔR is observed that is independent of metal width. Linear time dependence of ΔR is consistent with void growth that occurs as a result of stress relaxation in both the width and length directions of the metal line,[25] implying that the stress-relaxation front is identical over the range of widths examined. The mechanism of stress voiding in Cu/low-k vias appears to be independent of width.[27]

11.2.3 Void Nucleation

Although the process of void nucleation for SIV is not fully understood, it is interpreted by the competition of the tensile stress with the surface stress and surface energy of the void embryo,[28] leading to a critical stress threshold for void nucleation. Hydrostatic stress [Eq. (11.2)] is known to be the driving force for SIV void nucleation.[3] The most probable void-nucleation locations can be predicted based on the hydrostatic stress distribution, which can be obtained easily from finite-element modeling.[29] On the other hand, the von Mises stress [Eq. (11.3)] is usually used as a criterion for plastic deformation,

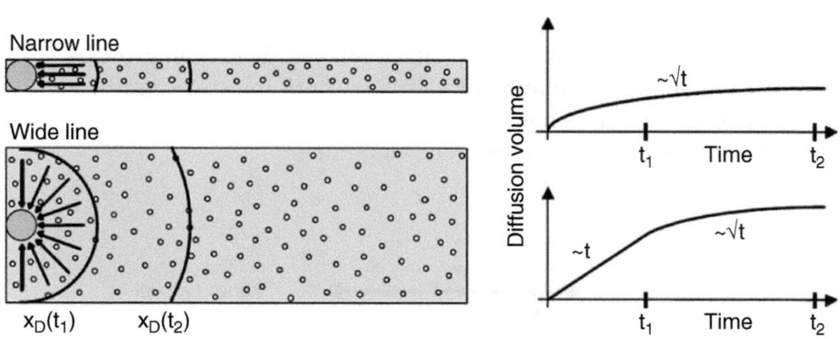

FIGURE 11.3 Diffusion area as a function of time calculated for narrow and wide lines, respectively. (*Used with permission from Cambridge University Press.*[26])

which is in a form of octahedral shear stress not affecting the volumetric changes of material. Processing variations,[30] microstructure,[31] surface-texture defects,[32] oxidation or "contamination patches,"[33–35] local delamination,[36] plastic deformation,[37] and stress concentration owing to triple-point grain-boundary intersections with interfaces are also believed to play roles in void nuclei.

$$\sigma_H = \frac{(\sigma_x + \sigma_y + \sigma_z)}{3} \tag{11.2}$$

$$\sigma_{\text{von-Mises}} = \frac{1}{\sqrt{2}}\left[(\sigma_1 - \sigma_2)^2 + (\sigma_2 - \sigma_3)^2 + (\sigma_3 - \sigma_1)^2\right]^{\frac{1}{2}} \tag{11.3}$$

where σ_x, σ_y, and σ_z are normal stress in the x, y, and z directions, respectively. σ_1, σ_2, and σ_3 are three principal stresses.

11.2.4 Stress Gradient

Once a void is nucleated, the stress gradient is believed to be the driving force for vacancy and/or atomic diffusion.[16] *The Stress-gradient region* is defined as the existence of a significant driving force that compels the vacancies to migrate toward a specific voiding site. This driving force is enabled by the relaxation of stress gradients that are present in the metallization system after the deposition process. Several other factors include thermal mismatch between interconnect and encapsulating oxide, changes in the microstructure as a result of grain growth or recrystallization, and geometric aspects of the connected metal lines, including the via size.

Figure 11.4 gives a schematic of the situation when a Cu atom is subjected to a higher tensile stress at the right-hand side. We can imagine that the Cu atom will be driven to the right (the side with higher tensile stress). In contrast, the vacancies will diffuse toward the more compressive (or lower tensile stress) region and coalescence or sink into voids where the stress will be relaxed. Voids can grow by the diffusion and accumulation of vacancies until the stress gradient is eliminated. When this happens, the interconnect will have severe reliability problems, such as open-circuit failures.

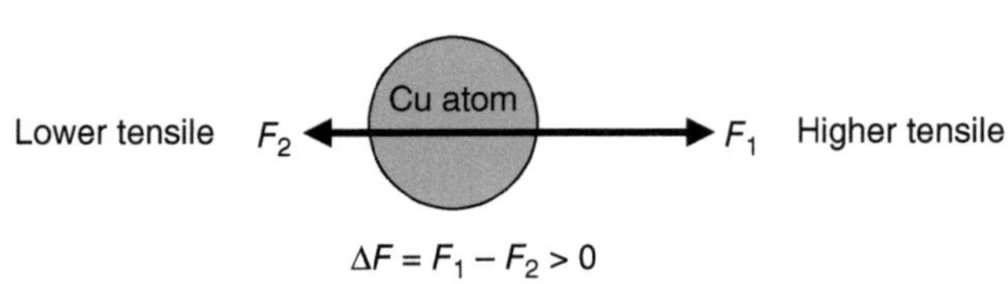

FIGURE 11.4 Schematic for the situation where a Cu atom is subjected to a higher tensile stress at the right-hand side.

The role of stress gradient is illustrated in the following manner.[38] Since stress-induced voids are the result of vacancy supersaturation, the time-averaged vacancy flux under a driving force is given by

$$J_{\mathrm{SIV}} = C(\vec{x},t)\cdot M\cdot F = C(\vec{x},t)\cdot\left[\frac{D_0\exp(-E_a/k_BT)}{k_BT}\right]\cdot\Omega\frac{\Delta\sigma}{\Delta x} \tag{11.4}$$

where J_{SIV} is the vacancy flux due to a stress gradient, $C(\vec{x},t)$ is the local concentration, M is a diffusional mobility, F is the driving force, D_0 is an effective diffusional prefactor, E_a is activation energy, k_B is the Boltzmann constant, T is the absolute temperature, Ω is the local atomic volume, and $\Delta\sigma/\Delta x$ is the local stress gradient. Equation (11.4) can be used to estimate the level of stress gradients required to cause a void. If the local vacancy concentration is assumed to be a constant and is approximately proportional to the inverse of the atomic volume, $f\,\Omega^{-1}$, where f is the number that can range between 0 and 1, Eq. (11.4) is expressed as

$$J_{\mathrm{SIV}} = f\cdot\left[\frac{D_0\exp(-E_a/k_BT)}{k_BT}\right]\cdot\frac{\Delta\sigma}{\Delta x} \tag{11.5}$$

Equation (11.5) can be used to calculate the estimated magnitude of stress gradients required to cause a void. For simplicity sake, the stress gradient is described in the following equation,[16] where D is the diffusion coefficient:

$$J_{\mathrm{SIV}} = \pm\frac{D}{k_BT}\left(\frac{\Delta\sigma}{\Delta x}\right) \tag{11.6}$$

The negative and positive signs in Eq. (11.6) indicate vacancy flux and atomic flux, respectively.

Based on Eq. (11.4), hydrostatic stress, vacancy concentration, and stress gradients that have significant effect on stress-induced voiding should be minimized to effectively suppress stress-induced-void failure in Cu interconnects.

11.2.5 New Failure Mechanism Observed for Stress Migration

The SM mechanism discussed earlier that exhibits a peak at an intermediate temperature is based on the two competing effects, diffusion and stress by thermal expansion mismatch when the test temperature is lower than 225°C. In contrast, when baking temperatures are higher than 225°C,[39] a new failure mechanism is observed, where oxidation of the barrier takes place, especially for the interconnects embedded in SiCOH low-k dielectric employed in advanced-technology nodes.[40] In this region, the resistance increase during HTS continued to increase with temperature. Such a resistance increase was prominent only in via-chain structures but negligible in line structures without vias.[39] By extending Fig. 11.1, Fig. 11.5 schematically shows the high-temperature effect when baking temperature is larger than 225°C, which shows a steady increase in degradation as a function of temperature (the dotted curve).

When comparing the relative resistance change (i.e., R_{stress}/R_0* 100%) after 1,000 hours of storage at 275°C for tetraethylorthosilicate (TEOS) and SiCOH stacks, Aubel et al.[40] found that the SiCOH stack showed a significant resistance, whereas the TEOS stack did not. The maximum resistance increase in the SiCOH stack can be about 30 percent. The activation energy for this mechanism is above 1.1 eV. In addition, about 20 percent of the tested samples have a resistance increase larger than 10 percent. On the other hand, the TEOS stack has a negligible resistance shift—only about 2 percent, which could be just measurement variation. It is interesting that after extensive failure analysis using focused ion beam (FIB)/scanning electron microscopy (SEM), no SM related voids could be found even for 30 percent *R*-shift samples with SiCOH dielectric stacks. This observation could be explained by the shape of the resistance trace over testing time. In conventional SM at a stressing temperature of ~200°C, after the void grows to a critical size at a certain time, the corresponding resistance will increase abruptly. However, with SiCOH dielectrics at high storage temperatures, only a slow creeping-resistance increase is seen, indicating that there is a unique mechanism taking place at high storage temperatures as explained by *the*

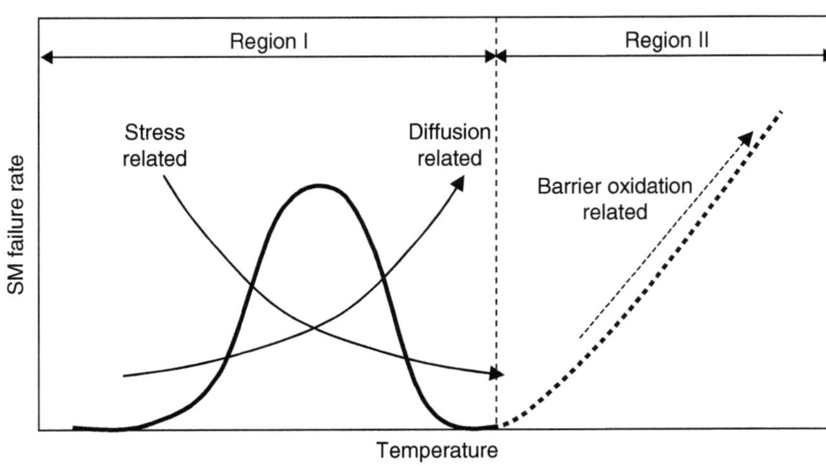

Figure 11.5 The dotted curve at the high temperature (>225°C) is added to Fig. 11.1, indicating the barrier-oxidation-related steady increase in degradation as a function of temperature.

partial oxidation of the Ta barrier in that region evidenced by the electron energy loss spectroscopy (EELS) analysis. The TaO_x has much higher resistance than Ta, which causes the continuous resistance increase during stressing. Given the various deposition and etch backsteps, no TaN remains under the via bottom but on the via sidewall. Therefore, there is a very thin Ta barrier layer (but no TaN) sandwiched rather between the via bottom and the lowermetal surface. It is likely that the lower metal surface is mechanically damaged by the etch-back process prior to Ta deposition. A rough interface between Ta and the lower galvanic Cu is seen in a high-resolution TEM image.[40] It is thus proposed that *it is the SiCOH layer that supports Ta oxidation by serving as an oxygen source,* leading to the barrier oxidation. Since the via sidewall is "protected" by TaN, only the via bottom is susceptible to oxygen owing to the open Ta interface between the barrier and the Cu. The rough lower interface makes Ta oxidation much easier to take place. It is thus necessary to perform this test at higher temperature to address potential process weaknesses in advanced technologies that uses low-*k* dielectrics.

11.2.6 Mathematical Model for Stress-Induced Voiding

The SM model that explains the SIV phenomenon exhibiting a peak at an intermediate temperature (Region I in Fig. 11.5) was developed by McPherson and Dunn[41] and is expressed as

$$R = C(T_0 - T)^N \exp\left(\frac{-E_a}{k_B T}\right) \tag{11.7}$$

where R is the creep rate, T_0 is the stress-free temperature for metal (approximate metal-deposition temperature for copper), T is the temperature for SM stressing, N is the *creep exponent* (2 to 4 for ductile metals), and C is a fitted constant. In this model, the *effective* or observed activation energy from an Arrhenius analysis will be temperature-dependent and is expressed as[42,43]

$$E_{a,eff} = E_a - Nk_B\left(\frac{T^2}{T_0 - T}\right) \tag{11.8}$$

where $E_{a,eff}$ is a temperature-dependent activation energy.

Based on the preceding understanding, the time-to-failure (TTF) can be written as[4]

$$\text{TTF} = C_0 \cdot (T_0 - T)^{-N} \cdot \exp\left(\frac{E_a}{k_B T}\right) \tag{11.9}$$

where C_0 is a fitted constant, and E_a is the activation energy.

The activation energy E_a is 0.8 eV extracted based on the $\Delta R/\Delta t$ data,[27] implying that vacancy diffusion along the Cu-dielectric barrier interface is the process responsible for stress relaxation and void growth. More specifically, E_a values ranging from 0.74 to 1.2 eV have been reported.[1,44] It is also seen that the E_a is independent of line width in the range 0.07 to 0.42 μm under study, indicating that the void-growth mechanism does not change over the whole stripe-width range examined.

A modification to the preceding diffusion-creep model was proposed by Croes et al.[15] in 2009 when considering baking temperatures greater than 225°C, where the degradation increases steadily with temperature as shown in Fig. 11.5. In this case, an additional Arrhenius component is added as follows[15]:

$$\Delta = A(T_0 - T)^{N/2}\exp\left(\frac{-E_a^I}{2k_BT}\right) + B\exp\left(\frac{-E_a^{II}}{2k_BT}\right) \tag{11.10}$$

where Δ is the drift/via, and E_a^{I} and E_a^{II} represent the activation energies of the diffusion mechanisms from regions I and II, respectively. The experimental activation energies in Regions I and II are 0.9 and 1.2 eV, respectively. These values are in good agreement with literature values for interface and grain-boundary diffusion.[22] Also, the fitted values of N (=2) and T_0 (=225°C) are in good agreement with literature values.[1] The activation energies obtained indicate that *the degradation mechanism dominating in Region I is due to interface diffusion, wheres the diffusion processes causing the drifts in Region II are driven by grain-boundary diffusion.*

11.3 Characterization of Stress Migration

11.3.1 Stress-Migration Test Structures

Typical SIV structures consist of big metal plates below (Fig. 11.6*a*) or above (Fig. 11.6*b*) a via or kelvin-like single via structure (Fig. 11.6*c*). These single-via structures are sensitive to the SIV resistance rises and have the benefit of easier assessment for failure analysis. They are designed to study different voiding mechanisms for voiding below a via (Fig. 11.6*a*), voiding in a via (Fig. 11.6*b*), and both mechanisms (Fig. 11.6*c*), respectively. The metal plate should be large enough (as defined in the design rules) as the vacancy source; for example, a 5 × 5 μm M_x-plane structure was designed with two 60-nm vias positioned 3 μm from each other, given in Fig. 11.6*a*, where each via was connected to a narrow M_{x+1} line that was connected to a bond pad. The wider the line below or above the via, the larger is the volume from which the vacancies can be collected to from a void. Therefore, wider lines yield a higher likelihood of voiding, which results in resistance increases during limited test times. In most investigations, extremely

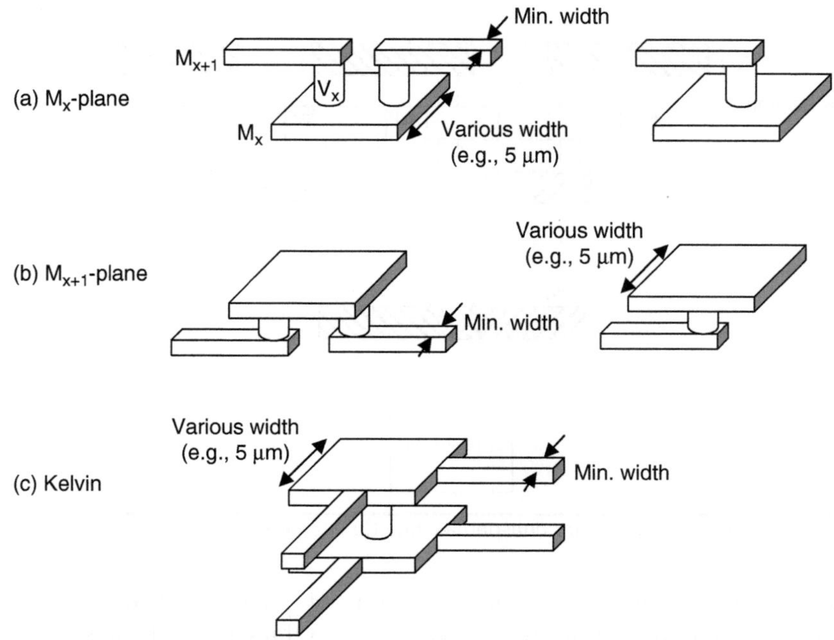

FIGURE 11.6 Test structures designed to study different voiding mechanisms: (*a*) below a via, (*b*) in a via, and (*c*) both below and in a via.

wide plates are used to evaluate the marginality of the SIV performance, but they are not allowed in regular designs.

In contrast, the via-chain (VC) structure is another SIV structure consisting of many vias connected through metal lines (either long or short), for which the resistance rises of vias are magnified and easy to detect, but it is relatively difficult to locate the failure location. Via-chain SIV structures can be a linear via chain or a stacked via chain (Fig. 11.7). Some more examples of special SIV structures are detailed below.

Figure 11.8 gives schematic stacked via-chain SM test structure for dual-damascene Cu interconnects.[45] Figure 11.8*a* is a cross-sectional

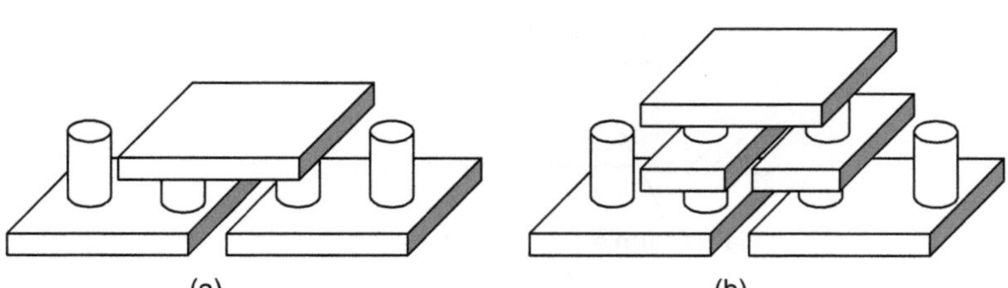

FIGURE 11.7 Via-chain (VC) SIV structures: (*a*) linear via chain and (*b*) stacked via chain.

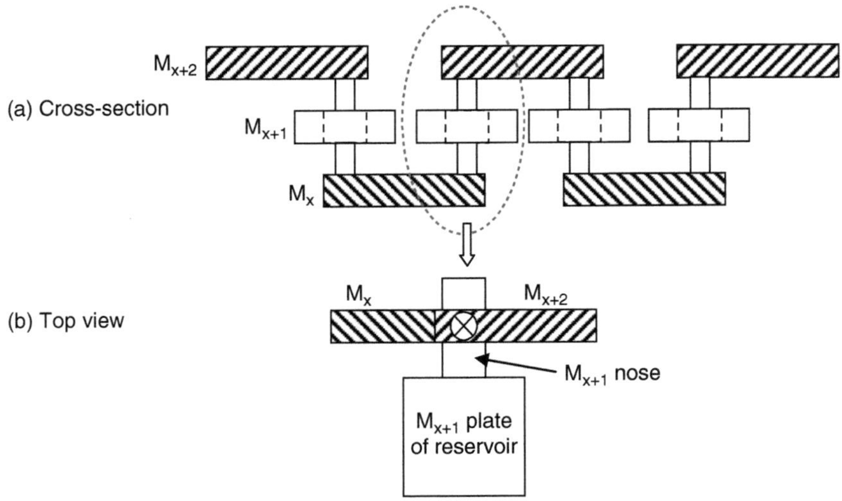

FIGURE 11.8 SM test structures for dual-damascene Cu interconnects.

view of the structure connecting from M_x to M_{x+2}. Each stacked via-chain segment connects to an M_{x+1}-plate reservoir through an M_{x+1} nose (Fig. 11.8*b*) of variying widths and lengths. An example of M_{x+1}-nose dimensions is 100 nm wide and 250 nm long.

Matsuyama et al.[46,47] studied some SIV structures, as illustrated in Fig. 11.9. The *wide pattern* is a chain pattern where the two metal levels

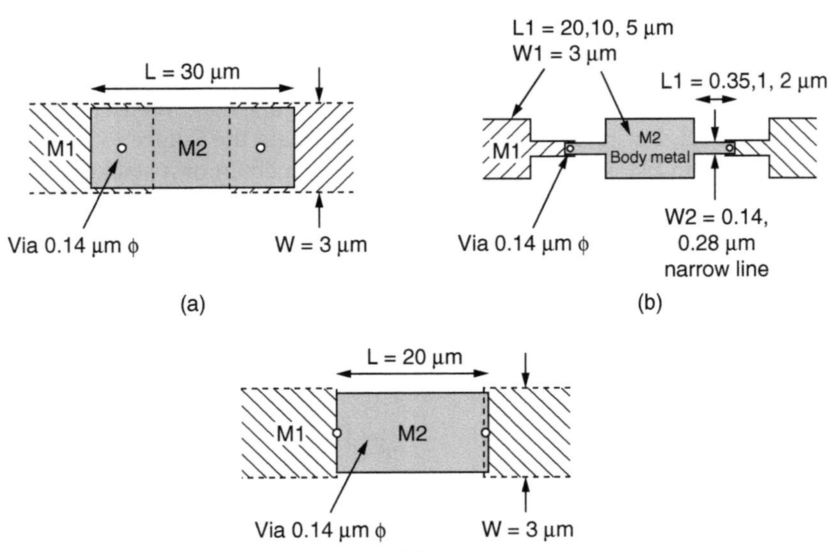

FIGURE 11.9 Schematic top-view drawing of (*a*) wide pattern, (*b*) extrusion pattern, and (c) zero-extrusion pattern. (*Used with permission from IEEE.*[47])

are connected in series by single vias. M_1 and M_2 lines have the same width (3 μm), and each has a length of 30 μm. The *extrusion pattern* is a chain connecting a narrow pattern to a wide pattern. In the *zero-extrusion pattern*, the length of narrow pattern is zero. The vias are located at the edge of the wide pattern.

11.3.2 Stress-Migration Characterization Methodology

SM phenomena are usually investigated by simple HTS tests at the wafer or package level.[48] Package level systems have the advantage of higher time resolution because the in situ resistance-measurement readouts can be performed continuously in real time, whereas the ex situ wafer-level test approach allows testing of large samples sizes or a wider variety of different test structure types simultaneously. The test itself is a simple resistance measurement over time. The stress temperatures range from 100 to 325°C. For an ex-situ wafer-level stress test, the initial resistance of the structure under evaluation is first measured as a prestress T_0 value. Then readouts will be taken at stress durations, for example, of 168, 500, 1000, and sometimes up to 2,000 hours to check the void-migration saturation/stability. A typical resistance-measurement current for small resistance is 0.3 mA.[40] During baking, no current is applied to the test structure to eliminate the effect of EM-induced voiding.

Lognormal plots of failure times resulting from changes in resistance and normal or lognormal distributions of percentage change in resistance of the test structure are methods used to study SM reliability statistics.[49,50]

There are also other ways to characterize SM, but they are less popular, for example, measurement of thermal stress in Cu lines and vias directly as a function of line width during the cooling down from passivation deposition temperature[14] or as a function of baking temperature using x-ray diffraction.[51]

11.4 Failure Modes of Stress Migration

Stress-induced voids are formed by the concentration of supersaturated vacancies inside interconnects. These supersaturated vacancies diffuse owing to stress gradients and accumulate at the stress-concentration sites. Once this void reaches a critical size (depending on the void location), the interconnect fails from either a high resistance increase or an open circuit.

There are mainly three failure modes induced by SM:

- The first is the pull-out-mode void at the via bottom shown as Fig. 11.10*a*. This failure mode is specific to dual-damascene structures when the upper line is wide (e.g., see Fig. 11.6*b*),[3,9,21,26,52,53] and is referred to as *upper-metal mode SIV*. This mode can be suppressed by optimizing the via shape[3,12,21] to improve barrier-metal coverage around the via and enhance adhesion between Cu and the barrier metal. There are two distinct characteristics in the via-line transition affected by local thinning.[12] One is in the upper part of the metal trench (owing to lower incoming mass flux); the other is in the transition area between the via and trench at the midvia sidewall (owing to stronger local resputter). Poor barrier-metal coverage brought a deficient Cu filling into the via. The insufficient adhesion between Cu and the underlayer causes the voids by agglomeration of Cu or "pullout" of Cu into the wide upper lines.

- The second is void formation in the middle of the via (Fig. 11.10*b*). This is another kind of *upper-metal-mode SIV*. This results from the combination of nonoptimized via processes[21] and a pronounced penetration of the via into the metal line underneath the copper-liner interface as the via bottom becomes obviously less critical for void nucleation.[12] As a result, the middle of the via is more prone to void nucleation when applying liner deposition techniques with resputter effect.
- The third mode is void formation at the metal line underneath the via (Fig. 11.10*c*),[53] which is referred as *lower-metal-mode SIV*.[54] To get voiding below a via (Fig. 11.10*c*), the main diffusion path of vacancies is the copper-cap interface and a reservoir of vacancies needs to be present in the metal layer under the via (e.g., see Fig. 11.6*a*), which is the main failure mode for SiO_2 as an intermetal dielectric (IMD). This mode of SIV can be well controlled by using multiple vias[28] or increasing the Cu grain size and improving the adhesion of the barrier metal with Cu.[21]

A focused ion beam (FIB) is used commonly to detect the location of the void. TEM is sometimes used for a high-resolution failure analysis to characterize the phenomenon. In addition, finite-element

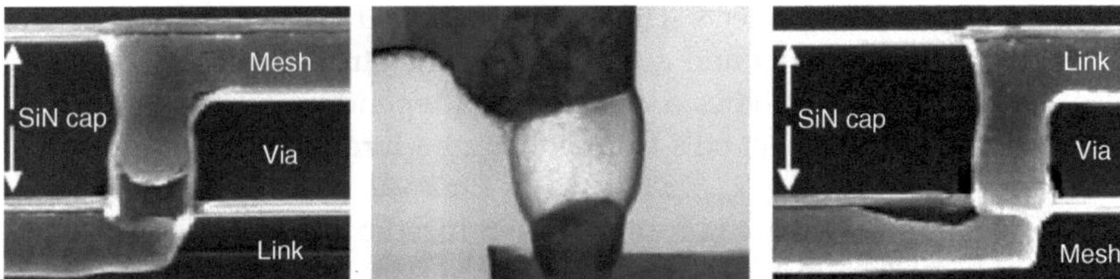

(a) (b) (c)

FIGURE 11.10 Typical stress-induced voids in damascene Cu interconnects. [(*a*) and (*c*) *used with permission from Cambridge University Press26, and* (*b*) (*used with permission from IEEE.*[47])]

analysis has been regarded as a necessary supplement for understanding the detailed mechanism of voiding in relation to the stress field, which will be elaborated in the following section.

11.5 Finite-Element Method for Stress Migration

Finite-element modeling (FEM) has been performed successfully to study the stress distribution and migration in a three-dimensional (3D) multilevel device structure for Al interconnection since the 1990s.[55] Given the complicated interconnect structure for Cu, finite-element-analysis simulations played a substantial role in assessing stress distribution and characterizing the relative SIV probability of Cu metallization.[56,57]

In addition, an understanding of the voiding mechanism certainly can shed light on the future design and process improvement of multilevel device structures. FEM operates on the principle that the analysis of a large and complex structure can be simplified by breaking it down into numerous parts or elements.[29] Each of these elements is then described by a set of relatively simpler equations. These element-specific sets of equations then are joined together to come up with an extremely large set of interrelated equations that describe the behavior of the entire structure. The dynamic void-growing process after nucleation can be simulated through FEM calculations with the aid of the element death technique.[58,59]

Simplified 2D axisymmetric models can be employed for symmetric reliability test structures. However, for more realistic narrow-to-wide chain asymmetric structures with a larger stress gradient that leads to faster void growth, the buildup of a 3D model was required for assessment of stress gradients.[60]

The most widely used commercial finite-element software for finite element analysis is ANSYS or ABAQUS. In three dimensions, the responses of interest were the normal stress components σ_x (along line length), σ_y (transverse to line), and σ_z (normal-to-line length), hydrostatic stress, von Mises stress, and the corresponding stress gradient. Details of the FEM modeling for SIV are given in this section.

11.5.1 Finite-Element-Method Model Description

Figure 11.11*a* shows a typical via-line structure with an M_1-V_1-M_2 configuration for the FEM model. The model geometry also includes M_1 and M_2 extensions (via-to-line overlap), which are widely used in practical layouts. The unit of the dimensions listed in Fig. 11.11*a* is the nanometer. The thickness of the barrier-layer Ta (tantalum) is 25 nm for the bottom and the sidewalls, and that of the etch stop layer undoped silicate glass (USG) 50 nm. The height and width of the via are 280 and 260 nm, respectively. The entire line length of the model is fixed at 2500 nm.

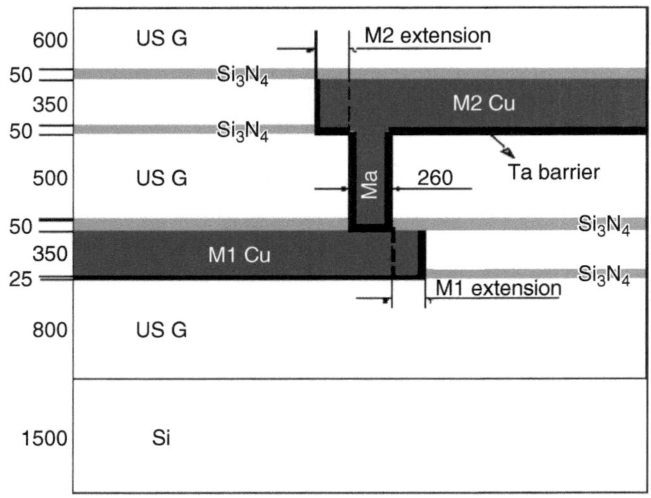

(a)

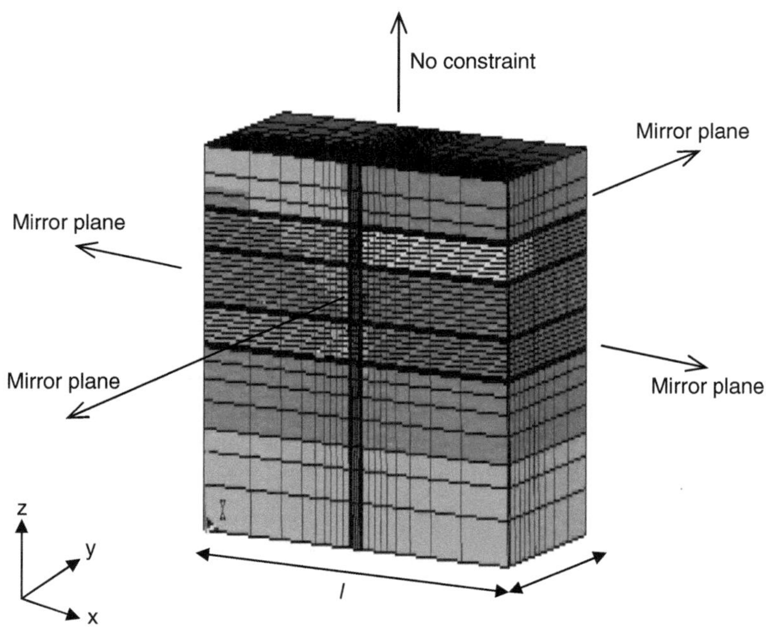

(b)

Figure 11.11 (*a*) Model used for the via-line structure with geometric description. (*b*) Model used for the via-line structure with a description of boundary conditions of finite-element mesh. (*Used with permission from Springer Science and Business Media.*[29])

Properties	Materials	CTE (ppm/°C)	Modulus (GPa)	Poisson's Ratio
Substrate	Si[63]	2.6	141	0.22
IMD	USG[64]	1.37	60	0.25
	CDO[65]	12.0	16.2	0.25
	TEOS[13]	1	59	0.16
	SiLK[13]	62	3.5	0.35
Etch stop	SiN[66]	3.2	220.8	0.27
Line	Cu[67]	17.7	104.2	0.352
Barrier	Ta[68]	6.5	185.7	0.342

TABLE 11.1 Thermomechanical Properties of the Materials Used in the Finite-Element Analysis

Figure 11.11*b* shows the finite-element mesh with the boundary conditions for the via-line structure given in Fig. 11.11*a*. The following boundary conditions are used: As planes $x = 0, l$ and $y = 0, w$ are mirror symmetric constraints owing to the repetitive structure, $u_x = 0$ in planes $x = 0$ and $x = w$; and $u_y = 0$ in planes $y = 0$ and $y = w$. Here u_x and u_y correspond to the displacements along the x and y directions, respectively. The element type used in this ANSYS simulation is the eight-node SOLID45, which is commonly used for 3D thermal-stress analysis.[61] A mesh refinement can be adapted in the copper interconnects because a high stress concentration is expected at these locations. Figure 11.12 gives a flowchart on using ANSYS for SM modeling.

Table 11.1 lists the properties of the materials used in the simulation. These properties are obtained from measurement of the thin film because the model dimensions are in micrometers. Each material is assumed to be an isotropic linear elastic solid without accounting for the microstructure-dependent nonlinear behavior of Cu. This assumption is reasonable because it is known that the width of the Cu lines is in the submicron range, and it will continue to display linear elastic behavior even at temperatures as high as 400°C.[62]

Thermomechanical calculations are performed for interconnect cooling from its initial stress-free temperature to the SIV test temperature of 200°C.[69] The stress-free temperature of 400°C is generally used because it is the deposition temperature of the dielectric material when the structure is supposed to be at the zero-stress state.[70] The final temperature of 200°C is selected to ensure the highest acceleration in SIV, and this temperature is termed the *critical temperature*.[69] This critical temperature is determined by the resistance changes of the Cu interconnects after HTS at various temperatures.

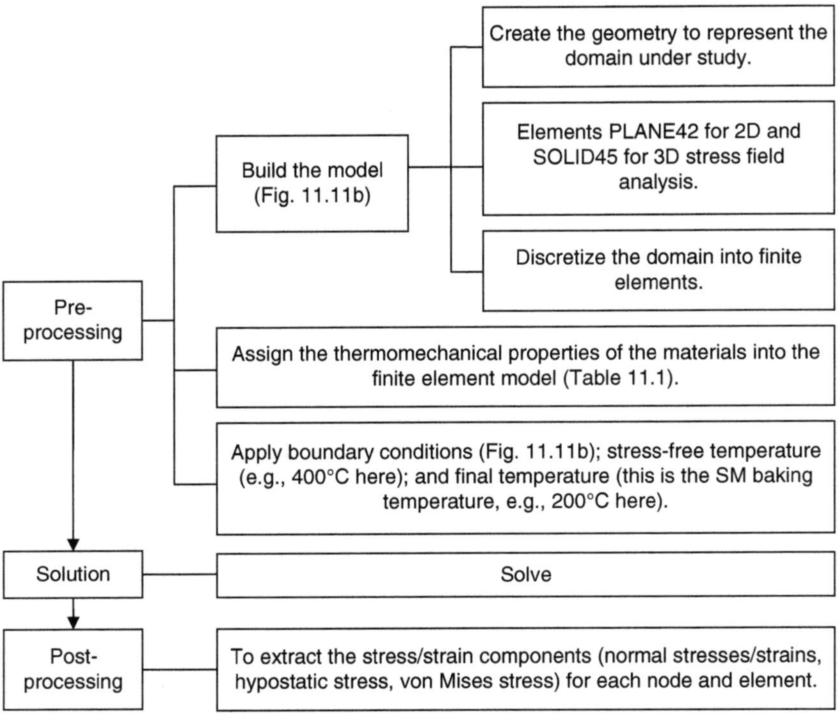

FIGURE 11.12 A flowchart for using ANSYS for SM modeling. (*Used with permission from Springer Science and Business Media.*[29])

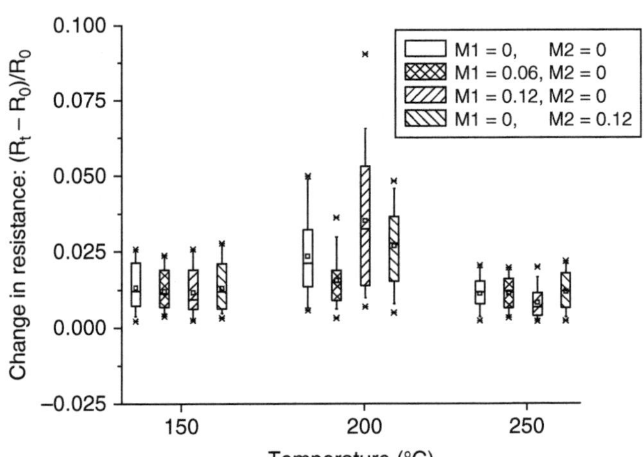

FIGURE 11.13 Normalized change in resistance after testing for 1,344 hours at various temperatures. (*Used with permission from Elsevier.*[69])

Figure 11.13 is an example of a normalized change in resistance with varied M_1 or M_2 extensions in USG dielectric after stressing for 1344 hours at 150/200/250°C. The maximum resistance change with the widest spread was observed at 200°C. This observation is also part of other reports found in the literature.[1,71] The results also indicate that SM is not very much sensitive to M_1 and M_2 extensions. Some other works used room temperature as the final temperature for stress simulation.[13,55]

11.5.2 Finte-Element-Method Parameters to Characterize Stress and Examples

Hydrostatic Stress and von Mises Stress

Different kinds of stress distributions in the copper interconnect such as von Mises stress, hydrostatic stress, and principal stresses can be simulated or extracted from FEM software such as ANSYS or ABAQUS. Hydrostatic stress is known to be the driving force for stress voiding or, in stricter sense, the driving force for stress-void nucleation.[3] Therefore, the hydrostatic stress distribution that is obtained from simulation is able to predict the highly probable void-nucleation location. Von Mises stress, which is usually used as a criterion for plastic deformation, is in a form of octahedral shear stress that does not affect volumetric changes in the material. The hydrostatic stress and von Mises stress can be formulated using Eqs. (11.2) and (11.3), respectively.

Examples of the hydrostatic and von Mises stress distributions in a TEOS-embedded via-line structure are shown in Fig. 11.14. The highest and the second-highest hydrostatic stresses are found at the bottom plane of the via and at the top plane of the lines, respectively, and the calculated stress value is about 600 to 700 MPa. Although a hydrostatic tensile stress on the order of 1 GPa normally would be needed to overcome the energy barrier for void nucleation,[72] it can be lowered when defects are present.[34] Thus, the bottom of the via can constitute a weak interface unless the barrier-deposition process is optimized. The top surface of the metal lines is likely to be contaminated during the chemical-mechanical polishing (CMP) process and have a lot of defects and/or high interfacial energy. In other words, either the bottom of the via or the top of the lines will be the dominant nucleation site owing to the high density of the interfacial defects as well as high stress values in these regions, and this agrees well with experiments.[26]

Stress Gradient

While the hydrostatic stress itself can explain the void nucleation sites, it may not explain the SIV behavior. For example, it was reported that the simulated hydrostatic stress decreases with

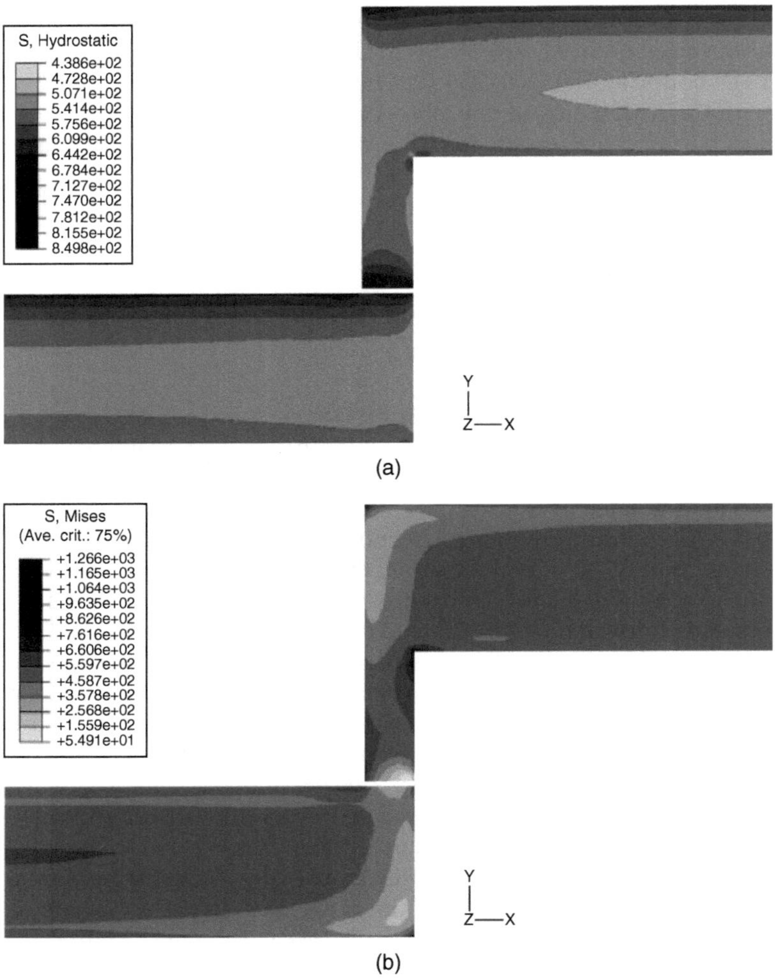

FIGURE 11.14 Contour plots of (*a*) hydrostatic and (*b*) von Mises stress of a TEOS-embedded via-line structure. (*Used with permission from Elsevier.*[13])

increasing line width.[7] However, experimental results indicated that wider metal lines have smaller mean-time-to-failure (MTTF); that is, they are more susceptible to SIV than narrower lines.[7] This phenomenon can be explained by the stress-gradient concept instead.

In fact, once a void is nucleated, the stress gradient is believed to be the driving force for vacancy and/or atomic diffusion, as described by Eq. (11.6), causing the void to grow, which is explained in Fig. 11.14. Vacancies diffuse toward the more compressive region and shrink near the voids where the stress is relaxed. Voids can grow by the diffusion and accumulation of vacancies, until the stress gradient is eliminated, and if this happens, it can induce severe reliability problems, such as open-circuit failures. An example is given below to combine the failure analysis and the stress gradient from FEM.

Figure 11.15 showed that the dielectric of carbon-doped oxide (CDO) is much worse than that of USG.[69] About 40 percent of the CDO samples showed open circuit failures after a 1344-hour test, whereas the maximum resistance change in the USG samples was only 10 percent. The data thus clearly indicate that the choice of dielectric materials has a very significant impact on SM lifetimes. Failure analysis using focused ion beam (FIB) indicated that failures in both CDO and USG were very similar in nature, with voids nucleating symmetrically at the bottom of the via[69] (Fig. 11.16).

As discussed earlier, the site of the most concentrated hydrostatic tension stress is believed to be the most probable void-nucleation site. Regardless of the dielectrics employed, the simulated concentration center of hydrostatic tensile stress was found situated at the edge of the via, as shown in Fig. 11.17, which is comparable with the actual failure behaviors of the via-line structure in Fig. 11.16. Indeed, the via bottom is a process-related weak interface that lowers SIV reliability, especially in narrow interconnects. The crack propagates rapidly owing to the large stress concentration at the void tip. However, Fig. 11.17 shows that the magnitude of the hydrostatic stress in CDO dielectric is lower than that in USG dielectric, which cannot explain the worse SIV resistance of the CDO dielectric shown in Fig. 11.15.

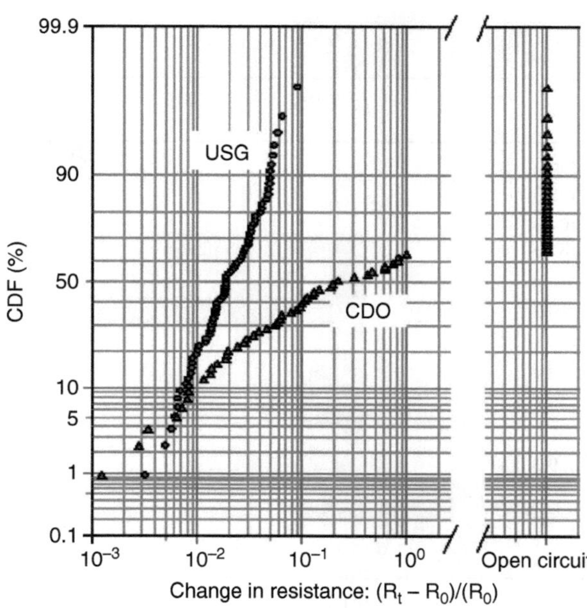

FIGURE 11.15 Changes in resistance of USG versus CDO intermetal-diaelectric (IMD) after 1344 hours (test temperature = 200°C). (*Used with permission from Elsevier.*[69])

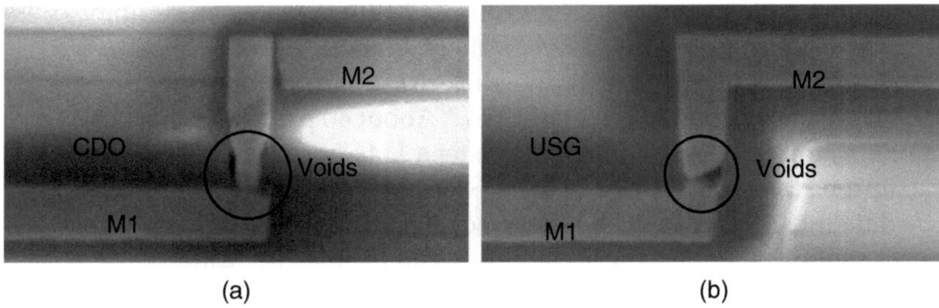

FIGURE 11.16 Postfailure FIB images: (*a*) dielectric = CDO; (*b*) dielectric = USG. (*Used with permission from Elsevier.*[69])

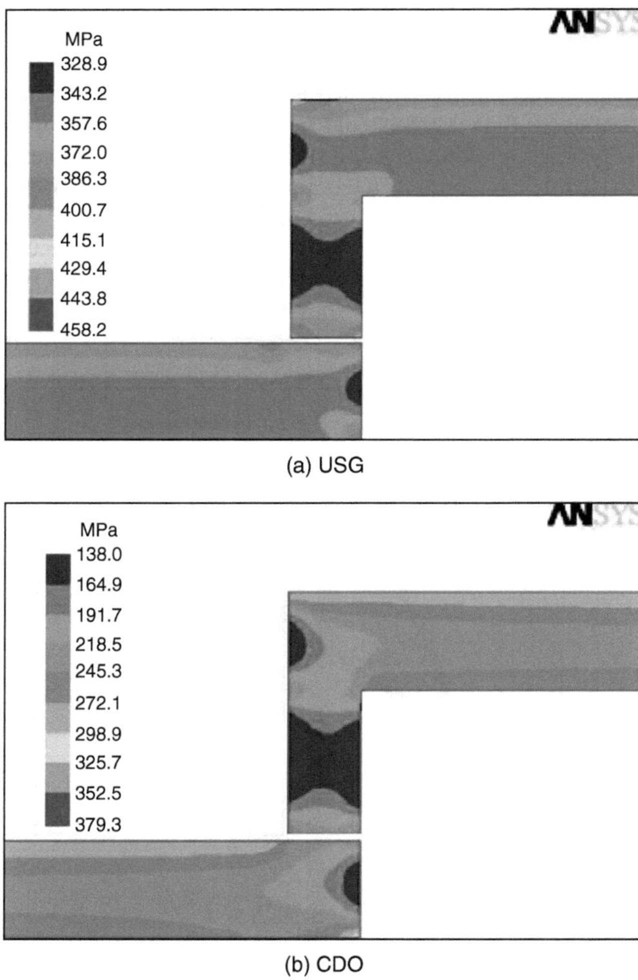

FIGURE 11.17 Finite-element-simulated contour plot of hydrostatic stress around via bottom (*a*) for USG and (*b*) for CDO dielectrics. (*Used with permission from Elsevier.*[69])

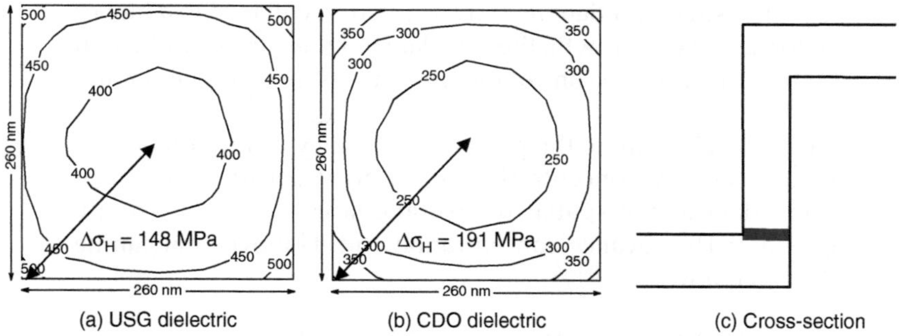

FIGURE 11.18 Hydrostatic stress contours at via bottom for (*a*) USG dielectric and (*b*) CDO dielectric. (*c*) Schematic cross-section of the metal-via structure highlighting the via bottom. (*Used with permission from Springer Science and Business Media.*[29])

To explain the experimental results, the hydrostatic stress gradients are computed from FEM and are shown in Figs. 11.18 and 11.19. Based on Fig. 11.18, the average stress gradient at the via bottom in a CDO structure along the diagonal is about 1.039 MPa/nm, which is about 30 percent higher than that in a USG structure (~0.805 MPa/nm). Similarly, based on Fig. 11.19, the average stress gradient at the via sidewall in a CDO structure along the z axis (~0.578 MPa/nm) is about 57 percent higher than that in USG structure (~0.368 MPa/nm).

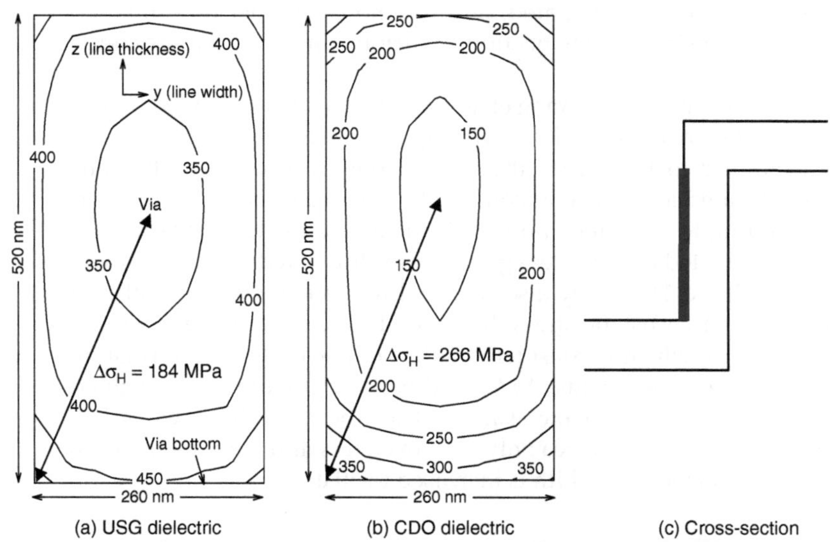

FIGURE 11.19 Hydrostatic stress contours at a via sidewall for (*a*) USG dielectric and (*b*) CDO dielectric. (*c*) Schematic cross-section of the metal-via structure highlighting the via sidewall. (*Used with permission from Springer Science and Business Media.*[29])

Since a higher stress gradient in a CDO structure will lead to a higher driving force for void growth, the calculated higher stress gradient in a CDO structure could be one of the key reasons for its higher failure rate.

To numerically mimic the physics of vacancy migration, a concept of the vacancy concentration or vacancy density was introduced to relate to the spatial hydrostatic stress gradient $\nabla\sigma_H$ by Huang et al.[73] The vacancy concentration and its movement can be described as follows[74]:

$$\frac{\partial C}{\partial t} = -\nabla\left[-\frac{D}{\Omega k_B T}(\Omega k_B T \nabla C - \Omega \nabla \sigma_H)\right] \tag{11.11}$$

where D is the diffusion constant, C is the atomic concentration, Ω is the atomic volume, k_B is the Boltzmann constant, T is the absolute temperature, and σ_H is the hydrostatic stress. Equation (11.11) indicates that the evolution of vacancy density (per unit volume) at different times equals the preexisting vacancy concentration plus its variation over various spatial distances while stressed by σ_H. In this modeling scheme, a numerical index reflecting the vacancy density then is developed to emulate the vacancy-concentration evolution within each material volume/element, in which each finite element of the constructed FEM model is considered to be a unit volume with a specified preexisting vacancy density. With the stress profiles simulated using FEM simulation, the vacancy movement along each spatial direction was calculated, and the index number or simulated vacancy density of each location at various time thus was obtained accordingly.

On the other hand, Wang et al.[53] devised an area-weighted hydrostatic stress, namely, $\Sigma_{i=1}^{N}[(\sigma_H)_i \times (area)]$ to evaluate the stress effect on SM. Similar to Fig. 11.18, they also reported a lower hydrostatic stress located at the via, in contrast to the area outside the via, corresponding to a vacancy flow from the area outside the via toward the via directed by the stress gradient, as illustrated schematically in Figs. 11.2 and 11.4. They also reported that a larger metal will produce more hydrostatic stress difference between the metal area and the via area, which is considered to be significant for SIV formation during bake experiments. At M_1 width of 1.4 μm, the area-weighted hydrostatic stress in the metal area outside the via is about two orders higher than that in via area only. However, this difference increases to four orders when M_1 width is increased to 30 μm.

11.6 Process Effects on Stress Migration

Controlling the SM phenomenon requires optimization of several processes that will be elaborated in this section.

11.6.1 Via-Gouging Effect

Via-gouging morphology at the via-line interface describes the via penetrating into the Cu metal line underneath (see Fig. 10.24*c*). It is determined by the depth of via penetration. Via gouging has been demonstrated to reduce contact resistance, EM, and SM.[45,75] One way to obtain via gouging is to use a plasma sputter etch during barrier deposition. More via gouging takes place with increasing resputter intensity.[12] An ex situ sputter etch in which deposition and sputter etch are processed in two different vacuum chambers can increase the gouging depth to 1.3× when compared with an in situ process with both deposition and sputter etch in the same chamber.[45] Deeper via gouging could reduce the number of early SM failures, which was attributed to a reduced stress gradient (hence the driving force of SM) for deeper via gauging through finite-element modeling.[12]

11.6.2 Metallization-Layer Dependence

The thermally induced stress in the vias may increase with increasing metallization layers.[50] Figure 11.20 shows the percentage change in the resistance of via 1 (V_1) to via 4 (V_4) chain test structures consisting of vias 0.2 μm in diameter with 100,000 links and 0.6-μm-wide metal interconnects from 0.13-μm technology after baking at various stress temperatures for 1,000 hours. The difference in the percentage change in the resistance was less prounced at the lower-level vias (V_1 and V_2) than at the higher-level vias (V_3 and V_4). Such findings look to be opposite to the intuition that the lowest-level via (i.e., V_1) should be

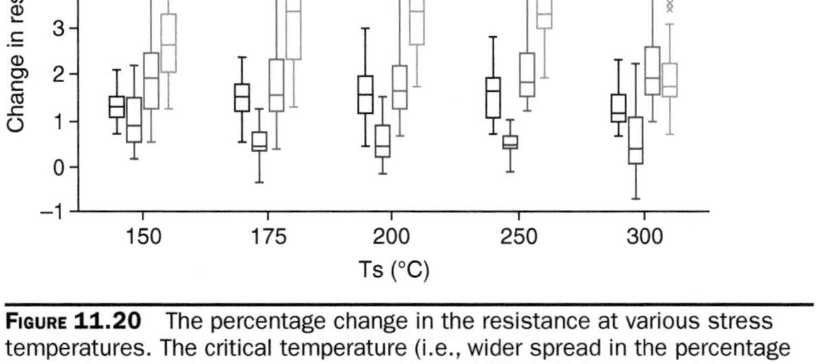

FIGURE 11.20 The percentage change in the resistance at various stress temperatures. The critical temperature (i.e., wider spread in the percentage change in the resistance) occurred between 150 and 200°C. In addition, the thermally induced stress in the vias increased with increasing metallization layers. (*Used with permission from IEEE.*[50])

the weakest structure because V_1 is subject to more thermal cycles during fabrication. Such as observation then was attributed to the stresses in the different metallization layers related to the radius of curvature of the wafer. Figure 11.21 shows a linear-like response of the radius of curvature of a 200-mm patterned wafer with respect to the metallization layer measured at different metallization layers after the respective Cu CMP process. Based on the well-known Stoney's equation[76] [Eq. (11.12)], it is clear that the stress increased correspondingly at the upper metallization layers:

$$\sigma = \frac{Et_{sub}^2}{6(1-\nu)t_{\text{film}}R} \tag{11.12}$$

where E is Young's modulus of the substrate, ν is Poisson's ratio of the substrate, t_{sub} is the thickness of the substrate, R is the net radius of curvature, and t_{film} is the thickness of the film. In short, there are a couple of suggestions as follows:

- For any meaningful comparison of SM performance, the study has to be performed at the same metallization layer.
- The stress-induced voiding effect in the Cu interconnect will become more challenging with the implementation of more metallization layers.

11.6.3 Barrier-Layer Effect

A thicker barrier is necessary to ensure overall reliability, including both SM and EM.[12] An unoptimized physical vapor deposition (PVD) barrier-layer deposition process may lead to a higher deposition rate

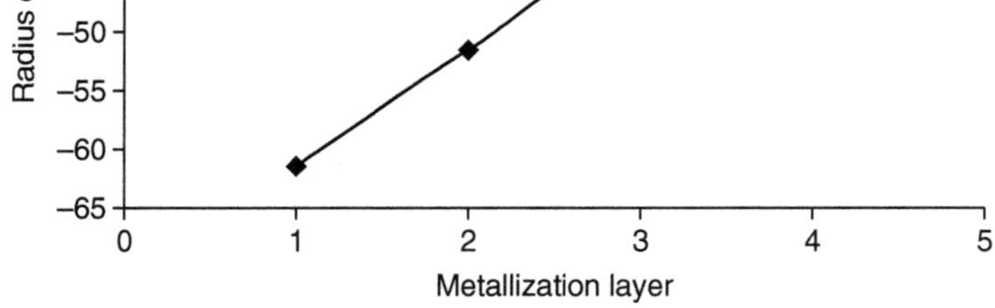

Figure 11.21 The linear-like response of the radius of curvature to the metallization layer indicated that the upper metallization layers experienced higher stress. (*Used with permission from IEEE.*[50])

at the center of the wafer. Therefore, there could be a strong correlation between the SM failure rate and the initial prebake resistance of the via-chain structure under study. Lim et al.[50] found that a better and thicker via bottom sidewall coverage (thus a higher initial resistance prior to the SM test) at the wafer center corresponds to a lower SM failure rate than that at wafer edge.

The typical voiding morphology induced by an unoptimized barrier process is given in Fig. 11.16. A schematic to illustrate the two modes of void formation at the via bottom is given in Fig. 11.22. In Mode I, the stress-induced voids will first nucleate at the via edge and stretch vertically (Fig. 11.22*a* and Fig. 11.16*a*). In Mode II, voids prefer to grow horizontally along the via bottom (Fig. 11.22*b* and Fig. 11.16*b*). Once the void nucleates in the via bottom, whether it will grow following Mode I or Mode II depends on the localized stress state as well as the integrity of the Ta barrier in the via bottom. Built-up tensile stress in the via bottom may be high enough to ensure void growth during SM, resulting in an open circuit. Figures 11.19 and 11.18 show the simulated stresses for Mode I and Mode II, respectively. Note that based on the simulation, the dielectric material does affect to the SM. Apparently, a void formed in the *x*-*y* plane (Mode II) will induce more pronounced resistance change than a void formed in *y*-*z* plane (Mode I).

One solution to reduce the stress-induced voiding caused by the small voids at the bottom corners of the via is to improve the

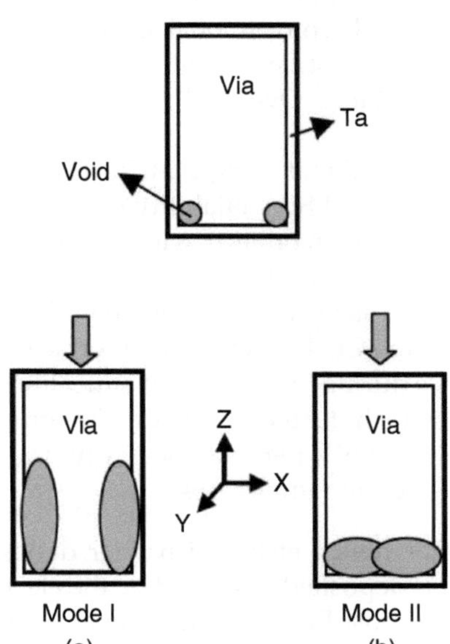

Figure 11.22 Schematic of two modes of void formation at the via bottom: (*a*) voids stretch vertically; and (*b*) voids grow horizontally, leading to a more detrimental situation. (*Used with permission from Elsevier.*[69])

PVD process of the diffusion barrier layer.[50,77,78] During the PVD process, the diffusion barrier layer is deposited first, followed by a resputtering step with the bias power turned on. This resputtering step enables part of the diffusion barrier layer, which is deposited earlier, to be redistributed at the sidewall of the via bottom. As such, the coverage can be improved.[50,79] Furthermore, resputtering also thins the diffusion barrier layer at the bottom of the via, thus leading to a lower via-chain resistance. A continuous diffusion barrier layer thus is achieved at the bottom of the via that results in a continuous Cu seed layer and void-free gap fill during the ECP process.

The preclean process prior to barrier-layer deposition also has much impact on SM performance.[80] The bias power of the argon plasma preclean should be kept as high as possible to reduce Cu recess to underlying metal, and the resputtering time should be kept as short as possible but sufficiently long to clean the via bottoms. The effects of the via diameter variation caused by the precleaning to SIV are also significant.

11.6.4 Cu Alloy Effect

Cu-Al alloy[81–83] is one of the effective ways to improve SIV reliability. Matsubara et al.[81] demonstrated that Cu-Al alloy interconnects using a solid-phase reaction between ECD Cu and PVD Al showed superior SIV resistance. In this process, after trench etching, followed by barrier TaN/Ta and seed Cu sputtering and Cu electroplating, two additional steps (i.e., Al sputtering and an anneal process) are added to form Cu-Al alloy, after which conventional Cu-CMP is performed. The annealing causes atomic scale mixing at the Al-Cu interface that improves the SIV resistance. Over 0.31 percent Al in Cu is required to eliminate SIV failure.

Iguchi et al.[82] reported that using Cu-0.5 wt% Al seed metal clearly can improve both EM and SIV reliability of 32-nm-node ultralow-k (k = 2.4) Cu multilevel interconnects with a bilayer low-k barrier cap (k = 3.9). The Cu-Al alloy seed technology does not require additional manufacturing process steps. In this technology, Al diffused from seed metal segregates at the Cu–barrier-metal interface and the Cu–low-k barrier cap interface during the optimized annealing condition (350°C for 30 seconds), which further reduces Cu ionic drift and thus improves both EM and SIV performance. A typical Cu-Al seed process includes the following three steps:

- *Step 1:* After trench etch and barrier deposition, the Cu-Al alloy seed is deposited, followed by the electroplated (EP) Cu film. During post-EP annealing, some of the Al diffuses into the EP Cu film, whereas the other elements remain near the barrier metal.
- *Step 2:* The excess EP Cu film, the seed metal, and the barrier metal on the dielectric interlayer then are removed by CMP.

- *Step 3:* The low-*k* cap is deposited. As the process continues, residual Al piles up from the neighbor barrier metal to the Cu-SiCN interface.

However, the negative effect of this technique is that the wiring resistivity increases about 10 percent when the Al concentration is 0.5 percent. On the other hand, the via resistance is not especially affected by the Al concentration. Increasing Cu grain size and optimizing the Al thermal diffusion to reduce the Al concentration in the Cu film are suggested to control the negative impact.

Tonegawa et al.[52] investigated the effect of impurity doping (Sn and Ti) to Cu from the Cu-alloy seed layer on SIV. Cu-alloy seed was deposited by ionized PVD after deposition of the Ta-based barrier metal. The concentration of Sn and Ti in the seed layer was 1.0 wt%. Given the higher diffisivity of Sn into Cu in comparison with Ti, the Cu-Sn alloy improved the SIV resistance more than the Cu-Ti alloy. Cu-Sn suppressed diffusion of vacancies through both the Cu grain boundaries and the interface of the Cu–barrier metal because Sn piled up to the Cu grain boundaries and Cu–barrier metal interface. On the other hand, Cu-Ti was effective in suppressing diffusion of vacancies through the interface of the Cu–barrier metal because Ti piled up only to the interface of the Cu–barrier metal, but did not adequately affect the suppression of vacancy diffusion through the Cu grain boundaries.

Ueki et al.[84] studied effects of Ti addition on SIV behaviors through three different methods of Ti addition (Fig. 11.23). The results are summarized below:

1. *Ti layer insertion under Ta-TaN stacked barrier metal.* The SIV under the via connected to a wide lower line is drastically suppressed. The Ti is added just to the via bottom of M_1 to improve adhesion between the Cu (M_1) and the Ta-TaN barrier (M_2), and void nucleation at the interface thus was suppressed. However, EM is not improved, probably because the Ti was not doped into all areas of the Cu lines.
2. *Ti layer insertion between a Ta-TaN barrier and Cu.* In this case, EM is improved owing to the Ti addition to the Cu lines. However, the SIV resistance is degraded. The Ti doping from the bottom surface of the Cu film resulted in a strong Cu(111) texture with high tensile stress. Cu(111) texture with incoherent twins was found to be prone to form SIVs.[31]
3. *Ti doping from the surface of the electrochemical-plated (ECP) Cu film.* This scheme is desirable because both SIV and EM resistance can be improved. The improvement adhesiveness at the interface between the via and the lower Cu line and the oxygen gettering from Cu by Ti during the ECP-Cu annealing play an important role in suppressing SIV and EM. This scheme also

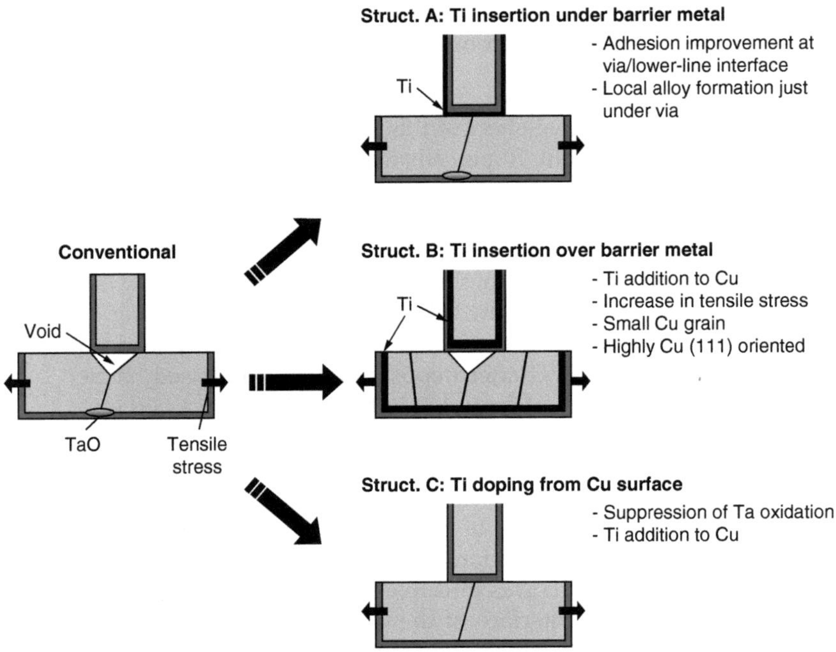

FIGURE 11.23 Illustration of the effects of each Ti addition on the interconnect properties. (*Used with permission from IEEE.*[84])

provides the least sheet-resistance increase of the structure. The via resistance is comparable with the conventional scheme without Ti doping. In short, structure C is the most promising scheme for SIV and EM performance improvement without sacrificing device performance.

Isobayashi et al.[85] developed Ag-doped Cu (Cu-Ag) alloy seed to drastically improve Cu interconnect SIV resistance. Cu-Ag alloy has been shown to have the properties of little drop of electrical conductivity in alloying and a high creep resistance even with the content about 0.1 wt% Ag. In this study, a 1 wt% Ag-doped Cu-alloy seed layer deposited by ionized PVD was introduced as the Cu alloy. In contrast, there is no distinction of EM between Cu-Ag and pure Cu because Ag has migrated to the Cu grain boundaries, and EM degradation dominated by the interface between the metal and capping-layer interface.

Shih et al.[86] studied the effect of ECP additive structures and concentrations on EM lifetime and SIV formation. An ECP additive (leveler A)—cationic polymeric amine with a small molecular weight—would scavenge chlorine, oxides of sulfur, and carbon in the plating solution. These impurities, such as C, S, and C1, could form an interstitial solid solution and then migrate from an interstitial position to a neighboring empty site during Cu grain growth and most likely accumulate in the grain boundary. The accumulation of impurities in the grain boundary could act as an effective barrier against vacancy diffusion to inhibit SIV formation. However, the ECP additive condition that produces highly impure Cu was found to increase the gap-filling pits on top of the submicron features that would reduce EM lifetime performance. On the other hand, by replacing the ECP additives with a larger-molecular-weight subtance (named *leveler B*), void-free gap filling can be achieved, which ensures the excellent EM performance while keeping the same SIV resistance as the leveler A additive.

11.6.5 Dielectric Dependence

For Fluoro silicate glass (FSG) as an intermetal dielectric, the stress-induced voiding behavior in Cu dual-damascene interconnects depends on the desorption amount of hydrogen, oxygen, and fluorine ions from the network structure of the FSG films at baking temperature.[87] Given the long heating time, the desorbed ions may diffuse out to the FSG-SiN and Cu-SiN interfaces and thus weaken the interfacial adhesion. Simultaneously, FSG film shrinks and increases tensile stress in adjacent Cu. As a result, Cu vacancies will move faster along the weakened interface, leading to void growth around via. A higher FSG stability at high temperatures (i.e., less desorption of ions) can be achieved through optimization of the CVD process.

The integration of Cu with dielectrics of lower constant values (low *k*) has proved more challenging, and its reliability robustness appears to be more process-dependent, attributed to the poor mechanical properties of the higher porosity of low-*k* dielectrics.[46,50,88] Figure 11.15 showed that the dielectric of carbon-doped oxide (CDO) is much worse than that of USG. Lim et al.[50] also reported that the SM performance of Cu interconnections with diffusion barrier layers deposited by the conventional method degraded when the dielectric constant was lowered from the range of 3.5 to 3.7 to that of 2.8 to 3.0. As discussed in Sec. 11.5.2 this observation cannot be explained by reduction of the biaxial modulus of the higher-porosity low-*k* dielectric film,[69,89] which leads to the intrinsic stress decreases correspondingly, but is attributed to the more severe stress gradients present in and above the via when integrated into porous low-*k* materials, making the voiding into the via more apparent.[15,46,69,88]

On the other hand, Zhai et al.[90] suggested that there is a stress threshold, that is, a critical stress σ_{crit}, for void formation. It was found by Hau-Riege et al.[91] that the σ_{crit} in Cu-TEOS is higher than for Cu–low-*k* material by a factor of 4× (100 MPa for Cu-SiO_2 versus 25 MPa for Cu-low-*k*). The critical stress can be considered as the resistance of the material/system, including the surface energy, plastic work in the metal, or other type of energy dissipation associated with a growing void, similar to the energy release rate in fracture mechanics. Thus an SM risk index associated with both simulated stress and critical stress is defined as follows[90]:

$$s(x,y,z) = \frac{\sigma(x,y,z)}{\sigma_{crit}} \tag{11.13}$$

The void is expected to nucleate first where the maximum value is located. The SM risk index, by considering both stress state and critical stress of a given Cu-IMD interconnect system, determines the probability of SIV from the stress perspective. It reveals that the SM risk index for Cu–low-*k* material is approximately 3× higher than that for Cu-TEOS, explaining why a Cu–low-*k* system is worse in SIV performance.

11.6.6 Copper-Microstructure Effect

Copper annealing has an impact on SM behavior. The voiding results from the supersaturation of vacancies that develops owing to grain growth when the Cu is not annealed properly prior to being fully constrained.[1] Oshima et al.[21] found that for the case of a via having wide metal below it, the voids were formed at the locations where grain boundaries existed and that SIV could be suppressed by enlarging the Cu grain (through optimizing the annealing conditions after ECP) to decrease the grain boundaries because the grain boundaries are a dominant diffusion path of the vacancies.

Figure 11.24 shows the effect of the postannealing condition on SIV for a structure with Ti doping from the surface of the Cu film

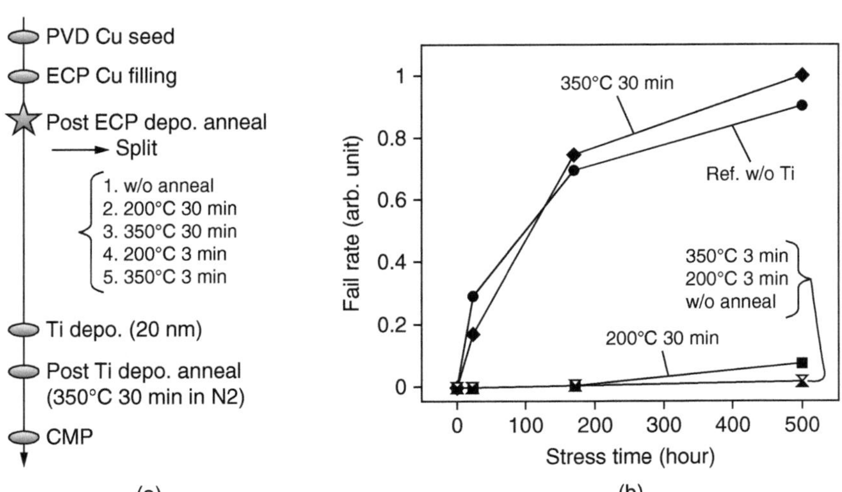

FIGURE 11.24 Effect of the postannealing condition on the SIV: (*a*) process flows; (*b*) SIV failure rate (baked at 150°C) for each postannealing condition. The Ti-deposition is shown in Fig. 11.23 (structure C). (*Used with permission from IEEE.*[84])

(see Fig. 11.23, structure C). The postannealing was performed in N_2 with different temperatures and times. It is seen that the sample with annealing at 350°C for 30 minutes did not show an improved SIV resistance. However, the shorter stress time and/or lower annealing temperature for the cases with Ti doping did improve SIV very much when compared with the reference without Ti doping. A near 0 percent failure rate can be achieved for the cases with postannealing of 3 minutes or without the postannealing.

After annealing, the Ti is found to diffuse through the Cu grain boundaries, but with a very limited amount of Ti addition (0.003 percent), which should not be the sole reason for the SIV resistance improvement. On the other hand, it is found that the oxygen has diffused into the Ta-Cu interface after annealing at 350°C for 30 minutes and formed TaO there that degraded SIV performance because Cu on TaO x is easier to diffuse than that on pure Ta.[92]

The oxygen at the Cu-Ta interface does not come either from the annealing ambient or from the inside of the Cu film, but rather from the surface oxide of the Cu. The oxygen on the Cu surface is supposed to diffuse along the grain boundary during annealing at 350°C for 30 minutes. When the annealing temperature is low or the time duration is short, the oxygen does not diffuse to Ta, so the Ta is not oxidized. When Ti is deposited on the Cu film without postannealing or with low-temperature or short annealing, the oxygen in Cu is incorporated into Ti during the following annealing because Ti has a higher affinity for oxygen than Cu.[84,93]

11.6.7 Quenching Effect

Kawano et al.[3] revealed that quenching treatment suppressed SIV owing to stress relaxation. In this method, the quenched sample was cooled to 77 K after the fabrication process. Quenching causes plastic deformation of the Cu line, resulting in a reduction in tensile stress. For example, a -75°C quench treatment reduced stress in ECD Cu from 280 to 110 MPa. Quenching treatment can achieve a 10× reduction in SIV failures.

11.6.8 Copper-Plating Chemistry

The concentration of organic additives in the copper-plating bath can be varied to modulate the microstructure and physical properties of the copper.[48] Via and metal SM performance then is affected by using different plating chemistries.

11.6.9 Cu-Capping-Layer Effect

When a cobalt-tungsten-phosphorus (CoWP) capping layer was employed in the copper (Cu) damascene interconnects, both EM and SIV resistance can be much improved.[94] However, one drawback of conventional electroless plating is that it introduces new contaminants into fabrication lines because the plating process includes palladium (Pd) for catalyst activation and alkaline-metal components in the plating solutions. Since the Cu surfaces after CMP are oxidized, and both the copper and dielectric surfaces are contaminated, Pd activation has been required to achieve uniform and selective metal capping. Ishigami et al.[94] developed a CoWP cap process without using Pd activation and also using alkaline-metal-free plating solution. Because no Pd or alkaline metals were used, contamination control in a fabrication line could be minimized.

Lin et al.[95] reported that when the cap layer (cap-b) has a lower modulus (67 versus 125 GPa) and a similar CTE (both ~16.5 ppm/°C) to Cu, the SIV degradation is much suppressed, whereas cap-a with a higher modulus (265 versus 125 GPa) and a large CTE mismatch of cap-a (1.5 versus 16.5 ppm/°C) with that of Cu, serious SIV degradation will occur. These observations were attributed to the stress-concentration and stress-gradient profile in the via structures through finite-element analysis.

11.6.10 Miscellaneous

There is a post–reactive-ion-etching (RIE) cleaning step to remove the damages at the SiCOH and Cu surfaces owing to various plasma processing steps. It was demonstrated that the cleaning chemistries do have an impact on SM performance.[45]

The sidewall profiles of both via and trench also affect SM performance because it is critical for formation of a good barrier and Cu seed coverage. Degraded barrier and Cu seed coverage could induce some with initial seams and voids along the sidewall and therefore cause some early SM failures. It should be noted that the profile of via and trench sidewall is controlled by multiple processes and many integration schemes, which makes optimization complicated and time-consuming.

11.7 Geometric Effect on Stress Migration (Stress-Migration Improvement Through Design)

Interconnect geometry has a large impact on SM behavior,[54,96] and this is elaborated in this section.

11.7.1 Effect of Metal Plate Geometry

Metal-Plate-Width Dependence

A typical comparison of SM performance for metal-plate-width dependence is given in Fig. 11.6*a* and *b*. In Fig. 11.6*a*, the metal on top of the via is with the minimum feature size, whereas the metal beneath the via has a very large width. In type B, it is the opposite.

The SIV failure mode of Fig. 11.6*a* was related to metal-x surface voiding underneath the via (see Fig. 11.10*c*),[1,21] and the failure mode of Fig. 11.6*b* tended to be associated with voiding inside the via (see Fig. 11.10*a* and *b*).[9,21] The two different modes of failure generally are believed to depend on the location of a reservoir of available vacancies or active diffusion volume for voiding. The results indicated that the SM resistance shift is quite sensitive to the metal lead width beneath the via Fig. 11.6*a*, whereas it is not for the case with a large metal lead over the via Fig. 11.6*b*, even though the baking time is up to 1000 hours.

Shao et al.[97] showed similar metal width dependence for the structures shown in Fig. 11.6*a* (Fig. 11.25*a*). It is clear that the median change in resistance increased with M_1 Cu line width, indicating that SIV reliability is better for narrower M_1 lines. However, the change in resistance may not make much difference when the M_1 line width is increased from 700 to 1,000 nm, which can be explained by the stress gradient, as detailed below.

The stress gradient acts as the driving force for vacancy movement, and the vacancies will diffuse toward the less concentrated stress regions. The volume-averaged stress gradient at the Cu-cap interface is plotted against line width (Fig. 11.25*b*). It is clear that the magnitude of the stress gradient increases with line width, indicating that wider lines are more prone to void growth and ultimately failure. However, for line widths larger than 700 nm, the amount of driving force tends to increase at a decreasing rate, resulting in saturation of the medium time to failure. This indicates that the change in resistance (which defines the failure times) for a 1,000-nm line width is comparable with the change in resistance for a 700-nm line width (Fig. 11.25*a*).

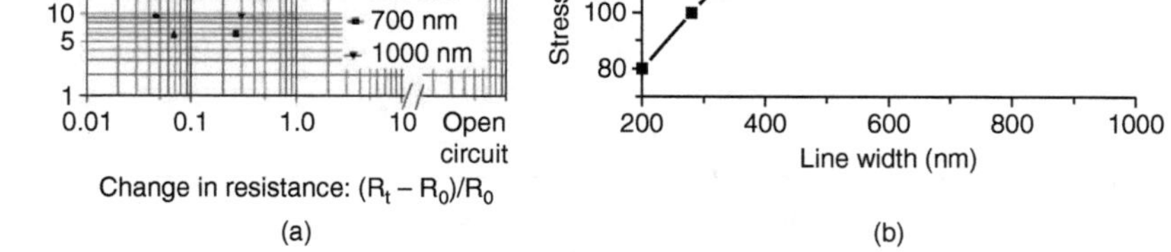

FIGURE 11.25 (*a*) Normalized change in resistance of M_1 Cu line for various line widths. (*b*) Volume-averaged stress gradients versus line width. (*Used with permission from Elsevier.*[97])

It was believed that the existence of an active-diffusion area played a role in void formation.[1,7,9,21] Again, the FEM simulation suggested that the differences in the stress gradients and hence the driving force of the stress induced a voiding mechanism as the root cause of the line-width effect.

Zhai et al.[90] pointed out that the hydrostatic stress increases clearly with line width in a Cu–low-*k* system, although it is hardly observed with line width in a Cu-TEOS system. Thus, as the line width increases, in addition to increasing active diffusion volume, the increasing stress in a Cu–low-*k* system serves as another acceleration factor that shortens the SM MTF. Von Mises stress also increases with line width for both Cu–low-*k* and Cu-TEOS systems, indicating increasing driving forces for plastic yielding in the metal line.

Metal-Plate Shape Dependence

For the three SIV structures shown in Fig. 11.9, Matsuyama et al.[47] reported that the extrusion pattern (see Fig. 11.9*b*) is much worse than the wide pattern (see Fig. 11.9*a*) and the zero-extrusion pattern (see Fig. 11.9*c*). Temperature dependence of the failure frequency observed for the three types of patterns is given in Fig. 11.26. The worst temperature is ~200°C, corresponding to the highest failure frequency. A further study of a series of extrusion patterns with the same body metal varied against the length of the narrow line shows that a longer narrow line will give a larger average resistance drift, indicating that the narrow line is important for vacancy diffusion. The copper and

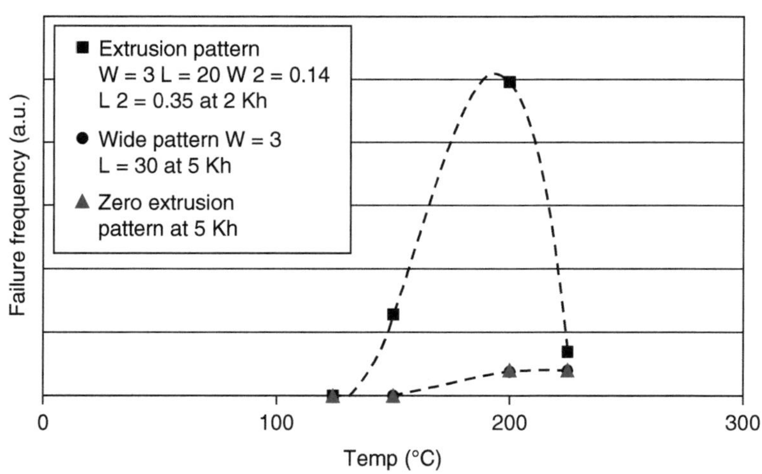

FIGURE 11.26 Temperature dependence of the failure frequency observed for three types of patterns. The worst temperature is ~200°C, corresponding to the highest failure frequency. (*Used with permission from IEEE.*[47])

barrier-metal interface, especially the side surface rather than the bottom surface, was postulated to be the main pathway for vacancy diffusion, resulting in a void in the via, as shown in Fig. 11.10*b*.

Via-Location Dependence

It was found that a single via landing on the edge of a large metal lead gives a higher resistance to SM than a via landing on the center. This phenomenon is well explained by the *active-diffusion-area model* (Sec. 11.2.2). As shown schematically in Fig. 11.27, case (*a*) with a via landed on the metal center gives a higher active diffusion volume than case (*b*) with a via landed on the metal edge, indicating that case (*a*) will supply more vacancy for stress-induced voids than case (*b*).

It also was generally found that the case with double vias results in better SM resistance than the test patterns having single vias after thermal stressing. There could be two reasons for the gain in SM resistance in the double-via case. One is attributed to the stress-gradient reduction in double vias leading to a reduced vacancy migration rate. Another possibility is that the second via effectively serves as another vacancy collector and shares the load of vacancy accumulation.

Narrow-Width Effect in Advanced Technology

According to the active-diffusion-area model (Sec. 11.2.2), a longer time is needed for void formation for vias connected to narrower Cu lines. If the Cu lines are sufficiently narrow, it is thought that voids cannot be formed owing to limited vacancies in the active area. However, this is not necessarily true, especially for the advanced-technology nodes when the line width is in the deep submicron region.

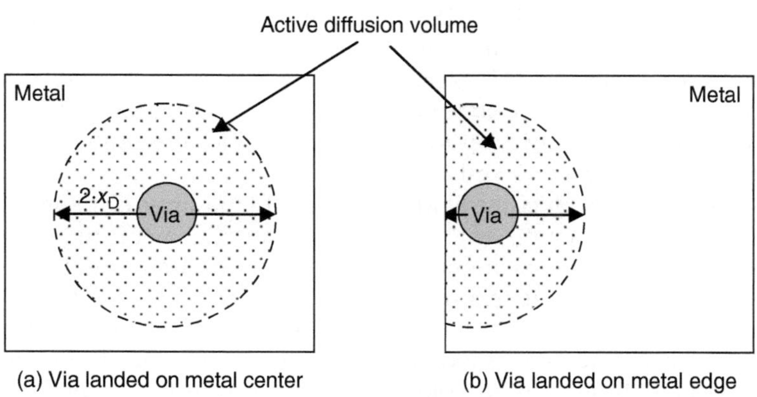

FIGURE 11.27 Schematic for active diffusion volume: In case (*a*), via landed on the metal center; in case (*b*), via landed on the metal edge. Case (*a*) gives a higher active diffusion volume than case (*b*).

Lin et al.[27] studied SIV behavior for advanced-technology nodes with 0.07-μm-diameter Cu–low-*k* ($k = 2.55$) via chains connected to metal widths in the range 0.07 to 0.42 μm. Basically, the wider width corresponds to a larger resistance shift, except for the minimum metal width 0.07 μm, which gives an anomalous large resistance increase compared with other metal widths. The authors attributed this abnormality to the nonoptimized via process because stress voiding at narrow widths is very sensitive to interconnect processing. In this case, the void is observed in the via for $w = 0.07$ μm rather than in the metal below the via.

Kouno et al.[98] also reported that the SIV mechanism of vias with narrow metal widths (SIV-N) is different from that of well-known stress-induced voiding under vias connected to wide Cu lines (SIV-W). As illustrated in Fig. 11.27*a*, if the line width of the Cu lines is wide enough, vacancies will be concentrated underneath the vias owing to conventional diffusion (see Fig. 11.2) in the active area consisting of diffusion along the grain boundary and the Cu–cap-layer interface. However, if the line width is as narrow as the diameter of vias, vacancy diffusion along the interfaces among the Ta-based barrier, Cu, and SiC*x*-based cap will dominate the degradation process during thermal stressing, making narrow-line SIV performance much worse.

11.7.2 Effect of Via Misalignment

Via-to-line misalignment was intentionally introduced through specially designed test structures.[26] Figure 11.28*a* shows a via chain of 10,000 W-plugs linked by 0.28×1.5 μm^2 lines in the aluminum and copper levels with via-to-line overlap varied from –0.06 to + 0.2 μm in both metal levels separately. The corresponding failure-time distributions with varying overlap between the W-plug and the copper level are summarized in Fig. 11.28*b*. It is seen that the distributions have similar shape factors, but failure times become steadily smaller with smaller overlap. The MTF is reduced by about 2× when the overlap is decreased from 0.00 to –0.06 μm. On the other hand, the failure time is increased when the overlap is enlarged from 0.00 to +0.20 μm, although the gain is less pronounced.

Figure 11.29 gives the FEM simulations of the stress components for via-line structures with and without misalignment under the elastic-deformation assumption. The regions of high stress gradients are denoted by the letters A, B, C, etc. A high stress gradient for σ_{xx} is located at the upper part of the metal line next to the via, whereas the high σ_{yy} gradient is located directly under the via. The corresponding stress-induced voids are also shown in Fig. 11.29. They can be attributed to different failure modes with different activation energies.[44]

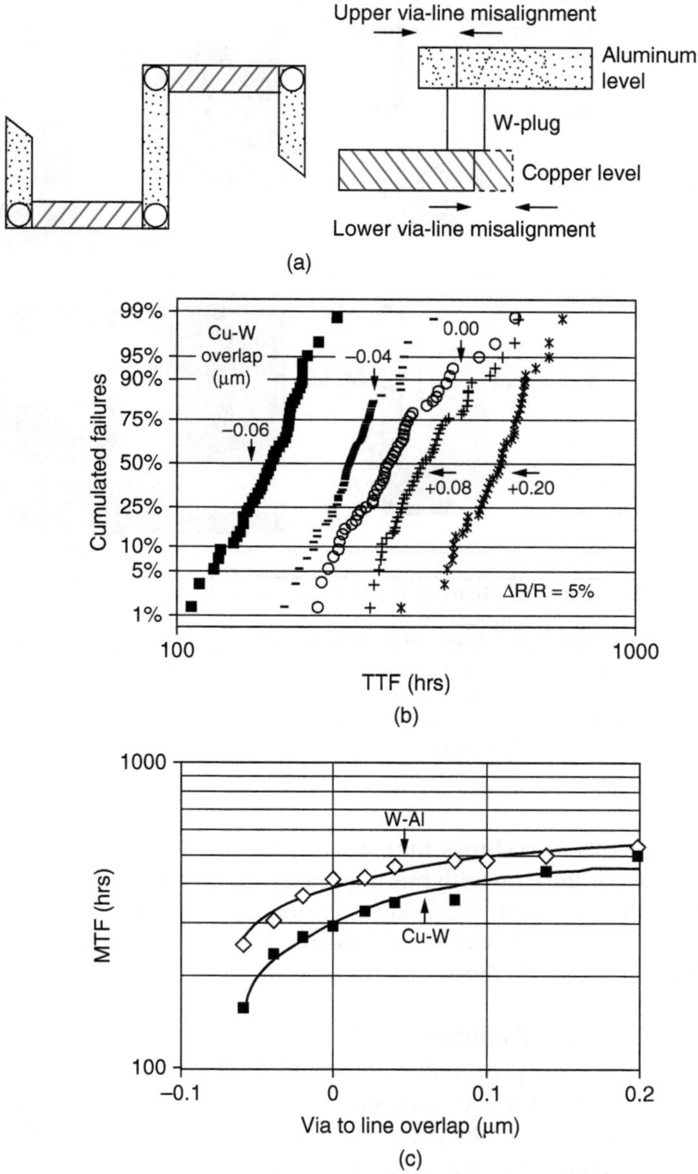

FIGURE 11.28 (*a*) Schematic of a via chain of 10,000 W-plugs linked by $0.28 \times 1.5\ \mu m^2$ lines in the aluminum and copper levels with via-to-line overlap. (*b*) Failure distributions obtained at 275°C on short via chains as a function of the Cu-to-W-via overlap using 5 percent relative failure criterion. The failure times become smaller with smaller overlaps between the copper line and the W-plug. (*c*) MTFs obtained at 275°C on short via chains as a function of the Cu-W and Al-W overlap, respectively. Increasing overlaps with respect to both copper or aluminum metal lines yield longer failure times. (*Used with permission from Cambridge University Press.*[26])

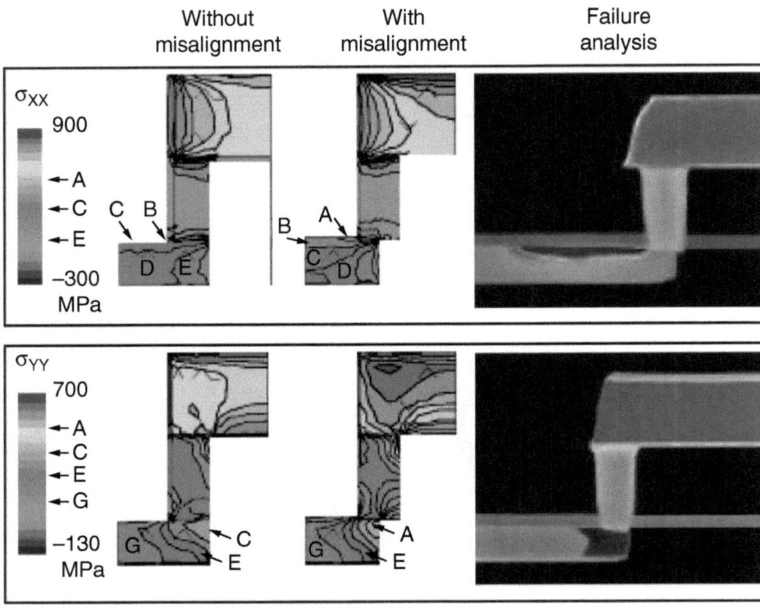

Figure 11.29 FEM simulations of the linear σ_{xx} (*top*) and σ_{yy} (*bottom*) stress components for via-line structures without and with misalignment of the line below the via, respectively. Regions of high σ_{xx} stress gradients are obtained in the upper part of the metal line next to the via; σ_{yy} gradients are directly under the via. In both regions, stress-induced voids can be found after HTS testing corresponding to two different SV failure modes. (*Used with permission from Cambridge University Press.* [26])

Figure 11.29 also shows that stress gradients in both the x and y directions for the case with misalignment is larger than for the case without misalignment. Consequently, higher driving forces are expected at a large negative via-to-line overlap leading to a reduction in SIV failure times on those structures, as observed experimentally.

11.7.3 Effect of dielectric slots

Dielectric slots were proposed to effectively suppress stress-induced void failure by the diffusive volume of vacancies and the hydrostatic stress around the via in the presence of the dielectric slot.[28,99,100] Note that no EM reliability degradation or improvement owing to the dielectric slot was found.[100] The idea that the dielectric slot could be retrospected to the introduction of a dielectric pillar structure into the metal lines was introduced by Suzuki et al.,[14] who showed that the SM failure rate became negligible or very low when dielectric pillars were present. The dielectric slots were incorporated into advanced Cu interconnects using a damascene process without additional masking steps. Some examples of interesting dielectric slotting design

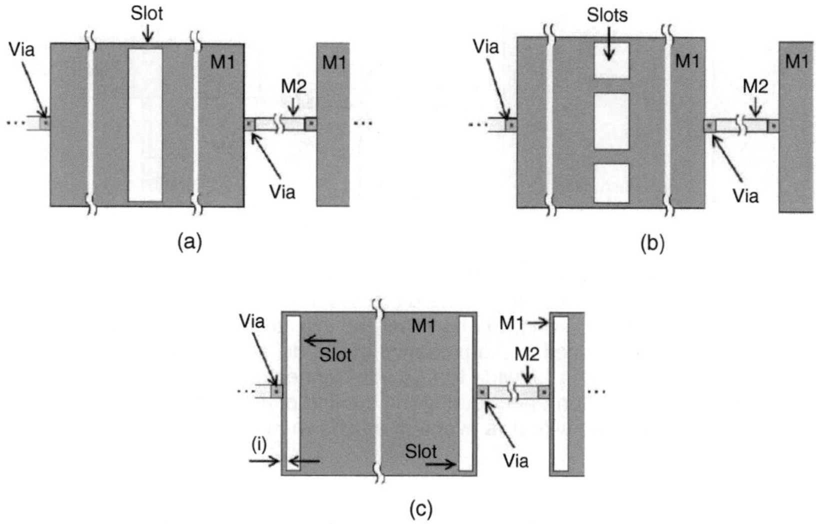

Figure 11.30 Dielectric slotting structures: (*a*) square segmented slots in the center of the block, (*b*) rectangular slots in the center of the block, (*c*) long rectangular slots in front of the active via. *(Used with permission from IEEE.*[28])

structures are shown in Fig. 11.30.[28] The location and shape of such dielectric slots were found to be very crucial in determining their effectiveness in SM reliability improvement. The dielectric slot must be sufficiently long to effectively impede vacancy coalescence in Cu interconnects for SM reliability improvement.[28]

The dielectric slot can improve SIV resistance drastically when compared with the conventional structure without dielectric slots. However, a negative impact is that the dielectric slot on the Cu interconnects can lead to approximately a 1.7 percent increment in the resistance of the structure. [99] The failure analysis result shows that in the absence of dielectric slots, a stress-induced void was formed beneath a failing via and extended significantly into the wide Cu metal lead. On the other hand, when dielectric slots were incorporated, no stress-induced void was formed beneath the via.

Two possible mechanisms contribute to the improved SM reliability. One mechanism is the hydrostatic stress reduction effect. It was observed from several stress-simulation works that when a via is located near the metal edge, the hydrostatic stress beneath the via would be lowered,[101,102] thus reducing the tendency of void nucleation. The other mechanism is a vacancy coalescence impedance effect. The dielectric slots could significantly reduce the effective diffusive volume of vacancies in the wide Cu metal leads, thus preventing the formation of a stress-induced void beneath the via (Fig. 11.31*a* and *b*). For example, assuming the effective diffusive length of the vacancies

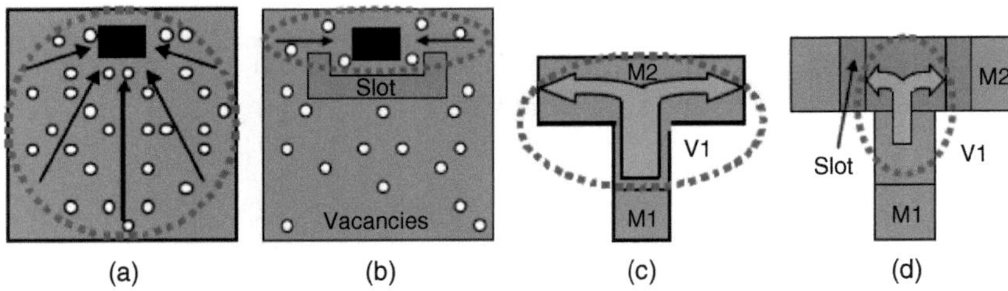

FIGURE 11.31 The possible mechanisms that resulted in improved SM reliability. Hydrostatic stress-reduction effect owing to the dielectric slot designed near the via. Moreover, dotted circles in (*a*) and (*b*) show that the effective diffusive volume of vacancies is significantly reduced in the presence of the dielectric slot. (*c*) and (*d*) show an additional mechanism for wide top Cu interconnects, which is the reduction in Cu volume-contraction effect during the cooling phase of the SM test. Thus the tendency of the via to pull out is avoided, as shown in (*d*). (*Used with permission from IEEE.*[99])

is 10 μm, a horseshoe-like dielectric slot in a 2-μm-wide Cu interconnects could decrease the effective diffusive volume of the vacancies by approximately 98 percent.[99]

Figure 11.32 shows the case for a single via connected to a wide-top Cu interconnect with a horseshoe-shaped dielectric slot. Again, it is seen that the dielectric slot can improve the SM robustness of wide-top Cu interconnects. The transmission electron microscopy

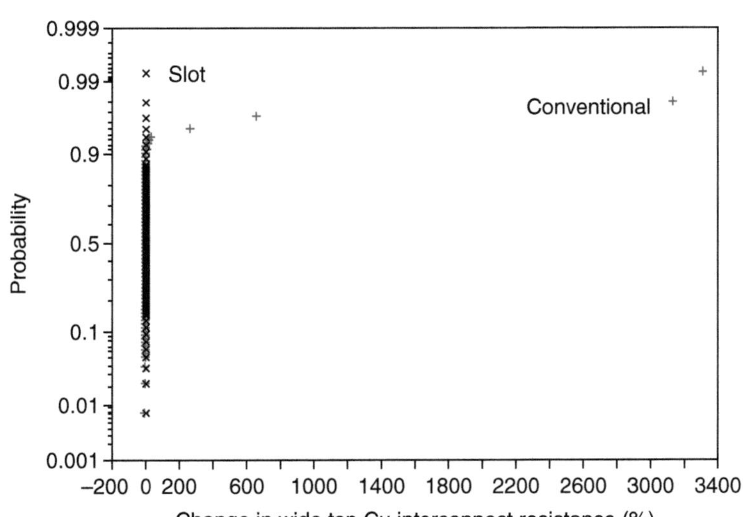

FIGURE 11.32 For wide-top Cu interconnects with a horseshoe-shaped dielectric slot, improved SM robustness has been seen. The outliers correspond to SM-damaged structures. (*Used with permission from IEEE.*[99])

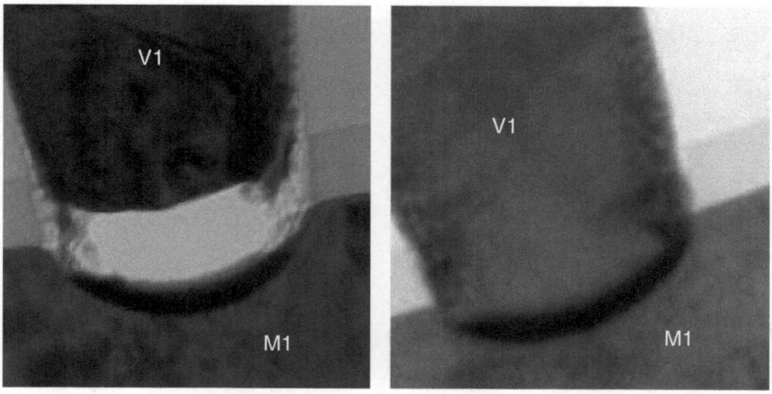

FIGURE 11.33 (*a*) A stress-induced void is formed within the via when a dielectric slot is absent. (*b*) No stress-induced void is found when a dielectric slot is incorporated. (*Used with permission from IEEE.*[99])

micrographs shown in Fig. 11.33 clearly demonstrate the formation of stress-induced voids within a via when there was no dielectric slot in the wide-top Cu metal lead, whereas no void was observed when the dielectric slots were incorporated. Given the slight difference in stress-induced voiding mechanisms[50] compared with the preceding case, in addition to the two mechanisms discussed earlier, a third mechanism was proposed for SM reliability improvement. It is believed that during the cooling phase of the SM test, the dielectric slots incorporated in the wide-top Cu metal leads could reduce Cu volume contraction, thus lowering the tendency of the via pull out (see Fig. 11.31*c* and *d*).

11.7.4 Dual (Multiple) Via Effect

Multiple redundant vias under Cu interconnects could be an effective solution to reduce the via pullback effect (see Fig. 11.22) during SM cooling owing to the mechanical properties of Cu.[50] The addition of a redundant via can act as a stress modifier of the original via,[73] which will reduce the vacancy diffusion to the via, as already discussed in preceding sections.

Yoshida et al.[54] revealed that a multivia structure in which two vias in parallel alignment survived the SIV test. Even though one via became open by SIV, the other via can survived owing to the stress relaxation by SIV formation in the opposite via. Lim et al.[50] also showed that SM performance was improved with the implementation of a dual-via interconnect. A larger area of the vias (i.e., two vias instead of one) could suppress the via pullback effect.

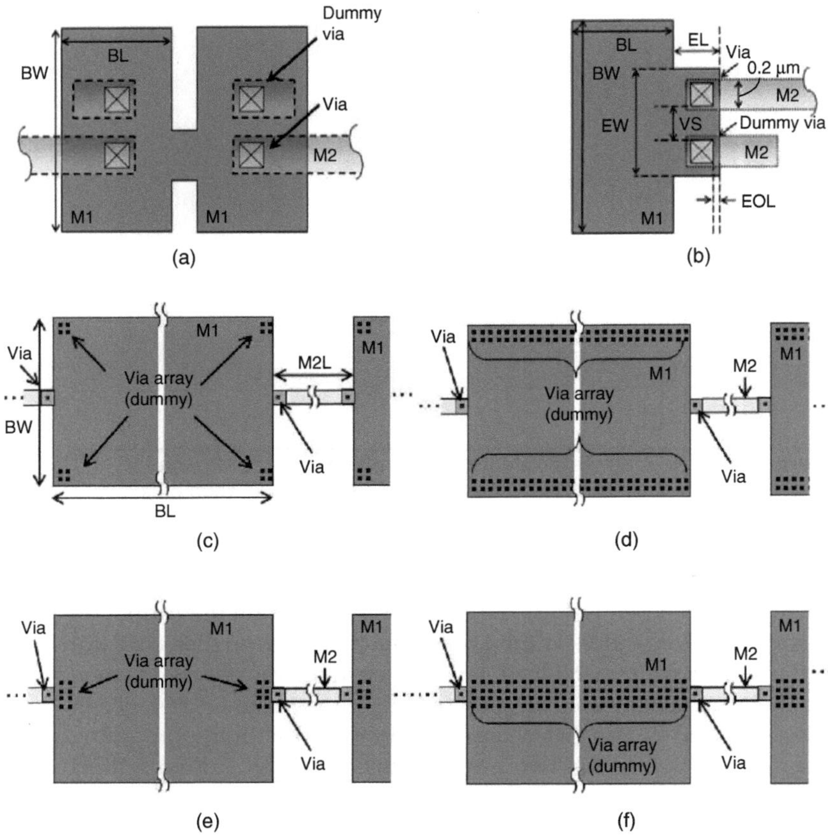

FIGURE 11.34 Stress-induced voiding dual-via structures: (*a*) via landing in center of large metal block and (*b*) via contained within an extension. Via array structures: (*c*) small four-via array at the outside edges, (*d*) large via array spanning the entire length of the block, (*e*) small eight-via array in front of the active via, (*f*) large via array in front of the active via.[28]

11.8 Process-Qualification Practices

For SM qualification, Joint Electron Device Engineering Council (JEDEC) JP-001[103] provides good guidance. However, the industry does not have an agreement on a standard SM test yet.[4] The more common practice is to stress the wafers containing test structures to elevated temperatures (150 to 300°C) for extended times (~1000 hours), periodically cooling the wafers to room temperature to be tested for any increases in resistance, to find out which temperature (termed as the *worst SM temperature*) is most sensitive to an SM resistance increase. Therefore, Table 11.2 only provides some guidelines for practice on the SM qualification.

Failure criteria	10% (or 100% or other percentage values) resistance shift as predefined per requirement from specific application.
Test structures	Predefined specified structures such as metal lines, kelvin via connected to large metal plates, via chains, stacked vias, and so on with varied metal width, as described in Sec. 11.3.1.
Sample size	Wafer mapping or half-mapping, two wafers, three lots, for each temperature and each metal/via level. Sample size should be large enough for statistical analysis.
Stress conditions	Stress (or baking) the wafers containing test structures to elevated temperatures (150 to 300°C) for extended times (e.g., 1,000 or 2,000 hours), periodically cooling the wafers to room temperature to be tested for any increases in resistance. In a typical practice is to separate the wafers into five groups, which are aged at five different temperatures (e.g., 150, 175, 200, 225,and 250°C), respectively. Readouts: Resistance measurements at selected intervals (e.g., 0, 168, 500, 1,000, and 2,000 hours). Then find out which temperature (termed as the *worst SM temperature*) is most sensitive to SM resistance increases.
Wafer/Package Level	Wafer-level or package-level.
Acceptance criteria	No sample has a 10% (or 100% or other percentage values) resistance shift as predefined per requirement from specific applications.
Model	Eq. (11.7): $\text{TTF} = C_0 \cdot (T_0 - T)^{-N} \cdot \exp\left(\frac{E_a}{k_B T}\right)$
Distribution	Lognormal

TABLE 11.2 Typical Process-Qualification Practices for Stress Migration

11.9 References

1. E. T. Ogawa, J. W. McPherson, J. A. Rosal, K. J. Dickerson, T. - C. Chiu, L. Y. Tsung, M. K. Jain, T. D. Bonifield, J. C. Ondrusek, and W. R. McKee, "Stress-induced voiding under vias connected to wide Cu metal leads," in *Proceedings of the 40th IEEE Annual International Reliability Physics Symposium*, 2002. pp. 312–321.
2. B. L. Park, S. R. Hah, C. G. Park, D. K. Jeong, H. S. Son, H. S. Oh, J. H. Chung, J. L. Nam, K. M. Park, and J. D. Byun, "Mechanisms of stress-induced voids in multi-level Cu interconnects," in *Proceedings of the IEEE International Interconnect Technology Conference (IITC)*, 2002, pp. 130–132.

3. M. Kawano, T. Fukase, Y. Yamamoto, T. Ito, S. Yokogawa, H. Tsuda, Y. Kunimune, T. Saitoh, K. Ueno, and M. Sekine, "Stress relaxation in dual-damascene Cu interconnects to suppress stress-induced voiding," in *Proceedings of the International Interconnect Technology Conference (IITC)*, 2003, pp. 210–212.
4. JEP122E, *Failure Mechanisms and Models for Semiconductor Devices*, JEDEC Solid State Technology Association, Arlington, VA, March 2009.
5. A. Heryanto, K. L. Pey, Y. K. Lim, W. Liu, J. Wei, N. Raghavan, J. B. Tan, and D. K. Sohn, "Study of stress migration and electromigration interaction in copper/low-κ interconnects," in *Proceedings of the 48th IEEE Annual International Reliability Physics Symposium*, 2010, pp. 586–590.
6. J. Curry, G. Fitzgibbon, Y. Guan, R. Muollo, G. Nelson, and A. Thomas, "New failure mechanisms in sputtered aluminum-silicon films," in *Proceedings of the 22nd IEEE IEEE Annual International Reliability Physics Symposium*, 1984, pp. 6–8.
7. A. Von Glasow, A. H. Fischer, and G. Steinlesberger, "Using the temperature coefficient of the resistance (TCR) as early reliability indicator for stress-voiding risks in Cu interconnects," in *Proceedings of the 41st IEEE Annual International Reliability Physics Symposium*, 2003, pp. 126–131.
8. Y. B. Park, and I. S. Jeon, "Effects of mechanical stress at no current stressed area on electromigration reliability of multilevel interconnects," *Microelectronics Engineering*, Vol. 71, 2004, pp. 76–89.
9. K. Y. Y. Doong, R. C. J. Wang, S. C. Lin, L. J. Hung, S. Y. Lee, C. C. Chiu, D. Su, K. Wu, K. L. Young and Y. K. Peng, "Stress-induced voiding and its geometry dependency characterization," in *Proceedings of the 41st IEEE Annual International Reliability Physics Symposium*, 2003, pp. 156–160.
10. S. -H. Rhee, Y. Du, and P. S. Ho, "Thermal stress characteristics of Cu/oxide and Cu/low-*k* submicron interconnect structures," *Journal of Applied Physics*, Vol. 93, No. 7, 2003, pp. 3926–3933.
11. D. Gan, G. Wang, and P. Ho, "Effects of dielectric material and line width on thermal stresses of Cu line structures," in *Proceedings of the International Interconnect Technology Conference (IITC)*, 2002, pp. 271–273.
12. A. H. Fischer, O. Aubel, J. Gill, T. C. Lee, B. Li, C. Christiansen, F. Chen, M. Angyal, T. Bolom, and E. Kaltalioglu, "Reliability challenges in copper metallizations arising with the PVD resputter liner engineering for 65 nm and beyond," in *Proceedings of the 45th IEEE Annual International Reliability Physics Symposium*, 2007, pp. 511–515.
13. J. M. Paik, H. Park, and Y. C. Joo, "Effect of low-*k* dielectric on stress and stress-induced damage in Cu interconnects," *Microelectronic Engineering*, 2004, No. 71, pp. 348–357.
14. T. Suzuki, S. Ohtsuka, A. Yamanoue, T. Hosoda, T. Khono, Y. Matsuoka, K. Yanai, H. Matsuyama, H. Mori, N. Shimizu, T. Nakamura, S. Sugatani, K. Shono, and H. Yagi, "Stress induced failure analysis by stress measurements in copper dual damascene interconnects," in *Proceedings of the International Interconnect Technology Conference (IITC)*, 2002, pp. 229–230.
15. K. Croes, C. J. Wilson, M. Lofrano, Y. Travaly, D. De Roest, Zs. Tökei, and G. P. Beyer, "Time and temperature dependence of early stage stress-induced-voiding in Cu/low-k interconnects," in *Proceedings of the 47th IEEE Annual International Reliability Physics Symposium*, 2009, pp. 457–463.
16. H. Okabayashi, "Stress-induced void formation in metallization for integrated circuits," *Material Science and Engineering R*, Vol. 11, 1993, pp. 191–241.
17. H. Matsuyama, H. Mori, and N. Shimizu, "Stress-induced failure analysis by stress measurements in copper dual damascene interconnects," in *Proceedings of the 40th IEEE Annual International Reliability Physics Symposium*, 2002, pp. 229–230.
18. D. Jawarani, H. Kawasaki, I. -S. Yeo, L. Rabenberg, J. P. Stark, and P. S. Ho, "In situ transmission electron microscopy study of plastic deformation and stress-induced voiding in Al-Cu interconnects," *Journal of Applied Physics*, Vol. 82, No. 4, 1997, pp. 1563–1577.

19. S. J. Lee, S. G. Lee, B. S. Suh, H. Shin, N. I. Lee, H. K. Kang, and G. Suh, "New insight into stress induced voiding mechanism in Cu interconnects," in *Proceedings of the IEEE International Interconnect Technology Conference (IITC)*, 2005, pp. 108–110.
20. E. Zschech, R. Hübner, O. Aubel, and P. S. Ho, "EM and SM induced degradation dynamics in copper interconnects studied using electron microscopy and x-ray microscopy," in *Proceedings of the 48th IEEE Annual International Reliability Physics Symposium*, 2010, pp. 574–580.
21. T. Oshima, K. Hinode, H. Yamaguchi, H. Aoki, K. Torii, T. Saito, K. Ishikawa, J. Noguchi, M. Fukui, T. Nakamura, S. Uno, K. Tsugane, J. Murata, K. Kikushima, H. Sekisaka, E. Murakami, K. Okuyama, and T. Iwasaki, "Suppression of stress-induced voiding in copper interconnects," in *Proceedings of the International Electron Devices Meeting (IEDM)*, 2002, pp. 757–760.
22. C. K. Hu, R. Rosenberg, and K. Y. Lee, "Electromigration path in Cu thin-film lines," *Applied Physics Letters*, Vol. 74, 1999, pp. 2945–2947.
23. T. Suzuki, A. Uedono, T. Nakamura, Y. Mizushima, H. Kitada, and Y. Koura, "Direct observation of vacancy defects in electroplated Cu films," in *Proceedings of the IEEE International Interconnect Technology Conference (IITC)*, 2004, pp. 87–89.
24. X. Federspiel, "Stress induced degradation of 90-nm node interconnects," in *Proceedings of the 42nd IEEE Annual International Reliability Physics Symposium*, 2004, pp. 623–624.
25. C. J. Zhai and R. C. Blish, II, "A physically based lifetime model for stress-induced voiding in interconnects," *Journal of Applied Physics*, Vol. 97, 2005, pp. 113503–113508.
26. A. von Glasow, A. H. Fischer, M. Hierlermann, S. Penka, and F. Ungar, "Electromigration failure mechanism studies on copper interconnects," in *Proceedings of the Advanced Metallization Conference in 2002*, AMC2002, MRS, Warrendale, PA, 2003, pp. 161–167.
27. H. Y Lin, S. C. Lee, and A. S. Oates, "Characterization of stress-voiding of Cu/low-*k* vias attached to narrow lines," in *Proceedings of the 46th IEEE Annual International Reliability Physics Symposium*, Phoenix, 2008, pp. 687–688.
28. G. D. R. Hall, D. D. J. Allman, and H. D. Bhatt, "Impact of via interactions and metal slotting on stress induced voiding," in *Proceedings of the 46th IEEE Annual International Reliability Physics Symposium*, 2008, pp. 392–398.
29. C. M. Tan, Z. H. Gan, W. Li, and Y. J. Hou, "Finite element method for stress induced voiding ," in *Applications of Finite Element Methods for Reliability Study of ULSI Interconnections*, London: Springer-Verlag 2011.
30. T. D. Sullivan, "Stress-induced voiding in microelectronic metallization: void growth models and refinements," *Annual Review of Materials Science*, Vol. 26, 1996, pp. 333–364.
31. A. Sekiguchi, J. Koike, S. Kamiya, M. Saka, and K. Maruyama, "Microstructural influences on stress migration in electroplated Cu metallization," *Applied Physics Letters*, Vol. 83, No. 10, 2003, pp. 1962–1964.
32. M. S. J. Koike, M. Wada, and K. Maruyama, "Effects of crystallographic texture on stress-migration resistance in copper thin films," *Applied Physics Letters*, Vol. 81, 2002, p. 1017.
33. R. J. Gleixner, B. M. Clemens, and W. D. Nix, "Void nucleation in passivated interconnect lines: effects of site geometries, interfaces, and interface flaws," *Journal of Materials Research*, Vol. 12, 1997, pp. 2081–2090.
34. B. M. Clemens, W. D. Nix, and R. J. Gleixner, "Void nucleation on a contaminated patch," *Journal of Materials Research*, Vol. 12, 1997, pp. 2038–2042.
35. P. A. Flinn, S. Lee, J. Doan, T. N. Marieb, J. C. Bravman, and M, Madden, "Void phenomena in passivated metal lines: recent observations and interpretation," in *Proceedings of the Fourth International Workshop on Stress Induced Phenomena in Metallization*, 1998, pp. 250–261.

36. Y.-L. Shen, "Stresses, deformation, and void nucleation in locally debonded metal interconnects," *Journal of Applied Physics*, Vol. 84, 1998, pp. 5525–5530.
37. M. A. Korhonen, W. R. LaFontaine, P. Børgesen, and C. -Y. Li, "Stress induced nucleation of voids in narrow aluminum based metallizations on silicon substrates," *Journal of Applied Physics*, Vol. 70, No. 11, 1991, pp. 6774–6781.
38. A. Gladkikh, M. Karpovski, and A. Palevski, "Microstructural and surface effects on electromigration failure mechanism in Cu interconnects," *Microelectronics Reliability*, Vol. 37, 1997, pp. 1557–1560.
39. W. C. Baek, J. P. Zhou, J. Im, P. S. Ho, J. G. Lee, S. B. Hwang, K. K. Choi, S. K. Park, O. J. Jung, L. Smith, and K. Pfeifer, "Oxidation of the Ta diffusion barrier and its effect on the reliability of Cu interconnects," in *Proceedings of the 44th IEEE Annual International Reliability Physics Symposium*, 2006, pp. 131–135.
40. O. Aubel, W. Yao, M. A. Meyer, H. J. Engelmann, J. Poppe, F. Feustel, and C. Witt, "New failure mechanism during high temperature storage testing and its application on SIV risk evaluation," in *IEEE International Integrated Reliability Workshop Final Report*, 2009, pp. 5–10.
41. W. McPherson and C. F. Dunn, "A model for stress-induced metal notching and voiding in very large-scale-integrated Al-Si(1%) metallization," *Journal of Vacuum Science and Technology B*, Vol. 5, 1987, pp. 1321–1325.
42. L. M. Klinger, E. E. Glickman, and V. E. Fradkov, "Extensions of thermal grooving for arbitrary grain-boundary flux," *Journal of Applied Physics*, Vol. 78, 1995, pp. 3833–3838.
43. E. Glickman and M. Nathan, "On the unusual electromigration behavior of copper interconnects," *Journal of Applied Physics*, Vol. 80, 1996, pp. 3782–3791.
44. A. von Glasow and A. H. Fischer, "New approaches for the assessment of stress-induced voiding in Cu Interconnects," in *Proceedings of the International Interconnect Technology Conference (IITC)*, 2002, pp. 274–276.
45. F. Chen, B. Li, T. Lee, C. Christiansen, J. Gill, M. Angyal, M. Shinosky, C. Burke, W. Hasting, R. Austin,T. Sullivan, D. Badami, and J. Aitken, "Technology reliability qualification of a 65nm CMOS Cu/low-k BEOL interconnect," in *Proceedings of the 13rd International Symposium on the Physical and Failure Analysis of Integrated Circuits (IPFA)*, 2006, pp. 97–105.
46. H. Matsuyama, T. Suzuki, H. Ehara, K. Yanai, T. Kouno, S. Otsuka, N. Misawa, T. Nakamura, Y. Mizushima, M. Shiozu, M. Miyajima, and K. Shono, "Investigation of stress-induced voiding inside and under VIA's in copper interconnect with wing pattern," in *Proceedings of the 46th IEEE Annual International Reliability Physics Symposium*, 2008, pp. 683–684.
47. H. Matsuyama, M. Shiozu, T. Kouno, T. Suzuki, H. Ehara, S. Otsuka, T. Hosoda, T. Nakamura, Y. Mizushima, M. Miyajima, and K. Shono, "New degradation phenomena of stress-induced voiding inside VIA in copper interconnects," in *Proceedings of the 45th IEEE Annual International Reliability Physics Symposium*, 2007, pp. 638–639.
48. G. B. Alers, J. Sukamto, P. Woytowitz, X. Lu, S. Kailasam, and J. Reid, "Stress migration and the mechanical properties of copper," in *Proceedings of the 43rd IEEE Annual International Reliability Physics Symposium*, 2005, pp. 36–40.
49. R. Lloyd, and J. J. Clement, "Electromigration in Cu conductors," *Thin Solid Films*, Vol. 262, 1995, pp. 135–141.
50. Y. K. Lim, Y. H. Lim, C. S. Seet, B. C. Zhang, K. L. Chok, K. H. See, T. J. Lee, L. C. Hsia, and K. L. Pey, "Stress-induced voiding in multi-level copper/low-*k* interconnects," in *Proceedings of the 42nd IEEE Annual International Reliability Physics Symposium*, 2004, pp. 240–245.
51. E. M. Zielinski, R. P. Vinci, and J. C. Bravman, "The influence of strain energy on abnormal grain growth in copper thin films," *Applied Physics Letters*, Vol. 67, 1995, pp. 1078–1080.
52. T. Tonegawa, M. Hiroi, K. Motoyama, K. Fujii, and H. Miyamoto, "Suppression of bimodal stress-induced voiding using high-diffusive dopant from Cu-alloy seed layer," in *Proceedings of the International Interconnect Technology Conference (IITC)*, 2003, pp. 216–218.

53. R. C. J. Wang, L. D. Chen, P. C. Yen, S. R. Lin, C. C. Chiu, K. Wu, and K. S. Chang-Liao, "Interfacial stress characterization for stress-induced voiding in Cu/low-*k* interconnects," in *Proceedings of the 12nd International Symposium on the Physical and Failure Analysis of Integrated Circuits (IPFA)*, 2005, pp. 96–99.
54. K. Yoshida, T. Fujimaki, K. Miyamoto, T. Honma, H. Kaneko, H. Nakazawa, and M. Morita, "Stress-induced voiding phenomena for an actual CMOS LSI interconnects," in *Proceedings of the International Electron Devices Meeting (IEDM)*, 2002, pp. 753–756.
55. L. T. Shi, and K. N. Tu, "Finite-element modeling of stress distribution and migration in interconnecting studs of a three-dimensional multilevel device structure," *Applied Physics Letters*, Vol. 65, No. 12, 1994, pp. 1516–1518.
56. C. H. Yao, T. C. Huang, K. S. Chi, W. K. Wan, H. H. Lin, Chin C. Hsia, and M. S. Liang, "Numerical characterization of the stress induced voiding inside via of various Cu/low *k* interconnects," in *Proceedings of the International Interconnect Technology Conference (IITC)*, 2004, pp. 24–26.
57. X. Federspiel and S. Orain, "90nm node damascene copper stress voiding model and lifetime extrapolation methodology," in *IEEE International Integrated Reliability Workshop Final Report*, 2004, pp. 64–70.
58. Y. Hou and C. M. Tan, "Stress-induced voiding study in integrated circuit interconnects," *Semiconductor Science and Technology*, Vol. 23, 2008, pp. 075023–075031.
59. K. Weide-Zaage, D. Dalleau, Y. Danto, and H. Fremont, "Void formation in a copper-via-structure depending on the stress free temperature and metallization geometry," in *Proceedings of the 5th International Conference on Thermal and Mechanical Simulation and Experiments in Micro-electronics and Micro-Systems*, 2004, pp. 367–372.
60. M. Lofrano, C J. Wilson, K. Croes , and B. Vandevelde, "Thermo-mechanical modeling of stress-induced-voiding in BEOL Cu interconnect structures," in *Proceedings of the 10th International Conference on Thermal, Mechanical and Multiphysics Simulation and Experiments in Micro-Electronics and Micro-Systems*, 2009, pp. 1–6.
61. ANSYS Elements Reference: ANSYS Release 5. 6, ANSYS, Inc. , Canonsburg, PA, 1999.
62. S. -H. Rhee, Y. Du, and P. S. Ho, "Thermal stress characteristics of Cu/oxide and Cu/low-*k* submicron interconnect structures," *Journal of Applied Physics*, Vol. 93, No. 7, 2003, pp. 3926–3933.
63. C. M. Tan, Z. H. Gan, and X. F. Gao, "Temperature and stress distribution in the SOI structure during fabrication," *IEEE Transactions on Semiconductor Manufacturing*, Vol. 16, No. 2, 2003, pp. 314–318.
64. J. H. Zhao, T. Ryan, P. S. Ho, A. J. Mckerrow, and W. -Y. Shih, "Measurement of elastic modulus, Poisson ratio, and coefficient of thermal expansion of on-wafer submicron films," *Journal of Applied Physics*, Vol. 85, No. 9, 1999, pp. 6421–6424.
65. A. Grill, "Plasma enhanced chemical vapor deposited SiCOH dielectrics: from low-k to extreme low-*k* interconnect materials," *Journal of Applied Physics*, Vol. 93, No. 3, 2003, pp. 1785–1790.
66. C. T. Lynch (ed.), *CRC Handbook of Materials Science*, Vol. II, Boca Raton, FL: CRC Press, 1975.
67. J. H. Zhao, Y. Du, M. Morgen, and P. S. Ho, "Simultaneous measurement of Young's modulus, Poisson ratio, and coefficient of thermal expansion of thin films on substrates," *Journal of Applied Physics*, Vol. 87, No. 3, 2000, pp. 1575–1577.
68. E. A. Brandes, and G. B. brook (eds.), *Smithell's Metals Reference Book*, 7th ed. Boston: Butterworth-Heinemann, 1999.
69. Z. H. Gan, W. Shao, S. G. Mhaisalkar, and Z. Chen, and H. Y. Li, "The influence of temperature and dielectric materials on stress induced voiding in Cu dual damascene interconnects," *Thin Solid Films*, Vol. 504, 2006, pp. 161–165.
70. L. T. Shi, and K. N. Tu, "Finite-element modeling of stress distribution and migration in interconnecting studs of a three-dimensional multilevel device structure," *Applied Physics Letters*, Vol. 65, No. 12, 1994, pp. 1516–1518.

71. B. Li, T. D. Sullivan, T. C. Lee, and D. Badami, "Reliability challenges for copper interconnects," *Microelectronics Reliability*, Vol. 44, No. 3, 2004, pp. 365–380.
72. W. D. Nix, and E. Arzt, "On void nucleation and growth in metal interconnect lines under electromigration conditions," *Metallurgical and Materials Transactions A*, Vol. 23, No. 7, 1992, pp. 2007–2013.
73. T. C. Huang, C. H. Yao, W. K. Wan, Chin C. Hsia, and M. S. Liang, "Numerical modeling and characterization of the stress migration behavior upon various 90 nanometer Cu/low-*k* interconnects," in *Proceedings of the International Interconnect Technology Conference (IITC)*, 2003, pp. 207–209.
74. S. Rzepka, M. A. Korhonen, E. R. Weber, and C. Y. Li, "Three-dimensional finite element simulation of electro and stress migration effects in interconnect lines," in *Proceedings of the Materials Research Society Symposium*, Vol. 473, 1997, pp. 329–335.
75. D. Edelstein, H. Rathore, C. Davis, L. Clevenger, A. Cowley, T. Nogami, B. Agarwala, S. Arai, A. Carbone, K. Chanda, F. Chen, S. Cohen, W. Cote, M. Cullinan, T. Dalton, S. Das, P. Davis, J. Demarest, D. Dunn, C. Dziobkowski, R. Filippi, J. Fitzsimmons, P. Flaitz, S. Gates, J. Gill, A. Grill, D. Hawken, K. Ida, D. Klaus, N. Klymko, M. Lane, S. Lane, J. Lee, W. Landers, W.-K. Li, Y. -H. Lin, E. Liniger, X.-H. Liu, A. Madan, S. Malhotra, J. Martin, S. Molis, C. Muzzy, D. Nguyen, S. Nguyen, M. Ono, C. Parks, D. Questad, D. Restaino, A. Sakamoto, T. Shaw, Y. Shimooka, A. Simon, E. Simonyi, A. Swift, T. Van Kleeck, S. Vogt, Y.-Y. Wang, W. Wille, J. Wright, C.-C. Yang, M. Yoon, and T. Ivers, "Comprehensive reliability evaluation of a 90-nm CMOS technology with Cu/PECVD low-k BEOL," in *Proceedings of the 42nd IEEE Annual International Reliability Physics Symposium*, 2004, pp. 316–319.
76. G. G., Stoney, "The tension of metallic films deposited by electrolysis," *Proceedings of the Royal Society*, London, A8, 1909, pp. 172–175.
77. K. Ishikawa, T. Iwasaki, T. Fujii, N. Nakajima, M. Miyauchi, T. Ohshima, J. Noguchi, H. Aoki, and T. Saito, "Impact of metal deposition process upon reliability of dual-damascene interconnects," in *Proceedings of the International Interconnect Technology Conference*, 2003, pp. 24–26.
78. A. H. Fisher, A. V. Glasow, S. Penka, and F. Ungar, "Process optimization: the key to obtain highly reliable Cu interconnects," *International Interconnect Technology Conference*, 2003, pp. 253–255.
79. G. B. Alers, R. T. Rozbicki, G. J. Harm, S. K. Kailasam, G. W. Ray and M. Danek, "Barrier first integration for improved reliability in copper dual damascene interconnects," in *Proceedings of the International Interconnect Technology Conference*, 2003, pp. 27–29.
80. J. -P. Wang, Y. -K. Su, and J. F. Chen, "Effects of surface cleaning on stressvoiding and electromigration of Cu-damascene interconnection," *IEEE Transaction on Device and Materials Reliability*, Vol. 8, No. 1, 2008, pp. 210–215.
81. Y. Matsubara, M. Komuro, T. Onodera, N. Ikarashi, Y. Hayashi, and M. Sekine, "Thermally robust 90 nm node Cu-Al wiring technology using solid phase reaction between Cu and Al," in *Symposium on VLSI Technology Digest of Technical Papers*, 2003, pp. 127–128.
82. M. Iguchi, S. Yokogawa, H. Aizawa, Y. Kakuhara, H. Tsuchiya, N. Okada, K. Imai, M. Tohara, K. Fujii, and T. Watanabe, "Optimization of metallization processes for 32-nm-node highly reliable ultralow-k (k = 2.4)/Cu multilevel interconnects incorporating a bilayer low-k barrier cap (k = 3.9)," in *Proceedings of the International Electron Devices Meeting (IEDM)*, 2009, pp. 871–874.
83. S. Yokogawa, K. Kikuta, H. Tsuchiya, T. Takewaki, M. Suzuki, H. Toyoshima, Y. Kakuhara, N. Kawahara, T. Usami, K. Ohto, K. Fujii, Y. Tsuchiya, K. Arita, K. Motoyama, M. Tohara, T. Taiji, T. Kurokawa, and M. Sekine, "Tradeoff characteristics between resistivity and reliability for scaled-down Cu-based interconnects," *IEEE Transactions on Electron Devices*, Vol. 55, No. 1, 2008, pp. 350–357.
84. M. Ueki, M. Hiroi, N. Ikarashi, T. Onodera, N. Furutake, N. Inoue, and Y. Hayashi, "Effects of Ti addition on via reliability in Cu dual damascene interconnects," *IEEE Transactions on Electron Devices*, Vol. 51, No. 11, 2004, pp. 1883–1891.

85. A. Isobayashi, Y. Enomoto, H. Yamada, S. Takahashi, and S. Kadomura, "Thermally robust Cu interconnects with Cu-Ag alloy for sub-45-nm node," in *Proceedings of the International Electron Devices Meeting (IEDM)*, 2004, pp. 953–956.
86. C. H. Shih, S. W. Chou, C. J. Lin, T. KO, H. W. Su, C. M. Wu, M. H. Tsai, Winston S. Shue, C. H. Yu, and M. S . Liang, "Design of ECP additive for 65-nm-node technology Cu BEOL reliability," in *Proceedings of the IEEE International Interconnect Technology Conference*, 2005, pp. 102–104.
87. H. S. Oh, J. H. Chung, J. W. Lee, K. H. Kang, D. G. Park, S. R. Hah, I. S. Cho, and K. M. Park, "The effect of FSG stability at high temperature on stress-induced voiding in Cu dual-damascene interconnects," in *Proceedings of the IEEE International Interconnect Technology Conference*, 2004, pp. 21–23.
88. R. C. J. Wang, C. C. Lee, L. D. Chen , K. Wu, and K. S. , Chang-Liao, "A study of Cu low-*k* stress-induced voiding at via bottom and its microstructure effect," *Microelectronics Reliability*, Vol. 46, 2006, p. 1673.
89. J. J. Liu, D. W. Gan, C. Hu, M. Kiene, and P. S. Ho, "Porosity effect on the dielectric constant and thermomechanical properties of organosilicate films," *Applied Physics Letters*, Vol. 81, No. 22, 2002, pp. 4180–4182.
90. C. J. Zhai, H. W. Yao, A. P. Marathe, P. R. Besser, and R. C. Blish, II, "Simulation and experiments of stress migration for Cu/low-*k* BEoL," *IEEE Transaction on Device and Materials Reliability*, Vol. 4, No. 3, 2004, pp. 523–529.
91. C. S. Hau-Riege, A. P. Marathe, and V. Pham, "The effect of low-*k* ILD on the electromigration reliability of Cu interconnects with different line lengths," in *Proceedings of the 41st IEEE Annual International Reliability Physics Symposium*, 2003, pp. 173–177.
92. L. Chen, N. Magtoto, B. Ekstrom, and J. Kelber, "Effect of surface impurities on the Cu-Ta interface," *Thin Solid Films*, Vol. 376, 2000, pp. 115–123.
93. *CRC Hand Book of Chemistry and Physics, 83rd ed., D. R. Lider.* Boca Raton, FL: CRC Press, 2002.
94. T. Ishigami, T. Kurokawa, Y. Kakuhara, B. Withers, J. Jacobs, A. Kolics, I. Ivanov, M. Sekine, and K. Ueno, "High reliability Cu interconnection utilizing a low contamination CoWP capping layer," in *Proceedings of the IEEE 2004 International Interconnect Technology Conference*, 2004, pp. 75–77.
95. M. Lin, J. W. Liang, and K. C. Su, "Stress characterization for stress-induced voiding in Cu/low-*k* interconnects with geometry and upper cap layer dependences," in *IEEE International Integrated Reliability Workshop Final Report*, 2008, pp. 32–35.
96. M. Ueki, M. Hiroi, N. Ikarashi, T. Onodera, N. Furatake, M. Yoshiki, and Y. Hayashi, "Suppression of stress induced open failures between via and Cu wide line by inserting Ti layer under Ta/TaN barrier," in *Proceedings of the International Electron Devices Meeting (IEDM)*, 2002, pp. 749–752.
97. W. Shao, Z. H. Gan, S. G. Mhaisalkar, Z. Chen, and H. Y. Li, "The effect of line width on stress-induced voiding in Cu dual damascene interconnects," *Thin Solid Films*, Vol. 504, 2006, pp. 298–301.
98. T. Kouno, T. Suzuki, S. Otsuka, T. Hosoda, T. Nakamura, Y. Mizushima, M. Shiozu, H. Matsuyama, K. Shono, H. Watatani, Y. Ohkura, M. Sato, S. Fukuyama, and M. Miyajima, "Stress-induced voiding under vias connected to "narrow" copper lines," in *Proceedings of the International Electron Devices Meeting (IEDM)*, 2005, pp. 187–190.
99. Y. K. Lim, K. L. Pey, J. B. Tan, T. J. Lee, D. Vigar, L. C. Hsia, Y. H. Lim, and N. R. Kamat, "Novel dielectric slots in Cu interconnects for suppressing stress-induced void failure," in *Proceedings of the International Electron Devices Meeting (IEDM)*, 2005, pp. 179–182.
100. A. Heryanto, Y. K. Lim, K. L. Pey, W. Liu, J. B. Tan, D. K. Sohn, and L. C. Hsia, "The effects of dielectric slots on copper/low-k interconnects reliability," in *Proceedings of the International Interconnect Technology Conference (IITC)*, 2009, pp. 92–94.
101. C. J. Zhai, H. W. Yao, P. R. Besser, A. Marathe, R. C. Blish II, D. Erb, C. Hau-Riege Sidharth, and K. O. Taylor, "Stress modeling of Cu/ low-*k* BEOL: application to stress migration," *42nd IEEE Annual International Reliability Physics Symposium*, 2004, pp. 234–238.

102. Y. K. Lim, S. C. Seet, T. J. Lee, and D. Vigar, "Stress migration reliability of wide Cu interconnects with gouging vias," in *Proceedings of the 43rd IEEE Annual International Reliability Physics Symposium* 2005, pp. 203–208.
103. JP-001, *Stress Migration (Stress-Induced Voiding): Foundry Process Qualification Guidelines* (Wafer Fabrication Manufacturing Sites, JEDEC/FSA Joint Publication, September 2002, pp. 9–11.

CHAPTER 12

Intermetal Dielectric Breakdown

12.1 Introduction

With the continuous shrinkage of the interconnection line width and device dimensions, the integration density increases. However, in the back-end-of-line (BEOL) integration, the capacitance between two neighboring metal lines at different voltages will increase when they become closer and closer. Therefore, the interconnect delay (i.e., *RC* delay) also increases with more power dissipation. According to Fig. 12.1,[1] the intrinsic gate delay decreases continuously with decreasing size of the transistors. However, the interconnect *RC* delay will go up abruptly when the feature size is reduced below 0.5 μm, where the interconnect delay becomes dominant over total delay.[1] Note that the plot in Fig. 12.1 is based on certain assumptions regarding the geometry of the interconnect array. However, the general trend can be seen clearly.

By assuming that the minimum metal pitch p equals twice the metal width W and assuming that the dielectric thickness above and below a metal line equals the thickness of the metal line, the *RC* delay is correlated with the interconnect length L, metal resistivity ρ, dielectric constant k, metal thickness d_m, and minimum metal pitch p as follows[2]:

$$RC = 2\rho k \varepsilon_0 \left(\frac{4L^2}{p^2} + \frac{L^2}{d_m} \right) \tag{12.1}$$

According to Eq. (12.1), there are several ways to reduce the *RC* delay: by using a metal line with lower resistivity (e.g., Cu to replace Al to reduce ρ), by using low-dielectric-constant (low-k) insulating materials (i.e., reduce k), or by using multilevel metallization (to reduce L) and/or changing the ratio of width to thickness of the metal lines. Note that *the use of multilevel metallization is intended to reduce the line length L*. If the integration density increases within only one level of

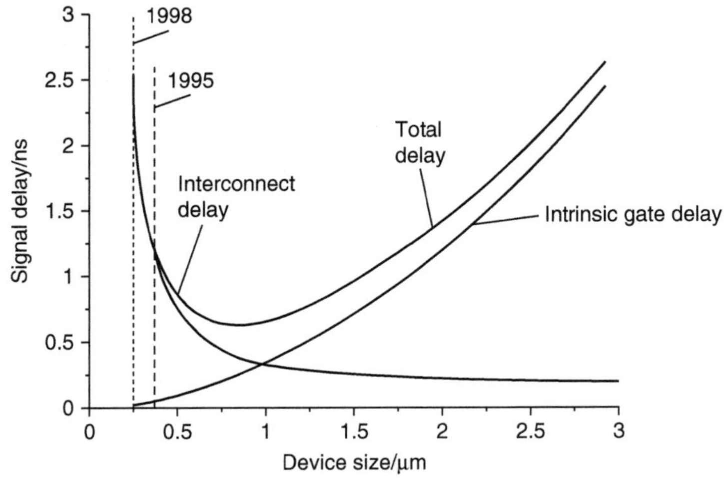

FIGURE 12.1 Dependence of signal delay on device size. (*Used with permission from Elsevier.*[1])

metallization, the total length will increase dramatically, leading to a larger *RC* delay because *RC* delay is proportional to L^2, as described in Eq. (12.1). A typical eight-level microprocessor BEOL of an IBM 90-nm CMOS technology with Cu/Plasma-Enhanced Chemical-Vapor Deposition (PECVD) Low-k[3] includes the Cu wiring having three hierarchical scaling factors (1×, 2×, and 6×) for pitches and cross-sectional dimensions, where 1× factor for M_1 to M_5 (in low-k SiCOH), 2× factor for M_6 (also in low-k SiCOH), and 6× factor for M_7 and M_8 [in fluorine-doped tetra-ethyl-ortho-silicate (FTEOS)]. The idea of this kind of design is analogous to the civil three-dimensional (3D) transport infrastructure to reduce the traffic loading on the ground.

In this chapter, the general characteristics of low-k dielectric insulating materials are introduced first. The test structures and test methodology for characterization of intermetal dielectric reliability then are presented, followed by a summary of the most up-to-date knowledge and understanding of the intermetal dielectric breakdown mechanisms and lifetime extrapolation models and on how to control reliability through design and process optimization, especially with a focus on low-k materials. The breakdown voltage V_{bd} obtained from a V-ramp test can be well correlated with the time-dependent dielectric breakdown (TDDB), which provides a fast wafer-level reliability assessment. The characteristics of the metal-via structure TDDB then are addressed. Finite-element modeling (FEM) as an auxiliary tool for dielectric reliability assessment is also introduced. Finally, the qualification requirements for BEOL dielectric reliability are summarized.

The general characteristics of low-k dielectric insulating materials are introduced below. SiO_2 is the most widely used insulating

Dielectric Material	Deposition Method	Type of Low-k	k_{eff}
SiO_2 (USG)[7]	Thermal	Original dielectric	3.9
TEOS[7]	CVD	Original dielectric	4.2
SiOF (FSG)[7]	CVD	Type I	3.5
SiOC (CDO)[7]	CVD	Type I	2.6–2.9
SiLK[14]	Spin-on	Type II	2.2
Porous MSQ[15]	Spin-on	Types II and III	2.3
Porous OSG[16]	PECVD	Type III	2.5

TABLE 12.1 A List of Several Low-*k* Dielectric Materials

materials in today's integrated circuits (ICs). The dielectric constant of *thermally grown* SiO_2 is $k = 3.9$ (Table 12.1). This number is the ratio of the permittivity of SiO_2 divided by permittivity of vacuum, that is, $k = \varepsilon_{SiO_2}/\varepsilon_0$, where $\varepsilon_0 = 8.854 \times 10^{-6}$pF/μm.[4] Sometimes it is referred to as undoped silicate glass (USG) with proven reliability. For process integration, SiO_2 has many good properties.[5,6] It is thermally and chemically stable and therefore does not degrade during processing. It is mechanically rigid (i.e., high elastic modulus) and is relatively impermeable to moisture (at least at the operating temperatures of ICs), which simplifies packaging and reduces cost.

High-quality SiO_2 films also can be deposited by plasma-enhanced chemical-vapor deposition (PECVD). However, for CVD of SiO_2 (TEOS in Table 12.1), the *k* value is closer to 4.2 because of the higher concentration of O−H bonds usually found in CVD films. As discussed earlier, to meet the *RC* delay-reduction requirement, lowering the dielectric constant of silica is necessary. There are many materials with lower dielectric constants, but few of them can be suitably integrated into a manufacturing process. Development efforts to reduce dielectric constants have focused primarily on three types of dielectric materials (Table 12.1 gives several low-*k* insulating materials) as detailed below:

1. *Type I:* In this type of dielectric, dielectric constant reduction is achieved through individual atom replacement by lowering the ionic and/or electronic contribution to the polarizability.[7] An example is fluorinated silica, sometimes called as *F-doped silica glass* (FSG), whose dielectric constant can be lowered from 3.9 to 3.5.[8,9] This reduction is attributed to the formation of Si−F bonds to replace the Si−O bonds. On the other hand, by doping SiO_2 with carbon instead of fluorine, the dielectric constant can be lowered to 2.7 to 3.0. The IBM group refers this kind of dielectric as SiCOH,[10] which

is used at the 90- and 65-nm technology nodes. Lower dielectric constants (2.2 or less) are possible by adding pores to the SiCOH, which is used at the 45-nm node and below. Other commercial products of carbon-doped silicon dioxide (CDO) include Black Diamond from Applied Materials,[11] and Aurora from ASM International N.V.[12] Aurora is the low-k material used in Intel 90-nm, 65- and 45-nm lines, whereas Black Diamond controls about 80 percent of the low-k material market.[8] Novellus Systems' Coral also belongs to this category.

2. *Type II:* The second practice for dielectric constant reduction is to employ the polymeric dielectrics,[1] which are generally deposited by a spin-on approach, such as those traditionally used to deposit photoresist, rather than CVD. Integration difficulties include low mechanical strength and thermal stability. SiLK from Dow Chemical is a well-known example of a low-k material in this category.[13] Other spin-on organic low-k materials includes polyimide, polynorbornenes, benzocyclobutene, and PTFE. In addition, there are two kinds of silicone-based polymeric dielectric materials deposited by spin-on, namely, hydrogen silsesquioxane (HSQ) and methylsilsesquioxane (MSQ). There is a good summary paper on the polymeric dielectrics applied to microelectronic low-k materials.[1]
3. *Type III:* The third practice is to introduce porosity, which is an effective way to decrease the dielectric constant.[7] A *pore* can be defined as any localized region in the silica matrix (may be a few angstroms in size) that contains low-polarizablity material and/or just a microvoid. A simple calculation shows that introducing 50 percent porosity ($k_1 = 1$ assumed as air for pores) into silica ($k_2 = 4.2$) will yield an effective k value of about 2. Dielectric constants of porous silicon dioxide lower than 2.0 have been reported. Integration difficulties related to porous silicon dioxide implementation include low mechanical strength and difficult integration with etch and polish processes. By ultraviolet (UV) curing, floating methyl groups in carbon-doped silicon dioxide can be eliminated, and pores can be introduced to the carbon-doped silicon dioxide low-k material. Products in this category include Black Diamond II,[11] Aurora 2.7, and Aurora ULK.[12] The reported k value can be as low as 2.5. By introducing pores into the SiLK resin, the dielectric constant value can be lowered to 2.2.[14]

However, the application of low-k dielectric insulating materials has raised some critical reliability concerns, such as the time-dependent dielectric breakdown (TDDB), because low-k materials generally have weaker intrinsic breakdown strengths than the traditional SiO_2 dielectrics. This issue becomes be more serious when the technology continues to shrink to a smaller dimension with finer interconnect pitches.[17–21] TDDB is a failure of a dielectric by breakdown at applied fields below the characteristic breakdown field of the dielectric material. Low-k TDDB is commonly characterized under accelerated conditions (normally at higher fields and higher temperature), and then the lifetimes are extrapolated back to lower fields and lower temperatures under use conditions.

Dielectric TDDB failure is analogous to the mechanical breakdown of solids.[22] Similar to the mechanical energy (characterized by $\frac{1}{2}Ye^2$, where e is strain and Y is Young's modulus) converted to heat during a mechanical stressing, the electrical energy (characterized by $\frac{1}{2}CV^2$ where C is the capacitance and V is the voltage) stored in the dielectric during electrical stressing will be converted into heat, too. Similar to work hardening inhibiting further deformation, dielectric trapping of injected charge prevents further injection through the evolution of space-charge fields. When the strains or injected charge levels exceed critical values, failure occurs.

12.2 Test Structures and Test Methodology

Several types of test structures have been used widely for intermetal dielectric (IMD) reliability study, including comb-comb (CC), comb-serpentine (CS), and metal-via (MV) structures, as detailed in this section.

12.2.1 Test Structures

Comb-Comb Structure

A comb-comb structure is composed of two metal combs inserted in each other (Fig. 12.2*a*). The line width and line-to-line spacing can be varied but typically are given by the design rules. To simulate the real situation in today's interconnects, the total length of the finger in the test structure is generally about 1 m or even longer. During reliability tests (either the voltage-ramped test or the constant-voltage stress test), one comb is grounded, whereas the other comb is positively biased, and the corresponding leakage current is recorded. Sometimes both anode and cathode currents are monitored separately to check whether there is Cu drift during stressing.[7]

Comb-Serpentine-Comb Structure

A comb-serpentine-comb structure is composed of a metal serpentine with two metal combs inserted from each side to cover both edges of the serpentine (Fig. 12.2*b*). The line width and the line-to-line spacing generally are given by the design rules. The total length of the serpentine in the test structure can be about 1 m or even longer. The effective capacitor area for the typical structure is roughly 1×10^{-3}cm^2.[7] During testing, the serpentine typically is grounded, whereas the comb-metal

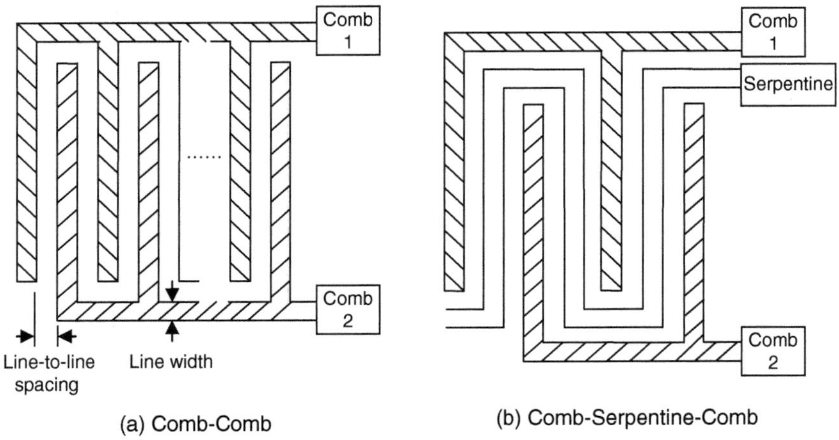

FIGURE 12.2 Schematic diagram of typical (*a*) comb-comb and (*b*) comb-serpentine-comb test structures.

is positively biased. The breakdown condition typically is defined as a sudden rise in leakage current of at least 2×. If a relatively high voltage is applied (e.g., > 50 V), the large amount of charge stored on the capacitor generally produces a hard breakdown, with the current reaching the compliance limit. Soft breakdown prior to hard breakdown is seldom observed for low-*k* dielectrics.

Metal-Via Structure

A metal-via structure is a kind of test structure with a metal-comb connected through numerous vias. Figure 12.3*a* and *b* gives the schematic layout top view of metal-via test structures with aligned vias and staggered vias, respectively. Although in layout the via is defined within the metal line, the distance between two adjacent vias or between a via and the adjacent metal is likely less than the distance between the layers (Fig. 12.3*c* and *d*), attributed to process control, even though such distance is considered the same in layout. Additionally, the vias are funnel-shaped so that the via distances in the top layer are less than those in the bottom layer. The interlevel dielectric (ILD) layer stack consists of the low-*k* layer and a polished cap layer and the top barrier of the interconnect. Thus, it is expected that the maximum electric field should be located in the top layer next to the capping layer, where there are several interfaces in parallel to the stress electric field. It can be assumed that there are microdefects at these interfaces that provide traps for current transport or charge trapping. In this edge region, the *E*-field can be twice (or more) as high as that in other regions. This is caused by the different dielectric constants of the capping layer ($k = 5$ to 7) and the low-*k* layer ($k = 2.5$-4), as well as because of the spike shape at the transition between sidewall barrier and capping layer.[23] Therefore, assessment of the metal-via

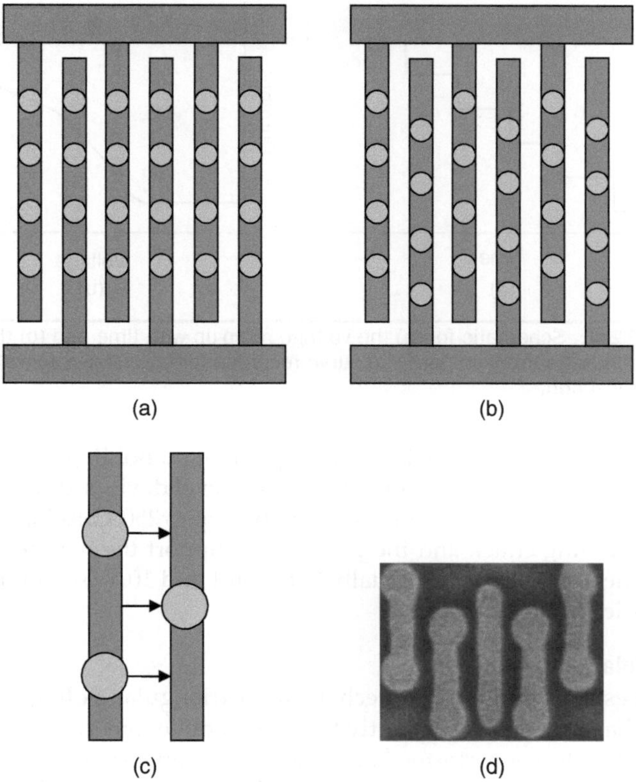

FIGURE 12.3 (*a*) Schematic layout top view of the metal-via test structure (vias aligned). (*b*) Schematic layout top view of the metal-via test structure (vias staggered). (*c*) Schematic top view of staggered vias with larger via size owing to process control. (*d*) SEM top view of staggered vias.

structure is quite necessary during process development and mass-production monitoring.

12.2.2 Test Methodology

Similar to the gate-oxide integrity (GOI) test detailed in Chap. 6, IMD reliability also involves V-ramp tests and TDDB tests, as specified in this section.

Stepped V-Ramp Tests

The V-ramp test as a reliability-evaluation technique has become increasing important for reliability engineers bacause it can be used to save time and resources during process development, qualification, and production monitoring as well. This method also can be used to support verification tools for reliability models.[24] A schematic of the stepped V-ramp test is given in Fig. 12.4. During a stepped voltage-ramp to a breakdown test, the field is stepped in uniform

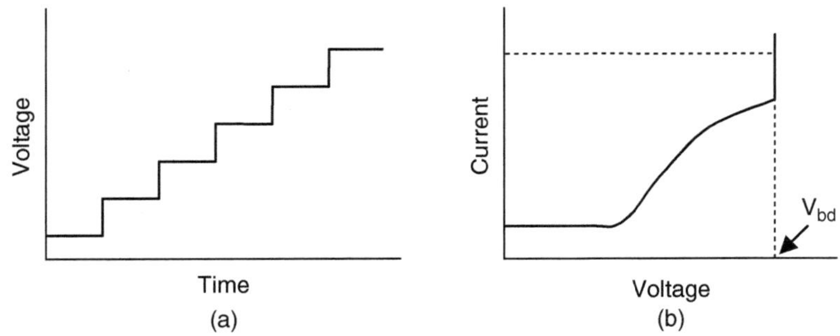

FIGURE 12.4 Schematic for (*a*) the voltage ramp-up with time and (*b*) the corresponding current-voltage (*I-V*) curve recorded for breakdown voltage V_{bd} characterization.

increments ΔE (or ΔV), with each step having a holding time of $\Delta\tau$. The test can be easily performed on wafer-level devices at room temperature or high temperature, for example, up to 250°C, as long as the wafer-holding chuck and the prober can support the high temperature. The ramp slopes are usually between 1 and 10 V/s, and the test time is less than 2 minutes.

Triangular Voltage Sweep

To investigate the TDDB mechanism, a triangular voltage sweep (TVS) has been used frequently to detect mobile ions such as Cu^+ or H^+.[25,26] This test is performed at an elevated temperature (such as 200 or 250°C) to greatly increase the mobility of the ions in the dielectric. The basic procedure is explained in Fig. 12.5. In this method, an initial voltage is applied to the capacitor structure under investigation (e.g., those given in Sec. 12.2.1) to generate an electric field of ~1 MV/cm, and it is maintained for some time (e.g., 20 seconds to 5 minutes depending on the dielectric thickness), suggested by

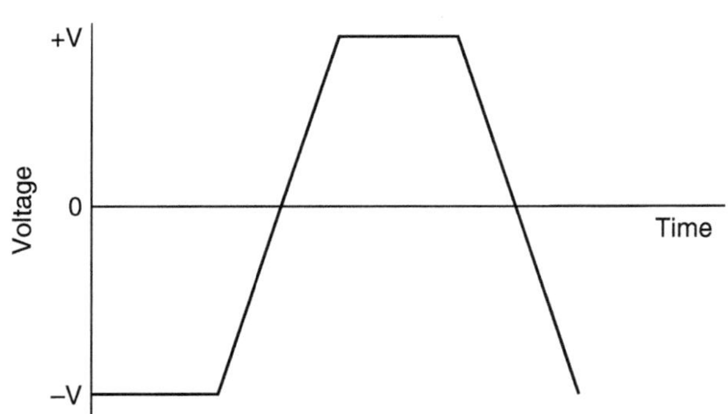

FIGURE 12.5 Schematic for the voltage versus time in a triangular voltage sweep (TVS). (*Used with permission from Ref. 26.*)

Chiron Technology[26] to ensure that the mobile ions reach a stable position on the plate with opposite polarity. The corresponding current is recorded. Then the applied voltage is ramped up at a slow linear rate (~1 MV/cm/s) through zero and up to a voltage that has the same absolute value of the initial stress condition but the opposite polarity. The voltage then is held for the same period of time as the initial stress condition. After this, the applied voltage is again ramped down through zero to the initial stress with the same sweep rate. TVS observation enables one to distinguish whether the stress-induced leakage current (SILC) comes from an intrinsic cause (e.g., Si–O bond breakage or dipole-dipole interaction of defects under the applied electric field) corresponding to a symmetric TVS hysteresis or an extrinsic degradation (e.g., mobile ions caused by moisture adsorption) corresponding to an asymmetric TVS hysteresis.[27] The typical *I-V* curve indicating Cu^+ diffusion induced by stress has a corresponding spike/peak.[25] The related Cu^+ concentration can be calculated from the area under the peak as follows[26]:

$$\text{Mobile ion density (ions/cm)} = \frac{I_p}{e^-} \cdot \frac{1}{T_{ox} A_{ox}} \cdot t_i \qquad (12.2)$$

where I_p is the measured peak current (coulombs/second or amps), e^- is the charge on an ion (coulombs), T_{ox} is the thickness of the dielectric (cm), A_{ox} is the area of the capacitor, and t_i is the integration time for the current measurement (seconds).

Constant-Voltage/Field Stress

Constant-voltage/field stress testing is the usual way to investigate the lifetime of the Cu/dielectric, which is in most case is referred as the *TDDB test*. The core of this test is a leakage-current measurement under accelerated conditions at elevated temperatures and much higher stress voltages than operating conditions. The resulting mean-time-to-failure (MTTF) could be used to extrapolate to operating conditions with the aid of area/length scaling and voltage acceleration and temperature acceleration factors. It is clear that different test structures with different areas/lengths are necessary, and the tests must be performed at different temperatures and different voltages (*E*-fields).

In general, the line-to-line leakage is recorded during a constant-voltage stress. Figure 12.6 gives a typical evolution of the leakage current with time under three different constant-voltage stresses at 125°C for BEOL IMD. A three-step degradation of the leakage-time curve is clearly seen. At the initial stressing stage (stage I), a clear reduction in the leakage current with time is observed, which is generally attributed to electron trapping. However, with longer stressing times, trap-assisted leakage makes the leakage current rise (stage II), which corresponds to a much longer relative time period until final breakdown (an abrupt

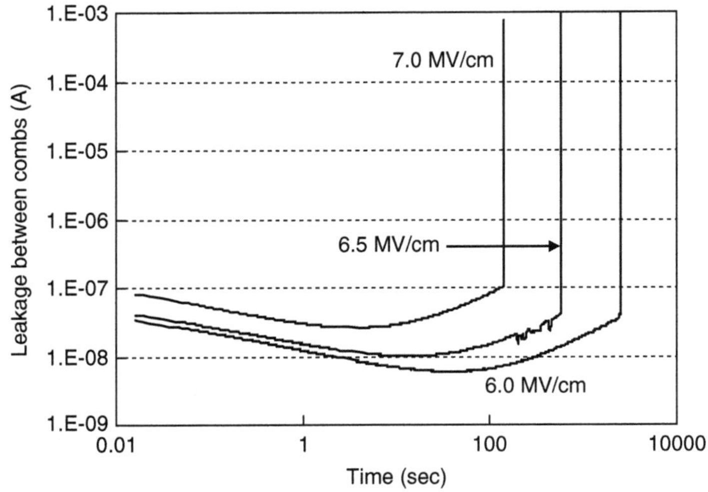

FIGURE 12.6 Typical line-to-line leakage current as a function of time during IMD TDDB measurement at 125°C.

current jump, stage III). As always expected, Fig. 12.6 shows that the time-to-failure (TTF) strongly depends on the applied electric field.

12.3 Failure Mechanisms/Modes for Intermetal Dielectric Breakdown

12.3.1 Failure Mechanisms

Low-*k* dielectric breakdown occurs owing to either bond damage or metal diffusion into the insulator as a result of the high electric field. Two mechanisms for IMDu breakdown are widely accepted: the thermochemical breakdown model (termed *intrinsic failure* owing to broken Si−O bonds or dipole-dipole interaction of defects under the applied electric field)[28] and the charge-injection model (termed as *extrinsic failure* induced by extrinsic defects such as Cu ions migrating from the Cu electrode).[29] These two models are elaborated in this section.

The *thermochemical breakdown model* assumes that the interaction of the dipole moment of the bond, for example, the Si−Si bond at an oxygen vacancy in SiO_2 (see Fig. 6.21 in Chap. 6), with the electric field weakens the bond, finally leading to the bond being broken either by thermal energy or by hole capture. The activation energy for bond breakage (i.e., damage) is reduced because of the electric field.[7] The natural logarithm of time-to-breakdown is directly proportional to the electric field, and thus the thermochemical model is the base for the *E*-model, which is a field-driven model. Since this model was already detailed in Chap. 6, more focus on the charge-injection model is given below.

The *charge-injection model* assumes the physical breakdown to be a copper diffusion into the dielectric. The diffusion of neutral Cu plays either no role or an insignificant role in TDDB in low-*k* dielectrics.[30] It is the *ionized Cu* that plays some role, followed by field-assisted diffusion kinetics. Recent investigations show the possibility of two stages in the breakdown behavior.[31] As indicated in Fig. 12.6, in the early part of the measurement, the dielectric system is charged with electrons, which leads to a field shielding effect and reduces the effective stress voltage. This results in a decrease in leakage current during the early part of the measurement. After a short time period, copper diffusion begins to be dominant, and the breakdown occurs.[31] *The charge-injection model is the base for the* $\sqrt{E}$ *model, which is a current-driven model.*

The SQRT(E)-model applicable to the SiCOH TDDB is a current-driven, Cu-catalyzed TDDB behavior.[32,33] It is generally assumed that accelerated electrons injected from the cathode follow interface Schottky emission (SE) or Poole-Frenkel (PF) conduction during their transport at the Cu-cap interface. SE is a thermionic emission process across the interface at low fields (<1.4 MV/cm), whereas PF emission is the field-enhanced thermal excitation of trapped electrons into the conduction band at high fields (>1.4 MV/cm).[32] Smaller line spacing can make the PF characteristic clearer.[34] SE conduction shows ln(I) linearity with $E^{1/2}$, whereas PF conduction exhibits ln(I/E) linearity with $E^{1/2}$ as follows[4,35]:

$$J_{\mathrm{SE}} = aT^2 \exp\left[\frac{q}{k_B T}\left(\sqrt{\frac{qE}{4\pi\varepsilon_0 k}} - \Phi_0\right)\right] \tag{12.3}$$

$$J_{\mathrm{PF}} = bE \exp\left[\frac{q}{k_B T}\left(\sqrt{\frac{qE}{4\pi\varepsilon_0 k}} - \Phi_1\right)\right] \tag{12.4}$$

where a and b are empirical constants, T is the ambient temperature, E ($=V/S$) is the electric field, V is the voltage applied to the two Cu electrodes with spacing S, q is the electron charge, ε_0 is the permittivity of free space, k is the dielectric constant, Φ_0 is the barrier height of electrons into the conduction band (Fig. 12.7*a*), $\sqrt{qE/4\pi\varepsilon_0 k}$ is the barrier lowering (band bending) owing to the applied electric field E, and Φ_1 is the depth of the trap potential well (Fig. 12.7*d*). During the V-ramp testing, when the applied voltage is low, the main contribution for the leakage current should be electron SE conduction (Fig. 12.7*a*), Fig. 12.7*b* is the case considering barrier lowering described by Eq. (12.3). With the applied voltage increasing, the conduction electrons will gain higher energy. Such energetic electrons could impact the Cu atoms at

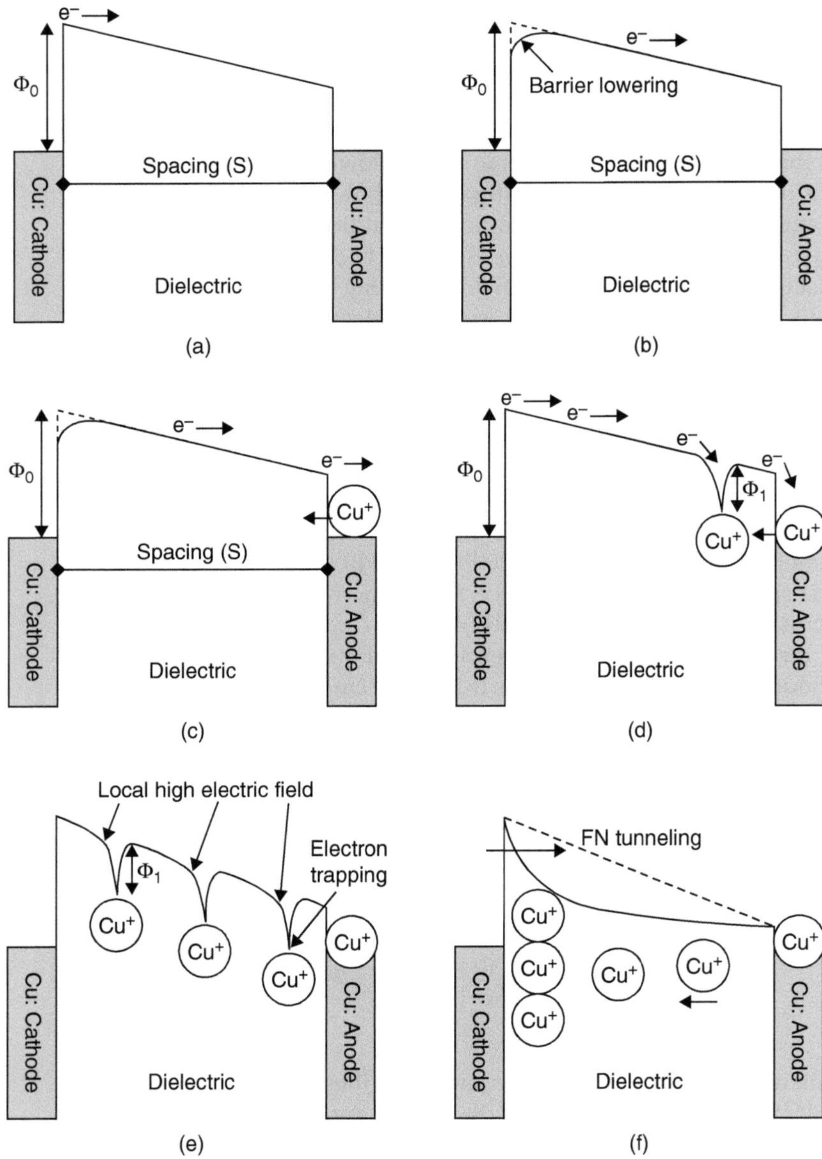

FIGURE 12.7 (*a*) Schottky emission (SE) conduction without barrier lowering. (*b*) SE conduction with barrier lowering. (*c*) Generation of Cu ions at the anode owing to energy release of accelerated electrons from cathode. The generated Cu ions are then injected into the dielectric. (*d*) Cu ions at the low-*k*—ESL interface could lead to electron traps. The trapped electrons could escape with the barrier height Φ_1, corresponding to the Poole-Frenkel condition. (*e*) More and Cu ions at the interface will lead to more electron traps, effective spacing narrowing, and an Cu ions at the interface will lead to more electron traps, effective spacing narrowing, and an increased effective barrier height. (*f*) Cu ion accumulation at the cathode will lead to FN tunneling.

the anode and accelerate the generation of positive Cu ions when they reach the anode. Those generated Cu ions then could inject into the low-*k*–cap-layer interface under the applied electric field (Fig. 12.7*b*). Some unlucky electrons will be trapped by these migrated Cu ions at the interface. The trapped electrons possibly could escape the trap potential well with depth Φ_1 (Fig. 12.7*d*), corresponding to the PF conduction. The unescaped ones may just be trapped or recombine with Cu ions, leading to neutral Cu atoms at the interface. Accumulated Cu atoms at the interface could form clusters of nanoparticles.[32] The Cu ions/atoms at the interface also would lead to a higher localized field and more band bending. In turn, like the trapping in gate oxide that leads to effective oxide thinning,[36] the migrated Cu atoms/ions also will result in a reduction in the effective Cu electrode spacing, referred as *effective spacing narrowing*. More and more Cu ions at interface will lead to a much larger electron trap and an increase in effective spacing narrowing and effective barrier height (Fig. 12.7*e*). Cu ions accumulating at the cathode will finally lead to FN tunneling and dielectric breakdown (Fig. 12.7*f*).

Conventionally, the TDDB test is performed under dc stressing. However, a real device is operated under ac conditions. The lifetime of ac TDDB is found to be larger than that of dc TDDB and is attributed to the backward migration during reverse bias.[37] However, owing to the increase in remaining Cu ions during repeated reverse biasing, backward migration of injected Cu ions was retarded. Therefore, the TTF of ac TDDB is not infinite but a few times larger than that of dc TDDB. The failure time increases as frequency increases because, under higher frequencies, the length of injection and depletion of Cu ions is short. Therefore, the number of injected and depleted Cu ions is small, and the failure time increases under higher frequencies. However, it is saturated at higher frequencies because of the contribution of diffusion.

12.3.2 Failure Modes

Dielectric breakdown between neighboring Cu lines generally occurs at the interface between the capping layer and the dielectric.[31] The electric field is highest at the corners of metal lines owing to the complicated geometric effect there, which initiates the leakage.[23] In addition, the Cu lines are generally tapered (wider at the top than at the bottom), so the space is smallest at the top of wires, leading to higher electric field. The interface is expected to have a higher trap density than the bulk dielectrics owing to bond mismatch between the different materials or to contaminants from the Cu Chemical-Mechanical Polish (CMP) process. It is expected that the interface between the capping layer and the dielectric is the preferred leakage path because of the combination of the high electric field and the high defect density.

Three failure modes have been observed[19,21] for the IMD breakdown as follows:

- *Mode A* (Fig. 12.8*a*) is the diffusion of copper at the low-*k*–dielectric barrier interface. This failure mode corresponds to copper diffusion through the weakest point of a comb structure, which is widely observed and accepted. *Improvement of the interface properties of the dielectric barrier is the way to eliminate mode A failures.*
- *Mode B* (Fig. 12.8*b*) has small cracks observed at the corner of the trench. These cracks could propagate into the upper passivation layer and sometimes join at neighboring lines, leading to a short. The cracks are sometimes observed in conjunction with interface diffusion identified as mode A.
- *Mode C* (Fig. 12.8*c*) has more severe cracks that continue along the upper passivation–cap-layer interface instead of continuing into the IMD. These cracks tend to appear on alternate lines that are positively biased.

Modes B and C dielectric breakdown failures are attributed to the mechanical forces from electrostatic attraction between the fingers of the comb. The high electric field and stress concentrated at the upper corner of a trench results in copper extrusion there. The mechanical

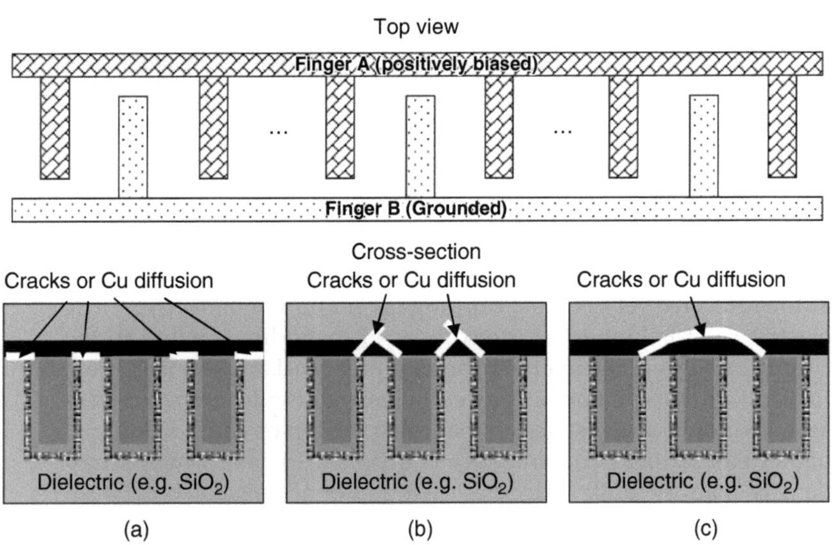

FIGURE 12.8 Schematic shows (*a*) mode A: diffusion of copper at the low-*k*–dielectric barrier interface. (*b*) mode B: small cracks at the corner of the trench; (*c*) mode C: severe cracks that continue along the upper passivation–cap-layer interface.

forces from electrostatic attraction can be estimated as follows[21]: The energy in a capacitor C with voltage V is $CV^2/2$. If we assume that this energy is distributed equally throughout the dielectric between the fingers of the comb, then the attractive force per unit length of capacitor would be

$$F = \frac{CV^2}{2ls} \tag{12.5}$$

where L is the length of the comb and s is comb separation. $C = k\varepsilon_0 d_m L/s$, with $k\varepsilon_0$ as the dielectric constant of the IMD and d_m the thickness of the Cu lines, Eq. (12.5) can be further simplified to

$$F = \frac{1}{2} k\varepsilon_0 d_m \cdot E^2 \tag{12.6}$$

where E is the electric field. As an example calculation, when the electric fields in the range of ~5 MV/cm, the energy density is ~2 J/m^2 between the fingers of the comb. This energy density is close to typical adhesion energies of 4 to 12 J/m^2 measured by four-point bending techniques for barrier adhesion to copper or low-k dielectric.[38]

12.4 Lifetime Models

Reliability engineers are routinely requested to predict the lifetime at which the dielectrics would reach a cumulative fraction failed on the order of a lower percentile (e.g., 0.1 percent or 1 ppm) for the entire chip.[39] However, rapid process-development cycles limit the number of samples, as well as the test time per sample, and the test-chip size limits the areas of the test structures. This section describes the models for such extrapolation from limited sample size toward lower percentiles based on the Weibull distribution from smaller chip areas (i.e., shorter metal length) toward larger chip areas (i.e., longer metal length) based on the Poisson area scaling transformation, from high-temperature toward lower operation temperature based on temperature related activation energy, and from large stress voltage toward lower operation voltage based on voltage/field acceleration models.

12.4.1 Weibull Distribution

Lognormal distributions tend to apply when gradual degradation occurs over time because of diffusion effects, corrosion processes, and chemical reactions. EM is generally modeled by lognormal statistics, as discussed in Chap. 10.[40] On the other hand, Weibull distributions appear to be applicable in cases where the weak link propagates to failure. Dielectric breakdown typically is well described by the Weibull distribution.[22] Lognormal distribution is proven to be not good at describing the dielectric breakdown behavior. It is found that after the lognormal area transformation, the new distribution will not be lognormal anymore.[41] As a matter of fact, using of the Weibull distribution usually gives a more conservative projection than the lognormal distribution, meaning that lognormal projection likely gives an overoptimistic projected lifetime.

The Weibull distribution is expressed as follows [note that this is another expression for Eq. (6.18)]:

$$\ln[1 - F(t)] = -(t/\eta)^{\beta} \tag{12.7}$$

where t is the variable under study, $F(t)$ is the cumulative distribution function (CDF), β is the Weibull slope, also called the *Weibull shape factor*, and η is the scale parameter of the Weibull distribution. The parameter η has the same unit as t, and its value is equal to that at CDF $F(t)$ = 63.2 percent. If Weibull statistics are used to analyze the times-to-failure distribution, $t_{63.2\%}$ generally is used as the characteristic time-to-failure. In contrast, in lognormal statistics, the characteristic time-to-failure is $t_{50\%}$. The Weibull slope, in general, becomes shallower with decreasing line-to-line spacing, which is similar to the smaller Weibull slope for thinner gate oxide, as given in Fig. 6.23 (Chap. 6). Chen et al.[41] reported that the Weibull slope is around 5 when the line to line spacing is 85 to 90 nm, whereas it decreases to ~4 when the spacing is 80 nm. Furthermore, the Weibull slope will decrease drastically to ~2 when the spacing is 70 nm. It is found that line-edge roughness increases with decreasing line-to-line spacing, leading to Weibull slope degradation for narrower spacing structures.

12.4.2 1/*E* Model, *E* Model, and SQRT(E) Model

Models are used to extrapolate the TDDB lifetime from the acceleration high-voltage/field-test conditions to the low-voltage/field-operation conditions. There are several remarkable models available, say, the $1/E$ model, E model, power law, and $\sqrt{E}$ [SQRT(E)] model for low-k material TDDB.[42]

Figure 12.9 compares the $1/E$ model, E model, and $\sqrt{E}$ model based on TDDB TTF data for low-k dielectrics. The TDDB data with electric fields above 24 V were stressed by wafer-level reliability (WLR) tests, and the others were stressed by package-level reliability (PLR) tests. For WLR data with high stress conditions and low TTF, all models can be well fitted, but the lifetime extrapolated to operating conditions can differ by tens of orders of magnitude using different models. However, when considering the PLR data, the SQRT(E) model is the best fitted. In other words, for lifetime extrapolation to operating conditions, the E model will be too conservative, whereas the $1/E$ model will give too optimistic predictions. This section will give the details of each model and compare their applications from a historical point of view.

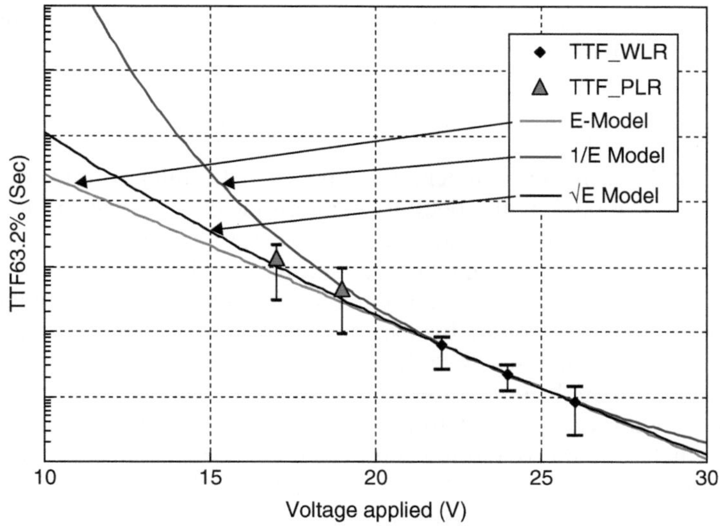

FIGURE 12.9 The experimental time-to-failure (TTF) in terms of stressing voltage by both WLR and PLR tests for comb-serpentine structure. The TTF values are fitted with the *E* model, 1/*E* model, and SQRT(*E*) model.

As detailed in Chap. 6, the 1/*E* model was found to be very useful in describing TDDB in SiO_2 gate dielectric films under high-field conditions, where the dominant conduction mechanism is Fowler-Nordheim tunneling.[43] The equation for the 1/*E* model is

$$T_{bd} = A \cdot \exp(G/E) \tag{12.8}$$

where T_{bd} is the time-to-breakdown, *A* is a fitted constant, *E* is the electric field, and *G* is the field acceleration factor in the 1/*E* model.

However, the breakdown strength in porous silica-based low-*k* films is usually relatively low, where the dominant conduction mechanism during TDDB testing is generally found to be either Schottky barrier emission or Poole-Frenkel hopping. A three-year TDDB study[44] published in 1998 showed that the *E* model gives a much better fit to long-term/low-field TDDB data, particularly at design electric fields of <5 MV/cm.

E Model

A commonly used reliability model (*E* model) to predict the T_{bd} for silica-based dielectrics with thicknesses greater than 4.0 nm was first introduced by Berman[45] and then explained based on a physical (thermochemical) foundation by McPherson.[46] The model assumes that $\ln(T_{bd})$ is directly proportional to the electric field in the dielectric.

Following the thermochemical theory, the time-to-failure equation can be written[7]

$$T_{bd} = A \cdot \exp(-\gamma \cdot E) \tag{12.9}$$

where A is a material-dependent constant with units of time, and γ (for the E model) is the field acceleration factor. γ is reported to be material independent[7] and line-to-line-spacing independent.[62] The reported γ is 4.13 ± 0.85 cm/MV for silica-based dielectrics.[62]

The electric field acceleration parameter γ is temperature-dependent, which is demonstrated by experiments[46] showing that the field acceleration increases with lower temperature. Therefore, TDDB stressing at high temperature to get a lower-voltage acceleration factor will mislead the lifetime extrapolation. TDDB tests should be conducted at the operating temperature T_{op} (e.g., 125°C) so as to eliminate the error caused by the extrapolation from high temperature to T_{op}.

$\sqrt{E}$ Model

The industry now tends to use the SQRT(E) formalism.[47,50] There are now at least three distinct versions of the SQRT(E) TDDB failure models[48,49] for interlevel dielectrics, and a simple formula can be expressed as follows:

$$T_{bd} = A \cdot \exp\left(-\gamma_{\sqrt{}} \cdot \sqrt{E}\right) = A \cdot \exp\left(-\gamma_{\sqrt{}} \cdot \sqrt{\frac{V}{s}}\right) \tag{12.10}$$

where A is a fitted constant, E is the electric field, V is the applied stress voltage, s is the nominal line-to-line spacing, and $\gamma_{\sqrt{}}$ is the field acceleration factor for the SQRT(E) model.

The $\sqrt{E}$ model assumes that damage in the dielectric is proportional to the total number of charge carriers injected into the material and that breakdown occurs at a critical charge level (i.e., the charge-to-breakdown). The amount of charge injected in the dielectric is proportional to the leakage current. Assuming a constant leakage current, the charge-to-breakdown Q_{bd} is related to the time-to-breakdown T_{bd} as follows

$$Q_{bd} = J \cdot T_{bd} \quad (\text{unit}: \text{C/cm}^2) \tag{12.11}$$

where J is the current density injected into the dielectric during stressing.

Assuming that leakage is due to Schottky emission (SE), the T_{bd} is obtained by combining Eqs. (12.11) and (12.3), and we get

$$T_{bd} = \frac{Q_{bd}}{J_{SE}} \propto \exp\left[-\frac{q}{k_B T}\left(\sqrt{\frac{qE}{4\pi\varepsilon_0 k}} - \Phi_0\right)\right] \tag{12.12}$$

From Eq. (12.12) it can be seen that the natural logarithm of time-to-breakdown is directly proportional to the square root of the electric field. Note that Eq. (12.12) also applies to PF conduction because the exponential terms are the same for both SE and PF emission.

The field acceleration factor is virtually not sensitive to the spacing change and area/length change in the SQRT(E) model. In contrast, the voltage acceleration factor tends to increase with a decrease in line-to-line spacing.[41]

According to Eq. (12.12), when taking derivative, we get the temperature-dependent $\gamma_{\sqrt{}}$:

$$\gamma_{\sqrt{}} = -\left(\frac{\partial \ln(T_{bd})}{\partial \sqrt{E}}\right)_T = \frac{q\sqrt{q/(4\pi\varepsilon_0 k)}}{k_B T} \tag{12.13}$$

12.4.3 Activation Energy

The temperature-dependent breakdown voltage V_{bd} from the V-ramp test can be evaluated for the same test structures and voltage acceleration based on the Arrhenius relation:

$$V_{bd} = A\exp\left(\frac{E_a}{k_B T}\right) \tag{12.14}$$

The temperature influence on V_{bd} could be quite low (Fig. 12.10*a*) and the corresponding activation energy is 0.007eV. A higher E_a for V_{bd} was reported at around 0.1 eV.[34] This clearly indicated that the kinetics was not diffusion dominated for V-ramp stressing.

For time-dependent dielectric breakdown (TDDB), the corresponding Arrhenius relation is

$$T_{bd} = A\exp\left(\frac{E_a}{k_B T}\right) \tag{12.15}$$

E_a is the activation energy for Si$-$O bond breakage, k_B is the Boltzmann constant, T is the temperature in kelvins.

The TDDB activation energies for three silica-based films (SiOF, SiOC, and porous MSQ) based on the comb-serpentine test structure[7] are all around 0.5 eV, independent of the film type. This observation suggests that the porosity in the silica matrix has little or no impact on diffusion kinetics during IMD TDDB stressing. On the other hand, an E_a of 0.3 eV was reported for $k = 2.7$ SiCOH dielectric films by Chen et al.[41], but 0.77 eV by Yiang et al.[50] The difference in the reported E_a values could be attributed to the different electric fields applied when extracting the activation energy by different authors, as explained below.

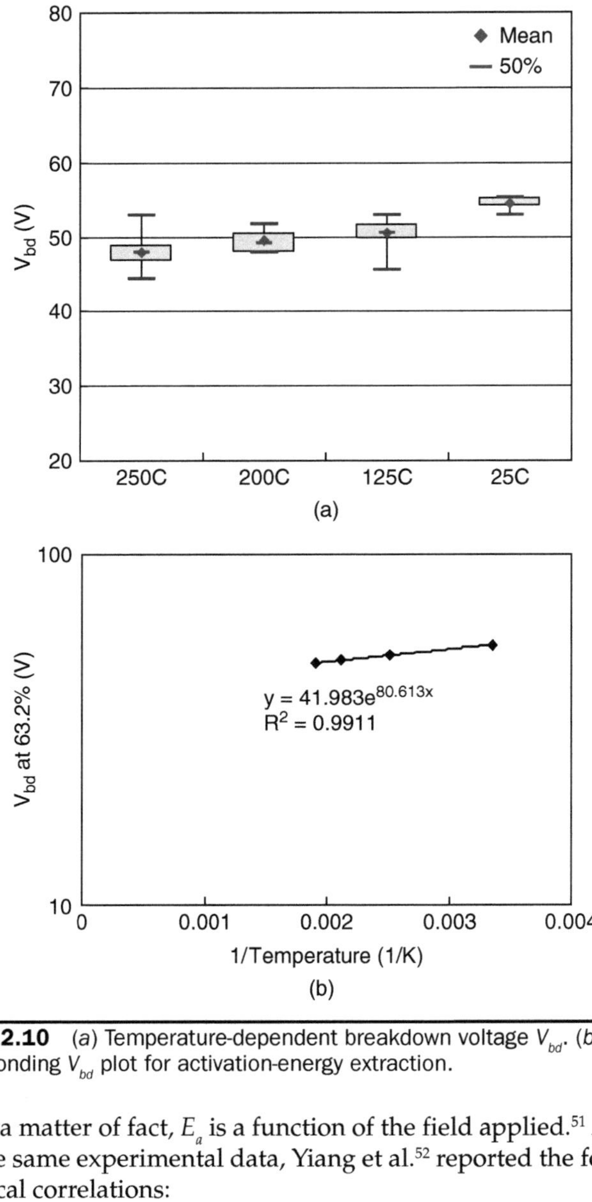

FIGURE 12.10 (*a*) Temperature-dependent breakdown voltage V_{bd}. (*b*) The corresponding V_{bd} plot for activation-energy extraction.

As a matter of fact, E_a is a function of the field applied.[51] After fitting the same experimental data, Yiang et al.[52] reported the following empirical correlations:

For the E model: $E_a = 3.20 - 0.97 * E$.
For the SQRT(E) model: $E_a = 5.71 - 3.14 * \mathrm{SQRT}(E)$.

It is clear that for both the E model and the SQRT(E) model, E_a increases as the applied field decreases, indicating that the E_a obtained under the higher-stress field is underestimated if we use it to extrapolate lifetime to the much lower operating field. Thus, for an accurate lifetime projection, TDDB tests should be conducted at T_{op} and accelerated based on stress voltage only. For the E model, the linear dependence of E_a on electric field E is explained by Eq. (12.16) as follows

$$E_a = E_{a0} - p_{eff} \cdot E \tag{12.16}$$

where E_{a0} is the zero-field activation energy for Si−O bond breakage, p_{eff} is the effective dipole moment for the molecule in the solid silica material (~13 eA).[100]

For the SQRT(E) model, according to Eq. (12.12), when taking the derivative, we get

$$E_a = \left[\frac{\partial \ln(T_{bd})}{\partial(1/k_B T)}\right]_E = B - \frac{q\sqrt{q}}{\sqrt{4\pi\varepsilon_0 k}} \cdot \sqrt{E} \tag{12.17}$$

where a linear dependence of E_a on SQRT(E) is shown.

12.4.4 Area/Length Scaling

In order to scale failure probability from small test structures to product-like metal-via, comb-comb, or vertical natural capacitor (VNCAP) size, the Poisson area scaling transformation[53,54] is generally employed. Following the area scaling rule described in Chap. 6, the length scaling can be expressed as follows:

$$\frac{\eta_1}{\eta_2} = \left(\frac{L_2}{L_1}\right)^{1/\beta} \tag{12.18}$$

or

$$\eta \approx L^{-1/\beta} \tag{12.19}$$

where η_1 and η_2 are the scale factors of two T_{bd} (or V_{bd}) Weibull distributions of the IMD structures with identical dielectric thicknesses but length L_1 and L_2, respectively. β is the Weibull shape factor (slope). Strictly speaking, a pure intrinsic failure of a perfectly homogeneous material should be area-independent and should occur at exactly the same time for all samples. In reality, the material has inhomogenities (even without any defects), that lead to a distribution of failure times.

It is noted that for V_{bd}, the Weibull slope is quite large (e.g., about 40 to 70). Therefore, the V_{bd} differences will be very small, even with a large area/length differences. The change in V_{bd} was indeed very small even for a 100× area difference.[41] On the other hand, the Weibull slope for the T_{bd} is much smaller (e.g., ~2 to 6) for the same structures,[41] which means that the T_{bd} is much more sensitive to the area/length of the test structures.

12.4.5 Defect Density (DD)

The defect density (DD) can be calculated based on the following equation[55]:

$$C \le 1 - \sum_{i=Y}^{N} \frac{N!}{i!(N-i)!}[\exp(-A \cdot DD)]^i [1 - \exp(-A \cdot DD)]^{N-i} \tag{12.20}$$

where C (e.g., 95 or 90 percent) is the confidence level that the actual defect density is less than or equal to DD, A is the area of test structure, and Y (i.e., yield) is the number of good samples among the total sample size N.

Based on Poisson's law for randomly distributed failures [Eq. (12.21)] and the formula for yield [Eq. (12.22)], the Weibull scale can be transferred into a defect-density graph by using Eq. (12.23):

$$P(m) = \frac{(A \cdot DD)^m}{m!} \cdot \exp(-A \cdot DD) \tag{12.21}$$

$$Y = 1 - \text{CDF} = P(0) \tag{12.22}$$

$$\text{weibit} = \ln[-\ln(1-\text{CDF})] = \ln\{-\ln[\text{P}(0)]\} = \ln(A \cdot DD) \tag{12.23}$$

where P is probability, Y is the yield, and CDF is the cumulative density function. For gate oxides, A is the active dielectric area between anode and cathode. For BEOL IMD, this area usually corresponds to the top-down layout area. Hence, the defect densities relate to a number of defects per $1/\text{cm}^2$ metal level. In this way, the voltage-to-breakdown and/or the time-to-failure can be transferred to defect density. The defect-density graph is independent on A, and according to Eq. (12.23), TDDB data obtained for larger areas will be plotted at lower DD values.[56]

12.5 Factors Affecting IMD Reliability

12.5.1 Material Dependence

The TDDB lifetime is very sensitive to the dielectric materials and related processes, especially for low-k materials. The TDDB lifetime typically decreases as the dielectric constant of the material decreases.[7,57] In particular, porous materials have lower TDDB lifetimes than nonporous materials. Possible reasons for the lower TDDB lifetimes are weaker bonds, higher trap densities, and lower barrier heights at the metal-insulator interface. Despite the lower TDDB lifetimes of porous materials compared with nonporous materials, the intrinsic reliability is still adequate for ICs.[6]

SiCOH is a low-k carbon doped oxide dielectric material comprised of Si, C, O, and H deposited by chemical-vapor deposition (CVD). Its TDDB degradation at the 65-nm technology node has been intensively investigated at IBM.[31] SiCOH TDDB is very sensitive to all aspects of integration, including barrier deposition, reactive-ion-etch (RIE), CMP, lithography, queue-time, rinse and cleaning, IMD deposition recipe, etc. Models for interlevel dielectric reliability in copper damascene structures vary considerably in terms of mechanism and location of the failures.

Some reports attribute the degradation of reliability in low-k structures to intrinsic properties of the IMD material.[58] It has been

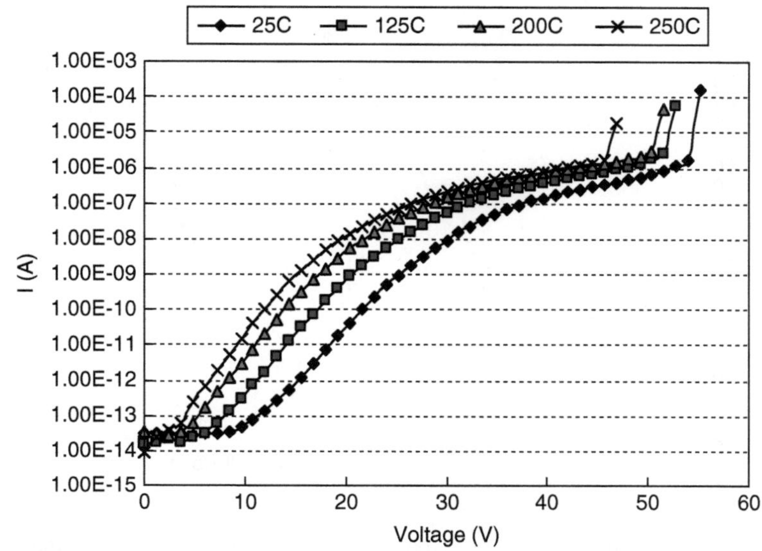

FIGURE 12.11 Typical temperature-dependent leakage-voltage curves for low-*k* dielectrics.

shown[7] that porous MSQ shows much lower breakdown strength than the other two dielectrics. Its current density at breakdown is also relatively low. Figure 12.11 shows typical leakage-voltage curves for low-*k* dielectrics at various temperatures. Higher temperatures will lead to larger leakage currents at the same voltage, although the breakdown strength does not much depend on the temperature shown in Fig. 12.10.

Figure 12.12 shows typical TDDB TTF Weibull distributions for low-*k* dielectrics. Ogawa et al.[7] have reported that the TTF depends on the low-*k* material. Low-*k* dielectrics with a higher porosity (porous MSQ and SiOC) have a shorter TTF than SiOF dielectric when subjected to the same stressing electric field. The Weibull shape factor β for nonporous SiOF ($k = 2.5$ to 2.8) is 2.8, which is much larger than that for the porous films: $\beta = 1.5$ for SiOC and $\beta = 1.3$ for porous MSQ, respectively. It is the different molecular bond configurations such as cross-linking and polarity that affect the drift rate of Cu ions in polymer low-*k* films because Cu drift usually reflects interactions between Cu ions and the local chemical environments of the low-*k* films.[31] Abrupt changes in the time to dielectric breakdown for spin-on porous low-*k* dielectrics were observed when the porosity of the low-*k* material was between 20 and 30 percent.[59] In summary, the pores have a large impact on the breakdown electric field and time-to-failure.

When the TDDB electric field dependence for the conventional thermal SiO_2, PE-TEOS, SiOF, SiOC, and porous MSQ were plotted based on the *E* model (Sec. 12.4.2), it was interesting that the

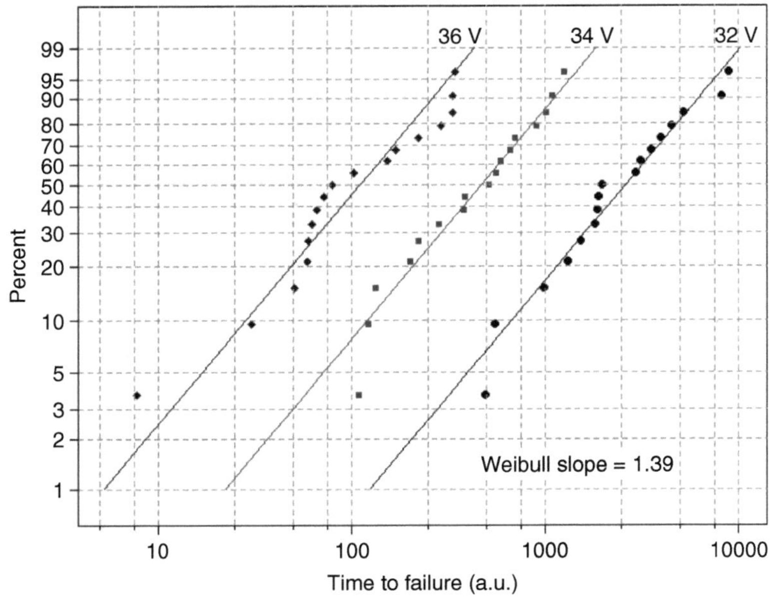

FIGURE 12.12 Typical TDDB TTF Weibull distributions for low-*k* dielectrics.

E-field-dependent T_{bd} was in parallel,[7] meaning that the field acceleration factor γ was similar regardless of the dielectric and how much porosity it had. The line spacing of the comb-serpentine test structure under study typically was 170 nm. γ was 4.5 ± 0.5 cm/MV for all these silica-based dielectrics. These TDDB observations suggest that the porosity in the silica matrix has little or no impact on failure kinetics (physics).

Some reports suggested that the mechanical strength of low-*k* materials was lower than USG and, therefore, cracks and interface delaminations were more likely take place, thus contributing to IMD failures.[19] The primary failure mode in comb capacitor structures with breakdown fields greater than 4 MV/cm was found not to be the diffusion of copper through the metal barrier layer or dielectric, but the mechanical cracking of the dielectric barrier or delamination at the barrier-IMD interface (see Fig. 12.8*c*),[21] where the *k* value of the low-*k* dielectric was 2.9.

12.5.2 Moisture Effect

Moisture Effect on Capacitance Stability

The presence of moisture may have a significant influence on the capacitance stability,[54] leading to increased leakage and capacitance and reduced TDDB lifetime as well, especially for ultralow-*k* materials.[60] It is known that the low-*k* SiCOH surface layer could be damaged and therefore absorb moisture during certain process steps involving exposure to moisture.[61] Chen et al.[54] reported that the capacitance measured at high temperature decreased nonlinearly within 1,000 seconds and eventually leveled off at a certain value. This phenomenon of temperature-dependent capacitance could be explained as follows: Moisture adsorption increases the leakage paths and film polarization, thus resulting in higher leakage current and a film k-value increase. As a result, a higher initial capacitance is expected. However, during the high-temperature aging, such trapped moisture will redistribute within different layers by diffusion until a new equilibrium state is achieved, which is related to a lower capacitance. Therefore, such capacitance reduction was found to be primarily temperature-dependent with little or no voltage dependence. An empirical formula was proposed to model such capacitance drift during temperature aging as follows[54]:

$$\frac{\Delta C}{C_0} = A \cdot \mathrm{Tanh}\left(-\sqrt{\frac{t}{D}}\right) \tag{12.24}$$

where

$$D = D_0 \cdot \exp\left(-\frac{E_a}{k_B T}\right) \tag{12.25}$$

and C_0 is the initial capacitance prior to temperature aging, ΔC is the capacitance change during aging, A is a fitted constant representing the final saturation value when time t approaches infinity, D is related to the moisture diffusivity, and E_a is the activation energy for the moisture diffusion. The experimentally obtained E_a is 0.83 eV.[54] According to Eq. (12.25), higher temperatures T correspond to a higher moisture diffusivity D, which further leads to a higher capacitance change ($\Delta C/C_0$) according to Eq. (12.24). Fortunately, with careful device layout design (e.g., using a seal ring as a moisture barrier), process control, and process optimization, the capacitance instability could be minimized successfully; that is, the constant A approached to 0 after optimization.

Moisture Effect on TDDB Performance

Moisture absorption also degrades IMD TDDB performance. For example, moisture likely penetrates into the IMD from the open area around the pad metal if TEOS rather than SiN passivation is used.[62] It was found that exposure to ambient conditions for several weeks (e.g, 25°C and 50 percent relative humidity) was sufficient to affect TDDB performance significantly.[60] Plasma treatments on the CVD SiOC:H film caused by ashing (N_2/H_2 plasma) also can introduce moisture into the SiOC:H low-k dielectric.[63] The thermal activation energy (E_a) in those samples with moisture adsorption even could be negative.[64] The TDDB field acceleration factor (and thermal activation energy) in those samples with moisture adsorption is significantly smaller than that in samples without moisture adsorption, suggesting that the failure kinetics for TDDB may be dramatically different between those two cases. Different finger-to-fingers spacings exhibited a similar moisture effect. Since SiCOH TDDB is attributed to Cu migration during stress,[31] it is likely that the trapped moisture could diffuse into the Cu line, oxidize the Cu, and accelerate Cu diffusion during high-temperature and high-bias stress. A surface layer containing cuprous oxide (Cu_2O) and cupric oxide (CuO) can be formed when a Cu surface is exposed to moisture. It is well known that either Cu_2O or CuO contains active Cu ions which could migrate into SiCOH along the SiCOH-cap interface. A large number of Cu ions could be generated with the help of moisture at the anode, allowing many Cu ions to inject into the SiCOH and form Cu nanoparticles and clusters at the interface. Such neutral Cu particles and clusters could further migrate along the interface following a diffusion-limited process caused by the concentration gradient. Therefore, the actual field and temperature influences on TDDB would be dramatically reduced, and both the voltage acceleration and thermal activation energy would be significantly smaller in those samples with moisture adsorption than those without moisture adsorption.

Seal *ring* is one of the solutions widely used to protect the chip from moisture.[27,60] High-temperature annealing (e.g., 12 hours at 150°C[60] or a long N_2 bake at 190°C[63]) also can remove the physisorbed moisture from the low-*k* dielectrics and thus improve TDDB performance.[65] Equation (12.26)[66] gives the so-called α bond breakage corresponding to the physisorbed water desorption. The activation energy E_a for α-type water desorption is 23 ± 2 kJ/mol. However, Li et al.[63] also demonstrated that there is another type of moisture determined by a more tightly bond OH group that cannot be removed by thermal annealing. This corresponds to a so-called β-type tightly hydrogen-bonded water for silica-based dielectrics, as shown in Eq. (12.27).[66] The E_a for β-type water desorption is 55 ± 17 kJ/mol. Thus, it is recommended to mitigate moisture uptake of IMDs from the beginning of the integration flow rather than relying on a postintegration annealing to restore the low-*k* characteristics.

$$\alpha:\ \equiv Si-OH\cdots H_2O \rightarrow \equiv Si-OH + H_2O \tag{12.26}$$

$$\beta:\ \equiv Si-OH\cdots H_2O\cdots OH-Si \rightarrow 2\equiv Si-OH + H_2O \tag{12.27}$$

12.5.3 Critical-Dimension (CD) Control

Within-wafer critical-dimension (CD) uniformity and line-width roughness (LWR) or line-edge roughness (LER) within the dielectric breakdown comb structure play key roles in the V_{bd} and TDDB distribution, which have drawn more and more attention especially in recent years, owing to dimension shrinkage.[41,67–72] This geometric effect leads not only to the narrowing of local spacing but also to the local concentration of high electric field, and both worsen the dielectric reliability performance.

LWR impact on low-k TDDB estimated on the basis of the E model [Eq. (12.9)] and the SQRT(E) model [Eq. (12.10)] is described in Eqs. (12.28)[69] and (12.29), [68] respectively:

$$\delta = \frac{t_{63-1}}{t_{63-2}} = \exp\left[-\gamma\left(\frac{V_1}{s_1 - w_{\text{rms}}} - \frac{V_2}{s_2 - w_{\text{rms}}}\right)\right] \tag{12.28}$$

$$\delta_{\sqrt{}} = \frac{t_{63-1}}{t_{63-2}} = \exp\left[-\gamma_{\sqrt{}}\left(\sqrt{\frac{V_1}{s_1 - w_{\text{rms}}}} - \sqrt{\frac{V_2}{s_2 - w_{\text{rms}}}}\right)\right] \tag{12.29}$$

where δ and $\delta_{\sqrt{}}$ are the ratio of two characteristic TDDB lifetimes (i.e., t_{63}) from the two different spacings (i.e., s_1 and s_2) under the same field ($E = V_1/s_1 = V_2/s_2$) condition for the E model and the SQRT(E) model, respectively. γ and $\gamma_{\sqrt{}}$ are field acceleration factor for the E model and the SQRT(E) model, respectively. w_{rms} is the effective LWR RMS (root mean square). From Eq. (12.28), it is seen that the LWR contribution becomes less with decreasing stress bias, and its contribution becomes larger with decreasing spacing.

The line-to-line spacing difference across a 300-mm wafer could be as large as 30 percent. More spacing variation is expected within test structures with smaller spacings and larger areas, which tend to degrade the Weibull slopes of both V_{bd} and TDDB.[68] LWR also can degrade chip performance owing to capacitance increase/variation, cross-talk, and resonance-induced speed delays. Therefore, LWR control has become critical for further dimension shrinkage in advanced-technology nodes.

LWR can be characterized by analyzing top-down SEM images (Fig. 12.13). In an SEM image of the pattern, a horizontal line is scanned from top to bottom for all the lines, and at least 100 scans are performed per each image. LWR then is obtained from the deviation of the line edge points from their mean line width. A useful parameter of LWR is the RMS because RMS depends not only on the mean but also on the standard deviation.

Spacing variation also can be determined experimentally from the measured capacitance distribution because line-to-line capacitance (measured through a nondestructive C-V measurement) normally has a good correlation with the line-to-line spacing.[41] A real spacing variation could be a skewed normal (non-Gaussian) distribution or even other types of distribution depending on process quality.

There are several sources including materials and BEOL processes that induce LWR.[68,73] Lithography is one of the sources for LWR, especially when critical dimensions shrink. Plasma etch processes often roughen the feature sidewalls, leading to anisotropic striations.

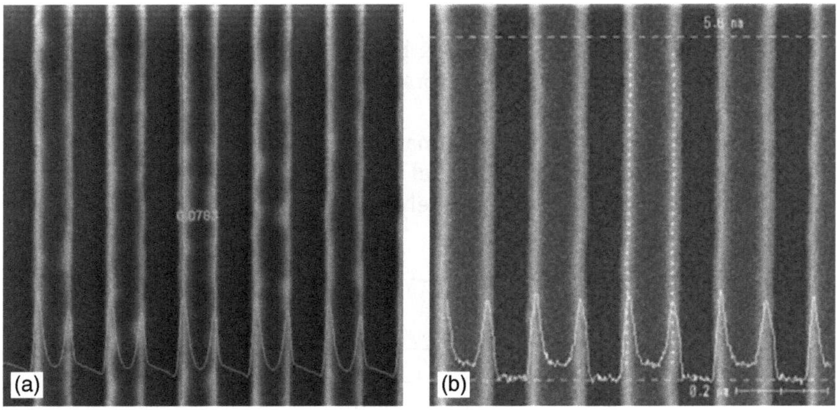

Figure 12.13 LWR was improved by a new trench etch recipe.The LWR is (*a*) 7.0 nm and (*b*) 5.2 nm before and after optimization of the trench etch process.

An unoptimized CMP process can induce edge corrosion of the copper line, which is another factor causing line edge roughness. The hydrophobic characteristic of low-*k* dielectrics may induce a high surface tension that could push the H_2O_2 molecules contained in the CMP slurry toward the edge of the Cu line, thus enhancing the corrosion reaction at the Cu line edge.

A descum step in the trench etch process was considered to improve the CD nonuniformity located at the wafer edge.[67] This method forces oxygen to flow toward the chamber-edge area by huge amounts of dilute chemical flux; thus, the transverse etch is realized. Power and power ratio of the trench etch process also can be adjusted for LWR improvement. Lower bias power results in a lower magnitude of striation on the dielectric line edge. With these efforts, inline CD measurement data showed that within the wafer, CD range was improved from about 15 to about 7 nm, whereas the LWR was improved from 7.0 to 5.2 nm, as shown in Fig. 12.13. Using this optimized trench etch recipe, IMD V_{bd} is clearly improved. Figure 12.14 is a V_{bd} comparison between unoptimized and optimized trench etch recipes, and the optimized etching recipe achieves dielectric breakdown performance with improved V_{bd} distribution and absolute values.

Based on Eqs. (12.3) and (12.4), it is clear that the leakage current will increase with an electrode spacing decrease, which will lead to a lower breakdown voltage V_{bd}. As a result, to improve the CD uniformity within a wafer or a V_{bd} test structure is one of the key factors to get a tighter V_{bd} Weibull distribution, as shown in Fig. 12.14.

12.5.4 Cu-Cap Interface Quality Control

Cu-cap interface quality control is one of the key factors to improve IMD reliability.[57,74] An NH_3-plasma treatment prior to SiN cap deposition on

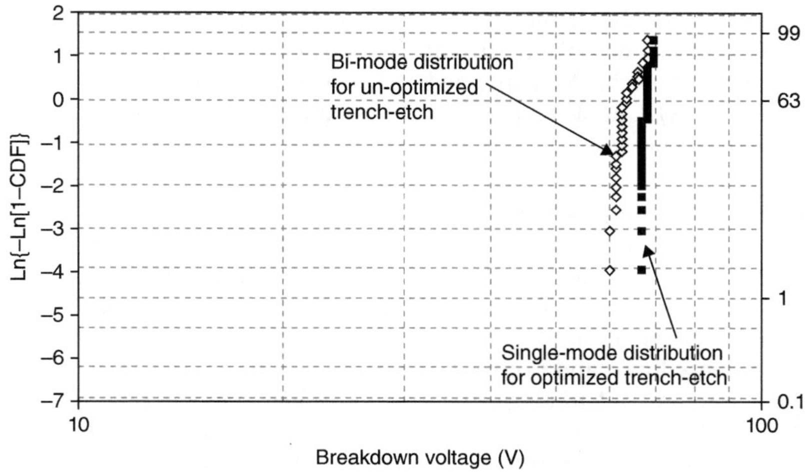

FIGURE 12.14 V_{bd} comparison between unoptimized and optimized trench etch recipes. With optimized trench etch, V_{bd} distribution turns to be single mode with larger values.

Cu interconnects improved TDDB attributed to the CuO reduction and CuN formation at the Cu-cap interface.[57] The addition of an SiH_4 treatment before capping SiN deposition led to more uniform breakdown fields throughout a 300-mm wafer and a higher Weibull slope and longer TDDB lifetime, attributed to the formation of a thin $CuSi_x/CuSi_xN_y$ layer at the Cu-capping SiN interface and a favorable morphology at the edge of the Cu-capping SiN interface.[75] The integration of a Ti-based barrier metal resulted in the best uniformity of breakdown fields and a dramatic enhancement in TDDB reliability related to the formation of Ti−Cu compound and Ti oxide.[75]

A high level of volatile organic compounds (VOCs) in manufacturing environment could result in a CMP-finished surface being contaminated and result in reduced dielectric breakdown voltages for worse interface adhesion. From correlation data of VOC and V_{bd} collected, the higher VOC content (>10 ppbv) corresponds to the worse V_{bd} performance.[67]

Furthermore, nitrogen doped carbide (NDC) tool faceplate lifetime needs to be controlled within an acceptable level.[67] When one faceplate is used for too long, it will be accumulatively deposited with Cu-containing substances during wafer processing. This material then is redeposited on the wafer surface in subsequent steps, which is not desirable. Off-line monitoring data showed that the whole-wafer refractive index (RI) clearly increased, and wafer-center RI increased more when the faceplate was used for a long time. The existence of redeposited Cu-containing material in between the dielectric surface and NDC cap layer will deteriorate the interface adhesion, causing a lower dielectric breakdown voltage.

Based on Eqs. (12.3) and (12.4), the barrier height Φ_0 is a critical factor that affects the leakage current. The leakage will increase with a lower barrier height. As mentioned earlier, a higher VOC level or a long-lifetime NDC tool faceplate will cause a lower barrier height, which results in worse IMD TDDB performance.

12.5.5 Novel Cap Layers

As detailed in Chap. 10, CoWP caps have been demonstrated to be promising materials for EM improvement for advanced-technology nodes. However, the CoWP cap could exacerbate the dielectric TDDB characteristics, including T_{bd}, voltage acceleration, and Weibull slopes for both dense low-*k* and porous ULK materials with comb-serpentine test structures of 1-m-long serpentine.[76] The degradation is caused by the defects such as metallic (Co) particles and the lateral growth of CoWP toward the low-*k* surface induced by the electroless CoWP deposition process. One solution path to minimize the detrimental impact of CoWP on TDDB is to selectively deposit the CoWP cap on the copper through careful selection of various chemistries and their sequence.[77,78]

Formation of a PECVD self-aligned barrier (PSAB) layer between the SiC cap and Cu line was proposed for improvement of both dielectric reliability and EM.[79] This improvement is attributed to the formation of Cu_xSi at the interface, as evidenced by X-ray Photoelectron Spectroscopy (XPS) composition analysis, which drastically enhances the adhesion between the SiC cap and Cu surface. This process is selective because the SiH_4 reacts only with the Cu surface to form Cu_xSi, but does not react with the low-*k* material. Optimization of both the self-aligned barrier and the SiC film can improve the adhesion between both the Cu and the dielectric layer on top, resulting in a 10 to 1,000× increase in TDDB lifetime.

12.5.6 Barrier Effect

The composition of the dielectric diffusion barrier also strongly influences IMD reliability.[57,80] It is well known that Cu can diffuse very fast in SiO_2, low-*k* materials, and Si and thus lead to serious current leakage and breakdown problems in inter- and intra-level dielectrics. To prevent Cu from diffusing into dielectrics and penetrating into Si, Cu must be fully sealed using a diffusion barrier (or sometimes referred as a *liner*). Both the leakage current and the dielectric breakdown strength are found to depend on the choice of the liner and liner integration.[81] Many physical aspects of the liner process can affect TDDB, in particular the post-CMP liner profile at the trench top. The quality of Ta-based liner owing to different liner processes and liner thicknesses not only influenced the SiCOH time-to-breakdown but also affected the TDDB distributions.[32] An H_2 plasma after etching can improve the breakdown voltage up to 34 percent.[82] In addition, a thick CVD TiN barrier could be very strong in protecting the porous ULK and thus may improve dielectric reliability.[82]

The barrier-integrity problem can be overcome by modification of the integration approach, termed *postetch porogen burnout* (PEBO).[83,84] This approach is demonstrated to be beneficial to a porous polymer dielectric. Porosity is created after patterning, and it was shown that the PEBO approach provides smooth sidewalls, resulting in a fully dense diffusion barrier on recess sidewalls and improved TDDB performance.

12.5.7 Self-Assembled Molecular Nanolayers as Diffusion Barrier

Typically, a layer of Ta, TaN, TiN, or a combination thereof is deposited at the trench bottom and sidewalls along with an insulating amorphous SiC or SiN cap layer on the top to inhibit Cu transport along the Cu–low-*k* interface and into the dielectric—either during processing or accelerated by local electric fields encountered during device operation—and thus improve EM resistance. Self-assembled molecular nanolayers (MNLs) have been demonstrated at the laboratory level to be attractive alternatives to conventional diffusion barriers and adhesion promoters if the bonds at both at the Cu-MNL and at the MNL-SiO_2 interfaces are strong.[85] MNLs with selected molecular termini at Cu-silica interfaces result in several orders of magnitude lower leakage currents and up to tenfold enhancement in failure time during aggressive electrothermal tests, indicative of decreased Cu transport across the Cu-silica interface. MNLs also have been shown to toughen Cu-SiO_2 interfaces by more than a factor of 3 by forming strong chemical bonds with the Cu overlayer and the silica underlayer.[86–88] Additionally, MNLs also can effectively passivate the Cu surface against oxidation and can reduce the surface oxide.[89] These attributes are attractive for minimizing surface- and interface-scattering induced resistivity increases in nanoscale Cu wiring.

12.5.8 Cu CMP Effect

CMP queue-time control: TDDB lifetime was found to deteriorate abruptly as the waiting time (queue time) between Cu CMP and Cu top capping-layer deposition in air increased.[31,90] This phenomenon can be attributed to the Cu ion migration along the CMP surface owing to the CuO_x formation at the Cu surface if the queue time is long enough. This queue-time impact on TDDB was believed to be a common problem for Cu technology regardless of IMD, line space, and liner.

CMP downforce control: A lower downforce during CMP can significantly reduced Cu corrosion and defects, thus improving TDDB performance.[31]

Post-CMP rinse/clean optimization: Post-CMP rinse/cleaning has a large impact TDDB performance.[91–93] Corrosion at the edge of the Cu lines is likely caused by oxygen diffusion from the air through the thin chemical layer during the post-CMP cleaning. Such corrosions enhance Cu ionization at the line edge, where the electric field concentrates so as to accelerate the dielectric breakdown. Using a lower wafer rotational speed (e.g., 750 rpm) creates a thicker chemical layer, which reduces Cu corrosions and thus improves TDDB.[94] High organic acid concentration of a benzotriazole (BTA)–containing slurry is necessary to prevent pittings as well as Cu dissolution.[94] Otherwise, some dissolved Cu atoms remain on the dielectric surface between adjacent Cu lines, which results in TDDB degradation.[93] By optimizing the in situ post-CMP rinse for a CMP process, a 10× improvement in TDDB was reported because the use of an optimized rinse could retard Cu oxidation and moisture adhesion processes.[31] During the post-CMP clean, it is important to control the redox potential to be less than -0.5 V versus NHE (normal hydrogen electrode) and to use a chelating agent on cleaning sequence. This is to effectively eliminate the small antioxidant-copper complex on the IMD without interacting with wiring and thus to improve TDDB lifetime.[73]

CMP-induced mechanical damage/scratch control: Cu CMP process-induced mechanical damage to the SiO_2 surface also can degrade TDDB. Adoption of a mechanical-damage-free slurry or a post-CMP HF treatment to remove the damaged layer from the surface can improve TDDB.[57] CMP polish scratch can result in embedded particles on Cu surface and liner damage, which severely affects TDDB reliability.[95] In situ CMP platen 3 (P_3) pad preclean with a kind of acidic chemical is found to reduce the polish-scratch density effectively and significantly improve the V-ramp/TDDB reliability performance. Liu et al.[95] reported that the TDDB TTF Weibull slope β and SQRT(E) model voltage acceleration factor γ values are 1.07 and 8.85 $V^{-1/2}$, respectively, when the polish scratch density is less than 1.0 cm^{-2}. β and γ values will reduce to 0.61 and 6.67 $V^{-1/2}$, respectively, if the polish scratch density increased to 5.21 cm^{-2}. What's more, when there is a very high polish-scratch density of 65.7 cm^{-2}, the corresponding β and γ values will be much lowered to 0.60 and 3.04 $V^{-1/2}$, repectively. However, inappropriate use of chemical preclean would cause Cu corrosion, leading to line-edge recession and thus worsen EM performance. Thus, a compromise between polish-scratch reduction and Cu corrosion associated with P_3 pad preclean needs to be achieved by controlling the acidic chemical concentration and clean time, as well as by increasing the postclean rinse time to effectively remove the remaining chemical on the polishing pad.

12.6 Correlation Between Voltage-Ramp (V_{bd}) and Time-Dependent Dielectric Breakdown (T_{bd})

The V-ramp test is used mainly as a reliability monitor during process development and production for qualitative assessment. However, there is a trend to enhance the utilization of the V_{bd} data from the V-ramp test to obtain quantitative information through correlation with the TDDB lifetime.[75] This is in one way to save cost and time, and the other way to more easily assess the decreased reliability margins originating either from introducing new materials and process generations or from the demands/forces of the market.[96,97]

Most of the work[98–100] corresponding to the conversion of V_{bd} to TDDB lifetime (T_{bd}) is based on the cumulative damage summation or integral method originally proposed by Berman.[45] In practice, during a stepped voltage-ramp (V-ramp) test (see Fig. 12.4), the voltage applied V (and the electric field $E = V/s$ as well with s as the local line-to-line spacing for two adjacent metal lines) is stepped in uniform increments ΔV (or $\Delta E = \Delta V/s$). In each step, the holding time is $\Delta\tau$, as shown in Fig. 12.15. The total step number is n when breakdown occurs, so

$$E_{bd} = n \cdot \Delta E \qquad (12.30)$$

where E_{bd} is the breakdown electric field during a V-ramp test. During the V-ramp test, at each step i (i = 1, 2, 3, …, n) the dielectric is subjected to a TDDB stress with stress field at E_i and time $\Delta\tau$, where

$$E_i = i \cdot \Delta E \quad (i = 1,2,3,\ldots,n) \qquad (12.31)$$

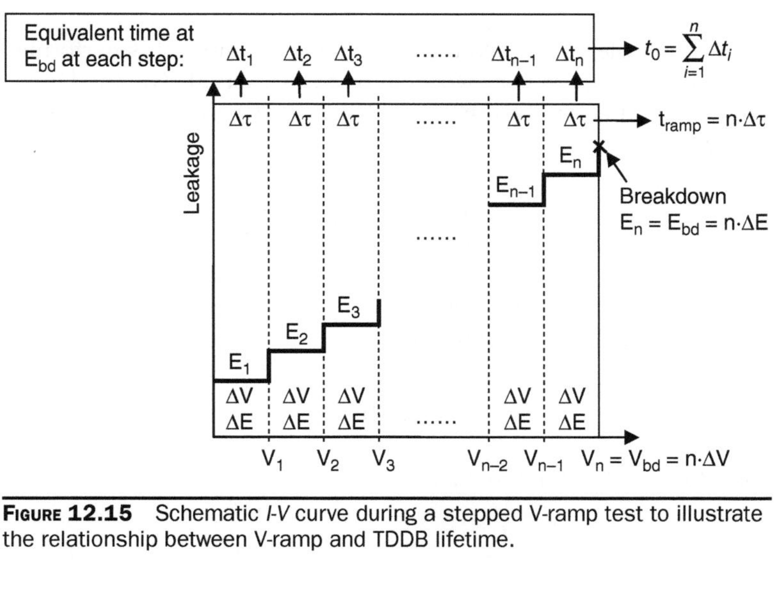

FIGURE 12.15 Schematic *I-V* curve during a stepped V-ramp test to illustrate the relationship between V-ramp and TDDB lifetime.

The time $\Delta\tau$ at step i^{th} can be converted to Δt_i at the breakdown field E_{bd} according to Eq. (12.9), as follows:

$$\Delta t_i = \Delta\tau \cdot e^{\gamma\cdot(E_i - E_{bd})} = \Delta\tau \cdot e^{\gamma\cdot(i\cdot\Delta E - E_{bd})} \qquad (12.32)$$

In other words, the total effective time t_0 spent at the breakdown field E_{bd} can be considered the summation of the preceding equation for each time step as follows:

$$t_0 = \sum_{i=1}^{n} \Delta t_i = \Delta\tau \cdot \sum_{n=0}^{n} e^{\gamma\cdot(i\cdot\Delta E - E_{bd})} = \frac{\Delta\tau}{1 - e^{-\gamma\cdot\Delta E}} = \frac{\Delta\tau}{1 - e^{-\frac{\gamma}{s}\cdot\Delta V}} \qquad (12.33)$$

Note that this relationship is also applicable to gate oxides, where s is the gate-oxide thickness. Using Eq. (12.9), the device under test's (DUT's) lifetime at any other electric field E from its breakdown electric field E_{bd} value can be obtained from the V-ramp test by the following relationship:

$$T_{bd}(E) = t_0 \cdot e^{\gamma\cdot(E_{bd} - E)} \qquad (12.34)$$

or expressed in terms of voltage V and breakdown voltage V_{bd} as

$$T_{bd}(V) = t_0 \cdot e^{\frac{\gamma}{s}\cdot(V_{bd} - V)} \qquad (12.35)$$

Let's then consider the case for a constant ramp rate RR, and Eq. (12.31) turns to be

$$E_i = RR \cdot t = \frac{\Delta E}{\Delta\tau} \cdot t \qquad (12.36)$$

As an example, $RR = 1$ volt/$\Delta\tau$ when the V-ramp test is performed with each 1-V increment within a $\Delta\tau$ time interval. The total effective time t_0 spent at the breakdown field E_{bd} [Eq. (12.33)] then can be expressed as:

$$t_0 = \int_0^{E_{bd}/RR} e^{\gamma(E_i - E_{bd})}\,dt = \int_0^{E_{bd}/RR} e^{\gamma(RR\cdot t - E_{bd})}\,dt = \frac{1 - e^{-\gamma\cdot E_{bd}}}{\gamma\cdot RR} = \frac{1 - e^{-\frac{\gamma}{s}\cdot V_{bd}}}{\gamma\cdot RR} \qquad (12.37)$$

If we assume that $\gamma = 4.1$ cm/MV[62] and $E_{bd} = 6$ MV/cm, we get $e^{-\gamma\cdot E_{bd}} \ll 1$. Then Eq. (12.37) can be simplified as:

$$t_0 = \frac{1}{\gamma\cdot RR} \qquad (12.38)$$

Inputting Eq. (12.38) into Eq. (12.35), we get

$$T_{bd} = \frac{1}{\gamma\cdot RR} \cdot e^{\gamma(E_{bd} - E)} = \frac{1}{\gamma\cdot RR} \cdot e^{\frac{\gamma}{s}(V_{bd} - V)} \qquad (12.39)$$

or in the other way,

$$V_{bd} = \frac{s}{\gamma}\ln(\gamma \cdot RR \cdot T_{bd} + V) \tag{12.40}$$

Equations (12.35) and (12.39) give the relationship between the time-to-dielectric-breakdown T_{bd} and the voltage applied V and V_{bd}, indicating that the T_{bd} exponentially depends on the V_{bd} and the voltage applied V. Equation (12.40) shows that the V_{bd} will be larger if the ramp rate RR is higher, which was already demonstrated experimentally.[100] Equations (12.40) and (12.39) require knowledge of the correct voltage acceleration, which is usually extracted from the intrinsic part of the TDDB TTF distribution.

During TDDB testing, we generally apply constant voltage at a higher stress voltage V_{str}, and then extrapolate the lifetime to operating conditions V_{op}. Based on Eq. (12.39), we get

$$T_{bd\_str} = \frac{1}{\gamma \cdot RR} \cdot e^{\frac{\gamma}{s}(V_{bd} - V_{str})} \tag{12.41}$$

$$T_{bd\_op} = \frac{1}{\gamma \cdot RR} \cdot e^{\frac{\gamma}{s}(V_{bd} - V_{op})} \tag{12.42}$$

Dividing Eq. (12.41) by Eq. (12.42), we get Eq. (12.43), which represents the E-model:

$$\frac{T_{bd\_op}}{T_{bd\_str}} = e^{\frac{\gamma}{s}(V_{op} - V_{str})} \tag{12.43}$$

This discussion is based on the E model. When a power-law model is used, the DUT's lifetime at any other voltage V can be extracted from its V_{bd} value obtained from the V-ramp test by using the following equation[98]:

$$T_{bd}(V) = \frac{V}{RR \cdot (n+1)}\left(\frac{V_{bd}}{V}\right)^{n+1} \tag{12.44}$$

with the ramp starting at V_{start}:

$$T_{bd}(V) = \frac{V^{-n}}{RR \cdot (n+1)}\left(V_{bd}^{n+1} - V_{start}^{n+1}\right) \tag{12.45}$$

It should be noted that calculations based on the preceding equations require the acceleration parameter (γ for the E model and n for the power-law model) to be known from previous long-term experiments. However, if it is not available, the parameter can be estimated using dual or multiple ramp rates.[100]

12.7 Time-Dependent Dielectric Breakdown Characteristics for Stacked-Via-Involved Comb Structures

BEOL stacked-via-involved comb structures (see Fig. 12.3) are attractive capacitors for radiofrequency and mixed-signal applications owing to their low cost, high area capacitance, excellent Q-factor, and linearity.[31] The high density that can be achieved by a stacked-via-involved comb structure makes them very much competitive with metal-insulator-metal (MIM) capacitors that usually require extra masks. However, owing to the special via-comb-involved configuration, an unoptimized process generally means that the distance between two adjacent vias or between a via and the adjacent metal is likely less than the distance between the layers (see Fig. 12.3). Therefore, long-term reliability becomes a real challenge. Among the reliability issues, TDDB and its defect density-related aspects are most concerned. These problems are further exacerbated by the integration of low-*k* or ultralow-*k* dielectrics and the fact that such dense structures may contain up to 100 million stacked vias and tens of meter metal lines.

A Weibull statistics model is also applicable to stacked-via-involved comb structure TDDB qualification to predict reliability lifetime. Poisson area scaling transformation also can be used to scale the failure probability from test structures to product-size stacked-via-involved combs. Table 12.2 provides a comparison between the metal-comb (no via involved) and stacked-via-involved comb structures. As a general observation, TDDB failure distributions obtained on complex via/line structures are found to be different from those of pure line patterns with the same layout metal width and line space. Typically, shorter failure times, larger voltage acceleration factors (γ in $1/V$), and smaller shape factors (β) below 1.0 are obtained.

Structure	Metal-Comb (No Via Involved) (Fig. 12.2)	Stacked-Via-Involved Comb (Fig. 12.3)
TTF	TTF better than stacked-via-involved comb	TTF shorter than metal-comb with the same layout metal width and line space
Shape factors (β)	>1	<1
Voltage acceleration factors (γ)	4.1 cm/MV[62]	5.5 cm/MV[54]

Table 12.2 A Comparison Between Metal-Comb (No Via Involved) and Stacked-Via-Involved Comb Structures.

In contrast, the shape factor (β) obtained for a metal-comb structure is >1.3, whereas it is about 0.7 for stacked-via-involved comb structure.[56] This is so because vias with oversized top diameters reduce the distance between the adjacent metal lines. Since the failure mechanism for metal lines both with and without vias remains the same (namely, breakdown at the low-*k*–cap interface), the smaller physical spacing at the via causes higher electric fields and hence shorter failure times. The *E*-field acceleration factor γ of a 65-nm SiCOH VNCap was determined to be about 5.5 cm/MV at 125°C. The temperature acceleration factor E_a based on the Arrhenius relation was determined to be about 0.35 eV at nominal 4.5 MV/cm stress field.[54]

12.8 Finite-Element Modeling for Dielectric Reliability Assessment

12.8.1 Finite-Element Modeling for Electric Field Simulation

Electric field simulation through the finite-element method (FEM) can provide insight for understanding dielectric breakdown.[101] As described in Sec. 12.5.3, LWR/LER is commonly observed, which worsens the reliability performance. Using FEM, it is found that the local enhanced electric field near the local extrusion can be 2.4 × that away from the extrusion.[70] A simple calculation is given below to illustrate the impact of this amount of electric field increase. For a 3.3-V application with 0.1-μm spacing, the local electric field E_0 can reach ~0.33 MV/cm, which is a significantly high electric field when one considers that the breakdown strength of some of the low-*k* dielectrics can be relatively low (<2 MV/cm). In the widely accepted *E* model for TDDB [Eq. (12.9)], a typical γ is 4.1 cm/MV for low-*k* dielectrics. Therefore, the TTF under the 2.4× electric field can be estimated as follows:

$$\frac{\mathrm{TTF}_{E,\max}}{\mathrm{TTF}_{E0}} = \exp[-\gamma(E_{\max} - E_0)] \qquad (12.46)$$

where the nominal electric field E_0 is assumed to be 0.33 MV/cm, the $\mathrm{TTF}_{E,\max}$ will be only 0.15 of TTF_{E0}. In other words, a 2.4× electric field will lead to ~85 percent time-to-failure reduction.

In a dual-damascene structure, misalignment sometimes is observed because the via and trench patterns are manufactured separately. In addition, surrounding corner structure is a common situation in a circuit layout where a metal-line terminal is surrounded by another metal line. For the low-*k* TDDB qualifications in advanced-technology nodes, the most challenging structures usually are the via-related structures such as via chains or via comb structures, as shown in

Fig. 12.3. A via bump with a large curvature shows much higher local electric field enhancement.[68]

12.8.2 Finite-Element Model for *k*-Value Extraction of Low-*k* Material

The effective dielectric constant K_{eff} is a concept to characterize the integrated working permittivity of a structure consisting of various dielectrics with different dielectric constants. K_{eff} of a specific structure depends on the ratio and number of different dielectric parts. Given the the complex structure of dual-damascene interconnections, this value is not practically measurable; it can, however, be determined using FEM simulation.

The *k*-value of a low-*k* material from the comb interline structure shown in Fig. 12.16 was extracted by using the commercially available FEM software (e.g., ANSYS[102]). In order to extract *k* from this kind of complex structure, the FEM simulation is incorporated into the analysis that involves the capacitance measurement and characterization of the actual dimensions of the dielectric spacing/dimensions from cross-sectional images [using focused ion beam (FIB), for example]. The procedure is shown in Fig. 12.17. In the FEM, an initial *k*-value of the low-*k* material under study is assumed as input for the model, which will, in turn, give a simulated capacitance of the interline structure. The simulated capacitance then will be compared with the measured value. If both capacitances fit well with each other, the initially assumed *k*-value will be considered as the real *k*-value for the low-*k* materials under study. On the other hand, if the simulated capacitance does not fit the measured one, a new *k*-value will be assumed until a good fit is achieved. In general, several iterations are necessary.

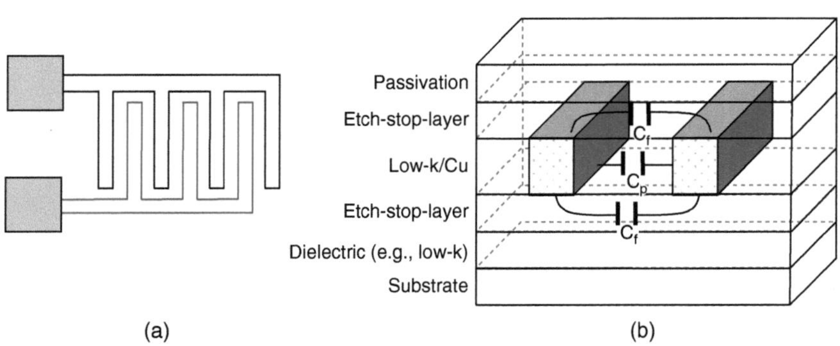

FIGURE 12.16 (*a*) Schematic top view of the comb structure (interline capacitor); (*b*) schematic cross-sectional view of the capacitance components involved in two metal fingers. The simplified capacitance components C_p and C_f refer to the parallel capacitance, and fringing capacitance respectively. (*Used with permission from Springer Science and Business Media.*[103])

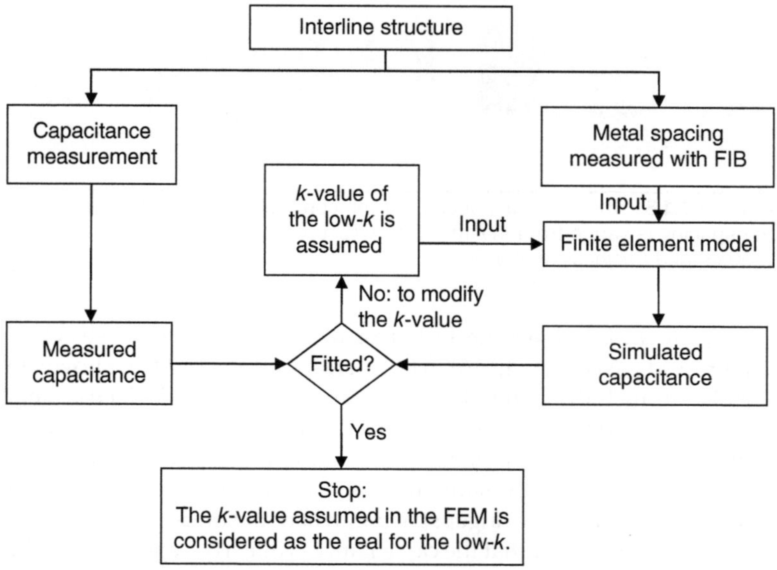

Figure 12.17 *k*-Value extraction methodology from interline capacitance structures based on physical and electric measurement and FEM simulation. (*Used with permission from Springer Science and Business Media.*[103])

ANSYS develops a special command macro (CMATRIX)[104] to extract self and mutual-capacitance terms for multiple-conductor systems. The element type available for 2D modelling of the structure is an eight-node electrostatic solid element PLANE121.[105]

12.8.3 Finite-Element Model for *k*-Drift of Low-*k* Dielectric Materials

Some low-*k* dielectric materials can degrade with time owing to chemical changes, moisture adsorption, vacuuming, and other reasons,[106,107] and the rate of degradation in *k* can be accelerated by high temperatures. As a result, low-*k* dielectric materials show a drift in the measured dielectric constant as a function of time and temperature. This *k*-drift can affect the speed of an IC, causing the device to fail at high speed some time in the future. This section provides a method to evaluate the *k*-drift in low-*k* dielectrics by combining the capacitance measurement and FEM. After *k*-drift is extracted as a function of time and temperature, it can be applied to the Simulation Program with Intagrated Circuit Emphasis (SPICE) model to evaluate the impact of *k*-drift of the low-*k* dielectrics on reliability and device performance.

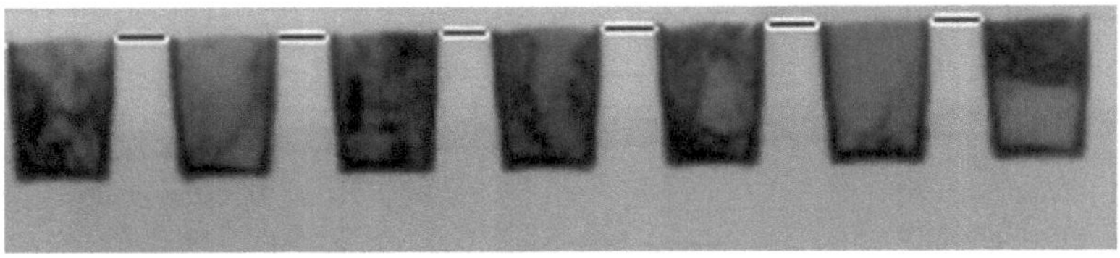

FIGURE 12.18 The FIB cross section of the comb structure, indicating that the spacing is varied from 4.7 to 5.2 nm. (*Used with permission from Springer Science* and *Business Media.*[103])

Figure 12.18 shows the FIB cross section of the comb structure under study (see Fig. 12.16). The corresponding FEM is shown in Fig. 12.19, including the 2D finite-element mesh and electric field vector obtained from ANSYS. The model has been verified by a case with a rectangular metal line over a conducting ground plane.[103]

Figure 12.20 gives the simulated capacitance in terms of *k* values of low-*k* materials and the finger spacing, where the finger width is fixed to be 120 nm. It is clear that the capacitance is linearly dependent on the *k* values of the low-*k* materials. This fitting methodology is similar to the previous report by Choudhury and Sangiovanni-Vincentelli.[108]

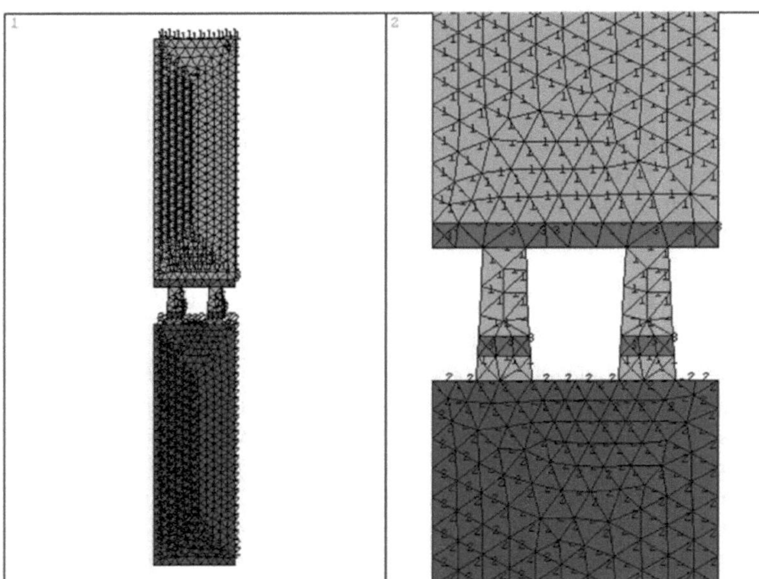

(a)

FIGURE 12.19 (*a*) A 2D finite-element mesh and (*b*) the electric field vector of the comb structure according to the dimensions from Fig. 12.18, including two electrodes only. (*Used with permission from Springer Science and Business Media.*[103])

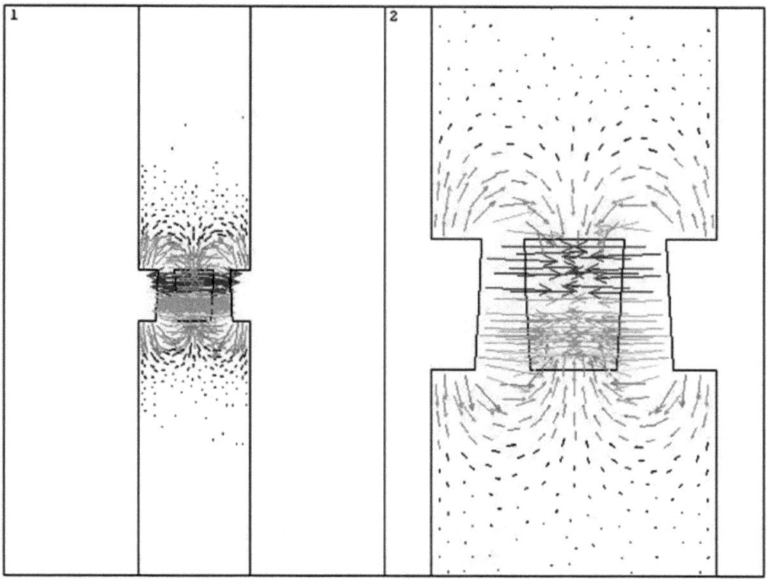

(b)

Figure 12.19 (*Continued*).

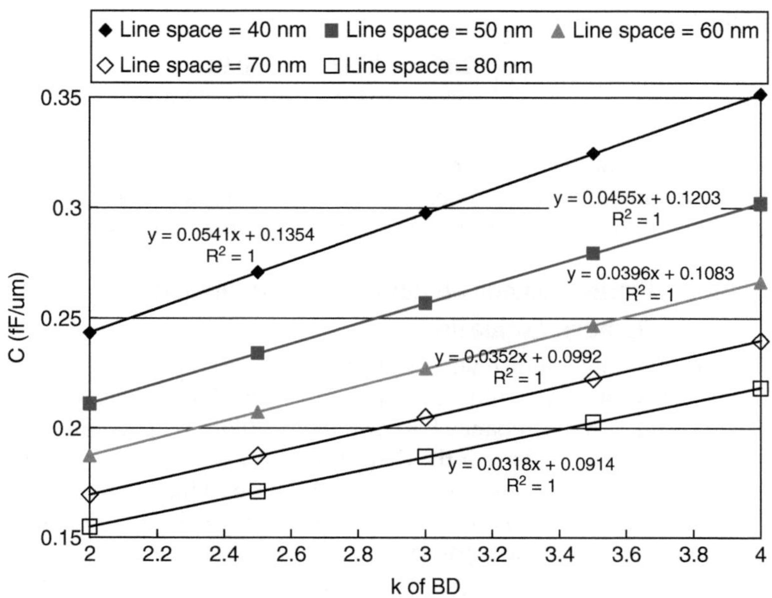

Figure 12.20 The calculated capacitance in terms of *k* of low-*k* material and finger spacing, where the finger width is 120 nm. The symbols represent the calculated capacitance, and the solid lines are the corresponding linear fit, which shows perfect linear fit with $R^2 = 1$. (*Used with permission from Springer Science and Business Media.*[103])

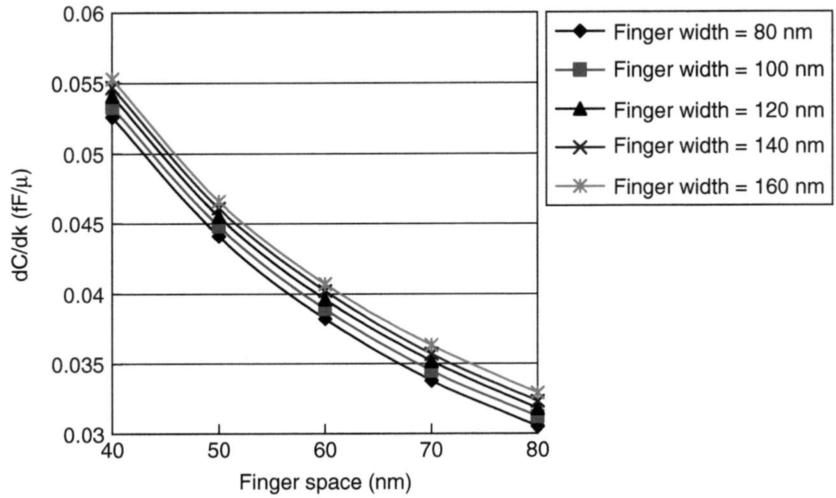

FIGURE 12.21 The gradient dC/dk is plotted in terms of finger space and finger width. (*Used with permission from Springer Science and Business Media.*[103])

Furthermore, the gradient dC/dk is plotted in terms of finger space and finger width in Fig. 12.21, and they fit nicely to a function given in Eq. (12.47):

$$\frac{dC}{dK} = 0.010447 + \frac{2.19}{\text{space}} - \frac{8.293}{\text{space}^2} - \frac{0.916}{\text{space} + \text{width}} \tag{12.47}$$

where space and width are in nanometers; dC/dK are in fF/μm. Hence, by measuring ΔC owing to degradation, it is straightforward to compare the k-drift (Δk) given that the space and width are known.

12.8.4 Finite-Element Model for Process-Induced Damage Evaluation

The interline capacitance will be changed owing to process-induced damage, such as dry etching, cleaning, film deposition, and CMP.[109,110] There could be a discrepancy between the simulated capacitance and the measured capacitance if the dielectric constant of the low-k material is assumed to be constant in the simulation, which is not necessarily true when there is process-induced damage. Study of Chikaki et al.[109] indicated that the capacitance discrepancy (i.e., process-induced damage from the process point of view) becomes more aggressive when low-k material is integrated, and the metal spacing is reduced to <800 nm. The effective dielectric constant can be considered as being modified when the spacing is narrower. The dependence of the dielectric constant of low-k film on Cu damascene spacing could be estimated by fitting the simulated result to the measured capacitance. The extracted dielectric constant of the low-k film increased from 2.1 to 4.5 for spacing from 800 to 140 nm.

Three possible damage modes, namely, interface degradation, void-like degradation, and sidewall degradation, were proposed by Chikaki et al.,[109] who studied the impacts of these degradations on capacitance changes using finite-element simulation. Their results showed that when an interfacial damaged layer has a thickness of 10 nm and a k value of 34.5 is added to the FEM, the simulated capacitance can fit very well with measured capacitance. It is supposed that this interfacial damage layer may be formed owing to wet chemicals, such as CMP, Cu electrochemical plating, and/or wet cleaning.

12.9 Process-Qualification Practice

The intermetal dielectric (IMD) TDDB is a required process-qualification item, especially for the case when the IMD is not SiO_2. The V-ramp test for breakdown-voltage characteristics serves as part of an engineering study. The dielectric between metal lines of one metal layer (either metal-comb or metal-serpentine-metal structures with spacing following the minimum design rule) (see Fig. 12.2) should be stressed. The IMD reliability between the metal layers connected with vias (see Fig. 12.3) also should be tested. Testing should be performed on all IMD layers, especially when the metal-layer scheme changes.[111] Long structures (e.g., 10 m long) with large sample size per wafer as a screen for quantitative defect density evaluation are necessary because short structures (i.e., 100 μm long) may not be effective in detecting the metallic contamination at the low-k surface.[77] For IMD reliability qualification, Joint Electron Device Engineering Council (JEDEC) JP-001[111] provides good guidance. Table 12.3 provides some guideline or practice for IMD reliability (V-ramp and TDDB) qualification.

Failure criteria	• V-ramp test: Failure (breakdown) is defined by the leakage current exceeding a specified value or a 10× (or other predefined number) leakage increase compared with the previous voltage step. • TDDB constant-voltage stress: Failure (breakdown) is defined by the leakage current exceeding a specified value at the stress voltage/temperature or 10× (or other predefined number) leakage increase compared with the previous time step.

TABLE 12.3 Typical Process-Qualification Practices for IMD Reliability (V-ramp and TDDB)

Test structures	1. Three different metal lengths should be tested for length-scaling purposes. 2. Three types of test structures are recommended for testing. The spacing between the fingers is defined as the minimum rule in the design rules: • A metal serpentine with two metal combs inserted from each side to cover both edges of the serpentine • Two metal combs inserted into each other (no serpentine) • Metal-via structures with two layers of metal serpentines or metal combs connected by vias
Sample size	• V-ramp test: Wafer mapping * 2 wafers * 3 lots • TDDB constant-voltage stress: Three stress voltages * 25 sites * 3 lots for field acceleration factor and three metal lengths * 25 sites * 1 lot for area/length scaling for all various combinations of IMD layers. The measured devices should be uniformly allocated across the wafer.
Stress conditions	Stress to be done at +125°C. Various temperatures are needed to get activation energy. 1. V-ramp test: Ramped voltage stress to failure 2. TDDB constant-voltage stress: Constant voltage stress at 2 * $V_{dd,max}$ or greater. Stress should continue until 50% or more of the samples exceed the failure criterion. Three stress voltages to be applied for voltage acceleration.
	Wafer level or package level
Acceptance criteria	TTF (0.1%) > 10 years @ 125°C/100 m-length/V_{op}
Model	1. V-ramp test: No model 2. TDDB constant-voltage stress: *E* model [Eq. (12.9)] or SQRT(*E*) model [Eq. (12.10)]
Distribution	Weibull distribution for both V-ramp V_{bd} and TDDB TTF.
Parameters	1. V-ramp test: • Weibull distribution of the voltage to failure V_{bd} • Record the complete extrinsic and intrinsic branches of the distribution 2. TDDB constant-voltage stress: • Weibull distribution of the time-to-failure (TTF) • Voltage/field acceleration factor γ, area/length scaling factor β for Poisson model, and activation energy (E_a)

Table 12.3 Typical Process-Qualification Practices for IMD Reliability (V-ramp and TDDB) (*Continued*)

12.10 References

1. G. Maier, "Low dielectric constant polymers for microelectronics," *Progress in Polymer Science*, Vol. 26, 2001, pp. 3–65.
2. M. T. Bohr, "Interconnect scaling—the real limiter to high performance ULSI," in *Proceedings of the International Electron Devices Meeting (IEDM)*, 1995, pp. 241–244.
3. D. Edelstein, H. Rathore, C. Davis, L. Clevenger, A. Cowley, T. Nogami, B. Agarwala, S. Arai, A. Carbone, K. Chanda, F. Chen, S. Cohen, W. Cote, M. Cullinan, T. Dalton, S. Das, P. Davis, J. Demarest, D. Dunn, C. Dziobkowski, R. Filippi, J. Fitzsimmons, P. Flaitz, S. Gates, J. Gill, A. Grill, D. Hawken, K. Ida, D. Klaus, N. Klymko, M. Lane, S. Lane, J. Lee, W. Landers, W. -K. Li, Y. H. Lin, E. Liniger, X. -H. Liu, A. Madan, S. Malhotra, J. Martin, S. Molis, C. Muzzy, D. Nguyen, S. Nguyen, M. Ono, C. Parks, D. Questad, D. Restaino, A. Sakamoto, T. Shaw, Y. Shimooka, A. Simon, E. Simonyi, A. Swift, T. Van Kleeck, S. Vogt, Y. -Y. Wang, W. Wille, J. Wright, C. -C. Yang, M. Yoon, and T. Ivers, "Comprehensive reliability evaluation of a 90 nm CMOS technology with Cu/PECVD low-*k* BEOL," in *Proceedings of the 42th IEEE Annual International Reliability Physics Symposium*, 2004, pp. 316–319.
4. S. M. Sze, *Physics of Semiconductor Device*. New York: John Wiley, 1986.
5. J. Gambino, A. Stamper, T. McDevitt, V. McGahay, S. Luce, T. Pricer, B. Porth, C. Senowitz, R. Kontra, M. Gibson, H. Wildman, A. Piper, C. Benson, T. Standaert, P. Biolsi, E. Cooney, E. Webster, R. Wistrom, A. Winslow, and E. White, "Integration of copper with low-*k* dielectrics for 0. 13-μm technology," in *Proceedings of the 9th International Symposium on the Physical and Failure Analysis of Integrated Circuits*, 2002, pp. 111–117.
6. J. Gambino, F. Chen, and J. He, "Copper interconnect technology for the 32 nm node and beyond," in *Proceedings of the IEEE 2009 Custom Intergrated Circuits Conference (CICC)*, 2009, pp. 141–148.
7. E. T. Ogawa, J. Kim, G. S. Haase, H. C. Mogul, and J. W. McPherson, "Leakage, breakdown, and TDDB characteristics of porous low-k silica-based interconnect dielectrics," in *Proceedings of the 41st IEEE Annual International Reliability Physics Symposium*, 2003, pp. 166–172.
8. http://en.wikipedia.org/wiki/Low-*k*, Access date: December 2010.
9. M. J. Shapiro, S. V. Nguyen, T. Matsuda, and D. Dobuzinsky, "CVD of fluorosilicate glass for ULSI applications," *Thin Solid Films*, Vol. 270, 1995, pp. 503–507.
10. A. Grill, "Plasma-enhanced chemical vapor deposited SiCOH dielectrics: from low-k to extreme low-*k* interconnect materials," *Journal of Applied Physics*, Vol. 93, No. 3, 2003, pp. 1785–1790.
11. http://www.appliedmaterials.com/products/producer_black_diamond_4.html, Access date: December 2010.
12. http://www.semiconductor.net/article/199167–Low_k_Bursts_Into_the_Mainstream_Incrementally.php, Access date: December 2010.
13. http://www.dow.com/silk, Access date: December 2010.
14. http://www.dow.com/silk/silky/feature.htm, Access date: December 2010.
15. N. Ohashi, K. Misawa, S. Sone, H. J. Shin, K. Inukai, E. Soda, S. Kondo, A. Furuya, H. Okamura, S. Ogawa and N. Kobayashi, "Robust porous MSQ (k=2. 3, E=12 GPa) for low-temperature (<350°C) Cu/low-*k* integration using ArF resist mask process," in *Proceedings of the International Electron Devices Meeting (IEDM)*, 2003, pp. 857–860.
16. Z. C. Wu, T. J. Chou, S. H. Lin, Y. L. Huang, C. H. Lin, L. P. Li, B. T. Chen, Y. C. Lu, C. C. Chiang, M. C. Chen, W. Chang, S. M. Jang, and M. S. Liang, "High performance 90/65-nm BEOL technology with CVD porous low-*k* dielectrics (k - 2. 5) and low-*k* etching stop (k - 3. 0)," in *Proceedings of the International Electron Devices Meeting (IEDM)*, 2003, pp. 849–852.
17. R. Tsu, J. W. McPherson, and W. R. McKee, "Leakage and breakdown reliability issues associated with low-k dielectrics in a dual-damascene Cu process," in *Proceedings of the 38th IEEE Annual International Reliability Physics Symposium*, 2000, pp. 348–353.

18. J. Noguchi, T. Saito, N. Ohashi, H. Ashihara, H. Maruyama, M. Kubo, H. Yamaguchi, D. Ryuzaki, K. Takeda, and K. Hinode, "Impact of low-*k* dielectrics and barrier metals on TDDB lifetime of Cu interconnects," in *Proceedings of the 39th IEEE Annual International Reliability Physics Symposium*, 2001, pp. 355–359.
19. W. S. Song, T. J. Kim, D. H. Lee, T. K. Kim, C. S. Lee, J. W. Kim, S. Y. Kim, D. K. Jeong, K. C. Park, Y. J. Wee, B. S. Suh, S. M. Choi, H. -K. Kang, K. P. Suh, and S. U. Kim, "Pseudobreakdown events induced by biased-thermal-stressing of intra-level Cu interconnects—reliability and performance impact," in *Proceedings of the 40th IEEE Annual International Reliability Physics Symposium*, 2002, pp. 305–311.
20. W. Wu, X. Duan, and J. S. Yuan, "A physical model of time dependent dielectric breakdown in Cu metallization," in *proceedings of the 41st IEEE Annual International Reliability Physics Symposium*, 2003, pp. 773–776.
21. G. B. Alers, K. Jow, R. Shaviv, G. Kooi, and G. W. Ray, "Interlevel dielectric failures in copper/ low-k structures," *IEEE Transactions on Device and Materials Reliability*, Vol. 4, No. 2, 2004, pp. 148–152.
22. M. Ohring, *Reliability and Failure of Electronic Materials and Devices*, New York: Academic Press, 1998.
23. K. Y. Yiang, T. S. Mok, W. J. Yoo, and A. Krishnamoorthy, "Reliability improvements using buried capping layer in advanced interconnects," in *Proceedings of the 42nd IEEE Annual International Reliability Physics Symposium*, 2004, pp. 333–337.
24. A. Aal, "Fast prediction of gate oxide reliability," in *IEEE International Integrated Reliability Workshop Final Report*, 2006, pp. 182–185.
25. I. Ciofi, Zs. Tőkei, D. Visalli, and M. Van Hove, "Water and copper contamination in SiOC:H damascene: novel characterization methodology based on triangular Voltage sweep measurements," in *Proceedings of the International Interconnects Technology Conference (IITC)*, 2006, pp. 181–183.
26. http://www.chironholdings.com/chirontechnology/references/App%20 Notes/TVS. pdf, Access date: December 2010.
27. H. Miyazaki, D. Kodama, and N. Suzumura, "The observation of stress-induced leakage current of damascene interconnects after bias temperature aging," in *Proceedings of the 46th IEEE Annual International Reliability Physics Symposium*, 2008, pp. 150–157.
28. J. W. Mcpherson and H. Mogul, "Underlying physics of the thermochemical E model in describing low-field time-dependent dielectric breakdown in SiO_2 thin films," *Journal of Applied Physics*, Vol. 84, No. 3, 1998, pp. 1513–1523.
29. G. B. Alers, K. Jow, R. Shaviv, G. Kooi, and G. W. Ray, "Interlevel dielectric failures in copper/low-*k* structures," *IEEE Transactions on Device and Materials Reliability*, Vol. 4, No. 2, 2004, pp. 148–152.
30. J. R. Lloyd, S. Ponoth, E. Liniger, and S. Cohen, "Role of Cu in TDDB of low-*k* dielectrics," in *Proceedings of the 45th IEEE Annual International Reliability Physics Symposium*, 2007, pp. 410–411.
31. F. Chen, K. Chanda, J. Gill, M. Angyal, J. Demarest, T. Sullivan, R. Kontra, M. Shinosky, J. Li, L. Economikos, M. Hoinkis, S. Lane, D. McHerron, M. Inohara, S. Boettcher, D. Dunn, M. Fukasawa, B. C. Zhang, K. Ida, T. Ema, G. Lembach, K. Kumar, Y. Lin, H. Maynard, K. Urata, T. Bolom, K. Inoue, J. Smith, Y. Ishikawa, M. Naujok, P. Ong, A. Sakamoto, D. Hunt, and J. Aitken, "Investigation of CVD SiCOH low-*k* time-dependent dielectric breakdown at 65-nm node technology," in *Proceedings of the 43rd IEEE Annual International Reliability Physics Symposium*, 2005, pp. 501–507.
32. F. Chen, B. Li, T. Lee, C. Christiansen, J. Gill, M. Angyal, M. Shinosky, C. Burke, W. Hasting, R. Austin,T. Sullivan, D. Badami, and J. Aitken, "Technology reliability qualification of a 65-nm CMOS Cu/low-*k* BEOL interconnect," in *Proceedings of the 13th International Symposium on the Physical and Failure Analysis of Integrated Circuits*, 2006, pp. 97–105.
33. J. P. Gambino, T. C. Lee, F. Chen, and T. D. Sullivan, "Reliability challenges for advanced copper interconnects: electromigration and time-dependent dielectric breakdown (TDDB)," in *Proceedings of the 16th International Symposium on the Physical and Failure Analysis of Integrated Circuits*, 2009, pp. 677–684.

34. O. Aubel, M. Kiene, and W. Yao, "New approach of 90-nm low-*k* interconnect evaluation using a Voltage ramp dielectric breakdown (VRDB) test," in *Proceedings of the 43rd IEEE Annual International Reliability Physics Symposium*, 2005, pp. 483–489.
35. Z. H. Gan, Y. J. Wu , K. Zheng, R. Guo, and C. C. Liao, "Fast method to identify the root cause for ILD V_{bd} fail," in *Proceedings of the 9th International Conference on Solid-State and Integrated-Circuit Technology Proceedings (ICSICT)*, 2008, pp. 254–257.
36. J. C. Lee, I. C. Chen, and C. M. Hu, "Modeling and characterization of oxide reliability," *IEEE Transactions on Electron Devices*, Vol. 35, No. 12, 1988, pp. 2268–2278.
37. S. -Y. Jung, B. -J. Kim, N. Y. Lee, B. -M. Kim, S. J. Yeom, N. J. Kwak, and Y. -C. Joo, "The characteristics of Cu-drift induced dielectric breakdown under alternating polarity bias temperature stress," in *Proceedings of the 47th IEEE Annual International Reliability Physics Symposium*, 2009, pp. 825–827.
38. M. W. Lane, E. G. Liniger, and J. R. Lloyd, "Relationship between interfacial adhesion and electromigration in Cu metallization," *Journal of Applied Physics*, Vol. 93, No. 3, 2003, pp. 1417–1421.
39. G. S. Haase and J. W. McPherson, "Modeling of interconnect dielectric lifetime under stress conditions and new extrapolation methodologies for time dependent dielectric breakdown," in *Proceedings of the 45th IEEE Annual International Reliability Physics Symposium*, 2007, pp. 390–398.
40. JESD 37, *Standard for Lognormal Analysis of Uncensored Data and of Singly Right-Censored Data Utilizing the Persson and Rootzen Method*, JEDEC Standard, October 1992.
41. F. Chen, P. McLaughlin, J. Gambino, E. Wu, J. Demarest, D. Meatyard, and M. Shinosky, "The effect of metal area and line spacing on TDDB characteristics of 45-nm low-*k* SiCOH dielectrics," in *Proceedings of the 45th IEEE Annual International Reliability Physics Symposium*, 2007, pp. 382–389.
42. J. Kim, E. T. Ogawa, and J. W. McPherson, "Time dependent dielectric breakdown characteristics of low-*k* dielectric (SiOC) over a wide range of test areas and electric fields," in *Proceedings of the 45th IEEE Annual International Reliability Physics Symposium*, 2007, pp. 399–404.
43. I. C. Chen, S. Holland, and C. Hu, "A quantitave physical modelfor time-dependent breakdown in SiO_2," in *Proceedings of the 23rd IEEE Annual International Reliability Physics Symposium*, 1985, pp. 24–31.
44. J. McPherson, V. Reddy, K. Banejee, and H. Le, "Comparison of E and I/E TDDB models for SiO_2 under long-term/low-field test conditions," in *Proceedings of the International Electron Devices Meeting (IEDM)*, 1998, pp. 171–174.
45. A. Berman, "Time zero dielectric reliability test by a ramp method," in *Proceedings of the 19th IEEE Annual International Reliability Physics Symposium*, 1981, pp. 204–209.
46. J. W. McPherson and D. A. Baglee, "Acceleration factors for thin gate oxide stressing," in *Proceedings of the 23rd IEEE Annual International Reliability Physics Symposium*, 1985, pp. 1–5.
47. J. R. Lloyd, "New models for interconnect failure in advanced IC technology," in *Proceedings of the 15th International Symposium on the Physical and Failure Analysis of Integrated Circuits*, 2008, pp. 1–7.
48. F. Chen, O. Bravo, K. Chanda, P. McLaughlin, T. Sullivan, J. Gill, J. R. Lloyd, E. Wu, R. Kontra, and J. Aitken, "A comprehensive study of low-*k* SiCOH TDDB phenomena and its reliability lifetime model development," in *Proceedings of the 44th IEEE Annual International Reliability Physics Symposium*, 2006, pp. 46–53.
49. N. Suzumara, S. Yamamoto, D. Kodama, K. Makabe, J. Komari, E. Murukami, S. Maegawa, and K. Kubota, "A new TDDB degradation model based on Cu ion drift in Cu interconnect dielectrics," in *Proceedings of the 44th IEEE Annual International Reliability Physics Symposium*, 2006, pp. 484–489.
50. K. -Y. Yiang, H. W. Yao, A. Marathe, and O. Aubel, "New perspectives of dielectric breakdown in low-k interconnects," in *Proceedings of the 47th IEEE Annual International Reliability Physics Symposium*, 2009, pp. 476–480.

51. N. Suzumura, S. Yamamoto, D. Kodama, H. Miyazaki, M. Ogasawara, J. Komori, and E. Murakami, "Electric-field and temperature dependencies of TDDB degradation in Cu/low-*k* damascene structures," in *Proceedings of the 46th IEEE Annual International Reliability Physics Symposium*, 2008, pp. 138–143.
52. K. -Y. Yiang, H. W. Yao, and A. Marathe, "TDDB kinetics and their relationship with the *E*- and SQRT(*E*)-models," in *Proceedings of the International Interconnect Technology Conference (IITC)*, 2008, pp. 168–170.
53. Y. -L. Lia, Zs. Tökei, Ph. Roussel, G. Groeseneken, and K. Maexa, "Layout dependency induced deviation from Poisson area scaling in BEOL dielectric reliability," *Microelectronics Reliability*, Vol. 45, No. 9–11, 2005, pp. 1299–1304.
54. F. Chen, F. Ungar, A. H. Fischer, J. Gill, A. Chinthakindi, T. Goebel, M. Shinosky, D. Coolbaugh, V. Ramachandran, Y. K. Siew, E. Kaltalioglu, S. O. Kim, and K. Park, "Reliability characterization of BEOL vertical natural capacitor using copper and low-*k* SiCOH dielectric for 65-nm RF and mixed-signal applications," in *Proceedings of the 44th IEEE Annual International Reliability Physics Symposium*, 2006, pp. 490–495.
55. JESD35-A, *Procedure for the Wafer-Level Testing of Thin Dielectric*, Electronics Industries Association, Washington D. C., 2001.
56. A. H. Fischer, Y. K. Lim, Ph. Riess, Th. Pompl, B. C. Zhang, E. C. Chua, W. W. Keller, J. B. Tan, V. Klee, Y. C. Tan, D. Souche, D. K. Sohn, and A. von Glasow, "TDDB robustness of highly dense 65-nm BEOL vertical natural capacitor with competitive area capacitance for RF and mixed-signal applications," in *Proceedings of the 46th IEEE Annual International Reliability Physics Symposium*, 2008, pp. 126–131.
57. J. Noguchi, N. Ohashi, J. Yasuda, T. Jimbo, H. Yamaguchi, N. Owada, K. Takeda, and K. Hinode, "TDDB improvement in Cu metallization under bias stress," in *Proceedings of the 38th IEEE Annual International Reliability Physics Symposium*, 2000, pp. 339–343.
58. R. Gonella, P. Motte, and J. Torres, "Assessment of copper contamination impact on inter-level dielectric reliability performed with time-dependent-dielectric-breakdown tests," *Microelectronics Reliability*, Vol. 40, 2000, pp. 1305–1309.
59. S. -S. Hwang, H. -C. Lee, H. W. Ro, D. Y. Yoon, and Y. -C. Joo, "Porosity content dependence of TDDB lifetime and flat band Voltage shift by Cu diffusion in porous spin-on low-*k*," in *Proceedings of the 43rd IEEE Annual International Reliability Physics Symposium*, 2005, pp. 474–477.
60 J. R. Lloyd, T. M. Shaw, and E. G. Liniger, "Effect of moisture on the time dependent dielectric breakdown (TDDB) behavior in an ultralow-*k* (ULK) dielectric," in *IEEE International Integrated Reliability Workshop Final Report*, 2005, pp. 39–43.
61. Y. H. Wang and R. Kumar, "Stability of carbon-doped silicon oxide low-*k* thin films," *Journal of The Electrochemical Society*, Vol. 151, No. 4, 2004, pp. F73–76.
62. J. Kim, E. T. Ogawa, and J. W. McPherson, "A statistical evaluation of the field acceleration parameter observed during time dependent dielectric breakdown testing of silica-based low-*k* interconnect dielectrics," in *Proceedings of the 44th IEEE Annual International Reliability Physics Symposium*, 2006, pp. 478–483.
63. Y. Li, I. Ciofi, L. Carbonell, G. Groeseneken, K. Maex, and Z. Tökei, "Moisture-related low-*k* dielectric reliability before and after thermal annealing," in *Proceedings of the 45th IEEE Annual International Reliability Physics Symposium*, 2007, pp. 405–409.
64. F. Chen and M. Shinosky, "Addressing Cu/low-*k* dielectric TDDB-reliability challenges for advanced CMOS technologies," *IEEE Transactions on Electron Devices*, Vol. 56, No. 1, 2009, pp. 2–12.
65. J. Michelon and R. J. O. M. Hoofman, "Moisture influence on porous low-k reliability," *IEEE Transactions on Device and Materials Reliability*, Vol. 6, No. 2, 2006, pp. 169–174.

66. J. Proost, M. Baknalov, K. Maex, and L. Delaey, "Compensation effect during water desorption from siloxanebased spin-on dielectric thin films," *Journal of Vacuum Science and Technology B*, Vol. 18, No. 1, 2000, pp. 303–306.
67. Q. Wang, Z. H. Gan, L. Zhao, K. Zheng, E. Bei, and J. Ning, "Low-*k* breakdown improvement in 65-nm dual-damascene Cu process," in *Proceedings of the 9th International Conference on Solid-State and Integrated-Circuit Technology Proceedings (ICSICT)*, 2008, pp. 1324–1327.
68. F. Chen, J. R. Lloyd, K. Chanda, R. Achanta, O. Bravo, A. Strong, P. S. McLaughlin, M. Shinosky, S. Sankaran, E. Gebreselasie, A. K. Stamper, and Z. X. He, "Line edge roughness and spacing effect on low-*k* TDDB characteristics," in *Proceedings of the 46th IEEE Annual International Reliability Physics Symposium*, 2008, pp. 132–137.
69. J. R. Lloyd, X-.H Liu, G. Bonilla, T. M. Shaw, E. Liniger, and A. Lisi, "On the contribution of line-edge roughness to intralevel TDDB lifetime in low-*k* dielectrics," in *Proceedings of the 47th IEEE Annual International Reliability Physics Symposium*, 2009, pp. 602–605.
70. A. Yamaguchi, D. Ryuzaki, K. Takeda, and H. Kawada, "Evaluation of line-edge roughness in Cu/low-*k* interconnect patterns with CD-SEM," in *Proceedings of the IEEE International Interconnect Technology Conference (IITC)*, 2009, pp. 225–227.
71. M. Vilmay, D. Roy, C. Monget, F. Volpi, and J. M. Chaix, "Copper line topology impact on the SiOCH low-*k* reliability in sub-45-nm technology node: from the time-dependent dielectric breakdown to the product lifetime," in *Proceedings of the 47th IEEE Annual International Reliability Physics Symposium*, 2009, pp. 606–612.
72. Zs. Tokei, Ph. Roussel, M. Stucchi, J. Versluijs, I. Ciofi, L. Carbonell, G. P. Beyer, A. Cockburn, M. Agustin, and K. Shah, "Impact of LER on BEOL dielectric reliability: a quantitative model and experimental validation," in *Proceedings of the IEEE International Interconnect Technology Conference (IITC)*, 2009, pp. 228–230.
73. D. Oshida, T. Takewaki, M. Iguchi, T. Taiji, T. Morita, Y. Tsuchiya, H. Tsuchiya, S. Yokogawa, H. Kunishima, H. Aizawa, and N. Okada, "Quantitative analysis of correlation between insulator surface copper contamination and TDDB lifetime based on actual measurement," in *Proceedings of the International Interconnect Technology Conference (IITC)*, 2008, pp. 222–224.
74. M. Tada, Y. Harada, H. Ohtake, S. Saito, T. Onodera, and Y. Hayashi, "Improvement of TDDB reliability in Cu damascene interconnect by using united hard-mask and Cap (UHC) structure," in *Proceedings of the International Interconnect Technology Conference (IITC)*, 2003, pp. 256–258.
75. H. Park, H. B. Lee, H. K. Jung, Z. S. Choi, J. Y. Bae, J. W. Hong, K. I. Choi, B. L. Park, E. J. Lee, J. W. Kim, J. M. Lee, G. H. Choi, and J. T. Moon, "Voltage ramp and time-dependent dielectric breakdown in ultra-narrow Cu/SiO_2 interconnects," in *Proceedings of the International Interconnect Technology Conference (IITC)*, 2008, pp. 49–51.
76. J. Gambino, F. Chen, S. Mongeon, D. Meatyard, T. Lee, B. Lee, H. Bamnolker, L. Hall, N. Li, M. Hernandez, P. Little, M. Hamed, and I. Ivanov, "Effect of CoWP capping layers on dielectric breakdown of SiO_2," in *Proceedings of the 14th International Symposium on the Physical and Failure Analysis of Integrated Circuits*, 2007, pp. 59–64.
77. F. Chen, M. Shinosky, B. Li, C. Christiansen, T. Lee, J. Aitken, D. Badami, E. Huang, G. Bonilla, T. -M. Ko, T. Kane, Y. Wang, M. Zaitz, L. Nicholson, M. Angyal, C. Truong, X. Chen, G. Yang, S. B. Law, T. J. Tang, S. Petitdidier, G. Ribes, M. Oh, C. Child, H. Sawada, A. Kolics, O. Rigoutat, and N. Gilbert, "Comprehensive investigations of CoWP metal-cap impacts on low-*k* TDDB for 32nm technology application," in *Proceedings of the 48th IEEE Annual International Reliability Physics Symposium*, 2010, pp. 566–573.
78. F. Chen, E. Huang, M. Shinosky, M. Angyal, T. Kane, Y. Wang, and A. Kolics, "A comparative study of ULK conduction mechanisms and TDDB characteristics for Cu interconnects with and without CoWP metal cap at 32-nm technology," in *Proceedings of the International Interconnect Technology Conference (IITC)*, 2010, pp. 1–3.

79. K. Chattopadhyay, B. van Schravendijk, T. W. Mountsier, G. B. Alers, M. Hornbeck, H. J. Wu, R. Shaviv, G. Harm, D. Vitkavage, E. Apen, Y. Yu, and R. Havemann, "In-situ formation of a copper silicide cap for TDDB and electromigration improvement," in *Proceedings of the 44th IEEE Annual International Reliability Physics Symposium*, 2006, pp. 128–130.
80. R. Gonella, "Key reliability issues for copper integration in damascene architecture," *Microelectronic Engineering*, Vol. 55, 2001, pp. 245–255.
81. Z. Chen, K. Prasad, C. Y. Li, P. W. Lu, S. S. Su, and L. J. Tang, "Highly reliable dielectric/metal bilayer sidewall diffusion barrier in Cu/porous organic ultra low-k interconnects," in *Proceedings of the 42nd IEEE Annual International Reliability Physics Symposium*, 2004, pp. 320–325.
82. C. Guedj, V. Arnal, J. F. Guillaumond, L. Amaud, J. P. Barnes, A. Toffoli, V. Jousseaume, A. Roule, S. Maitrejean, L. L. Chapelon, G. Reimbold, J. Torres, and G. Passemard, "Influence of the diffusion barriers on the dielectric reliability of ULK/Cu advanced interconnects," in *Proceedings of the International Interconnect Technology Conference (IITC)*, 2005, pp. 57–59.
83. Z. Tokei, V. Sutcliffe, S. Demuynck, F. Iacopi, P. Roussel, G. P. Beyer, R. J. O. M. Hoofman, K. Maex, "Impact of the barrier/dielectric interface quality on reliability of Cu porous-low-*k* interconnects," in *Proceedings of the 42nd IEEE Annual International Reliability Physics Symposium*, 2004, pp. 326–332.
84. R. Caluwaerts, M. Van Hove, G. Beyer, R. J. O. M. Hoofman, H. Struyf, G. J. A. M. Verheyden, J. Waeterloos, Zs. Tokei, F. Iacopi, L. Carbonell, Q. T. Le, A. Das, I. Vos, S. Demuynck, and K. Maex, "Post patterning meso porosity creation: a potential solution for pore sealing," in *Proceedings of the International Interconnect Technology Conference (IITC)*, 2003, pp. 242–244.
85. D. D. Gandhi, P. G. Ganesan, V. Chandrasekar, Z. Gan, S. G. Mhaisalkar, H. Li, and G. Ramanath, "Molecular-nanolayer-induced suppression of in-plane Cu transport at Cu-silica interfaces," *Applied Physics Letters*, Vol. 90, 2007, pp. 163507–163509.
86. A. Krishnamoorthy, K. Chanda, S. P. Murarka, G. Ramanath, and J. G. Ryan, "Self-assembled near-zero-thickness molecular layers as diffusion barriers for Cu metallization," *Applied Physics Letters*, Vol. 78, No. 17, 2001, pp. 2467–2469.
87. G. Ramanath, G. Cui, P. G. Ganesan, X. Guo, A. V. Ellis, M. Stukowski, K. Vijayamohanan, P. Doppelt, and M. Lane, "Polyelectrolyte nanolayers as diffusion barriers for Cu metallization," *Applied Physics Letters*, Vol. 83, 2003, p. 383.
88. N. Mikami, N. Hata, T. Kikkawa, and H. Machida, "Robust self-assembled monolayer as diffusion barrier for copper metallization," *Applied Physics Letters*, Vol. 83, No. 25, 2003, pp. 5181–5183.
89. P. G. Ganesan, A. Kumar, and G. Ramanath, "Surface oxide reduction and bilayer molecular assembly of a thiol-terminated organosilane on Cu," *Applied Physics Letters*, Vol. 87, No. 1, 2005, p. 011905.
90. J. Noguchi, N. Miura, M. Kubo, T. Tamaru, H. Yamaguchi, N. Hamada, K. Makabe, R. Tsuneda, and K. Takeda, "Cu-ion-migration phenomena and its influence on TDDB lifetime in Cu metallization," in *Proceedings of the 41st IEEE Annual International Reliability Physics Symposium*, 2003, pp. 287–292.
91. Y. Yamada, Y. Yagi, N. Konishi, N. Ogiso, K. Katsuyama, S. Asaka, J. Noguchi, and T. Miyazaki, "Analysis of post-chemical-mechanical-polishing cleaning mechanisms for improving time-dependent dielectric breakdown reliability," *Journal of the Electrochemical Society*, Vol. 155, 2008, pp. H301–H306.
92. S. Kondo, K. Fukaya, N. Ohashi, T. Miyazaki, H. Nagano, Y. Wada, T. Ishibashi, M. Kato, K. Yoneda, E. Soda, S. Nakao, K. Ishigami, and N. Kobayashi, "Direct CMP on porous low-k film for damage-less Cu integration," in *Proceedings of the International Interconnect Technology Conference (IITC)*, 2006, pp. 164–166.
93. J. Noguchi, N. Konishi, and Y. Yamada, "Influence of post-CMP cleaning on Cu interconnects and TDDB reliability," *IEEE Transactions on Electron Devices*, Vol. 52, No. 5, 2005, pp. 934–941.
94. N. Konishi, Y. Yamada, J. Noguchi, T. Jimbo, and O. Inoue, "Influence of CMP process on defects in SiOC films and TDDB reliability," in *Proceedings of the International Interconnect Technology Conference (IITC)*, 2005, pp. 123–125.

95. W. Liu, Y. K. Lim, F. Zhang, W. Y. Zhang, C. Q. Chen, B. C. Zhang, J. B. Tan, D. K. Sohn, and L. C. Hsia, "Effect of chemical mechanical polishing scratch on TDDB reliability and its reduction in 45-nm BEOL process," in *Proceedings of the 47th IEEE Annual International Reliability Physics Symposium*, 2009, pp. 613–618.
96. A. Aal, "A comparison between V-ramp TDDB techniques for reliability evaluation," in *IEEE International Integrated Reliability Workshop Final Report*, 2008, pp. 133–136.
97. F. Chen, P. McLaughlin, J. Gambino, and J. Gill, "A comparison of voltage ramp and time dependent dielectric breakdown tests for evaluation of 45-nm low-*k* SiCOH reliability," in *Proceedings of the International Interconnect Conference* 2007, pp. 120–122.
98. S. C. Fan, J. C. Lin, and A. S. Oates, "Accurate characterization on intrinsic gate oxide reliability using voltage ramp tests," in *Proceedings of the 44th IEEE Annual International Reliability Physics Symposium*, 2006, pp. 625–626.
99. A. Kerber, T. Pompl, M. Rohner, K. Mosig, and M. Kerber, "Impact of failure criteria on the reliability prediction of CMOS devices with ultrathin gate oxides based on Voltage ramp stress," *IEEE Electron Device Letters*, Vol. 27, No. 7, 2006, pp. 609–611.
100. G. S. Haase, E. T. Ogawa, and J. W. McPherson, "Breakdown characteristics of interconnect dielectrics," in *Proceedings of the 43rd IEEE Annual International Reliability Physics Symposium*, 2005, pp. 466–473.
101. H. Miyazaki and D. Kodama, "TDDB lifetime of asymmetric patterns and its comprehension from percolation theory," in *Proceedings of the 47th IEEE Annual International Reliability Physics Symposium*, 2009, pp. 814–818.
102. Webpage. http://www.ansys.com, Access date: May 2006.
103. C. M. Tan, Z. H. Gan, W. Li, and Y. J. Hou, *Applications of Finite Element Methods for Reliability Study of ULSI Interconnections*, London: Springer-Verlag 2011.
104. In *ANSYS Theroy Reference*, 5. 6 ed; Swanson Analysis System, Inc., (now ANSYS Inc.), Canosburg, USA, 1999.
105. In *ANSYS Element Reference*, 5. 6 ed; Swanson Analysis System, Inc., (now ANSYS Inc.), Canosburg, USA, 1999.
106. Webpage. http://www. electroline. com. au/elc/feature_article/item_042003a. asp, Access date: May 2006.
107. C. U. Kim. (May 2006), http://www.sematech.org/meetings/past/20031027/TRC%202003_03_Kim.pdf, Access date: May 2006.
108. U. Choudhury and A. Sangiovanni-Vincentelli, "Automatic generation of analytical models for interconnect capacitances," *IEEE Transactions of Computer-Aided Design of Integrated Circuits and Systems*, Vol. 14, 1995, pp. 470–480.
109. S. Chikaki, M. Shimoyama, R. Yagi, T. Yoshino, Y. Shishida, T. Ono, A. Ishikawa, N. Fujii, T. Nakayama, K. Kohmura, H. Tanaka, J. Kawahara, H. Matsuo, S. Takada, T. Yamanishi, S. Hishiya, N. Hata, K. Kinoshita, and T. Kikkawa, "Extraction of process-induced damage in low-*k*/Cu damascene structure," in *Proceedings of the IEEE International Symposium on Semiconductor Manufacturing (ISSM)*, 2005, pp. 422–425.
110. K. Maex, M. R. Baklanov, D. Shamiryan, F. Lacopi, S. H. Brongersma, and Z. S. Yanovitskaya, "Low dielectric constant materials for microelectronics," *Journal of Applied Physics*, Vol. 93, 2003, pp. 8793–8841.
111. JP-001, "Inter/intra-metal dielectric integrity," in *Foundry Process Qualification Guidelines* Wafer Fabrication Manufacturing Sites, JEDEC/FSA Joint Publication, September 2002u, pp. 12–13.

Index

Note: Page numbers followed by *f* denote figures; page numbers followed by *t* denote tables.

F

J

K

L

M

N

9 780071 754279

Printed by Libri Plureos GmbH in Hamburg,
Germany